Fuchs/Oppolzer/Rehme

Particle Beam Microanalysis

Distribution:

VCH, P.O. Box 10 11 61, D-6940 Weinheim (Federal Republic of Germany)

Schwitzerland: VCH, P.O. Box, CH-4020 Basel (Schwitzerland)

United Kingdom and Ireland: VCH, 8 Wellington Court, Wellington Street, Cambridge CB1 1HZ (England)

USA and Canada: VCH, Suite 909, 220 East 23rd Street, New York NY 10010-4606 (USA)

ISBN 3-527-26884-7 (VCH, Weinheim) ISBN 0-89573-505-9 (VCH, New York)

E. Fuchs/H. Oppolzer/H. Rehme

Particle Beam Microanalysis

Fundamentals, Methods and Applications

Weinheim · New York
Basel · Cambridge

Dr. Ekkehard Fuchs
Dr. Helmut Oppolzer
Dr. Hans Rehme
Siemens AG, Zentrale Forschung und Entwicklung
Otto-Hahn-Ring 6
D-8000 München 83

Published jointly by
VCH Verlagsgesellschaft mbH, Weinheim (Federal Republic of Germany)

Editorial Director: Walter Greulich
Production Manager: Dipl.-Ing. (FH) Hans Jörg Maier

Library of Congress Card No.: 90-12865

British Library Cataloguing-in-Publication Data
Fuchs, Ekkehard
Particle beam microanalysis.
1. Materials. Electron beam analysis
I. Title II. Oppolzer, Helmut III. Rehme, Hans
620.112

ISBN 0-89573-505-9
ISBN 3-527-26884-7 W. Germany

Deutsche Bibliothek Cataloguing-in-Publication Data

Fuchs, Ekkehard:
Particle beam microanalysis: fundamentals, methods and applications / E. Fuchs; H. Oppolzer; H. Rehme. – Weinheim; New York; Basel; Cambridge: VCH, 1990
ISBN 3-527-26884-7 (Weinheim ...)
ISBN 0-89573-505-9 (New York ...)
NE: Oppolzer, Helmut:; Rehme, Hans:

Printed on acid-free paper.

Composition: Filmsatz Unger & Sommer GmbH, D-6940 Weinheim. Printing: betz-druck gmbh, D-6100 Darmstadt 12
Printed in the Federal Republic of Germany

Preface

The microtechnologies, above all semiconductor technology, have made rapid advances in recent years. No end can currently be discerned to this development, for which the ability to reliably manufacture very thin layers and microscopically small structures is a vital precondition. Microanalytical methods allowing high lateral and vertical resolution, high sensitivity, low detection limits and high accuracy provide crucial support for these manufacturing processes. Analytical methods which use electron or ion beams as measuring probes are particularly well suited for this purpose. High lateral resolution is obtained by focusing the particle beam. High depth resolution is achieved by chosing appropriate signals stemming from the specimen surface. The strong interaction effects of electrons and ions with solid material also make these methods very sensitive and capable of attaining extremely low detection limits.

Microanalysis and trace analysis investigations are indispensable for

- developing and applying the relevant materials,
- mastering the process technologies involved, and
- manufacturing products of high quality and reliability.

Details in the microstructure such as second phases, interfaces, the type and arrangement of lattice defects as well as the elemental distribution (e. g. dopant profiles) can be analyzed with great sensitivity and good spatial resolution. Investigations of this type are used to obtain a knowledge and thus control of the material effects on the performance of miniaturized components. They are also an indispensable aid for determining the effects of microscopic and submicroscopic inhomogeneities on macroscopic material properties and thus provide the basis for synthesizing new materials. Microanalytical instruments also provide important backup for quality control. Modern microtechnology is particularly sensitive to effects from the environment (dust particles, corrosion, abrasion), which can cause serious faults during production or in product operation. Microanalytical techniques have proved particularly effective for rapidly diagnosing the causes of such faults. The relatively high acquisition costs for high-performance analysis instruments are often quickly recouped by such troubleshooting.

This monograph deals with particle beam methods of microanalysis which allow to identify substances (qualitative determination of the elements), measure element concentrations (quantitative determination of the elements) and localize their constituents (determination of the microstructure and distribution of the elements) as well as measuring electrical properties in microelectronic circuits. In selecting the methods treated, priority was given to those which complement each other and can be routinely applied in industrial laboratories: scanning and transmission electron microscopy, electron beam X-ray microanalysis, Auger electron microanalysis, and ion beam microanalysis as well as electron beam testing are described in detail. These methods possess a high lateral or vertical resolution, provide reliable results within a relatively short time and are suitable for dealing with a broad range of analytical problems.

The principal aim of this book is to support the analyst in his practical work. The theoretical basis is treated only to the extent required to obtain an understanding of the relevant physical relationships (such as the interaction of the measuring probe with the specimen or the dependence of the concentration on the intensity of the measured signal) and to allow effective use of the analytical instruments. The mode of operation of the instruments, the preparation of specimens, the evaluation of the measured signals as well as the detection limits are described in detail. A selection of practical examples drawn mainly from the field of semiconductor technology demonstrates the range of applications and the limitations of the various particle beam methods.

Publications concerned with the field treated here have grown tremendously in numbers over recent years. To keep the number of references within reasonable limits therefore, we have endeavoured to include overview publications such as books and review papers of recent date wherever possible. We were consequently able in such cases to dispense with sources of earlier date.

The symbols used in the formulae (listed on page ...) as well as the units are based essentially on the recommendations of the International Union of Pure and Applied Physics (IUPAP; Symbols, Units and Nomenclature in Physics, Document U.I.P. 20 (1978)). Only where the same formula symbol is recommended for different physical quantities and could lead to confusion did we deviate from this procedure.

The analytical work described in chapter 9 was performed principally by our co-workers in the Siemens Research Laboratories; we wish to extend our thanks to all of them. For fruitful discussions and valuable suggestions during the preparation of the manuscript we have to express our thanks to Dr. H. Cerva, R.v. Criegern, Dr. E. Demm, Dr. O. Eibl, Dr. W. Hönlein, Dr. W. Hösler, Dr. S. Görlich, Dr. H. Kabza, Dr. U. Knauer, Dr. J. Kölzer, R. Lemme, Dr. K. Lubitz, Dr. A. Mitwalsky, Dr. W. Pamler, Dr. E. Plies, Dr. E. Wolfgang, Dr. R. Treichler, I. Weitzel, and Dr. H. Zeininger. We further wish to thank I. Frosien, K. Lippoldt, and H. Weinel for typewriting, R. Michell for the English translation, Dr. A. Petford-Long and Dr. N. J. Long for proof-reading the manuscript, and R. Schwab for the drawings. We also wish to thank our company Siemens AG for their permission to publish this manuscript and the publishers for their tolerant cooperation. Finally we would like to express our gratitude to our wives for their understanding and patience.

München, May 1990

Ekkehard Fuchs
Helmut Oppolzer
Hans Rehme

Contents

List of frequently used symbols

Space, time:

a	acceleration
A	area
d	diameter, distance
r	radius
t	time, thickness
v	velocity
c	vacuum velocity of light
x, y, z	space coordinates
α, β, γ	plane angles
λ	wave length
Ω	solid angle

Mechanics:

E	energy of particles
$\mathbf{F}$	force
m	mass
p	momentum, pressure
W	energy
ϱ	density

Electricity:

$\mathbf{B}$	magnetic flux density
C	capacitance
$\mathbf{E}$	electric field strength
f	frequency
$\mathbf{H}$	magnetic field strength
I, i	electric current
j	electric current density
P	power
q	charge
R	resistance
U	electric potential, potential difference, voltage
ν	frequency
φ	phase
ω	circular frequency ($\omega = 2\pi f = 2\pi\nu$)

Optics, particle optics:

a	object distance
b	image distance
d_0	probe diameter
F	focal point
f	focal length
I	image size
M	image scale (magnification)
O	object size
t_F	depth of focus
w	working distance
α	divergence angle (aperture angle)
α_0	divergence angle (aperture angle) of the probe (in the object plane)
β	brightness
λ	wave length

Atomic physics, solid state physics:

$\boldsymbol{b}$	Burgers vector
d	lattice plane spacing
E_0	energy of primary particles
E_b	binding energy
E_c	energy of conduction band edge
E_{ex}	exit energy of particles
E_F	Fermi energy
E_g	band gap
E_v	energy of valence band edge
E_w	work function
$f(\Theta)$	atomic scattering amplitude
f_x	atomic scattering factor for X-rays
$\boldsymbol{g}$	reciprocal lattice vector
G	generation factor for electron-hole pairs
g	generation rate for electron-hole pairs
I	signal intensity
I_0	primary beam current, probe current, intensity of directly transmitted beam
I_g	intensity of diffracted beam
I_{BSE}	backscattered electron current
I_{cc}	charge collection current
I_{PE}	primary electron current
I_{SE}	secondary electron current
$\boldsymbol{k}$	wave vector
$\dot{N}$	sputter rate

$N(E)$	energy distribution of particles
$P^{\pm}$	ionization probability
R	range of particles in a solid, radius of diffraction rings
$\boldsymbol{R}$	displacement vector of a defect in a crystal lattice
$\boldsymbol{r}$	position vector in a crystal lattice
s	deviation parameter (from Bragg orientation) or excitation error
T	temperature
U_0	acceleration voltage
$\boldsymbol{u}$	line vector of dislocation
Y	sputter yield
Z	atomic number
z	depth coordinate
z_p	penetration depth
δ	secondary electron yield, spatial resolution
η	backscattering coefficient
η_{cc}	charge collection efficiency
η_{rr}	radiative recombination efficiency
Θ	scattering angle
ϑ	Bragg angle, diffraction angle
Λ	mean free path of electrons in a solid
λ_p	inelastic mean free path
λ_i	escape depth or information depth of particles
μ	absorption coefficient
ξ_g	extinction distance for electrons in a crystal
σ	cross-section, total yield of secondary and backscattered electrons ($\sigma=\delta+\eta$)
τ	life time of charge carriers, useful yield
Φ	dose, amplitude of scattered waves
$\Phi(\varrho z)$	depth distribution function of X-rays
Ψ	wave function
ψ	angle of incidence, take off angle, exit angle
ω_K	fluorescence yield for K transitions

Composition of compounds, quantitative analysis:

N_i	number of particles of species i
$N = \sum N_i$	total number of particles
$C_i = N_i/V$	concentration [cm^{-3}]
$X_i = N_i/N$	mole fraction
w_i	mass fraction
ε	relative mean error (relative mean standard deviation)
σ	statistical standard deviation

List of abbreviations

AC — Alternating current
AEM — Analytical electron microscope (microscopy)
AES — Auger electron spectrometer (spectrometry)

BF — Bright field
BIP — Bipolar
BSEs — Backscattered electrons

CBED — Convergent beam electron diffraction
CL — Cathodoluminescence
CMA — Cylindrical mirror analyzer
CMOS — Complementary metal on semiconductor
CRT — Cathode ray tube
CTEM — Conventional transmission electron microscope (microscopy)
CTF — Contrast transfer function
CVD — Chemical vapour deposition

DC — Direct current
DF — Dark field
DRAM — Dynamic random access memory

EBIC — Electron beam induced current
ECP — Electron channeling pattern
EDS — Energy dispersive spectrometer (spectrometry)
EDX — Energy dispersive X-ray spectrometer (spectrometry)
EELS — Electron energy loss spectrometer (spectrometry)
EM — Electron microscope (microscopy)
EPMA — Electron probe microanalyzer (analysis)
ESCA — Electron spectroscopy for chemical analysis

FEG — Field emission gun
FET — Field effect transistor
FWHM — Full width at half maximum

HF — High frequency
HOLZ — High order Laue zone
HREM — High resolution electron microscope (microscopy)
HVEM — High voltage electron microscope (microscopy)

IC — Integrated circuit
ICP — Inductive coupled plasma
IMFP — Inelastic mean free path
IMMA — Ion microprobe mass analyzer (analysis)
IR — Infrared

LEED — Low energy electron diffraction

MBE	Molecular beam epitaxy
MCA	Multi-channel analyzer
MFP	Mean free path
MOS	Metal on semiconductor
MOVPE	Metal organic vapour phase epitaxy
MQW	Multi quantum well
MS	Mass spectrometry
OES	Optical emission spectrometry
PEs	Primary electrons
PIs	Primary ions
PM	Photomultiplier
RBS	Rutherford back scattering spectrometer (spectrometry)
RF	Radio frequency
RHEED	Reflection high energy electron diffraction
RTA	Rapid thermal annealing
SAD	Selected area diffraction
SAM	Scanning Auger microscope (microscopy)
SAW	Surface acoustic wave
SEM	Scanning electron microscope (microscopy)
SEs	Secondary electrons
SIMS	Secondary ion mass spectrometer (spectrometry)
SIs	Secondary ions
SNMS	Sputtered neutrals mass spectrometer (spectrometry)
SQW	Single quantum well
STEM	Scanning transmission electron microscope (microscopy)
TEM	Transmission electron microscope (microscopy)
TEs	Transmitted electrons
UHV	Ultra high vacuum
VLSI	Very large scale integration
VPE	Vapour phase epitaxy
WDS	Wavelength dispersive spectrometer (spectrometry)
XMA	X-ray microanalysis

1 Introduction

1.1 What microtechnology requires of analysis

Microtechnology is characterized by
- extremely small functional elements,
- high structural complexity,
- great diversity of the materials involved, and
- sophisticated technological processes.

Semiconductor technology represents the predominant example of microtechnology. In comparison with other industrial procedures, therefore, microtechnology makes particularly high demands on analysis. Many different methods of analysis are required to characterize materials, processes, process agents, and ultimately complete components or systems. These methods range from simple physical and chemical techniques, such as optical microscopy and wet chemical analyses, to very sophisticated methods of analysis (such as electron microscopy, radiochemistry, secondary ion mass spectrometry). These analytical methods are required not only in research and development but also for failure analysis and quality control in production lines. They must thus also provide reliable results within a short time and be rapidly available. Rapid diagnosis of the causes is an important precondition for eliminating disturbances or failures during production.

Micro- and trace analysis methods are used for characterizing extremely fine structural details and inhomogeneities as well as extremely low admixtures and impurity levels. They provide qualitative and quantitative data about the material properties such as the microstructure, the chemical composition and its spatial distribution, the surface composition and morphology, as well as the crystal structure and atomic configuration.

The macroscopic properties of a material, such as its electrical or thermal conductivity, corrosion resistance or hardness, depend primarily on these microscopic properties, and this dependence becomes increasingly sophisticated as components, devices and systems become more complex, the stresses to which they are subjected increase and the size decreases. This applies especially to microelectronics components, whose electrical properties can be changed either intentionally or inadvertently by the incorporation of very small amounts of impurities into ultra-pure semiconductor material or by minor disturbances of the crystal lattice. Environmental influences ignored in other technologies, such as dust particles, trace contaminants in process media or supposedly negligible deviations from tolerance limits in thermal or chemical processes, can considerably reduce the yield of electronic components during manufacturing.

In general, the properties of materials depend greatly on the manufacturing process. The development of the materials and of their corresponding fabrication technology are therefore closely linked and are frequently inseparable. The task of analysis is, consequently, not only to characterize the material in line with the criteria mentioned above, but also to provide contributions to the correlation of physical and chemical properties with technological processes. Ultimately, analysis must supply results relevant to the quality and reliability of the end product. Analytical investigations must therefore accompany the research and development of materials and devices as well as the manufacturing processes all along the line. If the development is to advance smoothly, the results of these investigations must be rapidly available and be presented in a manner comprehensible to the technology engineer.

A characteristic change has taken place in the analytical work pursued in various subsectors. Previously, rapid analytical investigations were required to a larger extent, especially when unpredictable deviations from the planned route or failures occurred in the course of research and development or during production. Today, analysis must still contribute to solving such acute problems (trouble-shooting), but its main function consists increasingly in supporting research and development projects from an early stage. By starting with intensive material investigations, sidetracks in development, difficulties in production, and failures during application can be avoided. This necessitates close cooperation between analysts on the one hand and materials scientists, technology and production engineers on the other. The technology engineer learns to appreciate the potentials of analysis and the analyst becomes more familiar with the problems facing the technology engineer. In addition, this approach allows the analyst to adapt his methods in good time to the problems in hand and adopt new techniques so that he may be well equipped for future tasks.

1.2 Overview of microanalytical methods

Generally, there are two types of analytical problems posed by microtechnology (Fig. 1-1). In the first case, the available volume of material is very small, in the second case the point is to determine the chemical composition of minute admixtures in a larger volume of material.

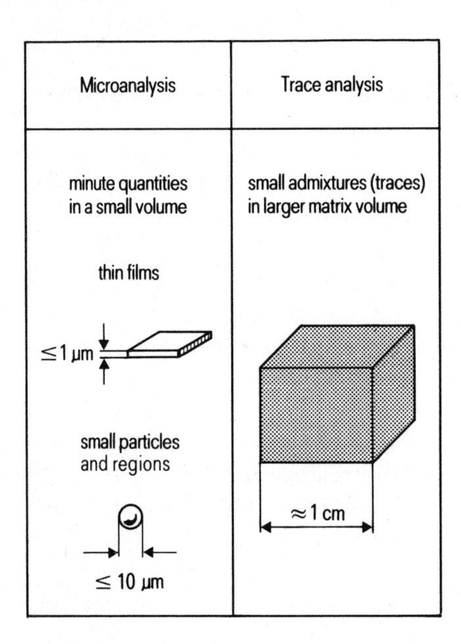

Fig. 1-1.
Definition of microanalysis and trace analysis.

Particularly effective solutions to these problems are offered by the highly-sensitive techniques of micro- and trace analysis.

Microanalysis involves the characterization of minute quantities in a small volume, such as thin films (film thickness $\leq$ 1 μm) or small particles (diameter $\leq$ 10 μm) which lie isolated on a surface or are embedded in a matrix. *Trace analysis*, in contrast, is used when a minute quantity of a very finely dispersed substance must be analyzed in a larger matrix volume (of the order of 1 cm^3).

The most important techniques of micro- and trace analysis which have proved their value in practical use are listed in Table 1-1 [1-1]. They are classified in line with the fundamental requirements of practical analytical work, namely

- identification,
- quantification, and
- localization.

Identification is the determination of the kind of elements present in the specimen independent of their amount.

Quantification specifies the fraction or concentration of the elements in the specimen and the uncertainty of their determination. In microanalysis the following definitions are used for the quantification:

- The atomic concentration C_i (SI) is the number N_i of atoms of an element i divided by the analyzed volume V:

$$C_i = N_i/V \,. \tag{1-1}$$

C_i has the dimension cm^{-3}.

- The mole fraction X_i is the number N_i of atoms of an element i divided by the total number $N = \sum N_i$ of all atoms in the analyzed volume:

$$X_i = N_i/N \,. \tag{1-2}$$

X_i is a ratio of two equally dimensioned quantities and therefore is dimensionless.

- The mass fraction w_i of an element i is the mass m_i of the atoms of the element i divided by the total mass $m = \sum m_i$ of all atoms in the analyzed volume:

$$w_i = m_i/m \,.$$

The following is obtained for the sum of all mass fractions in the specimen:

$$\sum w_i = 1 \,.$$

With the atomic concentration C_i *in* cm^{-3}, the mass A_i of N_a atoms in g, $N_a = 6.02 \cdot 10^{23}$ being the Avogadro number, and the mean density ρ in $g \cdot cm^{-3}$ it follows that:

$$w_i = \frac{C_i \cdot A_i}{\rho \cdot N_a} \,. \tag{1-3}$$

Table 1-1. Overview of important techniques of micro- and trace analysis.

Analytical problem	Method	Effect used for the analysis	Gives informations related to chemical composition with a detection limit[1] of		crystal structure		micro structure		Analysis error
			m_{min}/g	X_{min}	m_{min}/g	X_{min}	δ/m	$\Delta z/m$	%
Identification	X-ray diffraction	Elastic scattering at crystal lattice	yes		10^{-6}	10^{-2}			
	Electron diffraction	" " " " "	yes		10^{-15}				
Quantification	Gravimetry	Specific chemical reaction	10^{-5}	$>10^{-1}$ [2]					0.1... 1
	Volumetry	" " "	10^{-7}	$>10^{-2}$ [2]					0.2... 2
	Colorimetry	" " "	10^{-8}	10^{-6}					0.5... 5
	Polarography	Electrochemical potentials	10^{-10}	10^{-9}					2 ... 20
	Gas chromatography	Specific adsorption of gases	10^{-12}	10^{-9}					0.1... 1
	Atomic absorption spectrometry	Excitation of outer atomic levels	10^{-14}	10^{-12}					2 ... 10
Identification and Quantification	Infrared spectrometry	Excitation of molecular vibrations	10^{-6}	10^{-4}					5 ... 50
	Ion chromatography	Specific adsorption of ions	10^{-8}	10^{-9}					0.1... 20
	X-ray fluorescence spectrometry	Excitation of inner atomic levels	10^{-9}	10^{-5}					0.1... 10
	Optical emission spectrometry	Excitation of outer atomic levels	10^{-11}	10^{-8}					5 ... 10
	Mass spectrometry	Ratio of charge-mass of ions	10^{-14}	10^{-6}					1 ... 20
Localization	X-ray topography	Elastic scattering at crystal lattice			yes		10^{-5}		
	Light microscopy	Absorption and scattering of light			(yes)		$2 \cdot 10^{-7}$		
	Scanning electron microscopy	Emission of secondary electrons			(yes)		$2 \cdot 10^{-9}$		
	Transmission electron microscopy	Scattering of electrons			yes		$2 \cdot 10^{-10}$		
Identification Quantification and Localization	X-ray microanalysis	Excitation of inner atomic levels	10^{-15}	10^{-3}			10^{-6}	10^{-6}	1 ... 10
	Auger microanalysis	" " " " "	10^{-17}	10^{-3}			10^{-7}	10^{-9}	5 ... 20
	Secondary ion mass spectrometry	Emission of secondary ions	10^{-20}	10^{-9}			10^{-6}	10^{-9}	10 ...100
	X-ray microanalysis in the scanning transmission electron microscope	Excitation of inner atomic levels in thin foils	10^{-20}	10^{-3}			10^{-8}		2 ... 10

[1] m_{min} minimum detectable mass
X_{min} minimum detectable mole fraction
δ lateral resolution
Δz depth resolution

[2] analytical method for main components

w_i is also dimensionless. From equations (1-1), (1-2) and (1-3), the following applies for the relationship between mass fraction w_i and mole fraction X_i

$$w_i = \frac{X_i \cdot A_i \cdot N}{V \cdot \rho \cdot N_a} . \tag{1-4}$$

For the relation between the ratios of mass fractions and mole fractions of two elements (i, j) in a specimen we then obtain:

$$\frac{w_i}{w_j} = \frac{X_i \cdot A_i}{X_j \cdot A_j} . \tag{1-5}$$

Localization refers to the determination of the concentration or the mole fraction of an element or inhomogeneities in the microstructure (crystal defects, precipitates, etc.) as a function of position. The uncertainty of the position is characterized by the lateral or vertical (depth) resolution.

Some techniques provide only qualitative information and can therefore be used only for element identification, while other methods can only determine quantities of the elements, and require advance knowledge of which elements are present in the specimen. In a specimen of unknown composition, purely quantitative methods can therefore be used only when a previous qualitative analysis has revealed which chemical elements are present.

A good overview of the chemical composition of an unknown specimen is provided by data from optical emission spectroscopy or X-ray fluorescence spectrometry. This overview analysis can help to decide which techniques should be used for further analytical investigation. Optical emission spectroscopy also allows a quantitative determination of the gross composition down to the trace region, and X-ray fluorescence analysis permits very high compositional accuracy. Information about the chemical bonds in the material is supplied by infrared spectrometry, ion chromatography and extended X-ray absorption fine structure spectrometry (EXAFS). X-ray and electron diffraction are the classical methods used for investigating the crystal structure of materials. They can also be used to distinguish crystallographically different phases. Since every crystalline substance has a crystal lattice with specific lattice parameters, diffraction methods can provide an unambiguous identification in many cases. Crystal structure data are available for some several thousand substances and are continually being extended [1-2,3,4].

1.3 Particle beam methods

Of special interest in microtechnology are, quite naturally, the methods with high spatial resolution, in particular those which can identify unknown substances in small volumes and determine their composition with sufficient accuracy. The particle beam methods have proved to be outstanding for such applications [1-5, 6]. These methods use electrons or ions as measuring probes, because

- charged particles can be easily accelerated to the required energy by electric fields,
- the trajectories of charged particles are easily influenced by electric and/or magnetic fields and can therefore be focused to form small probes which can be directed onto defined regions, and because
- energetic particles interact strongly with the investigated material, resulting in signals which are useful for analysis and can be easily detected.

The particle beam methods mentioned here are all based on a common principle (Fig. 1-2). A beam of charged particles, the probe, is focused onto a small area of the specimen for analysis. The probe particles interact with the material at that spot and thus excite secondary radiation and particles such as X-rays, electrons or ions which are, to some extent, characteristic of the material composition. By measuring the energy and intensity of the secondary radiation or particles, the elements present in the specimen can be identified and quantified.

Every technique has its strengths and weaknesses. Techniques with high spatial resolution are in general less sensitive for measuring low concentrations. Techniques with high sensitivity for concentrations, on the other hand, require a larger volume for analysis or will be susceptible to a greater analysis error. Greater effort and experience are then required to obtain a reliable analytical result providing useful information. Considerable differences also exist with respect to the range of the periodic table covered by each analytical method. X-ray microanalysis is, for example, sensitive to elements of medium and high atomic number, whereas Auger elec-

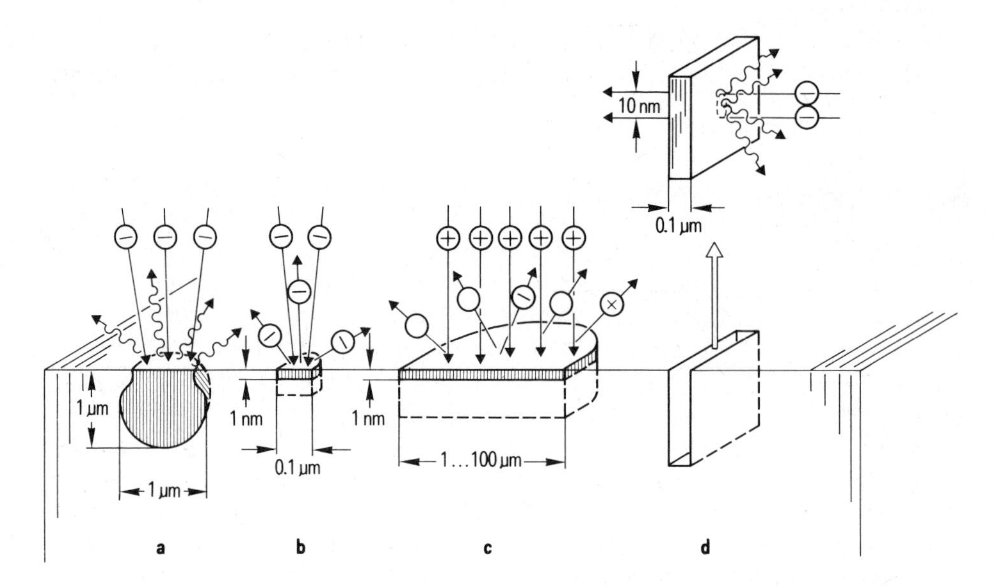

Fig. 1-2. Particle beam methods for solid material microanalysis: X-ray microanalysis provides analytical information from small volumes with a diameter of about 1 µm (a). In contrast, Auger microanalysis (b) and secondary ion mass spectrometry (c) are surface-specific methods with an information depth of a few nm. By means of successive abrasion of the surface by ion etching, the material composition as a function of depth can be determined (indicated by broken lines in b and c). For very thin foils which are cut from the object being investigated by preparing a thin cross-section (d), the microstructure can be imaged with high resolution in a transmission electron microscope and the element composition determined from the X-rays excited by the electrons.

tron microanalysis is more sensitive for the detection of light elements such as carbon and oxygen. This is why complex analytical problems often require several analysis techniques. A fundamental precondition for successful material research and development projects is therefore the support from an analytical laboratory capable of applying several analytical techniques which mutually support and complement each other. The very diverse analytical problems which arise in the field of microtechnology require careful selection of suitable analysis methods so that reliable results providing useful information can be obtained with minimum effort and outlay.

1.4 Analysis errors and detection limits

Like all measurements, every analytical measurement is subject to errors. The total analysis error is made up of various contributions, which are due to the following factors:

- the properties and composition of the specimen;
- the choice of measuring conditions and alignment of the analysis system;
- the statistical nature of the emitted radiation quanta or particles which provide the analytical information;
- separation of the element-specific signal from the measured signal;
- conversion of the element-specific signal to the sought-for element fraction in the specimen (quantification).

Properties and composition of the specimen

The specimen-specific analysis error is essentially due to

- the element-dependent signal intensity,
- the concentration of the elements, which differs from specimen to specimen, and
- the influence of other elements on the element-specific signal of interest.

Thus the X-ray yield, for instance, increases with increasing atomic number in the periodic table. Quantitative X-ray microanalysis for elements of middle and high atomic number is consequently burdened with smaller errors than the analysis of lighter elements. Further, stronger signals obtained at higher concentrations are measured with greater reliability than weaker signals at low concentrations. In unfavorable cases, the signal of the sought-for element can also be distorted by signals from other elements, for example by line overlap. Finally, the physical and chemical properties of the specimen material can disturb the analysis by charged particles (electrons, ions) in a sensitive way. Electrical charges, which are built up on the surface as a result of bombardment by the primary particles, can falsify results of the analysis or even make analysis quite impossible. Due to the relatively strong interaction of the primary particles with the specimen material, the specimen may decompose so that it loses its original composition or structure.

Selecting the measuring conditions and setting up the analysis system

The measurement conditions, e.g. acceleration voltage or probe current, must in general be selected so that an optimal signal is obtained. Even if great care is taken in this procedure, small alignment errors are unavoidable. In aligning the analysis system, two main conditions must be satisfied. The collection efficiency designates the fraction of the element-specific radiation which reaches the detector from the specimen. The greater this fraction is, the higher the signal. The analysis error is then smaller and the detection limit lower. On the other hand, the analysis system must permit sufficiently good separation of the element-specific signals so that the signal of the sought-for element is not falsified by the superimposed signal from another element present in the specimen. The resolution of the spectrometer system must thus be enhanced, but this is possible only at the expense of the signal intensity. The operating conditions chosen must therefore optimize the signal yield and the spectrometer resolution and adapt them to the analysis problem at hand.

Statistical error in the signal measurement

Even if low systematic errors are ensured by means of optimal instrument alignment, statistical errors must still be taken into account. Due to the quantum nature of radiation, the emission of radiation from the specimen follows statistical laws. The emitted quanta (X-ray quanta, electrons, ions) thus reach the detector in a statistical sequence. As a rule, the number of electrical pulses generated by the incident quanta in the detector is counted over the unit time.

If the same measurement is performed m times under identical conditions, the same number of pulse counts will not be measured m times. Instead, the measured values exhibit some variation, and a Poisson distribution is obtained. The best approximation to the unknown "true pulse number" is the arithmetical mean $\bar{n}$ of all m measured values. The individual values n_i ($i = 1, 2, \ldots m$) deviate more or less from the mean value $\bar{n}$. For a Poisson distribution, the mean deviation, known as the *standard deviation*, is given by

$$\sigma_P = \pm\sqrt{\bar{n}} \ . \tag{1-6}$$

In place of m measurements per unit time, particle beam methods perform only a single measurement over a time t. In this case, the N measured pulses are set equal to the arithmetic mean. The standard deviation is then

$$\sigma = \pm\sqrt{N} \tag{1-7}$$

and the relative mean error (or relative standard deviation) has the value

$$\varepsilon = \sqrt{N}/N = 1/\sqrt{N} \ .$$

In quantitative analysis, the intensity (counts per unit time, count rate) is generally used in place of the counts:

$$I = N/t \ . \tag{1-8}$$

If we assume that the time measurement is free of errors, the standard deviation of the intensity is given by:

$$\sigma_I = \pm\sqrt{N}/t = \pm\sqrt{(I/t)} \,, \tag{1-9}$$

and the relative standard deviation is

$$\varepsilon_I = \pm\frac{\sqrt{I/t}}{I} \,. \tag{1-10}$$

For a specified intensity, therefore, the relative error drops with $\sqrt{t}$.

With the aid of error theory and the use of the standard deviation, the probability P can be determined with which an error of size $k \cdot \sigma_I$ is not exceeded in measuring I. The true value I_w, which cannot be known in principle, then lies within the limits $I \pm k \cdot \sigma_I$ with this probability:

$$I - k \cdot \sigma_I \leq I_w \leq I + k \cdot \sigma_I \,.$$

In accordance with the error theory, a probability of $P = 68.26\%$ exists for $k = 1$. With this probability, the deviation from the true value is equal or smaller than $\pm\sigma_I$. If, for instance, an intensity of $I = 10^4\ s^{-1}$ has been measured during a period of 1 s, then in accordance with eq. (1-10) we can expect the true value not to deviate more than 1% from the measured value with a probability of 68.26%.

However, the International Union of Pure and Applied Chemistry (IUPAC) recommends the value $k = 3$ to be used in error considerations and when calculating detection limits [1-7, 8]. The probability that the measured result deviates by a maximum error of $3 \cdot \sigma$ or $3 \cdot \sigma_I$ from the true value is then 99.73%. However, this error is greater for the example mentioned above, namely 3%, but with a considerably higher reliability of over 99%. If a measurement with the same intensity ($I = 10^4\ s^{-1}$) is extended to 100 s, the statistical error is reduced to 0.3% with a probability of 99.73%. The measurement is thus more accurate with increasing measurement time. In practice, however, the measurement time must be limited. This is due to factors such as drift of the instrument parameters or changes in the specimen composition as a result of long radiation times. However, these considerations apply only to nondestructive methods of analysis. In secondary ion mass spectrometry, the specimen is consumed during the measurement, and every atom can be recorded only once (if at all). In this method of analysis, therefore, an increase in the measurement time does not lead to an improvement in the measurement accuracy (section 7.1).

Separation of the element-specific signal

As a rule, the signal I_{mA} measured in the spectrum, due to an element A, is composed of the desired element-specific intensity I_A and a background intensity I_b. The intensity I_A characteristic of the element alone is the difference between the measured intensity I_{mA} and the background intensity I_b:

$$I_A = I_{mA} - I_b \,. \tag{1-11}$$

Both intensities must be taken into account in calculating the statistical error. The deviations of both intensities from their true values are described by their standard deviations:

$$\sigma_{mA} = \sqrt{I_{mA}/t_{mA}} \; ; \quad \sigma_b = \sqrt{I_b/t_b} \; .$$

The squares of the standard deviations (known as variances) of superposed distributions are added in accordance with the error propagation law:

$$\sigma_A^2 = \sigma_{mA}^2 + \sigma_b^2 \; .$$

From this and with equations (1-9) and (1-10), we obtain the following expression for the relative statistical error of the element-specific intensity I_A:

$$\varepsilon_A = \frac{\sqrt{(I_{mA}/t_{mA})^2 + (I_b/t_b)^2}}{I_{mA} - I_b} \; . \tag{1-12}$$

It can be seen that the error becomes larger with increasing background intensity. It increases rapidly when the measured intensity differs only slightly from the background intensity.

An element is only just detectable when the signal intensity assigned to it can be distinguished with sufficient reliability from the background. That is the case when the number of the net pulses $N_A = I_A \cdot t_A$ is greater by a specific multiple k than the standard deviation $\sigma_b = \sqrt{N_b}$ of the background:

$$N_A = k \cdot \sqrt{N_b} \; . \tag{1-13}$$

When $k = 3$, there is a probability of 99.73% for the existence of an element assigned to the characteristic intensity I_A. The following relationship is therefore used for the *statistical etection limit*

$$N_{A\min} = 3 \cdot \sqrt{N_b} \; . \tag{1-14}$$

In microanalytical methods, however, the statistical error generally makes only a small contribution to the total analysis error. The separation of the measured signal and the quantification of the element-specific signal are burdened with considerably greater errors. The effects of these errors are described in the following chapters when considering the various analysis techniques.

The *limit of detection* in which all sources of error are taken into account is frequently used to characterize the efficiency of a method of analysis. This limit can be expressed as a relative quantity (i.e. minimum detectable mole fraction X_{min}) or as an absolute quantity (minimum detectable mass m_{min}). The detection sensitivity not only differs considerably from method to method, it also varies greatly for a specific method as a function of the material being analyzed. Therefore, the detection limits listed in Table 1-1 are only to be regarded as rough values for favorable examples of each case. The values for the minimum detectable mass refer in each case to an element which can be detected particularly easily with the relevant method and which makes up 100% of the analyzed volume. Today, minimum detectable masses down to the order of 10^{-20} g (X-ray microanalysis in the transmission electron microscope) are attained. This means that about 100 atoms of medium atomic weight can be identified. The attainable minimum detectable mole fractions are of the order of 10^{-12} (atomic absorption spectrometry). This means that one atom can still be analyzed among 10^{12} atoms (or ions).

1.5 References

[1-1] E. Fuchs, *Microelectronic Engng. 1*, 143 (1983)

[1-2] *Powder Diffraction File,* JCPDS, International Centre for Diffraction Data, Swarthmore, PA, USA 1985 (supplemented yearly)

[1-3] *Structure Reports,* International Union of Crystallography, D. Reidel Publishing Company, Dordrecht (Holland), Boston (USA) 1984

[1-4] *Powder Diffraction,* An international journal of materials characterization, JCPDS-International Centre for Diffraction Data, Swarthmore, PA, USA

[1-5] T. Ambridge, B. Wakefield, *Brit. Telecom Technol. J. 3,* 47 (1985)

[1-6] H.W. Werner, *Mat. Sci. Eng. 42,* 1 (1980)

[1-7] IUPAC: *Nomenclature, symbols, units and their usage in spectrochemical analysis -II, Spectrochim. Acta B 33B,* 242 (1978)

[1-8] IUPAC: *Guidelines for Data Acquisition and Data Quality Evaluation in Environmental Chemistry, Anal. Chem. 52,* 2242 (1980)

2 Fundamentals

2.1 Fundamental concepts of solid-state physics

2.1.1 Structure of solids, crystals

This chapter presents some refresher information on a number of important concepts of solid-state physics which will be needed to understand the methods of analysis described in this book [2-1,2,3].

Typically, the objects analyzed by the methods of particle beam microanalysis are solids. Matter which exists in a solid state is predominantly crystalline. Within certain regions, the elementary units (atoms, ions) of a solid are arranged periodically, with strict spatial regularity. The regular regions can vary in size. In a single crystal, the elementary units are uniformly arranged over the entire volume.

Today, single crystals with a diameter of up to 25 cm and more are produced for use in semiconductor technology. A polycrystalline substance has differently oriented regions (grains) separated from each other by grain boundaries. Even amorphous solids exhibit a certain order of their elementary building blocks, although this is found only in very small regions of atomic dimensions (short-range order). Liquids also exhibit just such a short-range order, which, however fluctuates with respect to space and time.

Crystal lattices

The smallest unit, the base of a crystal structure, can consist of a single atom or group of atoms. This base is translationally repeated along three non-coplanar directions throughout the crystal. The crystal volume built up by the three fundamental translation vectors ***a***, ***b***, ***c*** determines the unit cell of a given crystal structure. Every lattice point is accessible with the vector

$$\boldsymbol{r} = n_1 \cdot \boldsymbol{a} + n_2 \cdot \boldsymbol{b} + n_3 \cdot \boldsymbol{c} \tag{2-1}$$

(n_i are positive or negative integers, Fig. 2-1).

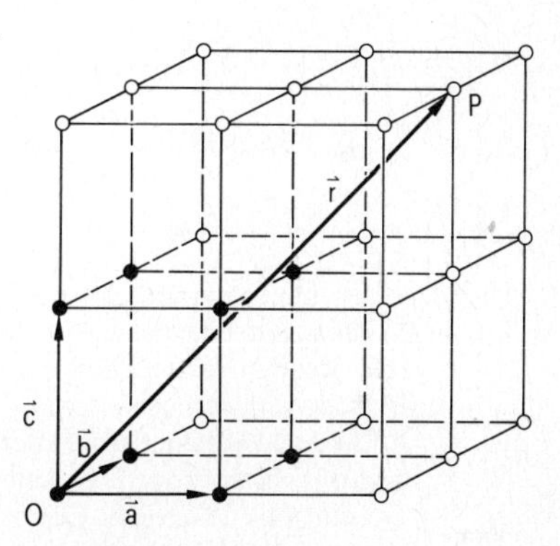

Fig. 2-1.
Point lattice with translation vectors ***a***, ***b***, ***c***. The lattice point P with lattice-specific coordinates $2a$, b and $2c$ is reached with the vector $\boldsymbol{r} = 2 \cdot \boldsymbol{a} + \boldsymbol{b} + 2 \cdot \boldsymbol{c}$. The cell defined by ***a***, ***b*** and ***c*** constitutes the primitive unit cell with lattice constants $a = |\boldsymbol{a}|$, $b = |\boldsymbol{b}|$ and $c = |\boldsymbol{c}|$. Crystal directions are defined by the coordinates of the lattice point closest to the origin O and are given in units of a, b and c: [212] for the direction defined by ***r***.

The lengths of the translation vectors are the lattice constants $a = |\boldsymbol{a}|$, $b = |\boldsymbol{b}|$, $c = |\boldsymbol{c}|$. Besides the translations, which describe the lattice, there are other symmetry operations (rotations, mirror planes) which transfer the lattice into itself. An in-depth consideration of the symmetry properties shows that 14 different fundamental types of point lattice (Bravais lattices) exist. They are classified into 7 crystal systems. Some lattice types of practical importance are shown in Fig. 2-2.

Every plane passing through three points of a crystal lattice is periodically populated at regular intervals by lattice units (atoms, ions). An infinitely large number of planes of the same kind run parallel to such a lattice plane with equal interplanar spacing *d*. They form a set of lattice planes. Such a set is characterized in terms of the lattice plane which is closest to the origin of the (lattice-specific) coordinate system. This lattice plane intersects the three coordinate axes at specific points. If the reciprocal values of the axial sections defined by these points are multiplied by their common denominator, a triplet of integers is obtained. These are the Miller indices (hkl) of the relevant lattice plane (Fig. 2-3). A set of lattice planes is unambiguously characterized by their interplanar spacing and these indices. For a cubic lattice with lattice constant a, the spacing d_{hkl} of the set of lattice planes is given by the Miller indices hkl through the relation

$$d_{hkl} = a/\sqrt{h^2 + k^2 + l^2} \, .$$

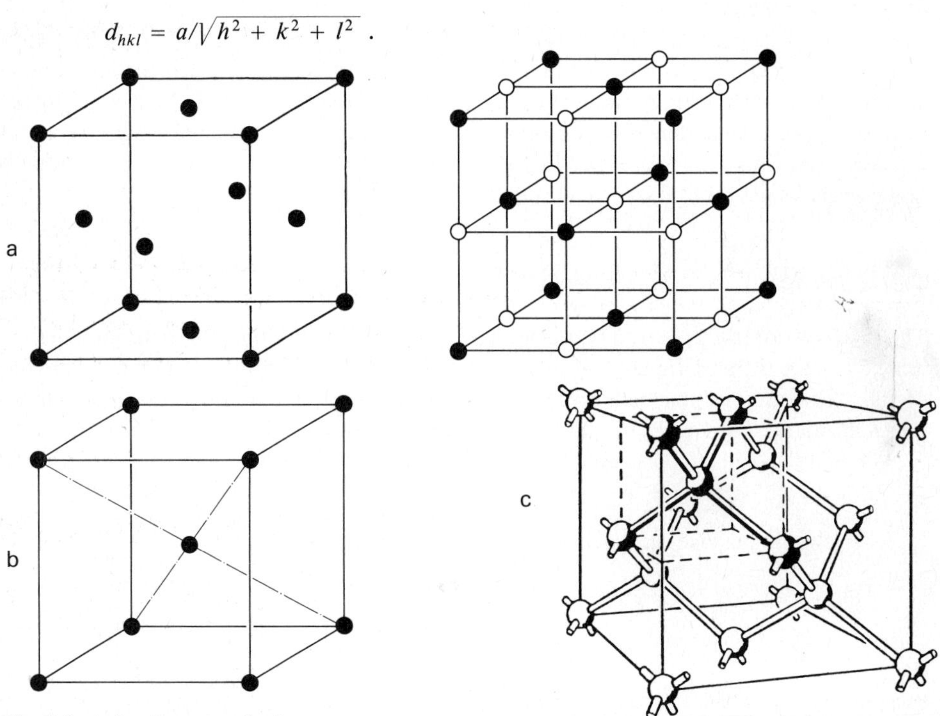

Fig. 2-2. A few important lattice types:
a) Face-centered cubic lattice. Next to it is the lattice of common salt (NaCl). Sodium and chlorine occupy one such lattice each.
b) Body-centered cubic lattice; the most common metal lattice.
c) Diamond lattice: each C atom is surrounded by 4 equidistant neighbors arranged at the corners of a tetrahedron. The materials important for semiconductor technology also crystallize in the diamond lattice (Si, Ge) or in the related sphalerite structure (GaAs, InP).

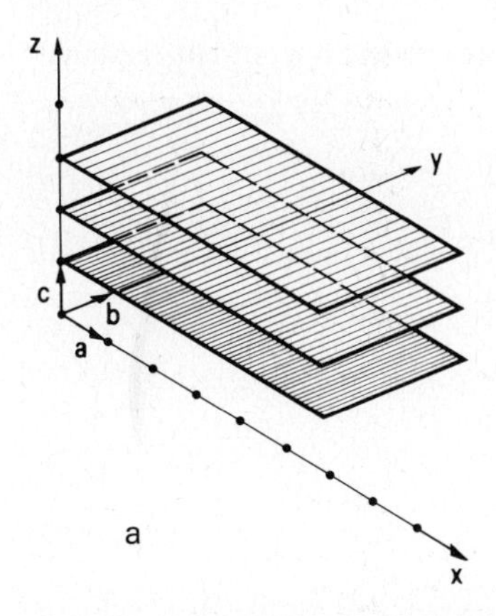

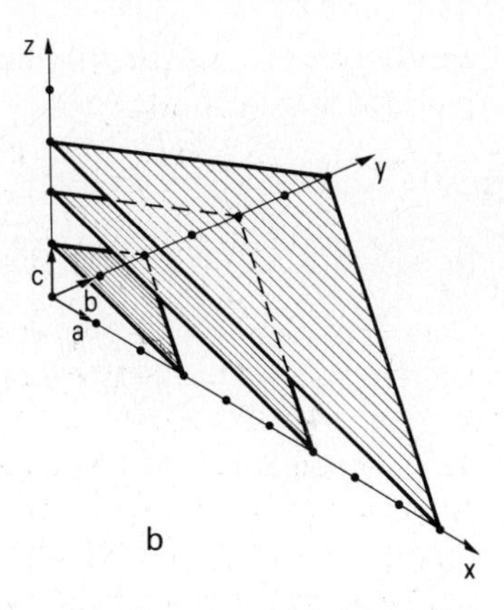

Fig. 2-3. Lattice planes in a cubic crystal lattice:
a) The set of planes normal to the z axis intersects this axis at the units 1, 2, 3,. The x and y axes are intersected at infinity. The Miller indices are $h = 1/\infty = 0$, $k = 1/\infty = 0$, $l = 1/1 = 1$. The set of lattice planes is correspondingly designated (001).
b) The axial sections of the first plane of this set are 3, 2 and 1; the common denominator is 6 and the corresponding Miller indices are: $h = (1/3) \cdot 6 = 2$, $k = (1/2) \cdot 6 = 3$ and $l = 1 \cdot 6 = 6$. This set of planes is therefore designated (236). The smaller the Miller indices, the greater becomes the interplanar spacing and the denser the population of the lattice planes with lattice points.

Diffraction at the crystal lattice

The elastic scattering of short-wave electromagnetic or particulate radiation at crystal lattices gives rise to diffraction phenomena in much the same way as the diffraction of light waves at diffraction gratings. The directions of maximum intensity may be easily explained with the aid of the lattice planes (Fig. 2-4). The diffraction of the wave radiation (wavelength λ) at the

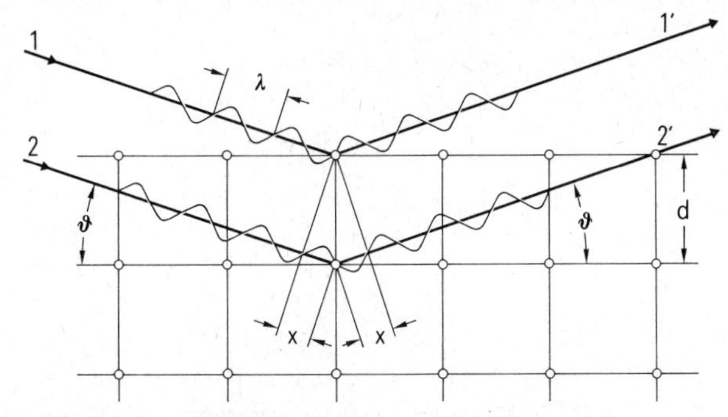

Fig. 2-4. Diffraction at the crystal lattice: the two incident waves 1 and 2 are initially in phase. After scattering at the atoms of the lattice, the waves 1′ and 2′ are of equal phase and lead to an intensity maximum only when the path difference $2x$ between them is equal to, or an integral multiple n of the wavelength λ. From this follows, with $x = d \cdot \sin \vartheta$, equation (2-2), known as Bragg's law.

crystal lattice can also be described as a "reflection" of this radiation at a set of lattice planes, provided that the Bragg condition

$$2 \cdot d \cdot \sin \vartheta = n \cdot \lambda \tag{2-2}$$

($n = 1, 2, 3, \ldots$) is met. By introducing the "wave vector" $\boldsymbol{k}$, which points in the direction of propagation of the wave and whose magnitude is $|\boldsymbol{k}| = 1/\lambda$, the Bragg equation becomes

$$k \cdot \sin \vartheta = n/2d\,,$$

or in vector form (Fig. 2-5a)

$$\boldsymbol{k} - \boldsymbol{k}_0 = \boldsymbol{g}\,, \quad \text{or} \quad \boldsymbol{k}_0 + \boldsymbol{g} = \boldsymbol{k}\,. \tag{2-3}$$

The vector $\boldsymbol{g}$ ($|\boldsymbol{g}| = 1/d_{hkl}$) is normal to the lattice plane (hkl) with interplanar spacing d. High-order reflections are taken into account by multiples of h, k and l. The vectors $\boldsymbol{g}$ may be used to set up the reciprocal lattice. The various points in the reciprocal lattice then correspond to the various planes of the direct crystal lattice. Analogous to the lattice vector $\boldsymbol{r}$ (equa-

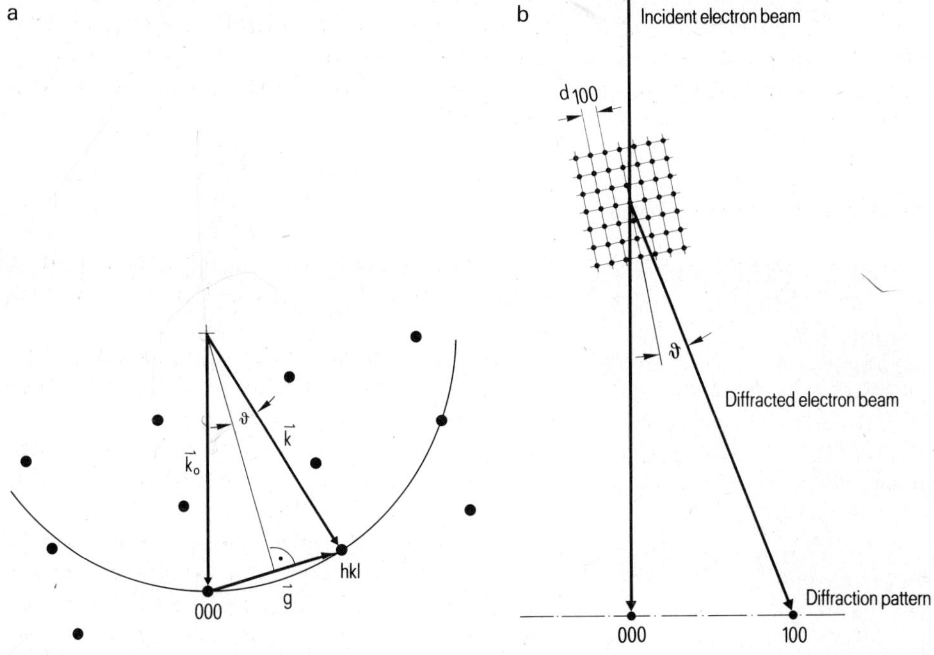

Fig. 2-5. a) The Bragg condition, represented by the wave vectors $\boldsymbol{k}_0$, $\boldsymbol{k}$, and the vector $\boldsymbol{g}$ (Ewald construction). From the Bragg equation (2-2) follows, with $|\boldsymbol{k}_0| = |\boldsymbol{k}| = 1/\lambda$, equation (2-3). The point hkl is the reciprocal lattice point assigned to the set of lattice planes (hkl). It is the end point of the vector $\boldsymbol{g}$ ($|\boldsymbol{g}| = 1/d_{\text{hkl}}$). A set of lattice planes fulfils the Bragg condition only if the reciprocal lattice point hkl assigned to it lies on the surface of the Ewald sphere (see text). b) The electron diffraction diagram represents an almost planar section through the reciprocal lattice. The undiffracted electron beam defines the point 000 in the diffraction pattern plane. This point corresponds to the origin of the reciprocal lattice. The diffraction points correspond to the points of the reciprocal lattice.

tion 2-1), the reciprocal lattice vector g is built up by the three reciprocal translation vectors a^*, b^* and c^*:

$$g = h \cdot a^* + k \cdot b^* + l \cdot c^* . \tag{2-4}$$

The vectors of the reciprocal lattice and the direct lattice vectors fulfil the relationships:

$$\begin{aligned} a^* \cdot a &= b^* \cdot b = c^* \cdot c = 1 , \\ a^* \cdot b &= a^* \cdot c = b^* \cdot a = b^* \cdot c = c^* \cdot a = c^* \cdot b = 0 . \end{aligned} \tag{2-5}$$

g allows every point in the reciprocal lattice to be accessed. In the reciprocal lattice, the unit of the linear dimensions is the reciprocal length (m^{-1}). From equation (2-3) and the reciprocal lattice, the possible diffraction maxima can be determined by a simple geometrical construction: a circle with radius $k = 1/\lambda$ (in the three-dimensional reciprocal space this is a sphere, known as the Ewald sphere) is drawn around the starting point of the vector k_0, the apex of the vector k_0 lying at the origin 000 of the reciprocal lattice (Fig. 2-5a). At the point where the circle (or spherical surface) passes through a point of the reciprocal lattice, the Bragg condition is fulfilled, i.e. a diffraction maximum occurs (Ewald construction). For very short-wave-length radiation as in electron beams with electron energies of the order of 10^4 to 10^5 eV, the radius of the Ewald sphere is very large. Consequently, the electron diffraction pattern in a recording plane normal to the incident beam will approximately represent a planar section through the reciprocal lattice (Fig. 2-5b).

Bonding forces in the crystal

The forces which keep the atoms, ions or molecules in a crystal lattice together are electrostatic Coulomb forces. Depending on the type of bonding, a distinction is made between ionic, atomic, metallic and molecular lattices.

In the *ionic lattice* the bonds are directly due to the different polarities of the lattice particles. Thus in NaCl sodium is the positive and chlorine the negative ion. A stable bond between the initially neutral sodium and chlorine atoms is set up because the energy gained by the Coulomb force of attraction is greater than that required to separate an electron from the Na atom and to attach it to the Cl atom. The net energy gain is the binding energy, which has a value of 6.4 eV per ion for NaCl.

Atoms of the same or similar type are also bound by electrostatic forces. The bonds in an *atomic lattice* are called covalent, homeopolar or valence bonds. The condition for this type of bond is that in adjacent atoms free electron orbits are present in an outer shell. The mutual approach of two atoms results in an overlap of the electron shells, which can proceed to the point where the electrons move in an orbit common to both atoms. The energy gain resulting from this arrangement is the binding energy. Typical valence crystals are formed by carbon and the semiconductors Si, Ge and SiC. The binding energies of covalently bound atomic crystals lie between 1 and 7 eV. In most cases bonds have both heteropolar and covalent character.

The *metallic bond* differs from the atomic bond in that the electrons cannot be permanently assigned to individual atomic pairs, but belong to the lattice as a whole. These electrons move more or less freely as an "electron gas" between the ions so that the overall system reaches an energy minimum. These free electrons are responsible for a number of features, including good electrical conductivity and other typically metallic properties. The bonds of the metal atoms result from the Coulomb interaction between the ions and the electrons moving freely between them. The binding energies have values between 1 and 8 eV.

Molecules with largely saturated electron shells are also held together in the crystal by – albeit very weak – Coulomb forces, provided that they have permanent or temporarily fluctuating dipole moments. The dipole moment of a lattice molecule induces an opposing moment in its neighbor. The resulting Van der Waals forces hold the crystal together. The molecular lattice is the form of crystallization of many organic compounds, and – at very low temperatures – of inert gases as well. The binding energies have values between 0.01 and 0.2 eV.

Lattice defects

Defect-free, ideal crystal lattices do not exist in nature. Real crystals exhibit irregularities in their lattice structures. Lattice defects may be point defects (vacancies, interstitials, impurities), linear defects (dislocations), planar defects (grain or twin boundaries, stacking faults) or three dimensional defects (precipitates, pores) (Fig. 2-6). They are responsible for many material properties such as conductivity (especially in semiconductors), colors of crystals, ductility of metals, or hardness of materials. Material properties can be modified in a desired manner by the deliberate incorporation of specific lattice defects.

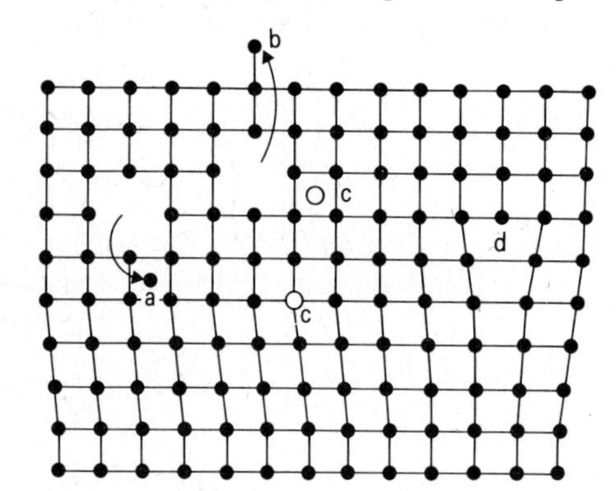

Fig. 2-6. Lattice defects:
a) Frenkel defect: an atom leaves its lattice site, leaving behind a vacancy and occupies an interstitial site.
b) Schottky defect: an atom leaves the lattice and migrates to the crystal surface.
c) Impurity atoms can also occupy interstitial or lattice sites.
d) Dislocations: the boundary of an additional lattice plane inserted in the crystal forms a dislocation line, which here runs normal to the paper. In the region surrounding the dislocation region, the crystal is elastically distorted.

2.1.2 Energy states in solids

The electrons of a single, free atom occupy specific, sharply defined energy states. These energy states can be represented graphically as orbital paths around the positive nucleus along which the electrons move (Bohr atomic model, Fig. 2-7a). Electrons in inner shells are more strongly bound to the nucleus than outer electrons, because the latter are further away from the nucleus and also partly shielded from the nuclear field by the electrons located more closely to the nucleus. The potential well model (Fig. 2-7b) is derived from this.

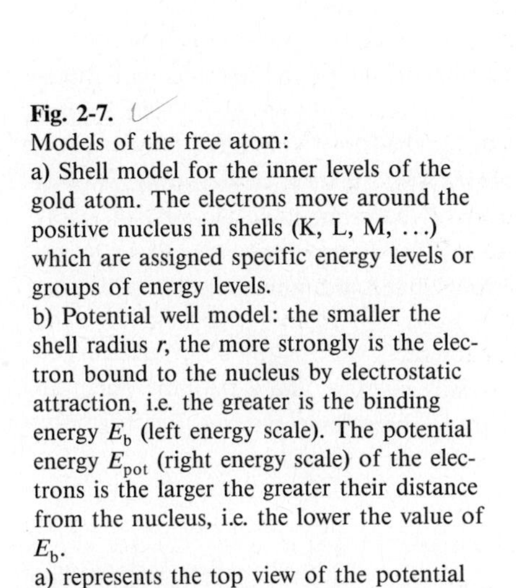

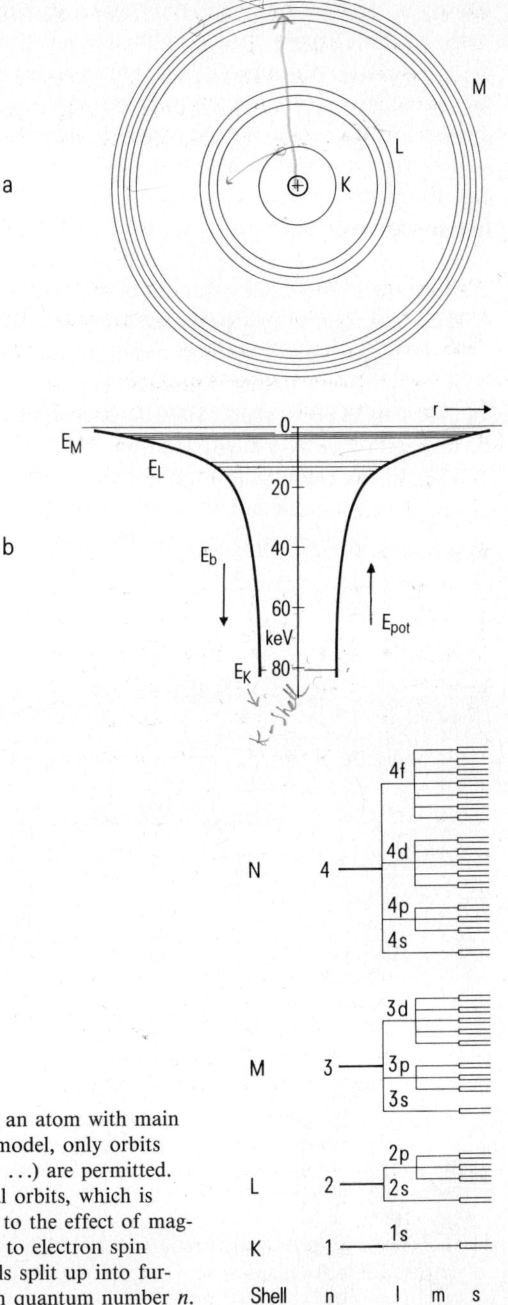

Fig. 2-7.
Models of the free atom:
a) Shell model for the inner levels of the gold atom. The electrons move around the positive nucleus in shells (K, L, M, ...) which are assigned specific energy levels or groups of energy levels.
b) Potential well model: the smaller the shell radius r, the more strongly is the electron bound to the nucleus by electrostatic attraction, i.e. the greater is the binding energy E_b (left energy scale). The potential energy E_{pot} (right energy scale) of the electrons is the larger the greater their distance from the nucleus, i.e. the lower the value of E_b.
a) represents the top view of the potential well shown in b).

Fig. 2-8.
Schematic representation of the energy levels in an atom with main quantum number $n = 4$. According to Bohr's model, only orbits with angular momentum $n \cdot h/2\pi$ ($n = 1, 2, 3, \ldots$) are permitted. Due to the eccentricity of the generally elliptical orbits, which is described by the secondary quantum number l, to the effect of magnetic fields (magnetic quantum number m) and to electron spin (spin quantum number s), the main energy levels split up into further levels. The splitting increases with the main quantum number n.

To remove an electron from its shell requires energy, the more of it the more strongly the electron is bound to the nucleus. Energy must therefore be supplied to an atom if an electron is to be "raised" from an inner shell to a shell lying further out. An atom "excited" in this way is in a higher energetic state than it was previously. It will have the tendency to relax as quickly as possible (in about 10^{-15} s), i.e. to return to its ground state again. This happens by an electron from a shell lying further out from the nucleus filling the "hole" created by the excitation. In this process, energy is released which was previously needed to generate the excitation. It is emitted in the form of radiation (X-ray or light radiation) or by emission of another electron from the atom (Auger electron, sections 2.4.3 and 6.1).

However, a complete explanation of these atomic processes can be given only with the methods of quantum physics. These describe the energy states which the electrons can occupy by four quantum numbers: n, l, m, s (Fig. 2-8). The electrons belonging to the same main quantum number n are combined to shells designated by the letters K, L, M, N, ... The K shell corresponds to the main quantum number $n = 1$, the L shell to $n = 2$ etc.

The occupation of the energy levels by electrons begins with the orbit closest to the nucleus ($n = 1$) and continues with increasing atomic number of the elements in the periodic table in accordance with specific rules. The number of electrons per energy level is regulated by the Pauli principle: every energy state designated by the four quantum numbers n, l, m and s can be occupied only by a single electron.

When atoms combine to form a lattice, every atom interacts with all the other atoms. The potential wells of the individual atoms (Fig. 2-7) partially overlap, and a periodic potential distribution is obtained (Fig. 2-9). The period is equal to the atomic spacing. Furthermore,

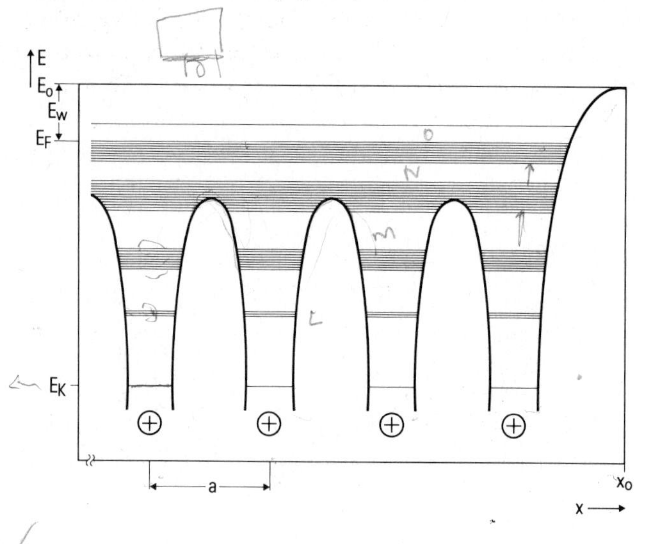

Fig. 2-9. Energy bands in the periodic potential distribution of an "unidimensional" crystal lattice. The energy levels occupied by electrons are represented by the series of lines. The upper limit of the occupied states is the Fermi level E_F. The uppermost band, which is here – as in a metal – only partly occupied, is hardly influenced by the periodic potential. The electrons in this band can move freely. Forbidden gaps exist between the bands, in analogy to the regions between the electron levels in the free atom. They cannot be occupied by electrons. At the crystal surface (right in the diagram at $x = x_0$) the potential reaches a value of E_0 (vacuum level), which exceeds the Fermi level by the amount of the work function E_W.

the close coupling of the atoms in the lattice leads, like the coupling of oscillating systems in mechanics, to the energy levels splitting into sublevels. Since every atom interacts with very many neighbors (there are about 10^{23} atoms in a cm^3), the sublevels are extremely numerous and they lie very close together. For the inner shells, however, this splitting is negligibly small. The results obtained by methods of analysis in which the inner levels are excited by energy transfer (e.g. X-ray microanalysis, Auger microanalysis) are therefore independent, to a first approximation, of the binding state of the investigated material.

In contrast to this, the energy states of the outer electrons are affected very strongly when the atoms combine to form lattices. The many sublevels which lie very close together form the energy bands. The width of these energy bands, in which the energy changes quasi-continuously, is the greater the lower the binding strength of the electrons to the nucleus. In the uppermost band, the electrons are hardly affected any more by the potential of the individual atom. This applies especially to the metals, in which the electrons are no longer assigned to the individual atom but to the entire crystal (crystal electrons) and can move more or less freely. For a description of many of a crystal's properties (e.g. electric and thermal conductivity), it is therefore often sufficient to consider only the energy states of the outermost bands.

Calculations using the methods of quantum mechanics show that the energy bands in the crystal can be very complex and are a function of the crystal structure and the type of atom. Many fundamental properties of solids can, however, already be clearly understood on the basis of the simplest models (potential well model of the crystal, Fig. 2-10; band diagram, Fig. 2-11). The electrons in the uppermost band (conduction band) surround the positive ions of the lattice like the particles of a gas. The density of this electron gas is, however, very high. If we assume as an approximation that for every metal atom one free electron contributes to the electron gas, the density of the electron gas is several thousand times greater than for a gas consisting of neutral particles. A "degenerated" gas of this kind is called a Fermi gas.

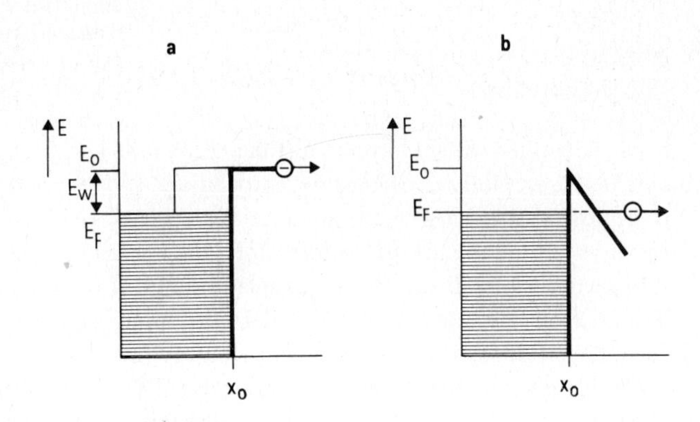

Fig. 2-10. Simplified potential well model of a crystal to explain electron emission.
a) Thermionic emission: by heating up the electron gas with thermal or photon energy, the energy of the crystal electrons can be increased to the point where they overcome the work function E_W and leave the crystal.
b) Field emission: an electric field applied to the crystal surface (at x_0) leads to a potential gradient. A potential barrier is set up. With a sufficiently high field, the quantum-mechanical tunnel effect enables the electrons to leave the crystal into the vacuum through the narrow potential barrier.

The permitted energy levels in the conduction band are occupied with electrons to an upper limit, which depends on the total number of electrons. The maximum energy of the occupied states at absolute zero temperature ($T = 0$ K) is the Fermi energy E_F. It can be calculated for all metals from the known electron concentration and electron mass data.

At temperatures $T > 0$ electrons slightly below E_F can be raised beyond E_F. The states below E_F become free, those above E_F are occupied. By supplying a sufficiently high energy (e.g. by heating the metal or irradiating it with high-energy photons), the electrons can be excited so much that they overcome the work function E_W (Fig. 2-10a) and leave the crystal. The result is thermionic emission or photo-emission. Crystal electrons can leave the solid also through the action of a sufficiently high electrostatic field (Fig. 2-10b). The external field gives rise to a potential gradient at the metal surface. The electrons can tunnel through the resulting potential barrier (quantum mechanical tunnel effect).

The two outermost bands are the most important ones for determining the properties of the solid: the conduction band and the valence band (Fig. 2-11). If the conduction band is only partially filled with electrons, i.e. if the Fermi level lies within this band, then states are available in close vicinity above E_F. At temperatures $T > 0$ electrons can occupy these levels and transport electric charge in the presence of an electric field. Solids with only partially occupied upper bands are the metals. In insulators, the conduction band is empty. The fully occupied valence band lying below this contains the states of the bound valence electrons. The energy gap between both bands which represents forbidden states for the electrons is relatively large. In a diamond, for instance, it has a value of 7 eV.

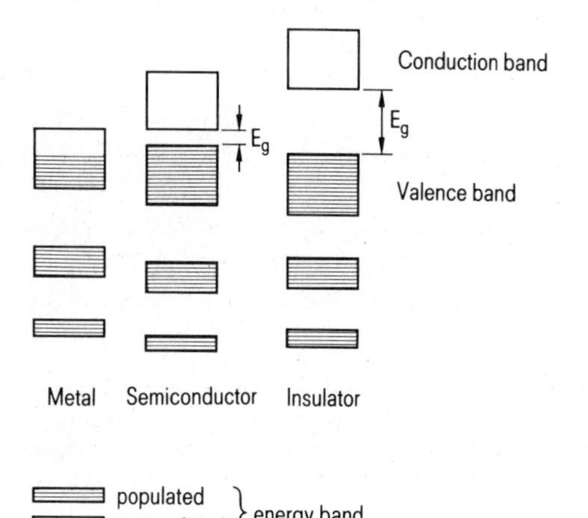

Fig. 2-11. Simplified band diagram for a conductor, a semiconductor and an insulator. The forbidden gap between the valence and conduction bands is marked by E_g (energy gap).

Semiconductors have a smaller band gap (Si 1.1 eV, Ge 0.6 eV). The transfer of sufficient thermal energy therefore allows valence electrons to reach the conduction band (Fig. 2-12a), this effect being equivalent to the ionization of atoms. The number of electrons reaching the conduction band increases exponentially with temperature. The conductivity (called *intrinsic conductivity*) therefore increases with temperature, a typical property of undoped semiconductors. In metals, however, the conductivity decreases with increasing temperature, since the mobility of the electrons is reduced by the thermal vibrations of the lattice atoms.

The prohibition against occupying energy states within the energy gap applies only to the electrons of the semiconductor atoms. Impurity atoms can occupy "forbidden states" within the gap and contribute to the conductivity; this is known as *extrinsic conductivity.* Impurity atoms which are positively ionized by transfer of energy act as electron donors and make the semiconductor n-conducting (Fig. 2-12b). Impurity atoms which are negatively ionized act as electron acceptors and make the semiconductor p-conducting (Fig. 2-12c). Both mechanisms are exploited for doping semiconductors. The energy levels of the doping atoms lie close to the band edges as shown in Fig. 2-12b,c. A low energy of only a few 0.01 eV is therefore required for ionizing the dopants. Thus doped semiconductors are already conductive at room temperature. Phosphorus, arsenic and antimony are examples of donors and boron of an acceptor in the case of silicon. Many more charge carriers are generated by doping than are present in the pure semiconductor crystal. In n-type material the electrons are called *majority carriers* and the holes *minority carriers.* In an analogous way, the holes are majority carriers and the electrons minority carriers in p-type material.

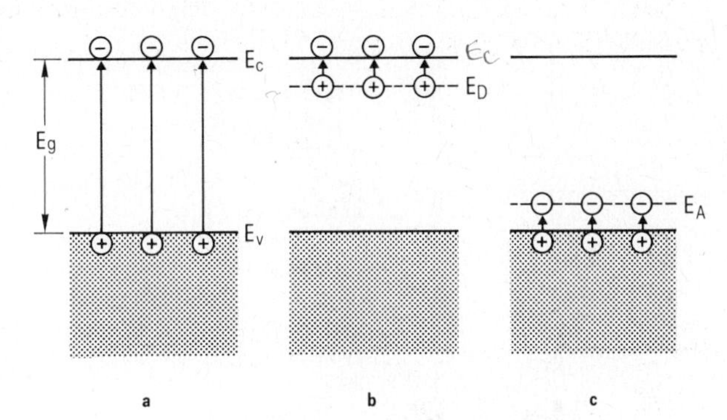

Fig. 2-12. Conductivity in a semiconductor, explained by the energy diagram (E_v upper edge of valence band, E_c lower edge of conduction band, E_g energy gap, E_D donor level, E_A acceptor level).
a) *Intrinsic conductivity:* with an energy transfer $\Delta E \geq E_g$, electrons are lifted from the valence band to the conduction band, where they move freely as do the empty states (holes) in the valence band.
b) *Extrinsic n-conductivity:* with a small energy transfer $\Delta E \geq E_C - E_D$, the donors are ionized and give electrons to the conduction band, where these move freely; the material is n-conducting.
c) *Extrinsic p-conductivity:* with a small energy transfer $\Delta E \geq E_A - E_v$, the acceptors take up electrons from the valence band. The holes (empty states) act as positive charge carriers and can move freely in the valence band; the material is p-conducting.

2.1.3 *pn-junctions in semiconductors*

The boundary between oppositely doped regions in a semiconductor is called a pn-junction; a diode is a device with one pn-junction. The principle of a pn-junction in thermodynamic equilibrium is shown schematically in Fig. 2-13: when p- and n-doped regions are in intimate contact, the majority carriers on both sides diffuse across the interface (Fig. 2-13c) due to the great concentration difference prevailing there: electrons move from the n side to the p side of the interface where they become minority carriers and as a rule will recombine with holes which are present there as majority carriers. Holes diffusing across the interface to the n-region behave analogously.

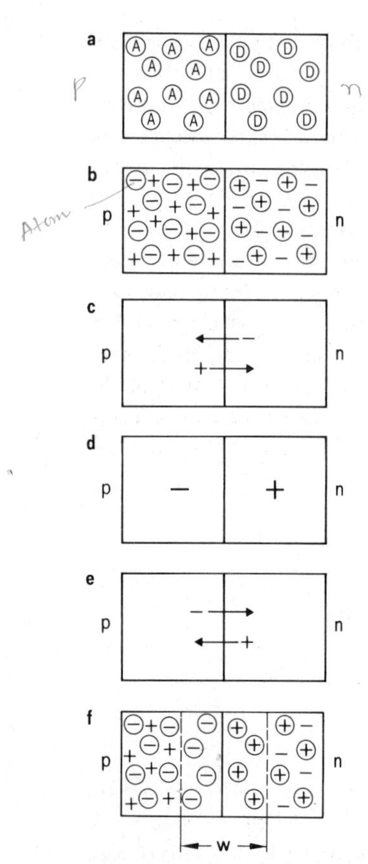

Fig. 2-13.
Principles of a pn-junction in a semiconductor (the various effects are illustrated here step by step; in reality they act together simultaneously):
a) A region doped with acceptors (A) borders abruptly on a region doped with donors (D).
b) The doping atoms are completely ionized: equal numbers of fixed negative ions (⊖) and of mobile holes (+) are produced on the left, this side becoming p-conducting. On the other side equal numbers of fixed positive ions (⊕) and of mobile electrons (−) are produced, resulting in n-conductivity. Both regions are still electrically neutral.
c) Diffusion currents flow due to the high concentration differences of the respective mobile (majority) charge carriers.
d) Because of the diffusion, the two regions are oppositely charged and an electric field therefore is built up around the interface.
e) Drift currents of minority carriers are generated by this electric field.
f) In equilibrium, drift and diffusion currents balance and no net current flows across the interface. On both sides of the junction the semiconductor suffers a depletion of majority carriers; a depletion zone of width w, which acts as a barrier layer, has been produced.

This diffusion process results in a potential difference (Fig. 2-13d) which drives drift currents of minority carriers in opposite directions (Fig. 2-13e). As soon as this potential difference has reached a specific value, the *diffusion voltage*, the diffusion and drift currents exactly balance; the diode is then in thermodynamic equilibrium. As a result of the diffusion and recombination processes, a zone on both sides of the interface is depleted of majority carriers (Fig. 2-13f). Within a *depletion zone* of width w, the fixed charges of ionized doping atoms are not compensated for by the respective majority carriers. A negative space charge is formed on the p side and a positive space charge on the n side of the interface. Integrating the electric field across the width w of the depletion or *space charge region* results in the diffusion voltage.

The diffusion voltage depends on the doping concentrations on both sides of the junction and assumes values between 0.5 V and 1.0 V at room temperature in the case of silicon. In thermodynamic equilibrium, no current flows (despite the diffusion voltage at the pn-junction) when the diode is connected to an external circuit. This is because contact voltage differences, which exactly compensate the diffusion voltage, arise at the contacts required for this circuit. If, however, the thermodynamic equilibrium is disturbed by irradiating the junction with electrons, ions or photons, a non-equilibrium voltage difference results at the pn-junction and an external current flows. This current can be used for analytical purposes, namely for characterizing semiconductor crystals (section 3.6.3) or detecting the radiation falling onto it (section 5.2.4).

Abrupt transitions, such as that assumed in Fig. 2-13, can be implemented only by means of special techniques. In most cases, pn-junctions are fabricated by diffusing or implanting acceptors into n-doped material or donors into p-doped material. The resulting junctions are continuous. This affects the width of the depletion zone and the magnitude of the electric field at the pn-junction, but not the fundamental principles illustrated in Fig. 2-13.

The geometrical position in the semiconductor at which the concentrations of donors and acceptors are identical is known as the *metallurgical* pn-junction (see Fig. 7-19). The *electrical* pn-junction is defined as the position at which the positive and negative free charge carriers have equal concentrations. Both junctions defined in this way are identical only in the case of symmetrical junctions (equal concentrations of acceptors and donors an both sides of the interface). In other cases they differ more or less depending on the difference between acceptor and donor concentrations and a careful distinction must therefore be made between the two types.

2.2 Charged particles

2.2.1 Forces acting on charged particles

Particles move in straight lines in a field-free vacuum. If we wish to use particles to set up measuring probes which can be guided to defined positions on a specimen, we need some means of influencing this rectilinear particle path. In the case of charged particles, electric and magnetic fields may be used for this purpose. Charged particles may be deflected or accelerated by the forces exerted by these fields.

In accordance with the fundamental relationship

$$\boldsymbol{F} = q \cdot \boldsymbol{E} \tag{2-6}$$

a charge q in an electric field of strength $\boldsymbol{E}$ experiences a force $\boldsymbol{F}$ resulting from the electrostatic interaction of charges (Coulomb force). In the case of positively charged particles, this force acts in the direction of the field, in the case of negative ones (e.g. electrons) in the opposite direction. The direction of a field, such as that existing between two electrodes of different polarity, is defined as running from the positive to the negative electrode. The variation of the energy W_{kin} and speed v of a charged particle (mass m, charge q) with the field strength E_0 may be easily explained with the aid of a model of a plate capacitor (plate distance d, potential difference U_0). Extending between the plates is the homogeneous field $E_0 = U_0/d$. A charge carrier starting at the electrode (anode) with a potential $U = 0$ is accelerated by a force $F = q \cdot E_0$. According to the law of conservation of energy, the work expended in traveling a distance d in the field (potential energy)

$$W_{\text{pot}} = F \cdot d = q \cdot E_0 \cdot d = q \cdot U_0$$

is equal to the kinetic energy of the charged particle

$$W_{\text{kin}} = (m/2) \cdot v_0^2 \, .$$

From this it follows that the velocity of the charge carrier after traversing the field is

$$v_0 = \sqrt{2 \cdot (q/m) \cdot U_0}$$

and its energy is

$$W_{\text{kin}} = q \cdot U_0 \, .$$

These relationships apply quite generally, i.e. they are independent of the path of the particle. After traversing a potential difference U, a charge q of mass m has a velocity

$$v = \sqrt{2 \cdot (q/m) \cdot U} \tag{2-7}$$

and an energy

$$W_{\text{kin}} = q \cdot U \, . \tag{2-8}$$

For charged particles, it is practical to express their kinetic energy as a function of the elementary charge

$$e = 1.602 \cdot 10^{-19} \text{ A s} \, .$$

Accordingly, a particle of charge e having traversed a potential difference of, say, 100 V has a kinetic energy

$$W_{\text{kin}} = 100 \text{ eV} \, .$$

It is customary to use the letter E instead of W for the energy of charged particles. Here, we have used the letter W to avoid confusion with the field strength E. At a later stage we will use E for the energy when no confusion is likely to occur.

Analytical instruments make use of electrons or ions with energies between 100 and 10^6 eV. 100 eV electrons travel at about 2% of the velocity of light ($c = 2.998 \cdot 10^8 \text{ m} \cdot \text{s}^{-1}$), whereas 1 MeV electrons already reach 98% of c (Fig. 2-14a). Electrons attain these very high velocities as a result of their very small mass

$$m_0 = 9.11 \cdot 10^{-28} \text{ g} \, .$$

Since these velocities become comparable to the speed of light at high voltages, the relativistic mass correction must be taken into account above about 10^4 V:

$$m = m_0 / \sqrt{1 - (v/c)^2}$$

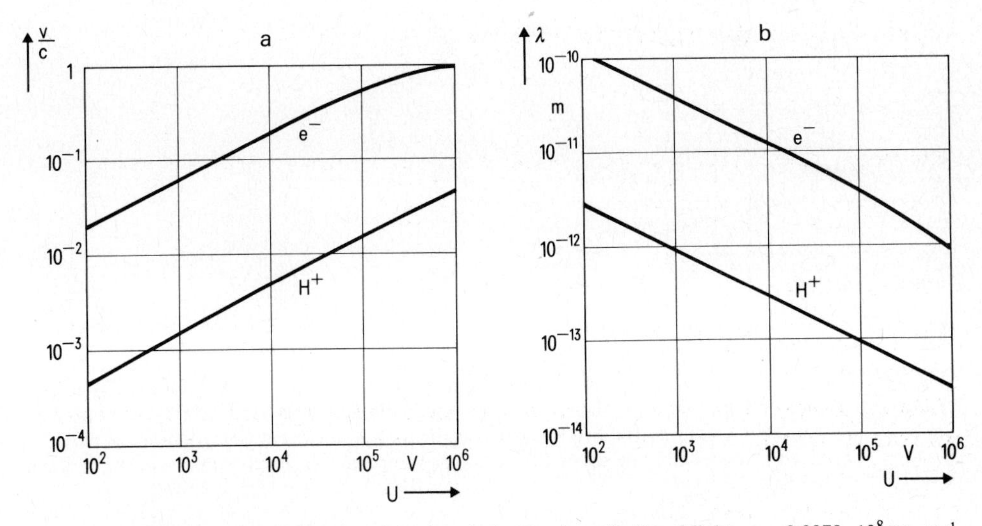

Fig. 2-14. a) Velocity v of charged particles in relation to the velocity of light $c = 2.9979 \cdot 10^8\ \mathrm{m \cdot s^{-1}}$ as a function of the traversed potential difference U.
b) Wavelengths λ of moving charge carriers (de Broglie wavelengths) as a function of the accelerating voltage U (e^-, electrons; H^+, protons).

where m_0 is the rest mass of the electron. Since even the lightest ion, the hydrogen nucleus, has a mass about 1800 times greater than an electron, ions are very much slower; thus 10^6 eV H^+ ions (protons) of mass

$$m_H = 1.67 \cdot 10^{-24}\ \mathrm{g}$$

travel at only about 5% of the velocity of light.

In order to guide moving charge carriers in specific directions, we require electric or magnetic deflection fields. Their effect may be easily explained by considering limited homogeneous fields which can be approximately realized in a plate capacitor or between two current-carrying coils.

If we neglect the peripheral fields of the deflection capacitor and limit ourselves to small angles, then the angle of deflection is given by (Fig. 2-15)

$$\gamma = \frac{U_c \cdot l}{2 \cdot U_0 \cdot d} \,. \tag{2-9}$$

Neither the charge nor the mass of the particle appears in this equation. Consequently, charged particles which have traversed the same accelerating voltage U_0 are deflected in the same way independently of their nature.

Different conditions exist in magnetic fields. A particle of charge q moving in a field of flux density B experiences the Lorentz force

$$\boldsymbol{F} = q \cdot (\boldsymbol{v} \times \boldsymbol{B}) \,. \tag{2-10}$$

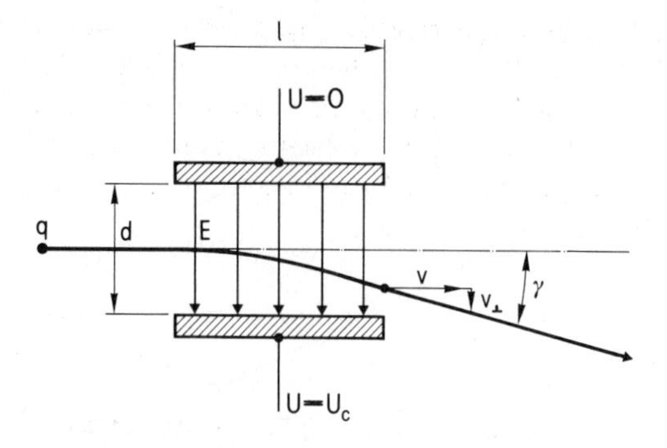

Fig. 2-15. Deflection capacitor: A charge carrier q entering a field $E = U_c/d$ with a velocity v is accelerated during time $t = l/v$ (when passing through a field of length l) by a force $F = q \cdot E = m \cdot a$ in the direction of the field (where m is the mass and a the acceleration of the particle). When leaving the field, the particle therefore picks up the velocity component in the direction of the field

$$v_\perp = a \cdot t = \frac{q}{m} \cdot E \cdot \frac{l}{v}.$$

From this it follows that the angle of deflection, which is independent of the nature of the particle, is given by

$$\tan \gamma = \frac{v_\perp}{v} = \frac{q}{m} \cdot E \cdot \frac{l}{v^2} = \frac{q \cdot U_c \cdot l}{m \cdot v^2 \cdot d} = \frac{U_c \cdot l}{2 \cdot U_0 \cdot d}$$

($v^2 = 2 \cdot (q/m) \cdot U_0$, see eq. (2-7)).

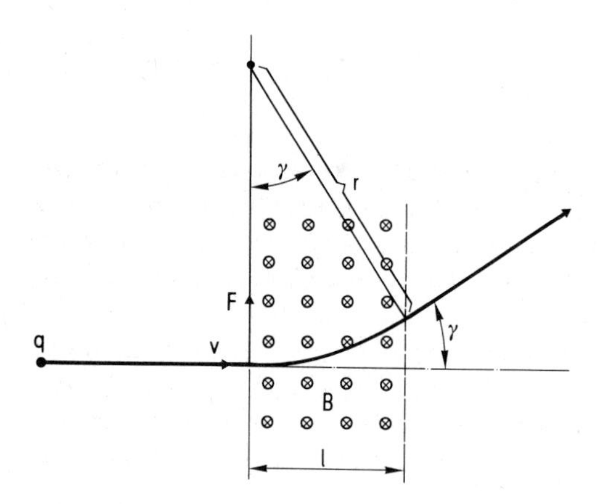

Fig. 2-16. Deflection in a magnetic field: In a field of flux density $\boldsymbol{B}$ running normal to $\boldsymbol{v}$, a charge carrier q with velocity v experiences a radial force $F = m \cdot (v^2/r)$, which makes it travel a circular path of radius r. With $F = q \cdot v \cdot B = m \cdot (v^2/r)$ the angle of deflection ($\gamma \ll 1$, field length l) is given by:

$$\gamma = \frac{l}{r} = \frac{q \cdot l \cdot B}{m \cdot v} = \sqrt{\frac{q}{m}} \cdot \frac{l \cdot B}{\sqrt{2 \cdot U_0}}$$

($v^2/r = a_r$ radial acceleration).

Accordingly, only the components $v_\perp$ and $B_\perp$ lying perpendicular to each other produce a force $F_\perp$ which acts perpendicular to both $v_\perp$ and $B_\perp$ (vector product). $\boldsymbol{F}$ is thus a radial force which causes only a change of direction, but not of velocity. The energy of a particle therefore remains unchanged in a magnetic field. A charged particle entering a homogeneous magnetic field perpendicularly moves in a circular path with constant velocity. In a field limited to a length l (Fig. 2-16), it experiences a deflection

$$\gamma = \sqrt{\frac{q}{m}} \cdot \frac{l \cdot B}{\sqrt{2 \cdot U_0}} \tag{2-11}$$

which depends on the ratio of its charge to its mass. As in the case of deflection in an electric field, the deflection angle is smaller the higher the energy of the particle.

Moving particles do not only exhibit properties which can be described in terms of mass points as in mechanics. Their properties also lead to phenomena such as diffraction and interference, which can only be explained by a wave propagation model. Thus, according to the laws of quantum mechanics, a particle with momentum $m \cdot v$ is assigned a wave of wavelength

$$\lambda = h/(m \cdot v)$$

(de Broglie wavelength, where h is Planck's constant). Together with equation (2-7), this yields

$$\lambda = h/\sqrt{2 \cdot q \cdot m \cdot U} \tag{2-12}$$

for the wavelength of the particle waves. Accordingly, electrons with an energy of 10 keV have a wavelength of 12 pm, protons with the same energy one of only 0.3 pm (Fig. 2-14b).

2.2.2 *Focusing of charged particles*

To produce a particle beam probe which can be concentrated onto an area of minimum size, the particles must be focused to a beam of smallest possible diameter. This can be done by using rotationally symmetric electric and magnetic fields. These fields act on beams of charged particles in a way similar to the action of glass lenses on rays of light, and the laws of optics can be applied also to particle beams.

An object point P_0 is imaged at an image point when the rays emerging from it at various angles α are recombined at the image point P_i by the optical system (the lens). This is obtained when the angle of deflection γ, due to the lens, increases proportional to the distance r of the rays from the optical axis. For sufficiently small angles, this relationship can be expressed by the imaging equation (Fig. 2-17)

$$\frac{1}{a} + \frac{1}{b} = \frac{1}{f}. \tag{2-13}$$

This is a fundamental equation in geometrical optics for the most important parameters of a lens: focal length f, object distance a and image distance b. In light optics, the imaging equation is valid for glass lenses with spherical surfaces. The image size for a given optical system

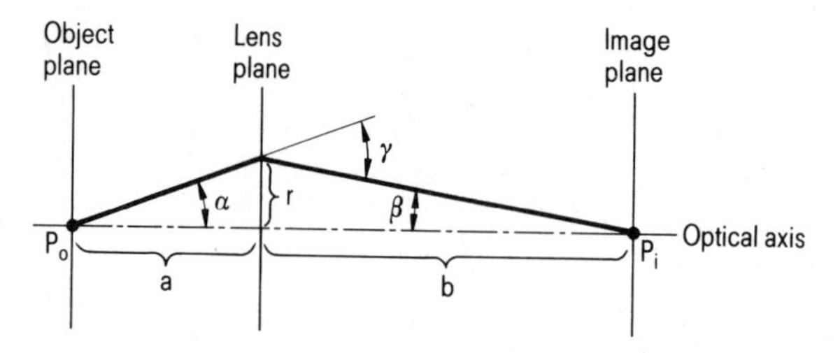

Fig. 2-17. Optical imaging: Rays emerging from the object point P_0 at various angles α are focused by the lens to an image point P_i when $\gamma = \text{const} \cdot r$. For small angles α and β this condition leads to the relationship

$$\gamma = \alpha + \beta = r \cdot \left[\frac{1}{a} + \frac{1}{b}\right].$$

This yields the imaging equation:

$$\frac{\gamma}{r} = \text{const.} = \frac{1}{a} + \frac{1}{b} + \frac{1}{f}$$

where f is the focal length of the lens. Further, from $\alpha = r/a$ and $\beta = r/b$ it follows that $\alpha : \beta = b : a$.

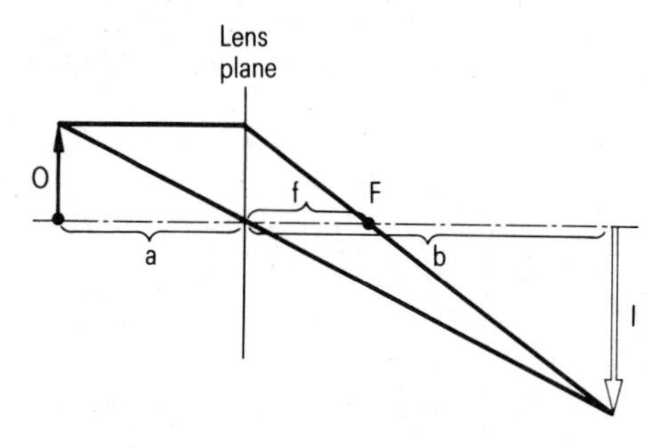

Fig. 2-18. Image construction: From $O/I = f/(b - f)$ and $O/I = a/b = 1/M$ where M is the imaging scale, it follows that $f/(b - f) = a/b$ which again yields the imaging equation (Fig. 2-17): $(1/a) + (1/b) = (1/f)$, where f is the focal length.

and object may be determined from this imaging equation and the construction shown in Fig. 2-18. The ratio

$$\frac{\text{object size}}{\text{image size}} = \frac{O}{I} = \frac{a}{b} = \frac{1}{M} \tag{2-14}$$

is defined as the image scale M. For $M > 1$, the imaging scale is also called the magnification, for $M < 1$ the demagnification.

The fundamental condition for these imaging or focusing properties

$$\gamma \sim r$$

is fulfilled by rotationally symmetrical electric and magnetic fields, provided that the deflection angles are sufficiently small and the rays run close to the optical axis (paraxial rays). This is illustrated in Fig. 2-19 with the example of a rotationally symmetrical electrical field extend-

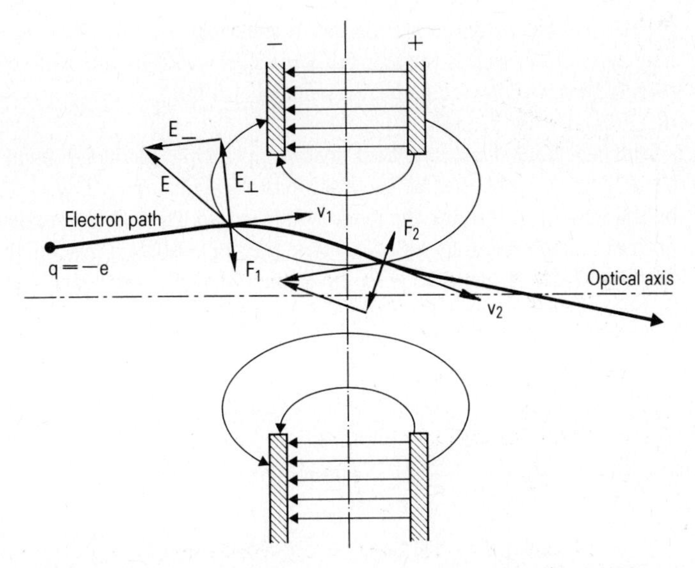

Fig. 2-19. Two-electrode electrostatic aperture lens: On the electron (charge e) a force $\boldsymbol{F} = -e \cdot \boldsymbol{E}$ is exerted. The component E_- acting in the direction of the electron velocity v_1 accelerates the electron and the component $E_\perp$ normal to it deflects it towards the axis. After the middle of the lens has been traversed, the deflection is now of opposite direction but is less effective due to the increased electron velocity v_2 so that the net result is a deflection directed towards the axis.

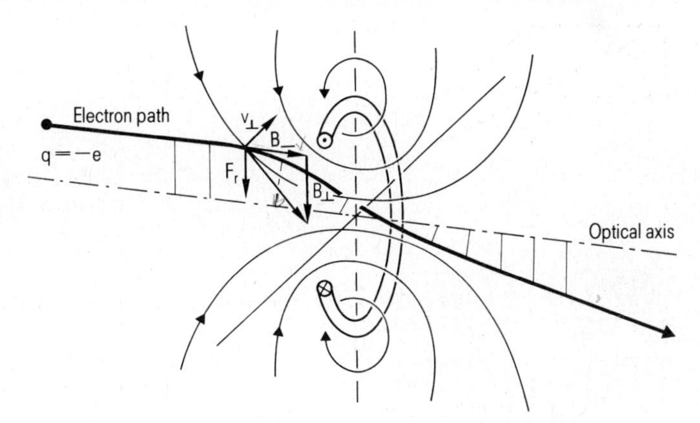

Fig. 2-20. Principle of the magnetic lens: On the electron entering the lens field from the left with a velocity $\boldsymbol{v}$ a force $\boldsymbol{F} = e \cdot (\boldsymbol{B} \times \boldsymbol{v})$ is exerted. The field component $B_\perp$ being perpendicular to $\boldsymbol{v}$ results in a velocity component $v_\perp$. The electron therefore moves in a spiral path. Since $v_\perp$ is also perpendicular to B_-, this additionally gives rise to a deflecting force F_r towards the axis. This force produces the lens effect.

ing between two circular apertures of equal size and with a common axis. It is generated by bringing the apertures to different potentials. The Coulomb force produces not only a deflection towards the axis which is proportional to the distance of the particle from the axis but also deceleration and acceleration of the particle within the lens field. Outside the lens field, the particle has different speeds before entering and after leaving the field. Fig. 2-20 shows how deflection towards the axis takes place in a rotationally symmetrical magnetic field. The field is generated by a circular current. Due to the Lorentz force, the particle moves through the field in a spiral path.

The "cardinal elements" focal position, principal plane and focal length can be defined for particle beam lenses just as for glass lenses in light optics (Fig. 2-21). The focal lengths may be changed by modifying the fields within broad limits. The practical designs of the most important particle beam lenses of the type used in analytical instruments are shown schematically in Fig. 2-22. Minimal focal lengths which lie within the order of magnitude of the electrode

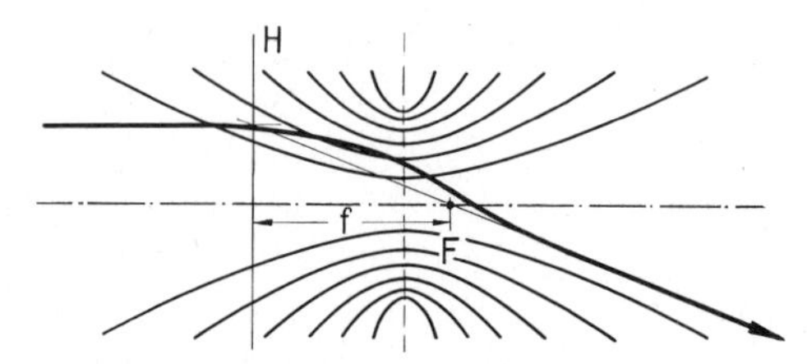

Fig. 2-21. Determination of focal point F and focal length f of a particle beam lens from the path of an electron incident parallel to the axis. The extension of the rectilinear part of the path after passing the lens field intersects the optical axis at the focal point F and the path parallel to the axis at the (front) principal plane H of the lens. The focal length f is the distance between the focal point and the principal plane.

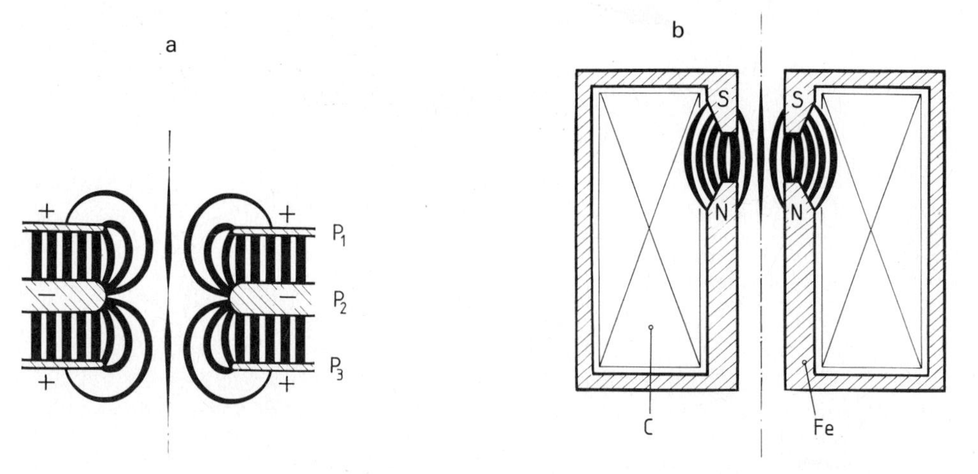

Fig. 2-22. Particle beam lenses of different design and their field distribution. The direction of the lines shows the local field direction, their thickness the local field strength. Short focal lengths and the resulting low equipment dimensions are obtained by compressing the imaging fields into a small volume.

a) Electrostatic einzel lens: the two outer electrodes P_1 and P_3 (diaphrams) are brought as close to the middle electrode P_2 as is permitted by the insulating strength of the constructional elements.

b) Magnetic pole piece lens: the field-generating current-carrying coil C is surrounded by an iron jacket Fe with only a narrow air gap (pole piece) in which the field is concentrated.

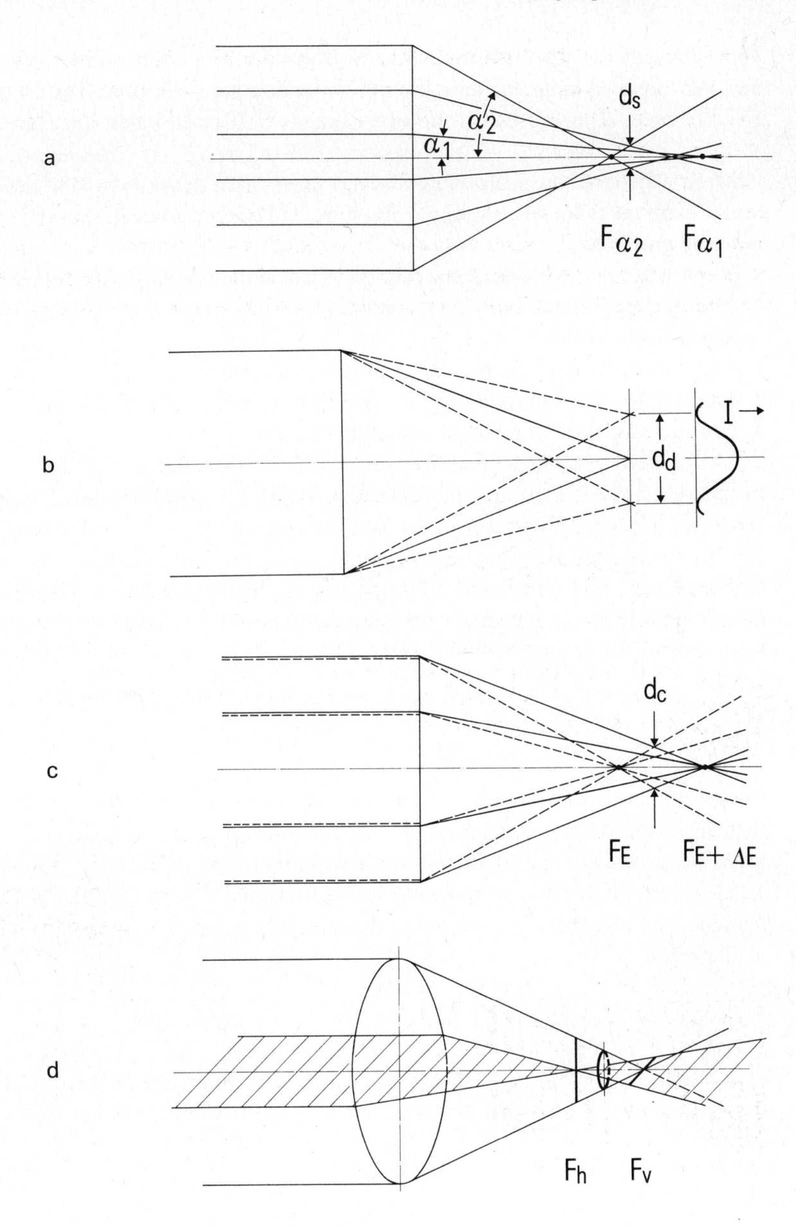

Fig. 2-23. The four most important imaging errors in particle beam lenses:
a) Spherical aberration: the focal point $F_{\alpha 2}$ of rays with larger angles of divergence α_2 (off-axis rays) is closer to the lens than that for paraxial rays ($F_{\alpha 1}$).
b) Diffraction error: the limitation of the wave front by the lens edge or apertures results in an intensity distribution I in the focal plane caused by diffraction.
c) Chromatic aberration: beams with different energies E and $E + \Delta E$, i.e. different wavelengths ("colors") are focused at different focal points F_E and $F_{E+\Delta E}$.
d) Astigmatism: due to the unavoidable deviation from rotational symmetry, two focal lines F_h and F_v perpendicular to each other are obtained instead of one focal point.

bore diameter can be obtained with the triple-electrode electrostatic einzel lens. With the iron-encapsulated magnetic lens, the minimum attainable focal lengths for 100 keV electrons lie in the order of magnitude of the bore diameter of the pole piece. The focal lengths are longer for ion beams due to the greater mass involved (cf. eq. (2-11)). These lenses act as converging lenses provided that no space charges occur. The current densities used in the lenses of particle beam instruments for analytical purposes are, in general, so small that the space charge does not play any role.

In our treatment of optical imaging up to this point, we made the following assumptions:
1. The imaging beams have a very small angle of divergence and run very close to the optical axis (paraxial rays).
2. The wave nature of the particle beams is neglected.
3. The energy of the particles is strictly uniform (monochromatic beams).
4. The imaging fields have ideal rotational symmetry.
5. The charge carriers do not affect each other by Coulomb interaction.

In reality, however, these assumptions were valid only approximately. As a result, the rays leaving an object point are not recombined at an image point but fill out an area of finite extension. This deviation from point-to-point imaging is called aberration. The limitations specified in points 1 to 4 above refer to properties of the imaging lenses. The aberrations relating to each point, namely spherical aberration, diffraction aberration, chromatic aberration and astigmatism will be treated below (Fig. 2-23).

Spherical aberration

With increasing distance from the axis, the deflection increases more than linearly. This means that the condition $\gamma \sim r$ is no longer fulfilled. This is an effect which occurs in light optics when the refractive surfaces are spherical, hence the origin of the term "spherical" aberration. The focal length of paraxial (on-axis) rays is thus greater than that of apaxial (off-axis) rays. An object point is therefore imaged by a beam with an angle of divergence α as a circular disc of diameter

$$d_s = C_s \cdot \alpha^3 \tag{2-15}$$

at the location of "minimal confusion", where C_s is the spherical aberration coefficient. The strong dependence of d_s on α means that the angle of divergence has to be kept very small.

Diffraction error

Since moving particles have the character of waves, diffraction phenomena occur at edges. The finite diameters of the lens apertures therefore lead to intensity minima and maxima in the wave field behind the lens. At the focal point, the intensity is not concentrated at a point but is distributed in the form of circular symmetrical maxima and minima, the major component of the intensity lying within the central maximum. The diameter of the first minimum is therefore defined as the diameter of the diffraction error disc

$$d_d = 1.22 \cdot (\lambda / \alpha)\,. \tag{2-16}$$

In order to make this error small, α should be selected as large as possible. But since the spherical aberration error increases with the cube of α, it is necessary to work with a very small angle of divergence and therefore an optimal angle must be sought. For electron microscopes, its value is of the order of 10^{-3} rad.

Chromatic aberration

It is impossible to generate particle beams of a specific energy, i.e. beams that are strictly monochromatic. The available beams always exhibit a certain energy spread ΔE. Particles with different energies are deflected by different degrees when traversing the lens. The more highly energetic, straighter beams have a greater focal length than those of lower energy. The error disc resulting from this has a diameter of

$$d_c = C_c \cdot (\Delta E/E) \cdot \alpha \,. \qquad \text{(2-17)}$$

To keep this error small, particle sources with low energy widths are to be used for particle probes.

Astigmatism

An astigmatic, i.e. non-punctiform, elongated image of a point is due to a lack of complete rotational symmetry of the imaging fields. The shape of the imaging fields is determined by the shape of the electrodes or the pole pieces. Since it is impossible to produce exactly circular apertures or pole pieces, deviations from rotational symmetry are unavoidable. A noncircular lens behaves like a spherical lens which is superimposed by a more or less weak cylindrical lens. The resulting error may be corrected by adding a low-power cylinder lens of the same magnification and opposite direction to compensate for the cylinder lens component. Stigmators are such compensation configurations which are used to adjust the strength and direction of the compensation field.

The assumption stated in point 5., namely that the charge carriers do not affect each other, is permissible as long as the current densities are not too high and the energies are not too low. We will discuss the influence of the space charge (Boersch effect) in more detail in section 2.3.

2.3 Electron beams

2.3.1 Electron sources

The intensity of the effects, excited in the target material by electron bombardment and exploitable for analysis, is greater the more electrons interact with the material per unit time. This means that electron probes are required which carry as large a current as possible. And since very small areas must often be analyzed in microtechnology, a small probe diameter is also required which means a high current per unit area. Electron probes with high current densities are obtained by using high-intensity electron emitters (cathodes) and suitable electron-optical focusing systems.

Electrons which move freely in a vacuum can be obtained by extraction from solids. Due to the high electron density in solids, emission from these provides high current densities. Such emission is excited by supplying energy. This can be done by

- high temperatures (thermionic emission),
- high electric fields (field emission),
- irradiation of light (photoemission),
- electron bombardment (secondary electron emission).

High current densities arc obtained by thermionic and by field emission. These are therefore the principal modes used for electron sources in analytical instruments.

In *thermionic emission* [2-4], a rise in temperature increases the kinetic energy of the electrons in the solid to the extent that they can overcome the forces keeping them within the solid. This allows them to overcome the work function (section 2.1.2). An emission current can be measured if the cathode is kept negative so that there is just enough potential difference between cathode and anode to draw away any emitted electrons. According to Richardson, the following relationship exists between temperature T, the work function E_W and the attainable saturation current density j

$$j = A \cdot T^2 \cdot \exp(-E_W / kT) \tag{2-18}$$

where k is the Boltzmann constant. According to the theory A consists of constants only. However, the experimental values vary with the kind of material (Table 2-1) and lie for metals between 50 and 100 $A \cdot cm^{-2} \cdot K^{-2}$. In this equation, the work function appears in the exponent and therefore has a particularly strong influence on the current density. The alkali earth metals have low work functions. They are, however, unstable in air even at room temperature and are consequently not used as electron sources in analytical instruments which have to be removable e.g. for cleaning purposes.

Tungsten and lanthanum hexaboride (LaB_6) are the principal materials used for thermionic emitters (Fig. 2-24). The tungsten cathode has proved to be a very stable electron gun for electron beam instruments (Fig. 2-25a). Although tungsten has a relatively high work function, tungsten cathodes can be operated at high temperatures due to the high melting point (3410 °C) of the metal, and saturation current densities of up to 10 $A \cdot cm^{-2}$ are obtained. A disadvantage of high operating temperatures is the relatively high vaporization rate which limits the lifetime to between 30 and 100 hours. Lanthanum hexaboride (m.p. 2210 °C) cath-

Table 2-1 Work function E_w and material constant A for thermionic emission from cesium Cs, lanthanum hexaboride LaB_6 und tungsten W:

	E_w/eV	$A/(A \cdot cm^{-2} \cdot K^{-2})$
Cs	1.8	162
LaB_6	2.6	25
W	4.2	≈ 100

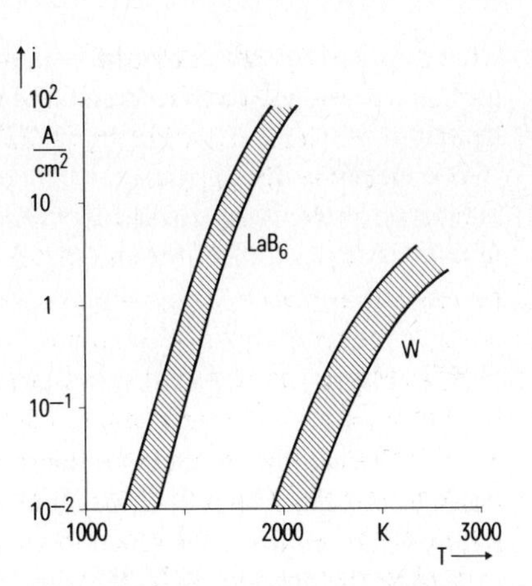

Fig. 2-24. Saturation current densities of thermionic electron sources made of LaB_6 and W: the thermionically excited electrons form a space charge cloud in front of the source. If a voltage is applied between the electron source (cathode) and an anode located opposite of it, then an electron current flows and approaches a limiting value, the saturation current, with increasing voltage. For a given cathode material, this saturation current depends only on the temperature.

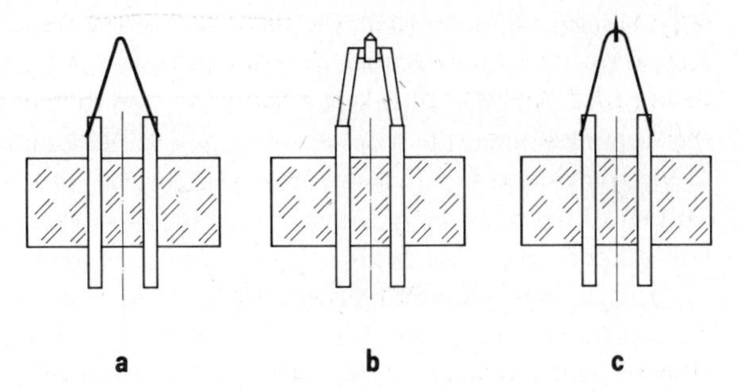

Fig. 2-25. Various types of cathode construction:
a) Tungsten hairpin cathode: a hairpin-shaped wire about 0.15 mm in thickness is welded to two metal pins which are embedded in ceramic material. The cathode is heated directly by current flow.
b) LaB_6 cathode: a single crystal rod with a conical point is held between two graphite blocks. Heating is also by direct current flow.
c) Field emission cathode: a fine tungsten tip is welded to the current-carrying tungsten wire.

odes can be operated at lower temperatures and thus have longer lifetimes (100 to 500 hours) [2-5,6]. Nevertheless, the lower work function allows a higher current density to be attained than with tungsten (up to 10^2 A · cm^{-2}). The preferred cathodic material consists of single crystal rods with conical points (Fig. 2-25b). The rod axis has a ⟨100⟩ or ⟨110⟩ orientation, since low-indexed crystal planes result in high current densities. This material is, however, highly reactive at high temperatures so that LaB_6 cathodes must be operated in a sufficiently high vacuum ($p < 10^{-4}$ Pa) [2-7].

Very high current densities in the order of 10^6 A · cm^{-2} are provided by field emission sources [2-8,9]. *Field emission* is based on the quantum mechanical tunnelling effect (see section 2.1.2). A high electric field causes a narrow potential barrier at the metal surface through which the crystal electrons can escape. To achieve sufficiently high fields fine tungsten wires with preferred crystal orientation and having a very sharp tip (radius of curvature of the tip 0.1 to 1 μm; Fig. 2-25c) are used as field emitters. The field strength at the tip must exceed 10^7 V · cm^{-1}. With such high fields, the fine tip can easily be damaged by bombardment from positive ions present in the residual gas. Field emission cathodes must therefore be operated in ultra-high vacuum (residual gas pressure below 10^{-7} Pa) in order to ensure stable emission.

The electrons do not leave the cathode with uniform energies. After emission they exhibit an energy distribution with a full width at half maximum (FWHM) of around 1 eV for thermionic cathodes and of around 0.2 eV for field emission cathodes.

2.3.2 Forming of electron probes

To use the electrons emitted into the vacuum from the cathode for forming a probe, they must be accelerated and focused. This requires an electrode system having the effect of a lens, consisting of the electron source and two rotationally symmetrical electrodes, the Wehnelt electrode and the anode (Fig. 2-26) (triode system). The heated cathode (heating voltage U_c) is at high negative potential. The potential U_w of the Wehnelt electrode is slightly negative with respect to the cathode potential and so a potential threshold for the electrons is built up in front of the cathode, which results in an electron cloud, producing a space charge. By varying the Wehnelt voltage, the space charge can be modified and so the beam current adjusted within wide limits. The higher the negative voltage, the higher the potential barrier which is created in front of the cathode. As a result, only the most energetic electrons emitted from the tip of the cathode, where the field is highest, can pass the anode aperture.

Cathode, Wehnelt electrode and anode act as an electron optical system which produces a small disk-like image (crossover) of the emitting area in front of the cathode. For a powerful electron probe, a large current density j_{c0} in the crossover and a small angle of divergence (aperture angle) are desirable. The brightness [2-10]

$$\beta_{c0} = j_{c0}/\Omega_{c0} = j_{c0}/(\pi \cdot \alpha_{c0}^2) \tag{2-19}$$

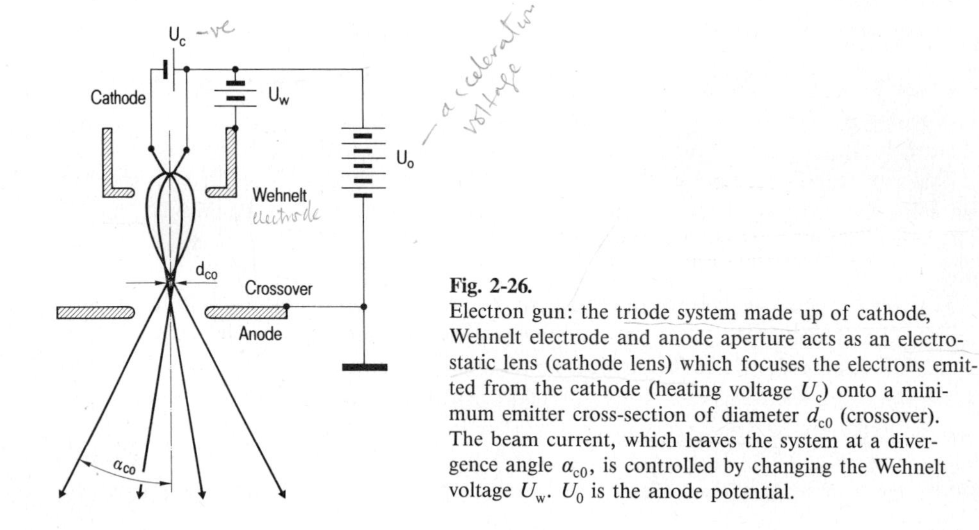

Fig. 2-26.
Electron gun: the triode system made up of cathode, Wehnelt electrode and anode aperture acts as an electrostatic lens (cathode lens) which focuses the electrons emitted from the cathode (heating voltage U_c) onto a minimum emitter cross-section of diameter d_{c0} (crossover). The beam current, which leaves the system at a divergence angle α_{c0}, is controlled by changing the Wehnelt voltage U_w. U_0 is the anode potential.

thus characterizes the performance of a particle beam probe ($\Omega_{c0} = \pi \cdot \alpha_{c0}^2$, solid angle). If no space effects occur, β is the same for all points on the axis of the optical system, independent of how many lenses and apertures are present in the beam path:

$$\beta = j/(\pi \cdot \alpha^2) = \beta_{c0} \,. \tag{2-20}$$

The current density in the crossover, j_{c0}, depends directly on the current density of the cathode, which in turn is a function of the cathode temperature T and the work function (eq. (2-18)) as well as the acceleration voltage U_0. For thermionic cathodes, an upper limit may be specified for the brightness [2-11]

$$\beta_{max} = j_c \cdot e \cdot U_0/(\pi \cdot k \cdot T) \tag{2-21}$$

where j_c is the current density at the cathode surface, k the Boltzmann constant, T the cathode temperature and U_0 the acceleration voltage. For an optimally adjusted electron gun therefore, the brightness is proportional to the acceleration voltage. In practice, brightness values between 10^4 and 10^5 $A \cdot cm^{-2} \cdot sr^{-1}$ are attained for tungsten cathodes at acceleration voltages between 20 and 100 kV. For LaB_6 cathodes, the brightness values are around one order of magnitude higher. The highest brightness values are attained by field emission cathodes (up to about 10^9 $A \cdot cm^{-2} \cdot sr^{-1}$), although at relatively small electron probe currents.

To create an electron probe with a very small diameter, the crossover must be imaged onto the object in a demagnified form. In Fig. 2-27, a two-stage demagnification system is shown schematically. To display the imaging characteristics of such a system more effectively, the an-

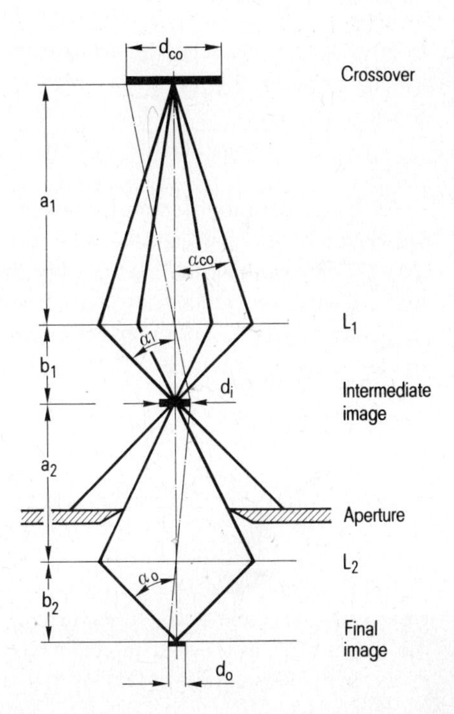

Fig. 2-27.
Two-stage demagnification of the crossover. The lens L_1 images the crossover with demagnified imaging scale $M_1 = b_1/a_1$ ($M_1 < 1$) onto the intermediate image plane. The intermediate image is once again demagnified onto the final image plane. There the image size (i.e. the diameter of the electron probe) is $d_0 = M_1 \cdot M_2 \cdot d_{c0}$. In practice the aperture is located close to the lens L_2, so that the divergence angle of the probe is approximately $\alpha_0 \approx D/2b_2 \approx D/2w$ (D diameter of the aperture; w working distance).

gles of divergence of the electron beams are enlarged in this drawing (the same applies to Fig. 2-26). In reality they are smaller than 10^{-1} rad. The rays which leave the crossover (diameter d_{c0}) with an angle of divergence α_{c0} are focused by the lens L_1 into an intermediate image. In accordance with the imaging scale $M_1 = b_1/a_1$ they propagate with a greater angle of divergence $\alpha_1 = \alpha_{c0}/M_1$ (cf. Figs. 2-17 and 2-18). In order to keep the aberrations small, (cf. section 2.2.2), an aperture is fitted at or a short distance before the second demagnification lens. This limits the angle of divergence behind L_2 to α_0. The diameter of the final image of the crossover, i.e. of the probe, is $d_0 = M_1 \cdot M_2 \cdot d_{c0}$ if we neglect the aberrations.

An important design parameter for analytical applications of the electron probe is the image distance b_2 of the second lens, which is approximately equal to the working distance w in the scanning electron microscope (section 3.2.2). This is the distance between the lower pole piece of the objective lens (here lens 2) and the specimen plane. The working distance must be sufficiently large to accommodate devices for collecting the signals generated by the probe such as detectors and spectrometers as well as to allow manipulation of the specimen. By modifying the focal length of L_2 and thus the angle of divergence α_0, the working distance can be varied within broad limits. However, as α_0 changes so do the probe current I_0 and the probe diameter d_0. The relationship between α_0, d_0 and I_0 can be easily determined with the aid of the brightness value (eq. (2-20)). Consequently, the current density in the probe is

$$j_0 = \beta \cdot \pi \cdot \alpha_0^2$$

and the probe current

$$I_0 = \beta \cdot (\pi/2)^2 \cdot d_0^2 \cdot \alpha_0^2 \tag{2-22}$$

from which we obtain the following expression for the probe diameter

$$d_0 = \frac{2}{\pi \cdot \alpha_0} \cdot \sqrt{\frac{I_0}{\beta}} \,. \tag{2-23}$$

The current density is, in reality, not constant over the probe diameter, but follows a Gaussian distribution. Equation (2-22) applies very well as an approximation, however, if we regard d_0 as the full width at half maximum of this distribution. It can be seen from equation (2-23) that, for small probe diameters, large brightness values and large angles α_0 are required. Due to the unavoidable aberrations, however, we are obliged to limit α_0. The aberrations increase the probe diameter (cf. eqs. (2-21) to (2-23)) to

$$\begin{aligned} d_0'^2 &= d_0^2 + d_d^2 + d_s^2 + d_c^2 \\ &= d_0^2 + \frac{(1.2 \cdot \lambda)^2}{\alpha_0^2} + C_s^2 \cdot \alpha_0^6 + \left[C_c \cdot \frac{\Delta E}{E}\right]^2 \cdot \alpha_0^2 \end{aligned} \tag{2-24}$$

the spherical aberration having the greatest effect, since the diameter of its error disk d_s increases with the cube of α_0. In the case of magnetic electron lenses, the spherical aberration coefficient C_s is, as a rough approximation, equal to the focal length. For a minimum probe diameter it follows from eq. (2-24) that the optimal angle of divergence lies between 10^{-3}

and 10^{-2} rad for focal lengths between 2 and 20 mm. It further follows from eq. (2-22) that with decreasing probe diameters d_0 the current I_0 in the probe decreases greatly (Fig. 2-28). This is of importance particularly for electron beam X-ray microanalysis and electron beam measuring techniques, since these generally require currents of the order of 10^{-7} A in order to obtain signals which may be successfully evaluated. For comparison purposes, Fig. 2-28 includes values measured at electron energies of 2.5 keV. With decreasing probe current they deviate increasingly from the theoretical curves. This is due to the increasing effect of the aberrations.

The space charge effects, which have hitherto not been taken into account, lead to an increase in probe size, and they become noticeable when the accelerating voltage is low and the current densities are high. Along the path from the crossover to the final image, Coulomb interactions occurring between the electrons increase their energy spread (energetic Boersch ef-

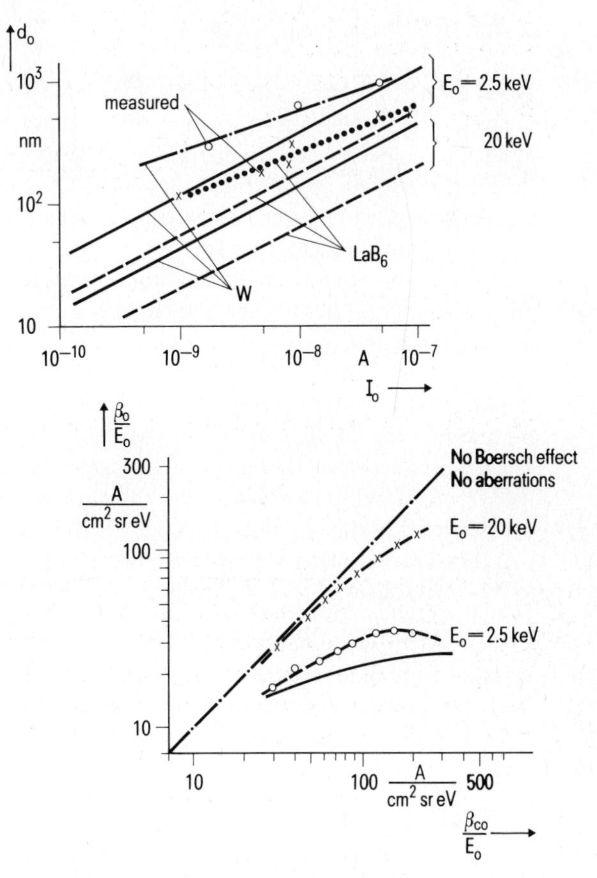

Fig. 2-28.
Probe diameter as a function of maximum probe current, estimated from eq. (2-23) with $\beta = \beta_{max}$ from eq. (2-21) for 2.5 keV and 20 keV beam energy. At 2.5 keV, values measured for W and LaB_6 cathodes are included for comparison.

Fig. 2-29. Brightness value β_0 in the probe as a function of the calculated brightness value β_{c0} in the crossover for electron energies $E_0 = 20$ keV and $E_0 = 2.5$ keV. To make the values comparable, they were standardized to the electron energy E_0. Without the Boersch effect and without aberrations, a linear relationship independent of the energy would be expected. However, the Boersch effect causes the increase in β_0/E_0 to keep below that of the linear relationship. The deviation is still relatively small at 20 keV, but attains a factor of 3 at 2.5 keV. (— theoretically calculated curve; $x, 0$ measured values). (Courtesy J. Frosien).

fect) and affect their paths (lateral Boersch effect) [2-12]. Both these effects lead to an increase in the probe diameter and thus to a decrease of the brightness value. With decreasing electron energies and increasing electron density the strength of the interaction increases and this is manifested as an overproportional decrease of the brightness value (Fig. 2-29). This is of importance especially for scanning electron microscopy of insulating objects and for electron beam measuring techniques where low electron energies are required with a sufficiently high current density.

2.4 Interaction between electrons and solids

2.4.1 Electron scattering

When an electron hits a solid surface, it penetrates into the microstructure of the solid and interacts with its atoms. The resulting effects allow the extraction of analytical information on various properties of the material. The elementary interaction process involves collision of the electron with an atom, either with the positive atomic nucleus or with the electrons of the atomic shell. Since the colliding particles are charged, the interaction is governed essentially by electrostatic Coulomb forces.

In a collision with an electron, the atomic nucleus hardly changes its position since it is very much heavier than the electron. The electron is therefore deflected at the nucleus without significant energy loss (elastic scattering); it merely changes its direction. Its deflection is greater the closer the electron passes to the nucleus and the lower its velocity. A simple approximation formula for the angle of deflection is obtained if the influence of the atomic shell is neglected, the nucleus is regarded as a positive point charge (Fig. 2-30), and only small angles of deflection are considered, such as are favored in reality. More rarely occuring large deflections originate from more direct collisions with the nucleus, which may even cause the electron to be backscattered in the opposite direction (Fig. 2-31a).

If the electron collides with the electrons of the atomic shell, then energy is transferred due to the equal masses of the colliding particles (inelastic interaction). The incident electron does

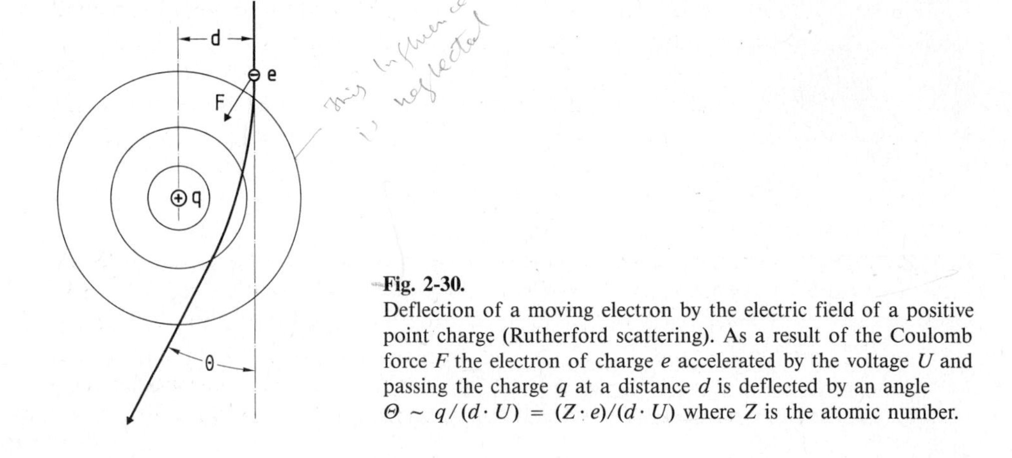

Fig. 2-30.
Deflection of a moving electron by the electric field of a positive point charge (Rutherford scattering). As a result of the Coulomb force F the electron of charge e accelerated by the voltage U and passing the charge q at a distance d is deflected by an angle $\Theta \sim q/(d \cdot U) = (Z \cdot e)/(d \cdot U)$ where Z is the atomic number.

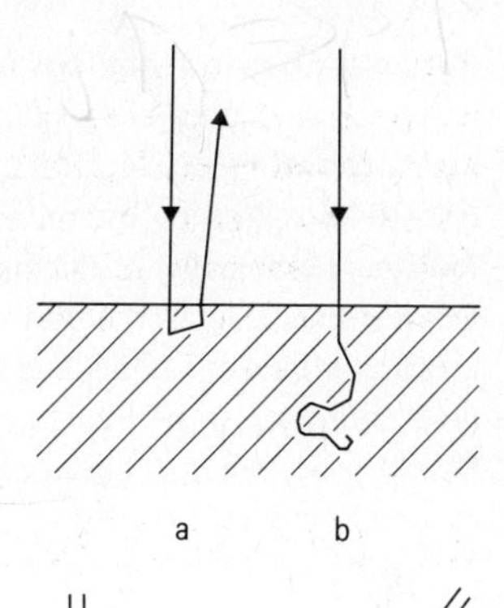

Fig. 2-31.
Scattering of electrons in a solid.
a) When elastic scattering occurs, few collisions with the atomic nuclei take place with large angles of deflection (and without significant energy losses). The electron can therefore leave the solid with almost unchanged energy.
b) Inelastic scattering results in many consecutive collisions, principally with the atomic electrons and with small scattering angles, leading to energy losses so that the electron finally remains in the solid.

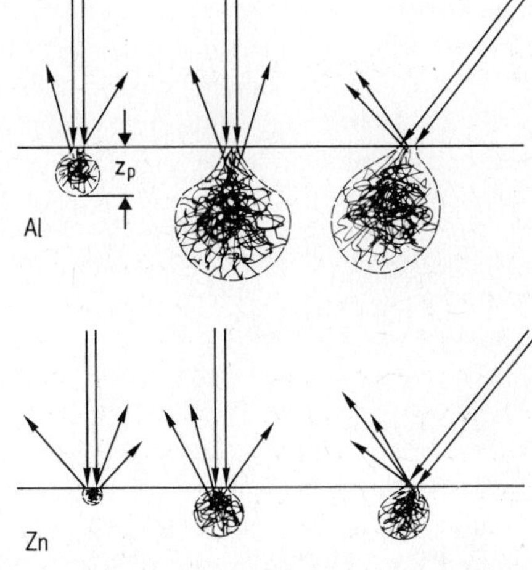

Fig. 2-32.
The shape and size of the volume containing the scattered electrons (scattering volume, energy dissipation volume) depend on material properties (density, atomic number) as well as on the energy and angle of incidence of the electrons. z_p is the penetration depth.

not only change its direction, it is also slowed down. During such inelastic interaction, however, scattering processes dominate which involve low energy losses and a low exchange of momentum resulting in a higher probability of small scattering angles (Fig. 2-31b).

Both processes operate side by side and simultaneously. As a consequence of these numerous different scattering events, an electron probe which originally was sharply focused in vacuum spreads over a greater volume after penetrating the solid (Fig. 2-32). The shape and size of this scattering volume depend upon the material as well as on the energy and angle of incidence of the probe electrons. In a material made up of elements of low atomic number and low density, there is little interaction and the electrons can penetrate more deeply into the material. With heavy elements, the electrons are scattered at an earlier stage, close to the surface. More electrons are therefore backscattered into the vacuum and the scattering volume is concentrated closer to the surface.

The diameter of the scattering volume and its extension normal to the surface are of significance for analytical applications. The diameter affects the spatial resolution attainable by electron beam microanalysis techniques. The penetration depth can be determined by measuring the transmission of electrons through thin films [2-13]. Theoretical calculations are usually based on an estimation of the energy losses along the electron path [2-14]. The most accurate theoretical values for extension and shape of the scattering volume are obtained with the aid of Monte Carlo calculations [2-15,16]. In these simulation calculations, the path of an electron is traced, assuming elementary scattering processes, until its energy has been reduced below a given small value, or until it leaves the solid again. The changes in direction caused by the individual scattering events and the path lengths between two scattering events are given by random numbers (as in the games of chance in the casino at Monte Carlo). To obtain realistic conditions, up to 10 000 individual electron paths must be traced in this way, which requires much computation time. These calculations, which are in good agreement with experimental measurements, show that the greatest diameter of the dissipation volume parallel to the surface is approximately equal to the penetration depth. Penetration depths mentioned in the literature, obtained from experimental data and theoretical calculations, are plotted in Fig. 2-33.

Scattering processes in thin films are of particular interest for transmission electron microscopy. They determine the image contrast and also the maximum film thickness for which there is transmission with sufficient image brightness and resolution. This critical film thickness is of the order of a few hundred nm for 100 keV electrons and elements of medium atomic number [2-22]. In such thin films, only relatively few scattering events take place. Most electrons penetrate the material in the forward direction without significant energy losses. Even for the critical film thickness the energy losses of 100 keV electrons are below 1%. Despite this, the

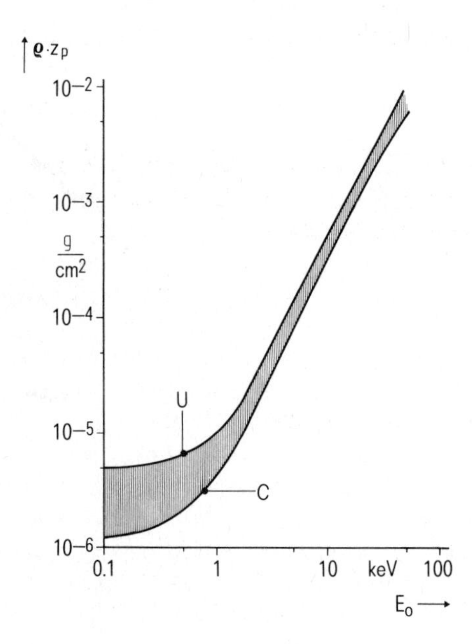

Fig. 2-33.
The penetration depth z_p of electrons into the material is conveniently measured in units of $\varrho \cdot z_p$ (ϱ density of the material). The values of $\varrho \cdot z_p$ are independent of the physical and chemical state of the material. They are scattered only slightly for the different elements and exhibit a similar dependence on the electron energy for all elements. Most of the published experimental values [2-17 to 21] lie between the theoretically calculated curves [2-14] for carbon (C) and uranium (U).

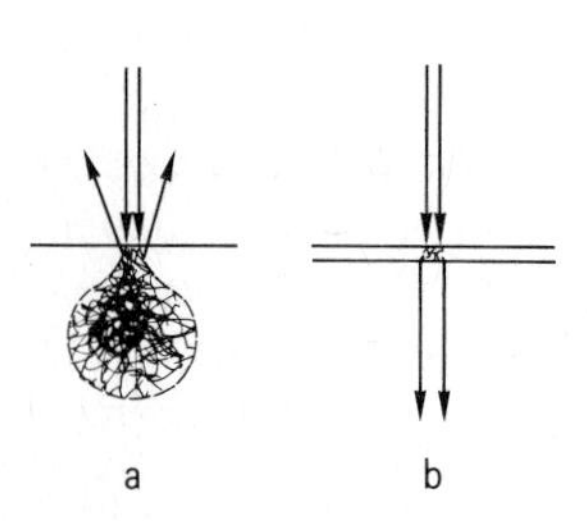

Fig. 2-34.
A comparison of electron scattering in a bulk material (a) and in a thin film (b). For a sufficiently thin film (around 100 nm), the broadening of the electron probe due to scattering is small. Only relatively few scattering events occur and the electron beam leaves the film without significant energy losses and with small scattering angles.

relatively few inelastic scattering processes suffice to generate measurable secondary effects which can be exploited for a material analysis of thin films (see section 5.6). The typical shape of the scattering volume for bulk materials cannot develop with the very thin films which are required for transmission electron microscopy (Fig. 2-34). The result is a better spatial resolution for analytical techniques using the transmission electron microscope.

The elastic interactions between electrons and atoms or between electrons and the crystal lattice lead to the following effects which are important for characterizing the material:
- electron backscattering,
- electron diffraction,
- deflection of electrons at ferromagnetic and ferroelectric domains.

The transfer of energy from the primary electrons to the solid, i.e. the inelastic interaction, leads to the following secondary processes which are important for analysis:
- emission of secondary electrons,
- excitation of oscillations in the electron plasma in the solid,
- emission of X-rays and Auger electrons,
- generation of electron-hole pairs.

2.4.2 Elastic electron scattering and electron diffraction

Electron backscattering

Backscattered electrons (BSE) have usually experienced only few scattering events with large scattering angles. Their number, summed over all scattering angles and referred to the number of primary electrons (PE), defines the backscattering coefficient:

$$\eta = I_{BSE}/I_{PE} \,. \tag{2-25}$$

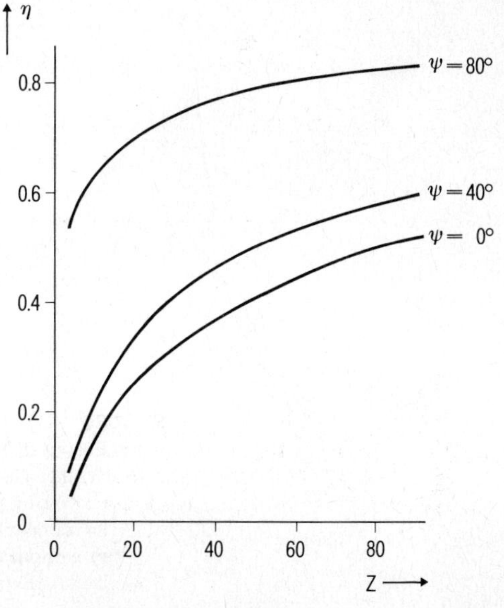

Fig. 2-35.
Variation of the electron backscattering coefficient $\eta = I_{BSE}/I_{PE}$ with the atomic number of the material and with the angle of incidence of the primary electrons ($\psi = 0$ means incidence normal to the surface) [2-24,25,26].

η depends only slightly on the electron energy, but strongly on the atomic number of the elements of the material [2-23,24]. Heavy atomic nuclei deflect the electrons more strongly, so that backscattering increases with atomic number (Fig. 2-35). The backscattering coefficient is also dependent on the angle of incidence of the primary electrons. Whereas in the case of

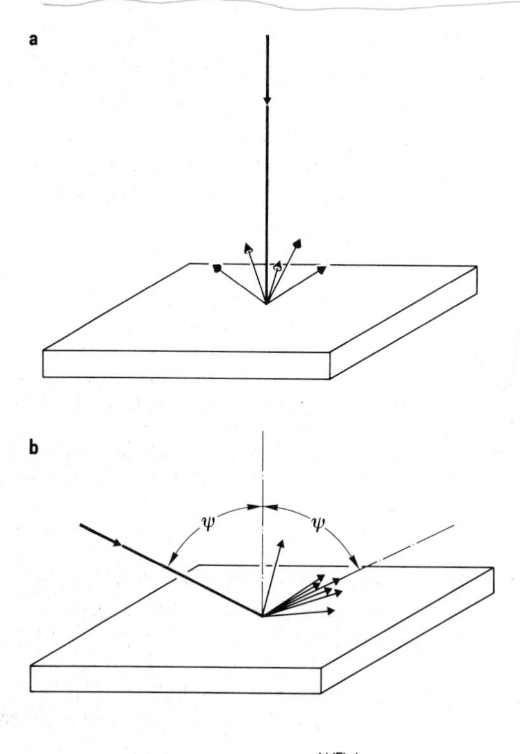

Fig. 2-36.
Electron backscattering as a function of the angle of incidence:
a) In the case of normal incidence, the primary electrons are backscattered in more or less opposite directions to the incident beam. The angular distribution of the backscattered electrons follows a cosine law $I(\psi) \sim \cos \psi$ with the maximum pointing normal to the specimen surface.
b) In the case of oblique incidence, the electrons are backscattered predominantly in the direction of the angle of reflection. This behaviour is more strongly pronounced for large ψ. Furthermore, the trajectories of the backscattered electrons lie mainly in the plane of the surface normal and the direction of the primary electrons.

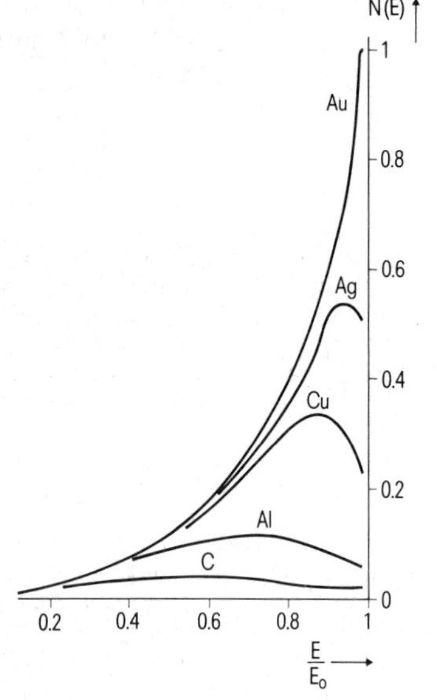

Fig. 2-37.
Energy distribution of the backscattered electrons. The parameter $N(E)$ denotes the number of backscattered electrons relative to the number of primary electrons in the energy interval between E and $E + \Delta E$. E_0 is the energy of the primary electrons. For lighter elements the energy maximum of the backscattered electrons is shifted towards smaller energies.

normal incidence less than half the primary electrons are backscattered, with grazing incidence only a few electrons remain in the solid, when the surface is smooth and planar. The angular distribution of the backscattered electrons also depends strongly on the direction of incidence of the primary electrons. The maximum of this angular distribution follows a reflection law (Fig. 2-36).

Not all backscattered electrons leave the solid without significant energy loss. A considerable number are backscattered into the vacuum after some inelastic collisions. As a consequence, with increasing atomic number the backscattered electrons will have an energy distribution whose maximum will move closer to the energy of the primary electrons [2-27]. In the case of very light elements the backscattered electrons can hardly be called elastically scattered. For carbon, their energy maximum lies at only half the primary electron energy (Fig. 2-37).

Electron diffraction

Just like electron backscattering, electron diffraction is based on Coulomb scattering at the atomic nucleus. However, to gain an understanding of the diffraction effects, the wave properties of moving electrons must be considered. If an electron wave hits a solid, secondary waves are emitted from every scattering center in all directions (Huygens' principle). Superposition of the scattered waves produces a wave field.

In amorphous materials, the atoms are arranged without long range order so that the scattered waves are superimposed irregularly and diffuse scattering occurs. In a crystal lattice the atoms are arranged regularly and the scattered waves are therefore also superimposed regularly. As a result diffraction occurs. The waves emerging from the individual atoms have regular path differences with respect to each other, which, in specific directions, amount to exactly one wavelength or a multiple of it. In these directions, the secondary waves are superimposed with equal phase and maxima in the scattered intensity, i.e. diffraction reflections, occur.

Diffraction may also be described as a selective reflection of the incident electrons at the lattice planes (section 2.1.1). Reflection takes place when the interplanar spacing d, the electron wavelength λ and the diffraction angle ϑ satisfy Bragg's law (Fig. 2-4)

$$n \cdot \lambda = 2 \cdot d \cdot \sin \vartheta \tag{2-26}$$

($n = 1, 2, 3, \ldots$). Due to the very small wavelengths of fast electrons (Fig. 2-14b: for 100 keV electrons $\lambda = 3.7$ pm), the diffraction angles are also very small. The reflection at the aluminum lattice with the largest Bragg angle corresponding to lattice planes with a spacing of $d = 0.234$ nm, occurs at $2 \cdot \vartheta = 0.91°$ for 100 keV electrons. For practical applications in electron microscopy therefore, the simplified form of Bragg's law can be used

$$n \cdot \lambda = 2 \cdot d \cdot \vartheta \, . \tag{2-27}$$

This very small diffraction angle has a further important consequence: essentially only those lattice planes which lie almost parallel to the incident beam contribute to the diffraction pattern. These planes are orientated perpendicular to the foil plane when electrons are transmitted trough thin foils with normal incidence (Fig. 2-38). In the case of a thin single crystal foil, diffraction at the various lattice planes produces a regular spot pattern (section 4.1.4).

Electron diffraction patterns of bulk material surfaces are obtained only for grazing incidence of the electron beam. At very low electron energies and correspondingly larger diffraction angles, larger angles of incidence are used (low energy electron diffraction, LEED).

Diffraction patterns of a different kind, namely Kikuchi patterns, are obtained when high energy electrons pass through relatively thick crystalline foils. In thick foils, inelastic scattering takes place in addition to elastic scattering. It occurs principally in the forward direction. The energy loss involved is small, so that the wavelength of the inelastically scattered electrons is still nearly equal to the wavelength of the incident electrons. These inelastically scattered electrons can now be diffracted according to Bragg's law at a set of lattice planes, with diffraction occuring on both sides of the planes. Due to the varying intensity of the electrons scattered in different directions, this also leads to different intensities of the diffracted electrons (Fig. 2-39). Since scattering takes place in all directions, the diffracted beams lie on conical envelopes; because of the very small diffraction angle the cone angle is almost 180°. In a recording plane normal to the drawing plane of Fig. 2-39 therefore, the diffracted electrons generate parallel bright (B) and dark (D) lines (Kikuchi lines) with a distance equal to the distance R of the diffraction spot from the directly transmitted beam (Fig. 2-38). Kikuchi patterns are obtained with transmission as well as with grazing incidence.

The backscattering of electrons from bulk single crystals produces pseudo Kikuchi or channeling patterns. Their origin is essentially due to the dependence of the backscattering factor on the crystallographic orientation (section 3.5).

Up to now, we have considered the influence only of atomic Coulomb fields on electrons. But moving electrons change their paths also in the electric and magnetic fields present in a material due to its ferroelectric [2-28] and ferromagnetic [2-29,30,31] properties. This leads to possible ways of investigating the domain structures in ferromagnetic and ferroelectric materials by means of electron probes.

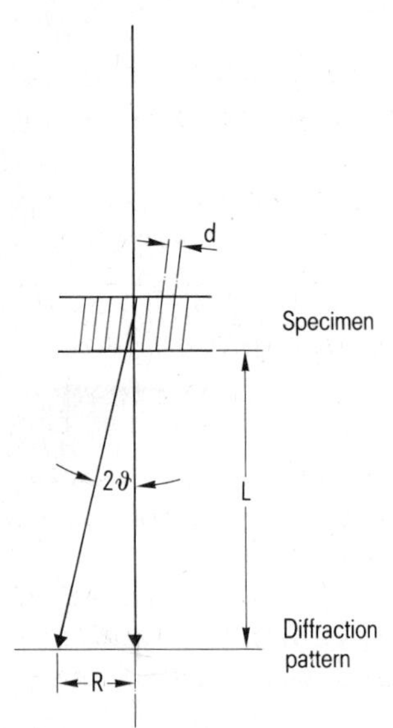

Fig. 2-38.
Transmission electron diffraction in thin foils. As a consequence of the very small diffraction angle ϑ, only those diffracted beams which originate from lattice planes orientated nearly perpendicular to the foil plane contribute to the diffraction pattern. The basic formula for electron diffraction $\lambda \cdot L = R \cdot d$ ($n = 1$ for first order reflections) follows from eq. (2-27), with L ("camera length") and R (distance of the diffraction spot from the undiffracted beam).

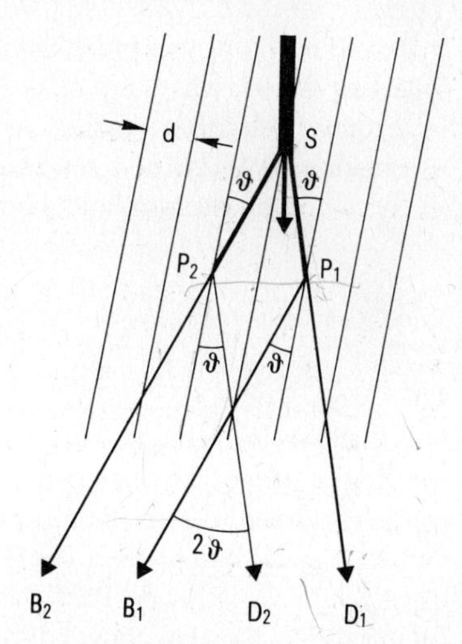

Fig. 2-39. Formation of Kikuchi patterns: The electrons scattered inelastically (at S) in thick single crystal foils (having lost only very little energy) can be diffracted at lattice planes if their direction meets the Bragg condition. This is the case for beams propagating in the directions B_2 and D_1. They are diffracted at the lattice planes at points P_1 and P_2. The intensity of the diffracted beams are missing in the incident beam direction in the general scattering background. Since at S more electrons are scattered close to the incident beam direction (the number of electrons in the various directions is indicated by the boldness of the arrows), more electrons in the direction D_1 are lost through diffraction than are gained through diffraction at P_2 into the direction D_2. The result is that the intensity with respect to the background is reduced in direction D (dark) and correspondingly increased in direction B (bright).

2.4.3 Inelastic electron scattering

The inelastic interaction between highly energetic electrons and matter leads to a transfer of energy to the atoms of the solid. This gives rise to secondary effects whose excitation require varyingly high amounts of energy.

Energies between a few and several 10 eV are required for exciting oscillations in the electron gas (plasmons). Energies of the same order of magnitude can detach outer shell electrons, which in insulators and semiconductors belong to the valence band. For such crystals, ionization leads to the formation of electrons and ions (electron-hole pairs). When these recombine, light may be emitted (from UV to IR). This effect is called cathodoluminescence.

Secondary electrons are emitted over a broad range of primary electron energies, the highest yield lying at energies of the order of 10^3 eV.

The ionization of inner shells requires energies between 10^2 and 10^5 eV. The electrons detached in this way from the atom reach the valence and conduction band of the solid. X-rays or Auger electrons are emitted during recombination after inner shell ionization.

By direct collision with electrons having energies above 10^4 eV, atoms can be displaced from their lattice sites. An essential part of the energy transfer during electron bombardment leads to excitation of lattice vibrations (phonons). Due to the small irradiation powers generally used in analytical instruments, however, the heating effect is low.

We will now turn to a more detailed treatment of the fundamental processes leading to the following effects, which are of importance for analysis:
- secondary electron emission,
- emission of X-rays and Auger electrons and
- formation of electron-hole pairs resulting in cathodoluminescence.

Secondary electron emission

The energy distribution of the electrons emitted from solid material during electron irradiation is shown in Fig. 2-40. It is marked by three regions. The electrons of region III, whose energies lie below 50 eV, are conventionally designated as secondary electrons, although this region also contains a small proportion of backscattered primary electrons. The energy distribution of the secondary electrons is practically independent of the energy of the primary electrons E_0, for $E_0 > 100$ eV. It exhibits a pronounced maximum, which lies between 1 and 5 eV for metals [2-33] and at about half this value for insulators. When plotted against the primary electron energy, the secondary electron yield, defined as

$$\delta = \frac{\text{Number of secondary electrons}}{\text{Number of primary electrons}} = \frac{I_{SE}}{I_{PE}}$$

follows a curve with a characteristic shape for all materials (Fig. 2-41). With increasing primary electron energy, the yield initially increases and reaches a maximum at which the number of secondary electrons exceeds that of the primary electrons for most materials. At higher primary electron energies the yield is reduced due to the generation of secondary electrons at greater depths from which they cannot reach the surface because of their low energy. For metals, the maximum yield lies between 0.4 and 1.6, for insulators and semiconductors it can be up to an order of magnitude higher. This can be explained by the energy losses suffered by

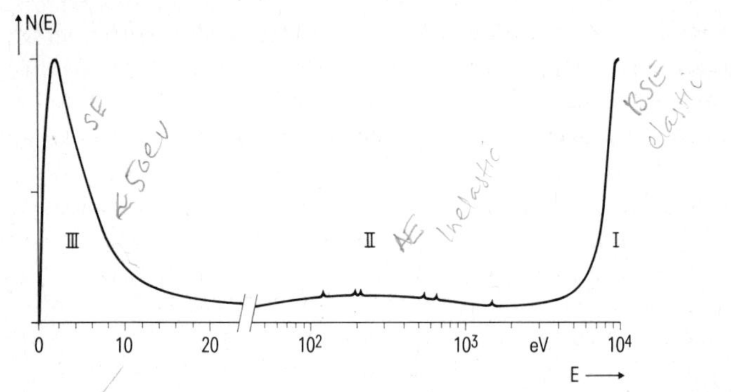

Fig. 2-40. Energy distribution of electrons emitted during bombardment of solid material by primary electrons with $E_0 = 10^4$ eV [2-32]. $N(E)$ is the number of electrons emitted in the energy range between E and $E + \Delta E$. Range I essentially contains elastically backscattered electrons, range II inelastically scattered ones and Auger electrons, the latter being indicated as small peaks. Range III contains the secondary electrons. The Auger peaks are shown very much higher than in reality. The secondary electron yield is up to 10^4 times greater than the yield of Auger electrons.

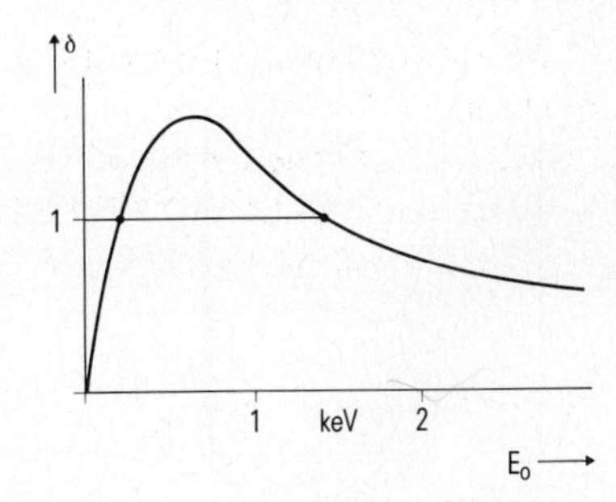

Fig. 2-41.
Secondary electron yield δ as a function of the energy E_0 of the primary electrons.

the secondary electrons on their way to the surface. In metals the secondary electrons are scattered by the high density of conduction electrons. In contrast, no such scattering occurs in insulators and semiconductors due to their lack of conduction electrons. For the same reason, the depth from which the secondary electrons escape, and which for metals lies between 2 and 10 nm, is up to ten times larger in insulators and semiconductors.

Apart from this, the yield of secondary electrons depends very strongly on the angle of incidence ψ of the primary electrons: $\delta \sim 1/\cos \psi$ where ψ is the angle between primary electron beam and surface normal. Since in the case of oblique incidence the secondaries are generated close to the surface, the yield increases with the angle of incidence. In the case of grazing incidence, it attains a value about six times as high as for normal incidence. This strong dependence, which applies in a similar way to the backscattered electrons, is to a great extent responsible for the contrast effects in the scanning electron microscope (section 3).

Emission of X-rays and Auger electrons

The energy distribution of X-rays, which are emitted from solids during bombardment by highly energetic electrons, consists of two contributions (Fig. 2-42). The characteristic spectrum containing sharp maxima is superimposed onto the bremsstrahlung spectrum which extends continuously over a large energy range up to a sharp limit (E_0). The continuous spectrum is due to the interaction with the atomic nuclei, which usually results in elastic scattering. With a low probability, however, the deceleration of the primary electrons in the field of the nucleus can lead to the emission of electromagnetic radiation. The energy of the emitted photons (X-ray quanta)

$$\Delta E = h \cdot \nu$$

stems from the kinetic energy of the primary electrons. Since there is a very large number of primary electrons suffering energy losses with arbitrary values a continuous spectrum is emitted. This has an upper sharp limit at the energy E_0 of the primary electrons, for no X-ray quanta with higher energies can be generated:

$$E_0 = e \cdot U_0 = h \cdot \nu_0 = h \cdot c / \lambda_{min} .$$

More important for analytical applications is the characteristic X-radiation which is produced in competition to the emission of Auger electrons as a result of inner shell ionizations

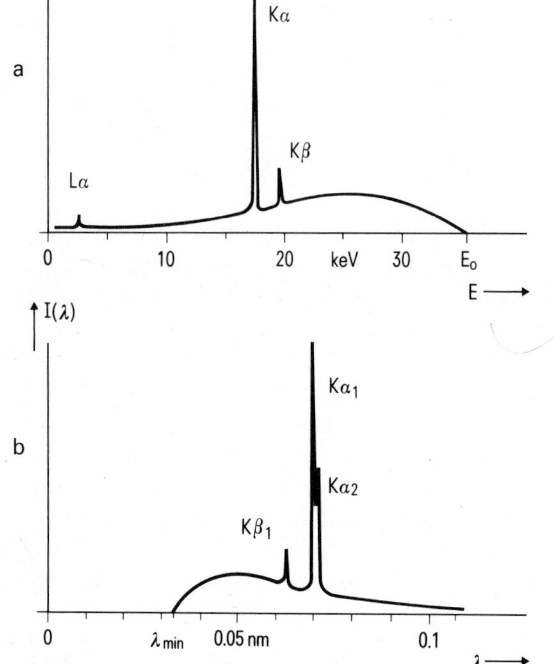

Fig. 2-42.
Energy distribution of X-rays excited from molybdenum by electrons of energy $E_0 = 35$ keV:
a) Intensity $I(E)$ as a function of the energy of the X-ray quanta. The bremsstrahlung spectrum has a sharp upper limit at the excitation energy E_0.
b) Intensity $I(\lambda)$ as a function of the wavelength of the X-rays.
The upper energy limit E_0 corresponds to a minimum wave length $\lambda_{min} = h \cdot c / E_0$. The relative intensities of the X-ray peaks for the K series are, for elements with medium atomic number, about $I_{K\alpha 1} : I_{K\alpha 2} : I_{K\beta 1} = 100:50:15$.

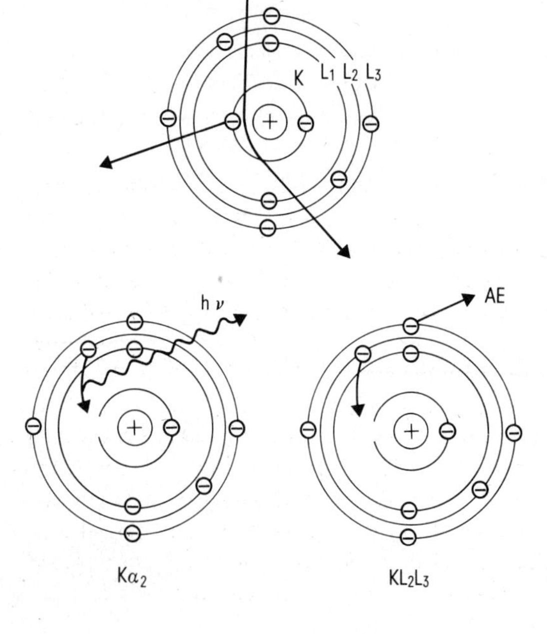

Fig. 2-43.
The emission of an X-ray quantum or Auger electron can be divided into two steps. The atom is initially excited by the removal of an electron from an inner shell (in this case the K shell) and thereby attains a higher energy state. After a very short time (about 10^{-15} s) it returns to its ground state when another electron from an outer shell (L shell) fills the vacancy. The energy released in this transition is used to emit an X-ray quantum or an Auger electron.

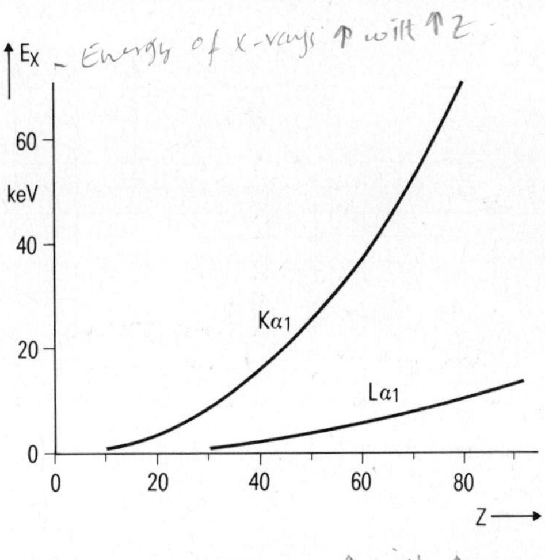

Fig. 2-44.
Energy of the X-ray lines $K\alpha_1$ and $L\alpha_1$ as a function of the atomic number. The quantum energy increases with atomic number Z according to Moseley's law (for $K\alpha_1$ lines):

$$E_{K\alpha 1} = \frac{3}{4} \cdot h \cdot R \cdot (Z - 1)^2$$

(Rydberg constant $R = 3.29 \cdot 10^{15}\ s^{-1}$; Planck constant $h = 4.135 \cdot 10^{-15}$ eV · s).

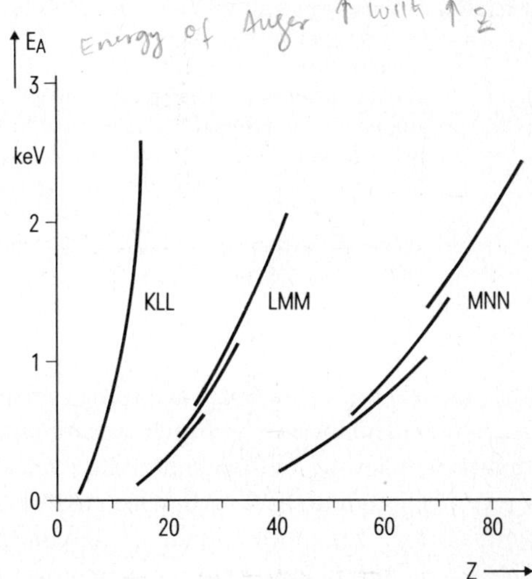

Fig. 2-45.
Energy of the Auger electrons as a function of the atomic number. Since the electrons are more strongly bound to the atomic nucleus as the atomic number increases, the energies of the Auger electrons and of the X-ray quanta increase with atomic number.

(Fig. 2-43). A primary electron incident with sufficiently high energy initially detaches an electron from an inner shell, for instance the K shell, and removes it from the atom. As a result, the atom attains a higher energy state. When the atom returns to its ground state, an electron from an outer shell (e.g. L shell) fills the vacancy in the K shell. The energy released in this process leads to the emission of either a photon (X-ray quantum) or another electron (Auger electron, for instance from the L shell). The vacancy now left in the L shell is filled by a further transition of an electron from an outer shell, with the associated emission of another X-ray quantum or Auger electron. This process is repeated in a similar sequence until the atom finally has returned to its neutral ground state. In this way, whole series (cascades) of X-ray lines and Auger electrons are produced. As a result of the element dependent electron configurations which are built up with increasing atomic numbers in line with simple rules, unambiguous relationships exist between the energies of the X-ray lines or the Auger electrons and the atomic numbers of the elements arranged in the periodic table (Figs. 2-44,45). The energy level diagram of the atom is more suitable for showing the possible electron transitions than the

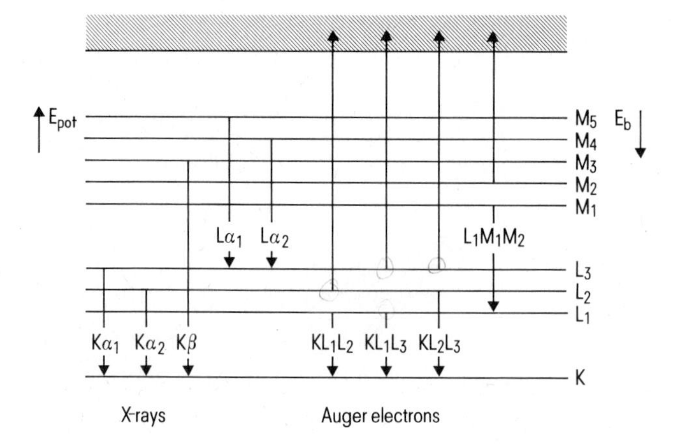

Fig. 2-46. Schematic energy level diagram of an atom with electron transitions leading to the emission of X-ray quanta and Auger electrons. The diagram contains two energy scales pointing in opposite directions. The potential energy E_{pot} is the energy difference between the level K and a level lying further out. The binding energy E_b is required for the complete removal of an electron from a shell (level) and is also referred to as ionization energy. In the designation for the X-ray lines, the capital letter stands for the final level and the small Greek letter for the initial level of the electron transition (e.g. Kα). In classifying the Auger lines, the first letter refers to the level of the original vacancy and the last two letters mark the levels of the vacancies which are generated after the emission of the Auger electron (e.g. KLL). Indices such as $K\alpha_1$ or KL_1L_2 characterize the sublevels involved.

shell model (Fig. 2-46). Some transitions which should be possible according to this diagram are, however, not observed. Which transitions are allowed or forbidden is determined by selection rules following from quantum mechanical considerations. The probability P_{XK} for electron transitions into the K shell, which lead to X-ray emission, increases greatly with the atomic number Z, in fact with about Z^4. In contrast, the K transition probability leading to the ejection of Auger electrons, P_{AK}, is approximately independent of Z. The ratio between the X-ray transitions and the total number of transitions, the fluorescence yield

$$\omega_K = \frac{P_{XK}}{P_{XK} + P_{AK}}$$

is thus approximately

$$\omega_K = \frac{A \cdot Z^4}{A \cdot Z^4 + B}$$

where A and $B = P_{AK}$ are constants [2-34]. In consequence, the emission of Auger electrons predominates for light elements, whereas with increasing atomic number the probability for the emission of X-ray quanta increases (Fig. 2-47).

Of great importance for microanalysis is the depth of the region below the specimen surface from which the X-ray quanta and Auger electrons are emitted. Due to their low energy and

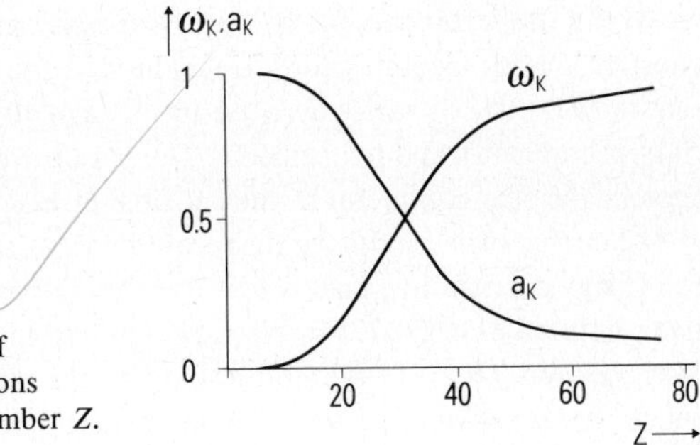

Fig. 2-47.
Yield of K transitions which lead to emission of X-rays (fluorescence yield ω_K) and Auger electrons ($a_K = 1 - \omega_K$) as a function of the atomic number Z.

strong interaction with the material, the escape depth is only about 0.5 to 3 nm for Auger electrons. In contrast, the characteristic X-radiation comes from depths of up to a few μm.

Generation of electron-hole pairs

The emission of X-rays and Auger electrons is based on effects which involve essentially only the inner electron levels. Since the environment of the atoms has hardly any effect on the inner electron shells, these levels are undisturbed, i.e. have well defined energies. In contrast, the outer levels are broadened as a result of such disturbances and form more or less wide bands which can no longer be assigned to individual atoms (section 2.1.2).

Due to the energy transfer from a high-energy electron beam, excitations and electron transitions also occur in the outer electron levels: an electron can be raised from the valence band into the conduction band. An empty state (hole) is left in the valence band acting just like a positve charge carrier. The mean excitation energy $\overline{E_i}$ for generation of such an electron-hole pair is small. But it is larger than the band gap energy E_g (Table 2-2). The difference $\overline{E_i} - E_g$ represents mainly the kinetic energy of the excited electron. Hence one 10 keV electron impinging on a silicon sample may produce nearly 3000 electron-hole pairs. This leads to a disturbance of the semiconductor's thermal equilibrium and causes the generated carriers to be "in excess". Therefore, in the case of a p-doped (n-doped) semiconductor the generated electrons (holes) are called "excess minority carriers".

Within a short time ($\approx 10^{-11}$ s) after excitation, electrons and holes reach energy states close to the band edges corresponding to thermal equilibrium (thermalization), the result of this being the same as for thermal excitation (Fig. 2-12a). They diffuse in random directions through the semiconductor before the electron recombines either directly or via localized sta-

Table 2-2 Band gap energy E_g and mean excitation energy $\overline{E_i}$ for the formation of one electron-hole pair for some important semiconductor materials:

Material	E_g / eV	$\overline{E_i}$ / eV
GaAs	1.43	4.6
Si	1.12	3.8
Ge	0.66	2.85
InSb	0.16	0.42

tes within the forbidden gap (traps), with a hole in the valence band (section 3.6.2). The time constant of this recombination mechanism is the lifetime of the excess minority carriers present in each case. It varies between 10^{-10} s and 10^{-3} s depending on the semiconductor material, its impurities and temperature. The path traversed by the charge carriers during their lifetime in the semiconductor is their diffusion length L.

The energy released during the recombination can be emitted in the form of light. The photon energy corresponds exactly to the energy difference between the states involved. This radiative recombination excited by electron bombardment is known as *cathodoluminescence* [2-35, 36b] (section 3.6.2). If electron-hole pairs are generated within an electric field, e.g. in the space charge region of a pn-junction of a semiconductor device, then the charge carriers are separated. If an electrical circuit is connected to the device, an *electron beam induced current* (EBIC) can be recorded [2-36a] (section 3.6.3).

2.5 Ion beams

2.5.1 Properties of accelerated ions

In the neutral atom, the nuclear charge is exactly compensated for by the number of negatively charged electrons in the atomic shell. The loss or gain of one or more electrons produces a positive or negative ion. Molecules may also be ionized.

The charge on an ion corresponds to the elementary charge or, in multiple ionization, to an – usually small – integral multiple of this charge. The ion mass is determined essentially by the nuclear mass and is, for elements of average atomic number, more than 10^5 times greater than the electron mass. The different physical effects of ions in comparison with electrons are thus largely due to their considerably greater mass.

It was shown in section 2.2 that after traversing an acceleration voltage of U_0, singly charged ions of mass m have the same kinetic energy $W_{\mathrm{kin}} = |q| \cdot U_0 = e \cdot U_0$ as electrons. According to eqn. (2-7), however, their velocity is $v \sim \sqrt{1/m}$, i.e. much smaller than the electron velocity (Fig. 2-14a). In an electrostatic field, a deflection of these ions by a given angle γ requires, in accordance with eq. (2-9), the same electric field strength $E = U_c/d$ as an equal deflection of electrons; the same applies to focusing. In a magnetic field, however, the same amount of deflection (or the same degree of focusing) as for electrons requires a far higher flux density because of $B \sim \sqrt{m}$ in accordance with eq. (2-11). The flux density attainable with rotationally symmetrical magnetic lenses is too low to focus high-energy ions effectively. Electrostatic systems are therefore commonly used for guiding and focusing ion beams.

Major differences between ions and electrons become apparent from the interaction of accelerated ions with matter (section 2.6). In penetrating matter, electrons can transfer their energy only to electrons in the atomic shells or in the energy bands of the solid, whereas ions, due to their large mass, also exchange energy with the atomic or molecular nuclei. The resulting effects are of interest for microanalysis applications and surface investigations. Their exploitation presupposes a mastery of ion sources of great brightness and the construction of ion probes with high current density.

2.5.2 Generation of free ions

There are many methods of generating free ions from gases, liquids or solids [2-37,38]. A number of mechanisms will now be described which are of significance for the generation of free ions for microanalysis applications.

Electron impact ionization

In moving through a gas-filled space, electrons lose their energy in inelastic collisions with the gas atoms (or molecules). Ions can be generated in this process if the kinetic energy of the electrons, E_0, is at least as high as the ionization energy E_i, i.e. the energy with which an electron is bound in the outermost shell (valence level) of a gas atom. Electron impact ionization occurs when this collision results in such an electron being emitted from the atom to leave behind a positively charged ion. The kinetic energy left over after the ionization ($E_0 - E_i$) is distributed over the striking (primary) electron and the emitted (secondary) electron. Both electrons can generate further ions by impact as long as they possess sufficient energy.

A particularly effective form of electron impact ionization occurs in an arc discharge. Here, the gas is contained between a cathode and an anode. The cathode is heated and thus excited to emit electrons. These electrons generate ions and secondary electrons on their way to the anode as described above. With suitable values of pressure and voltage, the discharge "ignites", and a quasineutral "plasma" is formed which contains equal numbers of electrons (primary and secondary ones) and ions. An extraction electrode draws the ions from the plasma to generate an ion beam.

A discharge of this type, which is initiated and maintained by electron emission from a thermionic cathode, is known as a "non-self-sustained" discharge. But discharges may also be produced without an additional electron source. A gas always contains a certain number of ions as a result of cosmic or radioactive radiation. These ions can generate secondary electrons on striking the cathode and thus initiate a "self-sustained" discharge if the field between the electrodes is sufficiently strong. To allow a plasma to be created and to ensure a stable discharge, every electron must produce as many ions as are required to generate just one electron at the cathode.

Irrespective of whether the electrons are generated thermionically or by ionic impact from the cathode, a high ionization efficiency can be ensured only if the electrons remain in the plasma long enough for as much of their energy as possible to be transferred to the gas by impact ionization. This may be achieved by artificially lengthening their path either by suitably shaping the electric field between cathode and anode or by employing a magnetic field. Both options are utilized in practice (section 2.5.3).

Surface ionization

On a hot metal surface which is hit by atoms (or molecules), a temperature-dependent equilibrium is set up between the arrival rate, the degree of coverage and the evaporation rate. At a low temperature, there is a high degree of coverage. Interaction occurs predominantly between the adsorbed atoms, and the atoms are evaporated as atoms. With the arrival rate remaining constant, the degree of coverage decreases as the temperature increases. At a critical temperature T_c (for some metals this may be as high as 1700 °C) the degree of coverage drops below a critical value, and interaction now predominates between the adsorbed atoms and the metal surface: electron exchange with the metal enables the atoms to be ionized and to be evaporated as positive ions. The ratio of the evaporation rate for positive ions to that for neutral atoms is (for $T > T_c$) given by [2-38]

$$\dot{N}_i/\dot{N}_a \sim \exp\left[(E_w - E_i)/kT\right] . \tag{2-28}$$

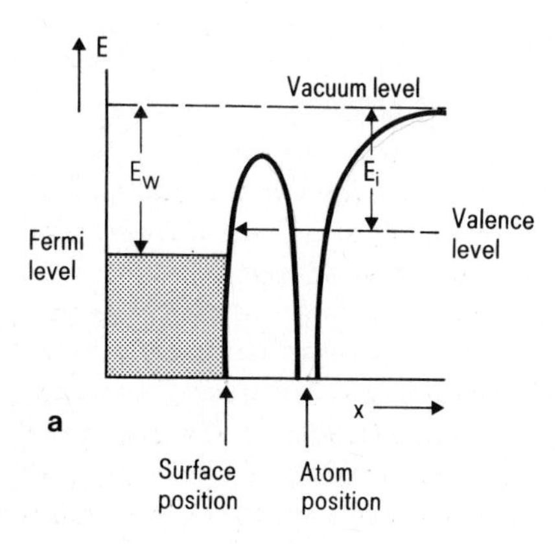

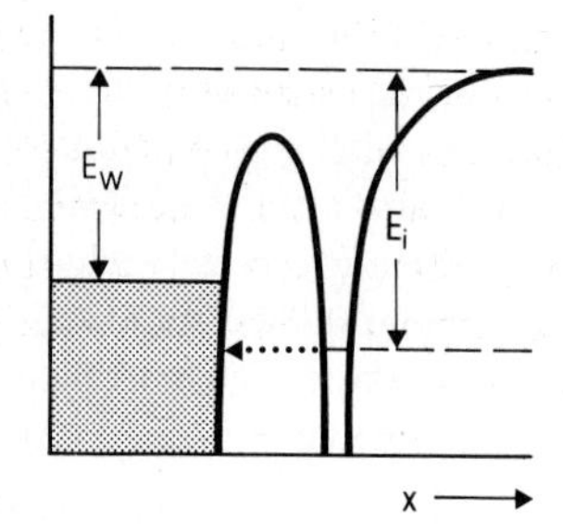

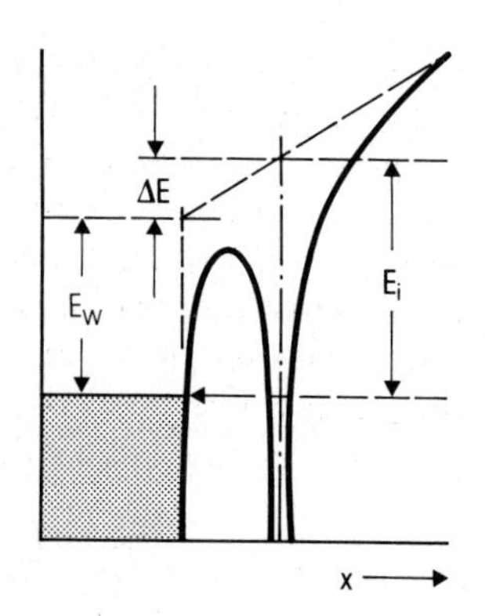

Fig. 2-48.
Potential energy E (x) in the environment of a solid metal surface and of an atom on this surface.
a) Surface ionization: the ionization energy E_i of the atom (which specifies the position of the highest occupied electron level) is smaller than the work function E_w of the metal. An electron transition from the atom into the metal is therefore possible to a site above the Fermi level (arrow).
b) The ionization energy E_i is greater than the work function E_w. In this case, no electron transition (dotted arrow) can take place.
c) Field ionization: by applying an electric field to the metal surface, the potential on the atom side is raised by $\Delta E = E_i - E_w$, so that an electron transition is possible.

A high ion emission thus requires the work function E_w of the metal surface to be greater than the ionization energy E_i of the adsorbed atom (Fig. 2-48a). In this case, an electron from the atom's valence level can tunnel through the potential barrier to the surface and occupy a vacant site above the Fermi level.

Metals with high work functions (they must also have a sufficiently high melting point $T_m > T_c$) include W, Re, Os, Ir and others. Low ionization energies are characteristic of the alkali metals such as Cs, Rb and Li. In practice (see section 2.5.3), for example in the surface ionization of cesium at about 1100 °C, use is made of porous tungsten, which possesses a large effective surface.

Field ionization

Fig. 2-48b shows the distribution of the potential energy for the case where an atom with ionization energy $E_i > E_w$ is located in front of (not necessarily on) a metal surface with work function E_w. Surface ionization through the transition of a valence electron to the surface is prevented by the non-availability of free sites on the equipotential level at the opposite side, i.e. in the metal's conduction band.

By applying a strong electric field of suitable direction (metal surface as anode), the potential is raised on the atom side (Fig. 2-48c). An electron transition from the atom to the surface can occur when the valence level reaches the Fermi level, i.e. when the energy level of the atom has been raised relative to the metal surface by the value $(E_i - E_w)$. After the electron transition, the atom is positively ionized.

This process, in contrast to surface ionization, does not require that the atom to be ionized be adsorbed on the metal surface. At a sufficiently high field strength, which must be of the order of 10^7 V/cm, atoms can be ionized even before they reach the surface. Such high electric fields can be realized with a moderate extraction voltage of about 10 kV only at extremely fine tips with a radius of curvature of less than 1 μm. This method may be used in microanalysis applications to generate ions from liquid metals which wet a pointed solid (liquid-metal ion sources).

Negative ion production

Negative ions are more difficult to generate than positive ions. A gas discharge produces not only positive but also negative ions, the latter being generated by the exchange of charges between ions and neutral particles, and by electron attachment. But these mechanisms become sufficiently probable only for electronegative gas atoms or molecules which easily accept an electron in their shell, e.g. O_2 and Cl_2. A further requirement is a low electric field strength in the discharge space, since only very slow electrons attach themselves to neutral particles. The generated ions must be rapidly extracted from the discharge space and transferred into a high vacuum, since they may otherwise be neutralized again by impact dissociation. A relatively efficient way of generating negative ions is the use of a duoplasmatron (see next section).

2.5.3 Ion sources

Different applications call for different ion sources, which vary in a number of important parameters such as beam current I_b (i.e. the total current extracted from the exit aperture of the ion source), energy spread ΔE (half-width value of the energy distribution), brightness β (see section 2.3.2) and beam purity.

Plasma ion source

Fig. 2-49 is a schematic diagram of a gas discharge ion source with a heated cathode [2-39]. The ionization chamber consists of a cylinder into which the gas is fed from the side (at G). The end faces of the cylinder form the two electrodes C_1 and C_2, both of which are at cathode potential. Electrode C_1 holds a thermionic cathode (filament F) while the other electrode (C_2) acts as a "repeller" for the electrons and contains an aperture for the ion extraction. The tubular anode (A) at positive voltage surrounds the discharge space. If no further arrangements were made, an electron emitted from the cathode would be directly accelerated to the anode (path 1), and the ion yield would be poor. Therefore a magnetic field coil (M) is provided to generate an axial magnetic field that forces the electrons to move along helical paths (path 2). They encounter a negative potential in the vicinity of the repeller electrode C_2 and are reflected. The same occurs on the return path at C_1. The electrons thus oscillate several times back and forth between C_1 and C_2 before losing all their energy, and finally strike the anode.

As a result of the ionizing impacts with the gas atoms, the discharge ignites at a suitable gas pressure and anode voltage to form a plasma as described in section 2.5.2. Positive ions, which reach the vicinity of the exit aperture of the electrode C_2, are affected by the electric field which is applied between C_2 and the extraction electrode E and extends into the aperture of C_2. Here they are accelerated and focused into a beam with the aid of further electrodes.

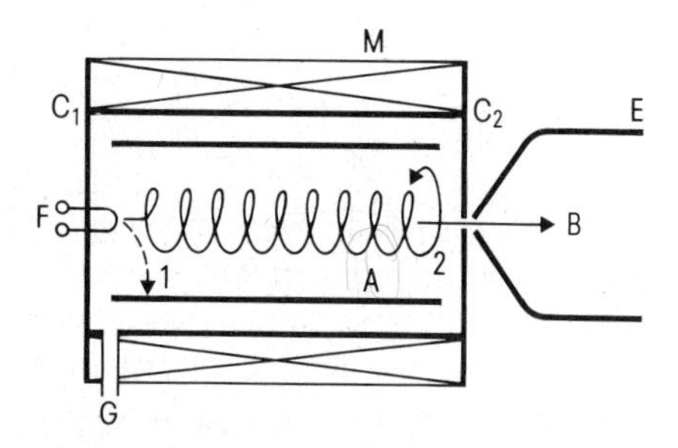

Fig. 2-49. Schematic diagram of a plasma ion source according to [2-39,40]. In the cylindrical discharge space, the electrons emitted from the cathode (C_1 with filament F) are guided by a magnetic field (coil M) along helical paths and reflected at the repeller electrode C_2. Along this path and before reaching the anode cylinder A, they generate positive ions by collisions with the gas atoms (gas inlet G). The ion beam B is extracted from the discharge space with the aid of the electrode E.

In principle, this setup can be used to ionize any gas. But reactive gases such as oxygen can attack the filament and greatly reduce its lifetime. In such cases therefore, self-sustained discharge with a cold cathode is preferred. For this purpose, the filament is replaced by a cold cathode arrangement, such as a massive aluminum tip. Higher values of magnetic coil current I_m, anode voltage U_a and gas pressure p are required to ignite the discharge in this case as compared to the use of a thermionic cathode. The energy spread ΔE is significantly greater than in thermionic cathode operation. The beam current I_b is made up predominantly of molecular ions; e.g. the yield of O_2^+, for instance, is almost 10 times greater than that of O^+.

The following are typical values for a plasma ion source with an exit aperture of 1 mm diameter:

Thermionic cathode mode (argon):
U_a = 40 to 60 V; $p \approx 10^{-1}$ Pa; $I_b \approx$ 10 μA [2-40]; ΔE = 1 to 10 eV [2-38].
Cold cathode mode (oxygen):
U_a = 400 to 500 V; $p \approx 5 \cdot 10^{-1}$ Pa; $I_b \approx$ 25 μA [2-40]; $\Delta E \approx$ 50 eV [2-38].

Duoplasmatron

The duoplasmatron is a further development of the plasma ion source [2-37]. Fig. 2-50 is a schematic diagram of a version with a hollow cathode [2-41]. Fitted between the cathode (C) and the anode (A) is an intermediate electrode (I) whose potential lies between those of cathode and anode; in some setups, the intermediate electrode voltage is kept floating. The anode and the intermediate electrode also constitute pole pieces of the electromagnetic lens (M).

The plasma is greatly constricted by the hole (approx. 5 mm dia.) in the intermediate electrode and the effect of the inhomogeneous magnetic field between the pole pieces. Intensive ionization occurs only in a very small volume of high atom and electron current density in the immediate vicinity of the exit aperture in the anode. This accounts for the high effective-

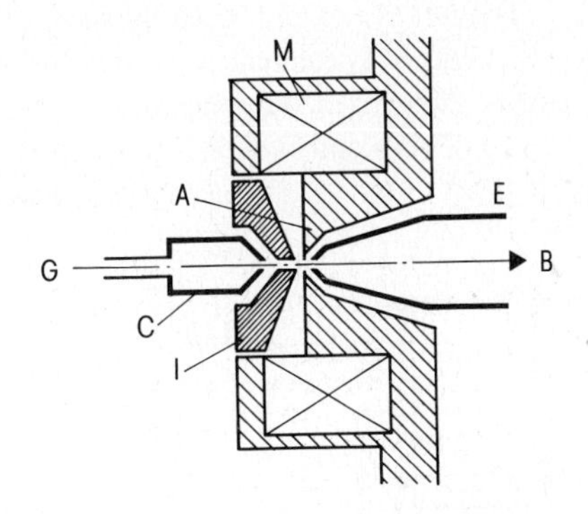

Fig. 2-50. Schematic diagram of duoplasmatron ion source according to [2-41]. The discharge gas is fed at G. An intermediate electrode I is placed between the cathode C and the anode A. By means of the aperture in I and the action of an inhomogeneous magnetic field (magnetic field coil M; I and A act as pole pieces), the plasma is concentrated in a narrow volume before the emission aperture in A. The emitted ions are drawn off with the aid of the extraction electrode E (B is the ion beam).

ness of this configuration. The ionization efficiency (number of ions emitted per number of gas atoms supplied) is between 50 and 95%; this is more than an order of magnitude greater than for the configuration shown in Fig. 2-49 [2-37]. The ions emitted through the anode aperture are drawn off with the aid of an extraction electrode (E) and focused into a beam by further ion-optical elements. Typical experimental parameters are:

U_a = 70 to 180 V; p = 1.5 to 5 Pa; I_b = 30 to 250 mA (exit aperture 1.2 mm dia.) [2-37];
β = 10^2 A/cm² sr; $\Delta E \approx$ 10 eV [2-38,43].

A duoplasmatron lends itself to the extraction of *negative* ions also – most effectively from the marginal regions of the discharge area. As far as the "hot core" of the plasma functions as a negative-ion source at all, the ions it emits are short-lived. Therefore, if negative ions are to be generated instead of positive ones, it is necessary not only to reverse the polarity, but also to shift the intermediate electrode about 1 mm off-center [2-42].

The following operating parameters have been specified for the generation of H^- ions:

U_a = 120 to 150 V; p = 5 to 10 Pa; I_b = 80 to 100 µA (exit aperture 0.9 mm dia.) [2-42].

Surface ionization source

Fig 2-51 shows a surface ionization source of the type used for the generation of cesium ions [2-44,45]. The heart of the setup is the ionizer, a porous frit (F) made of sintered tungsten. The Cs is stored in a reservoir (R) and is vaporized there with the aid of a heater (H_R). Cs atoms pass through the feed tube (T) to the tungsten frit (F) and diffuse through this to the front surface. The heater (H_F) maintains the frit at such a high temperature that the thermal surface ionization as discussed in section 2.5.2 produces Cs^+ ions. At a temperature of 1100 °C, about 99% of the Cs atoms diffusing into the surface are ionized [2-38]. They are drawn off and accelerated with the aid of the extraction electrode E.

The ion beams produced by such an arrangement are largely free of contaminants, since in practice only elements with very low work functions, but not the usual residual gas components, are ionized. The energy distribution of these elements is determined by the thermal energy (<1 eV) and is therefore narrower than in the other ion sources described.

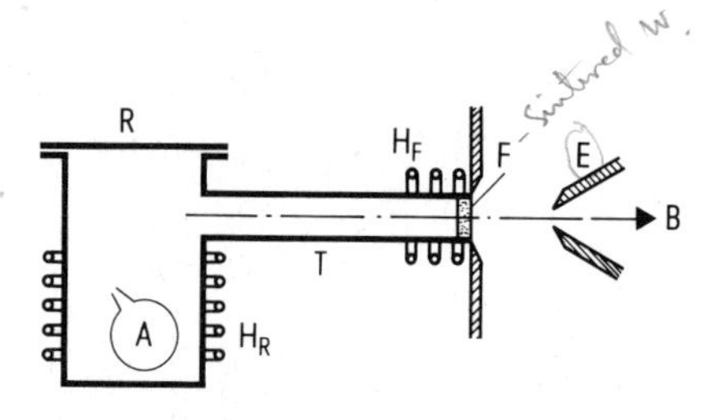

Fig. 2-51. Schematic diagram of surface ion source according to [2-45]. A reservoir R holds an ampoule A containing cesium, which is vaporized with the aid of the heater H_R. Cs atoms flow through the feed tube T and diffuse through the porous ionizer frit F. This is heated by H_F to such a high temperature that the atoms reaching the surface are desorbed as ions. These are accelerated and focused to form the beam B by the electrode E.

The following are typical operating parameters for a configuration with a frit diameter of 4 mm:

Temperature in the reservoir 290 °C; temperature of the ionizer frit 1100 °C; beam current I_b = 2.3 mA [2-45]; $\Delta E \approx 0.4$ eV; $\beta \approx 10^3$ A/cm² sr [2-43].

The configuration shown in Fig. 2-51 was further developed into a Cs microbeam source [2-46]. In this largely miniaturized version, the ionizer frit is replaced by a massive tungsten disk surrounded by the flow of vaporized Cs atoms. Electron bombardment is used to heat the reservoir and the ionizer, and the Cs ions generated at the front surface of the ionizer are drawn off through a narrow aperture of about 0.5 mm diameter.

Liquid metal ion source

Ion sources operating on the principle of field ionization are distinguished by their great brightness. They are therefore of interest for applications requiring high spatial resolution.

Fig. 2-52 shows two configurations for ionizing liquid metals [2-47]. In one case, molten metal is emitted from a heated storage container through a pointed capillary (Fig. 2-52a). In the other, an etched tungsten point is wetted by liquid metal (Fig. 2-52b). In both cases, the liquid metal is shaped into a cone under the action of the electric field, the cone shape being determined by the equilibrium between surface tension and electrostatic stress; the point radius may be smaller than 10 nm [2-48]. At a field strength of the order of 10^7 V/cm (extraction voltage of a few kV), positive ions are emitted from this point by the field ionization mechanism described above, and accelerated to the extraction electrode. Hitherto, liquid metal ion sources have been implemented preferentially with gallium, indium and cesium [2-49,50,51]. The beam current is small in comparison with other ion sources. Due to their great brightness, however, a very high beam current density may be attained. The energy width is due to the space-charge interactions in the ion beam and is therefore strongly dependent on the beam current.

Typical values for a gallium needle cathode [2-49] are:

I_b = 0.5 to 50 μA; ΔE = 5 to 50 eV; $\beta \approx 10^{-6}$ A/cm² sr.

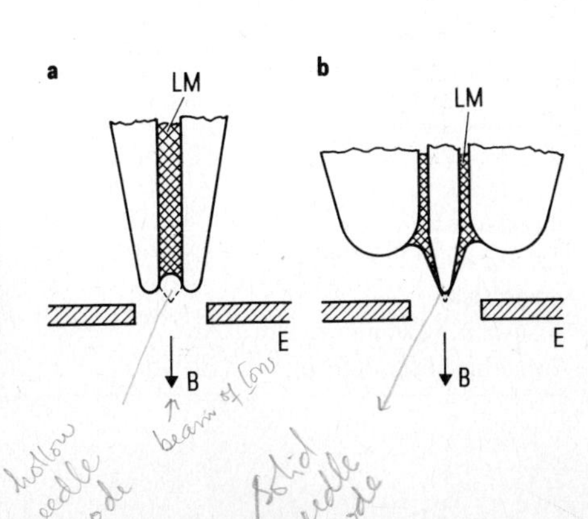

Fig. 2-52.
Liquid metal ion sources for field ionization [2-47]: a) hollow needle anode, b) solid needle anode. Under the action of an electric field, the liquid metal (LM) is drawn out to a point, shown by dotted lines, from which metal ions are emitted; E extraction electrode, B ion beam.

2.5.4 Forming of ion probes

Of outstanding importance for analytical applications are the properties of the ion beam incident on the specimen (i.e. of the ion probe). The most important parameters are: ion energy E_0, probe current I_0 (in secondary ion mass spectrometry also known as the primary ion current), probe diameter d_0 and beam purity.

The desired ion energy can be adjusted by the voltage at the extraction electrode of the ion source or at an additional acceleration electrode. Further ion-optical arrangements must be used to optimize the other parameters. They make use of beam-limiting apertures, which absorb a significant part of the emitted beam current I_b, so that $I_0 \ll I_b$ always results.

For the reasons mentioned in section 2.5.1, electrostatic lenses are used for focusing. As for electrons, the attainable probe diameter is defined to a first approximation by eq. (2-23), according to which

$$d_0 \sim \frac{1}{\alpha_0} \cdot \sqrt{\frac{I_0}{\beta}}.$$

The values of I_0 and α_0 cannot be freely selected due to the need to obtain a good signal-to-noise ratio and because of the aberrations of the lenses (spherical aberration as per eq. (2-15) $d_s \sim \alpha_0^3$). The critical parameter for the probe diameter is therefore the brightness β of the ion source. In most ion sources, the brightness is lower than for electron sources by between two and three orders of magnitude [2-41], i.e. $\beta < 10^3$ A/cm^2 sr. Only the liquid metal ion sources attain values of up to $\beta = 10^6$ A/cm^2 sr [2-49] and permit probe diameters in the submicron region. Due to the chromatic aberration (eq. (2-17)) it is not only the brightness but also the varying energy spread ΔE of the ions that affects the focusing. Other factors affecting the beam are particles of deviating charge and mass: multiply-charge ions, polymer ions, high-energy neutrals (which can be produced by charge exchange during or after acceleration) as well as ions from impurity elements (from the gas or the ion source components). These particles are focused either incorrectly or not at all, thus causing undesired interactions with the specimen. They must be eliminated from the beam path if well-defined and reproducible experimental conditions are aimed for. For "beam purification" (section 7.2.1), use is made of ion-optical filters, which allow the separation of ions with a defined charge-to-mass ratio (q/m) and a defined energy [2-38]. Table 2-3 lists the probe currents and probe diameters which can be realized on the specimen with different ion sources after focusing with ion lenses and, where required, after beam purification.

Table 2-3 Probe currents I_0 and probe diameters d_0, which may be realized on the specimen with different ion sources after focussing with ion lenses.

type of ion source	type of ions	E_0/keV	I_0/A	d_0/µm
plasma ion source	Ar^+, O_2^+	0.5 ... 15	10^{-10} ... 10^{-6}	1 ... 10^3
duoplasmatron	Ar^+, O_2^+	2 ... 15	10^{-9} ... $5 \cdot 10^{-6}$	1 ... 10^2
duoplasmatron	O^-	1 ... 15	... $5 \cdot 10^{-7}$	... 10^2
surface ionization source	Cs^+	3 ... 15	10^{-10} ... 10^{-6}	3 ... $5 \cdot 10^2$
microbeam surface ionization source	Cs^+	... 15	10^{-11} ... 10^{-6}	0.1 ... 10^2
liquid metal ion source	Ga^+, In^+	1 ... 40	10^{-12} ... 10^{-7}	$5 \cdot 10^{-2}$ ... 0.5

2.6 Interactions between ions and solids

2.6.1 General

The effects caused by interactions between accelerated ions and solids are exploited by a number of microanalytical techniques. Depth profiling based on Auger electron spectrometry (AES) makes use of ion beams to abrade the specimen by sputtering during measurement. The atomic layers which are successively revealed are analyzed to obtain a depth-dependent measurement signal. In secondary ion mass spectrometry (SIMS), the sputtered ions themselves constitute the measurement signal.

Sputtering is only *one* effect of the very complex processes occurring when a solid is irradiated with ions. Such effects can give rise to significant changes on the surface and in a layer close to the surface. The analyst must be aware of all these possible effects if he is to select optimal experimental conditions and avoid incorrect interpretation of the analytical results.

The processes treated here are limited to the energy region between 1 and 20 keV and to those effects, which are significant for the methods of analysis described in this book. More detailed presentations of this material can be found in textbooks dealing with ion implantation and sputtering [2-52 to 57] and in the original literature quoted there. In these sources, the specimens are usually called targets and the ions known as projectiles. Throughout this book, the object being analyzed will be called the *specimen* and the striking ions will be called *primary ions* (in line with the terminology usual for SIMS), to distinguish them from the secondary ions generated by sputtering (section 2.6.5).

2.6.2 Ion implantation

When ions of sufficiently high kinetic energy strike a solid, they penetrate it (with the exception of a small fraction of reflected ions, usually less than 1%) and are slowed down. The impacts distribute the kinetic energy of the primary ion over several atoms (primary recoil atoms), which in turn strike other atoms. The result is a collision cascade. The events occurring in such a cascade are shown clearly in Fig. 2-53 [2-58]:

a) As a result of successive collision processes, the primary ions traverse irregularly curved paths until they have lost all their energy and come to rest in the solid (ion implantation).
b) The specimen atoms struck by the collision cascade are displaced from their original sites and traverse paths exhibiting a more or less isotropic angular distribution. This results in a change of atomic arrangement (damage) and recoil mixing of the atoms in a layer close to the surface (see section 2.6.3).
c) Atoms in the uppermost atomic layers may also get struck by the collision cascade. They are emitted from the solid surface when the impulse transferred to them is great enough for them to overcome the surface binding forces (sputtering, see section 2.6.4).

These processes may be described with the aid of two theoretical approaches: the first is to determine the paths of individual particles by means of simulation calculations, as shown in Fig. 2-53. Generally valid statements about factors such as ranges, damage distribution etc.

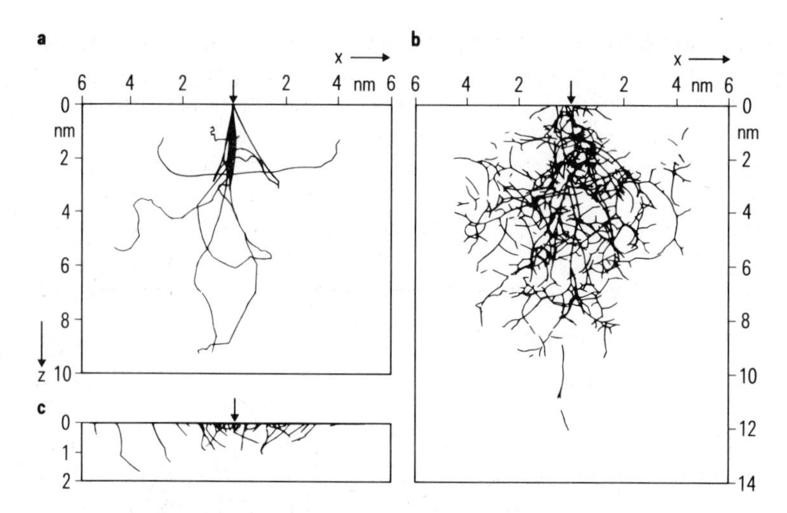

Fig. 2-53. Slowing down of 4 keV Ar^+ ions in copper [2-58,59]. Monte Carlo simulations for 10 (a,b) or 50 (c) events. The figure shows the paths of a) the penetrating ions, b) the recoil atoms and c) the sputtered atoms before leaving the target surface (at $z = 0$).

are then obtained by averaging over a sufficiently large number of single events. Computer programs for this procedure have recently been published [2-57].

The second approach consists of an analytical description of the penetration problem. The procedure is to consider, right from the outset, the total number of the particles and then to calculate their statistical behaviour in accordance with the laws of transport theory [2-57, 60, 61]. The analytical relations obtained by this method are advantageous in a discussion of the effects of different parameters (mass and nuclear charge of ion and specimen atoms, ion energy, bombardment angle etc.).

Amorphous single-element specimens

Let the simplest case be considered initially, namely an amorphous specimen consisting of only one kind of atom, and let attention be concentrated on the spatial distribution of the implanted ions.

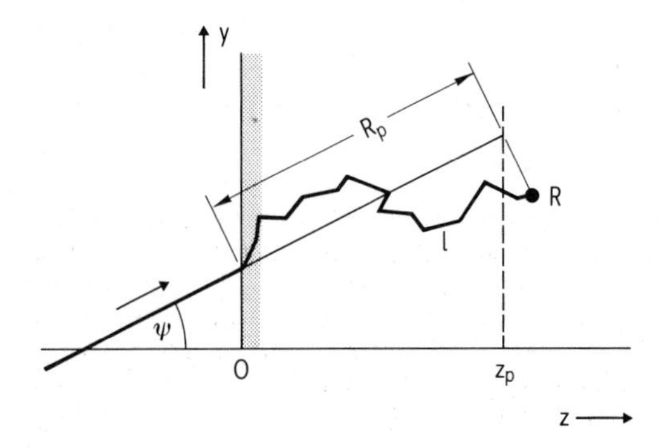

Fig. 2-54.
Path of a representative ion in the solid: total path length l, range R measured along the path l, projected range R_p, penetration depth $z_p = R_p \cdot \cos \psi$. Surface at $z = 0$.

The mean path length traversed by a single primary ion in the solid before it comes to rest is known as its *range R*, and its projection onto the direction of incidence as the *projected range R_p* (Fig. 2-54). For the sake of simplicity, the following treatment will cover the case of normal ion bombardment ($\psi = 0$), so that the penetration depth $z_p = R_p$.

Ions are retarded by two processes which are regarded as independent of each other: 1) elastic Coulomb interactions between the nuclear charges of the ion and the specimen atoms (nuclear collisions) retard and deflect the ion; 2) inelastic interaction with the electron shells causes the ion to lose energy without suffering any significant change of direction, the atoms thereby becoming excited or ionized. For the mean specific energy loss (i.e. the energy loss per path element) we obtain

$$\mathrm{d}E/\mathrm{d}l = (\mathrm{d}E/\mathrm{d}l)_\mathrm{n} + (\mathrm{d}E/\mathrm{d}l)_\mathrm{e} \,. \tag{2-29}$$

In this equation, E is the energy at location l (measured along the ion path), and the indices characterize the energy losses by nuclear collisions (n) and by electron excitation (e). Both are dependent on the ion energy. The mean range of an ion with an initial energy E_0 follows from eq. (2-29):

$$R = \int_{E_0}^{0} \frac{\mathrm{d}E}{(\mathrm{d}E/\mathrm{d}l)_\mathrm{n} + (\mathrm{d}E/\mathrm{d}l)_\mathrm{e}} \,. \tag{2-30}$$

Of greater interest than the range R measured along the path l is the penetration depth z_p, which is identical with the projected range R_p for normal ion bombardment. To obtain it, not only the energy losses but also the angular changes of the ion paths must be taken into consideration. This is done effectively by the projected-range theory of Lindhard, Scharff and

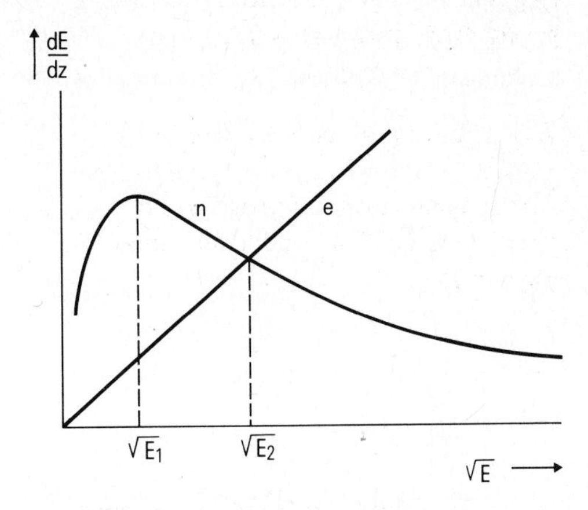

Fig. 2-55. Schematic showing the dependence on the ion energy E of the specific energy losses ($\mathrm{d}E/\mathrm{d}z$) caused by nuclear collisions (n) and electron excitations (e), according to [2-61]. At very high energies, $(\mathrm{d}E/\mathrm{d}z)_\mathrm{n}$ is relatively small, since at high velocities the time for the interaction between the colliding particles is small. $(\mathrm{d}E/\mathrm{d}z)_\mathrm{n}$ increases at medium energies and decreases again at low energies. This is because the reduction of the nuclear forces due to screening becomes increasingly effective with decreasing speed. Nevertheless $(\mathrm{d}E/\mathrm{d}z)_\mathrm{n} > (\mathrm{d}E/\mathrm{d}z)_\mathrm{e}$ still applies in this region. See Table 2-4 for values of E_1 and E_2.

Schiøtt (LSS theory) [2-61]. Due to their positive nuclear charges, the striking ion and the struck atom repel each other like billiard balls, so that the nuclear energy losses can be treated by the laws of classical collision theory. The "collision cross-section", which describes the closest possible approach of the colliding particles, is obtained with the aid of Coulomb's law, taking into account the partial screening of the nuclear charges by the electron shells. In estimating the electronic energy losses, it is assumed that a continuous retardation analogous to that in a viscous medium takes place, and that this retardation is proportional to the ion velocity (i.e. $\sqrt{E}$). Fig. 2-55 shows the fundamental dependence of the specific energy losses on the ion energy [2-61]. The characteristic energies E_1 and E_2 for a number of element combinations are specified in Table 2-4. According to this data, retardation by nuclear collisions predominates in all cases in the energy range treated here ($\leq$20 keV).

Table 2-4 Characteristic energies E_1 and E_2 (Fig. 2-55) in keV for implantation of O, Ar and Cs in Si and GaAs.

	E_1	E_2
O → Si	6	40
Ar → Si	22	230
Cs → Si	210	2300

	E_1	E_2
O → GaAs	12	32
Ar → GaAs	38	240
Cs → GaAs	250	2400

Since the retardation caused by the collisions is a statistical process, not all ions have the same range (even if they have the same mass and initial energy). A distribution is established around a mean projected range R_p (also known as the *1st moment of the range distribution*) and a standard deviation ΔR_p (also known as the *2nd moment or range straggling*). Values for these parameters, which were calculated in line with the LSS theory, can be found in [2-62,63].

The depth distribution of the implanted ions can be described to a first approximation by a symmetrical Gaussian distribution (Fig. 2-56):

$$C(z) \approx C_{max} \cdot \exp\left[-\frac{1}{2} \cdot \left[\frac{z - R_p}{\Delta R_p}\right]^2\right], \tag{2-31}$$

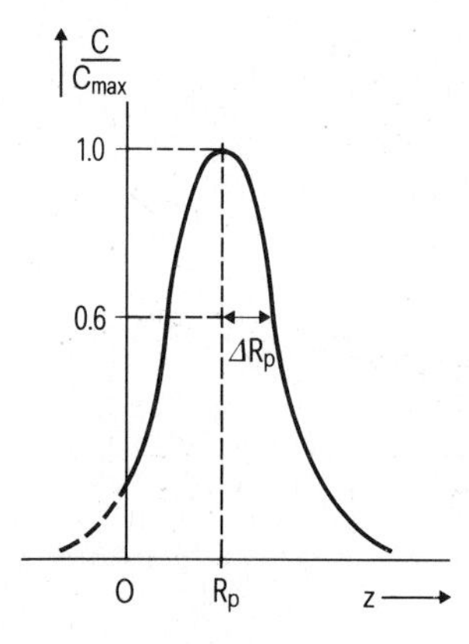

Fig. 2-56.
Theoretical Gaussian distribution curve for implanted ions: normalized concentration C as a function of the depth z. Surface at $z = 0$, mean projected range R_p; standard deviation (range straggling) ΔR_p. The broken part of the curve corresponds to reflected ions.

where $C(z)$ is the concentration of the implanted particles (number/volume) at a depth z. The maximum lies at a depth of $z = R_p$, and ΔR_p is the half width of the distribution at a concentration

$$C(R_p \pm \Delta R_p) = C_{max} \cdot \exp(-1/2) \approx 0.6 \cdot C_{max} .$$

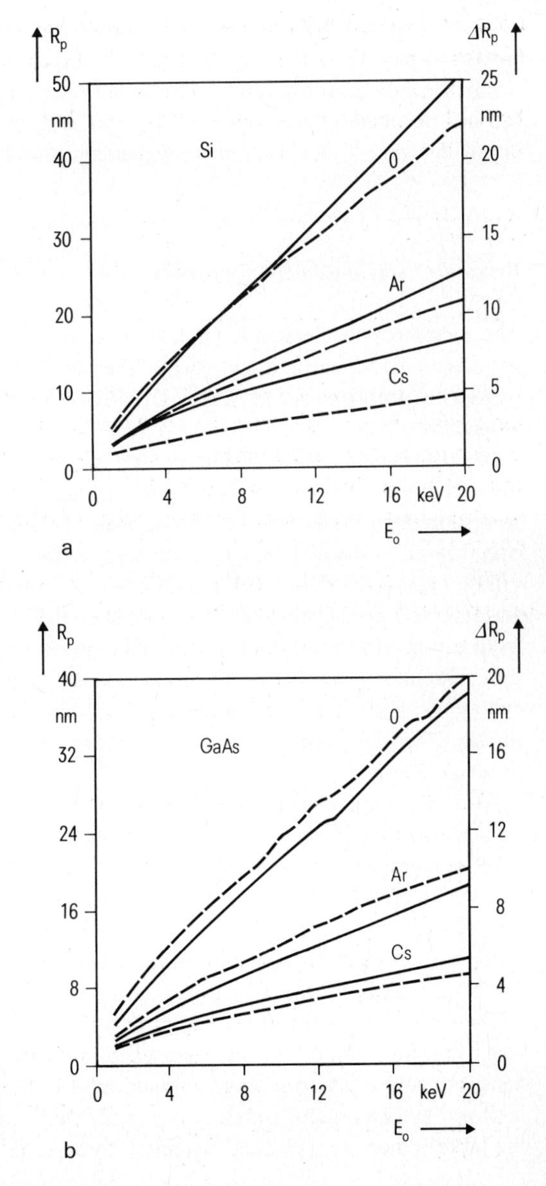

Fig. 2-57.
Projected ranges R_p (continuous curves) and range straggling (standard deviation) ΔR_p (broken curves) as a function of the ion energy for O, Ar and Cs ions: a) in silicon; b) in gallium arsenide. Mean values of 500 events were calculated using simulation program TRIM-2D [2-57].

The total number of all implanted ions (area under the curve of Fig. 2-56) corresponds to the incorporated ion dose ϕ (number/area). By integration of eq. (2-31) we obtain:

$$\phi = \int_0^\infty C(z)\,\mathrm{d}z \approx C_{max} \cdot \sqrt{2\pi} \cdot \Delta R_p ,$$

$$C_{max} \approx \phi/(\sqrt{2\pi} \cdot \Delta R_p) . \qquad (2\text{-}32)$$

The Gaussian distribution eq. (2-31) provides a good first approximation for implantation profiles, especially in the region of the low ion energies of interest here. Deviations from the symmetrical distribution can be described by approximations of higher order [2-62 to 64] but will not be treated here. Values of R_p and ΔR_p obtained from simulation calculations are shown in Fig. 2-57 for a number of element combinations [2-57].

Crystalline single-element specimens

The previous considerations applied to amorphous specimens. Polycrystalline solids with grain sizes $\leq$ 10 nm and statistically distributed orientation of the crystallites can also be treated like amorphous substances. Different conditions exist with respect to single crystalline specimens.

In single crystals, the collision cascade is influenced by the lattice structure [2-65]. According to the "transparency model", in which the atoms are regarded as spheres with a diameter corresponding to the effective cross-section, there exist directions of especially high transparency, known as *most open axis*. In these directions, which correspond to low-index axes, (e.g. ⟨100⟩, ⟨111⟩), the ions find "open channels" between densely populated rows of atoms. If an ion enters such a channel, it is guided in the electric field between the atoms as long as its direction of motion does not exceed a given angular deviation from the channel axis. It thus collides only by grazing the atomic shells and loses energy, if at all, only through electron excitations. It can traverse distances which are a multiple of the range which can be expected for any other directions in the lattice or for amorphous solids. This phenomenon is called the *channeling effect.*

As a rule, implantation profiles in single crystals have two maxima. One originates from the statistically scattered ions and corresponds to the distribution in amorphous material. The other maximum, located at greater depths, is produced by ions having overshot due to the channeling effect. How marked the two maxima are depends on the magnitude of the deviation between the direction of ion incidence and a low-index crystal axis. Even with bombardment in the direction of a high index axis, the experimental distribution curve contains a shoulder, which is formed by ions that have entered an open channel due to scattering.

The lattice structure influences not only the implantation, but also all other effects occurring in the interaction between crystal and ions (sections 2.6.3 and 2.6.4). As yet, no comprehensive theory exists to provide a quantitative description of these effects, but great promise is shown by computer simulation runs which take into account the position of the atoms in the lattice; they are, however, extremely complex [2-66].

Multi-element specimens

Alloys and compounds consist of atoms of different elements. The material is initially assumed to be of single phase. The ion implantation in this case does not take place in a significantly different way from that described so far. Except for a simple correction factor, the projected range R_p and its scattering ΔR_p correspond – at least for binary specimens of mean atomic number $\bar{Z}$ – to the values for a single element specimen of atomic number $Z = \bar{Z}$. The correction factor takes into account the atomic density (number/volume), to which R_p and ΔR_p are inversely proportional [2-62].

What has been said for crystalline single-element specimens applies analogously to crystalline specimens which contain several elements. Very unclear conditions prevail in multiphase systems. A mean range may at best be evaluated for very finely crystalline materials. Coarsely crystalline phase mixtures lead to locally diverse ranges and to selective effects which act especially on the surface topography (sec. 2.6.3) and the sputter yield (sec. 2.6.4).

Sputter influenced ion implantation

In the previous considerations dealing with the depth distribution of the implanted ions, the abrasion of the surface by sputtering during implantation was neglected. The distribution presented in Fig. 2-56 has in fact no fixed zero point. Instead the surface assumed at $z = 0$ moves into the solid with a speed $\dot{z}$, which is proportional to the sputter rate (section 2.6.4).

If we assume that in every infinitesimally short time interval a distribution according to eq. (2-31) from the actual surface is set up, then the distribution obtaining after a finite sputter time can be regarded as a superposition of all preceding distributions (neglecting the specimen modifications treated in the following section). The maximum of this distribution increases with the sputter time (i.e. with the dose ϕ) and at the same time approaches closer to the mo-

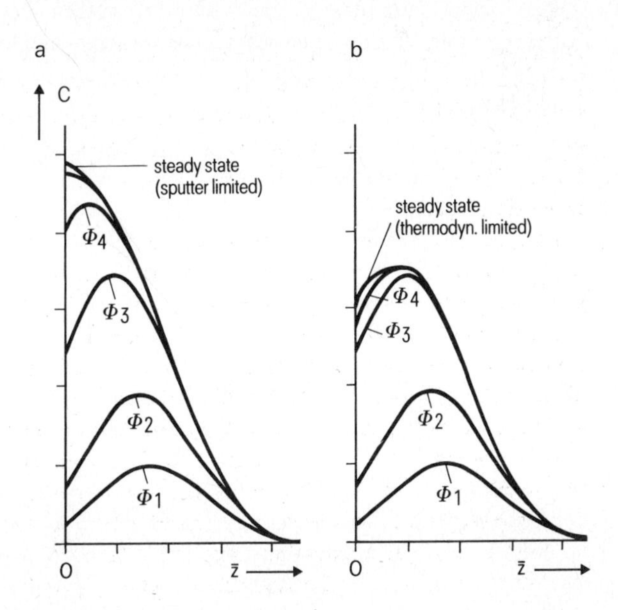

Fig. 2-58.
Schematic showing the dependence of the concentration C of implanted ions on the momentary depth $\bar{z}$ for simultaneous sputtering (momentary surface at $\bar{z} = 0$). The parameter used is the dose ϕ ($\phi_1 < \phi_2 < \phi_3 \ldots$), a) not assuming and b) assuming a thermodynamically defined concentration limit in the relevant matrix and out-diffusion of the excess ions. According to [2-68].

mentary surface (Fig. 2-58a) [2-64,67]. The steady-state concentration at the surface is attained as soon as the sputter rate of the previously implanted ions attains equilibrium with the implantation rate. The equilibrium concentration obtained from this simple model by calculation may well exceed the thermodynamically defined maximum concentration. This applies, to a large extent for the experimental conditions considered here, and the presentation in Fig. 2-58b is therefore more realistic [2-68]. Excess ions are in this case precipitated or transported away from the surface region by diffusion. The equilibrium condition is usually established after a layer with a thickness of the ion range R_p has been sputtered off.

2.6.3 Specimen modifications

The lattice structure, chemical composition and surface topography of a solid may be modified by ion bombardment. Also the ion implantation already discussed represents a modification of the specimen. It is exploited in processes such as the doping of semiconductors. In sputter depth profiling, ion implantation is a side-effect which can affect the sputter yield very significantly (section 2.6.4).

In addition to the effects caused directly by ion bombardment, secondary effects may occur due to a disturbance of the thermal equilibrium or to electric charges: diffusion, segregation and electromigration.

Structural modifications

On the interactions treated in the previous section, only the nuclear collisions lead to radiation damage, i.e. to a displacement of atoms. Here the excitation of electrons is of importance only insofar as the energy expended (E_e) is not available to the nuclear collisions ($E_n = E_0 - E_e$).

Recoil atoms push other specimen atoms from their sites and as a rule come to rest at interstitial sites. This leads to an enrichment of vacancies and interstitials in the region of the collision cascade. Further lattice disturbances are caused by clustering of vacancies, often in conjunction with impurity atoms.

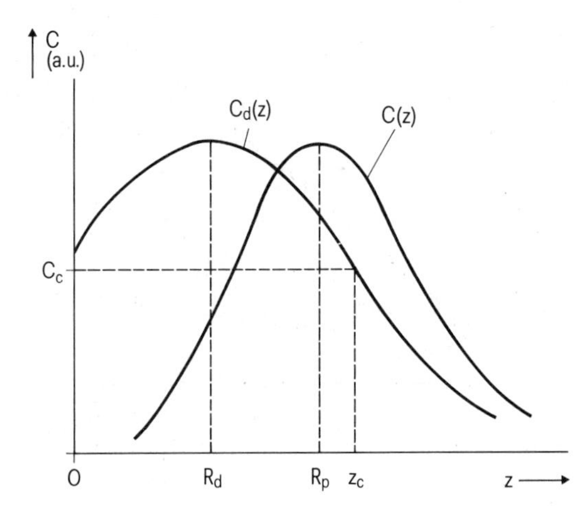

Fig. 2-59.
Damage profile $C_d(z)$ and implantation profile $C(z)$, profiles normalized to peak level. An amorphous zone is created in the range $0 \leq z \leq z_c$ when the critical concentration of point defects for the amorphization is at C_c.

The number of atoms directly or indirectly displaced by a primary ion is estimated to be [2-69]

$$N = E_n / 2 E_d \, , \tag{2-33}$$

where E_d is the displacement energy required to remove an atom permanently from its lattice site. For typical semiconductors E_d ranges between 8 and 30 eV and has a value of 14 eV for silicon. It can be seen from Fig. 2-55 and Table 2-4 that nuclear interactions predominate in the low energy region, so that here $E_e < E_n \approx E_0$. Under this assumption, eq. (2-33) can be

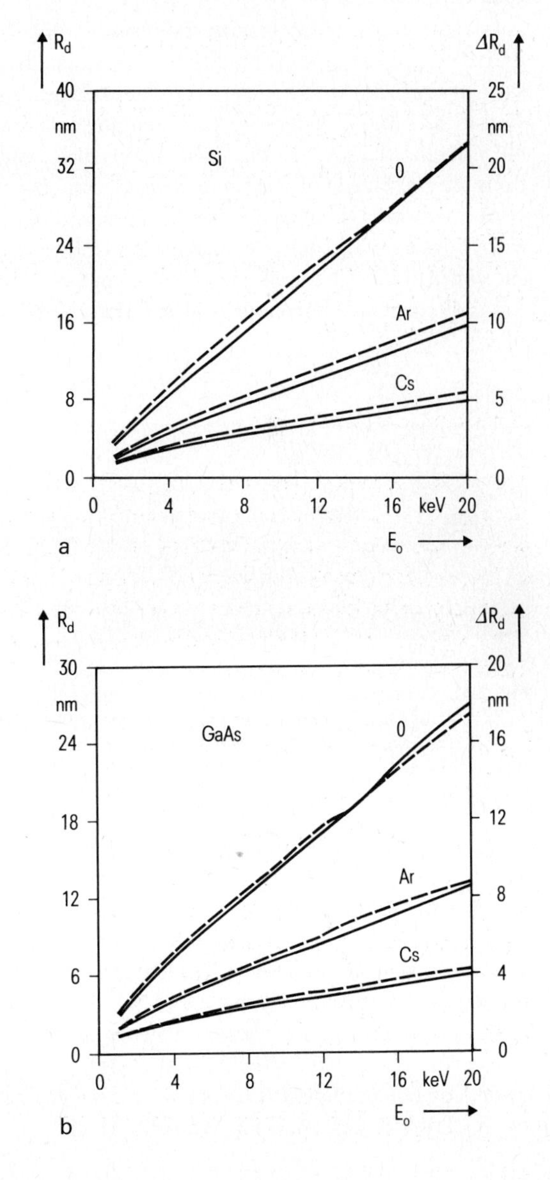

Fig. 2-60.
Mean radiation damage range R_d (continuous curves) and standard deviation ΔR_d (broken curves) as a function of the ion energy for O, Ar and Cs ions: a) in silicon; b) in gallium arsenide. Simulation calculations as indicated in Fig. 2-57.

used to show that an ion with an initial energy of a few keV can displace more than 100 atoms, i.e. generate more than 100 point defects.

At a low dose ($\phi < 10^{13}$ cm^{-2}) and as long as no multiple collisions occur, the local concentration C_d of the displaced atoms is proportional to the specific energy loss due to nuclear collisions: $C_d \sim (dE/dz)_n$. The depth dependence $C_d(z)$ is given by the energy dependence of $(dE/dz)_n$ (Fig. 2-55): during retardation, the ion energy values change from high to low, i.e. from right to left in Fig. 2-55. In the energy range under consideration here ($E < E_2$ in Fig. 2-55), the nuclear interaction dominates and the ions already cause radiation damage at the surface. As the energy decreases, $(dE/dz)_n$ initially increases to a maximum value, then decreases again.

Like the implantation profile, the resulting distribution of radiation damage $C_d(z)$ can be described to a first approximation by a Gaussian distribution (Fig. 2-59), i.e. by a relationship such as eq. (2-31) with a mean radiation damage range R_d and a corresponding standard deviation ΔR_d [2-64]. Values for these parameters are tabulated in [2-63]; R_d is between 10 and 40% smaller than R_p. In Fig. 2-60, values of R_d and ΔR_d obtained from simulation calculations are presented for the same element combinations as in Fig. 2-57 [2-57].

At a very high damage density, the specimen material can become amorphous. This is the case above a critical defect concentration C_c or a critical dose ϕ_c. If we assume that the critical concentration between the surface and $z = z_c$ is attained in the example shown in Fig. 2-59, then an amorphous zone is formed there, as can be seen on the electron micrograph of Fig. 2-61.

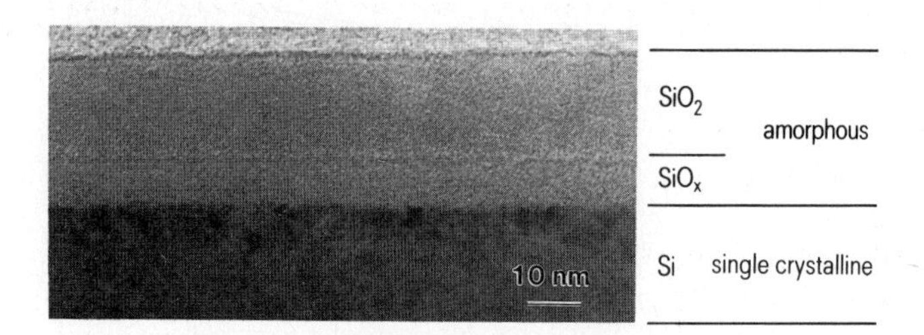

Fig. 2-61. Cross-section transmission electron micrograph of the crater bottom of a sputtered single crystal silicon specimen. Formation of amorphous layers of SiO_2 and SiO_x ($0 \leq x < 2$) as a result of normal bombardment with 12 keV O_2 ions [2-80].

Not all crystalline substances show this behavior. Some retain their crystalline structure under the same bombardment conditions or even become crystalline if they were previously amorphous. To this group belong most metals and their oxides. The semiconductors Si, Ge, GaAs, GaP, InAs as well as compounds such as SiO_2 and Si_3N_4 remain crystalline only above a transition temperature. Below this temperature, they become amorphous. A number of criteria exist for this disparate behavior [2-70], but as yet no generally valid explanation has been given. The transition temperature depends on the energy and mass of the primary ions and on the dose rate $d\phi/dt$, which indicates competing effects of radiation damage and annealing during ion bombardment. This behavior has a great effect on the sputter yield of semiconductor crystals (section 2.6.4, Fig. 2-64) [2-71].

Compositional modifications

In a multi-element specimen (alloy or compound) the atoms have different masses and binding energies. Under ion bombardment, this leads to a number of consequences, one being *preferential sputtering*, which means that one or other kind of atom is preferentially emitted from the surface. For almost equal binding energies, the atom species with the smaller mass (i.e. greater emission depth), and for almost equal masses the one with the lower binding energy is preferentially sputtered. This leads to a surface-zone in which the type of atom with the higher sputtering yield is depleted whereas the other one is enriched. It would be logical to expect that the depth of this layer corresponds more or less to the emission depth of the sputter particles of a few atomic layers (about 1 nm). But experiments show a zone of altered composition *(altered layer)* with a generally greater depth. Various material transport processes are responsible for these more extensive changes, which have a direct or indirect relationship to the ion bombardment [2-72,73]:

a) The large number of collisions and their isotropic directional distribution in a collision cascade (Fig. 2-53b) cause a total mixing of the atoms including the implanted ions *(cascade mixing)*. If the collision cascade strikes an interface, atoms are transported from both sides, i.e. in forward and reverse directions, across the interface. This leads to a symmetrically bilateral "blurring" of the interface.
b) Those atoms which are struck directly by the penetrating ion receive an impulse to move preferentially in the forward direction *(primary recoil mixing)*. Although the number of such direct impacts is small, the energy transfer is greater than otherwise in the cascade. This effect makes the blurring of an interface caused by cascade mixing unsymmetrical and extends it more deeply into the substrate. Annother result of this effect is that impurity atoms (e.g. adsorbates and contaminants) present on the surface are implanted in the target material *(recoil implantation)*.
c) The ion bombardment considerably disturbs the thermal equilibrium at the surface, either by sputtering off a component originally segregated at the surface or by altering an originally homogeneous surface layer by preferential sputtering. This disturbance is counteracted by transport processes which are additionally favored by the large number of lattice defects generated in the collision cascade. They are temperature-dependent and are known as *radiation-enhanced segregation or radiation-enhanced diffusion.*

The previously mentioned effects are often subsumed under the generic term *atomic mixing*. They all occur more or less simultaneously and reach down typically to a depth which corresponds approximately to the ion range (a few nm). In individual cases, however, much more extensive changes must be reckoned with:

d) At increased specimen temperatures, material from the specimen interior can reach the surface due to *thermal diffusion*.
e) Strongly insulating specimens can become charged under ion bombardment. This leads to the formation of an electric field which extends far into the interior of the specimen and can set easily mobile ions in motion, e.g. alkali ions in SiO_2 *(electromigration)*.

The compositional modifications depend on the ion dose and can go so far as to lead to the formation of precipitates (e.g. in the form of gas bubbles upon bombardment by noble gas ions) or new compounds (e.g. oxides of modified valence) in the altered layer.

Surface modifications

Another necessary concomitant of ion bombardment is the appearance of changes in the specimen surface topography [2-74,75]. If this was originally rough, it usually becomes even rougher (Fig. 2-62). But irregularities can be created even on very smooth surfaces; they take the form of steps, facets, pits, cones, pyramids etc. All these phenomena are caused by local differences in the sputter rate (section 2.6.4).

Surface impurities, contaminant particles and inclusions close to the surface, which are sputtered at a lower rate than their surroundings, screen the underlying material. As a result, continuous ionic bombardment leads to the formation of cones or pyramids, these latter being bounded by crystal growth surfaces. Precipitates which are due to the implantation of the impinging ions can also be the cause of such effects.

In the mechanical stress field associated with lattice defects, the binding energy of the atoms is reduced and the sputter rate correspondingly increased. At those sites at which lattice defects extend to the surface, this leads to the formation of etch pits similar to those resulting from chemical defect etching. For the same reason, grain boundaries are preferentially attacked and trenches formed there. This process however is superimposed by another effect: differently oriented crystallites of the same material also have different sputter rates, so that in polycrystalline specimens there can be considerable height differences between adjacent grains. The result is a step or terrace shaped surface topography (Fig. 2-62b). With increasing dose, the structures become coarser and in part overlap, often producing a very bizarre surface topography.

Similar phenomena are also observed in the depth profiling of layer structures. Sputtering through interfaces can lead to the creation of cones, needles, steps and similar deformations due to layer thickness fluctuations and local differences in sputter rate. These effects significantly impair the performance of sputter depth profiling (section 2.6.6).

The surface erosion induced by the ion beam is counteracted by the amorphizing process already discussed. Specimens which have a clean, even surface and become amorphous under

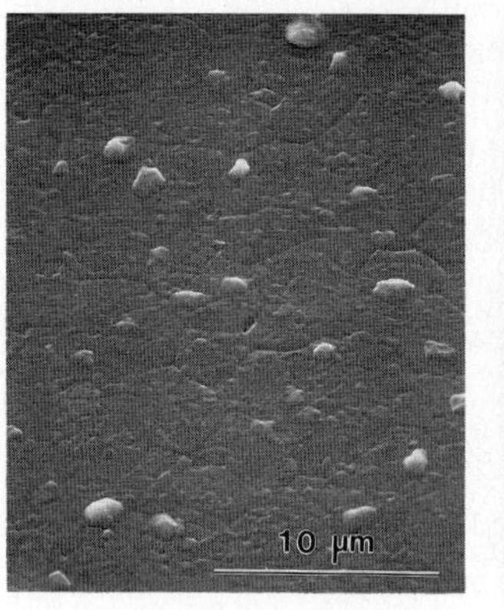

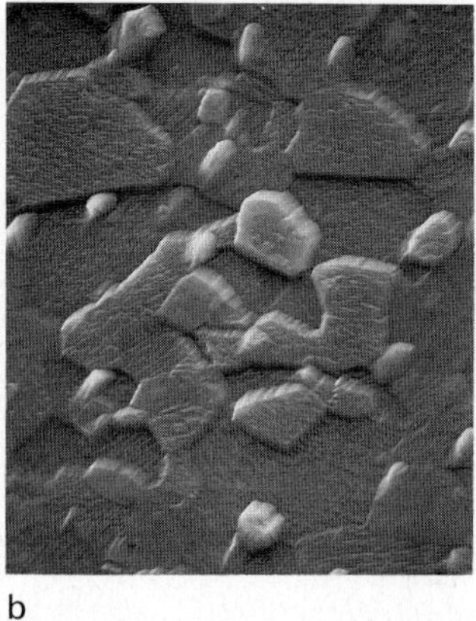

Fig. 2-62. Scanning electron micrographs of the surface of a polycrystalline aluminum sputter layer: (a) after sputtering and annealing; (b) after bombardment with 2 keV Ar^+ ions ($\phi = 1.2 \cdot 10^{18}$ cm^{-2}; $\psi = 55°$).

ion bombardment, such as most semiconductors, also retain their smooth surface during continuous ionic bombardment or show, at most, a slight waviness.

2.6.4 Sputtering

The preceding sections have given us the following information about sputtering: in a collision cascade, the primary ion energy is distributed among many atoms. Atoms are emitted from the uppermost atomic sites if the impulse transferred to them has a magnitude and direction allowing them to overcome the binding energy.

The sputter rate is defined as the number of sputtered atoms per unit time, $\dot{N} = \mathrm{d}N/\mathrm{d}t$. Within the range of validity of the collision cascade model, the sputter rate is proportional to the arrival rate of the primary ions (i.e. the primary ion flow $\dot{N}_0 = i_0/q$):

$$\dot{N} = Y \cdot \dot{N}_0 \,. \tag{2-34}$$

The proportionality factor is the *sputter yield* [2-76]. This varies, depending upon the ion-specimen combination and primary ion energy, between 10^{-3} and 10^2 atoms/ion; typical values are 1 to 5 atoms/ion.

Single-element sputtering

The sputtering of amorphous or fine-grained specimens will initially be considered. The yield depends upon how much energy the penetrating ions deposit close to the surface and how strongly the surface atoms are bound. For a given ion-specimen combination (and neglecting inelastic energy losses), the following applies [2-60, 76]

$$Y(E_0) \sim \frac{[(\mathrm{d}E/\mathrm{d}z)_\mathrm{n}]_{z=0}}{E_\mathrm{b}} \,. \tag{2-35}$$

Here the specific energy loss $(\mathrm{d}E/\mathrm{d}z)_\mathrm{n}$ at the surface ($z = 0$) must be inserted, where $E = E_0$. E_b is the surface binding energy; it has a value of several eV and is generally approximated by the sublimation energy [2-65].

Taking silicon as an example, Fig. 2-63 plots experimentally determined sputtering yields as a function of the ion energy [2-76]. The profile is similar for all ion-specimen combinations and can be read off qualitatively from the curve for $(\mathrm{d}E/\mathrm{d}z)_\mathrm{n}$ in Fig. 2-55 by setting $E = E_0$ for the energy plotted on the abscissa. With increasing energy, the yield initially increases and then reaches a shallow maximum (in Fig. 2-63 at about 10 keV for Ar^+). As the energy increases further, the collision cascade extends deeper into the interior of the specimen; the energy deposited close to the surface decreases, as does the sputter yield.

Like the specific energy loss, the proportionality factor in eq. (2-35) depends on the masses of the colliding particles. For a given specimen material, the yield increases with the ion mass and the yield maximum shifts towards higher energies (Fig. 2-63). Only for very light ions (H^+, He^+) will the maximum lie in the region below 10 keV.

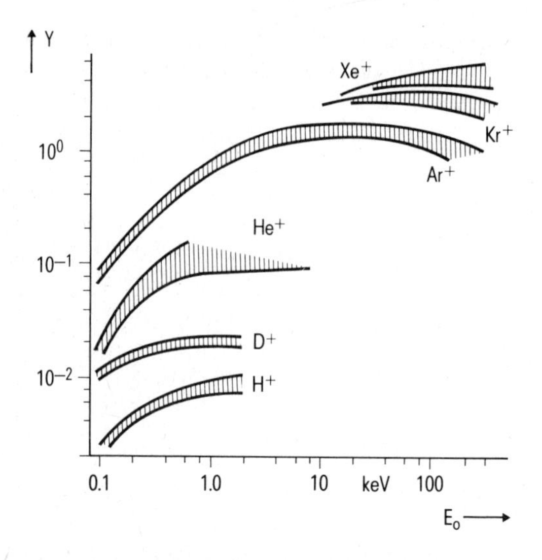

Fig. 2-63.
Sputter yield Y (atoms/ion) of silicon as a function of the primary ion energy E_0. Summary of experimental values for various primary ions; for references to the original works see [2-76].

When the ion bombardment changes from normal ($\psi = 0$) to oblique incidence, the yield initially increases monotonically with ψ. It attains a maximum, which can be a multiple of $Y(0°)$, between 60° and 80° [2-76]. As the angle of incidence increases further, the proportion of reflected ions (whose energy does not contribute to the sputtering process) increases and the sputter yield therefore declines steeply to $Y(90°) = 0$.

For *single crystals*, the lattice anisotropy influences the angular dependence of the sputter yield [2-71]. Although the yield curve has, to a first approximation, the form described, it always contains minima when the ions strike the crystal in the direction of a low-index lattice axis (see the curve for $T = 600°C$ in Fig. 2-64). This observation can be explained at least qualitatively by the channeling effect (section 2.6.2): the ions penetrating deeply into the crystal interior in the preferential directions deposit less energy close to the surface and thus generate fewer sputter particles than in the other directions.

In semiconductor crystals, the amorphization already discussed in section 2.6.3 has an effect on the sputter yield in the region below the transition temperature T_a: for $T < T_a$, the yield increases with ψ as for amorphous specimens; above this temperature the yield curve shows the minima characteristic of the single crystals [2-71,77] (Fig. 2-64). Under the experimental conditions to be considered here, semiconductors can be expected to show the same sputtering behavior as amorphous specimens.

The yield values discussed hitherto as well as equations (2-34) and (2-35) relate to the *total* yield, i.e. the mean number of all particles sputtered per ion, independently of their energy, their direction and their degree of excitation or ionization. Yield values which refer to sputter particles from a specific kinetic energy interval or solid-angle interval, or to particles in a specific ionization state (see section 2.6.5) are known as *differential yields*. Such yields are of interest for applications such as SIMS and are not proportional to the total yield.

The kinetic exit energy E_{ex} of the sputter particles ranges from zero to a few 100 eV. The distribution exhibits a broad maximum at about 10 eV and decreases at rising energies with approximately $1/E_{ex}^2$. These low energies are understandable since only a fraction of the primary ion energy is transferred to the individual, sputtered atom. This atom also suffers energy losses due to inelastic processes in the emission and in overcoming the binding energy.

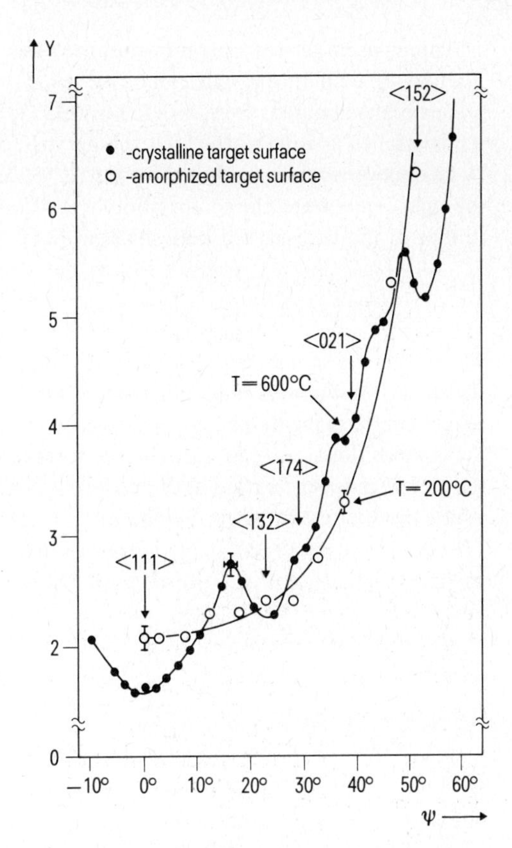

Fig. 2-64.
Sputter yield Y (atoms/ion) of an Si (111) surface as a function of the angle of incidence ψ for 30 keV Ar^+ ions; tilt axis ⟨112⟩ [2-77]. At the lower specimen temperature (200 °C) the single crystal surface is amorphized by the ion bombardment.

The distribution over the solid angle Ω resembles that of backscatter electrons: it can, for normal bombardment, be approximated by a cosine law (cf. Fig. 2-36a). With oblique primary ion bombardment, the maximum is deflected away from the incident beam (cf. Fig. 2-36b). Emission maxima can be observed for single crystals in the direction of low-index crystal axes. Collision sequences along the densely packed atomic rows as well as the directional properties of the surface bonds are among the mechanisms which are held responsible for this effect, but no concordant explanation currently exists [2-65].

In most cases the majority of the sputter particles are neutral atoms. A certain fraction may be emitted as atom clusters, or as atomic or molecular ions (section 2.6.5).

Multielement sputtering

A priori, the constituents of a specimen made up of several elements are not sputtered from its surface in proportion to their concentrations. This is because of the ion-beam-induced changes treated in the previous chapter. These are known as matrix effects and comprise ion implantation, preferential sputtering, atomic mixing, radiation enhanced diffusion, and segregation.

As long as no far-reaching transport processes are involved (thermal diffusion or electromigration), an equilibrium state can be set up: the components with the greater sputter yield are preferentially abraded, those with lower yield are enriched at the surface. The sputter equilibrium is reached as soon as the high sputter yield of a component is precisely compensated for by its impoverishment, and the low sputter yield of another component by its enrichment. In the *equilibrium state*, the composition of the flow of sputtered particles therefore corresponds to that of the undisturbed bulk material [2-60, 72]:

$$\dot{N}_i/\dot{N} = X_i\,, \tag{2-36}$$

where $\dot{N}_i$ is the sputter rate of element i, $\dot{N}$ the total sputter rate and X_i the mole fraction of the element in the bulk material. This always applies in the equilibrium case, even if some of the other effects mentioned above are active in addition to preferential sputtering. Because of the compositional modifications at the surface, a distinction must be made between X_i and the atomic number fraction at the surface X_i^{S} (i.e. in the approximately 1 nm thick layer from which the sputtered atoms originate).

According to eq. (2-34), the *total sputter yield* for a specimen made up of i elements is given by

$$Y = \dot{N}/\dot{N}_0 = \sum_i \dot{N}_i/\dot{N}_0 = \sum_i Y_i\,, \tag{2-37}$$

where Y_i is the *partial sputter yield* of the element i. Without matrix effects, this would be $Y_i = X_i^{\mathrm{S}} \cdot Y_i^0$ if Y_i^0 is the sputter yield of the pure i-element specimen. Due to the matrix effects which actually prevail, a component sputter yield Y_i^* is defined such that $Y_i = X_i^{\mathrm{S}} \cdot Y_i^*$ and thus

$$Y = \sum_i X_i^{\mathrm{S}} \cdot Y_i^*\,. \tag{2-38}$$

The difference between Y_i^0 and Y_i^* is then a measure of the matrix effect on the sputter yield. (In the literature, a clear distinction is not always made between the various yield values. Frequently, Y_i^0 is set in place of Y_i^* or Y_i^* is designated as the partial sputter yield.)

By comparison of eq. (2-37) with eq. (2-38) it follows that $X_i^{\mathrm{S}} \cdot Y_i^* = \dot{N}_i/\dot{N}_0$. Further, eq. (2-36) applies in the equilibrium case. The equation

$$\frac{X_{\mathrm{A}}^{\mathrm{S}}}{X_{\mathrm{B}}^{\mathrm{S}}} = \frac{X_{\mathrm{A}}/Y_{\mathrm{A}}^*}{X_{\mathrm{B}}/Y_{\mathrm{B}}^*} \tag{2-39}$$

can thus be derived for a specimen composed of the two elements A and B. It specifies the composition of the uppermost two to three atomic layers (exit depth of the atoms) during equilibrium sputtering. However, due to the effects described in section 2.6.3, the altered layer reaches down to a depth which approximately corresponds to the ion range. In order to attain the equilibrium state, a layer with a thickness between one and two ionic ranges must therefore be sputtered off.

The different sputter behavior of the components of multi-phase material is known as *selective sputtering.* It leads to the pronounced surface structures already treated in section 2.6.3.

Chemically influenced sputtering

All the above-described effects, which influence the sputter yield, are physical in nature. This is also true of the material transport processes which give rise to local changes of chemical composition in the specimen (section 2.6.3). They can be observed during bombardment with noble gas ions. In sputtering with other ions (e.g. oxygen, nitrogen), chemical reactions may occur between the implanted ions and the specimen atoms leading to the formation of compounds at the surface (oxides, nitrides) whose binding energy differs from that of undisturbed specimen material. According to eq. (2-35), a reduced binding energy causes an increase of the sputter yield and vice versa. This is known as chemically enhanced and chemically reduced physical sputtering, respectively [2-78]. (In individual cases, the binding energy can be so far reduced, that the resulting molecules are desorbed without further effects. This process is known as *chemical sputtering* [2-78]. It presupposes an increased specimen temperature of a few hundred °C and is of lesser importance for the methods of analysis treated here.)

These effects are also observed when sputtering takes place with non-reactive ions in an atmosphere of reactive gases. The precondition for this occurring is that the number of gas atoms incident per time interval on the specimen is comparable to or greater than the sputter rate. The gas atoms are absorbed at the specimen surface and incorporated by means of recoil implantation and radiation enhanced diffusion.

The degree of chemical change increases with the ion dose or the partial pressure of the reactive gas until sputter equilibrium is reached. Silicon, for example, can be completely oxidized at the surface by perpendicular bombardment with O_2 ions, an example is shown in Fig. 2-61. In the case of oblique bombardment, oxidation is incomplete [2-79].

In reactive sputtering, not only the sputter yield but also the nature of sputtered particles changes. The proportion of ions, in either atomic or molecular form, is dramatically increased. This is utilized in processes such as SIMS analysis: sputtering with oxygen or cesium ions can improve the secondary ion yield by several orders of magnitude, as compared to sputtering with noble gas ions.

2.6.5 *Ionization*

The number of positive or negative ions produced during sputtering of an element i is given by the *secondary ion yield*

$$Y_i^{\pm} = P_i^{\pm} \cdot Y_i \,, \tag{2-40}$$

where $P_i^{\pm}$ is the ionization probability for positive or negative ions and Y_i the partial sputter yield according to equation (2-37).

When a dimensionless, element-specific instrument constant β_i is introduced, the measurable *secondary ion intensity* has the value

$$I_i^{\pm} = \beta_i \cdot P_i^{\pm} \cdot Y_i \cdot \dot{N}_0 \,. \tag{2-41}$$

The secondary ion yield is referred to the number of primary ions and the secondary ion intensity to the time interval. In the literature, however, the quantity defined by equation (2-41) is also frequently known as the secondary ion yield.

Inert ion emission

Let us initially consider inert (or intrinsic) ion emission, i.e. the physical sputtering of pure elements or their alloys uninfluenced by chemical effects. A collision leading to sputter emission certainly involves electronic interactions by which the struck atom is excited or ionized (ionization is here considered as a special case of excitation). The lifetime of such excited states (10^{-16} s) is, however, much shorter than the time (10^{-13} s) which the excited particle needs to leave the specimen surface. Until the atom (ion) reaches the region beyond the influence of the surface, excitation and deexcitation processes take place at a high rate. The question of ionization probability $P^{\pm}$ thus involves the probability of survival of excited states. This is very low: only a small proportion of the sputtered particles ($<10^{-4}$) is emitted as ions. The great majority are atoms in their ground state.

To describe the process of ion emission during sputtering, numerous thermodynamic and quantum mechanical models have been developed; overviews of these can be found in [2-81 to 83]. Good agreement with many experimental results is shown by the quantum-mechanical tunnel model, which will now be described in simplified form.

Ion emission is regarded as the result of a two-stage process:

a) A surface atom receives an impulse from the collision cascade which enables it to leave the specimen.

b) When leaving the specimen, its charge condition is established by electron exchange with the specimen surface.

Fig. 2-65 shows a schematic profile of the potential energy between a (metallic) solid surface and an atom located directly in front of it. It is assumed that both of these are in their ground state. In leaving the surface, the atom can be ionized *positively* if an electron from the valence level (the highest occupied level of the free atom in the ground state) tunnels through the potential barrier (arrow 1). Since it can find no free site at the equipotential level on the opposite side, it must additionally be raised (at least) to the Fermi level (arrow 1′). For this to occur, an energy ($E_i - E_w$) is required.

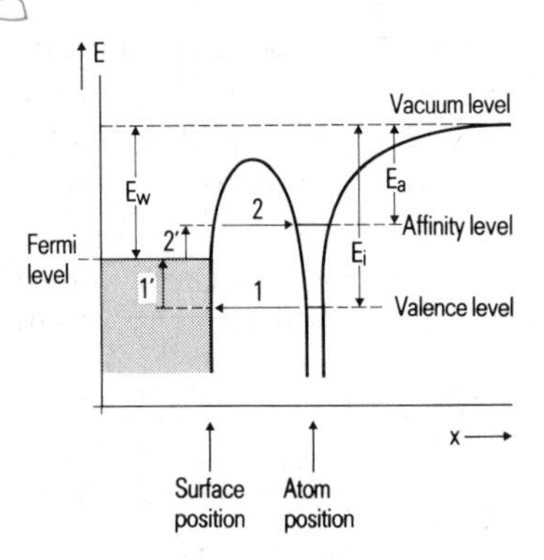

Fig. 2-65. Schematic profile of the potential energy $E(x)$ in the environment of a metallic solid surface and of a free atom in its ground state located directly in front of the surface. E_w work function, E_i ionization energy (position of the highest occupied electron level in the atom: valence level), E_a electron affinity energy (position of the lowest unoccupied electron level: affinity level). The atom can be ionized positively (arrows 1 and 1′) or negatively (arrows 2 and 2′) by way of electron exchange with the surface.

Negative ionization can take place if an electron from the solid tunnels through the potential barrier and reaches a free site in the atom's affinity level (arrow 2). To do so, it must previously be raised (at least) from the Fermi level to the affinity level (arrow 2'), for which an energy ($E_w - E_a$) must be expended.

The specimen surface and the atom remain in their ground state if the atom leaves very slowly (in the limiting case: infinitely slowly). The process then takes place without energy change (adiabatically) and a neutral atom is emitted. But the sputtering process is not adiabatic: due to the quantum-mechanical interaction with the surface, the movement of an atom at finite speed leads to the electrons having uncertain energy states. The position and width of the energy levels are modified when the atom leaves the specimen. As a result, there exists a finite probability for an electron interchange between specimen and atom and thus for an ionization. The probabilities are [2-83 to 85]:

$$P^+ \sim \exp\left[-(E_i - E_w)/cv\right] , \tag{2-42a}$$

$$P^- \sim \exp\left[-(E_w - E_a)/c'v\right] . \tag{2-42b}$$

It is here supposed that $E_i > E_w > E_a$ (Fig. 2-65); v is the normal component of the emission speed and c, c' are constants. On the basis of equations (2-42), the very low ion yield in inert sputtering is due to the low emission speed of the sputter particles (section 2.6.3: kinetic energy of the sputter particles at about 10 eV).

In the case of undoped semiconductors or insulators, the Fermi level lies within the band gap and (neglecting thermal excitation at room temperature) the electron states are occupied only up to the upper edge of the valence band (energy level E_v). In the equations (2-42) therefore, E_v must take over the role of E_w [2-84].

Chemically enhanced ion emission

Let us now return to the influence of reactive species on the ion yield in sputtering. It can be observed that the presence of electronegative atoms (e.g. oxygen) on the specimen surface greatly favors the formation of positive ions, whereas electropositive atoms (e.g. cesium) considerably increase the yield of negative ions [2-86]. It is here irrelevant whether the reactive constituents are implanted in the specimen during sputtering or whether they reached the specimen from the gas phase or via other measures or effects. No consistent model of ion emission in reactive sputtering with cesium and oxygen has yet been developed. Qualitatively different interpretations are thus given below [2-86].

The tunnel model already mentioned for inert sputtering has provided a valid description of the influence of *cesium* and the yield of negative ions (see Fig. 2-65, arrows 2). The incorporation of *cesium* (as of other electropositive elements) reduces the work function of the specimen and, according to equation (2-42b), increases the yield of negative ions. By experiment a connection has been established between the secondary ion intensity I^- and the work function E_w, which can be described by relations as those of equations (2-41) and (2-42b) [2-86].

The increase in positive ion yield in the presence of *oxygen* is due to the polar character of the metal-oxygen bond [2-87] and is described by the bond breaking model. The breaking

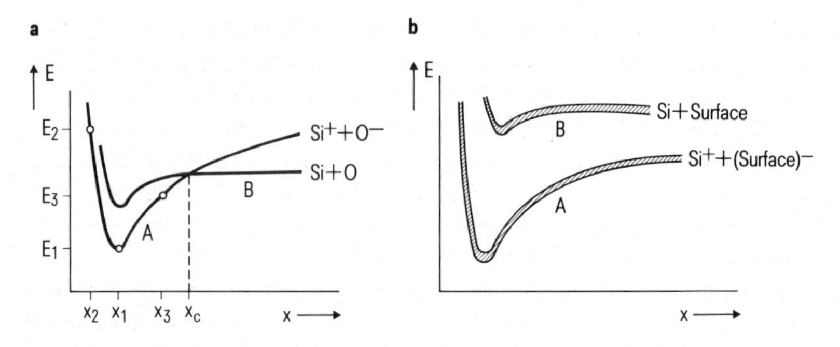

Fig. 2-66. Schematic profile of the potential energy E as a function of distance x a) between the components of an SiO molecule and b) between an Si atom or Si^+ ion and an SiO_2 solid surface. A indicates the curves for the ground state, B the curves for an excited state of the system; x_1, equilibrium distance. According to [2-84].

of the chemical bond with the oxygen can take place at the specimen surface [2-85] or in vacuum [2-88].

In the latter case it is assumed that neutral oxide molecules (e.g. SiO) are emitted and that these dissociate in vacuum (to Si^+ and O^-) only after leaving the surface. The energy for this process originates from the collisions during sputtering. Fig. 2-66a is a schematic representation of the potential energy of an SiO molecule as a function of the spacing between the two atoms. Curve A describes the ground state of the molecule. Due to the polar character of the bond, it largely reflects the Coulomb interaction of oppositely charged ions. Accordingly, the potential energy (starting from the equilibrium spacing at x_1) increases with the spacing and leads to a level at which both particles are ionized ($Si^+ + O^-$). Curve B refers to an excited state of the molecule with a lower polar fraction in the bond. Starting from this state, the separation of the particles leads to the neutral atoms (Si + O). Since these particles exert no Coulomb force on each other, no further energy need be expended after overcoming the largely covalent binding forces. The two curves intersect at $x = x_c$.

If the atoms in the ground state separate from each other starting at x_1, the energy of the system changes in accordance with curve A. At a spacing x_c (at the intersection with curve B) there exists a finite probability $(1 - P)$ that the system be transferred to curve B and later dissociates to Si + O. For the probability P that the system remains on the curve A and later dissociates into ions, an equation similar to (2-42) is obtained [2-84]:

$$P \sim \exp(-H^2/v)\,. \tag{2-43}$$

H describes the coupling between the two states and v is the relative velocity of the atoms at position x_c with respect to each other. Accordingly, $P = 0$ in the limiting case of infinitely slow (adiabatic) separation. But in the sputter process the atoms experience a sufficiently high relative speed so that the separation takes place nonadiabatically. The process is shown in Fig. 2-66a: the collision initially compresses the molecule, which assumes an energy E_2 (x_2) on the repelling branch of the energy curve. Subsequently (during the time that the molecule moves away from the surface), the atoms again move away from each other; see point E_3 (x_3). Their energy and relative velocity are great enough for the molecule to remain on curve A and finally to dissociate to $Si^+ + O^-$.

Now we assume that the bond is broken as soon as the particle leaves the specimen surface. In this case we have to consider no longer a free molecule but the coupling between the surface and the emitted particle. The energy curves for this case are less well known; a schematic representation is given in Fig. 2-66b [2-84]. The ground state is separated from the excited state in the solid by the band gap; the curves do not, therefore, intersect in the region $x > x_1$ (or only after a considerable distance). The system therefore has a great probability of remaining in the ground state (curve A) and the yield of positive ions is consequently very great for reactive sputtering with oxygen (or other electronegative elements).

The hitherto published measurement results, e.g. [2-86,88,89] are consistent with the models described here; they support either one or the other of the two models (bond breaking at the surface or in vacuum).

Molecular ion emission

It is known from SIMS spectra that a considerable part of the ionized species in reactive sputtering are molecules or atom clusters [2-78]. In sputtering of oxygen-doped silicon it is not only the yield of Si^+ ions that increases with the oxygen content, but also the yield of SiO^+ and Si_2O^+ ions [2-89]. Two models have been proposed to explain the emission of molecular ions [2-90]: according to the *fragmentation model*, these ions are emitted as complete fragments from the specimen surface, whereas the *recombination model* assumes the emission of atoms only and the formation of clusters by attractive interaction of adjacent atoms from the same sputter process. In neither case can it be expected that the sputtered molecules correctly

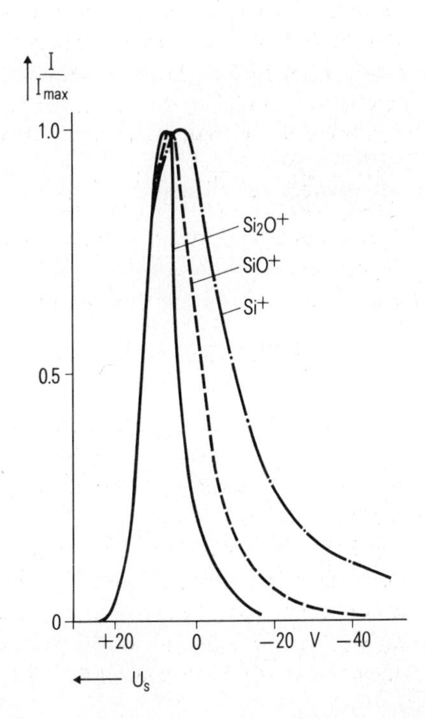

Fig. 2-67.
Energy distribution of atomic and molecular ions (Si^+, SiO^+, Si_2O^+) when sputtering silicon with oxygen ions. Signal intensity I in normalized form, measured with quadrupole mass spectrometer (sect. 7.2.2) through variation of the specimen voltage U_s. The kinetic energy of the secondary ions increases as U_s decreases.

represent the composition of the specimen. There is a high probability that they also contain primary ions or adsorbed atoms from the residual gas environment.

For a large number of elements it could be verified that the number of positive oxide molecule ions of the type MO^+ increases in relation to the number of positive atomic ions (M^+) as the dissociation energy increases; i.e. the stronger the bond, the greater is the probability for the emission of molecular ions.

The energy distribution of molecular ions differs distinctly from that of atomic ions (Fig. 2-67): the distribution curve is narrower and the maximum lies at a lower energy level. It must be assumed that at higher collision energies either the sputtered molecules disintegrate more easily or clustering is less probable than at lower energies.

2.6.6 Sputter depth profiling

In depth profiling by means of sputtering, a measurement signal is determined as a function of the sputter time. The analyst must then quantify the measured result, i.e. assign a depth scale to the sputter time and a concentration scale to the measured signal intensity [2-91 to 94]. The first part of this task is the same for all methods of sputter depth profiling. But the determination of the concentration scale depends on the method of analysis employed in each case; so different approaches are required for different methods, e.g. AES and SIMS. These are treated in detail in chapters 6 and 7, while in the following sections a number of fundamental problems occurring in depth and concentration calibration will be discussed.

Depth calibration

According to equation (2-34), the sputter rate has a value of $\dot{N} = Y \cdot i_0 / q$. If the primary ion current density $j_0 = i_0 / A$ and C is the particle number concentration in the specimen (number/volume), the erosion rate (also called sputter rate) is given by

$$\dot{z} = \frac{\dot{N}}{C \cdot A} = \frac{Y \cdot j_0}{C \cdot q} . \qquad (2\text{-}44)$$

The sputter depth has a value of $z = \dot{z} \cdot t$ in the case of a constant sputter rate. This case is fulfilled sufficiently closely, for instance in the depth profiling of dopants in semiconductor crystals where the dopant concentrations are low ($<10^{-3}$). In such cases, the sputter rate for the relevant experimental conditions can easily be determined by measuring the sputter time and the depth of the sputter craters. For shallow profiles, the deviation of the sputter yield within the pre-equilibrium stage of the sputter process must be taken into account.

In many cases, however, the sputter rate is not constant but changes with the depth, and thus with the sputter time. The sputter depth is then described by

$$z = \int_0^t \dot{z}(\tau) \, d\tau . \qquad (2\text{-}45)$$

For the case of sputtering through k different layers, the integral can be replaced by the sum $z = \sum \dot{z}_k \cdot \Delta t_k$. But this neglects the fact that $\dot{z}$ does not change abruptly at the boundary layers, even in the event of "sharp" transitions.

One parameter which characterizes the lack of accuracy in the depth profile is the *depth resolution* Δz. Various definitions exist for this parameter, it is most commonly defined as the depth region in which the signal intensity I at an infinitely steep concentration step changes from 16% to 84% of the step height. This definition is based on the assumption that the measured profile corresponds to the integral of a Gaussian distribution for which a Δz defined in this way corresponds to twice the standard deviation σ [2-92,95] (Fig. 2-68).

In general, several effects contribute to the depth resolution, so that

$$\Delta z_{tot} = \sqrt{\sum (\Delta z)^2} \ .$$

The effects described in sections 2.6.3 and 2.6.4. (primary ion implantation, atomic mixing, preferential sputtering) depend upon the penetration depth of the primary ions. Their contribution to the depth resolution is (like the effect of the electron exit depth for AES profiles; see section 6.5), independent of the sputter depth (Δz = const.). Other influences, which depend on the specimen or the instrument, cause an impairment of the depth resolution which increases with depth (e.g. $\Delta z \sim z$). Among these are a nonuniform ion current density as well as the surface modifications discussed in section 2.6.3, which are due to a rough or contaminated specimen surface or inhomogeneities in the specimen.

To minimize these latter influences, the analyst has to select carefully the area on the specimen so as to avoid local irregularities and to adjust accurately the ion beam to ensure a constant ion current density across the entire analyzed surface. The unavoidable sputter-produced influences on the depth resolution, which depend on the ion penetration depth z_p, may be reduced by working with low ion energy and with ions of high mass and by using a large bombardment angle (relative to the surface normal). This could be illustrated, for instance, by Auger electron depth profiling of a 28.4 nm anodically oxidized tantalum pentoxide layer on

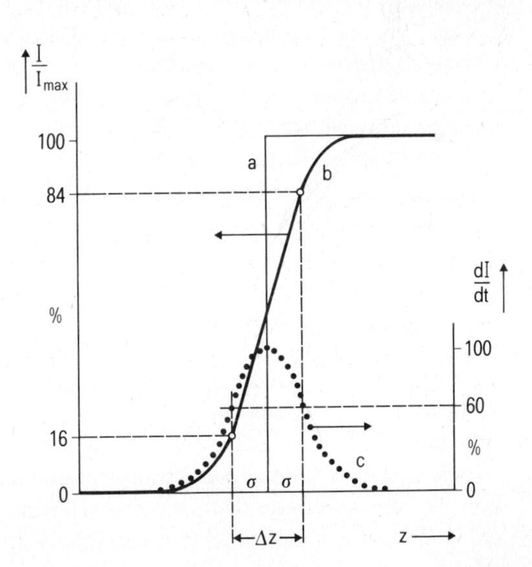

Fig. 2-68.
Determination of the depth resolution Δz in depth profiling: signal intensity I measured at a concentration step (a) as a function of the sputter depth z (curve b); associated Gaussian distribution dI/dt (curve c) with standard deviation $\pm\sigma$. According to [2-92].

tantalum, in which a depth resolution of $\Delta z = (1.41 \pm 0.07)$ nm was demonstrated (experimental conditions: O_{KLL} Auger electron signal; 2 keV Ar^+ ion bombardment at 35°) [2-96].

Concentration calibration

Let us first consider the case in which sputter equilibrium has been established. Equations (2-36) and (2-39) yield the following relationship for the ratio of the sputter rates of a two-element specimen:

$$\frac{\dot{N}_A}{\dot{N}_B} = \frac{X_A}{X_B} = \frac{Y_A^* \cdot X_A^S}{Y_B^* \cdot X_B^S} \,. \tag{2-46}$$

The meaning of this relationship varies with the different methods employed for analyzing sputter depth profiles. The left-hand side refers to the flow of sputtered particles, the flow composition corresponding to the sought-for concentration ratio in the bulk. It follows that a quantitative analysis free of matrix effects could be obtained if the sputter particles could be analyzed in their entirety or a constant matrix-independent fraction thereof. This may be done by means of subsequent ionization of the neutral particles in a high-frequency plasma [2-94] or by means of laser resonance excitation [2-97].

For SIMS, equation (2-41) indicates an intensity of positive or negative secondary ions of $I_i^\pm \sim P_i^\pm \cdot Y_i$. Due to $Y_i \sim N_i$ and equation (2-36), the following applies for equilibrium sputtering:

$$\frac{I_A^\pm}{I_B^\pm} = \frac{P_A^\pm \cdot X_A}{P_B^\pm \cdot X_B} \,. \tag{2-47}$$

Accordingly, the composition of the sputtered ion flow depends on the ionization probabilities, and thus to a high degree on the chemical effects discussed in section 2.6.5.

For quantitative analyses therefore, the best thing to do is to employ an empirical calibration with adapted standards. Where such standards are not available, another possibility is to use a semiquantitative method for calculating the ionization probabilities (the Andersen "LTE model" [2-98]; see section 7.4).

The right-hand side of equation (2-46) is relevant for those methods of analysis which investigate the surface modified by sputtering, e.g. AES. If the mole fraction X_i^S is successfully measured, a quantitative analysis could be possible if the component sputter yields Y_i^*(defined by equation (2-38)) were known. These yields are, however, matrix dependent and can be substituted by the total yield Y_i^0 of the pure *i*-element standards [2-99] only to a first approximation. In addition, this procedure assumes that the composition within the altered layer is constant and that the depth of this layer is greater than the analysed depth (which depends on the Auger electron energy). It is highly improbable, though, that both assumptions will be fulfilled in practice, so that even AES cannot always dispense with adapted standards.

Even greater difficulties occur when the equilibrium condition is not fulfilled, which is always the case within the first few nanometers at the surface after starting the measurement and in the surroundings of an interface layer when a new sputter equilibrium must be established in the transition from one layer to another. The difficulties involved here are explained on the basis of two examples.

Fig. 2-69 shows SIMS depth profiles of a 150 nm thick palladium layer on silicon [2-92]. Sputtering was done with Ar^+ ions in an oxygen atmosphere (O_2 jet). As the diagram shows, a rough sputter surface is formed within the polycrystalline Pd in region a (see surface modifications, section 2.6.3). As soon as the deepest valleys reach the Pd-Si interface, the Si signal increases steeply. In region b, Pd and Si are simultaneously recorded, until finally the coarsened sputter surface comes to lie completely within the Si substrate (region c). Two further artifacts should be noted in addition to the coarsening: first, the increased Pd signal in region b is due to increased oxygen incorporation as a result of the simultaneous presence of silicon (chemically enhanced ionization, section 2.6.5). Second, Pd reaches the Si substrate by means of cascade and recoil mixing (section 2.6.3) and a fraction of it is continually remixed into the silicon below the actual sputter level. This effect explains the very slow decay of the Pd signal in region c. It is also typical of other layer combinations.

The example of an Ag-Pd (80-20) alloy shown in Fig. 2-70 illustrates how the silver and palladium AES signals alter under simultaneous bombardment by Xenon ions if the ion energy is alternated between 0.5 and 5 keV [2-100]. During low-energy sputtering, Pd is enriched in a thin layer because Ag is preferentially sputtered away. A higher signal is therefore measured for Pd than for Ag. After switchover to 5 keV, the ions penetrate more deeply and mix the layer previously enriched with Pd with the bulk material which is poorer in Pd. An altered layer of greater depth with initially lower Pd content is formed. When the new sputter equilibrium is reached, the entire zone is again enriched with Pd, and to an even greater degree than before. Upon switching back again to 0.5 keV, the ions encounter a material strongly enriched with Pd. In the range of their again reduced penetration depth, preferential sputtering initially

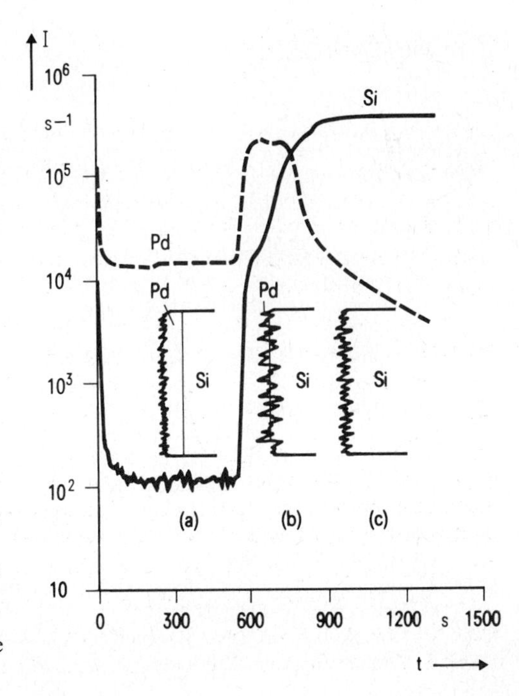

Fig. 2-69.
SIMS depth profile through a 150 nm Pd layer on Si: intensity I (counts per second) as a function of the sputter time t; 5 keV Ar^+ with oxygen jet [2-92]. Actual position of the sputter roughened surface: a) within the Pd layer, b) penetrating the Pd-Si interface, c) within the Si substrate.

causes an even stronger enrichment of Pd. Not until the altered layer previously formed at 5 keV has been completely sputtered away is the equilibrium that existed at the beginning of the experiment re-established.

Non equilibrium effects, such as those described in the above example, must always be expected when changing the sputter conditions, e.g. when sputtering through material interfaces. A knowledge of these effects is the most important precondition for avoiding erroneous interpretations of the measurement results. By varying and optimizing the experimental conditions, it should in most cases be possible to distinguish between reality and artefacts.

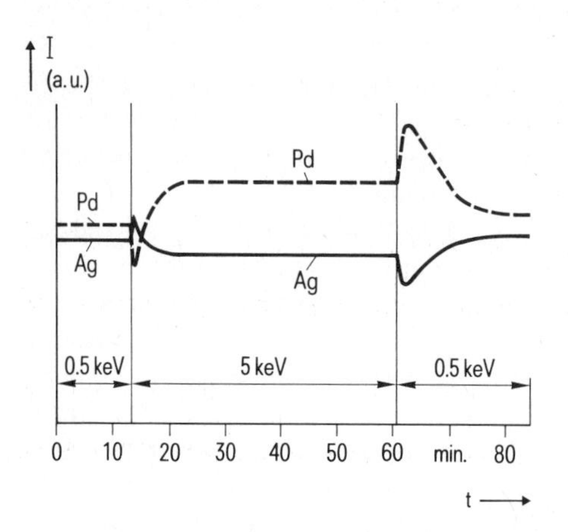

Fig. 2-70. Homogeneous Ag-Pd alloy (Ag atom fraction 80%): changes in the AES signal intensity I (peak to peak heights) with time t under Xe ion bombardment; disturbance of the sputter equilibrium through alteration of the ion energy between 0.5 and 5 keV [2-100].

2.7 References

[2-1] Ch. Kittel, *Introduction to Solid State Physics*, 5th ed. J. Wiley & Sons Inc., New York 1976
[2-2] W.A. Harrison, *Solid State Theory*. MacGraw-Hill, New York 1970
[2-3] A. Hart-Davis, *Solids: an introduction*. MacGraw-Hill (UK), Ltd., Maidenhead, Berkshire, England 1975
[2-4] R.O. Jenkins, *Vacuum 19*, 353 (1969)
[2-5] J.M. Lafferty, *J. Appl. Phys. 22*, 299 (1951)
[2-6] A.N. Broers in O. Johari (Ed.): *Scanning Electron Microscopy 1974*. SEM Inc. AMF O'Hare (Chicago) 1974, 9
[2-7] F.J. Hohn, T.P.H.Chang, A.N. Broers, G.S. Frankel, E.T. Peters, D.W. Lee, *J. Appl. Phys. 53*, 1283 (1982)
[2-8] E.W. Müller, *Z. Phys. 106*, 541 (1937)
[2-9] A.V. Crewe in E. Wolf (Ed.): *Progress in Optics*, XI. North Holland 1974, 225
[2-10] B. v.Borries, E. Ruska, *Z. Techn. Phys. 20*, 225 (1979)
[2-11] D.B. Langmuir, *Proc. IRE 25*, 977 (1937)
[2-12] H. Rose, R. Spehr, *Adv. Electronics and Electron Phys. Suppl. 13 C*, 475 (1983)
[2-13] V.E. Cosslett, R.N. Thomas in T.P. McKinley, K.F.J. Heinrich, D.B. Wittry (Eds.): *The Electron Microprobe*. Wiley, New York 1966, 243
[2-14] K. Kanaya, S. Ono in O. Johari (Ed.): *Scanning Electron Microscopy 1982*. SEM Inc., AMF O'Hare (Chigaco) 1982, 69
[2-15] K. Murata, T. Matsukawa, R. Shimizu, *Jap. J. Appl. Phys. 10*, 678 (1971)
[2-16] R. Shimizu, T. Ikuta, K. Murata, *J. Appl. Phys. 43*, 4233 (1972)

[2-17] L. Katz, A.S. Penfold, *Rev. Mod. Phys. 24*, 28 (1952)
[2-18] J.R. Young, *J. Appl. Phys. 27*, 524 (1956)
[2-19] J.E. Holliday, E.J. Sternglass, *J. Appl. Phys. 30*, 1428 (1959)
[2-20] M.J. Berger, S.M. Seltzer, NASA, SP-3012 (1964)
[2-21] V.E. Cosslett, R.N. Thomas, *Brit. J. Appl. Phys. 15*, 1283 (1964)
[2-22] F. Lenz, *Z. Naturforsch. 9a*, 185 (1954)
[2-23] L. Reimer, C. Tollkamp, *Scanning 3*, 35 (1980)
[2-24] H. Drescher, L. Reimer, H. Seidel, *Z. Angew. Phys. 29*, 331 (1970)
[2-25] S. Ono, K. Kanaya, *J. Phys. D 12*, 619 (1979)
[2-26] H. Seiler, *Z. Angew. Phys. 22*, 249 (1967)
[2-27] H.E. Bishop in R. Castaing et al. (Eds.): *X-ray Optics and Microanalysis*, IV. Int. Congr., Orsay 1965. Hermann, Paris 1966, 153
[2-28] W. Liesk, *Z. Angew. Phys. 21*, 25 (1966)
[2-29] H.W. Fuller, M.E. Hale, *J. Appl. Phys. 31*, 238 (1960)
[2-30] H. Boersch, H. Raith, D. Wohlleben, *Z. Phys. 159*, 388 (1960)
[2-31] E. Fuchs, *Z. Angew. Phys. 14*, 203 (1961)
[2-32] A.J. Dekker, *Solid State Physics 6*, 251 (1958)
[2-33] H. Seiler, *J. Appl. Phys. 54*, R1 (1983)
[2-34] M. Thompson, M.D. Baker, A. Christie, J.F. Tyson, *Auger Electron Spectrometry*. J. Wiley & Sons, New York 1985
[2-35] M.D. Muir, P.R. Grant in D.B. Holt et al. (Eds.): *Quantitative Scanning Electron Microscopy*. Academic Press, London, New York, San Franscisco 1974, 287
[2-36] D.B. Holt, *ibid.* a) 313, b) 335
[2-37] M. v.Ardenne, *Tabellen der Elektronenphysik, Ionenphysik und Übermikroskopie*, Band I. Deutscher Verlag der Wissenschaften, Berlin 1956
[2-38] R.G. Wilson, G.R. Brewer, *Ion Beams with Applications to Ion Implantation*. John Wiley & Sons, New York, London, Sidney, Toronto 1973
[2-39] K. Wittmaack, *Nucl. Instrum. Methods 118*, 99 (1974)
[2-40] *ATOMIKA Ion microprobe manuals*, Atomika, Technische Physik GmbH, Postfach 450135, D-8000 München 45
[2-41] H. Liebl, *Scanning 3*, 79 (1980)
[2-42] G.P. Lawrence, R.K. Beauchamp, J.L. McKibben, *Nucl. Instrum. Methods 32*, 357 (1965)
[2-43] H. Liebl, *Nucl. Instrum. Methods 187*, 143 (1981)
[2-44] P. Williams, R.K. Lewis, C.A. Evans, P.R. Hanley, *Anal. Chemie 49*, 1399 (1977)
[2-45] *General Ionex Model 133, Cesium Ion Source, Instruction Manual*. General Ionex Corp., 19 Graf Road, Newburyport, Mass. 01950, USA
[2-46] H.N. Migeon, C. Le Pipec, J.J. Le Goux in A. Benninghoven et al. (Eds.): *Secondary Ion Mass Spectrometry, SIMS V*. Springer, Berlin, Heidelberg, New York, Tokyo 1985, 155
[2-47] R. Clampitt, *Nucl. Instrum Methods 189*, 111 (1981)
[2-48] A.E. Bell, L.W. Swanson, *Nucl. Instrum. Methods in Phys. Res. B10/11*, 783 (1985)
[2-49] G.L.R. Mair, T. Mulvey, *Ultramicroscopy 15*, 255 (1984)
[2-50] R. Levi-Setti, G. Crow, Y.L. Wang, *Scanning Electron Microscopy 1985/II*, SEM Inc., AMF O'Hare 1985, 535
[2-51] A.R. Bayly, A.R. Waugh, K. Andersen, *Nucl. Instrum. Methods 218*, 375 (1983)
[2-52] R. Behrisch, *Sputtering by Particle Bombardment I, Topics in Applied Physics*. Springer, Berlin, Heidelberg, New York 1981
[2-53] R. Behrisch, *Sputtering by Particle Bombardment II, Topics in Applied Physics*. Springer, Berlin, Heidelberg, NewYork 1983
[2-54] R. Behrisch, *Sputtering by Particle Bombardment III, Topics in Applied Physics*. Springer, Berlin, Heidelberg, New York, to be published 1990
[2-55] O. Auciello, R. Kelly, *Ion Bombardment Modification of Surfaces; Fundamentals and Applications*. Elsevier, Amsterdam, Oxford, New York, Tokyo 1984
[2-56] H. Oechsner, *Thin Film and Depth Profile Analysis*. Springer, Berlin, Heidelberg, New York, Tokyo 1984
[2-57] J. F. Ziegler, J.P. Biersack, U. Littmark, *The Stopping and Range of Ions in Solids*. Pergamon Press, New York, Oxford, Toronto, Sydney, Frankfurt, Tokyo 1986

[2-58] T. Ishitani, R. Shimizu, *Phys. Letters 46A*, 487 (1974)
[2-59] K. Wittmaack, *Surf. Sci. 89*, 668 (1979)
[2-60] N. Sigmund, in [2-52], 9
[2-61] J. Lindhard, M. Scharff, H.E. Schiøtt, *K. Dan. Vidensk. Selsk. Mat. Fys. Medd. 33*, No. 14 (1963)
[2-62] J.F. Gibbons, W.S. Johnson, S.W. Mylroie, *Projected Range Statistics, Semiconductors and Related Materials*, 2nd ed. Dowden, Hutchinson and Ross Inc., Stroudsburg PA 1975
[2-63] K.B. Winterbon, *Ion Implantation Range and Energy Deposition Distributions, Vol. 2: Low Incident Ion Energies.* IFI/Plenum, New York, Washington, London 1975
[2-64] R. Kelly, in [2-55] 27
[2-65] M.T. Robinson, in [2-52] 73
[2-66] D.E. Harrison, *Radiat. Eff. 70*, 1 (1983)
[2-67] F. Schulz, K.Wittmaack, *Radiat. Eff. 29*, 31 (1976)
[2-68] K. Wittmaack, *Radiat. Eff. 63*, 205 (1982)
[2-69] G.H. Kinchin, R.S.Pease, *Rep. Progr. Phys. 18*, 1 (1955)
[2-70] H.M. Naguib, R. Kelly, *Radiat. Eff. 25*, 1 (1975)
[2-71] H.E. Roosendaal, in [2-52] 219
[2-72] G. Betz, G.K.Wehner, in [2-53] 11
[2-73] U. Littmark, W.O. Hofer, in [2-56] 159
[2-74] G. Carter, B.Navinsek, J.L. Whitton, in [2-53] 231
[2-75] R. Kelly, in [2-55] 79
[2-76] H.H. Andersen, H.L. Bay, in [2-52] 145
[2-77] H. Sommerfeldt, E.S. Mashkova, V.A. Malchanov, *Phys. Lett. 38A*, 237 (1972)
[2-78] J. Roth, in [2-53] 91
[2-79] W. Reuter, K. Wittmaack, *Appl. Surf. Sci. 5*, 221 (1980)
[2-80] R. v. Criegern, presented at the 5th Intern. Conf. Secondary Mass Spectrometry, Washington, D.C., 1985
[2-81] G. Blaise, A. Nourtier, *Surf. Sci. 90*, 495 (1979)
[2-82] P. Williams, *Surf. Sci 90*, 588 (1979)
[2-83] P. Williams, *Applied Atomic Collision Physics*, Vol. 4. Academic Press, London 1983, 327
[2-84] P. Williams, *Appl. Surf. Sci. 13*, 241 (1982)
[2-85] J.K. Norskov, B.J. Lundqvist, *Phys. Rev. B19*, 5661 (1979)
[2-86] M.L. Yu, *Nucl. Instr. Meth. Phys. Res. (B), B15*, 151 (1986)
[2-87] G. Blaise, G. Slodzian, *Surf. Sci. 40*, 708 (1973)
[2-88] W. Gerhard, C. Plog, *Z. Phys. B 54*, 59 and 71 (1983)
[2-89] K. Wittmaack, *Surf. Sci. 112*, 168 (1981)
[2-90] M.L. Yu, *Phys. Rev.B 24*, 1147 (1981)
[2-91] S. Hofmann, *Surf. Interface Anal. 2*, 148 (1980)
[2-92] C.W. Magee, R.E. Honig, *Surface and Interface Analysis 4*, 35, 1982,
[2-93] E. Zinner, *J. Electrochem. Soc. 130*, 199 C (1983).
[2-94] H. Oechsner, in [2-56] 63
[2-95] S. Hofmann, Le Vide-Les Couches Minces, No. Spec., 259 (1979)
[2-96] M.P. Seah, C.P. Hunt, *J. Appl. Phys. 56*, 2106 (1984)
[2-97] J.E. Parks, H.W. Schmitt, G.S. Hurst, W.M. Fairbanc, *Thin Solid Films 108*, 69 (1983)
[2-98] C.A. Andersen, J.R. Hinthorne, *Anal. Chem. 45*, 1421 (1973)
[2-99] H. Shimizu, M. Ono, K. Nakayama: *Surface Science 36*, 817 (1973)
[2-100] G. Betz, M. Opitz, P. Braun, *Nucl. Instrum. Methods 182/183*, 63 (1981)

3 Scanning electron microscopy

3.1 Principle

The scanning electron microscope (SEM) is used for imaging the surfaces of materials with the aid of signals derived from the interaction between the probe electrons and the specimen. Fig. 3-1 shows a schematic drawing of an SEM. The electrons emitted from the cathode are accelerated towards the anode. Acceleration voltages in the region of 2 to 30 kV are generally employed. The crossover (Fig. 2-26) produced in the electron gun is focused onto the specimen by the condenser lens and the probe-forming (objective) lens in a two-stage demagnification process as shown in Fig. 2-27. At sufficiently high acceleration voltages, an electron probe can be formed in this way having a diameter of, say, 5 nm and a current of 10^{-11} A. The electron

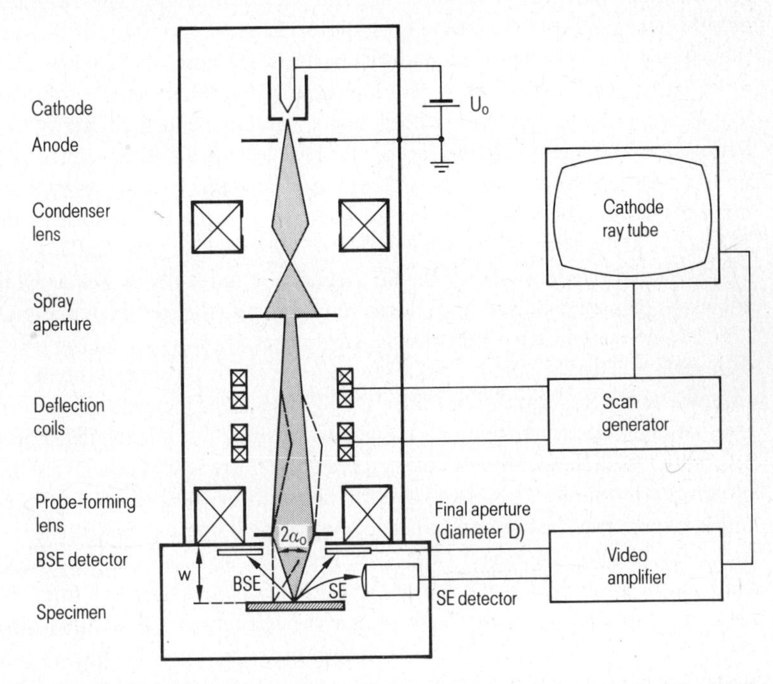

Fig. 3-1. Schematic drawing of a scanning electron microscope. An electron beam is generated in the gun of the electron-optical column and focused onto the specimen by the condenser and probe-forming lenses. The deflection coils scan the electron probe across the specimen. The signal supplied by the secondary (SE) or backscattered (BSE) electrons is amplified and controls the brightness of a cathode ray tube (CRT) whose electron beam is scanned synchronously with the electron probe. An approximate value for beam divergence $2\ \alpha_0 \approx D/w$ is obtained on the basis of working distance w, i.e. the free space between the specimen and the lower pole piece, and the diameter D of the final aperture (cf. Fig. 2-27).

probe is scanned across the specimen in raster mode with the aid of two pairs of deflection coils (one pair each for deflection in the x and y direction; only one pair is shown in Fig. 3-1). The upper and lower scan coils of the double deflection system deflect the electron beam in opposite directions. During scanning, the ratio of their deflections is kept constant so that the electron beam rocks around a fixed point. This pivot point has to lie in the plane of the final aperture. The scanning coils are driven such that the electron beam moves across the specimen continuously along a line (the line scan), e.g. in the x direction. By shifting the individual line scans slightly in the y direction (the frame scan) a raster pattern is formed. In synchronization with the deflection coil systems, the scan generator controls the deflection of the electron beam in the cathode ray tube (CRT). The intensity modulation of the CRT is controlled by one of the signals generated at the point of electron impact and recorded by suitable detectors, particularly the secondary electron (SE) and the backscattered electron (BSE) signal. The magnification M is adjusted by changing the excitation of the deflection coils and thus the raster amplitude. M is determined by the ratio of the constant width b of the CRT to the width a of the scanned specimen field, i.e. $M = b/a$. The smaller the scanned specimen field, the greater the magnification on the CRT.

The interaction between the primary electrons and the specimen gives rise to a series of effects in which particles or radiation are emitted (Section 2.4). These can be collected by suitable detectors and used for image formation. The most important of these effects for imaging purposes is the emission of secondary electrons. Because of their low escape depth of only a few nanometers, they allow a very high spatial resolution which, to a first approximation, is not influenced by scattering effects in the specimen and is primarily limited by the beam diameter. In the energy distribution of electrons leaving a bulk specimen (Fig. 2-40), electrons up to an exit energy of 50 eV are referred to as secondary electrons by definition. The maximum of the SE energy distribution, however, lies at a few eV. Because of this low energy, the SEs can be easily extracted by a detector to which a positive potential of a few 100 V is applied. The SE signal is determined by the SE yield δ (Section 2.4.3) as well as by the efficiency with which the SEs are collected by the detector. Although the specimen material also plays a part (in terms of material contrast), the SE yield depends primarily on the tilt angle of the specimen surface. The SE yield rises sharply with the tilt angle, due to the increasing surface area from which the SEs can escape. Since the SE detector is located on one side of the specimen, the secondary electrons generated at surfaces pointing towards the detector are collected more efficiently. As these surfaces appear brighter than those turned away from the detector, a shadow effect is produced. This, together with the strong dependence of the SE yield on the surface tilt, is responsible for the pronounced topographical contrast and the three-dimensional impression typical of SE images (Fig. 3-2). They look as if the specimen were illuminated by a light source positioned at the site of the SE detector and observed from the direction of the electron source. In the optical microscope the resolution is limited by diffraction effects to about 0.2 μm, i.e. about half the wavelength of visible light. Compared to an optical microscope, the SEM provides a much higher resolution (up to a factor of 100) in addition to a much greater depth of focus (Section 3.4.3).

Since the backscattered electrons have energies up to the primary electron energy (Fig. 2-40) they are not much influenced by electrostatic collection fields and leave the specimen along almost straight trajectories. They can therefore be effectively collected only by detectors covering a wide solid angle. The volume of the specimen from which BSEs are emitted is approximately that of a hemisphere with a radius equal to half the penetration depth (Section 2.4.1).

This limits the spatial resolution. On the other hand, the strong dependence of BSE emission on atomic number (Fig. 2-35) yields a high level of material contrast and thus allows the compositions of different phases in a material to be compared (Fig. 3-3).

Fig. 3-2.
Secondary electron image of the surface of an integrated circuit. Tilting the specimen in the SEM provides an oblique view which yields a three-dimensional impression of the surface topography. Interconnections run from top left to bottom right over a structured substrate. Round contact holes can also be seen near each edge of the image. The insulating layers were partially removed by selective etching. $U_0 = 25$ kV.

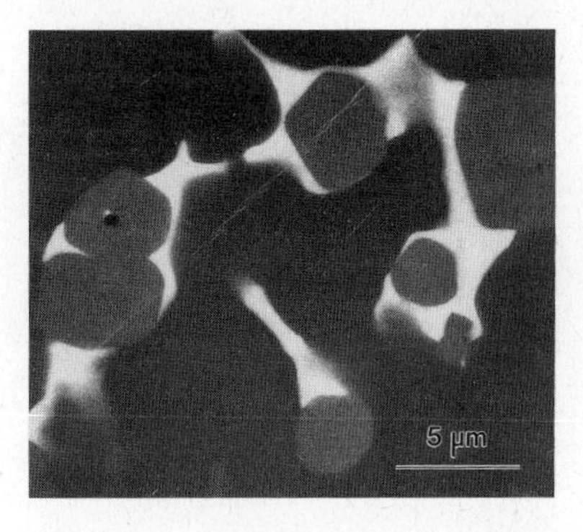

Fig. 3-3.
Backscattered electron image of a polished section through a ZnO varistor ceramic. Grains of the same composition and thus the same mean atomic number $\bar{Z}$, show the same brightness. The mean atomic number of a compound is given by $\bar{Z} = \sum w_i Z_i$ where w_i and Z_i are the mass fractions and atomic numbers of the various elements. In addition to the large dark ZnO grains ($\bar{Z} = 26$), some $Zn_7Sb_2O_{12}$ grains ($\bar{Z} = 31$) are present which appear slightly brighter. The Bi_2O_3 grains in the triple grain junctions are very bright because of their much larger mean atomic number of $\bar{Z} = 75$. $U_0 = 25$ kV.

Other effects generated by electron beam excitation which are of importance for analysis are explored later on in this book. They include the effects of the emission of characteristic X-rays and Auger electrons (Chapters 5 and 6) which can be exploited for elemental analysis, and the effect of surface potentials on SE emission (voltage contrast, Chapter 8). Some additional effects and the related techniques are discussed in Sections 3.6 and 3.7. The fundamentals of scanning electron microscopy and the various methods employed are described in detail in [3-1], with emphasis on the physics, while [3-2] and [3-3] stress the more practical aspects. Articles of general and specific interest on all aspects of the subject can also be found in the proceedings of the scanning electron microscopy conferences [3-4].

3.2 Instrumentation

3.2.1 Overall system

In many fields, scanning electron microscopes are used as routinely as optical microscopes. They are the most common of all particle-beam instruments and are commercially available in a broad range of models from the simple SEM to automatic linewidth measurement systems. However, they are all based on essentially the same design principles.

Fig. 3-4 shows an advanced SEM. The electron-optical column sits on top of the specimen chamber, the lower pole piece of the probe-forming lens being its upper cover. The specimen chamber contains the specimen stage for positioning, tilting and rotating the specimen under the beam (Section 3.2.3), and a number of openings for the various detectors (for SE, BSE and possibly X-rays etc.) (Section 3.2.4). The specimen chamber and column are connected to a vacuum system capable of producing a working pressure of about 10^{-3} Pa by means of a diffusion or turbomolecular pump. Turbomolecular pumps have the advantage of providing an oil-free high vacuum and thus prevent contamination of the specimen by hydrocarbons

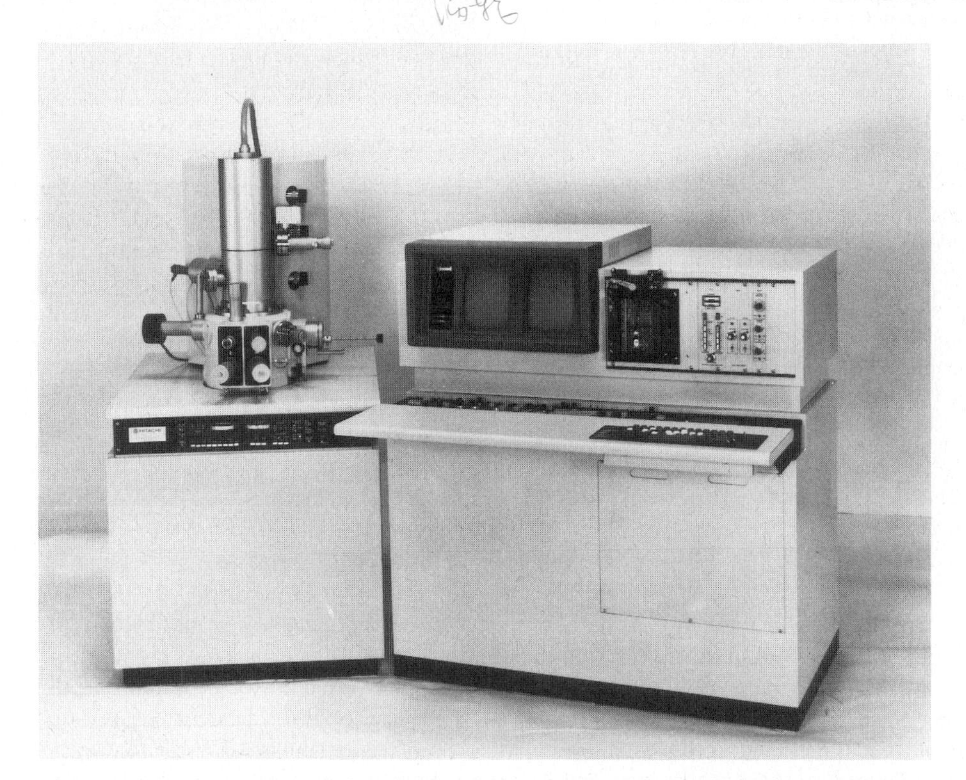

Fig. 3-4. View of an advanced scanning electron microscope equipped with a field emission gun: model S800 from Hitachi Ltd. In the part on the left, the electron-optical column sits atop the specimen chamber; the vacuum system is covered. The part on the right houses the operating controls, CRTs and all electric and electronic supplies.

from the residual gas. To avoid frequent venting of the specimen chamber, air-locks are usually employed for changing specimens. Use of LaB_6 and, in particular, field emission cathodes requires a higher vacuum (10^{-4} Pa and 10^{-7} Pa respectively), so that ion gettering pumps must be used for the electron gun. The various apertures in the electron optical column allow differential pumping of the gun vacuum with respect to the vacua in the column and the specimen chamber. The electron beam often passes through the column within a liner tube, allowing the lens and deflection coils to be located outside the high vacuum region. When cleaning the parts which come into contact with the electron beam, it is then sufficient to clean this tube in addition to the Wehnelt cylinder, the anode and the fixed apertures.

In addition to the high-voltage supply and the power supplies for the lenses and deflection coils, the electrical equipment comprises the electronics for signal amplification and processing. At least one display CRT is required for observation and a second one for photography. In the latest models, many of the instrument functions, from control of the vacuum systems and the stage to automatic focusing and automatic astigmatism correction, are controlled by a microcomputer.

3.2.2 Electron-optical column

The main components of the electron-optical column have already been described with reference to Fig. 3-1. LaB_6 and field-emission cathodes are increasingly replacing tungsten cathodes as electron sources because their higher brightness (Section 2.3.2) at comparable current levels allows smaller probe diameters. Unlike the setup shown in Fig. 3-1, two condenser lenses are generally employed to produce, together with the probe-forming lens, an image of the crossover on the specimen which is demagnified in three stages. The lower pole piece has only a small bore in order to keep the magnetic field at the position of the specimen small. The final aperture, which is usually variable and adjustable, is located just above the pole piece. Together with the working distance w, its diameter determines the divergence $2\alpha_0$ of the Probe (Fig. 3-1). For an aperture with a typical diameter of 100 µm and a working distance of 10 mm, a divergence angle of $2\alpha_0 = 5$ mrad is obtained. Since the diameter of the probe is inversely proportional to α_0 (Eq. 2-23), a better resolution can be achieved with a smaller working distance, which yields a larger value of α_0. A larger beam divergence, however, reduces the depth of focus (Section 3.4.3). The probe current is usually controlled by changing the condenser excitation. Focusing of the image is achieved by finely adjusting the probe-forming lens current or by changing the specimen height (z shift) mechanically. Astigmatism (Section 2.2.2) is compensated by a stigmator.

3.2.3 Specimen stage

The purpose of the specimen stage is to position the specimen under the electron beam and to tilt it so that the required specimen area can be imaged from the desired "viewing angle". In addition to the movement of the specimen along the axes x, y, z, tilting about an axis normal to the beam, as well as rotation, are required. The range of tilt should be as large as possible, although the ideal range up to an angle of 90° is only achieved with small objects. Many models are equipped with conical pole pieces which provide wide tilting angles also for inspecting large-area specimens such as silicon wafers. In the *x, y* plane, the object can usually be rotated through 360°. Eucentric stages have the advantage of preventing the area of the specimen under observation from being displaced when the specimen is tilted or rotated. The range of stage movement and hence the size of the specimen chamber should be selected according to the size of the specimens to be examined. Thus, inspection of silicon wafers with diameters up to 200 mm requires particularly large specimen chambers and stages (Section 3.4.4). Also, in the case of such large specimens, air-locks are used for specimen exchange. The stage may be driven manually or via motors which can be controlled by a computer.

3.2.4 Detectors

The most efficient means of secondary electron collection has proved to be a detector system of the type due to Everhart and Thornley [3-5]. In this system, the SEs emitted from the specimen are first extracted by a grid to which a 200 V positive potential is applied and are then accelerated onto a scintillator at high voltage (Fig. 3-5). Inorganic single crystals such as cerium-doped yttrium-aluminum garnet have started to replace plastics as the scintillator material since plastics have a reduced lifetime due to electron beam damage. Subsequent acceleration brings the secondary electrons up to a high energy level (10 keV) allowing a large number of electron-hole pairs to be generated in the scintillator, some of which recombine with the emission of photons. These photons are conducted via a light pipe to a photomultiplier which is located outside the high vacuum of the chamber. The photomultiplier delivers an electrical output signal which is then further amplified. This detector system, consisting of a combination of a scintillator and a photomultiplier, has the benefits of a very low noise level and a large bandwidth which is also suitable for image display at TV scan speeds.

The scintillator of the SEs detector is also struck by a small number of backscattered electrons (Fig. 3-5). If the positive grid voltage is switched off or a small negative voltage is applied, the SEs cannot reach the grid. In contrast, the fast BSEs can overcome the grid potential

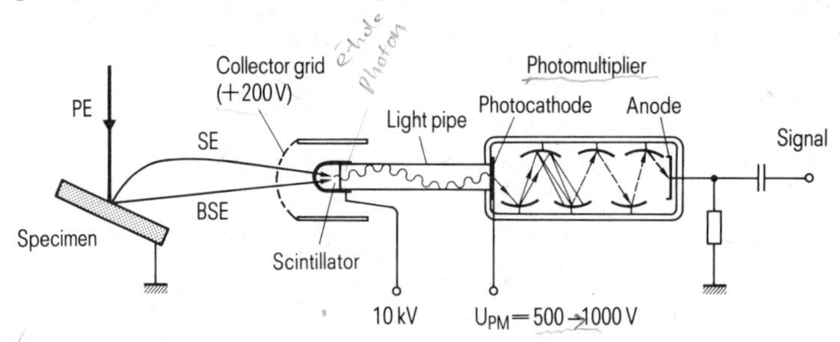

Fig. 3-5. Secondary electron detector due to Everhart and Thornley. The secondary electrons (SEs) generated by the primary electron (PE) beam are collected by a grid (which is biased at +200 V). They pass through the grid and are accelerated towards the scintillator which is biased at 10 kV. A thin metallic coating is vapor-deposited onto the scintillator so that a voltage can be applied to it. The 10 kV electrons easily penetrate the metal film and the photons generated in the scintillator are reflected back inside. The scintillator is coupled to a light pipe which transmits the photons to the photocathode of the photomultiplier. The individual dynodes of the photomultiplier consist of a material with high SE yield and together yield a multiplication factor of up to 10^6 for a single photoelectron. The SE signal amplification can be adjusted by varying the voltage U_{PM} applied to the photomultiplier.

and hit the detector. Unlike the slow SEs, they require no subsequent acceleration but have sufficient energy to generate numerous light quanta in the scintillator. In this mode, the Everhart Thornley detector acts as a BSE detector. However, as the emitted BSEs are collected over only a very small solid angle, they generate a weak signal. The detection efficiency is improved somewhat if a BSE detector is used whose scintillator is located close to the specimen (Fig. 3-6a). A large annular scintillator placed directly under the pole piece (Fig. 3-6b) will collect the emitted BSE over a wide solid angle.

Scintillators are not the only means of collecting backscattered electrons; semiconductor detectors can also be used for this purpose. They consist of Schottky diodes or diodes with pn-junctions which are reverse biased (Fig. 3-6c). The generation of a charge collection current in the space charge region of the semiconductor diode is due to the formation of electron-hole pairs (Section 2.4) and is identical to the process described in Section 3.6 as the electron-beam induced current (EBIC) method. In both types of detector – semiconductor and scintillator – the BSE signal increases with the energy of the primary electrons since backscattered electrons with higher energies generate more electron-hole pairs. Both detectors also exhibit an energy threshold of a few keV below which no signal is produced, because the BSEs must initially penetrate the metallization layer and, in diodes with pn-junctions, the p^+-doped region as well.

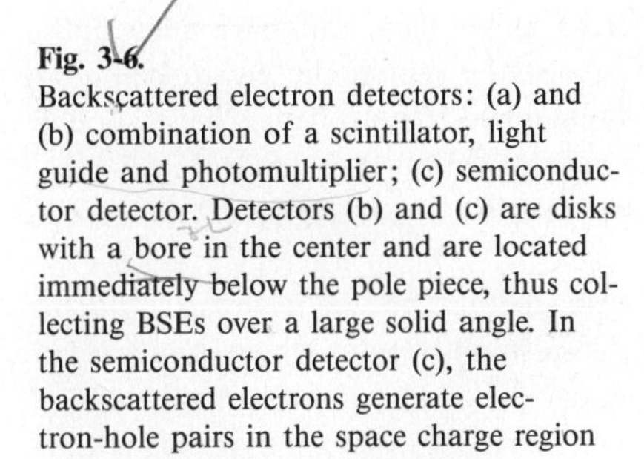

Fig. 3-6.
Backscattered electron detectors: (a) and (b) combination of a scintillator, light guide and photomultiplier; (c) semiconductor detector. Detectors (b) and (c) are disks with a bore in the center and are located immediately below the pole piece, thus collecting BSEs over a large solid angle. In the semiconductor detector (c), the backscattered electrons generate electron-hole pairs in the space charge region of the semiconductor diode. These are separated in the internal electric field of the diode and produce a charge separation current.

Backscattered electrons can also be detected by means of a converter plate coated with a material with high SE yield and placed below the pole piece. This plate transforms the BSEs into secondary electrons which can then be recorded by the SE detector [3-1]. By placing suitably biased grids around the specimen and in front of the converter plate, switchover is possible between the SE and the BSE signals, i.e. between topographical and compositional contrast.

Spectrometers are used as detector systems in special cases, such as for quantifying the voltage contrast of secondary electrons (Chapter 8) or to select low-loss, i.e. nearly elastically backscattered electrons. As these only come from a zone very close to the surface (≈ 10 nm deep) they provide high resolution and good contrast [3-6].

3.2.5 Signal and image processing

The signal at the SE or BSE detector output is proportional to the number of electrons recorded. It is initially amplified more or less linearly by the head and video amplifiers and subsequently controls the brightness of the CRT. However, the triode characteristic of the CRT means that the screen brightness is not a linear function of the video signal. This leads to further problems in image contrast quantification. Other nonlinearities arise during photographic recording of the CRT image and during photographic processing in the darkroom.

Optimal reproduction of the entire contrast range contained in the picture firstly requires adjustment of the signal amplification: higher signal amplification increases the contrast. Secondly, a constant background signal must be subtracted, a process known as adjusting the dark level. Thus a signal variation of only 1% can appear as a 10% variation when 90% of the average signal is deducted. In the latest instruments, contrast and dark level are adjusted automatically.

A variety of other signal processing methods are available to improve the reproduction of the relevant information contained in the picture. Thus an additional amplifier (γ control) can be used to selectively increase the contrast in regions of low or high signal level. Another means of contrast enhancement consists in differentiating the image signal. This results in an emphasis of those parts of the image in which the signal changes. Sometimes it is useful to display the image in reverse contrast, i.e. with light areas shown dark, and dark areas light. The Y modulation method can be used to obtain a quantitative representation of the image signal. In this method, the signal profile is recorded along one or several image lines as a line scan and, if desired, overlaid on the normal image made up of various gray levels (Fig. 3-14b). If many or all of the image lines are displayed in Y modulation, the intensity distribution is represented by a set of closely spaced curves.

Before digital image processing can be applied, the image must first be recorded in digital form. This is done by dividing each line of the raster into small sections where each section forms a picture element (pixel). Alternatively, the pixels can be addressed individually one by one. For this purpose, the deflection coils are controlled directly by the computer via a digital-to-analog converter (DAC). The signal measured for each pixel is integrated over a short time and stored. 256 kbytes of memory are required for a picture made up of 512×512 pixels with 256 gray levels. Once a picture is stored in the computer, all the procedures so far described for analog signal processing can be applied afterwards by means of appropriate programs without the need to scan the specimen over extended periods. A wide range of additional processing methods is available. Image analysis by stereology supplies values such as mean particle size, etc. Two-dimensional Fourier transformations allow certain periodiocities in the image to be better recognized. Linewidth measurement programs produce more accurate results independently of the operator (Section 3.4.4). Stereological methods can now often be applied using the computers of X-ray analysis systems (Section 5). An overview of computer-aided imaging is given in [3-3].

3.3 Specimen preparation

3.3.1 Changes of the specimen under electron bombardment

The specimen may be heated up or become electrically charged under electron bombardment. The electron beam also breaks up chemical bonds in organic molecules (hydrocarbons, etc.) diffusing at the specimen surface which may lead to the growth of contamination layers. Problems arising during imaging photoresist structures due to beam damage are described briefly in Section 3.4.4. The detrimental effects on SEM imaging caused by the above phenomena can largely be avoided by suitable measures regarding specimen preparation and/or instrumental conditions.

The low probe currents of < 1 nA used in scanning electron microscopy transfer so little energy into the specimen that it is in general easily dissipated. Heating of the specimen is then negligible [3-1]. The specimen may, however, heat up significantly in the case of unfavorable geometries such as fibers, or if the thermal conductivity of the specimen is particularly poor.

With high primary electron energies, the total electron yield is $\sigma = \delta + \eta < 1$, i.e. the number of emitted electrons is smaller than the number of incident ones (Sections 2.4 and 3.4.4). In such cases, electrically insulating materials are negatively charged under the electron beam. The potential at highly charged specimen areas may reach several kV. This leads to strong inhomogeneous electric fields above the specimen which may alter the trajectories of the primary electrons, resulting in image distortions or the surroundings of highly charged areas appearing dark [3-1]. The problem of charging during SE imaging of insulating specimens can be solved by coating the specimen with a thin electrically-conductive metallic film (Section 3.3.2) or by keeping the primary electron energy so low that $\sigma = 1$ (Section 3.4.4).

Hydrocarbon molecules are virtually always found on the surface of the specimen. They come from the laboratory air or may condense from the residual gas in the SEM onto the specimen. Possible sources in the SEM are the diffusion pump oil, the grease on the vacuum seals, and fingerprints. The hydrocarbon molecules diffuse along the surface of the specimen. Under electron bombardment their bonds are broken, i.e. they are cracked. Light atoms may escape and the fragments of the molecules cross-link in a process of carbonaceous polymerization. The cross-linked molecules settle, causing a contamination layer rich in carbon to build up. Since the thickness of this layer increases with electron dose, building up of a contamination layer is faster at high magnification where a smaller area of the specimen is scanned. If the probe is stationary, a contamination needle forms. If the contamination layer reaches a thickness of the order of 10 nm, the SE yield corresponds to the value of carbon, and the scanned area generally appears darker than its surroundings.

Contamination stemming from the specimen itself can be prevented or reduced by thorough cleaning (for instance by ultrasonic cleaning in ethanol). As far as the instrument itself is concerned, contamination depends on the vacuum system as well as on the cleanliness of the specimen chamber. The vacuum in the vicinity of the specimen can be improved by a cold trap surrounding the object area. Layers of contamination which have already formed on the specimen can subsequently be removed by etching in an oxygen plasma. This is especially advantageous when extensive measurements (by X-ray analysis, potential measurement techniques, etc.) have to be performed on a particular specimen site and contamination cannot be avoided.

3.3.2 Mounting the specimen

In the simplest case (small, electrically conductive specimens), virtually no specimen preparation is required and it is sufficient to attach the specimen to a support (stub) which is then mounted on the specimen stage of the SEM. However, the specimen is often, as in the case of semiconductor wafers, too large and must first be cut up unless the instrument is equipped with a large specimen chamber. Semiconductor wafers can easily be cleaved after scribing a line across them with a diamond tip. Unless the specimens are totally clean, it is always advisable to clean them before putting them in the SEM. Dust particles can be removed by blowing the specimens with dry air or nitrogen. A more thorough method is to use ultrasonic cleaning in, e.g., ethanol. This also removes residual fragments produced by scribing and fracture, which may stick to the cleaved pieces. The specimens should be blown dry after being taken out of the ultrasonic cleaner.

The specimens may be attached to the stubs by spring clamps, screws or adhesives. In all cases, a good electrical contact between the specimen and the stub is essential to prevent electrical charging. Adhesives providing electrical conductivity are silver paint or carbon paint. The paint should be completely dry to avoid subsequent contamination by the paint solvent. Drying can often take several hours but can be speeded up by heating the specimen in a vacuum furnace. When the specimen consists of electrically insulating material, it is advisable to apply the paint almost right up to the edge of the specimen site to be examined in addition to coating it with a conductive film. This ensures a conducting path to ground via all steps and edges. Various methods of specimen mounting using different types of stubs and adhesives, etc., are summarized in [3-7].

3.3.3 Coating with electrically conductive films

Various properties are required of a conductive coating besides good electrical conductivity, namely good adhesion, chemical stability (with respect to air, for example), formation of continuous films at low thickness having no structural features and showing uniform thickness regardless of specimen topography. Noble metals such as gold and platinum and alloys such as AuPd fulfill these material requirements best. Diode sputtering is the most commonly used method of deposition because it is quick and easy to apply and has numerous advantages over vapor deposition. An overview of various vapor deposition and sputtering techniques is given in [3-2]. Fig. 3-7 shows a schematic drawing of a diode sputtering unit. Under an argon pressure of 1 Pa, application of a high negative voltage of, say, 1 kV to the cathode starts a glow discharge. Ar^+ ions are accelerated towards the cathode, which represents the sputtering target (consisting of gold in Fig. 3-7), and knock metal atoms out of it. Repeated collisions with the gas atoms and ions lead to the formation of a "diffuse cloud" of metal atoms from which the anode carrying the specimen stubs is coated. Since the metal atoms impinge on the specimens from all directions, even rough and fissured surfaces are uniformly coated. The thickness of the deposited layer is proportional to the sputter rate, to the spacing between the cathode and anode, and to the sputtering time. The sputter rate increases with the glow current, which depends on the gas pressure and the high voltage. A good electrically and thermally conductive layer is continuous and is usually 10–20 nm thick.

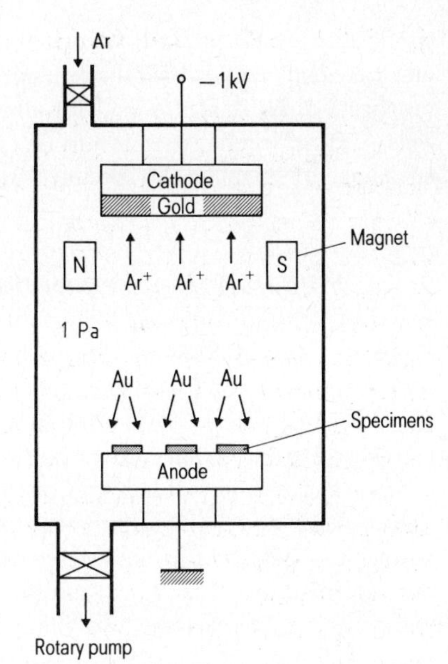

Fig. 3-7.
Schematic drawing of a diode sputtering unit. After the chamber is evacuated, argon is introduced up to a pressure of 1 Pa. During the glow discharge between the cathode (i.e. the gold sputter target) and the anode, gold atoms are knocked out of the target and are deposited on the specimens sitting on the anode from a "diffuse cloud". A permanent magnet deflects the electrons which would otherwise impinge on the anode. This prevents the heating of the specimen that would normally take place under electron bombardment.

Sputtering produces more uniform coatings with better adhesion than vapor deposition because the metal atoms impinge on the specimen with higher energies (several eV) and penetrate the first monolayers. This also leads to the formation of more growth nuclei, and sputtered coatings are therefore less prone than vapor-deposited layers to island formation. Nevertheless, under high magnification sputtered coatings also reveal structural features determined by the grain size of the film which depends mainly on the deposition conditions. Thus diode sputtering with high deposition rates can be used to produce particularly smooth coatings which are generally adequate for routine operation. To avoid masking fine details during imaging at high magnification, sputtered films should be kept as thin as possible, with very small grain size, but must still provide sufficient conductivity. An overview of specimen coating techniques resulting in fine-grained films is given in [3-3]: very fine-grained layers can be produced by using low cathode voltages (e.g. 500 V). Platinum or AuPd alloy sputter targets also yield more finely grained coatings than gold targets. Other methods such as ion beam sputtering or Penning sputtering permit deposition of even finer-grained layers than those obtained by diode sputtering. However, these methods are more expensive in terms of equipment and are so far not widely used in practice.

If an X-ray analysis is to be carried out in addition to SE imaging (Section 5), a gold sputter layer would result in X-ray lines of gold appearing in the X-ray spectrum of the specimen. This can be avoided by vapor-deposition of a conductive carbon layer. The carbon X-ray line has such a low energy that it is usually not recorded in the spectrum. Moreover, absorption of the incident electrons and emitted X-rays in a thin carbon layer is very low. The most commonly used method of vaporizing carbon is to press two carbon tips tightly together and then, by passing an electric current through them, heat them up to a point where the carbon sublimes at the tips. In another method, thin carbon fibers are vaporized by an electric current. Layer thicknesses are again generally in the 10–20 nm range. As carbon layers are brittle, thicker layers flake easily. If carbon is vaporized above a pressure of only 1 to 10^{-1} Pa, the carbon atoms frequently collide with the residual gas molecules and lose their preferential direction of propagation. As in sputtering, this allows rough specimen surfaces to be uniformly coated. This "diffuse" propagation means that the specimens can also be coated indirectly. It has the advantage of shielding the specimen from the heat radiated by the evaporation source.

3.3.4 Preparation of semiconductor devices

Process control by SEM in semiconductor technology precludes any specimen preparation steps since these would alter the semiconductor wafers and prevent their re-introduction into the production process. SEM inspection within the manufacturing line following the individual lithography and etching processes, etc., yields valuable information on process failures such as resist and etching residues and on critical structural dimensions (Section 3.4.4).

However, if additional information on complex semiconductor structures is required, such as thickness, edge profiles and step coverage of the various layers, cross sections must be made through these structures. Cross sections can be prepared by cleavage fracture or by grinding and polishing. For cleavage, two lines are scribed with a diamond tip on the wafer front surface on opposite sides of the area of interest. The scriber lines have to be aligned with respect to the crystal lattice of the wafer (parallel to ⟨110⟩ directions) and must penetrate through the processed layers of a device down into the silicon crystal. By careful placement of the scriber lines and a well-defined application of bending moments, the cleavage plane can be located within about 10 μm [3-8]. Polished cross sections allow much better positioning accuracy for the desired section plane than cleavage when the specimen is not embedded. In this case, not only the cross section but the surface of the chip as well can be inspected under an optical microscope. Fig. 3-8 shows a simple grinding setup with the polishing done on a roughened glass plate [3-9]. This setup is also commercially available [3-10]. Because of the constant polishing rate, an accuracy of about 0.2 μm can be obtained for the position of the section plane with careful monitoring in an optical microscope. Combined observation of the polished section plane and the surface of the chip provides positional orientation. Polished cross sections are a valuable aid in failure analysis as well as for technology characterization (Chapter 9).

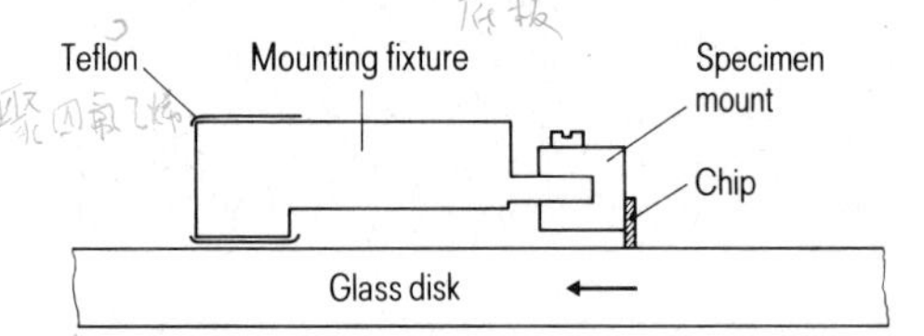

Fig. 3-8. Mounting fixture and specimen mount for producing high-precision cross sections of integrated circuits by polishing on a roughened glass disk [3-9, 10]. The chip or wafer fragment is first glued on the mount with wax. After initial grinding with grinding paper, the rear section of the mounting fixture is wrapped in Teflon tape for polishing on the glass disk to prevent metal from being abraded. The glass plate is rinsed with water during polishing. The polishing pressure is produced by the weight of the mounting fixture itself, which is held in a fixed position by a bar. With the glass disk rotating at 150 min^{-1} and a grinding path radius of 7 cm, resulting in a peripheral velocity of about 1 ms^{-1}, the polishing rate for silicon chips a few mm in width is approximately 1 μm/min.

General practice is to selectively etch the cleaved or polished cross sections. This delineates the individual layers more clearly in the SE image by creating contrast-enhancing steps at the interfaces between the layers. Fig. 3-9 shows as an example an etched cross section through a 4 Mbit memory chip. A variety of etching agents are used for selective etching [3-11, 3-12]. An oxide etch (7 parts HF to 1 part NH_4F) preferentially attacks only the SiO_2 layers. Other etching solutions, like a 20:1 mixture of HNO_3 and HF, attack the silicon and not SiO_2 and can also be used to selectively etch the electrically active regions in the silicon substrate, thereby delineating the pn-junctions.

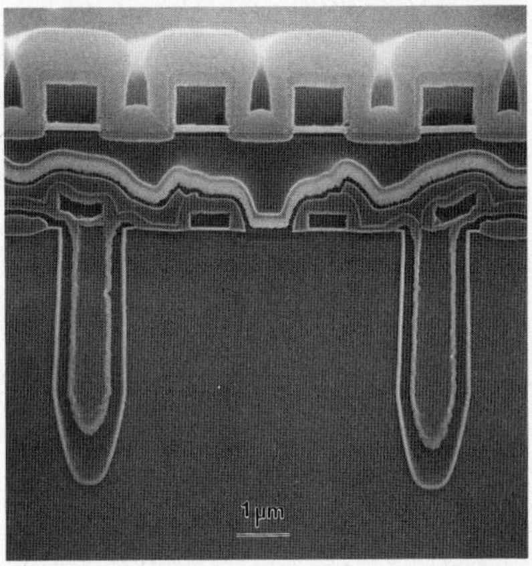

Fig. 3-9.
Polished cross section through a 4-Mbit memory chip (SE image) showing the trench capacitors. Selective etching delineates the individual layers as well as the pn-junctions clearly. As the chip contains several dielectric layers which would otherwise charge under the electron beam (particularly the top Si_3N_4 passivation layer), the specimen was coated with gold. $U_0 = 15\,kV$.

SEM-based failure analysis of semiconductor devices generally requires more extensive preparation [3-13]. In mounted devices, the chip must first be detached from its package. This is particularly difficult with plastic packages, but it can be done, for example by using hot sulphuric acid or by plasma etching. Since the causes of failures generally lie in buried layers, the overlying layers have to be removed. Usually, the passivation must be etched off first. Plasma etching is recommended for nitride passivation layers. The various etching agents suitable for the different material of the layers are summarized in [3-12,14].

Cross sections also provide more information on GaAs devices. GaAs is more brittle than silicon and is easy to cleave along the (110) planes. This is quite sufficient for optoelectronic components such as laser diodes, as the layers to be investigated are grown epitaxially, i.e. with the same crystal orientation as the GaAs substrate. The problem in preparing cross sections through integrated circuits on GaAs is that the gold metallization layers smear easily. However, these smeared gold residues can be removed by polishing on cloth using very fine alumina powder.

3.4 *Imaging with secondary electrons*

3.4.1 *Contributions to the secondary electron signal*

The signal recorded by the SE detector (Section 3.2.4) is made up of several contributions (Fig. 3-10a). The secondary electrons excited directly by the primary electrons and generated within the escape depth λ_{SE} are called type 1 SE (SE 1). Furthermore, the backscattered electrons release secondary electrons (SE 2) before leaving the surface. These two contributions cannot be separated and determine the secondary electron yield δ defined in Section 2.4.3. The SE 2 component is proportional to the backscattering coefficient, which increases with the atomic number (Fig. 2-35) and is therefore responsible for the slight increase of the SE yield with atomic number. The ratio of the SE 2 to SE 1 contributions is between 0.5 and 1, depending on the atomic number and the angle of incidence. The backscattered electrons emitted by the specimen strike various parts of the specimen chamber (mainly the pole piece of the probe-forming lens) and there, in turn, again excite secondary electrons (SE 3) which are extracted by the SE detector grid. Finally, backscattered electrons may also impinge on the SE detector scintillator directly, thus increasing the SE signal.

Fig. 3-10b shows the spatial distribution of the secondary electrons leaving the surface. Only the SE 1 excited by the primary electrons contain the desired information with the high spatial resolution determined primarily by the diameter of the electron probe (Section 3.4.3). This component accounts for only 25–50% of the total SE signal [3-1]. The residual component contains the information provided by the backscattered electrons, which originate at much greater depths in the specimen, i.e. from a hemisphere with a radius corresponding to half the penetration depth (Section 3.5). This residual component represents a background signal which degrades the contrast and also, to some extent, the resolution. Its contribution to the image can, however, be reduced by suitable signal processing (dark-level subtraction, Section 3.2.5).

a

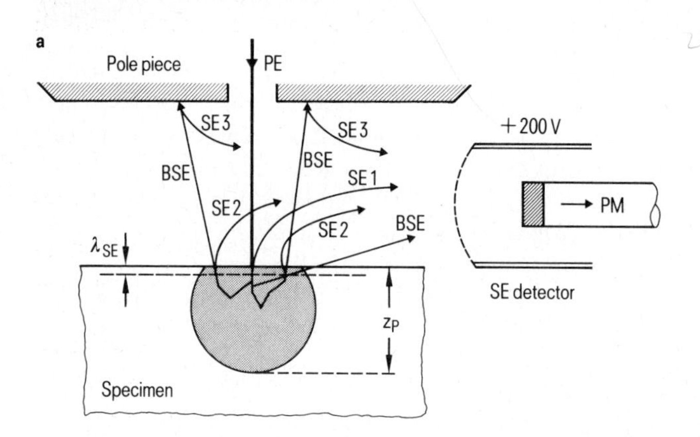

b

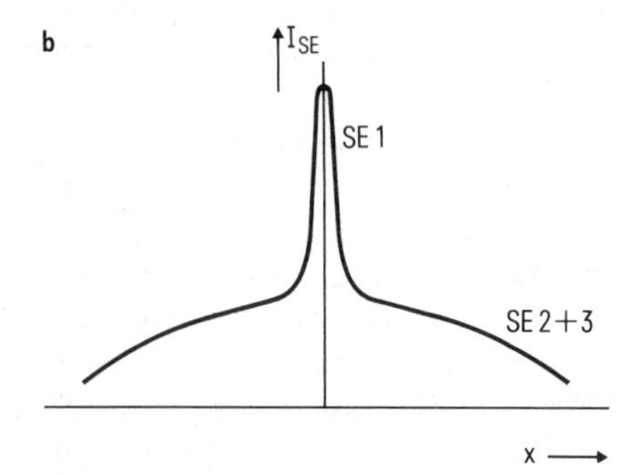

Fig. 3-10.
Contributions to the secondary electron signal.
a) Schematic diagram showing type of electrons collected by the SE detector: the primary electrons PE release only the SE 1 electrons; SE 2 are released by the backscattered electrons in the specimen. In addition, BSE striking the pole piece, etc., excite SE 3 electrons and BSE may also strike the SE detector directly. b) Exit distribution of secondary electrons showing schematically the SE 1 and SE 2 + 3 components.

Not all secondary electrons produced in these different ways are collected by the SE detector. Figs. 3-10a and 3-1 show that the SE detector is generally mounted on one side of the specimen in the specimen chamber because of the limited working distance. The number of SEs which actually enter the detector depends on the design of the specimen chamber and detector, on the working distance and the specimen tilt angle and also on the grid voltage. Thus, with a plane, untilted specimen and a short working distance, a larger number of SEs strike the pole piece if they are generated on the side pointing away from the detector. This results in an intensity gradient in the image at low magnification.

The relationship between SE yield δ and angle of incidence ψ of the primary electrons is given by $\delta \sim 1/\cos \psi$ (normal incidence at $\psi = 0$). This results in a strong increase of the SE yield at shallow incidence (Section 2.4.3). The angular dependence of the SE signal is also affected by the BSE component directly entering the detector. This component exhibits a different angular dependence than the SE yield and reaches a maximum at those angles of incidence for which the backscattered electrons, in accordance with the law of reflection (Fig. 2-36), are preferentially scattered in the direction of the SE detector.

3.4.2 Image contrast

The contrast in SE images of the specimen surface results from the dependence of the individual contributions to the SE signal on parameters such as angle of incidence and PE energy, but also on the detector geometry. The following contrast mechanisms can be distinguished:
- surface tilt contrast
- shadow contrast
- diffusion contrast
- material contrast
- voltage contrast.

The greatest contribution to the SE contrast is formed by the strong dependence of the SE yield, and hence the SE signal, on the tilt angle of the imaged specimen area. This results in the *surface tilt contrast.* Because the SE detector is located to the side of the specimen, a *shadow contrast* is also produced: surfaces of the specimen facing away from the SE detector appear dark because only a small proportion of the secondary electrons excited there can be extracted by the detector. The surface tilt and shadow contrasts are responsible for the high topographical contrast in SE images providing a three-dimensional impression of the surface topography. Several examples are shown in Fig. 3-11. In the large crystallites in Fig. 3-11a, the individual crystal faces appear with different brightness according to their tilt and azimuth angles (the latter determine their orientation with respect to the detector). In Figs. 3-11a to 3-11d, the SE detector is located above the upper edges of the micrographs. Knowing the position of the detector, elevations and depressions can be distinguished unambiguously (Fig. 3-11b). A pair of stereo images directly provides a three-dimensional view of the specimen surface. Since the secondary electrons generated in the pores and cavities of a specimen are very difficult to extract, pores and cavities always appear dark (Fig. 3-11c). If an image is to be taken of shallow surface steps or fine surface roughness, it is advisable to tilt the specimen very steeply (Fig. 3-11d). The dimension of interest (normal to the surface) can then be recorded directly, and contrast is particularly enhanced because of the great change in SE yield at a high angle of incidence .

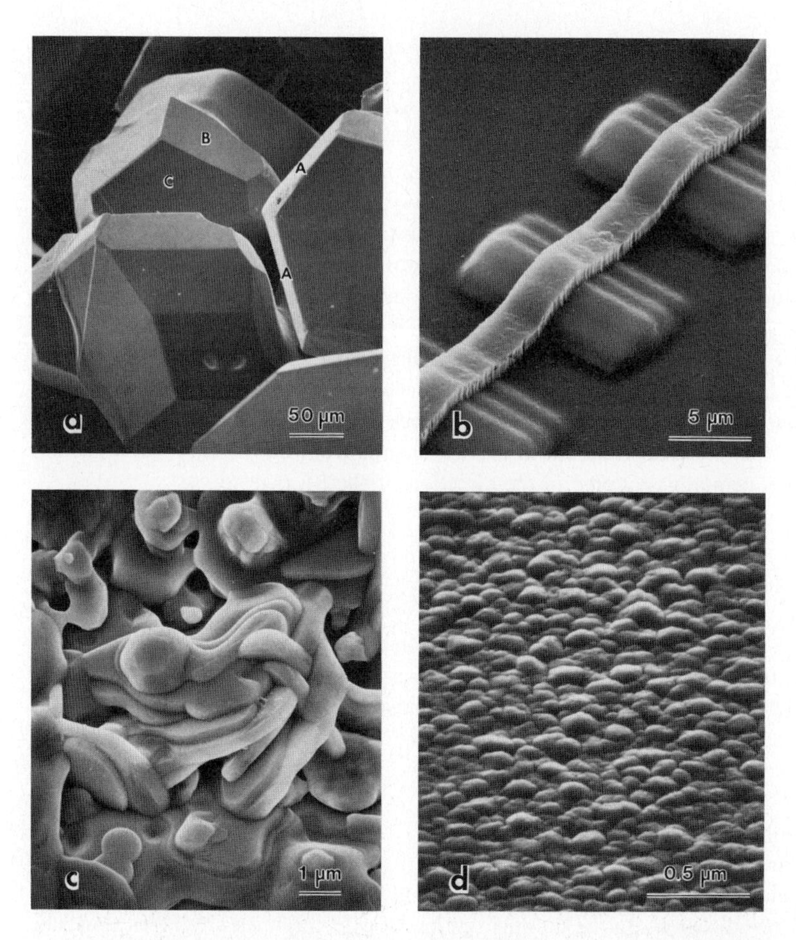

Fig. 3-11. Topographical contrast in secondary electron images. a) The individual crystal faces of the erbium crystallites appear with different brightness according to their tilt angles and their orientation with respect to the detector. The two faces marked A have a very large tilt angle relative to the PE beam and are therefore very bright. The brightness difference between faces B and C has a different origin. The face marked B is pointing towards the detector, which is located above the top edge of the image, and therefore appears brighter than face C which is turned away from it. b) The aluminum interconnection runs over elevations on a structured silicon wafer. c) Surface of a $Ba(Ti,Ni,Zn,Ta)O_3$ ceramic with plate-like crystallites. The pores appear dark because the SEs released within them cannot be extracted. d) Polysilicon surface. The hillocks formed by the individual crystallites are seen more clearly by tilting the specimen at a large angle (60°). (a) to (d): $U_0 = 15$ to 25 kV.

In surface details with a radius of curvature less than about half the penetration depth of the primary electrons, as is the case with edges and small particles, the surface area from which the diffusely scattered electrons can leave the specimen is increased. This increases the contribution of the SE 2 and SE 3 electrons (Fig. 3-12). Since the diffusely scattered electrons are responsible for this type of contrast, it is referred to as the *diffusion contrast*. Examples are given in Fig. 3-13. The diameter of the needles in Fig. 3-13a (about 50 nm) is much smaller

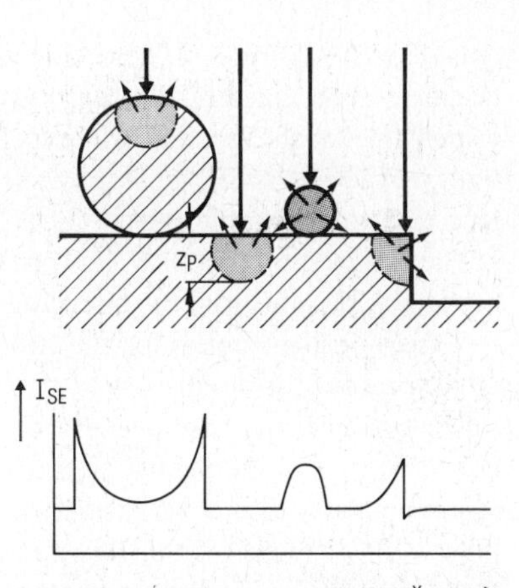

Fig. 3-12.
Diffusion contrast of small particles and edges:
a) spheres with a diameter greater and smaller than the penetration depth z_p of the primary electrons, and edge; b) SE signal profile across spheres and edge. The large sphere appears brighter along the perimeter because of surface tilt contrast. In the case of the small sphere, on the other hand, the diffusely scattered electrons release secondary electrons (SE 2 + SE 3) over the entire surface, and the whole sphere therefore appears brighter with an intensity maximum in the center. In the same way, emission of diffusely scattered electrons is favored at edges, producing a higher SE signal. (According to [3-1]).

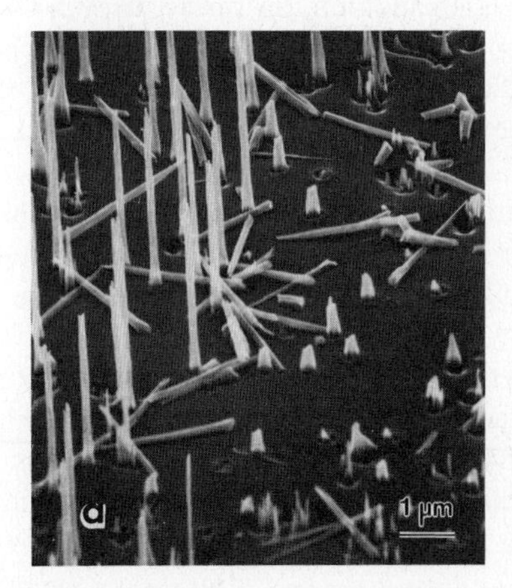

Fig. 3-13. Diffusion contrast in secondary electron images. a) The silicon needles resulted from an etching experiment using reactive ions. They either rise up vertically on the silicon surface or are broken down. Since their diameter is much smaller than the PE penetration depth, they appear bright overall. U_0 = 25 keV. b) The surface of a thermally etched lithium niobate single crystal with (001) orientation shows a terraced structure. The terrace edges appear bright because SE emission is enhançed. U_0 = 15 kV.

than the penetration depth of the PEs (about 5 µm). Thus, when PEs strike a needle, SEs are – as described above – emitted from the entire surface of the needle, which then appears bright as a whole. Because the SEs released on the needles pointing upwards can be extracted more easily, these are brighter than the needles lying on the ground. In the terraced structure shown in Fig. 3-13b, more SEs can leave the specimen at the edges of the terraces. Consequently, these edges appear as bright lines largely irrespective of their orientation. The width of the bright zone at the edges increases with the PE energy because the penetration depth increases approximately in proportion to $E_0^{1.6}$ (Section 2.4.1). The fact that edges yield a clear SE signal has been exploited during selective etching of cleaved or polished cross sections through integrated circuits (Fig. 3-9). The individual layers can be distinguished clearly thanks to the bright lines provided by the finely etched steps at their interfaces. Diffusion contrast is also responsible for the fact that a fine surface roughness, which is imperceptible at low magnifications, enhances the SE signal from such surfaces far beyond the level typical for a smooth surface with the same tilt angle.

The marked topographical contrast makes SE images look as though the specimen were illuminated with light from the direction of the SE detector. But there are, however, no comparable effects in light optics for the strong increase in the SE signal with the tilt angle and the diffusion contrast at edges.

The principal factor responsible for *material contrast* is the large contribution of backscattered electrons to the SE signal, a contribution which increases with atomic number. For PE energies greater than 5 keV, the backscattering coefficient rises sharply with atomic number (Fig. 2-35). Regions with different mean atomic number therefore show different contrast in the SE image. Fig. 3-14 compares the SE image of a cleaved cross section through a dielectric mirror with its BSE image. In the SE image shown in Fig. 3-14a, the individual layers can be distinguished as bright and dark stripes. In addition, the surface irregularities (fracture lines) show up. Because of the greater mean atomic number, the Ta_2O_5 layers appear brighter than the SiO_2 layers. In the BSE image shown in Fig. 3-14b, on the other hand, the surface details disappear and the material contrast between the layers is more pronounced. It must be borne in mind with SE images that the BSE-induced contributions to the SE signal originate from a specimen depth up to half the penetration depth and do not possess the high spatial resolution of the SE 1 component. At lower PE energies, material differences in the SE yield itself come into play. Use of such low energies is of particular value in semiconductor technology applications and is described in more detail in Section 3.4.4.

In addition to the mechanisms described above, differences in surface potential also affect image contrast. This phenomenon, known as *voltage contrast*, is dealt with in greater detail in Chapter 8. Differences in local surface potential also arise during charging (Sections 3.3.1 and 3.4.4). Furthermore, magnetic fields extending outside the specimen may also influence SE contrast (Section 3.7.3).

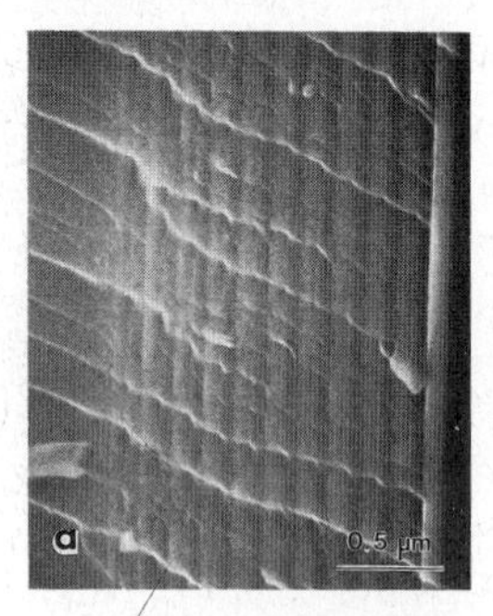

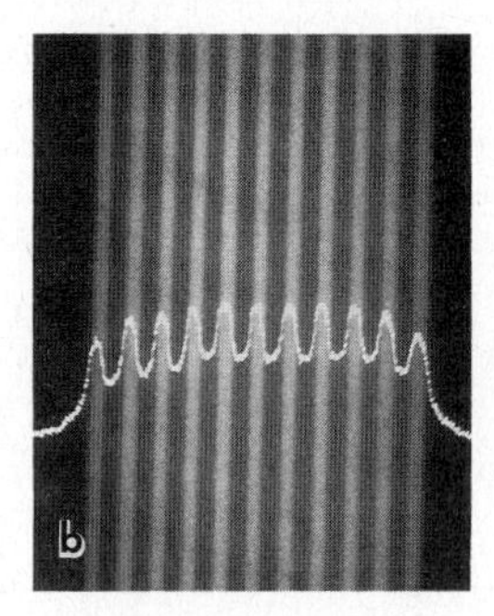

Fig. 3-14. Material constrast: secondary electron image (a) compared with backscattered electron image (b). The images show a cleaved cross section through a dielectric mirror consisting of thin, alternating layers of SiO_2 and Ta_2O_5. Because of their different mean atomic numbers, the layers are visible due to material contrast in both images. Whereas the SE image (a) clearly reveals surface irregularities in addition to the layers, the BSE image (b), recorded by an annular detector, shows only the stripe pattern of the layers. In image (b), the BSE signal profile of a line scan (in *Y* modulation) is overlaid on the BSE image. $U_0 = 25$ kV.

A sophisticated detector strategy allows more quantitative information to be obtained. When the detector extraction field is suppressed by placing a hemispherical grid over the specimen, the exit directions of the secondary electrons are not changed. Then, the signal difference of two SE detectors lying on opposite sides (in the x direction) is proportional to dz/dx if z (x, y) describes the surface topography [3-1]. The surface profile can be reconstructed by integration.

3.4.3 *Spatial resolution and depth of focus*

The limit of *spatial resolution* of SE images is approximately equal to the escape depth λ_{SE} of the SEs. This gives a minimum information volume of about λ_{SE}^3. In general, this volume has a diameter of only a few nm (Section 2.4.3). In addition, the diameter d_0 of the probe must be considered, yielding an exit diameter for the SE 1 contribution of $(d_0^2 + \lambda_{SE}^2)^{1/2}$. This contribution alone provides the desired high spatial resolution. The other contributions of the SE signal derive from the information volume of the backscattered electrons. The latest instruments achieve probe diameters of around 4 nm for LaB_6 cathodes and 1 to 2 nm with field-emission cathodes at acceleration voltages of 20 kV. However, in order to demonstrate the high resolution, special test specimens with a low λ_{SE} and correspondingly fine surface details are necessary. At lower voltages of only a few kV, the resolution is about one order of magnitude worse because of the larger probe diameter (Sections 3.4.4). Small probe diameters can be achieved only with large divergence angles, i.e. a small working distance, and high brightness electron guns (Equation 2.23). There is always a trade-off between the diameter of the probe and the probe current, a smaller-diameter probe with less current adversely affecting the signal-to-noise ratio and hence the contrast. Resolution and contrast are thus intimately coupled and cannot be selected independently.

However, the resolution limit becomes a matter of practical concern only at high magnifications. At low magnifications, the resolution of an SE image is limited by the raster line spacing. On high-resolution CRTs with a 10 cm × 10 cm screen having, say, 1,000 lines, the line spacing is 0.1 mm. This is only half the resolution of the naked eye (0.2 mm), and there is no point in magnifying Polaroid negatives, which are about the same size as the CRT, by more than twice. The probe diameter limits the resolution only at high magnifications (greater than 10,000, for example).

The *depth of focus* is determined by the divergence angle of the electron probe α_0 and is defined as that depth range which still appears sharp, i.e. in which loss of resolution resulting from probe divergence is smaller than the line spacing during scanning. The geometrical situation is explained in Fig. 3-15. Relating the focus depth t_F to the width a of the scanned area on the specimen, gives $t_F/a = 1/(\alpha_0 \cdot 10^3)$. At high magnifications, α_0 must be approximately 10 mrad. This gives $t_F/a = 10^{-1}$, i.e. the depth of focus is always about 10% of the image width, irrespective of the magnification. At small magnifications, α_0 can be reduced to values of around 1 mrad by using smaller-diameter apertures and increasing the working distance (Fig. 3-1). The depth of focus is then about the same as the width of the imaged specimen area, so all details within a cube whose sides are equal to the image width appear sharp. With the same resolution, depth of focus in an SEM is always at least one to two orders of magnitude greater than in an optical microscope. This is the great advantage of the SEM over the optical microscope, even at low magnifications.

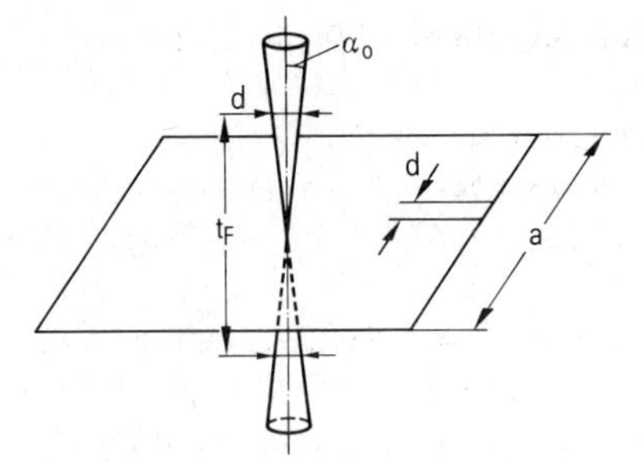

Fig. 3-15.
Depth of focus in SE images. The electron beam with divergence angle α_0 scans a specimen area of width a with a line spacing $d = a/1000$ (1000 lines). Depth of focus t_F is that depth range in which beam broadening due to divergence of the electron beam is at most equal to the line spacing. This gives $d = a/10^3 = 2\alpha_0 \cdot t_F/2$, or $t_F = a/(\alpha_0 \cdot 10^3)$.

3.4.4 Low-voltage microscopy and linewidth metrology

The improvements brought about in scanning electron microscopes operated at low acceleration voltages are mainly based on the requirements of semiconductor technology. Now that the minimum structural dimensions of VLSI circuits have shrunk to ≤ 1 μm and are approaching 0.5 μm, optical inspection methods are no longer adequate. SE imaging provides the necessary resolution, but at conventional PE energies of between 10 and 30 keV beam damage and charging represent a problem since the structures to be investigated often consist of photoresist or insultating layers (e.g. SiO_2). Both beam damage and charging can be avoided by using low acceleration voltages in the 0.5 to 2 keV range. Low voltage operation, however, imposes instrumental problems which have to be considered. The design of the electron-optical column has to be optimized since the image aberrations increase dramatically at low beam energy (especially the chromatic aberration) and the Boersch effect also increases. Enhanced sensitivity to stray magnetic fields requires additional shielding. Furthermore, the extraction field of the SE detector influences the primary beam. This can be avoided by using, e.g. a detector above the lens which collects the secondary electrons through the bore of the pole piece [3-15]. Since the gun brightness decreases with acceleration voltage ($\beta \sim U_0$, Section 2.3.2) the probe current is reduced. The poorer signal-to-noise ratio results in a loss of image information. This can be balanced by employing high-brightness electron sources, image storage techniques and signal integration. Using a field emission gun, for $E_0 = 1$ keV a resolution in the region of 10 nm and even below can be achieved.

Beam damage occurs mainly during imaging of photoresist structures when a high E_0 is used. The high-energy primary electrons break up bonds in the bulk of the photoresist, releasing light atoms. If these form gas bubbles, the volume increases and the surface becomes rippled. If the light atoms diffuse out, the photoresist shrinks. The change in volume alters the shape and dimensions of the structures. It has been found during irradiation with various electron energies that the thickness of PMMA photoresist is reduced as the electron dose increases [3-16]. The reduction in thickness is greatest for those E_0 values with a penetration depth approximately equal to the resist thickness. On the other hand, only a slight reduction in thickness has been measured at a low energy of 1 keV. At such low energies the damage is confined to a shallow surface region and is less disastrous to the bulk of the photoresist. Such low energies can therefore be employed for imaging purposes with minimum change of the resist structure. However, the electron dose must always be carefully chosen to provide an optimum compromise between resolution and radiation damage. A further hazard during the inspection of integrated circuits with high energy PE is that radiation damage may alter their electrical device properties [3-17] (Chapter 8).

In any discussion about *charging*, the dependence of total electron yield ($\sigma = \delta + \eta$) on primary electron energy must be invoked (Fig. 3-16). At low energies, η has a value of between 0.2 and 0.3 and σ is determined primarily by the SE yield δ. In the diagram shown in Fig. 3-16, there are two values of E_0 (E_{01} and E_{02}) for which $\sigma = 1$. The current of incident PEs is then equal to that of the escaping SEs and BSEs and charge equilibrium is obtained, i.e. no charging takes place. For $E_0 > E_{02}$ ($\sigma < 1$), fewer electrons leave the specimen than strike it, resulting in negative charging. As this negative charge can only be eliminated by a residual specimen current, it reaches high levels for specimens with good insulation properties. The negative charging reduces the effective potential difference between the cathode and the anode, and hence also the PE impact energy (arrow 1). Charging continues until the impact energy reaches a value of E_{02}. If, for example, a specimen made from a material with an E_{02} value of 2 keV is irradiated by primary electrons with an energy of 5 keV, it charges up to -3 kV; the potential difference between the specimen and the cathode is then 2 kV and $\sigma = 1$. For PE energies between E_{01} and E_{02} ($\sigma > 1$) a positive charge is built up. The impact energy increases (arrow 2) and approaches E_{02}. In the region $E_0 < E_{01}$ ($\sigma < 1$) charging is negative, and the effective impact energy tends toward zero (arrow 3). It can easily be seen from the direction of the arrows in Fig. 3-16 that E_{01} is an unstable point of equilibrium. Only for E_{02} can a stable charge equilibrium be attained and used for practical purposes.

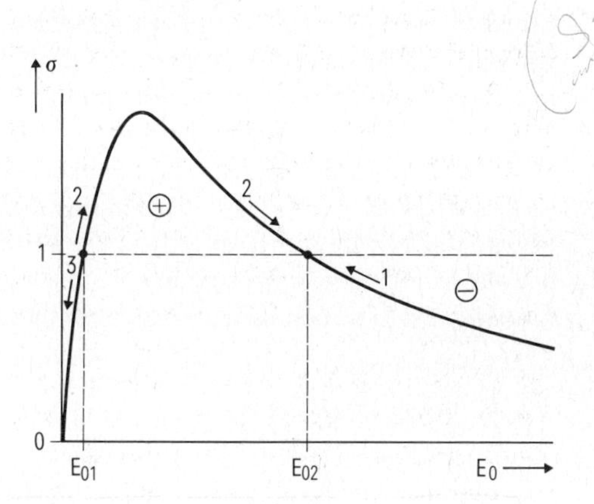

Fig. 3-16.
Total electron yield $\sigma = \delta + \eta$ as a function of PE energy E_0. For the two values E_{01} and E_{02}, $\sigma = 1$, and no charging takes place. Positive charging occurs in the range between E_{01} and E_{02} ($\sigma > 1$); negative charging occurs for $E_0 < E_{01}$ and $E_0 > E_{02}$. The arrows indicate the changes of the effective impact energy of the PEs.

In insulators, however, positive charging in the region $E_{01} < E_0 < E_{02}$ is limited to only a few eV since a positively charged object point can attract low-energy secondary electrons excited at adjacent points and thereby discharge. However, this way of stabilizing the surface potential is achieved at the cost of the SE yield and results in changes in image contrast.

E_{02} generally has a value of between 1 and 3 keV. These values are given in Table 3-1 for some materials relevant to semiconductor technology and apply for normal incidence; they increase with angle of incidence ψ of the PEs, as both δ and η increase with ψ. With tilted specimens, charge equilibrium can therefore be reached at higher PE energies. The E_{02} values are also influenced by microroughness and adsorbed surface layers. In real specimens with an uneven surface and composed of different materials, each surface area has a different σ vs. E_0 curve and charge equilibrium cannot be achieved for all areas at the same time. If E_0 is

Table 3-1. Values of primary electron energy for charge equilibrium (E_{02} in Fig. 3-16) for various materials measured in an SEM for normal incidence under standard high vacuum conditions.

Material	E_{02} [keV]	Reference
gold	8	[3-18, 19]
copper	2.8	[3-18, 19]
aluminum	2.8	[3-19]
GaAs	4.3	[3-19]
silicon	1.6	[3-18, 19]
SiO_2	3.0	[3-20]
passivation	2.0	[3-20]
polyimide	0.6	[3-19]
AZ1470 resist	1.0	[3-20]
epoxy	1.1	[3-19]
ceramics	2	[3-18, 19]

selected smaller than the lowest value of E_{02} necessary for local charge equilibrium, charging will have only a positive sign. However, as mentioned above, positive charging is limited to a few eV by redistribution of secondary electrons. Accordingly, no deviations have been observed in the region of positive charging when measuring the linewidth (see below) of photoresist patterns on GaAs [3-21]. A negative surface charge is already building up on the photoresist when E_0 is greater than 0.8 keV, leading to the image distortion indicated schematically in Fig. 3-17 and an overestimated linewidth.

Because of the low penetration depth – which at 2kV is only about 2% of the value at 20 kV – with the same probe current, more secondary electrons are generated near the specimen surface where they can escape. This means that the fraction of SE 1 (Fig. 3-10) in the overall SE signal which are excited directly by the PEs is increased at the expense of the other contributions stemming from greater depths. The higher proportion of SE 1, which contain the information with the desired high spatial resolution, should improve the *SE contrast.* At PE en-

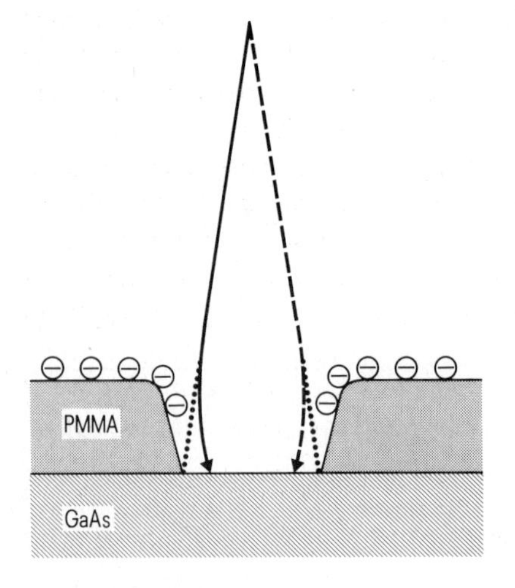

Fig. 3-17.
Linewidth measurement of insulating structures: photoresist (PMMA) on GaAs (schematic cross section through a window in the photoresist). The negative charging of the photoresist deflects the electron beam inwards. A larger deflection by the scan coils is therefore necessary for the electron beam to reach the edges of the window. This makes the window width appear larger [3-21].

ergies as low as E_{02}, however, the nature of image contrast formation changes [3-20]. At energies above 5 keV the main cause of the SE contrast arises from the variation of the SE yield δ with the angle of incidence (surface tilt contrast). This variation, however, becomes less with decreasing E_0 and disappears at about E_{02}. The residual topographic contrast that remains comes from variations in the detector collection efficiency as a function of emission angle (shadow contrast). This lack of strong topographical contrast makes low voltage images more difficult to interpret than images recorded at high voltages. Because of the reduced scattering volume diffusion contrast is diminished at low energies, and edges therefore appear sharper.

In addition to a column optimized for low voltage operation, scanning electron microscopes used for *in-process control* in semiconductor technology require a large specimen chamber and a stage with a sufficiently large range of movements to allow examination of wafers up to 200 mm in diameter. Besides clean-room compatibility, a clean, oilfree vacuum system is required to guard against contamination of the specimens, as mentioned above. A conical pole piece of the probe forming lens allows tilting of the wafers. SEM inspection is used to qualitatively determine pattern fidelity and to detect defects on wafers and masks. Evaluation of pattern fidelity comprises, e.g., checking for resist and etching residues and monitoring of the angle and shape of sidewalls. Apart from pattern inspection, critical dimensions, e.g. linewidth and pitch as well as placement accuracy of patterns are to be examined. Critical dimensions are measured either by analyzing the stored frames or by recording and measuring an SE signal (or line scan) profile. Reviews on critical dimension metrology in the SEM are given in [3-22] and [3-23]. Inaccuracies during SEM metrology are due partly to the instrument and partly to the specimen. Apart from environmental influences such as vibration and stray magnetic fields, instrument-induced sources of inaccuracy include errors in magnification reading, nonlinearity of the scan, lens hysteresis and resolution constraints due to the limited number of pixels. It is most vital to calibrate the magnification accurately against a calibration standard, as neither the magnification values generally provided nor the micron markers displayed on the CRT are sufficiently precise. The parameters affecting the magnification (such as the working distance or the lens current) should also be checked before starting work. Measurement results also depend on the position of the detector. The two sidewalls of a line structure usually do not furnish the same contrast. Although this type of contrast conveys a three dimensional impression, it interferes with precise measurement. To obtain a symmetrical contrast, the edges of the structure must be aligned with the axis of the SE detector. The edges can then be set parallel again on the screen by the raster rotation. In dedicated linewidth metrology systems (see below) no specimen tilt and rotation is necessary since symmetrical contrast is obtained by means of a special SE detector which sits above the probe-forming lens and extracts the SEs through the bore of the pole piece [3-15]. As mentioned above, specimen-related inaccuracies are caused by charging and beam damage effects, but they may be also due to inadequate understanding of how the electron beam interacts with complex specimen geometries. The line scan profiles and hence the results of linewidth measurements thus vary with the electron energy. Monte Carlo modeling of line scan profiles can help to clarify the edge definition from the SE signal [3-22].

Linewidth measurement in a manual mode involves the positioning of cursors at the points on the image or the linescan, between which the distance is to be computed. This is time-consuming and the measurement accuracy depends on operator skills. Analysis by digital image processing (Section 3.2.5) provides quicker and more objective results. The signal-to-noise ratio is usually improved by image integration in a digital frame store. If the edges of line structures are set parallel to the y axis, the signal-to-noise ratio can also be improved by means of a low-pass filter acting only in the y direction [3-24] (Fig. 3-18). The pattern edges can then be located by applying additional image processing methods. Fig. 3-18 shows how numerical differentiation, discrimination and a thinning procedure finally produce a binary image of the pattern containing only the edge information. Linewidth distribution can be depicted in the form of a histogram.

When recording line scan profiles, it is also good practice to apply signal averaging to reduce noise. In Fig. 3-19a the SE signal profiles between the two horizontal lines A and B was averaged to form the line scan profile shown in the lower part of the image. At the edges of the polysilicon lines the SE signal rises sharply. Various algorithms can be applied for automatic

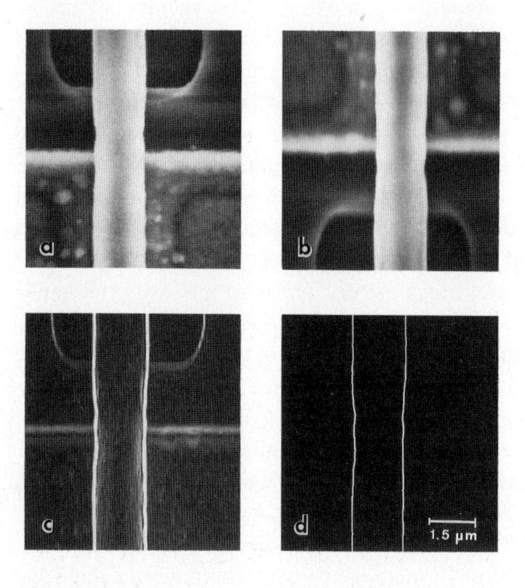

Fig. 3-18.
Edge definition by digital image processing. The SE image of a photoresist pattern (a) is first passed through a low-pass filter (b) and then numerically differentiated (c), thus emphasizing the edges. Discrimination and thinning yields a binary image (d) which now contains only the edge information [3-24]. $U_0 = 1$ kV.

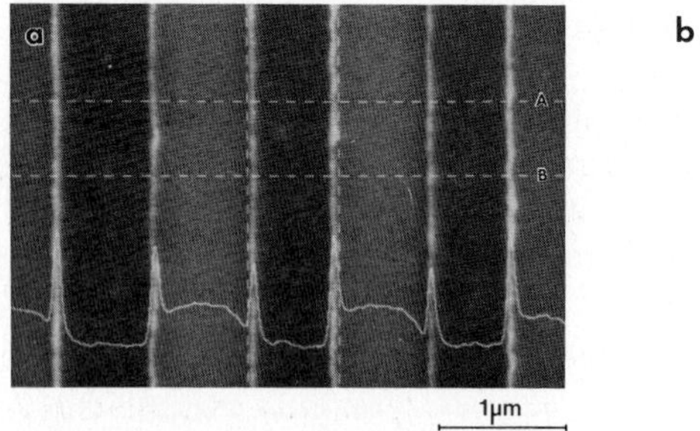

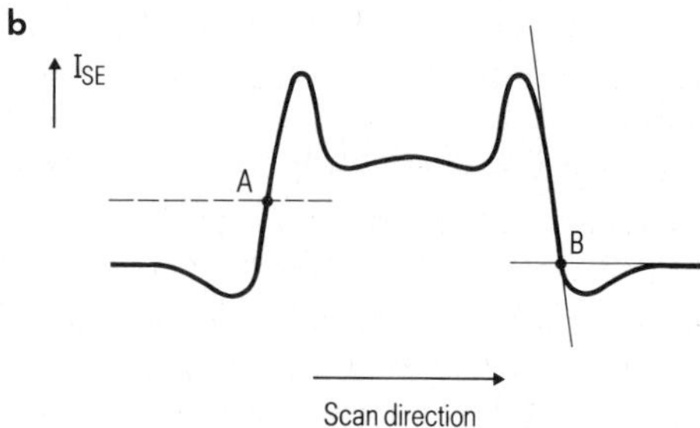

Fig. 3-19. Linewidth measurement by using a line scan profile of the SE signal. a) Pattern of polysilicon lines on SiO_2. The line scan profile shown in the lower part is the average of the SE signal between the lines A and B. $U_0 = 0.8$ kV. (Courtesy E.Demm and T.Harder). b) Schematic line scan profile showing algorithms for automatic edge location: threshold crossing (point A, left) and linear regression (point B, right).

signal evaluation. In automatic threshold crossing the threshold level is set to a preselected percentage of the total signal (Fig. 3-19b, left). The points where the signal crosses the threshold define the position of the edges and the computer will automatically measure and display the distance between the points. In the linear regression method both the base line and the profile edge are approximated by a straight line [3-25]. The point of intersection of these two lines is then assumed to correspond to the position of the bottom edge of the true pattern with high accuracy (Fig. 3-19b, right). This method is relatively insensitive to signal variations and changes of slope angles. Dedicated pattern linewidth metrology systems have been developed for in-process control in IC production. In combination with a computer-controlled positioning system, these systems allow the desired patterns to be localized on 150 mm diameter wafers [3-15]. In addition, wafer handling can be automated by the use of a cassette system, putting the entire operation on an automatic basis.

3.5 *Imaging with backscattered electrons*

3.5.1 *Image contrast and resolution*

The BSE signal and hence also the image contrast are strongly influenced by the position and type of the detector. A number of special BSE detector configurations and types were described in Section 3.2.4 (Fig. 3-6). Irrespective of the type of detector, the following contrast mechanisms may be distinguished:

- topographical contrast
- material contrast
- crystal orientation contrast.

In BSE images, *topographical contrast* is due to masking of the BSEs by elevations forming a shadow and also by the dependence of the backscattering coefficient on the tilt angle of the individual surface areas. Topographical contrast is enhanced by positioning the BSE detector at the side of the specimen, as shown in Fig. 3-6a. Because of the sharp shadows produced in this configuration, a better impression of the surface topography can be obtained at low magnifications than in SE images. The influence of surface tilt on contrast results from the angular distribution of the emitted BSEs (Fig. 2-36) and the detector position. As in SE imaging, diffusion contrast is also to be expected with backscattered electrons, as diffusely scattered electrons can escape preferentially at edges, for example. The extent to which this process generates a higher BSE signal depends on the direction of BSE collection.

Annular BSE detectors below the pole piece (Figs. 3-6 b,c) result in low topographical contrast because hardly any shadows are produced and the influence of surface tilt on the BSE signal is small with tilt angles of less than 40°. In such configurations, *material contrast* predominates (Fig. 3-14b). Variations in the BSE signal are due mainly to the sharp increase in the backscattering coefficient with atomic number Z (Fig. 2-35). For this reason, material contrast is also called Z contrast; it is so sensitive that differences in the mean atomic number of $\Delta Z < 1$ can still be detected. With the help of a calibration curve, the individual phases in multiphase specimens can be determined from the mean atomic number. Fig. 3-3 shows a section through a ZnO varistor ceramic also containing $Zn_7Sb_2O_{12}$ and Bi_2O_3 phases. The grains of the various phases appear with different brightness depending on their mean atomic number. BSE pictures of this kind are frequently used in X-ray microanalysis of metallographic sections to distinguish the individual phases, which can then be analyzed in detail (Chapter 5). As the backscattering coefficient of thin films – either unsupported or on a substrate – increases linearly with the film thickness, the BSE signal of such films can therefore be used to determine their thickness [3-26].

Because of channeling effects, the backscattering coefficient of single crystals also depends on the direction of incidence of the electrons with respect to the crystal lattice. In flat, polycrystalline specimens, the different grains appear with different brightness depending on their orientation (*crystal orientation contrast,* Fig. 3-20). Grains for which the electron beam lies close to a low-index crystal direction appear bright (Section 3.5.2). Tilting the specimen a few degrees alters the orientation contrast substantially, making it easy to distinguish from a material or topographical contrast. To obtain good orientation contrast, the divergence angle of the electron probe must not be larger than 1 to 10 mrad. The maximum change in backscattering coefficient with crystal orientation is 1 to 10% for electron energies between 10 and 20 keV, and it decreases with increasing energy. As signal changes are small, the BSE signal must be highly amplified and a high probe current is required to obtain low-noise images.

Resolution and contrast in BSE images are largely subject to the same factors as described in Section 3.4.3 with respect to SE images. One difference, however, is that the diameter of the probe is irrelevant, provided it is small, as the resolution is determined essentially by the diameter of the BSE exit area. This diameter is approximately equal to the penetration depth and diminishes with decreasing PE energy. This provides for better resolution, but the BSE signal is reduced at the same time because in both scintillators and semiconductor detectors (leaving aside any threshold energy) the BSE signal is proportional to the energy of the collected BSEs. PE energies smaller than about 5 keV produce only very poor BSE signals. BSEs that have suffered heavy energy losses make only a very minor contribution to the total signal, unlike high-energy BSEs which penetrate the material only slightly and constitute a very strong signal component. As a result, the resolution is often better than the penetration depth. Under favorable conditions, resolutions in the 10 nm range are possible: a resolution of 20 nm can be estimated from the BSE signal profile shown in Fig. 3-14b, for example.

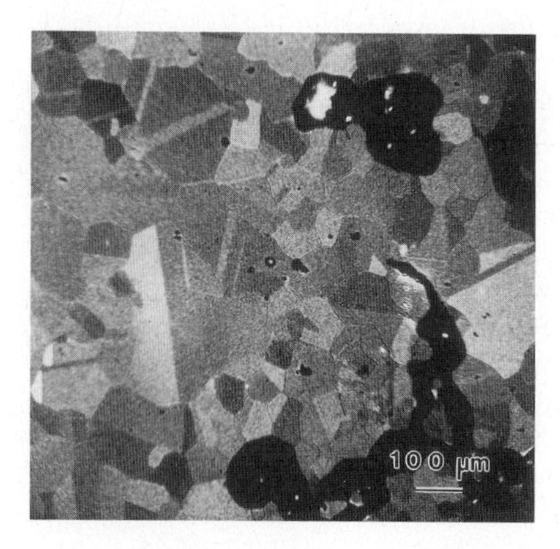

Fig. 3-20.
Backscattered electron image of polycrystalline silicon. The orientation contrast between the different grains is due to the dependence of the backscattering coefficient on the angle of incidence of the electrons with respect to the crystal lattice. $U_0 = 20$ kV.

3.5.2 Electron channeling patterns

In polycrystalline specimens, orientation contrast is generated by the different orientations of the individual grains. However, contrast can also be observed in single crystals because the angle of incidence of the PEs (or rocking angle) varies when the specimen is scanned. This is of little consequence at high magnifications, i.e. with only slight deflection of the beam on the specimen. At low magnifications, however, the deflection and the resulting variation in rocking angle is large enough to produce contrasts characteristic of the particular crystal and its orientation (Fig. 3-21). Such a BSE image is called an electron channeling pattern (ECP) by analogy with the anisotropy effects occurring during ion scattering in crystals (channeling effect, Section 2.6.2). Its pattern is identical to that of Kikuchi patterns described in Sections 2.4.2 (Fig. 2-39) and 4.4 observed in the transmission electron microscope. For this reason, ECPs are also called pseudo-Kikuchi patterns. The theoretical and experimental aspects of ECPs are described in detail in [3-27] and [3-1].

Fig. 3-21. Electron channeling pattern of a silicon single crystal with an orientation close to the [100] direction. The bright bands and bright-dark lines can be associated with specific lattice planes.

However, ECPs differ from the Kikuchi patterns by the way in which the contrast is produced. To understand the change in the BSE signal with the direction of electron incidence shown in Fig. 3-22, some results of dynamical diffraction theory must be invoked (Section 4.5). This theory states that electrons in the crystal propagate in the form of Bloch waves in a direction parallel to the diffracting lattice planes. There are two types of Bloch waves with different intensity distributions (Fig. 3-23). Which of them are excited more strongly depends on the ratio of the angle of incidence ψ and the Bragg angle ϑ. Type I waves are preferentially excited when $\psi < \vartheta$, and type II waves when $\psi > \vartheta$. The local intensity of the Bloch wave is a measure of the electron charge density. For type I waves, the intensity maxima are located at the site of the lattice planes, and for type II waves the maxima are located in between them. Type I waves are characterized by pronounced attenuation due to increased backscattering at the atomic nuclei. Hence a higher BSE signal is observed in the $\psi < \vartheta$ range (A to B in Fig. 3-22) where mainly type I waves occur. Outside this range, type II Bloch waves predominate; they can penetrate deeper into the crystal and therefore generate a weaker BSE signal.

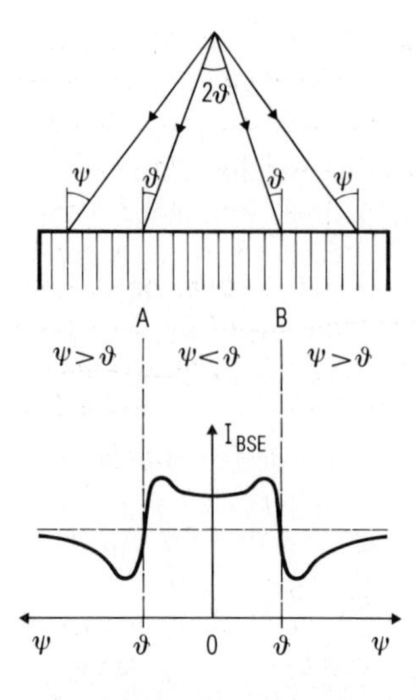

Fig. 3-22.
Variation in backscattering signal I_{BSE} with the direction of incidence of the electrons on a crystal. During a line scan, the angle of incidence (rocking angle) varies from $\psi > \vartheta$ via $\psi < \vartheta$ back to $\psi > \vartheta$. At positions A and B, where the Bragg condition $\psi = \vartheta$ is satisfied, the backscattering signal I_{BSE} changes strongly. In the range $\psi < \vartheta$, I_{BSE} is increased for low-index lattice planes.

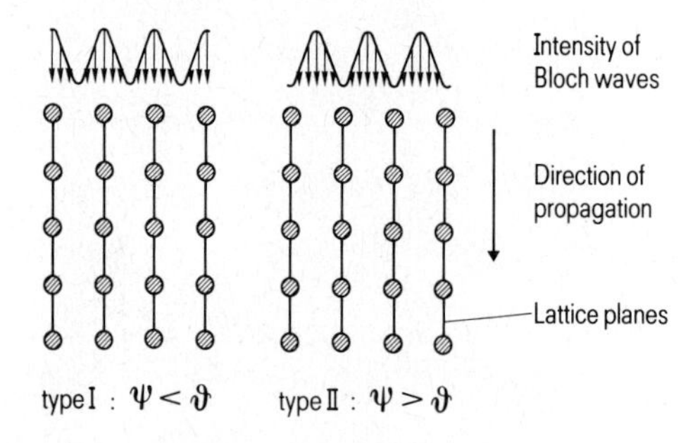

Fig. 3-23.
Intensity distribution of the two types of Bloch waves, which propagate along the diffracting lattice planes. For Type I waves, the intensity maxima are at the site of the lattice planes, in Type II waves between the lattice planes.

Lattice planes with a small Bragg angle (i.e. large interplanar spacing) produce bright bands with dark edges at the outsides (Fig. 3-21). For planes with greater Bragg angle, the edges of these bands appear as bright-dark lines, the width of which diminishes with increasing ϑ. The bands and lines are assigned to specific lattice planes in the same way as in the Kikuchi patterns, as explained in Section 4.4 (Fig. 4-24).

In order to obtain an ECP using the standard method described at the beginning, i.e. to image a flat single crystal with low magnification, a very large area of the specimen must be scanned. To achieve a maximum rocking angle of, say, $\pm 10°$, a specimen area of about 5 mm × 5 mm has to be scanned when the working distance is 15 mm. For a 10 cm screen, a magnification of only 20 times must be used in this case. For 20 kV electrons, the Bragg angle ϑ for the (220) reflection of silicon is 22 mrad (or 1.26°). Good angular resolution in ECP requires a beam divergence which is much smaller than ϑ, i.e $\alpha_0 \leqq 1$ mrad. This is about one order of magnitude smaller than the divergence used for normal imaging. In the standard method therefore, a very small aperture diameter must be employed, or the beam must be sufficiently defocused to strike the crystal with a small divergence. Annular BSE detectors provide the largest signals because of their high collection angle.

Specimen regions several millimeters in size are generally available only on semiconductor single crystals. However, in practice ECPs must be recorded from small regions. This can be done by shifting the pivot point of the electron beam from the plane of the probe-forming lens to the specimen plane. One way of doing this is to decrease the excitation of the lower deflection coils. Because of the large aberration coefficients of deflection systems, however, it is better to use the probe-forming lens to deflect the electron beam back to the specimen point of interest (Fig. 3-24). This deflection focusing method allows ECPs to be recorded from regions less than 10 μm in diameter. To obtain sufficiently large rocking angles, the final aperture must be removed, i.e. the opening must have a diameter of several millimeters. Beam divergence can be restricted by means of the spray aperture located higher up in the electron-optical column.

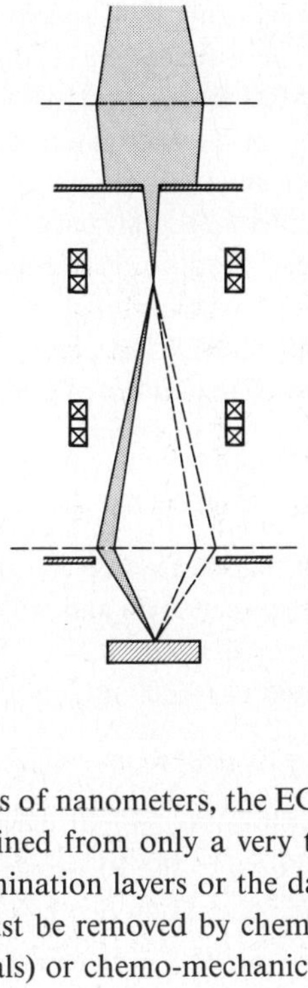

Fig. 3-24.
Deflection focusing method for recording electron channelling patterns (ECP) of small regions. Condenser 2 first focuses the electron beam in the plane of the upper deflection coils; the probe-forming lens then focuses it onto the specimen. The lower deflection coils are switched off and the probe-forming lens deflects the beam back to the specimen point. The system can be changed over from "ECP" to "image" mode by switching on the lower deflection coils.

As Bloch waves are absorbed within a depth of a few tens of nanometers, the ECP information, like the orientation contrast in BSE images, is obtained from only a very thin surface layer. ECPs can therefore be seriously impaired by contamination layers or the damage layer caused by mechanical polishing. Such a damage layer must be removed by chemical etching (for semiconductors), electro-chemical polishing (for metals) or chemo-mechanical polishing (for semiconductors and ceramics).

ECPs are used above all to determine the crystal orientation of grains in polycrystalline materials. Panoramic maps make it easier to identify specific orientations. If fine lines originating from lattice planes with a small interplanar spacing (large Bragg angle) are used for evaluation, lattice parameter differences as small as $\Delta a/a \geqq 5 \cdot 10^{-4}$ can be determined locally. The broadening of the lines provides information about high defect densities.

3.6 Cathodoluminescence and electron beam induced current mode

3.6.1 Charge carrier generation

Electron-hole pairs can be generated in a semiconductor or insulator by electron bombardment (Section 2.4.3). This induces electrical conductivity in the specimen. If internal electric fields are present, an electromotive force (Section 3.6.3) may be produced. Furthermore, luminescence may occur as a result of the energy released during charge carrier recombination (Section 3.6.2). Similar effects – photoconductivity, the photovoltaic effect, and photoluminescence – are observed when the material is irradiated with light (photons). In contrast to electrons, when using light specific energy states can be excited in the specimen by selecting the appropriate wavelength, i.e. the photon energy. On the other hand, the focused electron beam of an SEM makes it easier to generate images with high spatial resolution and to localize and characterize material inhomogeneities in microregions by the effects mentioned above.

To generate an electron-hole pair by electron bombardment, a mean excitation energy $\overline{E_i}$ is necessary (Table 2-2). A primary electron which is absorbed in the solid and loses all its kinetic energy E_0, can produce an average of $N = E_0/\overline{E_i}$ charge carrier pairs. Backscattered electrons also play a part in charge carrier generation: it can be assumed as a good approximation that on average a BSE loses half its energy (i.e. $E_0/2$) in the solid [3-28]. The *generation factor* (i.e. the average number of electron-hole pairs generated for each PE) can therefore be described by

$$G \approx (1 - \eta/2) \cdot E_0/\overline{E_i} \tag{3-1}$$

where η is the backscattering coefficient. For PE energies between 10 keV and 30 keV, generation factors between 2000 and 50000 are obtained for the semiconductors listed in Table 2-2 ($\overline{E_i} = 0.4 \ldots 5$ eV; $\eta = 0.2 \ldots 0.4$). The *generation rate* (i.e. the number of electron-hole pairs generated per unit time) is

$$g = G \cdot I_0/e \tag{3-2}$$

where I_0 is the probe current. The *generation volume* in which the electron-hole pairs are produced is equal to the PE scattering volume whose lateral extension corresponds approximately to the penetration depth (Fig. 2-32). For many practical purposes, it can be regarded as a sphere touching the surface with the penetration depth as its diameter. For aluminum, silicon, SiO_2 and $E_0 = 10$ keV, for example, this diameter is approximately 1 μm. However, as a result of diffusion, the excess charge carriers generated also reach regions outside the generation volume, where they may contribute to the processes described in the following.

3.6.2 Cathodoluminescence

Recombination processes

Generation of electron-hole pairs by electron excitation has been treated in Section 2.4.3. *Recombination* of an electron from the conduction band with a hole in the valence band is possible in a variety of ways: it may take place by a band-to-band transition or step by step via localized energy states within the forbidden gap. Most of these transitions occur without any light emission. In these cases the energy released is transferred to the crystal by way of excitation of electrons (Auger process) and phonons. Light emission (cathodoluminescence CL) [2-35, 36b, 3-29, 30] may result from those transitions in which the energy released lies within the range of the photon energy of visible light. Examples of such recombination mechanisms are shown in Fig. 3-25.

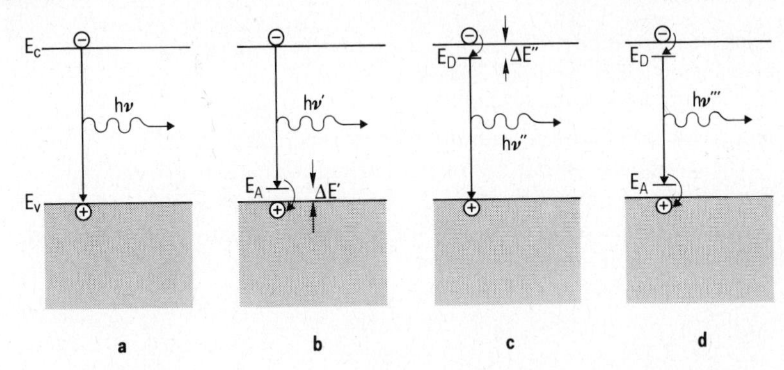

Fig. 3-25. Mechanisms of cathodoluminescence due to recombination processes in semiconductors or insulators: a) recombination by direct band-to-band transition leads to "intrinsic cathodoluminescence"; b) to d) recombination via localized states in the forbidden gap results in "extrinsic cathodoluminescence". E_A = acceptor level, E_D = donor level.

In a band-to-band transition as shown in Fig. 3-25a, light is emitted with a photon energy $h\nu \geqq E_g = E_c - E_v$, in which the equals sign applies to the case of low doping. This type of CL is called *intrinsic* luminescence, and it provides local information on crystal properties influencing the energy gap or the occupation of states near the band edges. Such properties are temperature, doping concentration and composition of compound semiconductors (e.g. x in $Ga_x Al_{1-x} As$) [3-31]. However, intrinsic CL is only likely to occur in semiconductors in which recombination can take place "directly", i.e. without changes in momentum. Semiconductors of this kind such as GaAs and ZnS are called "direct semiconductors". In "indirect semiconductors" like Si, Ge and GaP, the special band structure requires a phonon to take part in the recombination process so that momentum is conserved. The probability of this kind of process taking place is very small, and the yield – if CL occurs at all – is low.

In *extrinsic CL*, recombination takes place via localized energy states within the forbidden gap (Fig. 3-25b, c, d). If these states have an energy difference ΔE from the band edges, the photon energy $h\nu = E_g - \Delta E$. However, no distinction can be made between the cases indicated in Figs. 3-25b to d on the basis of the CL spectrum alone. They may be even more complex if the recombination process involves additional states lying deeper inside the forbidden gap. Such states can be due to point defects (impurities or lattice defects) and act as "traps" for electrons or holes. The variety of possible transitions and the thermally induced broadening of the energy states make extrinsic CL spectra difficult to interpret.

At low temperatures, the CL bands split up into sharp lines resulting from transitions between clearly defined energy levels. Excitons (loosely bound electron-hole pairs with energy states within the forbidden gap) give rise to additional, very sharp CL lines. Furthermore, the probability of radiative recombination at the expense of non-radiative transitions increases at low temperatures. This results in more and sharper spectral lines of greater intensity. CL spectra of cold specimens therefore yield much more information than spectra recorded at room temperature.

Cathodoluminescence intensity

Each recombination process can be characterized by the lifetime τ of the excess minority carriers (Section 2.4.3) involved. If several competitive recombination mechanisms are operating, the total lifetime of the excited state concerned is given by $1/\tau = \sum 1/\tau_i$. If one radiative (r) and one non-radiative (n) transition is involved, $1/\tau = 1/\tau_r + 1/\tau_n$ applies accordingly.

The recombination rate $\dot{N}_r$ (number of recombinations per unit time) is inversely proportional to the lifetime of the excited state concerned. Thus $\dot{N}_r \sim 1/\tau$ applies for the total recombination rate, and $\dot{N}_{rr} \sim 1/\tau_r$ for the rate of radiative recombination.

Radiative recombination efficiency η_{rr} is defined as the ratio of radiative recombinations to the total number of all possible recombinations. Thus $\eta_{rr} = \dot{N}_{rr}/\dot{N}_r = \tau/\tau_r$, giving [3-29]

$$J_{CL} = \eta_{rr} \cdot g = (\tau/\tau_r) \cdot G \cdot I_0/e \tag{3-3}$$

for the CL intensity (number of photons per unit time) produced in the specimen with the generation rate g from Eq. (3-2). Accordingly, all effects influencing the lifetime of the generated charge carriers – electric fields (Section 3.6.3), doping inhomogeneities, lattice defects, surface recombination – affect the CL intensity. As a result of this, the CL light may vary not only in intensity but also in its spectral distribution. These variations can be used to study the effects mentioned above. An example is given in Fig. 3-26 [3-32].

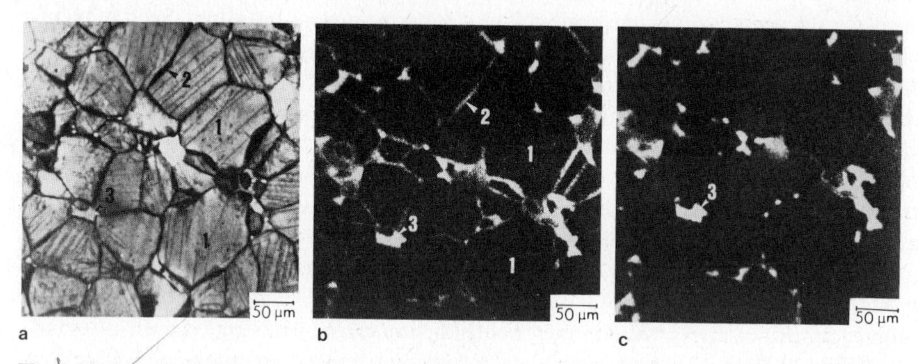

Fig. 3-26. Cathodoluminescence micrographs of barium titanate ($BaTiO_3$) ceramics of the same specimen area using (a) the integral CL intensity (λ = 300 nm to 800 nm) and (b), (c) monochromatic intensity: (b) λ = 856 nm, (c) λ = 770 nm (U_0 = 20 kV; T = 300 K) [3-32]. $BaTiO_3$ has a maximum in the CL spectrum at λ = 500 nm. The $BaTiO_3$ grains (1) therefore appear bright in image (a). The grain boundary zones (2) exhibit a high barium vacancy concentration. A competitive recombination process in which the vacancies are involved leads to a CL maximum at λ > 850 nm. The grain boundary zones are therefore dark in image (a) and bright in image (b). The local distribution of a second phase $Ba_6Ti_{17}O_{40}$ (3), which has a CL maximum at $\lambda \approx$ 770 nm, is shown in Fig. c.

Cathodoluminescence measuring technique

Despite the high generation factors for electron-hole pairs (Section 3.6.1), the external photon yield is small: to avoid any damage to the specimen and allow high spatial resolution, the probe current must not be set too high ($I_0 \leqq 10^{-7}$ A). Furthermore, the intensity generated inside the specimen, described by Eq. (3-3), is considerably reduced by absorption in the crystal and by total reflection at the specimen surface. For example, in GaAs (because of the high refractive index and low critical angle (with respect to the surface normal) for total reflection) only 1.5% of the photons produced can leave the crystal [3-1]. One of the fundamental problems in CL measuring technique is therefore how to collect the emitted radiation over the largest possible solid angle with high efficiency.

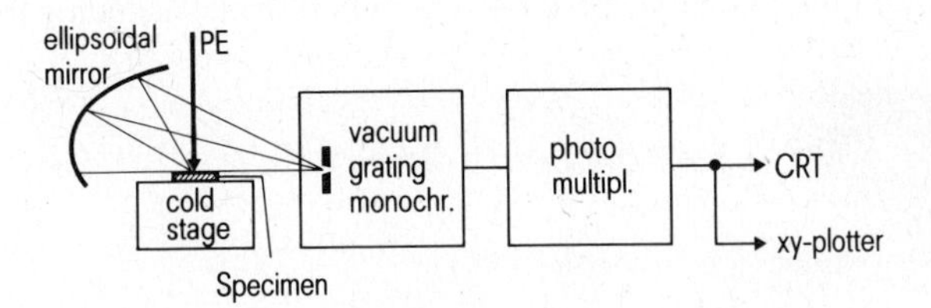

Fig. 3-27. Cathodoluminescence detection system consisting of an ellipsoidal mirror, vacuum grating monochromator and photomultiplier (according to [3-25]). The photomultiplier signal is used either to control the brightness of the CRT or to record a spectrum with an *xy* plotter. For recording an integral image, the monochromator is adjusted such that the direct, undiffracted beam (containing all wavelengths) enters the photomultiplier. A continuous-flow liquid helium cryostat can be used for the cold stage.

Various setups have been proposed employing lenses [3-33] or parabolic or ellipsoidal mirrors [3-34, 35] as light collection systems. Fig. 3-27 shows a configuration using an ellipsoidal mirror in which the point of PE impact on the specimen is at one of its focal points. The emitted CL radiation is focused onto the second focal point of the ellipsoid where the entrance slit for an optical spectrometer is located. Light quanta emitted into a solid angle of $0.8\,\pi$ can be collected.

If desired, the radiation can be dispersed according to wavelength using a spectrometer and detected with the help of a photomultiplier. Just as in X-ray microanalysis (Chapter 5), this kind of setup can be used to record a spectrum or to produce images using the light of a selected spectral line *(spectral CL)*. When the entire spectrum is used, *integral CL* images are formed.

Prism or grating monochromators are used as spectrometers. They operate in vacuum to reduce absorption losses. It must be ensured that the spectral sensitivity of the photomultiplier matches the particular wavelength range of the CL light to be analyzed. For very low intensities the photomultipliers are operated in single-photon counting mode, in which the signal-to-background ratio is better than in the more common analog technique.

Phase-selective amplification can be used to eliminate disturbing light radiation from the thermionic cathode and other sources. This is done by periodically switching the PE beam on and off (beam blanking, see Section 8.2) and measuring the CL signal by means of a lock-in amplifier tuned to the PE beam pulse frequency. This configuration can also be used for time-resolved CL measurements, for example to determine the lifetime of the generated excess minority carriers from the signal decay during the time interval between consecutive PE pulses: by varying the delay of the measurement window in the amplifier with respect to the beam blanking pulse, the entire exponential profile of the decay curve can be measured. The lifetime to be determined is the time the CL signal needs to decay to 1/e of its original value. One precondition is that the time resolution Δt of the setup must be significantly smaller than the lifetime to be measured, e.g. $\Delta t < 1$ ns, if lifetimes of the order of a few ns in highly doped GaAs are to be measured [3-36]. Measurement times can be substantially reduced by using sensitive, integrating detection systems [3-36, 37].

Specimen *cooling stages* are employed to take full advantage of the greater sharpness and intensity of spectral lines at low temperatures. They achieve temperatures close to that of liquid helium. This enhances the information content of CL spectra substantially. A summary of the commonly used CL detection systems, including liquid helium stages, is given in [3-38].

3.6.3 Electron beam induced current (EBIC)

Charge collection current

In an electric field, the electrons and holes are accelerated in opposite directions, i.e. positive and negative charges are separated. In the case of a specimen showing radiative recombination, CL intensity is at a minimum at the site of an electric field because charge separation inhibits recombination. However, charge separation can itself be utilized as a signal by collecting the electrons at one of the specimen contacts and the holes at another. In a closed circuit, a charge collection current I_{CC} flows [2-36a, 3-28, 39].

Charge separation can be caused by an electric field extending inside the specimen. Such "built-in" fields exist in the depletion zone of pn-junctions (Fig. 2-13) and Schottky barriers or occur as a result of inhomogeneous dopant distribution (e.g. pp^+-, nn^+-junctions). The electric field in a pn-junction, for example, causes electrons generated in the p-region to be attracted to the n-side and the holes to be pushed back. Exactly the opposite takes place with the charge carriers produced in the n-region. So what counts is the movement of the respective minority carriers, which become majority carriers on the other side of the pn-junction. This current changes the equilibrium potential difference at the pn-junction (i.e. the diffusion voltage). A non-equilibrium potential difference results which acts as an electromotive force. In a closed circuit, it drives the charge collection current. This phenomenon is called the electron-voltaic effect, by analogy with the photovoltaic effect.

In the absence of internal electric fields, a charge collection current only flows if a bias is applied to the outside of the specimen. This corresponds to the photoconductivity effect, and some authors therefore call it the β conductivity effect (in which β ray stands for electron beam) [2-36a].

Charge collection efficiency

With 100% charge collection, the charge collection current may, according to Eq. (3-2), reach a maximum value of

$$I_{cc}^{max} = G \cdot I_0 \,. \tag{3-4}$$

The amount of current actually flowing depends on the properties of the specimen and the experimental conditions. Fig. 3-28a shows a setup for measuring the charge collection current at a pn-junction. The pn-junction is normal to the specimen surface, and the depletion zone has a width w. The generation volume is located ouside the depletion zone. In such cases, only minority carriers which diffuse far enough toward the pn-junction can enter the zone of influence of the electric field. Under simplified assumptions (no field outside the depletion zone, no recombination inside the depletion zone, no additional surface recombination), the laws of diffusion yield an exponential decrease of charge collection current with distance ξ from the depletion zone ($\xi = |x - x_n|$ and $\xi = |x - x_p|$ respectively):

$$I_{cc} = I_{cc}^{max} \cdot \exp(-\xi/L) \tag{3-5}$$

where L is the diffusion length of the particular minority carriers. If the PE beam is scanned across the specimen surface in the x direction, the generation volume and the space charge region will overlap at some point. Charge separation is then no longer a pure diffusion problem and the charge collection current can no longer be described by Eq. (3-5): it now depends on the width of the depletion zone, the diameter of the generation volume and the degree of overlap.

Only when the depletion zone is wider than the diameter of the entire generation volume are all the electrons and holes affected by the electric field, so that charge collection is 100%. In this case, $I_{cc} = I_{cc}^{max}$. However, if the depletion zone is narrower and the generation volume

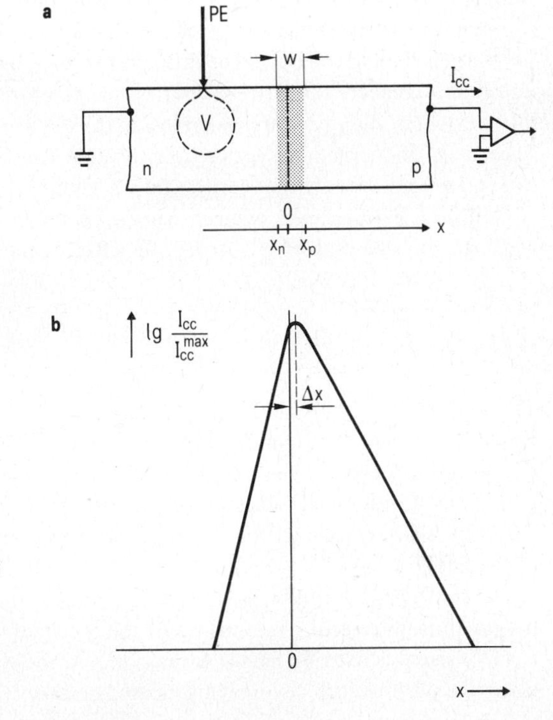

Fig. 3-28.
Charge collection at a pn-junction (schematic): a) experimental setup (PE primary electron beam, V generation volume, w width and x_n, x_p boundaries of depletion zone, I_{cc} charge collection current); b) line scan profile of $I_{cc}(x)$ normal to the pn-junction normalized to its maximum value ("EBIC profile"). Δx denotes the distance between the EBIC maximum and the location of the pn-junction.

and depletion zone only partly overlap, I_{cc} remains smaller than I_{cc}^{max}. The situation can generally be described by

$$I_{cc} = \eta_{cc} \cdot I_{cc}^{max} \tag{3-6}$$

where the charge collection efficiency $\eta_{cc} \leqq 1$. The equals sign applies for 100% charge separation.

In regions with a very high electric field strength such as weak points in a biased pn-junction, the electron beam induced charge carriers may be accelerated to such high energies that they generate other charge carrier pairs through impact ionization (charge carrier multiplication). In this case $I_{cc} > I_{cc}^{max}$ is measured and the charge collection efficiency seems to be $\eta_{cc} > 1$.

Charge collection measuring technique

Fig. 3-28a indicates how the specimen must be connected to measure the charge collection current externally. Eqs. (3-4) to (3-6) specify the current the specimen can deliver. The value actually measured depends on the ratio of the resistance of the measurement setup (external resistance R_{ex}) to that of the specimen (internal resistance, defined by $R_{in} = dU/dI$ in the case of a pn-junction).

If $R_{ex} \ll R_{in}$ (short-circuit case), the charge collection current I_{cc} given by Eqs. (3-5) and (3-6) flows through the outer circuit. No externally detectable voltage is produced. This mode of operation is called the EBIC (electron beam induced current) method. However, if $R_{ex} \gg R_{in}$ (open circuit case), the charge carriers separated at the pn-junction cannot flow off. They produce an inverse voltage to the diffusion voltage which – acting like a bias in the forward direction – reduces the potential barrier at the pn-junction. As a result, there is an internal drift current of majority carriers opposing the electron-beam generated current of the minority carriers. The inverse voltage assumes a value such that these two currents just compensate each other and the resulting current flow is zero ($I_{cc} = 0$). Externally, with this mode of operation, an electron beam induced voltage (EBIV) is measured [3-40]. In the following, we consider only the case of $R_{ex} \ll R_{in}$, as EBIC is the most widely used method.

Device characterization

The configuration shown in Fig. 3-28a is suitable for measuring the diffusion length of minority charge carriers, for example. To do this, the PE beam is scanned across the vertical pn-junction in the x direction and $I_{cc}(x)$ is plotted on a semilogarithmic scale (Fig. 3-28b). The diagram then shows straight lines at a sufficient distance on both sides of the pn-junction, the slopes of which can be used to determine the diffusion lengths L_n and L_p according to Eq. (3-5) [2-36a].

The configuration in Fig. 3-28a can also be used to visualize the two-dimensional shape of pn-junctions intersecting the surface of the specimen normally or obliquely. In this case, the $I_{cc}(x, y)$ signal is used for image formation and the resulting EBIC image shows bright lines along the pn-junctions. By imaging cleaved cross-sections, both the depth and the lateral extension of pn-junctions can be determined (Fig. 3-29). The width of the bright lines depends on the EBIC profile (Fig. 3-28b) and the amplifier setting. Digital image processing permits the maximum of the EBIC profile to be localized with much greater accuracy (Fig. 3-30) [3-41]. However, the EBIC maximum does not always coincide with the electrical pn-junction: the deviation (Δx in Fig. 3-28b) is only negligible if the depletion zone is very narrow (i.e. with a high dopant concentration $>10^{18}$ cm^{-3} on both sides) or if the pn-junction is approximately symmetrical (i.e. with approximately the same dopant concentrations and diffusion lengths on both sides). In other cases, deviations of a few 0.1 μm [3-41] must be expected, with signifi-

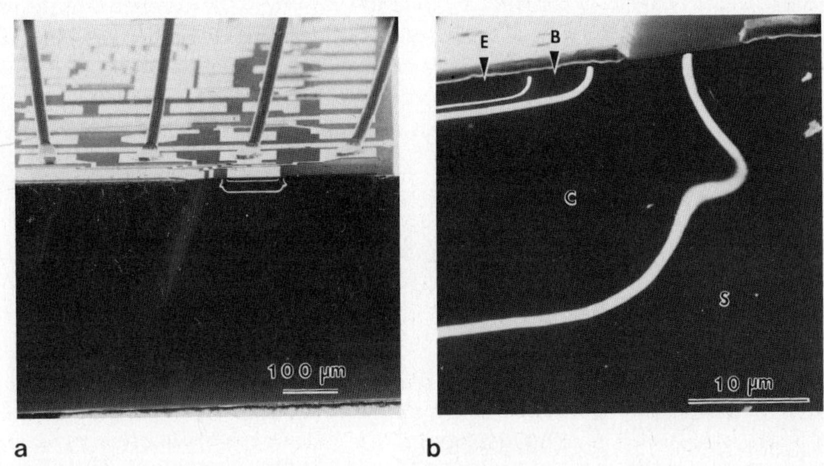

Fig. 3-29. Superimposed SE and EBIC images of an integrated bipolar circuit showing both the cleaved cross section and the highly-tilted surface of the device (U_0 = 5 kV). In the overview (a) the wires bonded to the pads of the circuit are seen. Due to the EBIC signal the pn-junctions appear as bright lines. At higher magnification (b) all pn-junctions between emitter (E), base (B), collector (C) and substrate (S) are visible. [3-41].

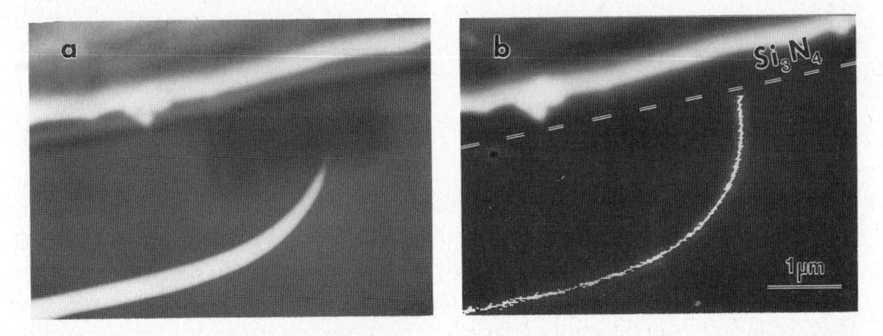

Fig. 3-30. Cross section through integrated circuit with pn-junction; superimposed SE and EBIC micrographs (the dashed line indicates the interface between the nitride passivation and the Si substrate). a) Conventional EBIC imaging: in the vicinity of the passivation layer the EBIC contour vanishes because the signal falls below the amplifier dark level due to enhanced recombination at the interface. b) Computer mapping of the EBIC maximum: the pn-junction is delineated uniformly along its total length. [3-41].

cantly greater deviations in the case of very wide depletion layers (e.g. in radiation detectors) [3-43].

Independently of these fundamental effects, the measurement result can also be seriously falsified by contamination and/or surface states due to preparation. Great care must therefore be taken in the preparation and treatment of the specimen (preferentially: fracturing, annealing, desorption of contaminants by electron bombardment). This is the more imperative the lower the depth of the pn-junctions to be imaged and the higher the consequent demands on spatial resolution [3-42].

If the PE beam is scanned across the surface of a device, the two-dimensional extent of the doped region lying below the surface can be imaged (Fig. 3-31) provided that the pn-junctions lie within the generation volume or are reached by diffusing charge carriers. The image brightness is generally modified by the different PE absorption in the overlying device structures (edges of oxide layers, interconnections, etc.), which therefore become visible (Fig. 3-31b). Information on the depth of the junctions is obtained by varying the acceleration voltage.

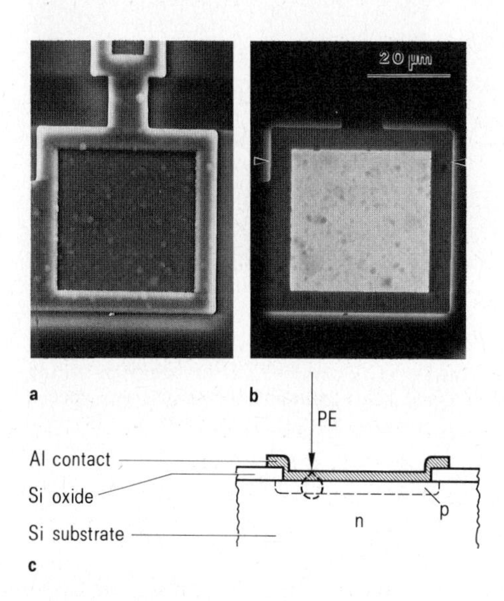

Fig. 3-31.
Diode in a silicon circuit. a) SE image, b) EBIC image (U_0 = 20 keV), c) schematic cross section along the line marked by arrows. The pn-junction lying parallel to the device surface produces an EBIC signal within the total junction area. The contrast is modified as a result of the varying PE absorption in the device layers. The bright fringes outside the aluminum contact are due to those carriers which diffuse from their generation point to the pn-junction. (Courtesy A. Dallmann).

Failure analysis

Disturbances in the vicinity of a pn-junction may be reflected in an increased or reduced EBIC signal [3-44, 45]. For example, defects in the layers above a pn-junction (e.g. interconnect interruptions, pores) result in a larger PE current reaching the space charge region at these points, which therefore appear brighter than their surroundings in the EBIC image. On the other hand, local defects in the pn-junction itself (short-circuits, pipes, etc.) are sites with a reduced signal because the electric field and hence charge separation is disrupted. Weak points in which charge carrier multiplication may lead to avalanche breakthrough can be localized by an increased EBIC signal.

Many device failures can be traced to crystal lattice defects which may cause higher recombination rates or charge carrier multiplication. What effects they have depends on their location

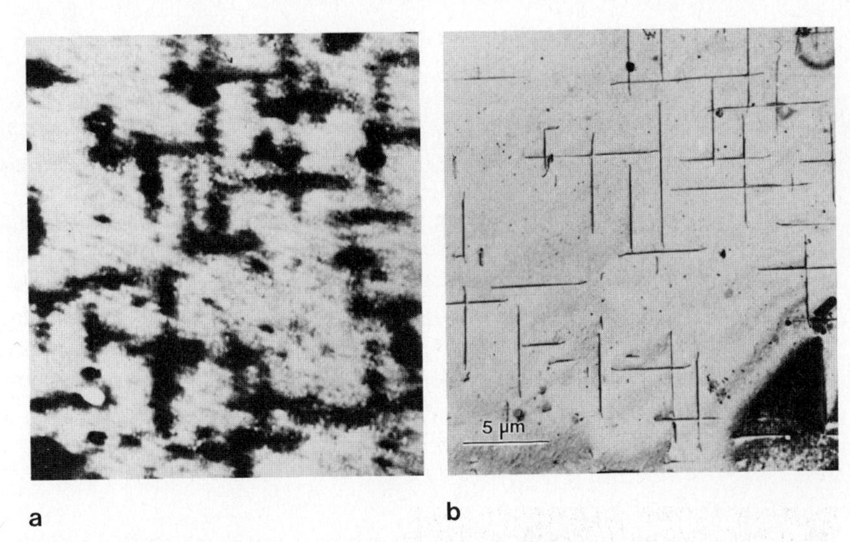

Fig. 3-32. Near-surface dislocations in silicon after phosphorus diffusion forming a pn-junction at 1 µm depth. a) EBIC micrograph taken in an SEM (U_0 = 5 kV), b) comparative TEM micrograph of the same specimen area (U_0 = 1000 kV). [3-46].

with respect to the pn-junction [3-46]. If the defects lie in the field-free region they merely enhance recombination. Fewer induced charge carriers then diffuse to the pn-junction and the lattice defects appear dark in the EBIC image (Fig. 3-32a).

The effects are different if lattice defects extend into the electric field of the space charge region or even penetrate the pn-junction. They then give rise to leakage currents, charge carrier multiplication and finally avalanche breakthrough. If a reverse bias is applied to the specimen, a transition from "recombination contrast" to "multiplication contrast" can be observed in the EBIC image at a certain bias threshold: the defective spots first appear darker and then, because of the charge carrier multiplication, distinctly brighter than the rest of the pn-junction area.

Numerous studies of lattice defects (dislocation networks, single dislocations, stacking faults, grain boundaries) by EBIC analysis have been reported [e.g. 3-47, 48, 49]. Schottky barrier layers are often deposited onto the semiconductor surface for the specific purpose of imaging near-surface lattice defects by the EBIC signal [3-50]. After the EBIC study, the electrically active defects can be identified by diffraction contrast analysis in comparative examinations in a transmission electron microscope (Fig. 3-32b) [3-46, 51]. Lattice defects resulting in charge carrier multiplication generally consist of metal precipitates or process-induced stacking faults decorated by metal impurities.

3.7 Other methods

3.7.1 Overview

For some methods which can also be employed in SEMs equipped with suitable attachments, specialized instruments have been developed – X-ray microprobes for X-ray microanalysis and Auger microprobes for Auger microanalysis. These instruments are described in Chapters 5 and 6. The voltage contrast method is particularly suitable for testing integrated circuits (Chapter 8). The following pages provide a brief description of some additional methods: the specimen current mode, imaging of magnetic fields and thermal wave microscopy.

3.7.2 Specimen current mode

The current of the electron probe striking the specimen is partially absorbed by the specimen and then discharged to ground. It is therefore called the specimen current or absorbed current. It is needed to provide a current balance between the primary electron current I_0 and the currents of the secondary and backscattered electrons leaving the specimen. With SE yield δ and backscattering coefficient η, the specimen current I_S is therefore

$$I_S = I_0 \cdot [1 - (\delta + \eta)] \ .$$

The specimen current depends on the angle of incidence ψ via δ and η and may change its sign if $\sigma = \delta + \eta > 1$. In flat specimens where δ varies only slightly, the specimen current yields the opposite contrast to the BSE signal and like it can be used to represent material contrast. To measure the specimen current as an image signal using an electrometer, the specimen must be insulated when mounted. The capacitances of the mount with respect to the chamber, cable, etc., result in a time constant that limits the scan speed which can be used to record specimen current images.

3.7.3 Imaging of magnetic fields

Two types of contrast formation can be used for imaging magnetic fields. Overviews are given in [3-1,3,52]. Type 1 magnetic contrast is based on the effect of external stray fields on the secondary electron trajectories [3-53]. Such stray fields are found, for example, in uniaxially magnetic materials such as cobalt, where they complete the magnetic flux between the differently oriented ferromagnetic domains (Fig. 3-33), and also in magnetic tapes. As these stray fields only extend a few μm above the surface of the specimen, they have a negligible effect on the primary electrons. Although the stray fields do not affect the number of emitted SEs, they deflect the secondary electrons when passing through them (Fig. 3-33). This results in a tilt of the entire angular distribution of the emitted SEs. To obtain optimum contrast, the specimen must be rotated so that the tilt of the angular distribution points either towards the SE detector or away from it. Rotation of the specimen around 180° inverts the contrast.

Fig. 3-34 shows the domain structure in a barium ferrite crystal as an example [3-54]. When applying type 1 magnetic contrast it has to be borne in mind that diffuse stray fields rather than the domains are imaged.

In magnetic materials such as iron where the magnetic flux remains in the specimen, the domains cannot be observed by type 1 contrast. Such magnetic fields can, however, be visualized by their effect on the electrons scattered inside the specimen [3-55]. This type 2 magnetic contrast mechanism is shown schematically in Fig. 3-35. However, since the scattered electrons are subject to only minimal deflection by the internal magnetic fields, contrast in the BSE image is very weak (a few 0.1%). Here again, optimal contrast depends on correct tilt and rotation of the specimen. In the diagram of Fig. 3-35, the tilt axis is parallel to ***B***.

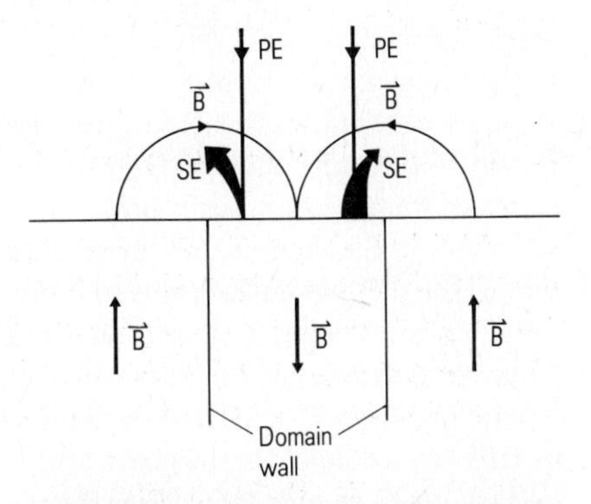

Fig. 3-33.
Type 1 magnetic contrast. Due to the horizontal components of stray fields ***B***, a Lorentz force acts on the emitted SE normal to the projection plane ($\boldsymbol{F} = -e\,(\boldsymbol{v} \times \boldsymbol{B})$). Depending on the direction of ***B***, SEs are deflected out of the projection plane (left) or into it (right). A contrast is produced if the SE detector is located on an axis normal to the projection plane. Stray fields of different directions show up with different brightness and thus display the domain structure.

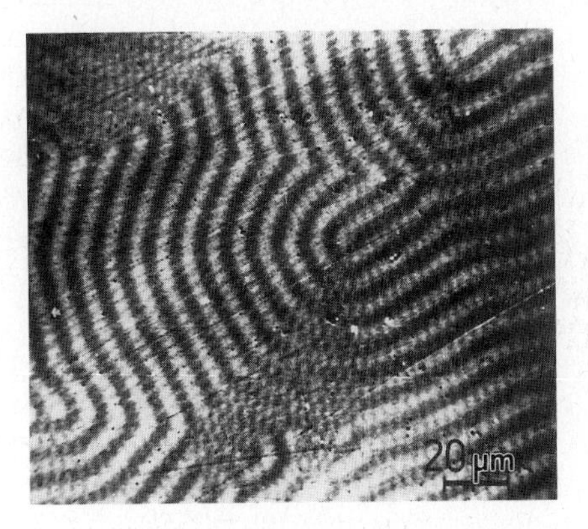

Fig. 3-34.
Ferromagnetic domain structure in a barium ferrite ($BaFe_{12}O_9$) single crystal revealed by type 1 magnetic contrast [3-54].

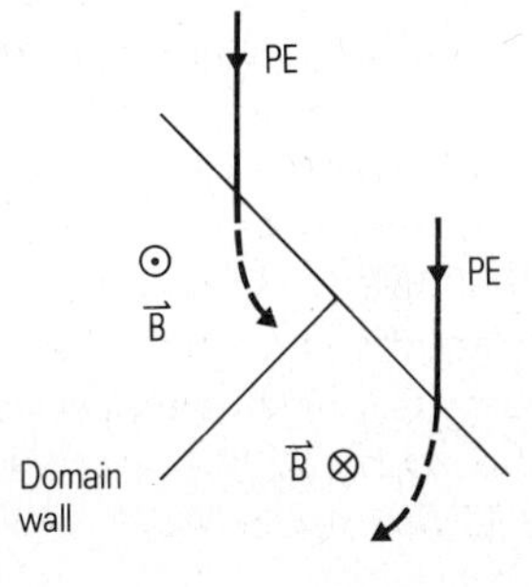

Fig. 3-35.
Type 2 magnetic contrast. The electrons scattered in the specimen are deflected by the magnetic field within the specimen. If the specimen is tilted, deflection is either toward the surface (left) or away from it (right). More electrons are backscattered from the left domain than from the right one because the electron paths are closer to the surface there. In the BSE image, the left domain therefore appears bright and right one dark.

3.7.4 Thermal wave microscopy

In this method [3-56] a pulsed electron beam is scanned across the specimen and heats it up periodically. This produces a thermal wave which propagates into the specimen from the point of electron impact as a spherical wave (Fig. 3-36), though it is so strongly attenuated that its amplitude decreases to e^{-1} over a distance of a few μm. However, the temperature oscillations give rise to a periodic expansion and contraction of the specimen which excites an acoustic (elastic) wave. This explains the other name by which the method is known – scanning electron acoustic microscopy (by analogy with laser acoustic microscopy, in which a pulsed laser beam is used to periodically heat up the specimen). The acoustic wave can be recorded with the help of a piezoelectric transducer (made, say, of lead zirconate titanate, PZT) to which the specimen has been securely glued. The acoustic wave has many times the wavelength of the thermal wave. Its sole purpose is to transmit the information contained in the thermal wave. Because of the low efficiency of the thermal to mechanical energy conversion – generally only in the order of 10^{-8} [3-3] – the acoustic signal is of very low intensity, although it increases linearly with beam power and pulse frequency. Image contrast results from local variations in the thermal and mechanical properties of the specimen in the propagation volume of the thermal wave. Such variations can be caused by pores, cracks and plastic deformations, etc., and by ion-implantation damage in semiconductors. The spatial resolution is limited by the range and attenuation constant of the thermal wave, and amounts to several μm at a pulse frequency of 1 MHz [3-3]. Values of less than 1 μm would appear possible. In the case of III-V semiconductors, the piezoelectric properties of these crystals allow the electron acoustic signal to be recorded directly without the need of a transducer: the acoustic waves generate charge displacements which can be measured by electrodes connected to the specimen [3-57]. Fig. 3-37 shows an InP wafer with Zn diffusion regions and Ohmic contacts. In the SE image (Fig. 3-37a) only the pattern of the contacts is seen. The thermal wave image (Fig. 3-37b), however, shows mainly the Zn diffusion regions in the InP substrate which are not visible in Fig. 3-37a. Thermal wave microscopy is still in its infancy and a lot of basic work is necessary before it can be put to more quantitative use.

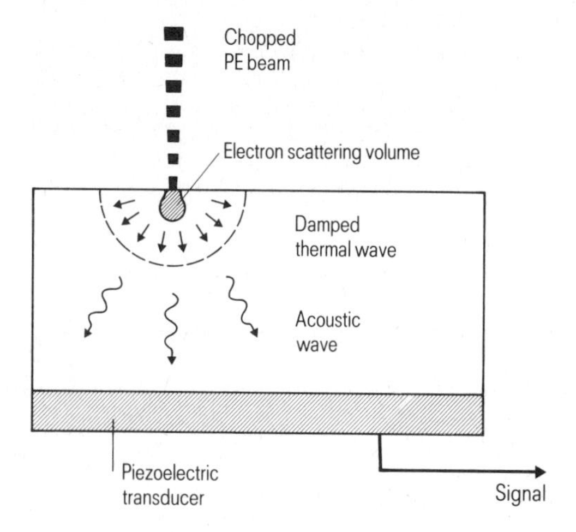

Fig. 3-36.
Principle of thermal wave microscopy. A pulsed electron beam periodically heats the specimen, producing a thermal spherical wave. Due to the periodic expansion and contraction of the specimen, this wave in turn excites a high-frequency acoustic (elastic) wave which can be recorded by a piezoelectric transducer.

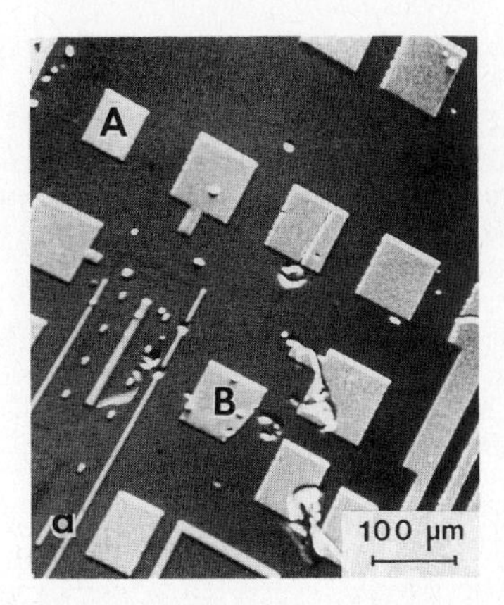

Fig. 3-37. SE image (a) and thermal wave image (b) of processed InP wafer. The Zn diffusion regions and the Ohmic contacts were fabricated with different masks. In (a) only the contacts are seen whereas (b) shows mainly the Zn diffusion regions. These are also visible when they lie below the contacts: the diffusion region A displays a different pattern than the contact. Contacts alone (B) produce only slight contrast in the thermal wave image. U_0 = 30 kV. [3-57].

3.8 References

[3-1] L. Reimer, *Scanning Electron Microscopy.* Springer, Berlin 1985

[3-2] J.I. Goldstein, D.E. Newbury, P. Echlin, D.C. Joy, C. Fiori, E. Lifshin, *Scanning Electron Microscopy and X-ray Microanalysis.* Plenum Press, New York 1981

[3-3] D.E. Newbury, D.C. Joy, P. Echlin, C.E. Fiori, J.I. Goldstein, *Advanced Scanning Electron Microscopy and X-ray Microanalysis.* Plenum Press, New York 1986

[3-4] *Scanning Electron Microscopy* (Ed. Om Johari). 1968-1977: IITRI, Chicago; 1978 and following years: SEM Inc., AMF O'Hare

[3-5] T.E. Everhart, R.F.M. Thornley, *J. Sci. Instr. 37*, 246 (1960)

[3-6] O.C. Wells, *Appl. Phys. Lett. 19*, 232 (1971)

[3-7] J.A. Murphy, *Scanning Electron Microsc. 1982/II,* SEM Inc., AMF O'Hare 1982, 657

[3-8] S.W. Joens, K.K. Shah, H.T. Shishido, *Scanning Electron Microsc. 1982/II,* SEM Inc., AMF O'Hare 1982, 573

[3-9] B.R. Hammond, T.R. Vogel, *Proc. Reliability Physics Symp.,* IEEE, New York 1982, 221

[3-10] Technology Associates, 51 Hillbrook Drive, Portorola Valley, CA 94025, USA

[3-11] D. S. Koellen, D.I. Saxon, K.E. Wendel, *Scanning Electron Microsc. 1985/I,* SEM Inc., AMF O'Hare 1985, 43

[3-12] T. Mills, *Proc. Reliability Physics Symp.,* IEEE, New York 1983, 324

[3-13] F. Beck, *Präparationstechniken in der Konstruktions- und Fehleranalyse von integrierten Halbleiterschaltungen.* Physik-Verlag, Weinheim 1988

[3-14] R.W. Belcher, G.P. Hart, W.R. Wade, *Scanning Electron Microsc. 1984/II,* SEM Inc., AMF O'Hare 1984, 613

[3-15] T. Ohtaka, S. Saito, T. Furuya, O. Yamada, *Proc. SPIE, Vol. 565,* 1984, 205

[3-16] S. Erasmus, *J. Vac. Sci. Technol. B 5,* 409 (1987)

[3-17] M. Miyoshi, M. Ishikawa, K. Okumura, *Scanning Electron Microsc. 1984/IV,* SEM Inc., AMF O'Hare 1982, 1507

[3-18] M. Brunner, N. Menzel, *J. Vac. Sci. Technol. B 1,* 1344 (1983)

[3-19] M. Brunner, D. Winkler, B. Lischke in A. Heuberger, H. Beneking (Eds.): *Microcircuit Engineering.* Academic Press, London 1985, 399
[3-20] D.C. Joy in L.M. Brown (Ed.): *Electron Microscopy and Analysis 1987.* Inst. Phys. Conf. Ser. No. 90: Chapt. 7. Bristol 1987, 175
[3-21] M. Brunner, R. Schmid, *Scanning Electron Microsc. 1986/II,* SEM Inc., AMF O'Hare 1986, 377
[3-22] M.T. Postek, D.C. Joy, *Solid State Technology,* November 145 (1986) and December 77 (1986)
[3-23] M. Brunner, R. Schmid, *Microelectronic Engineering 7,* 41 (1987)
[3-24] J. Frosien, *J. Vac. Sci. Technol. B 4,* 261 (1986)
[3-25] H. Yamaji, M. Miyoshi, M. Kano, K. Okumura, *Scanning Electron Microsc. 1985/I,* SEM Inc., AMF O'Hare 1985, 97
[3-26] H. Niedrig, *Scanning Electron Microsc. 1978/I,* SEM Inc., AMF O'Hare 1978, 841
[3-27] D.C. Joy, D.E. Newbury, D.L. Davidson, *J. Appl. Phys. 53,* R81 (1982)
[3-28] H.J. Leamy, *J. Appl. Phys. 53,* R51 (1982)
[3-29] D.B. Holt, S. Datta, *Scanning Electron Microsc. 1980/I,* SEM Inc., AMF O'Hare 1980, 259
[3-30] G. Pfefferkorn, W. Bröcker, M. Hastenrath, ibid., 251
[3-31] S.M. Davidson, *J. Microscopy 110,* 177 (1977)
[3-32] G. Koschek, D. Köhler, E. Kubalek, in P. Vincenzini (Ed.): *High Tech Ceramics.* Elsevier, Amsterdam 1987, 1591
[3-33] A.D. Yoffee, K.J. Howlett, *Scanning Electron Microsc. 1973,* IITRI Chicago 1973, 301
[3-34] E.M. Hörl, E. Mügschl, *Proc. Vth European Congr. Electron Microscopy 1972,* 502
[3-35] L.J. Balk, E. Kubalek, *Scanning Electron Microsc. 1977/I,* IITRI Chicago 1977, 739
[3-36] M. Hastenrath, L.J. Balk, K. Löhnert, E. Kubalek, *Journal of Microscopy 118,* 303 (1980)
[3-37] A. Steckenborn, H. Münzel, D. Bimberg, *Inst. Phys. Conf. Ser. 60,* 185 (1981)
[3-38] D.B. Holt, ibid., 165
[3-39] D.B. Holt, M. Lesniak, *Scanning Electron Microsc. 1985/I,* SEM Inc., AMF O'Hare 1985, 67
[3-40] L.J. Balk, E. Kubalek, E. Menzel, *Scanning Electron Microsc. 1975/I,* IITRI, Chicago 1975, 447
[3-41] H.K. Schink, H. Rehme, *Electronics Letters 19,* 383 (1983)
[3-42] R. Kuhnert, Thesis, Techn. Univ. München, 1989
[3-43] H.W. Marten, O. Hildebrand, *Scanning Electron Microsc. 1983/III,* SEM Inc., AMF O'Hare 1983, 1197
[3-44] J.D. Schick, *Scanning Electron Microsc. 1981/I,* SEM Inc., AMF O'Hare 1981, 295
[3-45] J.D. Schick, ibid. 1985/I, 55
[3-46] J. Heydenreich, H. Blumtritt, R. Gleichmann, ibid. 1981/I, 351
[3-47] K.V. Ravi, C.J. Varker, C.E. Volk, *J. Electrochem. Soc. 120,* 533 (1973)
[3-48] C.J. Bull, P. Ashburn, J.P. Gowers, *Solid-State Electronics 23,* 953 (1980)
[3-49] J.M. Dishman, S.E. Haszko, R.B. Marcus, S.P. Murarka, T.T. Sheng, *J. Appl. Phys. 50,* 2689 (1979)
[3-50] L.C. Kimerling, H.J. Leamy, J.L. Benton, S.D. Ferris, P.E. Freeland, J.J. Rubin in H.R. Huff, E. Sirtl (Eds.): *Semiconductor Silicon 1977,* The Electrochem. Soc. Inc., Princeton, N.J., 468
[3-51] R.B. Marcus, M. Robinson, T.T. Sheng, S.E. Haszko, S.P. Murarka, *J. Electrochem. Soc. 124,* 425 (1977)
[3-52] D. Hesse, K.-P. Meyer, in H. Bethge, J. Heydenreich (Eds.): *Electron Microscopy in Solid State Physics,* Elsevier, Amsterdam 1987, 496
[3-53] D.C. Joy, J.P. Jakubovics, *J. Phys. D 2,* 1367 (1969)
[3-54] L.J. Balk, J.B. Elsbrock, *Scanning Electron Microsc. 1984/I,* SEM Inc., AMF O'Hare 1984, 141
[3-55] J. Philibert, R. Tixier, *Micron 1,* 174 (1969)
[3-56] G. Davies, *Scanning Electron Microsc. 1983/III,* SEM Inc., AMF O'Hare 1983, 1163
[3-57] N. Kultscher, K. Steiner, L.J. Balk, in P.Hess, J. Pelzel (Eds.): *Photoacoustic and Photothermal Phenomena,* Springer Series in Optical Sciences 58, Berlin 1988, 237

4 Transmission Electron Microscopy

4.1 Principle

4.1.1 Basic layout

In contrast to the SEM, in which the electrons probe bulk specimens, the transmission electron microscope (TEM) makes use of high-energy electrons to irradiate thin specimens around 0.1 μm or less in thickness. The electrons are transmitted and form an image of the specimen magnified in several steps with the aid of electron-optical lenses. The entire image can be observed directly on a fluorescent screen and is not built up line by line on a CRT as in the SEM.

Fig. 4-1 shows the layout of a TEM in the form of a ray diagram. The electrons emitted from the electron gun are accelerated to an energy of, say, 100 keV. The condenser lens then shapes them to an approximately parallel beam which illuminates the specimen uniformly. The objective lens produces a first image, which is then further magnified by the intermediate and projector lenses and finally projected onto the fluorescent screen (Fig. 4-1a). If a film is inserted into the beam path instead of the fluorescent screen, the image can be recorded by exposing the photographic emulsion.

The image contrast is generated in the following way: the incident electrons are elastically scattered or diffracted in the specimen by small angles. In most cases, only those electrons which penetrate the specimen without scattering (directly transmitted beam) are used for image formation. To allow this, an adjustable aperture is inserted into the back focal plane of the objective lens (Fig. 4-1a). This objective aperture limits the scattering angle of the electrons scattered at the specimen to the order of $\Theta_0 = 1$ mrad and allows essentially only the direct beam to pass through. An image produced in this way is therefore called the bright-field image (see below) in analogy to light optics. Since the scattered or diffracted electrons are excluded from this image, any changes in the intensity of the transmitted beam due to inhomogeneities in the specimen with respect to, e.g., density, thickness, and orientation, create an image contrast (section 4.1.6).

Provision of thin specimens with a thickness in the region of 0.1 μm requires suitable specimen preparation techniques (section 4.3) which are significantly more complex than those of SEM specimens. Compared with the SEM, however, the TEM offers a higher resolution which extends down to atomic dimensions. Beyond this, the transmission of thin specimens yields a wealth of information about the material structure, such as grain size, crystal defects etc.

4.1.2 Scattering of fast electrons

The various elastic and inelastic scattering processes of the electrons were described in section 2.4. To interpret the image contrast in the TEM, a knowledge of the angular distribution of the elastically and inelastically scattered high-energy electrons is additionally required. The contrast is produced largely by the elastically scattered electrons. To obtain a good image reso-

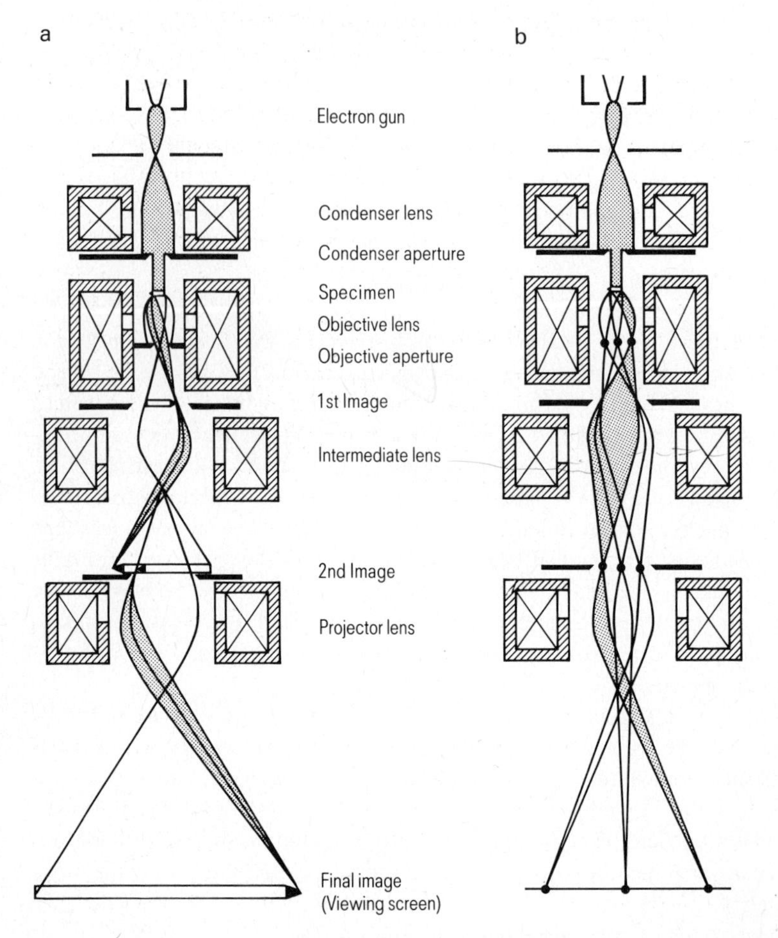

Fig. 4-1. Schematic layout and ray diagrams of a transmission electron microscope. a) Ray diagram for bright-field imaging: the aperture angle of the bundle of rays leaving each object point is limited to approx. 1 mrad by the objective aperture. One bundle is marked by shading, but its aperture angle is exaggerated. Since this angle is in fact extremely small, the directly transmitted beam contributes primarily to image formation. b) Ray diagram for selected area diffraction: the back focal plane of the objective lens in which the diffraction pattern is formed is imaged on the viewing screen by changing the intermediate lens excitation. One ray bundle is marked by shading. The selector aperture in the first (intermediate) image plane is used to select the area contributing to the diffraction pattern.

lution, the specimens must be thin enough so that each incident electron is scattered only once. The principles of *elastic scattering* at a single atom can then be used to explain the image contrast in terms of scattering at the specimen atoms. The amplitude of a wave scattered elastically by a single atom, i.e. the atomic scattering amplitude $f(\Theta)$ is, for small scattering angles Θ:

$$f(\Theta) = \frac{m_0 \cdot e^2}{2 \cdot h^2} \cdot \left(\frac{\lambda}{\Theta}\right)^2 \cdot (Z - f_x) \tag{4-1}$$

(where m_0, e and λ are the rest mass, charge and wavelength of the electron, h is Planck's constant, Z the atomic number and f_x the atomic scattering factor for X-rays.) At the high energies used in the TEM, m_0 and λ must be corrected for relativistic effects (section 2.2.1). For $f_x \to 0$, equation (4-1) describes the Rutherford scattering at the nucleus (Fig. 2-30), which dominates at large scattering angles. The term with f_x takes into account the contribution of the electron shell. It is important only at small scattering angles. In this case, the distance d of the electron trajectory from the nucleus (Fig. 2-30) is large and the nucleus is screened by the electron shell. In Fig. 4-2, the atomic scattering amplitude $f(\Theta)$ is plotted as a function of Θ for several elements. $f(\Theta)$ remains finite for $\Theta \to 0$, since f_x tends to Z. As Θ increases, $f(\Theta)$ decreases strongly. At $\Theta = 20$ mrad $\approx 1°$, $f(\Theta)$ is already equal to or smaller than a quarter of its value for $\Theta = 0$. This means that the contribution of the electrons which are elastically scattered with large angles, is low. However, this contribution is important since it is responsible for the contrast formation in the TEM (section 4.1.6) and for the reduction of the spatial resolution in the X-ray microanalysis of thin specimens (section 5.6.2).

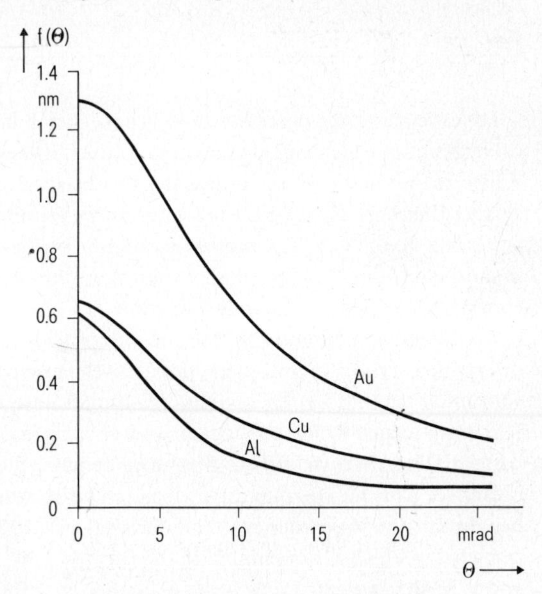

Fig. 4-2. Atomic scattering amplitude $f(\Theta)$ as a function of Θ for 100 keV electrons. $f(\Theta)$ was calculated for aluminum, copper and gold using eq. (4-1). [4-9].

The mean free path Λ between two scattering processes can be taken as a measure of the "strength" of the interaction between the electrons and the specimen atoms. Λ is inversely proportional to the cross section σ of the scattering process:

$$\Lambda = \frac{1}{C \cdot \sigma} \tag{4-2}$$

where C is the concentration (number of particles per cm^3). The total scattering cross section of a single atom σ_{tot} is made up of the sum of the cross sections for elastic and inelastic scattering $\sigma_{tot} = \sigma_{el} + \sigma_{inel}$. The ratio $d\sigma/d\Omega$ (Ω solid angle) is known as the differential cross section. The elastic differential cross section is closely related to $f(\Theta)$ in eq. (4-1):

$$d\sigma_{el}/d\Omega = |f(\Theta)|^2 \tag{4-3}$$

The increase of $f(\Theta)$ with atomic number (eq. (4-1) manifests in a strong reduction of the mean free path for elastic scattering (Table (4-1)). In the energy range between 100 and 200 keV, these mean free paths extend from 10 nm up to several 100 nm and are approximately proportional to the energy. In crystals, the elastic scattering is strongly modified by the crystal periodicity: in specific directions given by Bragg's law (eq. 2-2), beams of high intensity are diffracted.

Table 4-1. Mean free paths for the elastic scattering of high-energy electrons for several solid materials [4-1]

Target element	Mean free path in nm	
	100 keV	200 keV
carbon	200	400
silicon	120	240
copper	21	42
gold	11	22

Inelastic scattering processes such as core shell ionization form the basis for various analytical techniques which can also be used in the TEM (section 4.1.7). At low atomic numbers inelastic scattering processes dominate, i.e. the total mean free path for inelastic scattering is smaller than that for elastic scattering. At high atomic numbers the reverse is true, and elastic scattering dominates. The scattering angles for the various inelastic scattering processes are in most cases significantly smaller than those for elastic scattering, i.e. inelastic scattering occurs principally in the forward direction.

The largest contribution to inelastic scattering is caused by the thermal motion of the atoms or ions in a crystal. In this interaction between electrons and lattice vibrations (phonons), only very small amounts of energy (several 10 meV) are generally transferred. However, the large scattering angles which also occur lead to a diffuse background intensity in the diffraction pattern. This *thermal diffuse scattering* reduces the intensity of the diffracted beams. The mean free path for thermal diffuse scattering decreases strongly with increasing atomic number: from 1 μm for carbon to 20 nm for gold with 100 keV electrons. The fraction of the excitation energy which is not emitted in the form of particles or radiation remains in the specimen and is finally converted into heat. A further contribution stems from the excitation of collective oscillations of the valence and conduction electrons *(plasmons)*. The mean free path for plasmon excitation ($\Delta E \approx 20$ eV) by 100 keV electrons lies between 50 and 150 nm. The high-energy incident electrons can also transfer energy to single electrons. Depending on the transferred energy, secondary electrons are emitted or inner shells ionized. The contribution of these *single electron scattering* processes to the inelastic scattering is, however, small.

Since the scattering angles are small, electrons which have undergone energy losses due to inelastic scattering processes can also pass through the objective aperture and contribute to the image. However, they traverse the TEM imaging lenses with reduced energy. The chromatic aberration (section 2.2; Fig. 2-23c) of the objective lens then leads to blurring of the image and thus limits the specimen thickness which can be usefully transmitted in the TEM. The maximum specimen thickness which can be used depends on the desired resolution. For high-resolution electron microscopy of atomic dimensions, specimen thicknesses in the region of 10 nm are required. When investigating microstructure and crystal defects, specimen thicknesses between 0.1 and 1 μm can be examined, depending on the atomic number. These values increase up to several μm in high-voltage electron microscopy with electron energies of, say, 1 MeV.

4.1.3 Relationship between imaging and diffraction

Before we go on to describe the various imaging modes and electron diffraction in the TEM, the relationship between imaging and diffraction will be explained in more detail. As in light optical imaging, electron imaging initially also makes use of the laws of geometrical optics. The section of the ray path between the specimen and the first image is shown in Fig. 4-3. A thin specimen is illuminated with a parallel beam. All rays leaving one *specimen point* are deflected by the lens so that they meet at one image point (cf. Fig. 2-17), producing a magnified image of the specimen. Of special importance is the back focal plane. Those rays which leave the specimen in a specific *direction* meet in this plane, so that a specific point on the focal plane corresponds to a particular direction of radiation. The direction parallel to the axis corresponds to the focal point lying on the optical axis. With a diffracting specimen, the directions drawn in Fig. 4-3 correspond to diffraction maxima of orders 0, 1 and −1. The radiation diffracted into the various directions is focused by the lens at different points on the focal plane, thus producing a focused diffraction pattern with sharp diffraction spots. The focal and image planes represent two distinct planes in the imaging ray path which in principle contain all the information coming from the specimen, only in different forms. With an ideal lens (without aberrations) the imaging process can be described by Fourier transforms. Thus a Fourier transform of the wave function in the specimen plane provides the wave function in the back focal plane which represents the diffraction pattern. From this, an inverse Fourier transform then yields the wave function in the image plane.

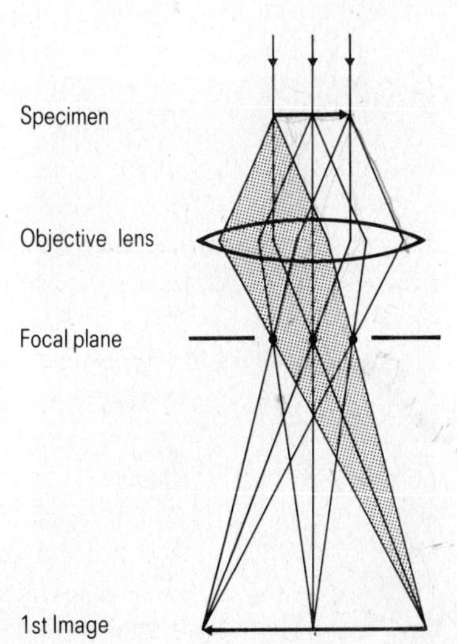

Fig. 4-3.
Imaging with an objective lens: all the rays coming from one point on the specimen meet at one point in the first image. In contrast, the rays which leave the specimen in the same direction are focused in the back focal plane. This is where the diffraction pattern is formed.

4.1.4 Electron diffraction

As described above, the back focal plane of the objective lens contains the specimen's diffraction information. If the excitation of the intermediate lens is reduced, this back focal plane can be imaged on the fluorescent screen instead of the first image plane (Fig. 4-1b) to obtain an electron diffraction pattern of the specimen. By inserting an additional aperture into the first (intermediate) image plane (selector aperture) the specimen area from which the diffraction information originates can be selected. This method is therefore called *selected area diffraction* (SAD). Since the selector aperture lies in the plane of the first image, which already has a magnification of perhaps 20:1, a specimen region of 1 µm diameter is selected with an aperture diameter of 20 µm. The smallest specimen areas which can be selected are about 0.5 µm in size (section 4.4). By changing the intermediate lens excitation, the TEM can be switched over between operation in the imaging mode and the diffraction mode.

Fig. 2-38 shows how a diffraction maximum (reflection) is created by diffraction at a set of lattice planes which lies almost parallel to the electron beam because of the small Bragg angles. If an electron beam falls onto a thin single crystal parallel to a low-index crystal direction, it is diffracted by several sets of lattice planes, all of which are parallel to the electron beam, and the diffraction pattern consequently has the form of a spot pattern (Fig. 4-4a). This pattern is characteristic for the structure and orientation of the crystal. For thicker single crystals, Kikuchi lines occur in addition to the diffraction spots (section 4.4.3). Their occurrence due to the diffraction of inelastically scattered electrons was explained in section 2.4.3 (Fig. 2-39). In the case of fine-grained crystalline specimens whose grain size is much smaller than the area from which the diffraction pattern originates, ring diffraction patterns are produced (Fig. 4-4b). Among the large number of randomly oriented grains, some of them are always oriented such that electrons are diffracted from a specific set of lattice planes so that individual diffraction spots are produced. Many such spots sum up to form a diffraction ring. The radii of these rings are inversely proportional to the interplanar spacings (Fig. 2-38). Both spot patterns and ring patterns allow unknown materials to be identified. Section 4.4 deals with the interpretation of diffraction patterns as well as with the additional information which can be derived from them.

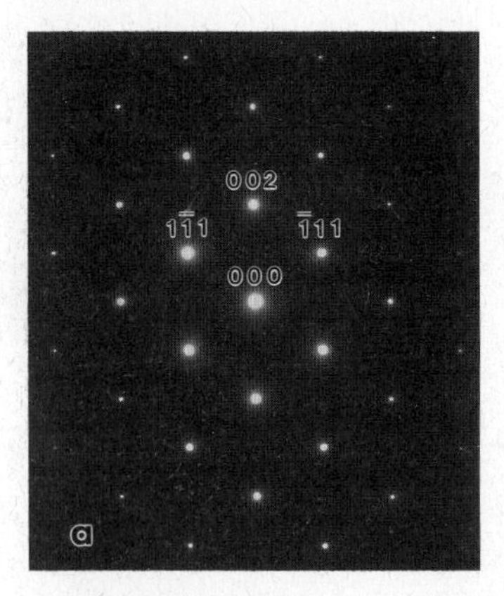

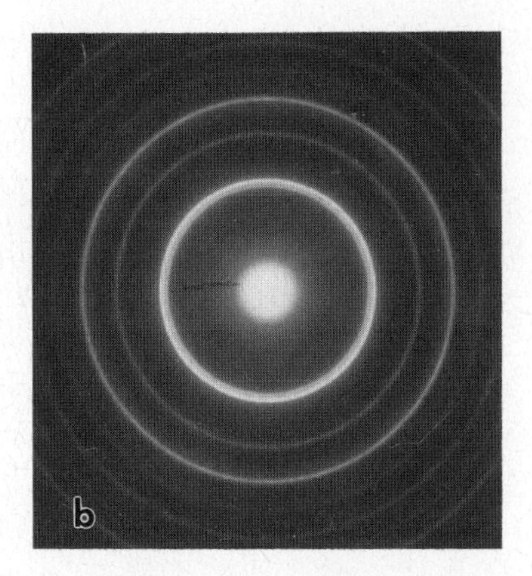

Fig. 4-4. Electron diffraction patterns. a) Spot pattern from single crystalline silicon in [110] orientation. b) Ring pattern from polycrystalline silicon.

4.1.5 Imaging modes

Imaging is impaired by aberrations (section 2.2.2). Since most of the image aberrations increase with the angle of divergence this angle has to be limited by an aperture in the back focal plane of the objective lens. The objective aperture then intercepts parts of the scattered or diffracted electrons and contributes to the contrast formation. By suitable selection of the size and position of the objective aperture, different imaging modes can be used to obtain a good image contrast, depending on the type of specimen and the information to be obtained from a TEM investigation. Different parts of the unscattered and/or scattered electrons then contribute to the formation of the image (Fig. 4-5). In *bright-field (BF) imaging* (Fig. 4-5a), as already described above, a small aperture is employed and centered onto the optical axis so that in effect only the directly transmitted beam passes through. The electrons scattered or diffracted at the specimen by angles larger than Θ_0 are intercepted by the objective aperture and cannot contribute to the image. Depending on the crystalline state of the specimen, a distinction is made between scattering contrast in amorphous objects and diffraction contrast in crystalline specimens (sections 4.1.6 and 4.5).

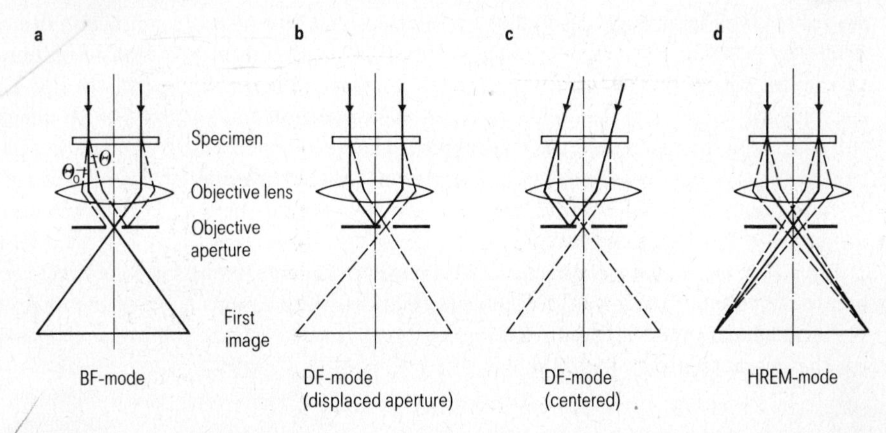

Fig. 4-5. Imaging modes. a) Bright-field mode: only the unscattered electrons and electrons scattered up to the small objective aperture angle Θ_0 contribute to the image. b) Dark-field mode (displaced aperture) and c) dark-field mode (centered): only part of the electrons scattered with a large angle ($\Theta > \Theta_0$) are used for imaging. The direct beam is intercepted. d) High-resolution mode: both directly transmitted and scattered (diffracted) electrons are used for image formation.

In *dark-field (DF) images*, the image is generated not by the direct beam but by the intensity of diffracted or scattered electrons. This can be achieved most simply by displacing the objective aperture as shown in Fig. 4-5b. To position this aperture the TEM is switched to SAD mode. The objective aperture, which is located in the plane of the diffraction pattern, is then imaged together with the diffraction pattern and can be adjusted at the desired position (e.g. on a diffraction spot). Displacing the aperture has the drawback that the diffracted electrons pass through the objective lens along off-axis trajectories at which the spherical aberration (Fig. 2-23a) degrades the image more strongly. This can be avoided by tilting the incident beam so that the diffracted electrons pass through the centered aperture and traverse the microscope along the optical axis (Fig. 4-5c). DF images require a longer exposure time than BF images since the intensities of the diffracted beams, composed of scattered electrons, are smaller than those of the direct beam.

In very thin and weakly scattering specimens, the interaction between the electrons and the specimen is weak and only the phase of the incident electron wave is changed. In order to obtain image contrast in spite of this, both the electrons in the direct beam and the scattered or diffracted electrons must be included in image production (Fig. 4-5d). A phase contrast is then created by allowing the scattered or diffracted electrons to interfere with the direct beam (sections 4.1.6 and 4.6). In thin crystals, the crystal lattice can thus be directly imaged in a specific projection. Since the lattice spacings to be imaged mostly lie at the resolution limit of TEMs, this type of imaging is known as the *high-resolution mode*.

4.1.6 Image contrast

The image contrast observed from a specimen is closely linked to the imaging mode. In all cases, the image contrast is due essentially to the elastically scattered electrons: these electrons are scattered by large angles as well as small angles and thus generate contrast in bright-field images by being excluded from the image plane (scattering contrast). In high-resolution images, they interfere with the direct beam (phase contrast). In contrast, the inelastically scattered electrons are scattered preferentially by small angles (section 4.1.2) so that for BF images they still pass through the objective aperture in considerable numbers. However, since they have lost energy, they are not focused onto the image plane due to chromatic aberrations but generate a scattering background which reduces the contrast. This is the main reason why the specimens in the TEM must be so thin.

The *scattering contrast* produced in BF images from amorphous specimens is determined by the unscattered fraction of the incident electrons. With I_{in} and I_{tr} being the intensities of the incident and transmitted (unscattered) electrons respectively, the following relation applies for the unscattered part [4-2, 3]:

$$\frac{I_{tr}}{I_{in}} = \exp\left(-\frac{N_a}{A} \cdot \sigma_{tot} \cdot \varrho \cdot t\right) \tag{4-4}$$

where N_a is Avogadro's number, A the atomic weight and σ_{tot} the total scattering cross section of an atom. The product of density ϱ and specimen thickness t is known as the mass thickness. Equation (4-4) implies that the intensity of the unscattered part of the beam declines exponentially with the mass thickness. The same applies to the dependence on $\sigma_{tot} = \sigma_{el} + \sigma_{inel}$, whereby σ_{el} is closely associated with $f(\Theta)$ of eq. (4-1) (eq. (4-3)) and increases strongly with atomic number and wavelength. An analogous dependence applies to σ_{inel}. It follows from this that the unscattered electrons decrease in number strongly both with increasing atomic number and increasing wavelength, i.e. decreasing energy. In the above case, the transmitted fraction was set approximately equal to the unscattered fraction. However, the objective aperture can be passed by electrons up to a scattering angle of Θ_0. The transmitted fraction therefore increases with Θ_0. In equation (4-4), this is accounted for by the total scattering cross section used being a function of Θ_0. The exact relations between σ_{tot} and Z, E_0 and Θ_0 are given in [4-2], for example. Scattering contrast is produced in a specimen when different regions differ in thickness t, density ϱ or mean atomic number Z. In darker regions of a BF image, at least one of these three parameters is increased. Fig. 4-6a shows, as an example, a cross section through a layer structure SiO_2-Si_3N_4-SiO_2 in which the Si_3N_4 layer appears darker than the adjacent SiO_2 layers due to its greater density.

In crystalline specimens, the image contrast results from differences in the diffracted intensity, so that we refer to it as *diffraction contrast*. If a thin crystal is oriented such that a specific lattice plane is in Bragg orientation, a diffracted beam is produced (Fig. 2-38). In a BF image, the diffracted electrons are intercepted by the objective aperture and the intensity of the transmitted electrons is that of the incident intensity reduced by the diffracted intensity. Specimen regions at which electrons are more strongly diffracted therefore appear darker in the BF image. In a DF image, in which the objective aperture is positioned at the diffracted beam (reflection), these regions would appear brighter than the surroundings. If only one diffracted beam is present, the DF image is complimentary to the BF image for thin crystals (section 4.5). Figs. 4-6b and c show, as an example, a layer of polycrystalline silicon. In BF image 4-6b, the great differences in contrast between the individual grains are due only to the incident electrons being diffracted with differing intensities depending on the respective grain orientation. If the specimen is slightly tilted, other grains will be in Bragg orientation and thus appear dark. To record the DF image 4-6c, the objective aperture was positioned over an arc segment of the most intense diffraction ring. Those grains whose diffracted intensities pass through the aperture appear bright. Specific details of contrast – e.g. the fringes at the grain boundaries of dark grains in the BF image or of bright grains in the DF image – depend on interference effects of the electron waves in the crystal. These must be taken into account in order to describe the dependence of the diffracted intensity on factors such as the specimen thickness and the exact orientation of thin single crystals with respect to the incident electron beam. This is done in section 4.5, where the diffraction contrast of crystal defects is also explained.

When passing through *extremely thin* objects, the incident electron wave is only weakly scattered. To obtain image information, the phase change of the scattered waves must be exploited, as in light optics. In light microscopy, phase contrast is converted into amplitude contrast through a $\lambda/4$ phase plate. By this additional phase shift of the diffracted rays by 90°, phase contrast becomes visible. So far no ideal phase plates exist for electron waves. In the TEM, however, *phase contrast* can be made visible by utilizing the spherical aberration and defocusing of the objective lens. This is because these two effects cause scattered electrons to undergo an additional phase shift compared with the direct beam. An optimum phase shift can be adjusted by means of suitable defocusing. This phase shift then acts as a $\lambda/4$ plate and converts the phase contrast of the specimen into an amplitude contrast. In contrast to the light optical case, however, the phase shift is not the same for all scattering angles. The exact relationships are treated in section 4.6. The image contrast results from interference of the unscattered and scattered electron waves. To allow both of these to contribute to the image, a large objective aperture must be used, as was previously described for the high-resolution mode (cf. Fig. 4-5d). Phase contrast is used mainly to image object details whose spacings lie close to the resolution limit. If the beams diffracted by thin crystals are included in the image formation, the crystal lattice can be imaged. Fig. 4-6d shows a silicon crystal which is projected along the [110] direction. Apart from the direct beam, the four reflections of type {111} and the two of type {200} were used to produce this image. The {111} and {200} planes corresponding to the reflections are marked in the image. The dot pattern of the high-resolution image reflects the lattice periodicity. Whether the bright or dark dots correspond to the atomic posi-

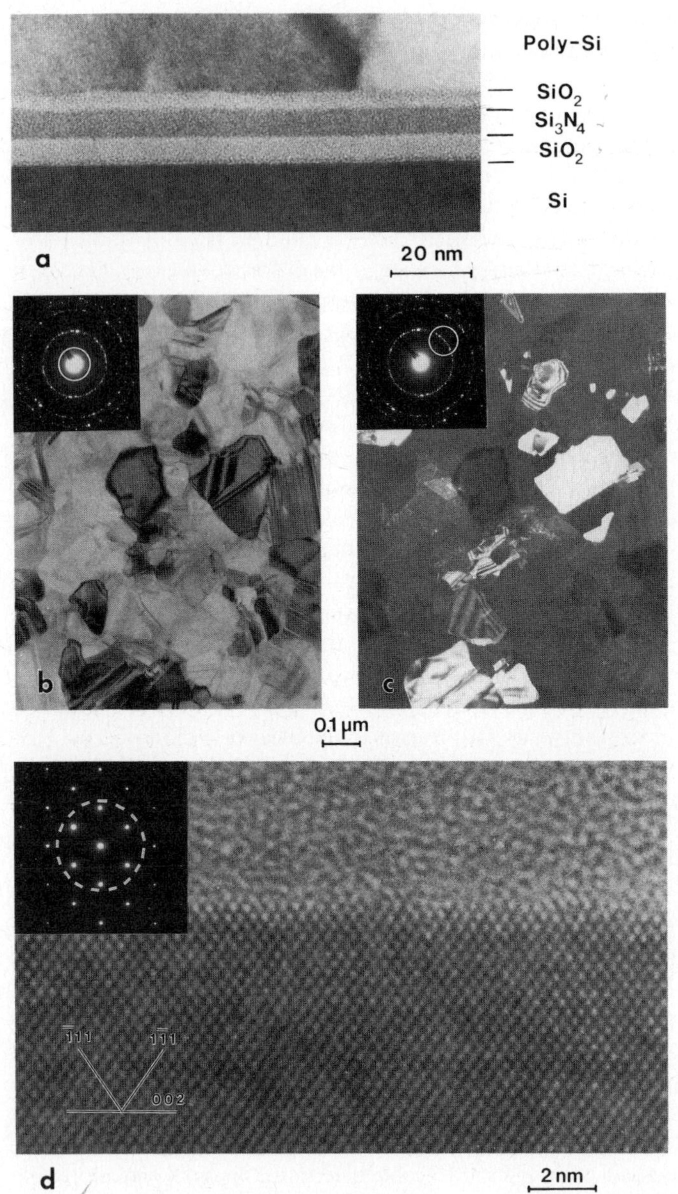

Fig. 4-6. Examples of various types of image contrast. a) *Scattering contrast:* in the bright-field image of a thin cross section through a structure consisting of amorphous insulating layers (SiO_2-Si_3N_4-SiO_2), the Si_3N_4 layer appears darker due to its greater density. b) and c) *diffraction contrast* in a layer of polycrystalline silicon caused by differences in the intensity diffracted by the individual grains: in the bright-field image (b), grains at which strong diffraction occurs appear dark, whereas they are bright in the dark-field image (c) if their diffracted intensity contributes to the image. The inserts show diffraction patterns (from a larger area of the specimen) in which the position of the objective aperture is marked. d) *High-resolution* electron micrograph of silicon lattice in [110] projection. The various lattice planes corresponding to the individual diffraction spots are indicated. The diffraction pattern also shows the size and position of the objective aperture.

tions (or in this case, to pairs of atoms), can be determined only by making a comparison with theoretically calculated images (section 4.6).

4.1.7 Analytical electron microscopy

When irradiating a thin specimen in the TEM, the same scattering and excitation processes occur as when irradiating a bulk specimen in the SEM. In addition to the transmitted electrons, which do not occur in bulk specimens, secondary (SE) and backscattered electrons (BSE) are also produced as well as X-rays and Auger electrons. These other emitted particles or quanta can be used for additional analytical methods, as in the SEM, if they are recorded by suitable detectors.

Selected area electron diffraction has already been mentioned as an analytical method, as the diffraction patterns provide information about the crystal structure of the specimen. Since every crystalline substance has a specific crystal structure, it can be identified with the aid of electron diffraction patterns. By this identification an analysis of the chemical composition of the specimen is obtained. However, it is common practice to use the term analytical electron microscopy only for those methods in which a focused electron probe is directed onto a specimen and used to generate various signals. The advantage of using such analytical techniques in the TEM is that the spatial resolution is one or two orders of magnitude better than for SAD or when analysing bulk specimens due to the reduced interaction volume (Fig. 2-34). The specimen region used for the analysis can be selected in the highly magnified image by gradually reducing the electron beam diameter while observing the image. If the TEM is equipped with a scanning attachment, the specimen can be scanned by a focused electron probe as in the SEM. The TEM is then operated as a scanning TEM (STEM). To record the image, the transmitted electrons (TE) are collected by a detector. The signal is amplified and then used to control the intensity modulation of a CRT. Instead of the TEs, the SEs or BSEs can also be used as image signals. The specimen point selected for the analysis is correlated to the image as in the SEM.

In Fig. 4-7, the three most important methods of analytical electron microscopy are schematically shown: X-ray microanalysis (XMA), electron energy-loss spectroscopy (EELS) and electron microdiffraction. A full presentation of these various methods is found in [4-4], for example. Energy-dispersive spectrometers with semiconductor detectors are particularly suited for *X-ray microanalysis* in the TEM. The operation of these detectors is described together with the special features of XMA of thin specimens in Section 5.6.2. One important point is that the quantitative composition of thin specimens can be determined relatively simply from the intensity ratios of the characteristic X-ray lines from the various elements.

In *electron energy-loss spectroscopy*, the transmitted electrons are passed through a spectrometer. In most cases a magnetic sector field spectrometer is used. It operates for electrons with different energies like a glass prism for white light: electrons leaving an object point are focused to different image points lying along a line depending on their energy (Fig. 4-7). A spectrum is obtained when those electrons which pass through the exit slit of the spectrometer are recorded as a function of the spectrometer magnetic field. The spectrum represents the energy distribution of the transmitted electrons, showing absorption edges at those energy losses above which more electrons lose energy due to excitation processes within the specimen. Apart from energy losses due to plasmon excitation, the energy absorption edges particularly

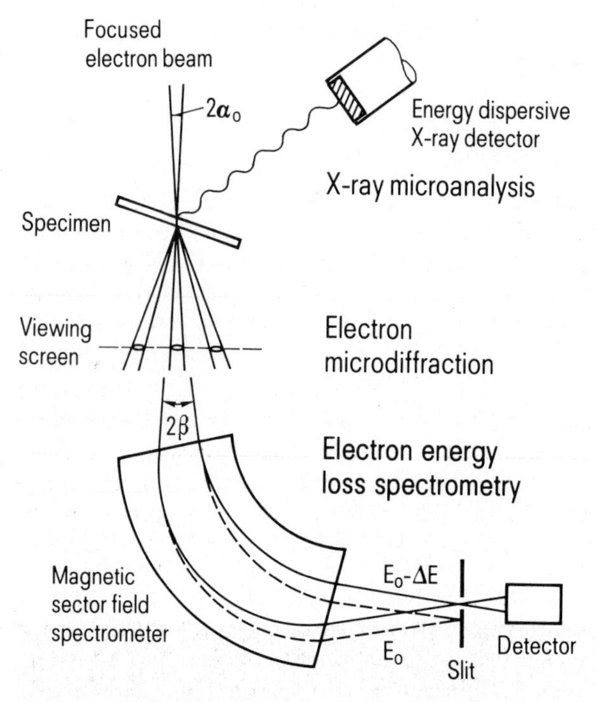

Fig. 4-7. Methods of analytical electron microscopy. Energy-dispersive detectors are used for X-ray microanalysis. In electron energy loss spectroscopy, the electrons are dispersed according to their energy (E_0 and $E_0 - \Delta E$) in a magnetic sector-field spectrometer. In electron microdiffraction, a diffraction pattern is formed with a focused electron beam directed onto a defined specimen point.

comprise the inner shell ionizations of the specimen atoms and thus permit a chemical analysis to be made (section 4.7.2). EELS is particularly suited for analysing light elements, which is more difficult with XMA. Apart from the need of specially designed detectors the fluorescence yield for light elements ($Z < 11$) is low. This is not the case for EELS, where the primary ionization process is detected. From the fine structure of the spectra, information can additionally be obtained about the chemical binding states of the specimen atoms. However, easily interpretable spectra are obtained only when the analyzed specimen areas are so thin that multiple scattering processes do not occur, i.e. the specimen thickness must be of the order of the mean free path for inelastic scattering. For elements of medium atomic number, these thicknesses are of the order of several 10 nm. A more detailed description of EELS is given in section 4.7.2.

In *electron microdiffraction*, a diffraction pattern is recorded with a focused incident electron beam. Diffraction information can thus be obtained from small regions with a diameter down to 10 nm. Since the electron beam loses its parallelism by focusing, the diffraction spots become disks. In convergent beam diffraction, the angle of convergence of the incident electron beam (angle $2\alpha_0$ in Fig. 4-7) is increased until the diffraction disks touch. Intensity modulations and fine lines appear within the disks containing a wealth of crystallographical information. The various methods of electron microdiffraction are treated in section 4.4.4.

4.2 Instrumentation

4.2.1 Overall system

The *electron-optical column* in a TEM consists not only of the (condenser) lenses for illuminating the specimen as in the SEM but also comprises a series of additional lenses below the specimen for image formation (Fig. 4-1). The use of higher electron energies compared with the SEM, (generally 100 to 200 keV) means that larger coils, which generate more powerful magnetic fields, are required for the lenses. TEM columns (section 4.2.2) are therefore significantly larger than SEM columns.

Fig. 4-8 shows a view of an advanced TEM with 200 kV acceleration voltage. At the lower end of the column is the viewing chamber with a glass window through which the image can be observed on the fluorescent screen. Below this is the plate camera which contains magazines with films. To record an image, a film is transported under the fluorescent screen and exposed with the screen folded upwards. To the right and left of the viewing chamber are the control panels. The high-voltage generator and the power supply are generally accommodated in separate units.

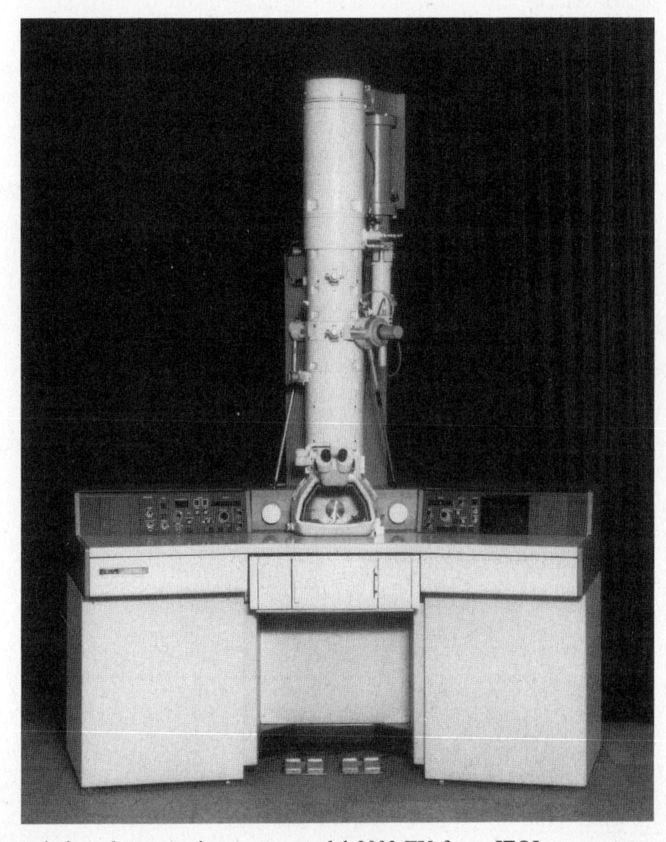

Fig. 4-8. View of advanced transmission electron microscope, model 2000 FX from JEOL.

As regards the requirements for the *vacuum system*, similar criteria apply as for the SEM: specimen contamination with hydrocarbons from the residual gas must be avoided. This means that ion gettering pumps are used for the specimen chamber and the electron gun, which is operated mostly with LaB_6 cathodes. When using cold traps, a vacuum of around 10^{-5} Pa is attained in the specimen environment, the partial pressure of hydrocarbons being particularly low ($<10^{-7}$ Pa). The vacuum in the viewing chamber is significantly poorer due to the films in the plate camera lying below it. The specimen chamber and the viewing chamber are therefore differentially pumped: an aperture located at the lower end of the column maintains a pressure difference of a factor of 100. The viewing chamber and plate camera are usually connected to a high-capacity diffusion pump. They are separated by a valve to allow rapid film exchange without breaking the column vacuum. Since the photographic emulsion of the films contains moisture, the films must be dried out in a desiccator before being loaded into the microscope.

In general, TEM specimens take the form of thin disks with a diameter of 3 mm and a thickness of a few tenths of a millimeter. Frequently, only small regions at the edge of a small hole close to the specimen center can be examined in transmission (section 4.3). The specimens are mounted in a specimen holder and inserted together with it into the microscope via an airlock. They can then be moved and tilted by means of a *specimen stage* (section 4.2.3).

For STEM operation and additional methods of analysis, suitable *detectors* are required (section 4.2.5). Such detectors have two possible locations: in the specimen surroundings like the X-ray detector and the detectors for secondary and backscattered electrons, and below the plate camera like the detector for transmitted electrons and the EELS spectrometer.

Many of the functions of modern instruments are controlled by a computer or microprocessor. This applies above all to the vacuum system and the control of the lenses. Consequently, where analytical investigations require the use of different methods with various excitation conditions for the illumination and the imaging lens system or for various deflection and alignment coils, these can be stored in a memory and called up again when required.

4.2.2 Electron-optical column

Electron gun

The electron gun (section 2.3.2, Fig. 2-26) consists of a cathode, a Wehnelt cup and an anode. In modern instruments, only LaB_6 cathodes are used because of their higher brightness which yields a greater image intensity at high magnification and a fine electron probe carrying more current. The requirements for the insulation between cathode and anode are larger than in the SEM due to the higher voltages in the TEM.

Illumination lens system

The schematic layout of the TEM in Fig. 4-1 shows only one condenser lens. But the illumination systems of commercial TEMs contain at least two condenser lenses (Fig. 4-9) so that various illumination conditions can be selected. An *illumination system with two condenser lenses* of this kind does not, unlike the SEM (cf. Fig. 2-27), have the function of focusing the electron

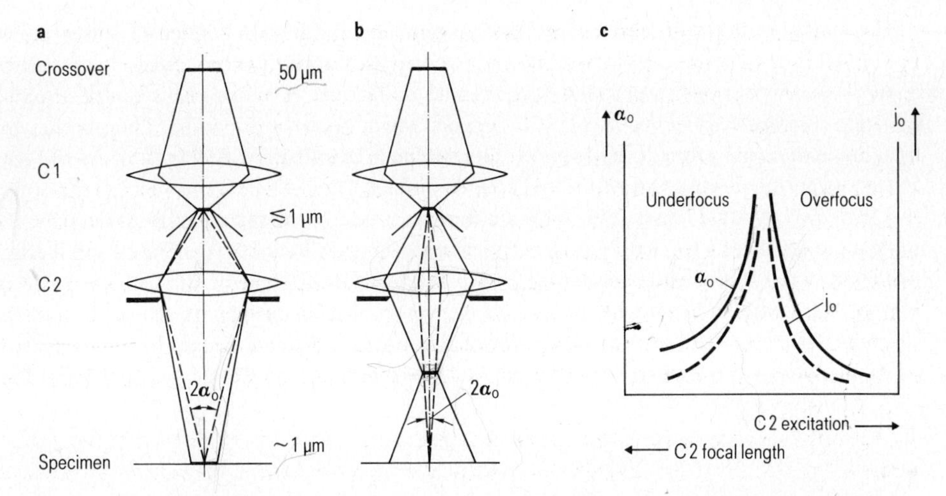

Fig. 4-9. a), b): Ray diagrams (full lines) for an illumination system consisting of two condenser lenses (C1 and C2). The beam divergence is indicated by broken lines. a) If condenser 2 is focused onto the specimen plane, the beam divergence α_0 is determined by the C2 aperture. b) Like underfocusing (not shown here), overfocusing of C2 enlarges the illuminated specimen area and reduces α_0. c) Relationship between C2 excitation and beam divergence α_0 or current density j_0.

beam as finely as possible, but should above all match the size of the illuminated specimen area to the specimen region observed on the fluorescent screen at different magnifications. Specimen areas which lie outside this region are then not illuminated and therefore cannot be heated or contaminated either. In addition, the image intensity and the beam divergence must be adjusted as required. However, since these parameters are closely related, they cannot be selected independently of each other.

The condenser lens 1 (C1) is always strongly excited and reduces the crossover of the gun with a diameter of say, 50 μm, to somewhat below 1 μm. If the condenser 2 (C2) is excited to in-focus operation (Fig. 4-9a), this image of the crossover is projected onto the specimen with slight magnification. The beam diameter there is about 1 μm. The beam divergence α_0, i.e. the divergence of radiation incident onto a specimen point, is given by the diameter of the C2 aperture. The image intensity is proportional to the current density j_0, which is directly related to α_0 via the brightness β (eq. (2-20): $j_0 = \beta \pi \alpha_0^2$). Current density and beam divergence are a maximum when condenser 2 is focused (Fig. 4-9a). The diameter of the illuminated object region can be enlarged by means of over- or underfocusing of the condenser lens. Fig. 4-9b shows the case of overfocusing. The beam divergence is now determined by the second image of the crossover. Since this image is much smaller than the C2 aperture, the beam divergence is strongly decreased. At the same time the current density and thus the image intensity are reduced. The same happens for underfocusing of C2. This dependence of beam divergence and current density on the C2 excitation is shown schematically in Fig. 4-9c. Small beam divergences of < 1 mrad are required to obtain sharp diffraction spots. For recording SAD patterns therefore, condenser 2 is always strongly defocused. Higher current densities can be attained by a smaller excitation of condenser 1. However, this means that α_0 also increases, since the crossover of the gun is demagnified less strongly.

The minimum beam diameter which can be generated by the two condenser lenses shown in Fig. 4-9 is of the order of 1 μm. However, for analytical investigations – e.g. X-ray microanalysis – it would be useful to have probes with a diameter ≪ 0.1 μm. This can be attained by using *condenser objective lenses* (CO lenses), which allow both parallel illumination for imaging, and a fine probe to be formed. The specimen is immersed deeply into the CO lens so that its pre-field acts as an additional condenser lens. The post-field then acts as the objective lens. Two separate lenses (Fig. 4-10) are therefore used in the ray diagram to describe the optical properties of a CO lens. Condenser 1 is again strongly excited to generate a small effective source size. By focusing condenser 2 onto the front focal plane of the CO lens (FFP in Fig. 4-10a), a parallel beam whose diameter is given by the C2 aperture is projected onto the specimen. The post-field of the CO lens forms the first intermediate image. To produce a fine probe, condenser 2 is excited so that a parallel beam falls on the CO lens (Fig. 4-10b). This

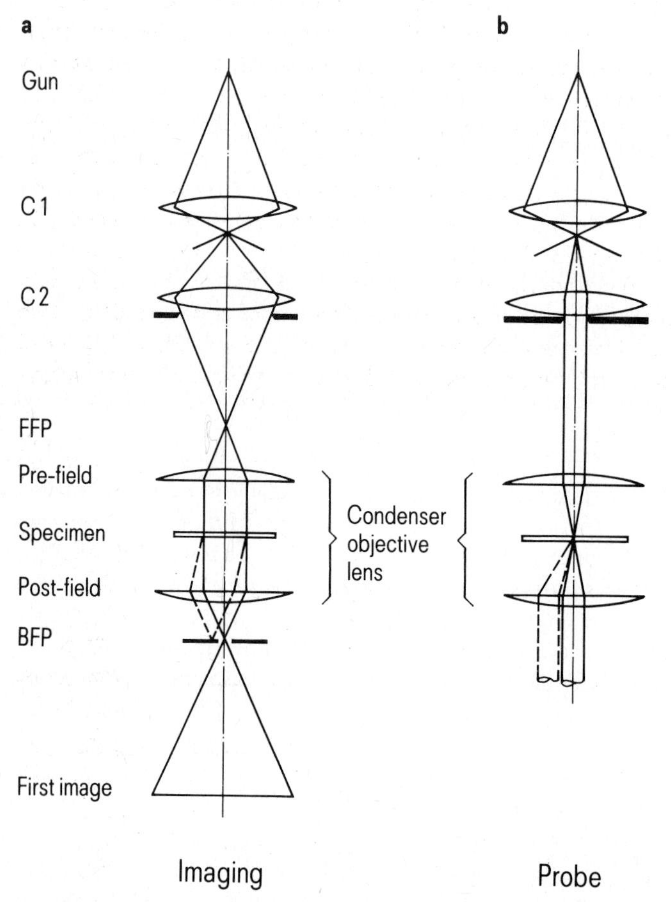

Fig. 4-10. Specimen illumination by the pre-field of a condenser objective (CO) lens. a) For imaging condenser 2 (C2) is focused onto the front focal plane (FFP) and the pre-field consequently produces a parallel incident beam; BFP = back focal plane. b) To produce a fine probe, a parallel beam formed by C2 is focused by the pre-field onto the specimen.

beam is then focused by the pre-field to a probe whose angle of divergence is determined by the C2 aperture. Probe diameters of a few nm can be attained in this way.

Modern microscopes, which require many different illumination conditions for analytical purposes, are equipped with a third, additional condenser lens which is located between lenses C2 and CO. By use of this lens, the size of the specimen area illuminated during imaging is no longer defined by the C2 aperture as in Fig. 4-10a, but can be continuously varied by changing the C2 excitation (as in the illumination system shown in Fig. 4-9). An analogous situation applies to the divergence angle α_0 of the probe, which in Fig. 4-10b is also defined by the C2 aperture: α_0 can now be continuously varied by this third condenser lens. This is of great importance, especially for convergent beam electron diffraction (section 4.4.4).

A still finer electron probe than that described above is often required for STEM operation. It is generated by switching off condenser 2 and the third condenser lens and focusing the beam passing through a small C2 aperture only with the aid of the pre-field of the CO lens (Fig. 4-11). The probe diameter can be estimated from the demagnification by the pre-field. As described above, the crossover image generated by the strongly excited condenser 1 has a diameter of about 0.5 μm. The demagnification M is given by the ratio of image distance to object distance (eq. (2-14)). In this case $M = f_0 / a$, where f_0 is the focal length of the CO lens and a the distance of the CO lens from the crossover image (cf. Fig. 4-11). For $f_0 = 1$ mm and $a = 250$ mm, $M = 1/250$ and we obtain a probe diameter of $d_0 = 2$ nm. In reality, values of 1 nm are attained. For scanning operation a double deflection system which consists of two pairs of deflection coils is used in the same way as in the SEM. The coils are driven such that the beam is rocked about the front focal point of the CO lens. The approximately parallel beam is then focused independently of its direction of incidence on the CO lens and the probe scans across the specimen in the x and y directions. Due to the symmetrical structure of the CO lens at whose center the specimen is located, the post-field again forms a parallel beam, which rocks about the back focal point of the CO lens during scanning. This means

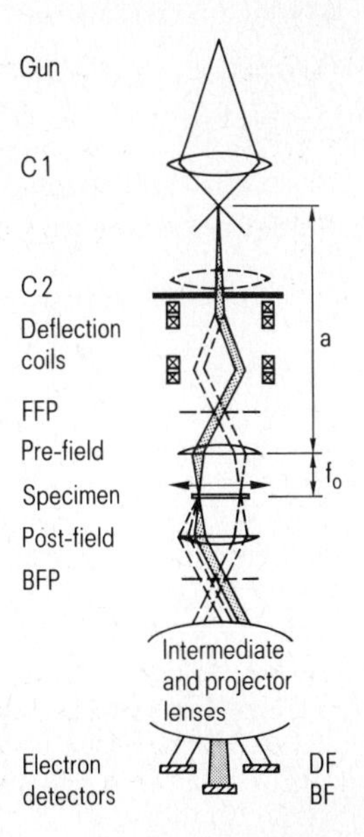

Fig. 4-11.
Ray diagram for STEM operation. With condenser 2 switched off, the probe is generated only by focusing with the pre-field of the CO lens. During scanning, the beam is rocked around the focal point in the front focal plane (FFP). In the back focal plane (BFP), a stationary diffraction pattern is formed which is then transferred to the detector plane with the aid of the intermediate and projector lenses. The directly transmitted electrons are collected by the bright-field detector (BF detector). The annular dark-field detector (DF detector) collects the electrons scattered or diffracted in various directions.

that despite the scanning, a stationary diffraction pattern is produced in the back focal plane in which the direct beam is separated from the diffracted beams. If the diffraction pattern is transferred by the succeeding intermediate and projector lenses to the detector plane, which is generally located below the plate camera, the various signals can be collected there (Fig. 4-11, section 4.2.5).

Imaging lens system

The rays deflected by the objective lens traverse the following lenses along trajectories which are almost parallel to the axis, since their angles with the optical axis decrease with increasing magnification (cf. Fig. 2-17). The deviation from the paraxial ray is greatest in the objective lens. The resolution of the TEM is consequently limited by the aberrations of the objective lens, which must therefore satisfy the highest requirements.

From these lens aberrations, the spherical aberration above all must be taken into account. Furthermore, the aperture employed to reduce the spherical aberration produces a defraction error in turn. The influence of the chromatic aberration is small (section 4.6) and is neglected here, since both the high voltage and the lens currents can be very well stabilized ($\Delta U/U$ and $\Delta I/I$ are around 10^{-6}). Mechanical defects (material inhomogeneities and the finite production tolerances of the objective pole piece) result in nonperfect rotational symmetry of the magnetic field. This aberration (astigmatism) can be compensated with the aid of the stigmator, i.e. an electron-optical cylinder lens with both variable focal length and direction.

To estimate the *resolution* therefore, we need to consider merely the diameters of the error disks of the spherical aberration d_s and of the diffraction error d_d. Their dependence on the divergence angle α (eq. (2-15, 16) is shown schematically in Fig. 4-12. The error disk which results by superposition of d_s and d_d is obtained by adding the squares of d_s and d_d (cf. eq. 2-24): $d^2 = d_s^2 + d_d^2$. At the optimum aperture angle α_{opt}, a minimum value is obtained for d, which represents the theoretical resolution limit [4-2, 3]:

$$d_{min} = A\,(C_s \cdot \lambda^3)^{1/4} \tag{4-5}$$

where λ is the electron wavelength and C_s the spherical aberration coefficient. The value of A must be determined from wave-mechanical calculations. A has a value of 0.66 for the point

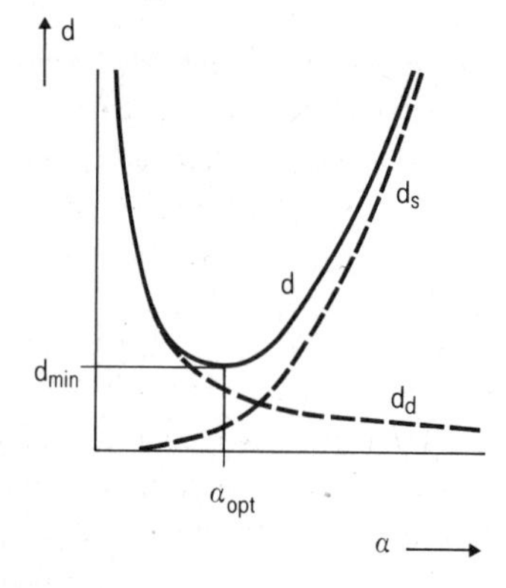

Fig. 4-12.
Relationship between resolution and divergence angle α. The diameters of the error disks for spherical aberration d_s and for the diffraction error d_d depend on α in different ways: $d_s \sim \alpha^3$ and $d_d \sim 1/\alpha$. The superposition of d_s and d_d yields the theoretical resolution limit d_{min} at α_{opt}.

resolution (see section 4.6). A high resolution, i.e. small d_{min}, is obtained when C_s and especially λ are as small as possible. Objective lenses which are optimized for high resolution operation exhibit C_s values $\leq$ 1 mm and have a particularly small pole piece gap, which limits the tilt. If higher voltages (and thus smaller values of λ) are used – in the region of 200 to 400 kV – a point resolution below 0.2 nm can be obtained.

The first intermediate image of the objective lens has a magnification of say, $M = 25:1$, and is further magnified by a series of additional lenses in the *magnification lens system.* In contrast to Fig. 4-1, three intermediate lenses and a projector lens are generally used today. The overall magnification then results from the product of the magnifications of all individual lenses. Whereas the excitations of the objective and projector lenses remain almost constant, the magnification can be varied by the three intermediate lenses over a wide range. Thus the maximum magnification possible on the screen is up to $10^6:1$. To attain very small magnifications of, say, 50:1, the objective lens must be switched off. The first intermediate lens or an additional small lens is then used as the objective lens having a long focal length. Using several intermediate lenses offers the additional advantage that when changing the magnification, the image rotation which is unavoidable in magnetic lenses (cf. Fig. 2-20) can be reduced to low values. In addition, different camera lengths (cf. Fig. 2-38) can be chosen for the diffraction mode. The magnification values displayed usually have errors. If we want to know the magnification exactly, a calibration must be performed with suitable test specimens. Reproducible magnifications can be ensured by setting the specimen always to the same z position (e.g. the eucentric point, section 4.2.3). This position can also be checked with the aid of the objective lens current.

To ensure optimum performance of the TEM, the optical axes of the various lenses or lens systems must be carefully aligned. In the first place, the axis of the electron gun must be aligned to that of the illumination system. This is then aligned with the axis of the objective lens. The electron beam must be shifted and tilted for each individual alignment step. Double deflection systems, which are well known in scanning operation (Fig. 4-11), are used for this purpose. A more detailed description of the components of the electron-optical column is given, e.g., in [4-2,5].

4.2.3 Specimen stage

The mechanical stability of the specimen stage must be very high so that images do not become blurred at high magnification due to specimen drift or vibrations during exposure. In addition, high precision is required to allow the specimens to be moved in a controlled way at high magnification. With TEM specimens of 3mm diameter, a movement range of ± 1 mm in the x and y directions is usually sufficient. Rods are mainly used as specimen holders due to their simpler construction; they are introduced from the side into the gap between the upper and lower pole pieces of the objective lens (side-entry stage). In addition, space for the objcctive aperture must be available within the gap of the pole pieces.

When investigating crystalline specimens it is in general necessary to tilt the specimen so that the electrons impinge along the required crystal direction. The angular range of tilting depends on the size of the pole piece gap. This is especially small when a high resolution is to be attained (section 4.2.2). For TEMs, which are designed especially for operation in the high-resolution mode close to the resolution limit, another type of specimen holder is used which is introduced through the bore of the objective lens pole piece from above (top-entry stage). The range of tilt of such specimen holders is limited up to about $\pm 25°$. Pole pieces with a large gap allow tilt ranges of up to $\pm 60°$ with side-entry goniometers. These goniometers are designed such that the tilt around the rod axis occurs eucentrically after correct alignment. This requires the specimen to be moved in the z direction, so that a selected specimen position lies at the eucentric point. The tilt axis then runs through this specimen position so that it can be observed during tilting with almost no lateral shift.

4.2.4 Image recording and processing

For recording images in the TEM, the photographic emulsion of the films (or plates) is directly exposed. To allow sufficient absorption of the high-energy electrons, thick emulsions (e.g. 50 μm) must be used. An electron dose of about 10^{-11} C cm^{-2} is required together with a sufficiently long developing time to obtain a photographic density of unity. The resolution of the emulsion is limited by its grain size and by electron scattering to about 30 μm. A high-resolution image with 0.2 nm resolution must therefore be recorded with a magnification of at least 150000 : 1. To allow the image details to be recognized with the naked eye (about 0.2 mm resolution), the negative must be subsequently magnified at least 7 times. A film with a size of 6×9 cm contains $6 \cdot 10^6$ image points. This corresponds to a very large amount of information.

At the high magnifications required for high resolution, the image intensity is low and the alignment (above all the correction for the objective astigmatism) as well as visual observation are therefore impaired. These problems may be reduced by using a TV system equipped with an image intensifier. The electron image falls on a transparent fluorescent screen mounted on a glass-fibre plate. This plate is located below the plate camera and separates the microscope vacuum from the outside. It is coupled optically to a second glass-fibre plate which forms the entrance to the TV system.

Recording images with a TV system allows various methods of analog image processing such as contrast enhancement (by greater signal amplification) and background subtraction. Beyond this, all methods of digital image processing are available when the TV system is linked up to a computer [4-6]. Image integration in a memory is of great importance, since it allows the signal-to-noise ratio and the sensitivity to be greatly increased. To evaluate high-resolution images, optical diffractograms are generally recorded (section 4.6). The same information can be obtained by performing a two-dimensional Fourier transformation on the stored image in the computer. Since the time required for this is of the order of only 1 s, the information is available on-line. This approach is very helpful when working in the high-resolution mode. In addition, the stored images can be further processed by applying filters to the two-dimensional Fourier space. Only specific regions of the Fourier transform are then used for calculating the image. This allows periodic structures, for example, to be emphasized.

4.2.5 Electron detectors and analytical attachments

In order to collect the various types of electron signal in STEM operation, suitable *electron detectors* are required. To detect the transmitted electrons, the stationary diffraction pattern is imaged onto the detector plane – as described in Section 4.2.2 (Fig. 4-11). As for the backscattered electrons in the SEM, a combination of scintillator and photomultiplier is generally used as a detector. The detector for the directly transmitted electrons (bright-field detector) lies in the optical axis (Fig. 4-11). The collection angle is defined by the detector diameter. It can be simply varied by changing the effective camera length, i.e. the magnification in the diffraction pattern. Frequently, additional use is made of an annular dark-field detector which detects all electrons scattered or diffracted in a specific angular range (Fig. 4-11). To collect a high fraction of the backscattered electrons, it is also helpful to use an annular detector. Since very little room is available in the pole piece gap, semiconductor detectors (Fig. 3-6c) are especially suited for this purpose. The secondary electrons moving along spiral trajectories along the magnetic field lines are extracted upwards through the bore of the pole piece, and are finally collected by a detector of the Everhart-Thornly type, as in the SEM (Fig. 3-5).

Due to their small space requirement, the *X-ray detectors* used are almost exclusively energy-dispersive spectrometers. The mode of operation of these detectors is described in section 5.2. Detectors with ultra-thin windows allow analysis of light elements down to boron, as in XMA of bulk specimens. If detectors with high-purity germanium crystals are used instead of Si (Li) detectors the energy resolution is improved to about 125 eV for the Mn-Kα line. The higher energy resolution is especially helpful in the low-energy region, since it increases the signal-to-background ratio. Furthermore, the higher X-ray absorption of germanium allows also to use the high-energy Kα lines of heavy elements for analysis. At this point, some instrumentation problems still remain to be discussed; they must be taken into account when attaching an X-ray detector to a TEM or STEM.

A high collection efficiency is obtained when the detector is positioned as close as possible to the specimen and detects a large solid angle of the X-rays emitted from the specimen. By arranging the detector above the specimen, a large take-off angle is obtained for the X-rays (Fig. 4-13), which allows analysis without additional tilting of the specimen.

The aim of every measurement is to position the electron probe onto the specimen point of interest and collect the X-rays emitted from it. However, various processes can also lead

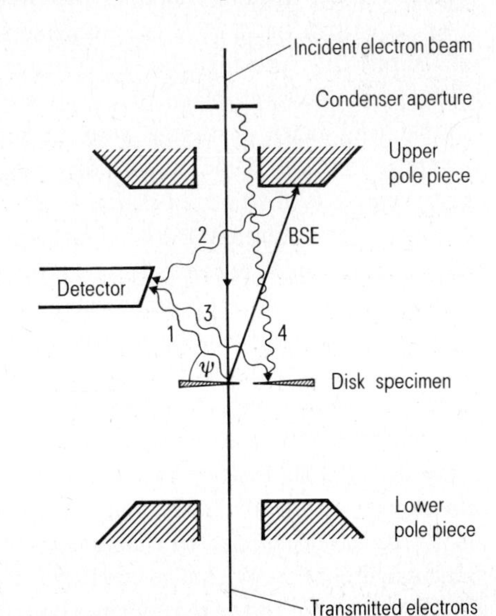

Fig. 4-13.
X-ray detector for X-ray microanalysis of thin specimens in the TEM. The take-off angle ψ changes when the specimen is tilted. In addition to the X-rays (1) excited by the incident electron probe on the specimen, the detector also collects spurious radiation (2 and 3) produced by the backscattered electrons (BSE) and hard X-rays (4) from the illumination system. The spurious radiation must be minimized by suitable measures (see text).

to X-rays being generated at other points of the specimen and the specimen environment and reaching the detector as spurious radiation (Fig. 4-13) [4-7]. Firstly, the electrons backscattered from the specimen strike the specimen holder and pole pieces and excite characteristic radiation there. This simulates the presence of additional elements (such as the iron of the pole piece). This type of spurious radiation can be minimized by a collimator in front of the detector and the use of low-atomic-number materials (such as beryllium whose X-rays are not detected) in the specimen environment, e.g. for the specimen mounting. Secondly, hard X-rays from the illumination system (above all bremsstrahlung) as well as uncollimated electrons (not shown in Fig. 4-13) hit the specimen away from the point of interest and excite X-rays there. This type of spurious radiation can be checked by measuring the "hole counts". This is done by passing the electron probe through a hole in the specimen so that no direct X-ray excitation takes place. X-rays, which are nevertheless collected by the X-ray detector, are excited by the spurious radiation. They can be minimized by fitting additional apertures in the second condenser.

The principle of *electron energy-loss spectroscopy* (EELS) with the aid of a magnetic sector field (or magnetic prism) has already been described when discussing Fig. 4-7. The mode of operation of such a spectrometer, which is used together with an electric sector field for secondary ion mass spectrometry, is treated in more detail in section 7.2 (Fig. 7-6). In contrast to SIMS, in EELS the electron mass is constant and the magnetic prism disperses the electron beam only according to different energies (cf. eq. (7-9)): all electrons leaving from one object point with equal energy come together at one image point. Two rays with an energy difference δE are separated in the image plane by a distance δx. The parameter $D = \delta E / \delta x$ is known as the dispersion of the spectrometer. The following relation applies to magnetic prisms [4-8]

$$D = 2R/E_0 \tag{4-6}$$

where R is the bending radius of the circular electron trajectories in the prism and E_0 the electron energy. In an ideal spectrometer, the energy resolution ΔE_R is given by the width S of the selecting slit which lies in the image plane of the spectrometer: $\Delta E_R = S/D$. Like all other lenses, however, magnetic prisms also have aberrations which limit their performance, so that ΔE_R increases strongly with the acceptance angle β [4-8]: $\Delta E_R \approx \pi \beta^2 E_0$. To collect a high proportion of the inelastically scattered electrons, electrons must be detected up to a scattering angle of 10 mrad (section 4.7.2). Since such a large acceptance angle would produce a poor energy resolution, the imaging lenses are used to reduce the angular divergence. A lens of magnification M reduces the divergence angle to β/M (Fig. 2-17). In this way both a high collection efficiency and a good energy resolution in the region of $\Delta E_R \approx 1$ eV can be attained. Apart from the angular divergence, the energy resolution also depends on the object size in the spectrometer object plane and naturally on the quality of the spectrometer. The aberrations of magnetic prisms can be reduced by curved entrance and exit faces.

There are two ways to arrange the electron-optical coupling of the EEL spectrometer to the TEM [4-9]. In the first, operation is in normal TEM mode, the specimen point of interest being aligned with a hole in the center of the fluorescent screen. The size of the analyzed specimen area can be selected by changing the magnification. If small areas are to be analyzed at high magnification, the image intensity is reduced and so is the number of collected electrons, and the signal-to-noise ratio becomes poor. In this case it is better to use the second approach, i.e. to operate in STEM mode, where the focused electron probe is directed onto the point of interest.

In the configuration shown in Fig. 4-7, the spectrum is serially recorded by a detector (scintillator and photomultiplier) by varying the spectrometer excitation, for example. If a position-sensitive detector is introduced into the image plane of the spectrometer, the spectrum can be simultaneously recorded (parallel detection). The measuring time is then reduced by about two orders of magnitude. A thin scintillator crystal which is coupled via a glass-fiber plate to a photodiode array, forms such a parallel detection system [4-10]. By using a system of three quadrupole lenses between the spectrometer and the parallel detector, the dispersion and thus the analysed energy-loss range can also be changed. In both cases (serial or parallel detection) the data is stored or integrated in a computer (or a multichannel analyzer) and is then available for further evaluation, e.g. a quantitative analysis (section 4.7.2).

4.2.6 Types of microscope

A way of distinguishing between different types of transmission electron microscope is by their application: high-resolution or analytical electron microscopy. Since it is often of advantage to use high-resolution and analytical methods side by side when investigating solid materials, this criterion applies only to a limited extent. However, differences exist because instruments designed to attain maximum possible spatial resolution have a limited specimen tilt and analytical capability. A parameter for distinguishing between various types of TEM and which also directly affects the cost is the maximum acceleration voltage. Standard instruments operate with maximum voltages in the region of 100 to 200 kV. Higher voltages offer the advantages of superior penetrating power and, due to the lower electron wavelength, of higher resolution (section 4.2.2). However, increasing the voltage also means that the size of the column increases, as does the outlay for the high-voltage generation and the insulation. For material investigations, TEMs operating with medium-range high voltages of up to 300 or 400 kV are increasingly used, since they can exploit these two advantages. In addition, they are not yet large enough to require their own building, which is usually the case for high-voltage microscopes with acceleration voltages ≥ 1 MV. These latter allow information about volume properties to be obtained from the transmission of very thick (several μm) specimens and, due to their greater pole piece gaps, also facilitate in-situ experiments.

For scanning operation, TEMs are generally equipped with a scanning attachment. In addition, dedicated STEMs are available which operate only in scanning mode and have no further imaging lenses below the objective lens used to focus the probe. These instruments usually have a maximum acceleration voltage of 100 kV and are equipped with a field emission gun. Due to its high brightness, the field emission gun provides an electron probe which has a small diameter of 1-2 nm in conjunction with a high current of 1 nA. A probe of this kind allows analytical investigations to be made with a spatial resolution of around a few nm.

4.3 Specimen preparation techniques

4.3.1 Introduction

The preparation of specimens often accounts for a major part of the work involved in transmission electron microscopy, so that particular attention must be devoted to this work to ensure successful TEM application. Various methods of preparation have been developed for different materials and types of specimen. Scanning electron microscopy, which images the surface directly, has largely superseded replica techniques as a means of revealing fine surface structures. The following description will concentrate on the preparation of bulk specimens for TEM [4-11, 12, 13]. These specimens are commonly of such large dimensions that they do not transmit electrons directly in the TEM. Exceptions include thin layers that can be removed from the substrate under favourable conditions and can then be examined without further thinning (section 4.3.6).

When preparing bulk specimens for TEM, the thinning process involves a number of steps. The first of these is to prepare a specimen small enough to fit into the TEM specimen holder. Most specimen holders are designed to accommodate disks measuring 3 mm in diameter and a few tenths of a millimetre in thickness (Fig. 4-14a). The methods subsequently used for thinning will depend on the material in question. In the main, electrolytic thinning is used for metals and chemical etching for semiconductors (section 4.3.3). Ion-beam thinning is a universal method which is particularly suitable for specimens that are difficult to thin in any other way (e.g. ceramics, section 4.3.4). It is usual to protect the edge of the disk during final thinning, so that in time a hole appears in the middle of the disk (Fig. 4-14b). Around this hole thin wedge-shaped regions are formed which will be electron transparent in the TEM. The thick periphery makes the specimen self-supporting and hence easy to handle.

Throughout the preparation process it is important not to change the properties of the specimen and in particular the material structure – such as by causing additional deformation of a metal. Any contamination layers must also be removed. The desired thickness of the specimen depends on the information of interest, ranging from several μm for examining single crystal specimens with low defect densities under a high-voltage microscope to about 10 nm for high-resolution microscopy. The preparation of metallic materials will merely be outlined here. Considerably more attention will be devoted to semiconductor materials (and ceramics)

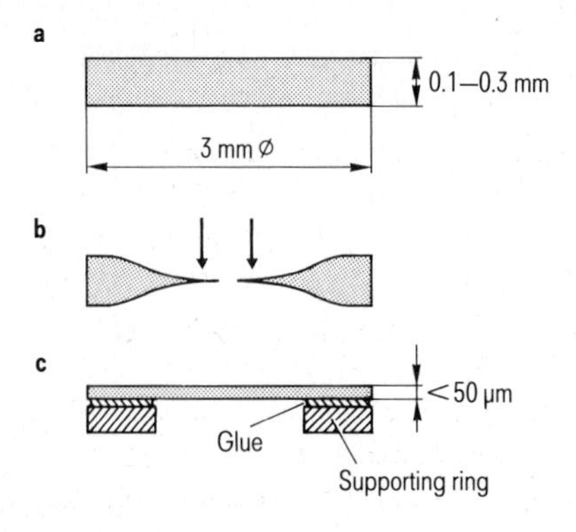

Fig. 4-14. Different types of specimens (schematic cross sections): a), b) Self-supporting disk specimens after initial preparation from bulk material (a), after chemical or electro-chemical polishing (b): the thin edges of the hole in the center are electron-transparent (arrows); c) After initial preparation down to a thickness of <50 μm for final thinning by ion milling; the specimen is glued onto a supporting ring.

and in particular to the preparation of cross-sectional specimens (section 4.3.5), because this method is especially important for microtechnology.

4.3.2 Initial preparation of bulk specimens

The first step in preparing the 3 mm disks is to cut slices from a piece of material. A *low-speed diamond-wheel saw* with a diamond-impregnated slicing wheel is suitable for this purpose. However, saws of this type produce a damage layer which may be as thick as 100 μm. The slices should not be cut too thin so that this damage layer can be removed afterwards. *Wire saws* produce considerably less damage. This method involves drawing a long, thin metal wire impregnated with diamond powder across the material so that it is cut along a line. The cutting speed depends on the load applied and on the velocity of the wire, amongst other factors. Wires wetted with a grinding suspension (or acids if the material is a metal) can also be used for cutting.

Spark machining is a universal method of cutting and initially thinning metal specimens and producing disks. In this process, the specimen and tool are immersed in an oil bath and a high voltage is applied. The tool is then brought closer to the specimen until sparks are generated. The sparks erode the material. The oil cools the specimen and carries away the eroded material. A control mechanism keeps the distance between tool and specimen constant. A wire or a knife is used to cut specimens; a fine tube will produce disks.

To cut disks from brittle or hard materials that are not infrequently poor conductors as well – such as ceramics and semiconductors – it is advisable to use an *ultrasonic drill.* Vibrations are generated by an ultrasonic source and transmitted to a tool. A grinding suspension (e.g. silicon carbide in water) between tool and specimen allows the material to be removed. In this way, the shape of the tool is transferred to the specimen. To cut disks, the tool must have the shape of a fine tube.

As a rule, *mechanical grinding and polishing* is required to reduce the specimen to the desired thickness prior to final thinning. Depending on the thickness and brittleness of the specimen, grinding and polishing may either precede or follow the cutting of the disk. Special grinding holders allow adjustment to the final thickness for initial thinning. This thickness depends on the method selected for final thinning. Initial thinning to a thickness of 100 to 300 μm is sufficient for electrolytic or chemical polishing (section 4.3.3). For ion thinning (section 4.3.4), it is advisable to thin the disks as much as possible at the initial-thinning stage, since ion thinning rates are very low. Semiconductor single crystals can be polished down to thicknesses of < 20 μm. At such thicknesses, these specimens are no longer self-supporting and must be glued to support rings (Fig. 4-14c).

Since very thin, brittle specimens such as ceramics readily break, they cannot be ground as thinly as this. It is better to thin them to an initial thickness of no less than say, 100 μm, and then grind a dimple in the center of the disk. Special machines known as *dimple grinders* have been developed for this purpose – their method of operation is outlined in Fig. 4-15. A revolving grinding wheel made of metal is lowered onto the specimen which is glued to a mount. Material is removed with the aid of a grinding suspension. A spherical depression or dimple is ground by rotating the specimen about the axis of the disk. The residual thickness at the center of the dimple can be set as required. Thicknesses of less than 10 μm can be achieved by careful adjustment. In order to reduce grinding damage, it is advisable to remove the last

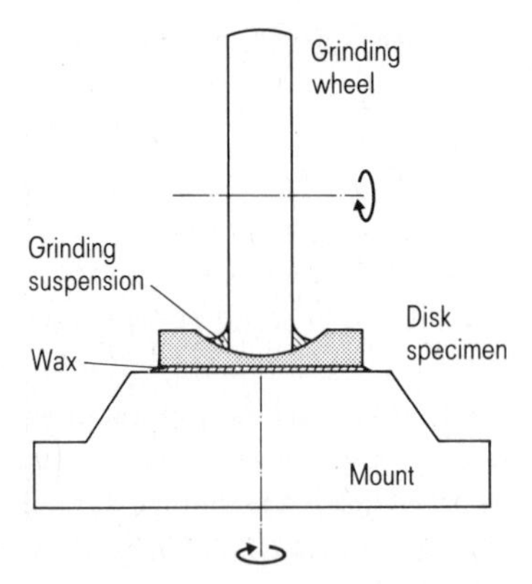

Fig. 4-15.
Principle of dimple grinder. Rotation of both grinding wheel and specimen around the indicated axes produces a spherical dimple.

few μm with a polishing wheel. Silicon specimens become transparent to light at thicknesses of < 10 μm. If the mount used is transparent, the residual thickness can also be checked on the basis of optical transparency.

4.3.3 Chemical and electrochemical polishing

Chemical polishing is mainly used for insulators and semiconductors, electrochemical polishing for metals. In both cases the polishing solution consists of an oxidizing agent and an agent which dissolves the oxide. For example, a mixture of nitric acid and hydrofluoric acid is used for silicon. The nitric acid (HNO_3) oxidizes the silicon and the hydrofluoric acid (HF) then dissolves the oxide. Agitation of the polishing solution makes the removal of material faster and more uniform in both chemical and electrochemical polishing. Polishing solutions and conditions for a number of materials are summarized in [4-11, 12].

Quite simple equipment is adequate for the *chemical polishing* of semiconductor materials. The specimen is glued to a mount with wax or lacquer and the edge of the disk masked with wax to protect it from etching attack. When immersed in a beaker of polishing solution, the specimen is thinned from one or both sides, depending on the geometry of the mount. The polishing solution is agitated with a magnetic stirrer. A lamp positioned behind the specimen allows polishing to be terminated when perforation occurs. Jet polishers (see below) of the type used for metals are also suitable for thinning semiconductors. A 5:1:1 mixture of $HNO_3:HF:CH_3COOH$ has proved effective for polishing silicon; highly diluted solutions of bromine or chlorine in methanol are among those suitable for GaAs. With the aid of special apparatus, large areas of a Si wafer can be thinned to a uniform thickness of only a few μm [4-14]. Photolithography and etching are used to produce disks from this thin foil which of course have to be glued to support rings. Areas several square millimetres in size can then be examined in a high-voltage microscope. This method is particularly attractive for examining lattice defects in specimens with very low defect densities.

When a metal is *electrochemically polished*, the specimen assumes the function of the anode in an electrolytic cell. In early apparatus, the specimen was placed opposite a plate-type or pointed cathode. Today it is common practice to use jet polishers to produce self-supporting disk specimens (Fig. 4-16). The specimen is mounted in a holder in such a way that the

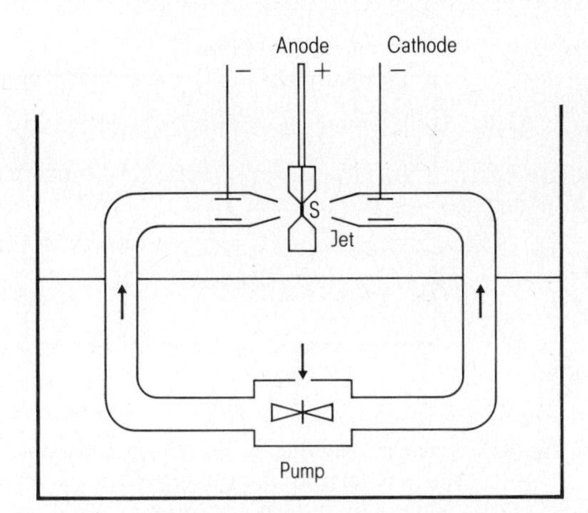

Fig. 4-16.
Schematic drawing of double jet polisher. The specimen S is polished by two jets of electrolyte.

edge is covered and contact is made with an electrode. A pump forces electrolyte through one or two nozzles (as shown in Fig. 4-16), depending on the configuration. The nozzles may or may not be immersed in the electrolyte along with the specimen. The equipment is also provided with systems for specimen illumination and observation. If the light which penetrates the hole when the disk is perforated is collected by a detector, this signal can be used to terminate polishing automatically. Rapid termination is required because the thin edges of the hole would soon be etched away if polishing were to continue. Recording of the current-voltage characteristic allows the optimum polishing conditions to be determined. Voltage is optimum along the plateau of this characteristic. Provision must also be made for cooling or heating the electrolyte, because the polishing conditions are strongly dependent on temperature.

4.3.4 Ion-beam thinning

In ion-beam thinning or ion milling, gas ions are accelerated to an energy of several keV and directed onto the specimen under oblique incidence. By this ion bombardment, material is removed from the specimen by sputtering (section 2.6.4). Fig. 4-17 is a schematic illustration of an *ion thinning apparatus.* Two ion guns are arranged in a vacuum chamber such that the specimen is bombarded from both sides at the same time. Argon is the gas most commonly used for the ion guns. A glow discharge is ignited when a high voltage is applied between anode and cathode (section 2.5.3), and a beam of particles is emitted from the hole in the cathode; the beam mostly comprises high-energy neutrals, and only a small fraction of ions [4-15]. At a fixed gun voltage, the particle flux is controlled by changing the flowrate of the gas.

The specimen is mounted in a holder designed to cover the edge of the disk. To create a symmetrical dimple under ion bombardment, the specimen holder is rotated about an axis normal to the specimen. As soon as a hole appears in the center of the specimen, the current of those ions passing through can be measured – light emitted from a light source can also be collected to the same effect. The thinning process can be automatically terminated in this way. A modern apparatus will feature an airlock which allows fast specimen exchange. This is particularly useful if a specimen has to be inspected at frequent intervals in a high-magnification optical microscope. Mechanical polishing and ion milling of semiconductor wafers

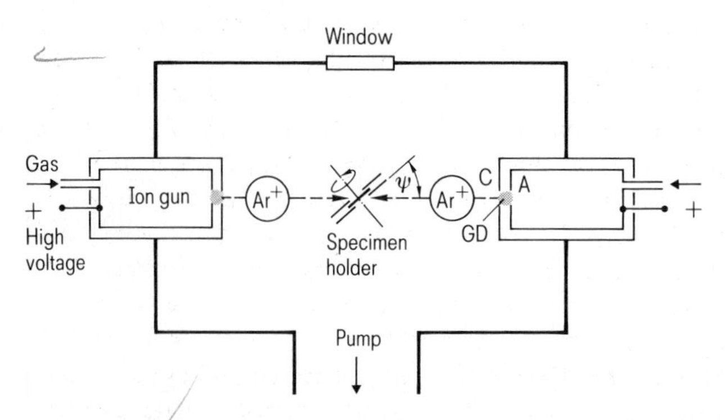

Fig. 4-17. Schematic drawing of ion milling apparatus. A beam of Ar ions is emitted from a glow discharge (GD) in the ion gun between anode (A) and cathode (C). The specimen is rotated during etching with incidence angle ψ.

from the back surface will provide plan-view specimens from the top region of the wafers. In addition, ion milling allows the preparation of cross-sectional specimens (next section).

Invariably, selection of the *thinning parameters* represents a compromise between the highest possible thinning rate and an acceptable damage level due to ion bombardment. The choice depends on the material to be thinned (see below). Since ion milling is more than one order of magnitude slower than chemical or electrochemical polishing, the aim is generally to achieve the highest possible thinning rates. The thinning or sputtering rate is directly proportional to the flux of particles incident on the specimen. In order to maximize this flux, the ion guns must be aligned so that the center of the beam strikes the center of the specimen. With increasing operation time of the ion guns, the particle flux decreases because the holes in the cathodes grow larger and the beam becomes less focused. The cathode plates must therefore be replaced after about hundred hours of operation. Given a constant flux of particles, the sputtering rate increases linearly with ion energy up to about 10 keV. However, since the ions are more sharply focused at high energies, the increase in sputtering rate with gun voltage becomes more than linear. At high energies, the ions penetrate the specimen more deeply, causing ion implantation and damage. Tolerable specimen damage determines the maximum energy to be selected. The relationship between sputtering rate and angle of incidence ψ (to surface) shows a maximum at $\psi = 20° - 25°$. At higher angles more ions are implanted into the specimen. In practice, smaller angles of incidence between 10° and 15° are frequently used along with specimen rotation in order to prevent surface structures forming during ion thinning. At angles of incidence as low as this, the depth of damage (see below) is less, and larger transparent areas are created. Momentum transfer and hence the sputtering rate are also dependent on the mass ratio between the gas atoms in the incident beam and the target atoms. Both variables are maximum when the masses are equal and decline as the difference between the masses increases. In practice, however, the outlay involved in using different gases is prohibitive, and virtually exclusive use is made of argon. In special cases, the residual gas in the ion milling unit may affect the sputtering rate. Thus, chemisorbed oxygen layers reduce the thinning rate of aluminium because the binding energy of the surface atoms is increased.

Specimen modifications due to ion thinning (section 2.6.3) are caused by specimen heating on the one hand and irradiation damage on the other. An increase in specimen temperature can be avoided by reducing the ion currents and ion energies and/or cooling the specimen with liquid nitrogen. Irradiation damage during ion milling of semiconductors results in the formation of a thin amorphous zone at the surface. The thickness of this zone increases with ion energy and angle of incidence. It is approximately 8 nm when silicon is thinned with argon ions at 5 keV energy and an angle of $\psi = 14°$, dropping to approximately 3 nm at an energy of 3 keV. This means that final thinning with low energies of between 2 and 3 keV is absolutely essential for the preparation of specimens in the 10 to 20 nm thickness range which is required for HREM inspection. Other artefacts may be created when compound semiconductors are thinned: if InP is milled by an argon-ion beam under standard conditions (5 keV, room temperature), small indium islands form on the surface. Alternatively, high densities of small crystal defects appear when II-VI semiconductors such as CdTe are thinned. Use of lower ion energies and currents as well as specimen cooling (with liquid nitrogen) will reduce both artefacts. These artefacts, however, disappear almost completely when iodine ions are used instead of argon for thinning [4-16]. In addition, this option will also considerably reduce the thickness of the amorphous surface layer – from approx. 3 nm for GaAs thinned by argon ions (3 keV) to less than 1 nm by iodine ions. In some ceramics too, defects are produced by ion bombardment. This means that high-temperature superconductors such as $YBa_2Cu_3O_7$ can be thinned satisfactorily only with low energies (3 keV) and specimen cooling.

4.3.5 Preparation of thin cross sections

In microtechnology, particular significance attaches to the preparation of cross-sectional specimens, these being most suitable for TEM characterization of layer structures. The vertical direction is displayed along with the lateral one and interfaces can thus be directly imaged. Furthermore, this technique also permits direct study of depth-related material effects such as interface reactions. In semiconductor technology, the possibility of cross-sectional specimen preparation has opened up a new and broad spectrum of applications for transmission electron microscopy [4-17].

The method commonly used in the preparation of cross-sectional specimens involves bonding two specimens face to face. The method described below proved very worthwhile as a routine technique in the authors' laboratory [4-18]. The individual steps are summarized in Fig. 4-18. To begin with, pieces measuring 3 mm × 3 mm are cut or cleaved from semiconductor wafers. Two of these pieces are then bonded together face to face with a two-component epoxy resin (Fig. 4-18a). The lines on the front of the specimen in Fig. 4-18 represent device structures which will be imaged in cross section. By clamping the two pieces together with the force of a spring the excess epoxy is squeezed out between the pieces. Clamping is maintained during curing of the epoxy at an elevated temperature of about 120 °C.

Since subsequent ion milling will remove the epoxy at a faster rate than the specimen material, it is advisable to keep the epoxy film as thin as possible (less than 1 μm). To this end, the pieces must be meticulously cleaned – by ultrasonics in ethanol, for example – in order to remove crystal fragments and other particles prior to bonding. Next, the sandwich-like specimen is attached to a mount with wax and cut into slices measuring approx. 150 μm in width by a high-speed diamond dicing saw of the type used to cut wafers into individual chips (Fig. 4-18b). By this means, a number of starting specimens are produced for the grinding and polishing processes which now follow. The slices are first polished on one side, then turned over on the lapping block with the other side upward and finally polished down to a thickness of approx. 20 μm (Fig. 4-18c). Silicon specimens can be polished down to less than 10 μm [4-19]. Final thinning is accomplished by ion milling (Fig. 4-18d). After ion milling – or beforehand in the case of extremely thin specimens – the specimens are glued to supporting rings to facilitate handling (Fig. 4-18e). Where the thin edge around the hole meets the bond, the device structures of interest can be imaged in cross section (Fig. 4-19). It is advisable to polish the specimens mechanically to as low a thickness as possible. This will shorten the ion milling time and also reduce the effect of differences in the sputtering rate during thinning of layer structures with different materials. Furthermore, the specimens will have larger transparent areas.

The desired information can be obtained from any specimen with a full-surface layer or periodic stripe pattern such as is shown in Fig. 4-18. Getting the sectional plane to pass through

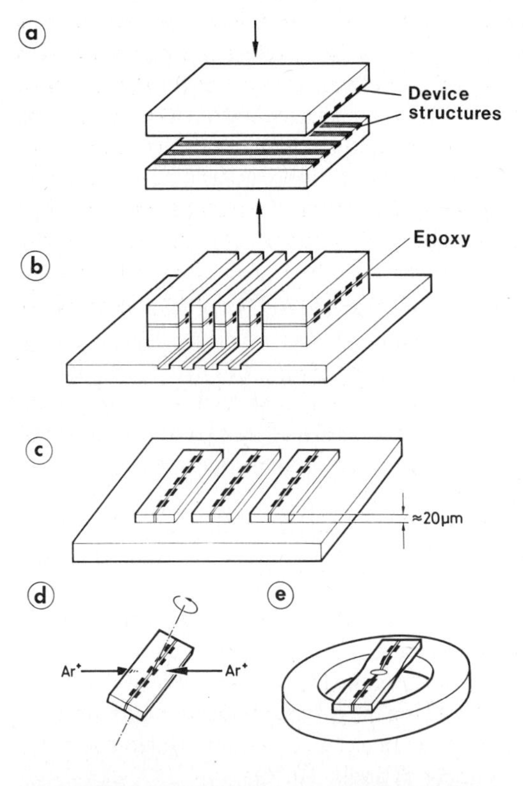

Fig. 4-18.
Preparation of cross-sectional specimens by the face-to-face method. a) Bonding together two pieces (3 mm × 3 mm) face to face; b) Cutting of slices (thickness of about 150 μm) with a dicing saw; c) Lapping and polishing the slices down to a thickness of about 20 μm; d) Argon ion thinning at a shallow incidence angle; e) The specimen is etched until a hole appears in the center, and is glued onto a supporting ring for easier handling. The thin edges of the hole are transparent to electrons (thickness < 0.5 μm).

Fig. 4-19.
TEM image showing both pieces of cross-sectional specimen. Only the bottom piece contains the device structures of interest.

a specified area causes considerable difficulty unless the area measures at least 20 μm × 20 μm. Smaller regions can only be targeted if they are arranged in a periodic array, as in the cell field of a memory chip, for example. All test devices also have periodic structures such as multi-transistors or fields of contact holes. To begin with, one has to make sure that the region of interest or test structure lies within the 150 μm slice cut off with the diamond saw. The dicing saw allows the cutting path to be positioned with an accuracy of 10 μm. To provide an orientation for positioning of the cutting path, only a *tiny* piece of a silicon wafer is bonded to the IC specimen [4-20]. The pattern in the surroundings of the test structure then provides the required orientation. Identification of the desired region is even easier when a thin piece of glass is glued to the IC specimen. If the sectional plane is to pass through a 20 μm area, a mechanical gauge must be used to check the amount of material removed from each face of the cross-sectional specimen during the grinding and polishing processes. Precise knowledge of the circuit layout is a prerequisite for targeting specific elements within a periodic structure. The position of the sectional plane within the period of the structure can then be determined on both sides of the cross-sectional specimen with the aid of a high-magnification optical microscope. When, during final thinning, the sectional plane passes through a desired element in the periodic structure, thinning is terminated on this side. On the other side of the cross-sectional specimen, thinning continues until one of the desired elements becomes transparent to electrons. Frequent checks in an optical microscope make this procedure time-consuming, but it allows the sectional plane to be set to an accuracy of better than 1 μm. This accuracy is required to target the trench capacitors of 4 Mbit memory chips, for example, elements which measure only about 1 μm × 1 μm (Chapter 9). In an alternative method, positioning of the sectional plane through a specific point of the device is achieved by polishing

a

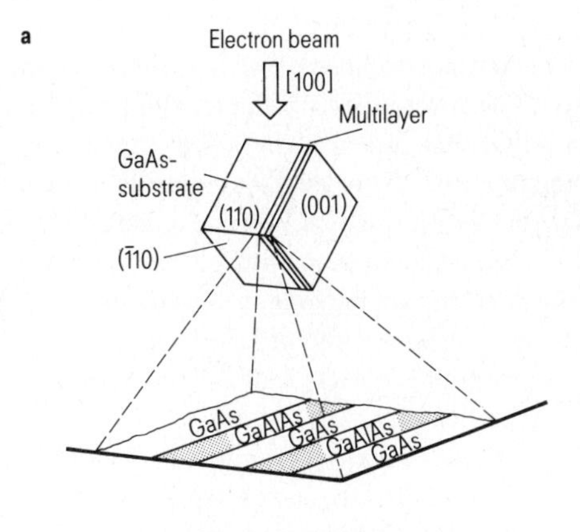

b

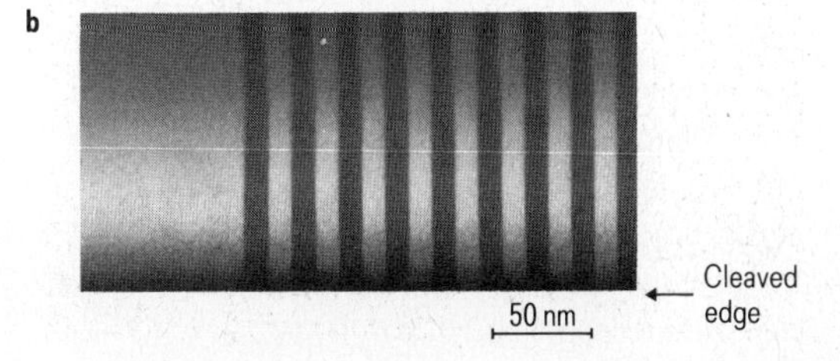

Fig. 4-20.
Cleavage method for preparation of wedge-shaped specimens from hetero-epitaxial layers of III-V semiconductors. a) Schematic illustration showing geometry of cleaved specimen and image formation, b) (200) dark-field image of cleaved edge from GaAs-GaAlAs multilayer on GaAs substrate. The GaAlAs layers appear bright (Section 9.3.2).

the first side of the cross-sectional specimen up to a few 0.1 μm from the point of interest [4-19]. When a thin piece of glass is glued to the IC specimen this polishing procedure can be controlled in an optical microscope in the same way as described in Fig. 3-8 for SEM cross sections.

The methods of preparing cross-sectional specimens described above are suitable for all types of layer structures on any substrate materials (Si, GaAs, glass, etc.) but require various preparation steps. In contrast to this, a very simple method for cross-sectional imaging of heteroepitaxial layers of III-V semiconductors was reported [4-21]. This method exploits the property of these semiconductors such as GaAs and InP to cleave precisely along {110} planes. A wedge-shaped specimen with a sharp 90° edge (Fig. 4-20a) can be produced by cleaving a {100}-oriented wafer along two {110} planes normal to each other. If the cleaved edge is viewed in transmission after orienting it normal to the electron beam, a cross-sectional image of the layer structure is obtained. By way of example, Fig. 4-20b shows a GaAs-GaAlAs multilayer structure.

4.3.6 Thin films and small particles

If the specimens of interest are thin films or small particles, they will be electron transparent without any further thinning. Specimens of this type need only be mounted on supporting grids and in some cases on supporting films as well, before they can be imaged in the TEM. To fit the specimen holders, a typical *supporting grid* will measure 3 mm in diameter; the size and shape of the mesh may vary. Usually made of copper, special-purpose grids are also available in platinum (chemically stable) or molybdenum (thermally stable). *Supporting films* must be thin and strong and must exhibit no structural features to ensure that the TEM image of the specimen is not impaired. Polymer films (formvar or collodion) or carbon films are generally used. The latter exhibit greater stability under electron bombardment. In some instances SiO_2 or Al_2O_3 films are also employed. Preparation of carbon films is as follows: a thin film of carbon (approx. 20 nm thick) is vapor-deposited on a glass substrate, as described in section 3.3. The film can be stripped from the glass by oblique immersion in water. The film floats on the surface and can be picked up on a grid. If a continuous supporting film would have a detrimental effect on imaging, it is advisable to prepare holey carbon films [4-22]. The diameters of the holes are of the order of 1 μm. If the film or a particle lies above a hole, it can be imaged without the background of the supporting film, which becomes particularly visible at high magnifications.

Thin films are generally fabricated by vapor deposition or sputter deposition. It is easiest to deposit the film directly onto a supporting film on a grid – a procedure which is, however, not always possible. Furthermore, differences in nucleation may result in different microstructures of films deposited in this way compared to films deposited on a bulk substrate. A film deposited on a bulk substrate must subsequently be removed from it. If the adhesive bond is not excessively strong, the film can be removed with the aid of an adhesive tape, which can be subsequently dissolved in a solvent. If the adhesion between film and substrate is strong, an alternative is to dissolve the substrate chemically. Films are frequently deposited on NaCl single crystals which are then dissolved in water. If the substrate is insoluble, the final option is conventional preparation with mechanical polishing and ion thinning from the substrate side of the specimen.

There are several ways of depositing *small particles* on a supporting film [4-23]. Ultrasonic agitation can be used to produce a suspension of a powder in a solvent (e.g. ethanol). If a drop of this suspension is placed on a supporting film, the particles are left on the film once the solvent evaporates. If the particles in question are precipitates in a metal, they can be isolated by an extraction replica. This technique is described in Fig. 4-21.

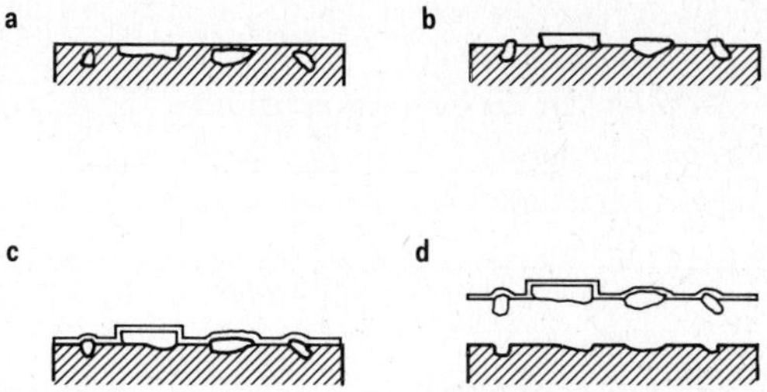

Fig. 4-21.
Extraction replica technique. a) Polished surface of two-phase metal (matrix and precipitates), b) Etching of the matrix exposes the precipitates, c) Carbon coating, d) After a second etching step, the replica is floated off.

4.4 Electron diffraction patterns

4.4.1 Introduction

The diffraction of particle beams by a crystal lattice was described in section 2.1 with the aid of the reciprocal lattice and the construction of the Ewald sphere (Fig. 2-5a): diffraction maxima are produced – or excited – when the Ewald sphere passes through a reciprocal lattice point. Since the electron wavelength is very small for the acceleration voltages used in transmission electron microscopy, the radius of the Ewald sphere $|\boldsymbol{k}| = 1/\lambda$ is very large and an electron diffraction pattern thus approximately represents a planar cross section through the reciprocal lattice (Fig. 2-5b). However, if the lattice is not primitive (has more than one atom in the unit cell), not all the points of the reciprocal lattice correspond to permitted reflections. The structure factor then determines which reflections are allowed and which not; this is described in greater detail in section 4.4.2.

Electron diffraction patterns are obtained in the TEM by imaging the back focal plane of the objective lens (Fig. 4-1b). All the electrons diffracted by the specimen in a given direction are focused at a specific point on this plane (Fig. 4-3). To obtain diffraction patterns from small, defined areas, either selected area diffraction (SAD) or microdiffraction must be used. In SAD, the area of the specimen from which the diffraction information originates is selected by means of an aperture in the first image plane; in microdiffraction, this area is determined by the probe diameter. The divergence angle $2\,\alpha_0$ of the probe is an additional parameter in microdiffraction. At small divergence angles, microdiffraction patterns are identical to SAD patterns. At larger angles, convergent-beam electron diffraction patterns are generated. The information contained in these patterns is explained in greater detail in section 4.4.4.

When relating the diffraction pattern to the image, one has to take into account that there is always a 180° rotation between diffraction pattern and image (Fig. 4-3); the direction in the diffraction pattern is the same as in the specimen plane but is inverted in the first image. Additional rotations caused by the different lens excitations in imaging and diffraction modes are avoided largely by the imaging lens system of modern microscopes comprising several intermediate lenses.

Different types of specimen (varying crystallinity, thickness, etc.) produce different types of electron diffraction pattern; amorphous substances produce diffuse rings and fine-grained specimens create ring patterns. Also, very thin single crystals produce spot patterns and thicker single crystals produce Kikuchi patterns. An analysis of various diffraction patterns is described in section 4.4.3.

By means of the diffraction patterns, the crystal system and the type of lattice of a material can be determined. Unknown materials can be identified in this way. In addition, the patterns yield crystallographic information such as the precise direction of the electron beam with respect to the crystal. This is of critical importance in interpreting images of crystal defects (section 4.5). If several different grains or phases are present, their orientation relationship can be determined.

In previous sections, the *geometrical theory of diffraction* was used to describe the relationship between crystal structure and diffraction patterns. It describes the direction in which electron waves are scattered by the crystal lattice with maximum intensity, thus giving the possible diffraction directions of beams from an infinitely-extended single crystal. As explained in section 2.1, and above, the geometrical theory leads to the construction of the reciprocal lattice but allows only a geometrical interpretation of diffraction patterns. More comprehensive theories, namely the kinematical theory and the dynamical theory, are needed to calculate the intensity diffracted at a real crystal of finite dimensions.

In the *kinematical theory* the elastic scattering contributions of the individual crystal atoms are simply summed (section 4.4.2). The assumption made is that only single scattering processes occur. The total diffracted intensity should also be small, so that the incident intensity is not changed significantly, and every atom of the crystal is irradiated with the same intensity. These assumptions are satisfied for very thin specimens of the kind often used in TEM. The kinematical theory explains the intensity of diffraction spots and can also describe the diffraction contrast of single crystals and crystal defects. For thicker crystals, however, the theory generally only provides qualitative results. To obtain a quantitative description in this case, the *dynamical theory* is required, which takes into account multiple scattering and thus explains the diffracted intensities from thicker crystals correctly (section 4.5.1).

4.4.2 Structure factor and shape of diffraction maxima

In the following treatment, the kinematical theory will be used to calculate the diffracted intensities. Since it describes elastic and coherent scattering, we may use the Huygens' principle for its mathematical formulation. This states that every scattering center (atom) can be regarded as the source of a secondary wavelet propagating in all directions with the same wavelength as the incident wave. The amplitude ϕ of the wave scattered in a given direction $\boldsymbol{k}$ with an angle Θ is calculated by superposing all the secondary wavelets scattered in that direction:

$$\phi = \sum_i f_i(\Theta) \cdot \exp\left[-\mathrm{i}\,\Delta\varphi_i(\Theta)\right] \tag{4-7}$$

where $f_i(\Theta)$ is the amplitude (cf. equation (4-1)) and $\Delta\varphi_i(\Theta)$ the phase difference, of the ith secondary wavelet in a direction $\boldsymbol{k}$ with respect to the wavelet scattered at the origin 0. The amplitude of the incident wave is assumed to be unity. The phase difference is based on the different paths a and b traversed by the scattered waves in the direction $\boldsymbol{k}$. As Fig. 4-22 shows, this phase difference is $\Delta\varphi = 2\pi(\boldsymbol{k} - \boldsymbol{k}_0)\cdot\boldsymbol{r}$, whereby eq. (4-7) becomes

$$\phi = \sum_i f_i(\Theta)\cdot\exp[-2\pi\mathrm{i}(\boldsymbol{k} - \boldsymbol{k}_0)\cdot\boldsymbol{r}]\,. \tag{4-8}$$

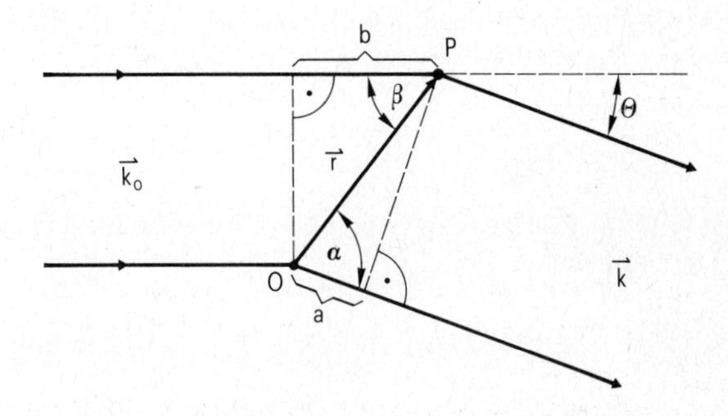

Fig. 4-22. The path difference Δl between the two waves emitted from the lattice points O and P in the scattering direction $\boldsymbol{k}$ is the difference between distances a and b: $\Delta l = b - a$ where $a = r\cdot\cos\alpha$ and $b = r\cdot\cos\beta$. Then, $\Delta l = r(\cos\beta - \cos\alpha)$ or in vector notation

$$\Delta l = r\left[\frac{\boldsymbol{k}\cdot\boldsymbol{r}}{|\boldsymbol{k}|\cdot|\boldsymbol{r}|} - \frac{\boldsymbol{k}_0\cdot\boldsymbol{r}}{|\boldsymbol{k}_0|\cdot|\boldsymbol{r}|}\right].$$

With the relations $\Delta l/\lambda = \Delta\varphi/2\pi$ and $|\boldsymbol{k}| = |\boldsymbol{k}_0| = 1/\lambda$ the phase difference is: $\Delta\varphi = 2\pi(\boldsymbol{k} - \boldsymbol{k}_0)\cdot\boldsymbol{r}$.

If the coordinates of the atoms are broken down into the coordinates $\boldsymbol{r}_l$ of the unit cells of the lattice and those of the atom positions $\boldsymbol{r}_i$ within the unit cell ($\boldsymbol{r} = \boldsymbol{r}_l + \boldsymbol{r}_i$), eq. (4-8) can be split into two factors

$$\phi = \underbrace{\sum_{i=1}^{n} f_i(\Theta)\cdot\exp[-2\pi\mathrm{i}(\boldsymbol{k} - \boldsymbol{k}_0)\cdot\boldsymbol{r}_i]}_{F}\cdot\underbrace{\sum_{l=1}^{N}\exp[-2\pi\mathrm{i}(\boldsymbol{k} - \boldsymbol{k}_0)\cdot\boldsymbol{r}_l]}_{G} \tag{4-9}$$

The first factor F is known as the *structure amplitude* and depends only on the atom positions $\boldsymbol{r}_i$ within the unit cell. The summation includes all n atoms of the unit cell. The second factor G is the *lattice amplitude* and is determined by the external shape of the crystal (see below). This summation includes all N unit cells of the crystal. The scattering amplitude ϕ is a maximum when the expression $(\boldsymbol{k} - \boldsymbol{k}_0)\cdot\boldsymbol{r}_l$ in factor G is an integer. Under these conditions, the exponential function $\exp(-2\pi\mathrm{i}n) = \cos 2\pi n - \mathrm{i}\cdot\sin 2\pi n$ has its greatest possible value, namely unity. This is the case when $\boldsymbol{k} - \boldsymbol{k}_0 = \boldsymbol{g}$, in other words, when the Bragg condition (eq. 2-4) is satisfied. This follows from equations (2-1) and (2-5), because the scalar product of a real lattice vector and a reciprocal lattice vector is always an integer. The structure amplitude for the reflection g is thus

$$F_g = \sum_{i=1}^{n} f_i(\vartheta) \cdot \exp(-2\pi i g \cdot r) . \tag{4-10}$$

The Bragg angle ϑ is used instead of the scattering angle Θ. If the structure amplitude is squared, the intensity is obtained: $|F_g|^2$ is known as the *structure factor*.

In a worked example, the structure amplitude will be calculated for a body-centered cubic lattice (Fig. 2-2b) in which the atoms in the unit cell are positioned at the coordinates (0, 0, 0) and (1/2, 1/2, 1/2) (cf. eqs. (2-1) and (2-4)):

$$F_{hkl} = f(\vartheta) \cdot [1 + \exp(-\pi i (h + k + l))]$$

With the relations $\exp(-\pi i n) = 1$ for even integers n and $\exp(-\pi i n) = -1$ for odd values of n, it follows that

$$F_{hkl} = 2f(\vartheta) \quad \text{if} \quad (h + k + l) \text{ is even and}$$

$$F_{hkl} = 0 \quad \text{if} \quad (h + k + l) \text{ is odd.}$$

Thus, only reflections for which $(h + k + l)$ is even are permitted. Analogous selection rules apply to other lattices. Thus, for a face-centered cubic lattice (Fig 2-2a), reflections are only permitted whose (hkl) are all even or all odd. $h + k + l = 4n$ applies as an additional condition for reflections of the diamond cubic lattice (Fig. 2-2c) when all (hkl) are even.

Strictly speaking, eq. (4-8) is valid only in the kinematical approximation, in other words when the intensity of the diffracted beam is significantly smaller than that of the direct beam: $I_g \ll I_0$. This is true only for extremely thin crystals. Electrons diffracted by thick crystals are reflected back into the direction of the direct beam (section 4.5). This modifies the intensities of the diffracted beams which then can no longer be calculated directly from the structure factors.

Double diffraction occurs when a diffracted beam is diffracted again by another set of lattice planes. The reflection g_3 produced in this way is the sum of the first two reflections: $g_3 = g_1 + g_2$. In this way, diffracted beams may occur whose structure factor is zero and which are thus forbidden according to kinematical theory. One example can be seen in the diffraction pattern of the [110] pole of silicon (Fig. 4-4a). The selection rules for the diamond lattice (see above) state that a (002) reflection is not allowed ($h + k + l \neq 4n$). Double diffraction (or "Umweganregung") via the $(1\bar{1}1)$ and $(\bar{1}11)$ reflections, however, generate a (002) reflection with strong intensity. Double diffraction also accounts for the presence of small satellite diffraction spots which surround the main spots of crystals containing microtwins or precipitates.

The diffraction pattern of a crystal contains more than just the reflections whose reciprocal lattice points lie on the plane intersecting the origin of the reciprocal lattice. The Ewald sphere also intersects the adjacent planes, and a diffraction maximum occurs where a reciprocal lattice point in one of these planes coincides with the Ewald sphere (Fig. 4-23a). If the diffraction pattern is collected over a large angle (with a short camera length) and/or the reciprocal lattice plane spacing normal to the incident beam direction is small, these reflections can be observed. When the incident beam is parallel to a low-index crystal direction, they are located on circles whose center is the direct beam. The circular zones containing such reflections are known as *higher-order Laue zones* (HOLZ). Besides the pattern of the zero Laue zone, Fig 4-23b also shows the ring of reflections of the first Laue zone. The radius of the first Laue zone R_L is given approximately by: $R_L = (2H/\lambda)^{1/2}$, where H is the spacing between the reciprocal lattice planes along the beam direction and $1/\lambda$ is the radius of the Ewald sphere.

Only in the case of infinitely extended single crystals would the intensity of the diffracted radiation meet at mathematical points. According to eq. (4-9), however, the diffraction maxima of finite crystals have an intensity distribution of varying width, with the *shape of the diffraction maxima* depending on the shape of the crystal. Hence, diffracted intensity is also observed if the Ewald sphere, instead of coinciding exactly with a point in the reciprocal lattice, merely passes in the vicinity of such a point. In order to use the Ewald construction also in this instance for determining the possible diffraction maxima, a vector *s* is introduced which

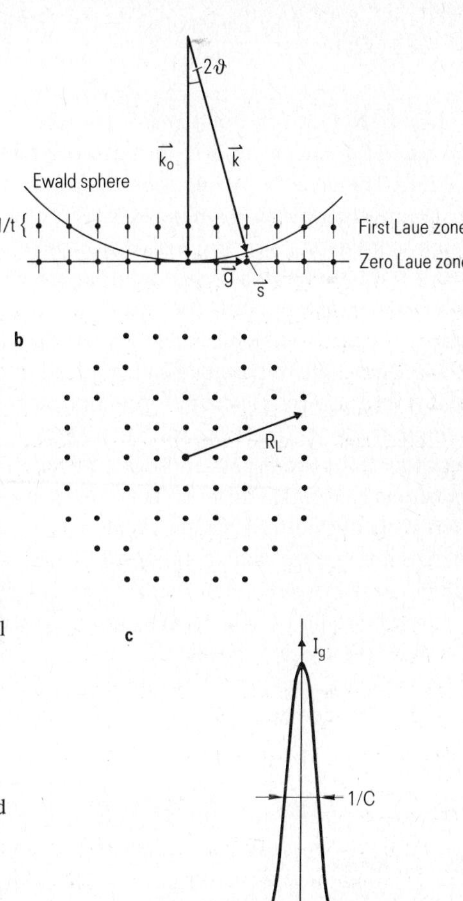

Fig. 4-23.
(a) Diffraction at a thin crystal. For thin crystals, the diffraction maxima are spikes extending normal to the foil plane. Reflections which do not lie in the plane containing the origin of the reciprocal lattice belong to higher-order Laue zones. Only the zero and first zone are drawn. The deviation from the Bragg orientation is described by the excitation error *s* which is a vector in reciprocal space. b) Schematic diffraction pattern showing zero-order and first-order Laue zones. Only those reflections appear whose spikes intersect the Ewald sphere in Fig. (a). The central part corresponds to the zero-order zone. Reflections of the first-order Laue zone form a ring with radius R_L. c) The intensity I_g of a diffracted beam from a thin crystal of thickness $t = C$ varies with the excitation error (or deviation parameter) s_z according to the third term of equation (4-12).

defines the distance between the reciprocal lattice point and the Ewald sphere. $\boldsymbol{s}$ thus connects the reciprocal lattice vector $\boldsymbol{g}$ to the Ewald sphere (Fig. 4-23a). Since $\boldsymbol{s}$ describes the deviation from the Bragg orientation, the Bragg condition is then: $\boldsymbol{k} - \boldsymbol{k}_0 = \boldsymbol{g} + \boldsymbol{s}$. If we substitute this in the expression for G, we obtain $(\boldsymbol{g} + \boldsymbol{s}) \cdot \boldsymbol{r}_l = \boldsymbol{g} \cdot \boldsymbol{r}_l + \boldsymbol{s} \cdot \boldsymbol{r}_i$ in the exponent. It was already shown that $\boldsymbol{g} \cdot \boldsymbol{r}_l$ is an integer. If the sum of equation (4-9) is replaced by an integral, the following relation is obtained for a crystal having the shape of a parallelepiped of edge lengths A, B and C:

$$G = \frac{1}{V_c} \int_0^A \int_0^B \int_0^C \exp\left[-2\pi \mathrm{i}\,(s_x x + s_y y + s_z z)\right] \mathrm{d}x\,\mathrm{d}y\,\mathrm{d}z \qquad (4\text{-}11)$$

V_c is the volume of the unit cell. When eq. (4-11) is integrated and squared, we obtain $|G|^2$, which describes the intensity distribution in reciprocal space:

$$I_g = \frac{1}{V_c^2} \cdot \frac{\sin^2(\pi A s_x)}{(\pi s_x)^2} \cdot \frac{\sin^2(\pi B s_y)}{(\pi s_y)^2} \cdot \frac{\sin^2(\pi C s_z)}{(\pi s_z)^2}\,. \qquad (4\text{-}12)$$

Fig. 4-23c shows the relationship between I_g and s_z, where $s_x = s_y$ were set zero. The greatest part of the intensity is in the central maximum which has a full width at half maximum of $1/C$. The intensity of the subsidiary maxima decreases rapidly as s_z increases.

The intensity distribution in reciprocal space varies with the shape of the crystal. Let us begin by considering a crystal with the shape of a thin foil of thickness $t = C$. This is the usual shape of the specimens used in transmission electron microscopy. Since the dimensions A and B are extremely large, only the third term in eq. (4-12) is of significance. The intensity distribution in the reciprocal space therefore has the shape of a spike (or rod) lying normal to the foil plane. The length of the spike is determined by the width of the central maximum in Fig. 4-23c and is inversely proportional to the thickness t of the foil. If the incident beam direction is normal to the foil plane as in Fig. 4-23a, the spikes are parallel to the beam direction. The resulting diffraction spots are sharp, because the spikes penetrate the Ewald sphere in an approximately normal direction. The occurrence of spikes accounts for the fact that the diffraction spots of thin foils remain visible even at considerable deviation from the Bragg orientation. Since s_z is the main component of $\boldsymbol{s}$, $s_z \approx s$ can be set where s is also referred to as the deviation parameter.

If the crystal is of a different shape, the intensity distribution in reciprocal space is also different. If the crystal is needle-shaped (e.g. for precipitates), the distribution will take the form of disks lying normally to the needles. In general, the intensity distribution always extends along a direction parallel to the smallest dimension of the crystal.

4.4.3 Analysis of diffraction patterns

Ring patterns

When electrons are diffracted by polycrystalline samples consisting of very fine grains, ring patterns are generated (Fig. 4-4b). In order to obtain sharply focused diffraction rings, the divergence of the incident beam must be reduced by overfocusing condenser 2 (section 4.2.2).

The same applics to spot patterns. Broadening of the rings despite reduction of beam divergence is indicative of an extremely small grain size (<5 nm). The reason for this effect is that as crystal size decreases, the diffraction maxima become less clearly defined (the width at half maximum shown in Fig. 4-23c increases for all three dimensions). At a grain size of approx. 2 nm, the diffraction rings are so broad that they resemble the diffuse rings from amorphous materials. With the aid of lattice fringes, nanocrystallites measuring less than 2 nm in size can be identified in high-resolution images. The position and intensity distribution of the diffuse rings produced by amorphous materials depend on the arrangement of the nearest neighbor atoms.

Ring patterns are evaluated by measuring the radii R of the individual diffraction rings. These radii are related to the associated interplanar spacings d_{hkl} as follows (Fig. 2-38):

$$d_{hkl} = \lambda \cdot L/R \tag{4-13}$$

The product of λ and the camera length L can be calibrated with the aid of the diffraction pattern produced by a known material. If the radii R on the diffraction pattern are measured, the values of d_{hkl} can be determined using eq. (4-13) and then used to identify unknown materials. The powder-diffraction files [4-24] compiled for the evaluation of X-ray diffractograms from powder specimens are helpful in this context. These files contain the d values and the relative X-ray intensities of the reflections for all known crystalline materials, together with the associated crystallographic data. In the accompanying search manuals, the substances are classified by the d values of the three most intense X-ray lines. The relative intensities of the diffraction rings observed in electron diffraction do not agree with those of the X-ray lines because, unlike the case of X-ray diffraction, the former diminish rapidly as the Bragg angle increases (cf. Fig. 4-2), but comparisons can still be made. The error of roughly 1% in the evaluation of electron diffraction patterns is approximately one order of magnitude higher than that for X-ray diffractograms. Hence, difficulties may be encountered in unambiguously identifying rarely occurring materials. In such a case, it is useful to begin by identifying the elements contained in the material. Compared with X-ray diffraction, however, electron diffraction allows much smaller volumes to be analysed.

The diffraction rings change in intensity if the crystallites in the specimen exhibit a preferred orientation (texture). In thin films produced by vapor-deposition or sputtering, this preferred orientation is frequently normal to the film plane and is rendered visible in the diffraction pattern only by the discrepancy between the intensity of some rings (or absence of rings) when compared with the ring pattern from a specimen with no texture. The preferred orientation can be detected directly by tilting the layer at a steep angle or using reflection high-energy electron diffraction (section 4.4.5). In such diffraction patterns the preferred orientation shows up as a nonuniform intensity distribution along the diffraction rings.

Spot patterns

When an electron beam is transmitted through a thin single crystal parallel to a given crystal direction, a spot diffraction pattern is observed (Fig. 4-4a). The patterns of low-index crystal directions (zone axes) contain numerous spots (with short **g** vectors) and exhibit a high degree of symmetry. The spot in the center of the pattern (and with the greatest intensity) corresponds to the directly transmitted beam. The other spots arise by diffraction at lattice planes lying approximately parallel to the electron beam. There is no difficulty in indexing diffraction patterns of a known material. With the aid of eq. (4-13), it is then a simple matter to determine the type of the individual (hkl) reflections on the basis of the distance between them and the direct beam. Three reflections with the shortest possible **g** vectors forming a parallelogram will suffice. In assigning specific indices to the reflections, care must be taken to ensure that the angles between the **g** vectors correspond to those in the diffraction pattern. The choice of specific indices is always arbitrary, but must be self-consistent. This means that vector addition of two reflections must give the correct indices for the third (cf. Fig. 4-4a). The beam direction is finally obtained from the cross product of two reflections whose **g** vectors are non-parallel.

It is particularly easy to index diffraction patterns from cubic materials because the cubic symmetry is maintained in reciprocal space and the axes of the real and the reciprocal lattice are parallel. As the lattice symmetry of non-cubic materials decreases, indexing becomes more and more difficult, and it is common practice to make use of tables of interplanar spacings and angles. Detailed instructions can be found e.g. in [4-25, 26]. The standard spot patterns for the most prominent zone axes have been summarised in [4-25, 26, 9], among others, for the most important lattice types.

If a specimen contains an unknown material, diffraction patterns must be recorded at different orientations. The crystal has to be tilted through large angles so that various low-index crystal directions can be examined. Kikuchi patterns (see below) are of assistance when a crystal of sufficient thickness is to be tilted. It is difficult to tilt the specimen through a large angle if the crystallite to be examined is small, because the goniometers are not fully eucentric and the region of interest is therefore displaced during the tilting process. One remedy is to defocus the diffraction image, so that the diffraction spots appear as small disks in which image contrast is visible. With their aid, the specimen can be moved to compensate the displacement. The tilt angle between the poles can be read from the goniometer or derived from the diffraction patterns of adjacent crystallites or the matrix, where an orientation relationship exists.

If eq. (4-13) is used to determine the d_{hkl} values for the various reflections, an attempt can be made to identify the material from the powder diffraction files in the same way as for a ring pattern. The chances of success are especially promising if additional information is available (e.g. the chemical elements concerned) which greatly restricts the number of possible compounds. In the absence of such information, the crystal system can be determined from the diffraction patterns. In this process, the symmetry properties of the diffraction patterns are evaluated (e.g. fourfold or sixfold axes). By using a trial and error approach, a start is made by indexing the pattern with the highest level of symmetry, then continuing through the other patterns until a consistent solution is found.

The presence of an ordered structure in an alloy results in additional diffracted intensities between the lattice reflections. In the case of a superlattice, i.e. long-range order, sharp superlattice reflections are produced. The degree of order can be determined from their intensity. In contrast, short-range order produces regions of diffusely scattered intensity. If the crystal matrix contains precipitates with a particular orientation relationship, the diffraction pattern is a superposition of the patterns of the matrix and precipitate. Additional reflections may be produced by double diffraction (a beam diffracted by the matrix is diffracted a second time by a precipitate).

Kikuchi patterns

Section 2.4 describes how Kikuchi lines are created (Fig. 2-39). In thick crystals, the Bragg diffraction of inelastically scattered electrons produces a pair of Kikuchi lines consisting of one bright and one dark line, with the dark line always lying closer to the direct beam. The lines are normal to the $\boldsymbol{g}$ vector of the lattice planes and their spacing is equal to $|\boldsymbol{g}|$, i.e. the distance between the reflection and the direct beam. From these properties, it follows that Kikuchi patterns can be indexed in a manner exactly analogous to that for spot patterns. By way of example, Fig. 4-24a shows the diffraction pattern of a silicon crystal with the incident electron beam exactly along the [001] direction. If the specimen were even thicker, the intensity of the diffraction spots would diminish further, leaving only the Kikuchi pattern visible. Fig. 4-24b is a schematic illustration of the Kikuchi-line pattern in which the associated reflection ($+\boldsymbol{g}$ or $-\boldsymbol{g}$) has been assigned to each line.

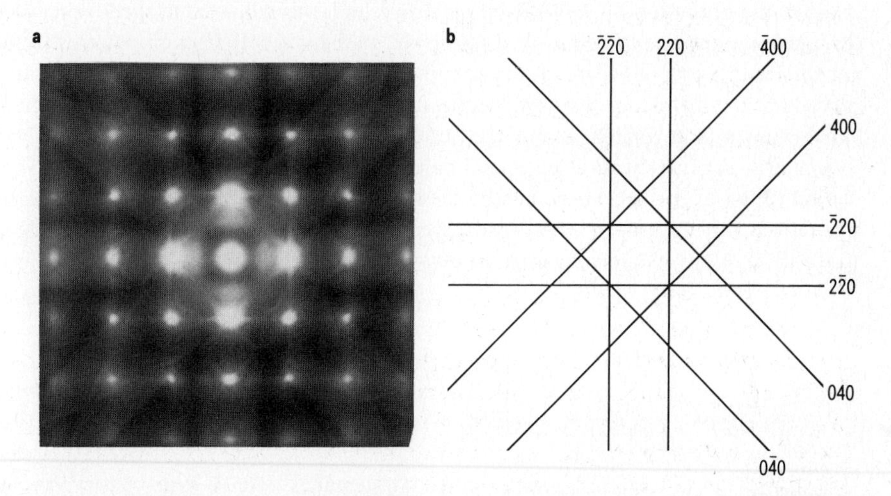

Fig. 4-24. (a) Selected area diffraction pattern of silicon crystal viewed exactly along the [001] orientation. Both diffraction spots and Kikuchi lines show up. (b) Schematic drawing of Kikuchi pattern with indexing of lines. Only lines corresponding to low-index lattice planes are shown.

The explanations in section 2.4 deal with the position of the Kikuchi lines, but ignore their intensity. Rather than a pair consisting of one bright and one dark line, low-index lattice planes produce a Kikuchi band within which the scattering background in the diffraction pattern is reduced. The explanation requires results from the dynamical theory of diffraction and is analogous to that given in section 3.5 for the formation of channelling patterns (Figs. 3-22, 23). Enhanced absorption of electron waves within the bands reduces TEM transmission (cf. Fig. 4-36) or increases backscattering in the SEM respectively. This explains why the Kikuchi bands are dark in the TEM whereas bright bands are observed in the SEM (Fig. 3-21).

For practical purposes, the special value of Kikuchi lines is that they can be used to determine and even adjust the exact crystal orientation. In the following, we assume only one set of diffracting lattice planes, i.e. that there is only one strong reflection. If the electron beam is incident exactly parallel to the diffracting lattice planes, the two Kikuchi lines lie symmetrically on each side of the direct beam (as seen in Fig. 4-24 for each set of lattice planes). If the crystal lies in an exact Bragg orientation, the excitation error (or deviation parameter) s is zero. In this case the dark Kikuchi line passes through the direct beam and the bright one through the reflection (Fig. 4-25a, c). If the crystal is tilted such that the angle of incidence on the lattice planes becomes larger than the Bragg angle, s becomes greater than zero and the Kikuchi lines move away from the direct beam. This situation is illustrated in Figs. 4-25b and d for an additional tilt through an angle ε. Electrons inelastically scattered in other directions then satisfy the Bragg condition and produce Kikuchi lines. The whole Kikuchi pattern moves as the crystal is tilted as if the two were rigidly connected. In contrast, the position of the diffraction spots changes only slightly during tilting, but if their spikes no longer intersect the Ewald sphere, their intensity tends to zero. Since the Kikuchi patterns seem to be directly coupled to the crystal, the incident beam direction can be determined more accurately from them than from the spot patterns. Furthermore, the excitation error s can be obtained simply from the distance of the bright Kikuchi line from its associated diffraction spot (Fig. 4-25b). Another important use of the Kikuchi patterns is that they allow controlled tilting of crystals through large angles. This controlled tilting is required when diffraction patterns from different zone axes are to be recorded, for example. As the means of orientation during the tilting procedure, the diffraction pattern is observed and compared with schematic Kikuchi maps which show a large angular range (e.g. $\pm 30°$) [4-26, 9].

Similarly, the investigation of crystal defects by diffraction contrast (section 4.5) requires careful tilting of the specimen. Firstly, the specimen must be tilted in such a way that only one reflection is excited, in other words, there is only one strong diffracted beam in addition to the direct beam (two-beam case). To obtain the desired reflection, the specimen is initially

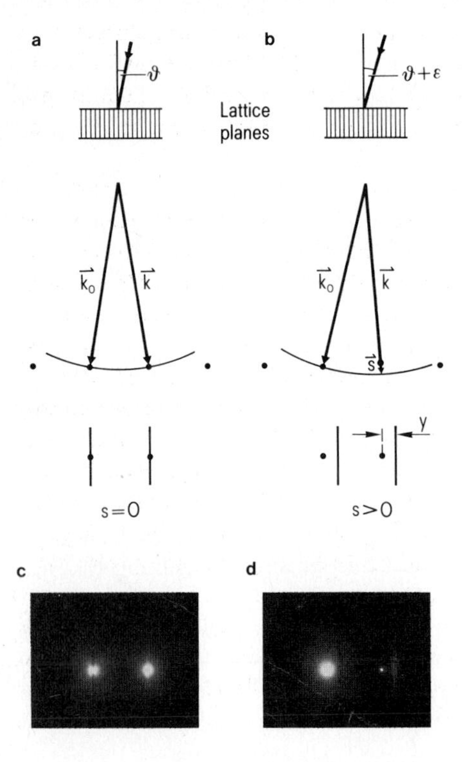

Fig. 4-25.
Relation between Kikuchi lines and diffraction spots. a) For a crystal in exact Bragg orientation ($s = 0$) the Kikuchi lines pass through the spots of the direct and diffracted beams; b) For an additional tilt ε ($s > 0$) the Kikuchi lines are shifted. The shift y is proportional to s: $s = y/(dL)$ (d lattice spacing, L camera length). Since s is a vector in reciprocal space, it has the dimension of reciprocal length. c), d) Diffraction patterns for $s = 0$ (c) and $s > 0$ (d).

tilted onto a pole which contains the desired reflection, using a Kikuchi map for reference. The next step is to tilt the specimen along the Kikuchi lines of the reflection away from this pole until only the desired reflection appears bright and all others disappear or become weak. Secondly, the deviation from the Bragg orientation must be set to some value, generally $s > 0$, because the image contrast depends sensitively on the deviation parameter s. As indicated above, for $s > 0$ the bright Kikuchi line lies beyond the associated diffraction spot and the distance between the two is proportional to s.

4.4.4 Electron microdiffraction

Diffraction modes

In *selected area diffraction*, the area of the specimen is selected by an aperture located in the image plane of the objective lens. The minimum selectable area is approx. 0.5 μm in size. With an objective-lens magnification of $M = 20:1$, this requires a selector aperture measuring 10 μm in diameter. The restriction in the size of the specimen area is due to the spherical aberration of the objective lens. This aberration causes the specimen area shown in the dark-field images formed by the Bragg reflections to be shifted with respect to the area shown in the bright-field image. This shift increases as ϑ^3, so that reflections with a large Bragg angle ϑ do not originate from the area selected in the bright-field image.

In order to obtain diffraction patterns from areas smaller than those possible with SAD, a fine electron probe is used in *electron microdiffraction*. If the probe is stationary with respect to the specimen, the specimen area from which the diffraction pattern originates is determined mainly by the probe diameter. The divergence angle of probes with small diameters (of the order of 10 nm and less) is large. It was explained in section 4.2.2 how condenser objective pole pieces can be used to produce a fine probe and still allow conventional TEM imaging (Fig. 4-10). If an additional condenser lens is employed, the angle of divergence can be varied continuously within certain limits.

Fig. 4-26 shows the types of diffraction patterns obtained as a function of the divergence angle of the electron probe. If the divergence is significantly smaller than the Bragg angle, *spot patterns* are formed (Fig. 4-26a) in which the spots are small diffraction disks. To obtain small disks, an extremely small second condenser aperture must be used (e.g. 20 μm). This considerably reduces the intensity both of the image and the diffraction pattern, which in turn makes practical work difficult. However, spot patterns are more suitable than convergent-beam patterns (see below) for determining the separation of, and angles between, the reflections. If beam divergence is increased without overlapping of the diffraction disks, *convergent-beam electron diffraction (CBED) patterns* are obtained (Fig. 4-26b). The individual disks display the intensity distribution as a function of the deviation parameter s. The information contained in these disks is described in greater detail below. If the angle of divergence is significantly larger than the Bragg angle, a *Kossel-Möllenstedt pattern* is produced (Fig. 4-26c). This pattern may be regarded as the superposition of several, extremely large diffraction disks. Kossel-Möllenstedt patterns are similar to Kikuchi patterns in appearance. Whereas in Kikuchi patterns the angular divergence of the electrons which are subsequently diffracted derives from inelastic scattering in the specimen itself, in a Kossel-Möllenstedt pattern the divergence is defined by the divergence of the incident probe. Kossel-Möllenstedt patterns are not much used in TEM for analytical purposes.

Besides the modes using a stationary probe as shown in Fig. 4-26, microdiffraction patterns can also be produced by rocking the beam about a given point on the specimen (analogous to the formation of electron channelling patterns from bulk specimens, section 3.5 – see [4-27], for example). A more detailed description of these methods will not be given here.

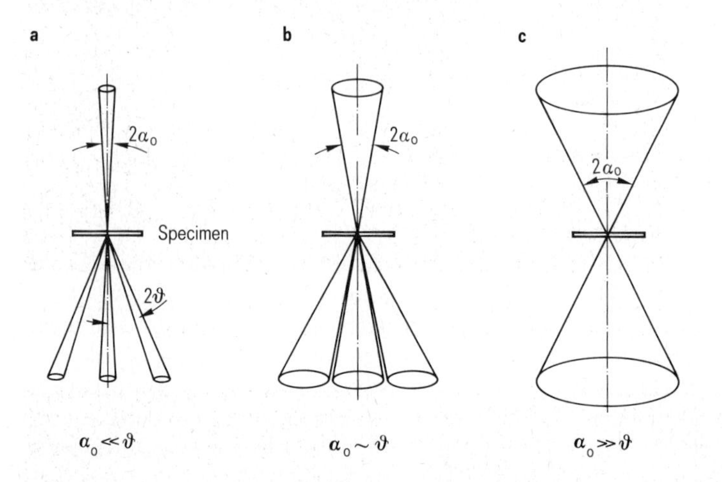

Fig. 4-26. Modes of electron microdiffraction with a stationary probe. Depending on the ratio of the divergence angle α_0 of the probe to the Bragg angle ϑ, spot patterns (a), convergent-beam patterns (b), or Kossel-Möllenstedt patterns (c) are formed.

Convergent-beam electron diffraction (CBED)

Every point within a diffraction disk corresponds to a particular direction of incidence from the illumination cone. The intensity distribution within the disks of the direct beam and the diffracted beams therefore reflects the change in intensity with deviation parameter s (i.e. the rocking curve). The dependence of the diffracted intensity I_g on s was calculated from the lattice amplitude G in eq. (4-12) and is shown in Fig. 4-23c. Fig. 4-27 shows a CBED pattern in which only one reflection is strongly excited. The fringes in the disk of the diffracted beam correspond to the subsidiary maxima in Fig. 4-23c. The spacing of the minima is inversely proportional to the thickness of the specimen. CBED patterns of this type are thus particularly suitable for determining the local specimen thickness, a procedure often required in X-ray microanalysis, for example. Quantitative evaluation of the fringe spacing, however, requires the use of dynamical diffraction theory (see section 4.5.1) instead of eq. (4-12) or (4-15) derived from kinematical theory [4-28]. In contrast to Fig. 4-23c, the disk intensity profile is also correctly reproduced by the dynamical theory (Fig. 4-36).

If a diffraction pattern is imaged with a large collection angle (i.e. electrons leaving the specimen in a large angular range are recorded) along a low-index zone axis, reflections of the higher-order Laue zones (HOLZ) become visible. Instead of disks, the HOLZ reflections in CBED patterns have the shape of lines, because the angular width of reflections with a large Bragg angle becomes small. For every bright line in a HOLZ ring, a dark line appears in the disk of the direct beam. Fig. 4-28a shows a CBED pattern of GaAs in the [111] pole with the first Laue zone also visible. The central disk containing the dark HOLZ lines is shown at a

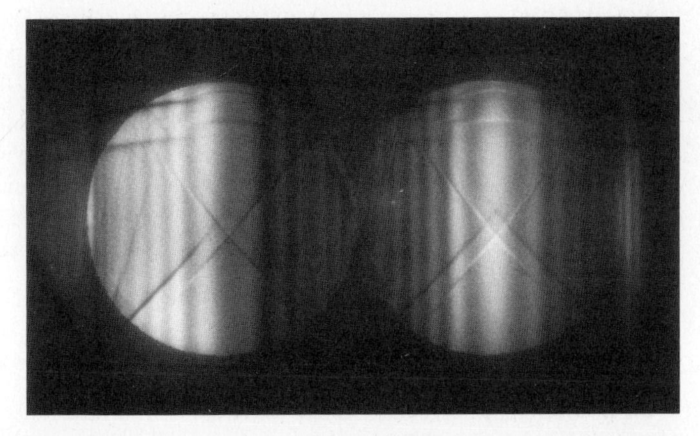

Fig. 4-27. Convergent-beam electron diffraction pattern with one strong diffracted beam (right). The local specimen thickness can be derived from the fringe spacing in the diffracted beam.

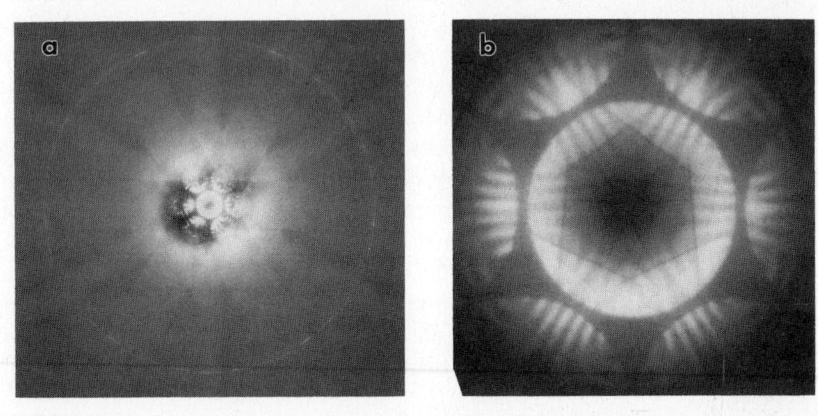

Fig. 4-28. Convergent-beam electron diffraction patterns of GaAs in [111] orientation. a) Large-angle pattern showing first-order Laue zone, b) Central disk with dark HOLZ lines.

higher magnification (larger camera length) in Fig. 4-28b. The HOLZ lines are equivalent to the bright and dark Kikuchi lines in the same way as already described for Kossel-Möllenstedt patterns. In contrast to the Kikuchi lines, however, HOLZ lines also appear in thinner specimens and are very sharp. The dark lines in the central disk can be indexed with the aid of the corresponding bright lines in the HOLZ rings. As the result of dynamic diffraction effects, pairs of dark lines may be produced and interactions may occur at the intersections. Since the HOLZ lines in the central disk are very sharp and correspond to large Bragg angles, their positions and hence their patterns are sensitive to small changes in interplanar spacings. Thus, changes in lattice spacings $\Delta d/d$ can be measured to an accuracy of a few times 10^{-4}. Moreover, the HOLZ lines contain three-dimensional information about the crystal structure. For example, multibeam effects which influence the intensity of particular HOLZ lines can be used to determine the polarity of non-centrosymmetric crystals [4-29] (e.g. in InP).

The CBED patterns of low-index zone axes depend strongly on the thickness of the specimen and its orientation. To avoid the information in the disks being averaged over areas of different thickness or orientation, a probe of small diameter has to be used. In addition, it is important to avoid the growth of contamination spots (Section 4.7.3). If CBED patterns are recorded precisely along the zone axis, the symmetry of the pattern reflects that of the crystal in each individual projection – thus providing detailed information on crystal structure. The point group of the crystal can be determined from the symmetry elements of CBED patterns imaged along different zone axes [4-30]. Furthermore, symmetry properties of forbidden reflections provide information concerning the space group [4-31]. Surveys on CBED are to be found in [4-32, 33, 27, 9]; numerous examples have been compiled in [4-34, 35].

4.4.5 Reflection high-energy electron diffraction

If the electron beam strikes a specimen with a smooth surface under grazing incidence, diffraction patterns can be recorded in a reflection geometry. Hence, this method is known as reflection high-energy electron diffraction or RHEED (in contrast to low-energy electron diffraction (LEED) which uses energies of less than several 100 eV). It supplies information on crystallinity and the crystal structure of a zone close to the surface of the specimen and has the major advantage of obviating any specimen preparation for thinning. Since the Bragg angles are of the order of 1°, the incidence angle of the electron beam must also be of this order.

For RHEED patterns to be recorded, a small specimen can be mounted in the TEM specimen holder in such a way that the incident electron beam forms a very low angle with the specimen surface. Because of this grazing incidence, areas of the specimen in widely varying z-positions contribute to the diffraction pattern. Well-focused diffraction patterns are obtained only when an extremely small specimen area is selected with the selector aperture. It is easier to use a special attachment to produce the diffraction pattern without the aid of a lens. This attachment is inserted in the TEM in place of the projector lens and the lenses above it serve only to produce a parallel electron beam. Using the first technique reflection electron microscopy is also possible, whereas the RHEED attachment permits only diffraction patterns to be recorded.

The patterns obtained may be diffuse diffraction rings, ring patterns or spot patterns with Kikuchi lines, depending on the crystallinity of the specimen. Fig. 4-29a shows a RHEED pattern from a polysilicon layer. Half of the diffraction pattern and the direct beam lie in the shadow of the specimen. Because of a well-developed preferred orientation (texture) of the grains, the intensities are not uniformly distributed along the diffraction rings. Only the (220) ring exhibits a strong intensity (arrowed in Fig. 4-29a) at the intersection with the surface normal, which runs from bottom to top in the center of the image. This is due to a preferred orientation in the ⟨110⟩ direction. A RHEED pattern of a planar single crystal surface will mainly exhibit a Kikuchi pattern (Fig. 4-29b). The diffraction spots are elongated to form streaks, because the diffraction information stems only from a region close to the surface, to a depth of a few nm. In contrast to diffraction in transmission (Fig. 4-23a), the spikes normal to the specimen surface are in the plane of the diffraction pattern and therefore appear as streaks. The penetration depth is significantly greater in a specimen with a rough surface because the peaks of the specimen topography are transmitted.

RHEED investigations are suitable for applications which require a rapid evaluation of the crystallinity of layers. In addition, fine-grained surface layers can be identified if the ring diagram is successfully indexed. Since the interaction volume is a shallow zone with a depth of

<10 nm, RHEED represents a sensitive method of surface analysis. If RHEED is implemented in a TEM, a beam diffracted in reflection can be used to produce a dark-field image. This method of reflection electron microscopy (REM) also allows atomic surface structures to be investigated [4-36].

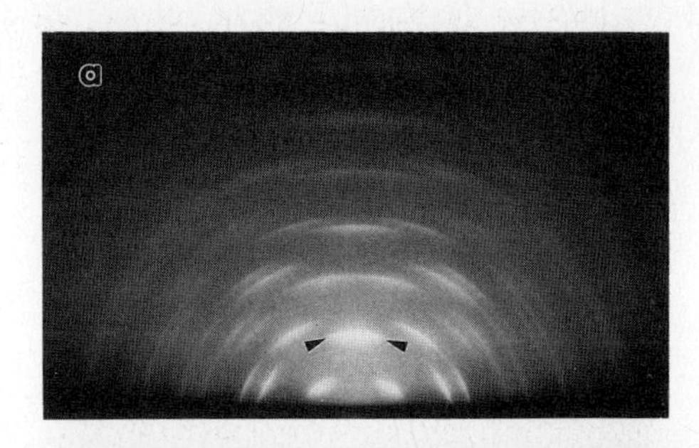

Fig. 4-29.
RHEED patterns from polysilicon film with ⟨110⟩ texture (a) and single-crystal silicon surface (b).

4.5 Diffraction contrast from perfect crystals and crystal defects

4.5.1 Diffraction contrast from perfect crystals

It was explained in section 4.1 that in a bright-field image the contrast arises from the intensity distribution of the directly transmitted electrons at the lower specimen surface. In the case of crystalline material, the transmitted intensity I_0 depends on how strongly electrons are diffracted by the crystal lattice. To obtain specific diffraction contrast, the crystal must be tilted such that electrons are essentially diffracted only at a single set of lattice planes. In addition to the direct beam, only one strongly diffracted beam is then present, i.e. only a single reflection is strongly excited (two-beam case). In reality, since diffraction maxima have the shape of spikes (Fig. 4-23a), several reflections are frequently excited in addition to $\boldsymbol{g}$ (reflections of higher order (systematic reflections) $2\boldsymbol{g}$, $3\boldsymbol{g}$ etc.). In the following section, however, we assume that only two beams are invariably present: the direct and the diffracted beam *(two-beam approximation).* If the incident beam has unit intensity, then $I_0 = 1 - I_g$ applies for the intensity of the direct beam. The diffracted intensity I_g depends on the exact diffraction conditions (section 4.4.2), i.e. both on the operating reflection $\boldsymbol{g}$ and the deviation from the Bragg orientation (parameter s). Both can be determined or adjusted in addition to the beam direction with the aid of the diffraction pattern (section 4.4.3): observation of the Kikuchi patterns allows the specimen to be tilted in a defined way to excite a particular reflection, and the spacing between the bright Kikuchi line and the associated diffraction spot is proportional to s (Fig. 4-25b).

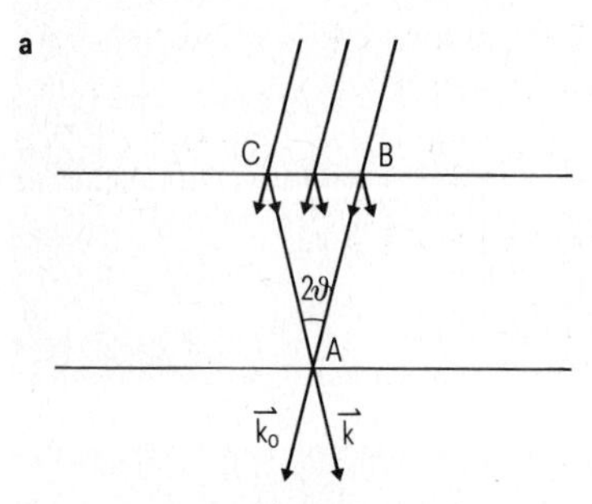

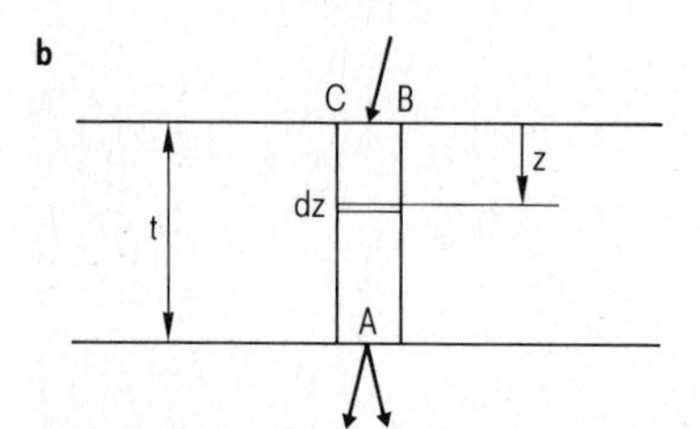

Fig. 4-30.
Column approximation: a) In the two-beam case, only electron waves within the narrow triangle ABC formed by the direct beam $\boldsymbol{k}_0$ and the diffracted beam $\boldsymbol{k}$ contribute to the image intensity at point A of the exit surface. For a specimen thickness of $t = 100$ nm the distance BC is only 1 to 2 nm. b) To calculate the intensities of the direct and diffracted beams, the triangle ABC can therefore be replaced by a column along the z-direction.

In order to calculate the image contrast, we consider the intensities of the direct and diffracted beams at a point on the lower specimen surface (point A in Fig. 4-30a). Only the undiffracted and diffracted electron waves which interact with the specimen within the triangle ABC contribute to these intensities. Due to the small Bragg angle ϑ the acute triangle ABC can be replaced by a narrow strip (a column in three-dimensional terms) (*column approximation,* Fig. 4-30b). To a good approximation, the intensities of adjacent columns are independent of each other. If we use the *kinematical* diffraction theory, the diffracted amplitude ϕ_g at the lower specimen surface is obtained by summing or integrating the scattering contributions of each thickness element dz along the column over the specimen thickness t:

$$\phi_g \sim \int_0^t \exp(-2\pi i s z)\, dz\,. \tag{4-14}$$

This also follows from eq. (4-11) when only the integration over the layer thickness $C = t$ is considered. Since in a perfect crystal $s_Z = s$ is constant along a column, we obtain the following expression for the intensity of the diffracted beam in analogy with eq. (4-12):

$$I_g \sim \frac{\sin^2(\pi s t)}{(\pi s)^2}\,. \tag{4-15}$$

A perfect single crystal of constant thickness yields a uniform image intensity. Intensity differences, i.e. image contrast, do not arise until the thin crystal exhibits differences in specimen thickness t, is bent or contains defects. In the case of a bent crystal foil, the deviation parameter s changes in the lateral direction. If the change in s in such a direction is uniform, an intensity distribution is obtained as shown in Fig. 4-23c. In the dark-field image, a strong bright contour with weak fringes at the margins appears, corresponding to the central and subsidiary maxima of Fig. 4-23c. In its complementary bright-field image ($I_0 = 1 - I_g$) the contour is dark. These *bend contours* specify points of identical specimen orientation in view of the diffracting lattice plane. This is shown schematically in Fig. 4-31. In a cylindrically bent crystal, both the $+\boldsymbol{g}$ and $-\boldsymbol{g}$ reflections generate a bend contour. The crystal in Fig. 4-32 has an orientation around ⟨100⟩ and is bent approximately spherically. The pattern of the numerous bend contours therefore reflects the symmetry of the ⟨100⟩ pole. Bend contours can be observed particularly in thin regions of the specimen since these are frequently bent.

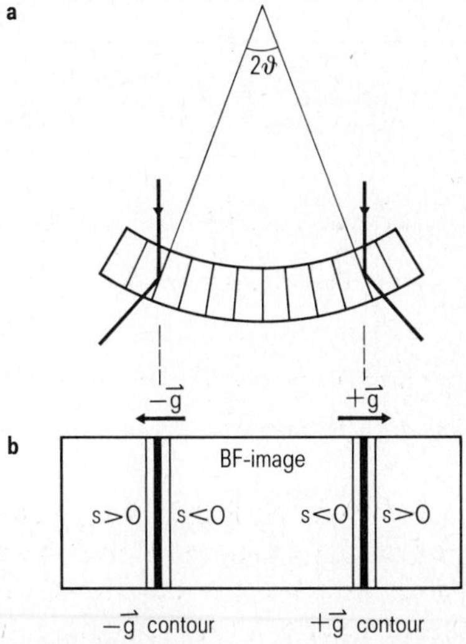

Fig. 4-31.
Formation of bend contours in a cylindrically-bent crystal for both $+\boldsymbol{g}$ and $-\boldsymbol{g}$ reflections. a) At those locations of the crystal at which the Bragg condition is fulfilled, bend contours are produced. b) Central and first subsidiary maxima of Fig. 4-23c (corresponding to dark lines in the BF image) are drawn in the schematic bright-field image.

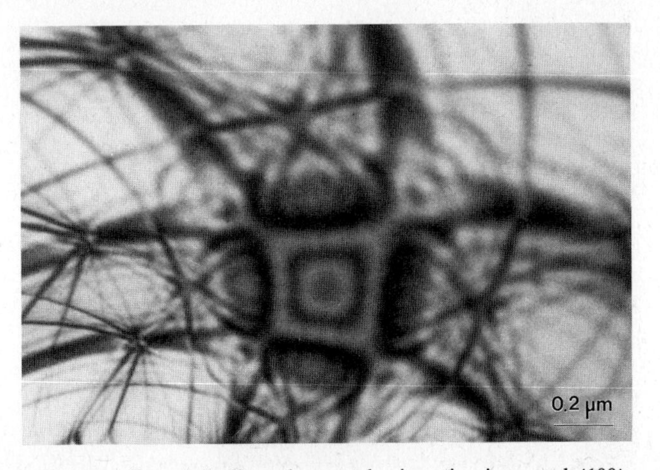

Fig. 4-32. Bend Contours in spherically-bent silicon foil. Since the crystal orientation is around ⟨100⟩, numerous bend contours appear reflecting the symmetry of the ⟨100⟩ pole.

The dependence of the diffracted intensity I_g on the crystal thickness t at a given orientation (s = const) follows from equation (4-15). At crystal thicknesses $t_n = n/s$ ($n = 1, 2, 3 \ldots$) $I_g = 0$, i.e. extinction of the intensity occurs. The parameter $t_g = 1/s$ specifies the difference in thickness between two extinctions and corresponds in the case of $s \gg 0$ to the effective extinction distance, which is explained below. The undiffracted intensity I_0 and diffracted intensity I_g accordingly oscillate with thickness, a minimum of I_g corresponding to a maximum of I_0 (Fig. 4-33a). In the case of wedge-shaped specimens (Fig. 4-33a) *thickness fringes* then appear in the image (Fig. 4-33b). The specimen thickness is constant along a fringe, and in-

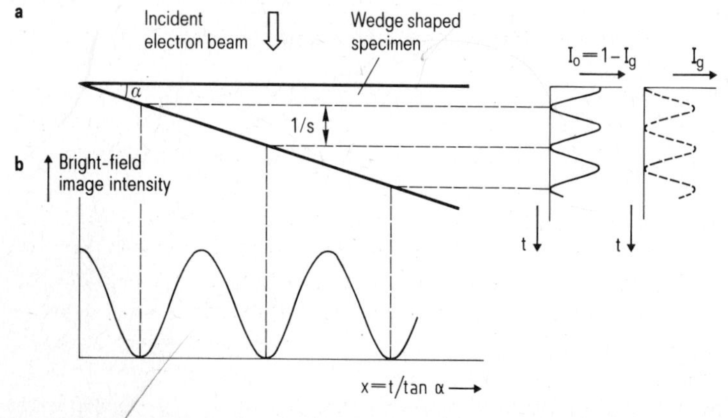

Fig. 4-33. Formation of thickness fringes. a) The intensities of the direct and diffracted beams I_0 and I_g oscillate with increasing specimen thickness t (right). This results in thickness fringes in a wedge-shaped crystal. b) Bright-field image intensity I_0 along the direction perpendicular to the edge of the wedge ($x = t/\tan\alpha$): each minimum in the curve I_0 versus t corresponds to a dark fringe.

creases by $t_g = 1/s$ from one fringe to the next. In Fig. 4-34, the thickness fringes can be seen at the wedge-shaped edge of a silicon specimen. If the shape of the specimen deviates from that of a wedge, the fringes are no longer parallel. Apart from edges along the hole in thinned foils, thickness fringes can also be seen at grain boundaries intersecting the specimen obliquely if a reflection is strongly excited only in one grain (Fig. 4-6b,c).

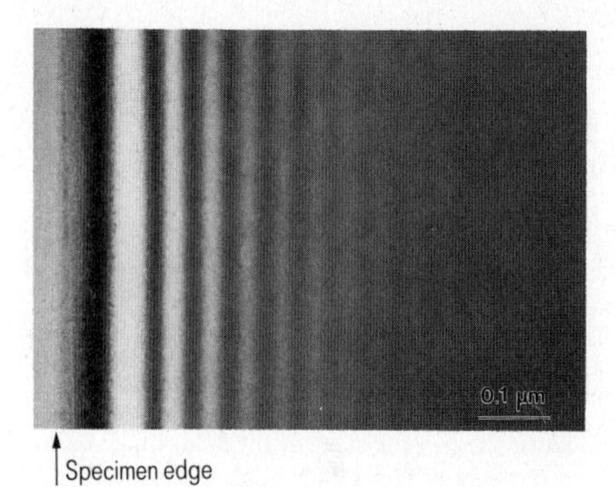

Fig. 4-34.
Bright-field image of a silicon crystal showing thickness fringes at the wedge-shaped specimen edge.

A comparison of the intensity profile in Fig. 4-33 with the experimental TEM image in Fig. 4-34 shows that no quantitative agreement exists. In the first place, the thickness fringes in Fig 4-34 disappear with increasing specimen thickness, and secondly their spacing is not reproduced correctly by equation (4-15). The precondition for the kinematical theory, namely that the diffracted intensity is small ($I_g \ll 1$), is no longer satisfied for thicker specimens, since if $s = 0$ is set in equation (4-15), it follows that $I_g \sim t^2$. For increasing thickness, the intensity of the diffracted beam would then become greater than that of the incident beam. In addition, for $s = 0$, the period t_g of the depth oscillations would become infinite, which is not the case in reality. The kinematical theory therefore applies only for very thin crystals. It also applies when $s \gg 0$ is chosen. In both cases, $I_g \ll 1$. When imaging crystal defects, $s \gg 0$ is frequently used, and in these cases the kinematical theory is applicable. To obtain a correct description of diffraction at thicker crystals and also for $s = 0$, however, the kinematical theory must be extended in two ways. In the first place we must bear in mind that the diffracted beam can be rediffracted again into the direction of the direct beam. This leads to the dynamical theory. Secondly, inelastic scattering must also be taken into account by the introduction of absorption.

The *dynamical theory of electron diffraction* will not be treated in more detail at this point. A derivation and description can be found in [4-25, 37], for example. Here we merely wish briefly to explain the assumptions involved and present some results which are important for interpreting image contrast. For the kinematical theory, the scattering contributions of the individual thickness elements of the form $d\phi_g \sim \phi_0 \exp(-2\pi i s z)\, dz$ can be simply integrated (eq. 4-14), since it was assumed that $\phi_0 = \text{const} = 1$. In contrast to this, the dynamical theory takes into account the fact that ϕ_0 does not remain constant but changes with depth in the crystal due to repeated scattering processes. Every thickness element therefore generates not only the scattering contribution $d\phi_g$ in the direction of the diffracted beam, but also a contribution $d\phi_0$ in the direction of the direct beam. If we solve the system of equations for $d\phi_g$ and $d\phi_0$ (Howie-Whelan equations), we obtain the following relation in place of eq. (4-15) for the intensity of the diffracted beam:

$$I_g \sim \frac{\sin^2(\pi \bar{s} t)}{(\pi \bar{s})^2}\,. \tag{4-16}$$

This equation has the same form as equation (4-15), except that s is replaced here by $\bar{s}$:

$$\bar{s} = (s^2 + 1/\xi_g^2)^{1/2} \tag{4-17}$$

where

$$\xi_g = \frac{\pi V_c}{\lambda \cdot F_g(\vartheta)} \tag{4-18}$$

(V_c is the volume of the unit cell, λ the electron wavelength, and $F_g(\vartheta)$ the structure amplitude of eq. (4-10)). For $s = 0$, we obtain, in agreement with the experimental results, a finite value for the periodicity of the thickness oscillations of I_g: $1/\bar{s} = \xi_g$. Since ξ_g specifies the thickness difference between two extinctions, it is referred to as the *extinction distance*. For electrons with energies between 100 and 200 kV, the values of ξ_g are usually in the range of

20 nm to 200 nm. For $s \neq 0$, it makes sense to define an effective extinction distance $\xi_{g,\mathrm{eff}}$, where:

$$\xi_{g,\mathrm{eff}} = \frac{1}{\bar{s}} = \frac{\xi_g}{(1 + w^2)^{1/2}} \tag{4-19}$$

The dimensionless parameter $w = s \cdot \xi_g$ (excitation error) was used here. At $s = 0$ (exact Bragg orientation of the crystal), $\xi_{g,\mathrm{eff}}$ has a maximum (ξ_g) and decreases with increasing deviation parameter s. For large s values ($s \gg 0$), $\xi_{g,\mathrm{eff}} \approx 1/s$, i.e. $\bar{s} \rightarrow s$ and eq. (4-16) becomes the analogous eq. (4-15) in the kinematical theory.

Eq. (4-16) correctly describes the thickness dependence of the intensity oscillations, i.e. the spacing of thickness fringes. To explain the decrease in intensity with crystal thickness, the influence of the inelastically scattered electrons must be taken into account. This is done formally by introducing an absorption term, and the wave functions then contain factors of the type $\exp(-\mu z)$ [4-25]. In reality, however, the inelastically scattered electrons are not absorbed physically in the specimen, but are prevented from contributing to the image formation by the objective aperture, which is equivalent to an absorption process. Taking into account this absorption, we then obtain the dependence shown in Fig. 4-35 for the undiffracted intensity I_0 and the diffracted intensity I_g of the film thickness t. Fig. 4-35a shows $I_0(t)$ and $I_g(t)$ for $s = 0$. With increasing values of s or w (Fig. 4-35b) the amplitude of the oscillations decreases, i.e. the contrast of the thickness fringes becomes weaker. In addition, their spacing is smaller than for $s = 0$ in Fig. 4-35a, according to eq. (4-19).

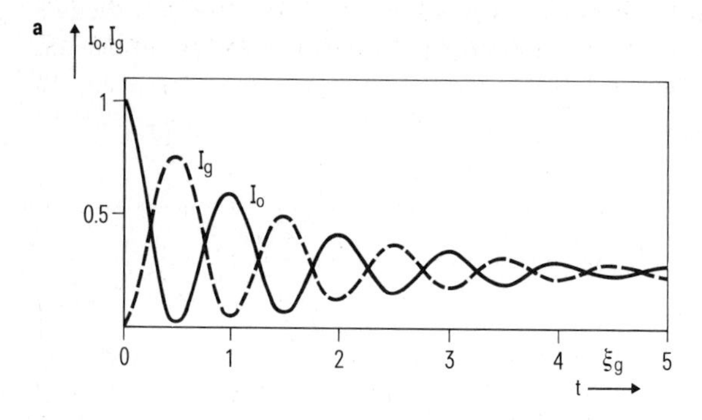

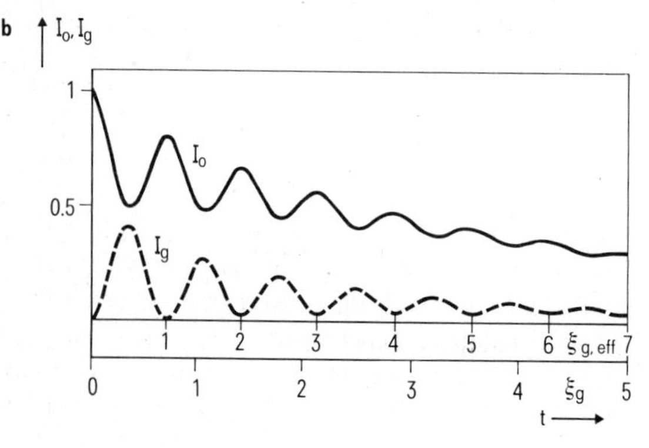

Fig. 4-35.
Bright-field and dark-field image intensities I_0 (solid line) and I_g (dashed line) as a function of specimen thickness t calculated by dynamical theory for different deviation parameters: a) $s = 0$, b) $s = 1/\xi_g$ ($w = 1$) [4-38]. As in Fig. 4-33 the curves represent intensity profiles of thickness fringes.

The results of the dynamical theory described above can also be obtained by solving the time-independent Schrödinger equation for a periodic potential in terms of *Bloch waves*. These are waves whose amplitudes vary in line with the lattice periodicity. For a two-beam case, two Bloch waves are obtained as the solution. Their intensity distributions are shown schematically in Fig. 3-23. Each Bloch wave forms a wave field which propagates in the crystal along the diffracting lattice planes. Waves of type I are preferentially excited when the angle of incidence on the lattice planes is $\psi < \vartheta$, i.e. $s < 0$, and waves of type II when $\psi > \vartheta$ ($s > 0$). The local intensity of the Bloch waves is a measure of the electron charge density of the waves. For Bloch waves of type I, the intensity maxima are located at the lattice plane sites, for waves of type II they are located between them. Waves of type I are thus attenuated more strongly, since increased scattering takes place at the lattice atoms. In contrast, Bloch waves of type II can propagate through the crystal with only low attenuation. The different properties of the two Bloch waves were already used in section 3.5 to explain the contrast of electron channeling patterns.

The curves of Fig. 4-35 can be interpreted with the aid of the Bloch waves in the following way. At low thickness, both Bloch waves are present, and their interference generates the thickness oscillations. With increasing thickness, the Bloch waves of type I are absorbed. Since only waves of type II remain, the oscillations disappear. The differential absorption of the Bloch waves also results in the dependence of I_0 on s (or w) being an asymmetrical function (Fig. 4-36). For $s < 0$ ($w < 0$), I_0 is small, since the preferentially excited waves of type I are greatly attenuated. In contrast, for $s > 0$ ($w > 0$), I_0 and thus the transmission is high, since the preferentially excited waves of type II are only slightly attenuated. The effect of this strongly differing attenuation for $s < 0$ and $s > 0$ is also known as *anomalous absorption*. The reduction of I_0 for $s < 0$ results in the bright-field image region within the bend contours of the reflections $+\boldsymbol{g}$ and $-\boldsymbol{g}$ appearing darker than the regions outside. In contrast, the diffracted intensity I_g is a symmetrical function of w.

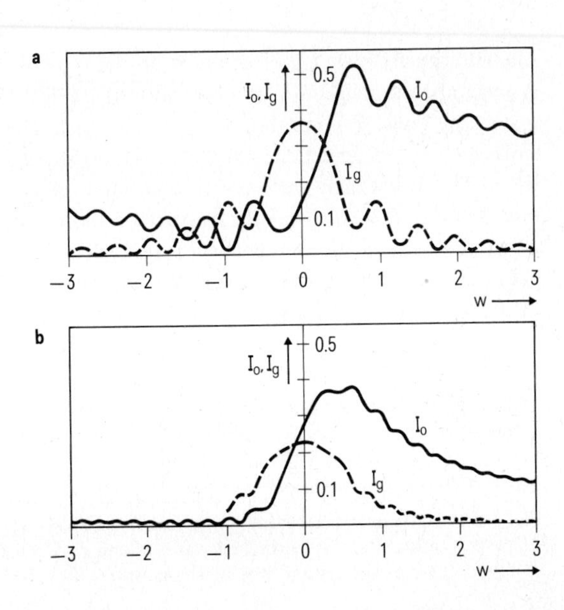

Fig. 4-36.
Bright-field and dark-field image intensities I_0 (solid line) and I_g (dashed line) as a function of excitation error $w = s \cdot \xi_g$ calculated by the dynamical theory for $t = 2.5\ \xi_g$ (a) and $t = 5\ \xi_g$ (b) [4-38]. The curves represent intensity profiles of bend contours.

4.5.2 Diffraction contrast for imperfect crystals

Imperfect crystals contain crystal defects. These can have various origins. In metals, defects are generated largely by plastic deformation, whereas in the case of semiconductor crystals, some defects are produced during crystal growth, and others are induced by various technological processes (section 9.2.2). Crystal defects may be classified in line with their dimensional extension: one-dimensional or line defects are dislocations, two-dimensional defects include stacking faults or grain boundaries and three-dimensional defects include precipitates. Disturbances of the crystal lattice by individual point defects, e.g. vacancies, are so small that they cannot be imaged directly in the TEM.

In the presence of a crystal defect, the positions of atoms in the defect environment are displaced by the vector $\boldsymbol{R}$ with respect to the positions in a perfect crystal. The position $\boldsymbol{r}$ in a perfect crystal is therefore changed by the defect into a position $(\boldsymbol{r} + \boldsymbol{R})$. The *displacements* $\boldsymbol{R}$ are smaller than or equal to the interatomic spacings. In the case of dislocations, for example, they decrease with increasing distance from the dislocation core, but change only slowly from one unit cell to another. The structure amplitude (eq. (4-10) then remains unchanged by the defect.

To calculate the diffracted amplitude we again use the column approximation. This is very exact if the displacements do not change greatly across the width of the column, which is generally the case. If we use the kinematical theory as an approximation, then the individual scattering contributions along the column are integrated as in eq (4-14). Since we have to replace $\boldsymbol{r}$ by $(\boldsymbol{r} + \boldsymbol{R})$ in the case of defects, however, the phase factor (see explanation to eq. (4-11)) in the exponent of the integral is now $(\boldsymbol{g} + \boldsymbol{s}) \cdot (\boldsymbol{r} + \boldsymbol{R})$. Remembering that $\boldsymbol{g} \cdot \boldsymbol{r}$ is an integer (and $\exp(-2\pi i n) = 1$) and neglecting $\boldsymbol{s} \cdot \boldsymbol{R}$, the phase factor becomes $(\boldsymbol{s} \cdot \boldsymbol{r} + \boldsymbol{g} \cdot \boldsymbol{R})$. We then obtain

$$\phi_g \sim \int_0^t \exp\left[-2\pi i (sz + \boldsymbol{g} \cdot \boldsymbol{R})\right] dz \tag{4-20}$$

The additional phase factor depends on the type of defect involved. As for a dislocation, $\boldsymbol{R}$ is generally a function of $\boldsymbol{r}$ and thus also of z (Fig. 4-37, centre). In special cases, such as for a stacking fault, $\boldsymbol{R}$ is constant (Fig. 4-37, right). The case of $\boldsymbol{g} \cdot \boldsymbol{R} = 0$ is of special interest since such displacements lie parallel to the diffracting lattice planes and do not bend or shift them. The defect then produces no additional phase shift and is invisible. The condition for invisibility $\boldsymbol{g} \cdot \boldsymbol{R} = 0$ forms the basis for a more precise characterization of the defects, e.g. when determining the Burgers vector of dislocations.

For an exact description of the image contrast of defects, the dynamical theory must be applied. Analytical solutions can be obtained in simple cases, such as for stacking faults. In the

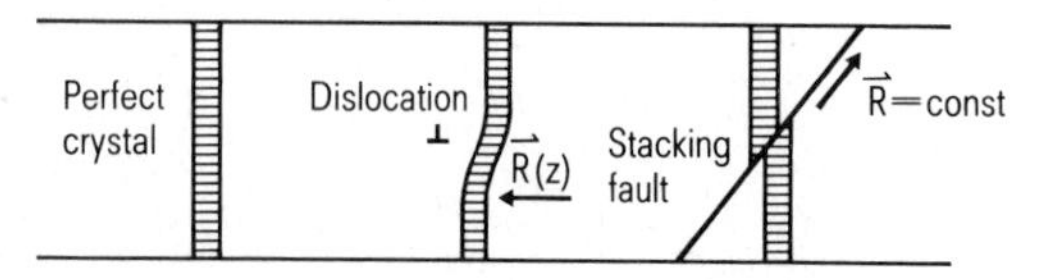

Fig. 4-37. Columns in perfect and imperfect crystals used for calculating the image intensity. For a dislocation, the displacement $\boldsymbol{R}$ varies along z. A stacking fault displaces the column in the lower part of the crystal by a constant value.

case of complex displacement fields, numerical integration can be used, for instance to calculate contrast profiles of defects. Beyond this, entire images of defects can be calculated by computer simulation [4-39]. In the following sections, the results of numerical calculations will be used to describe the contrast properties of the most important defects. Detailed descriptions of defect contrast can be found in [4-25, 37, 40]. In [4-26, 9] emphasis is placed on the practical work.

4.5.3 Dislocations

Dislocations are line defects in whose close vicinity the lattice planes are severely bent. By way of example, Fig. 4-38a shows an edge dislocation which may be considered as the termination line of an extra lattice plane inserted into the crystal. Dislocations are characterized by their *Burgers vector* (Fig. 4-38). The Burgers vector ***b*** of an *edge dislocation* is perpendicular to the dislocation direction which is normal to the drawing plane in Fig. 4-38a. The Burgers vector for a *screw dislocation* is parallel to the dislocation line and the set of parallel lattice planes normal to its direction is transformed to a single helicoid surface (Fig. 4-38c). For a mixed dislocation, the Burgers vector and the dislocation direction characterized by a unit vector ***u*** along the dislocation line enclose an arbitrary angle. The displacements ***R*** have only one component parallel to ***b*** if the dislocation is of the screw type, whereas edge dislocations have two components: a larger component also parallel to ***b*** and a smaller component parallel to $\boldsymbol{b} \times \boldsymbol{u}$.

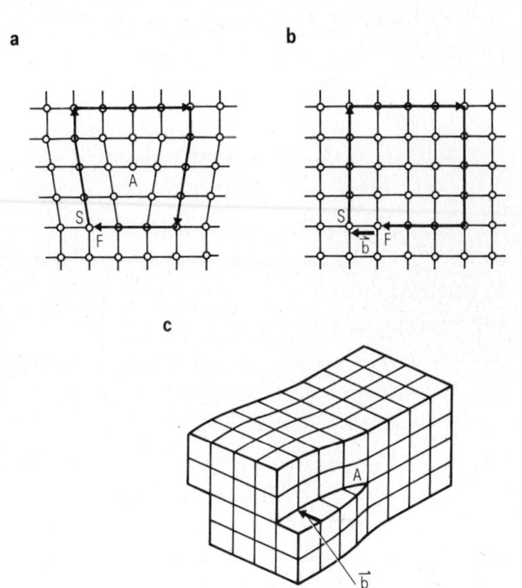

Fig. 4-38.
a), b) Definition of Burgers vector. a) An edge dislocation is formed by inserting an additional lattice plane into the crystal. A closed circuit is drawn around the dislocation. (Starting point S and finish F coincide). b) If the same circuit as in Fig. (a) (with an equal number of steps in the various directions) is drawn in a perfect crystal, the circuit is no longer closed. The closing vector $\overline{FS}$ is called the Burgers vector ***b***. For an edge dislocation, ***b*** is normal to the dislocation line at A in Fig. (a) which terminates the inserted lattice plane and runs normal to the drawing plane. c) For a screw dislocation, the Burgers vector ***b*** is parallel to the dislocation line at A.

Contrast of dislocations

In general, the bright-field mode will be chosen for imaging dislocations and the specimen will be tilted to obtain a two-beam condition with a deviation parameter of $s > 0$. Transmission is then optimized. $s > 0$ means that the Bragg condition is not fully satisfied in the regions of the perfect crystal. However, since the diffracting lattice planes in the vicinity of dislocations are bent (Fig. 4-39), there are regions on one side of each dislocation where the Bragg condition is exactly satisfied. The intensity diffracted from such regions is greater than that in the surroundings and the dislocations therefore appear as dark lines in the bright-field image. The center of the contrast profile is shifted with respect to the position of the dislocation. The direction of this lateral shift depends on the sign of $(\boldsymbol{g} \cdot \boldsymbol{b})\, s$. Since their Burgers vectors have different signs, the images of the two dislocations in Fig. 4-39 are shifted in opposite directions.

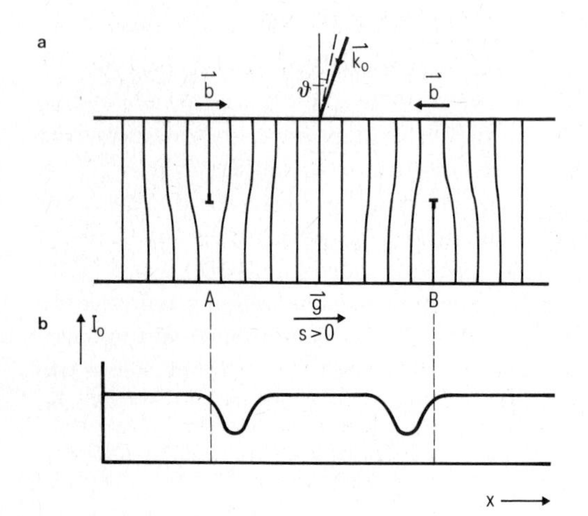

Fig. 4-39. Diffraction contrast of edge dislocations. a) For $s > 0$ the incidence angle of $\boldsymbol{k}_0$ is larger than the Bragg angle ϑ. Due to local bending of the diffracting lattice planes, the Bragg condition is satisfied on the right side of dislocation A and on the left side of dislocation B. b) Profile of bright-field image intensity showing the resulting shift of dislocation images with respect to their positions.

The width of the dislocation contrast is approximately $\xi_{g,\mathrm{eff}}/3$ and is thus dependent on s [4-41]. The contrast of dislocations is strongest when s is almost zero, its width then being several 10 nm. Fig. 4-40a shows dislocations in an epitaxial GaAs layer with $s \approx 0$. When the excitation error is increased, the contrast width decreases (Fig. 4-40b). This behavior is exploited particularly in the weak-beam technique (see below), where the value of s is especially large and contrast widths of a few nm are consequently obtained. Edge dislocations appear wider than screw dislocations. For $|\boldsymbol{g} \cdot \boldsymbol{b}| = 1$, the contrast of a dislocation always consists of a single line; double lines may occur for $|\boldsymbol{g} \cdot \boldsymbol{b}| = 2$.

If a dislocation runs at an angle from top to bottom through the specimen, an oscillatory contrast is produced when $s \approx 0$, the depth periodicity corresponding to the effective extinction length of eq. (4-19) (Fig. 4-40a). This contrast diminishes as s increases (Fig. 4-40b). In a thick specimen the oscillatory contrast can be observed only at those parts of the dislocation that are close to the top and bottom surfaces.

Since all the displacement vectors of screw dislocations are parallel to the Burgers vector, screw dislocations are invisible when $\boldsymbol{g} \cdot \boldsymbol{b} = 0$. The additional phase factor in eq. (4-20) then vanishes. In addition to $\boldsymbol{g} \cdot \boldsymbol{b} = 0$, $\boldsymbol{g} \cdot (\boldsymbol{b} \times \boldsymbol{u}) = 0$ must also be satisfied for edge dislocations if they are to be invisible. If $\boldsymbol{g} \cdot \boldsymbol{b} = 0$ and $\boldsymbol{g} \cdot (\boldsymbol{b} \times \boldsymbol{u}) \neq 0$, a slight residual contrast remains that is symmetrical with respect to the position of the dislocation line (dislocations at A in Fig. 4-40). This contrast is usually weaker than the contrast for $\boldsymbol{g} \cdot \boldsymbol{b} \neq 0$ and can thus be distinguished from the latter.

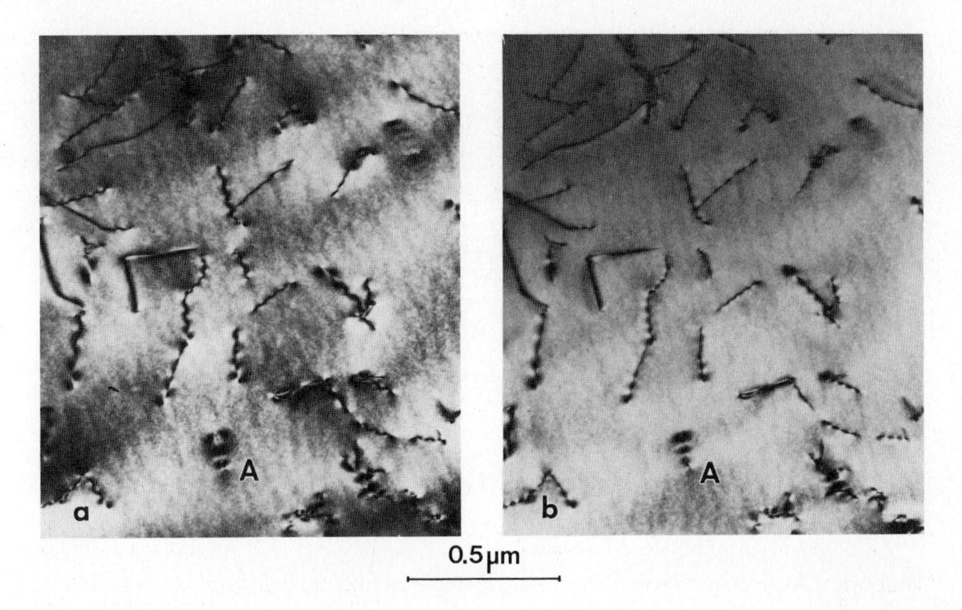

Fig. 4-40. Bright-field images of dislocations in GaAs with $s \approx 0$ (a) and $s > 0$ (b). The contrast width decreases with increasing excitation error s. Dislocations which are inclined to the foil plane exhibit an oscillatory contrast. For dislocations A $\boldsymbol{g} \cdot \boldsymbol{b} = 0$, and the residual $\boldsymbol{g} \cdot (\boldsymbol{b} \times \boldsymbol{u})$ contrast shows "symmetry along line". The high dislocation density in the epitaxial layer was generated by lattice mismatch due to deposition onto an InP substrate.

Burgers vector analysis

In order to determine Burgers vectors, a number of bright-field images are recorded under two-beam conditions and reflections sought for which $\boldsymbol{g} \cdot \boldsymbol{b} = 0$ is satisfied. The dislocations are then invisible or exhibit only a residual contrast. Once two such reflections, $\boldsymbol{g}_1$ and $\boldsymbol{g}_2$, have been found, the direction of $\boldsymbol{b}$ is given by $\boldsymbol{g}_1 \times \boldsymbol{g}_2$. The possible directions of $\boldsymbol{b}$ are frequently known, because only certain lattice vectors of minimum length are feasible for $\boldsymbol{b}$, for example $\boldsymbol{b} = 1/2 \langle 110 \rangle$ for the face-centered cubic (fcc) lattice as well as the diamond lattice. In this case, three reflections of the $\{111\}$ type suffice to distinguish between the six different $\langle 110 \rangle$ directions [4-25]. Fig. 4-41 shows three dislocations in a GaAs substrate crystal which have interacted. In Figs. 4-41 b, c, and d, $\boldsymbol{g} \cdot \boldsymbol{b} = 0$ is satisfied for one dislocation each. These dislocations still exhibit a weak $\boldsymbol{g} \cdot (\boldsymbol{b} \times \boldsymbol{u})$ contrast. In order to determine the dislocation type, the dislocation direction $\boldsymbol{u}$ must be determined as well as the Burgers vector $\boldsymbol{b}$. $\boldsymbol{b}$ and $\boldsymbol{u}$ define the glide plane of a dislocation. Dislocations 2 and 3 in Fig. 4-41 have glide planes of the $\{111\}$ type, dislocation 1 lies on a $\{100\}$ plane. Since dislocations can not easily glide on $\{100\}$ planes, it follows that this dislocation can only have been produced by a reaction between the other two.

The contrast properties discussed above apply only to elastically isotropic crystals. They are not applicable to strongly anisotropic materials, for which computer-simulated images are required in order to determine the Burgers vectors [4-39].

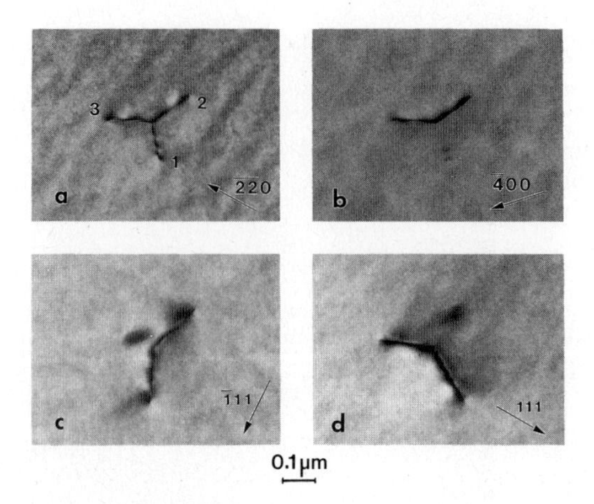

Fig. 4-41.
Reacting dislocations in GaAs single crystal. a) $\boldsymbol{g} \cdot \boldsymbol{b} \neq 0$ for all dislocations, b), c), d) $\boldsymbol{g} \cdot \boldsymbol{b} = 0$ for one dislocation each.

Dislocation loops

Dislocation loops may be created by the agglomeration of point defects on close-packed lattice planes. If vacancies or interstitials condense in an fcc crystal on a {111} plane, a stacking fault is produced (section 4.5.4.). This stacking fault is enclosed by a dislocation having a Burgers vector $\boldsymbol{b} = 1/3 \langle 111 \rangle$ normal to the stacking fault (Frank loop). This is a partial dislocation, because its Burgers vector is not a lattice vector. A Frank loop of this type can be converted to a perfect dislocation loop by a dislocation reaction. The sign of the Burgers vector indicates whether the loop was created by the condensation of vacancies or interstitials. In determining this sign, use is made of the property that the contrast is always to one side of the dislocation (cf. Fig. 4-39). If two images are produced with the reflections $+\boldsymbol{g}$ and $-\boldsymbol{g}$ but the same s (e.g. $s > 0$), the contrast is outside the loop in one case and inside the loop in the other. As the result of this inside-outside contrast, the loops appear to be of different size (Fig. 4-42). If the direction of $\boldsymbol{g}$ and the plane of the loop are determined, the difference in contrast will permit vacancy- and interstitial-type loops to be distinguished [4-26, 9]. The inside-outside

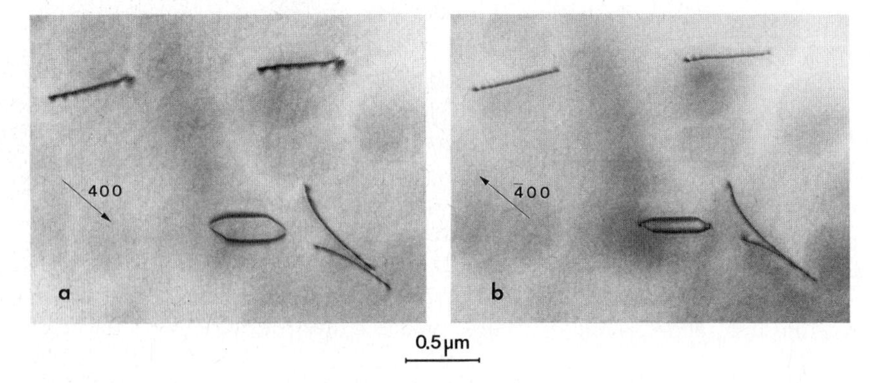

Fig. 4-42. Inside-outside contrast from dislocation loops in silicon. For $\boldsymbol{g} = (400)$ the dislocation contrast lies outside of the dislocation core position (a), for $\boldsymbol{g} = (\bar{4}00)$ inside (b). The dislocation loop is of the vacancy type.

contrast can be analyzed only if the loops are of sufficient size, in other words, if their diameters are larger than the contrast width of a dislocation ($\xi_{g,eff}/3$). If the loops are smaller, they appear as dark spots. Under weak-beam conditions, the limit at which the inside-outside contrast can be applied is approx. 10 nm. Even smaller loops produce a characteristic black-white contrast at $s = 0$ with the aid of which they can be analyzed [4-42].

Weak-beam method

In bright-field images, the contrast width of dislocations is often so large that complex configurations are no longer clearly visible. Although this width decreases as s increases, the dislocation contrast also diminishes because the background intensity in the bright-field image remains high. In contrast, the background intensity for large values of s is very low in a dark-field image (weak beam) and the dislocations appear as fine bright lines with low intensity but high contrast. Contrast along the dislocations is very narrow since only in regions very close to the dislocation core are the diffracting lattice planes bent sufficiently for the Bragg condition to be satisfied locally and electrons to be diffracted. As already indicated, contrast widths of a few nm are attainable. This means an improvement in resolution by a factor of between 5 and 10. By way of example, Fig. 4-43 shows that a high density of small dislocation loops can be seen much more clearly in the weak-beam image than in its bright-field counterpart. In experimental work, the fine bright lines of the dislocations are difficult to identify. To facilitate focusing, the $\boldsymbol{g}$ reflection in the bright-field image is initially set at $s = 0$. If the incident beam is then tilted such that the beam of the $\boldsymbol{g}$ reflection runs along the optical axis of the microscope (Fig. 4-5c), the $3\boldsymbol{g}$ reflection is excited in the dark-field image with $s = 0$. A larger excitation error, however, results for the $\boldsymbol{g}$ reflection which is used for imaging. When focusing is completed by use of the bright-field image, the weak-beam image can be recorded after changing to the dark-field mode.

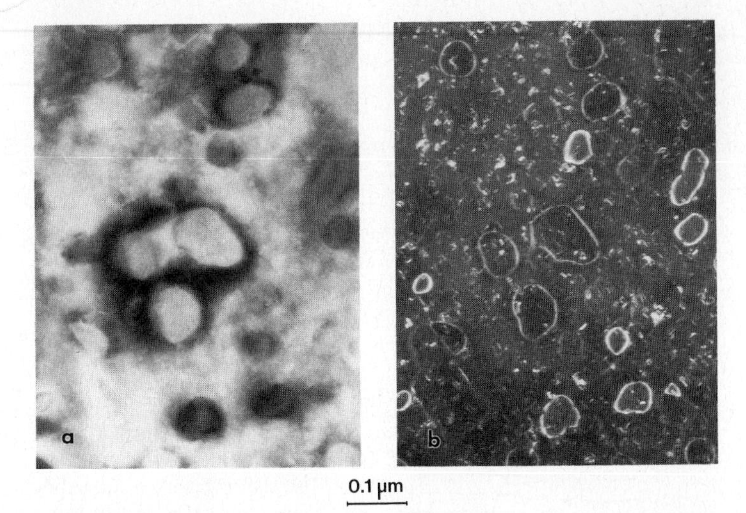

Fig. 4-43. Comparison of bright-field (a) and weak-beam image (b) from small dislocation loops in silicon produced by BF_2 ion implantation and thermal annealing. In addition to the loops, small defect clusters appear in the weak-beam image.

4.5.4 Planar defects

Planar defects split a crystal into two parts which may differ in position and/or orientation. In translation boundaries, the two parts are merely displaced relative to each other. The translation may be characterized by a constant displacement vector $\boldsymbol{R}$ (Fig. 4-37, right). In orientation boundaries, the two parts are rotated with respect to each other and the displacement $\boldsymbol{R}$ increases with distance from the boundary.

Translation boundaries

The most important translation boundaries are *stacking faults* in fcc crystals. These faults lie on the close-packed lattice planes of {111} type. The atoms on successive {111} planes assume three different types of position: A, B, and C (Fig. 4-44a). The stacking sequence in a perfect single crystal is thus: ABCABCABC... A stacking fault is created by the removal or insertion of a lattice plane. This does not destroy the close-packed arrangement and the nearest-neighbor order. If a lattice plane is removed (e.g. B), the stacking sequence is ABCA|CABC... and an *intrinsic* stacking fault is formed (Fig. 4-44b). If an additional lattice plane is inserted (e.g. A), an *extrinsic* stacking fault is produced with a sequence ABCAB|A|CABC... (Fig. 4-44c).

a

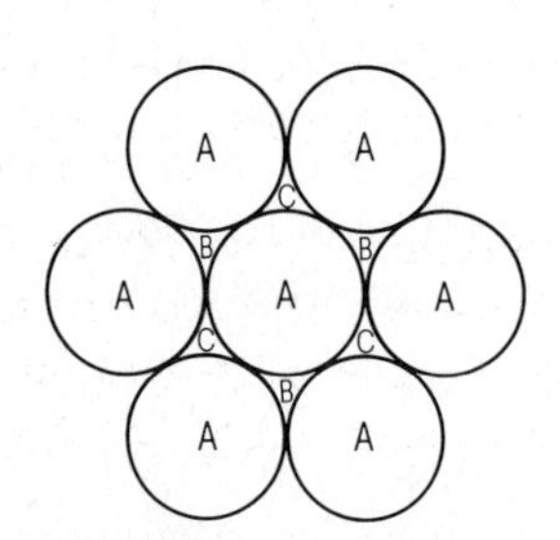

b

Perfect crystal — Intrinsic stacking fault

A B C A B C A B C — A B C A C A B C

c

Perfect crystal — Extrinsic stacking fault

A B C A B C A B C — A B C A B A C A B C

Fig. 4-44.
Stacking faults in fcc crystals. a) Close-packed layer of atoms at positions A on the (111) plane. The adjacent layers above or below the drawing plane have atoms at positions B and C. For a perfect fcc crystal, the stacking sequence is ABCABCABC. b) An intrinsic stacking fault is formed by removing one layer of atoms; c) An extrinsic stacking fault is formed by inserting an additional layer. Since the distance between two layers of the same type is given by the length of the vector ⟨111⟩, the displacement vector for inserting or removing one layer is $\boldsymbol{R} = \pm 1/3\ \langle 111 \rangle$.

The lines in the sequence of letters indicate the position at which the stacking sequence is disrupted, i.e. the position of the stacking fault. Depending on the type of stacking fault, the displacement vector is $\boldsymbol{R} = \pm 1/3 \langle 111 \rangle$. An intrinsic stacking fault is also produced when the atoms beneath the stacking-fault plane are displaced, e.g. from position B to position C. This type $1/6 \langle 112 \rangle$ displacement is equivalent to $\boldsymbol{R} = 1/3 \langle 111 \rangle$, since it differs from the latter only by a full lattice vector, and does not – by definition – change the lattice ($1/6 \langle 112 \rangle = -1/3 \langle 111 \rangle + 1/2 \langle 110 \rangle$). However, differences occur in the partial dislocation at the boundary of the stacking fault.

A stacking fault produces a phase shift of $\alpha = 2\pi \boldsymbol{g} \cdot \boldsymbol{R}$ in the electron waves (cf. eq. (4-20)). Stacking faults in fcc crystals produce only values of $\alpha = 0$ or $\pm 2\pi/3$. When $\alpha = 0$, the stacking fault is invisible. Since a stacking fault generally intersects the specimen obliquely, the phase shift occurs at different depths. This produces a characteristic contrast of bright and dark fringes parallel to the line of intersection between the stacking fault and the specimen surface (Fig. 4-45). Since the depth periodicity is equal to the effective extinction distance, the spacing between the fringes decreases as s increases. In a thick specimen, the fringe contrast is attenuated at the center of the foil. In the bright-field image, the fringe pattern is symmetrical about the center of the foil (Fig. 4-45a). In contrast, the dark-field image is asymmetrical, resembling the bright-field image at the top surface of the specimen and forming a complementary image at the bottom (Fig. 4-45b). The sign of α and hence the type of stacking fault involved can be derived from the nature of the outermost fringes (bright or dark) in a pair of bright-field and dark-field images. Precise instructions are to be found in [4-25, 26, 9], among others. If the stacking sequence is inverted within a narrow zone, a microtwin is produced with a stacking sequence such as ABCA|CBACB|CABC. Thin microtwins exhibit a fringe contrast analogous to that of stacking faults.

Antiphase domain boundaries occur in AB alloys with an ordered structure. Different domains are displaced with respect to each other such that the lattice sites of A and B are interchanged on one side of a domain boundary compared with the other side. Such domain boundaries can generate a phase shift of $\alpha = \pm\pi$. The fringe contrast generated in this case

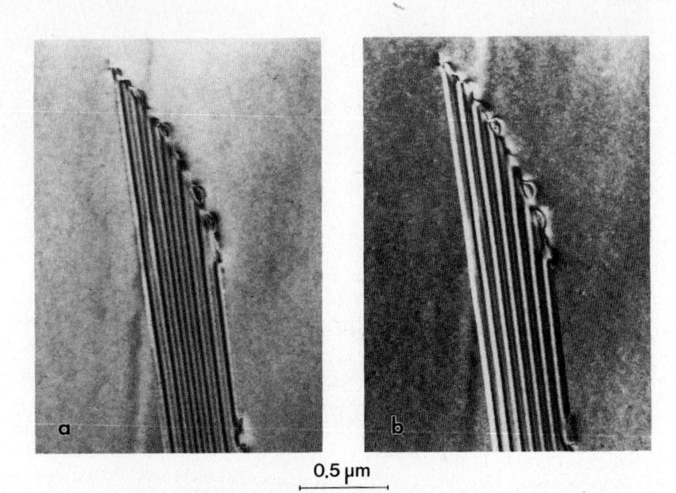

Fig. 4-45. Fringe contrast from an inclined stacking fault in silicon for $s \approx 0$. a) bright-field image, b) dark-field image. The stacking fault is of extrinsic type.

has different properties than that produced by stacking faults: for $s = 0$, the fringes in the bright-field and dark-field images are complementary and the fringe spacing corresponds to $\xi_g/2$ (in contrast to ξ_g for stacking faults).

Orientation boundaries

A simple case of an orientation boundary is a grain boundary. Generally, a reflection is excited in only one of the two crystals and thickness fringes appear at the grain boundary, as in a wedge-shaped specimen. Means of obtaining information on the intrinsic structure of a grain boundary include the investigation of grain-boundary dislocations by contrast experiments, or the study of the atomic configuration by high-resolution electron microscopy [4-38].

If the two parts of the crystal (or their reflection vectors) are tilted only slightly with respect to each other, the difference in the diffraction conditions is small. It may be described by the difference in the excitation error: $\Delta s = s_1 - s_2$ or $\delta = \Delta s \cdot \xi_g$. The fringes observed at boundaries of this type are thus known as δ-fringes in contrast to the α-fringes observed at stacking faults. Such slight tilts occur at ferroelectric 90° domain walls in $BaTi0_3$, for example. The tetragonal structure of $BaTi0_3$ differs only slightly from the cubic: $c/a \approx 1.01$. Since the c-axis changes its direction by 90° from one side of the domain wall to the other, a slight tilt results. The fringe pattern which is observed at a domain wall running obliquely through the specimen (Fig. 4-46) depends only on the difference Δs. Moreover, the fringe pattern is symmetrical about the centre of the foil in the bright-field image but symmetrical in the dark-field image (Fig. 4-46).

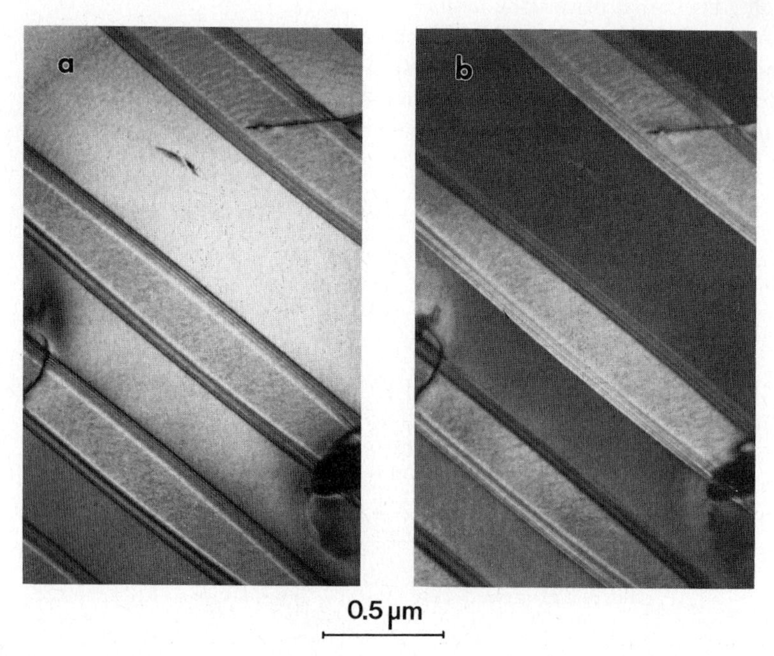

Fig. 4-46. Ferroelectric 90° domain boundaries in $BaTi0_3$ ceramic showing δ-fringe contrast. (a) bright-field image, (b) dark-field image.

4.5.5 *Precipitates*

If the solubility limit is exceeded in a two-component system, precipitates of a second phase may be produced in the matrix (of the first phase). A distinction is made between coherent, partially coherent and non-coherent precipitates, depending on their boundaries with the matrix. Coherent precipitates have the same crystal structure as the matrix and the lattice planes are continuous through precipitate and matrix (Fig. 4-47a). However, differences in the lattice parameter produce strains (Fig. 4-47b). Precipitates are partially coherent when they have at least one non-coherent boundary while the others are coherent (Fig. 4-47c). At the non-coherent boundary, the lattice planes no longer match due to differences in structure, for example. Coherent and partially coherent precipitates exhibit an orientation relationship between particle and matrix. In contrast, non-coherent precipitates generally exhibit no such relationship, because as a rule the precipitate and matrix have different crystal structures (Fig. 4-47d). The TEM contrast of a precipitate has two causes: the matrix may be distorted in the vicinity of the precipitate and a strain contrast produced *(matrix contrast)*; alternatively, the particle itself may influence the diffracted intensity *(precipitate contrast)*.

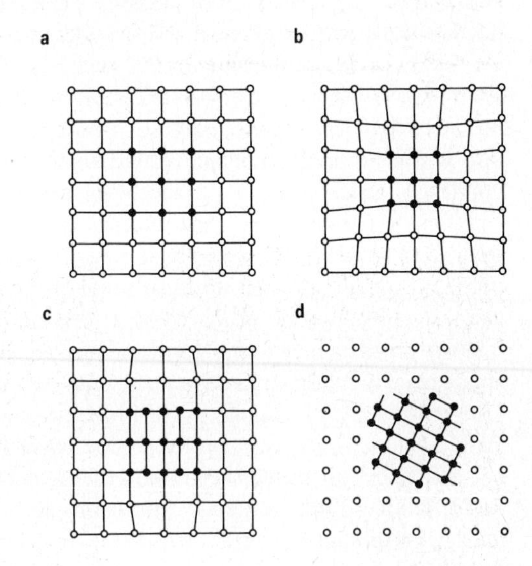

Fig. 4-47.
Types of precipitate: a), b) Coherent precipitates without (a) and with (b) lattice misfit; c) Partially coherent precipitate with lattice misfit; d) Non-coherent precipitate.

The *matrix contrast* depends on the size and geometry of the displacement field around the particle. This field is radially symmetric about a spherical particle. Since $\boldsymbol{g} \cdot \boldsymbol{R} = 0$ holds for the direction normal to the diffraction vector, larger particles exhibit a "line of no contrast" along this direction. Dark lobes appear on both sides of this line and the typical coffee-bean contrast is observed. Fig. 4-48 shows a thin oxide platelet in silicon; the platelet exerts high strains on the surrounding silicon matrix and was imaged end-on. Here, too, lines of no contrast can be seen: according to the respective diffraction vectors, this line lies along the platelet in Fig. 4-48a whereas it is normal to it in Fig. 4-48b. Extremely small spherical particles produce a black-white contrast similar to that exhibited by very small dislocation loops.

Precipitate contrast may be due to a number of causes. If the atoms of the particle are significantly heavier than those of the matrix, scattering contrast (section 4.1) makes the particles

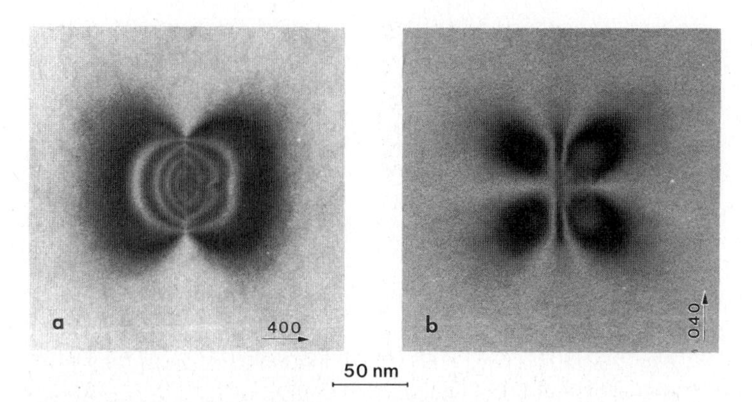

Fig. 4-48. Strain contrast around SiO_2 platelet in silicon. The platelet is imaged end-on using the (400) reflection (a) and the (040) reflection (b). The "line of no contrast" is normal to g.

appear dark (Fig. 4-49a). If the precipitate and matrix have different structure factors, the extinction distance ξ_g is different. This results in a change in the effective specimen thickness. The structure-factor contrast is thus particularly large at the edge of thickness fringes for $s = 0$. Fig. 4-49c shows the special case of amorphous precipitates ($Si0_2$ in Si) which are visible in the weak-beam image because of thickness fringes at the inclined boundaries. Orientation contrast is produced by partially coherent or non-coherent particles if certain lattice planes of the particle strongly diffract the incident electrons while the matrix is not in Bragg orientation. In the dark-field image formed using a reflection from the particles, the particles show up bright. Plate-like precipitates may produce a displacement of the matrix and thus a phase shift of the electron waves. If these precipitates lie obliquely in the specimen, a fringe contrast similar to that produced by a stacking fault becomes visible.

Moiré patterns may also be produced by precipitates. They occur when two crystals with approximately the same interplanar spacings or orientation lie on top of each other (Fig. 4-49b). Below, only the simple case of two crystals with the same orientation but slightly different interplanar spacings, d_1 and d_2, is discussed. If the reflection diffracted at the upper part of the crystal is diffracted back toward the direct beam by the lower part, an additional reflection is produced due to double diffraction. Since this reflection is close to the direct beam, it may pass through the objective aperture. Interference with the direct beam produces a fringe pattern (Moiré fringes) normal to g and with a spacing given by:

$$D = \frac{d_1 \cdot d_2}{d_1 - d_2} \,. \tag{4-21}$$

The formation of Moiré fringes may also be understood in terms of differences in the projection of the lattice planes of the two crystals on top of each other. Due to the difference in lattice spacings, areas in which the lattice planes of the upper crystal lie exactly above those of the lower one alternate with others in which the planes of the upper crystal lie above the spaces between the lattice planes of the lower crystal. Moiré patterns are also produced when two crystals with the same interplanar spacing are rotated slightly with respect to each other. If several reflections produce double diffraction, a two-dimensional Moiré pattern is formed.

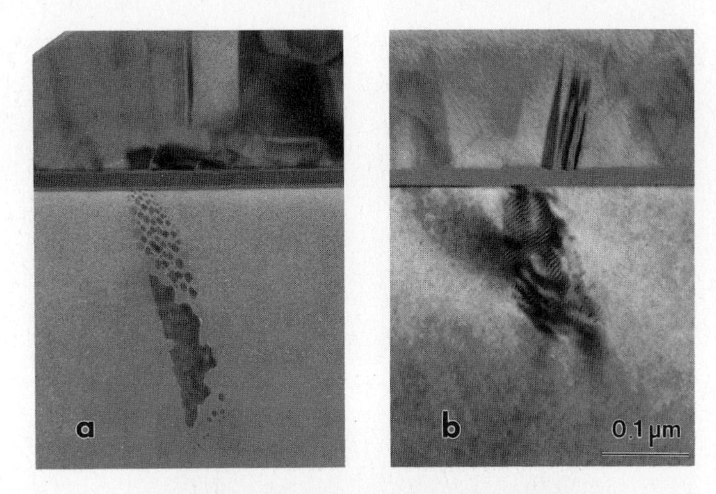

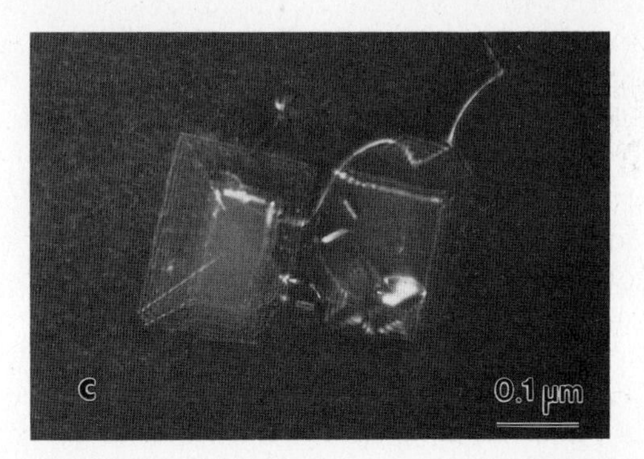

Fig. 4-49. Precipitate contrast: a) b) Copper silicide precipitates in silicon below the gate dielectric (cross sections through device structure). In Fig. (a) no reflection is strongly excited and the precipitates are visible due to scattering contrast. The precipitate of Fig. (b) exhibits Moiré fringes which are roughly normal to the excited (220) reflection (beam orientation close to [001]. c) Amorphous SiO_2 precipitates in silicon with the shape of truncated octahedra. Due to the small extinction distance in the weak beam image (large s) very narrow thickness fringes appear at the inclined interfaces.

Voids or *bubbles*, such as may be formed by the agglomeration of gas atoms in a material, represent a special type of precipitate. Large voids change the effective specimen thickness and thus become visible. Analogous to the Fresnel fringes at the edge of a specimen (section 4.8.1), small voids produce a bright outer fringe along their perimeter and therefore appear bright in an underfocused image and dark in an overfocused one. By way of example, Fig. 4-50 shows the result of an implantation of BF_2 ions in silicon by which small bubbles filled with fluorine are formed at a depth of approximately 0.1 μm.

The various causes of the precipitate contrasts and the effects to which they give rise in the diffraction pattern are discussed in detail in [4-25].

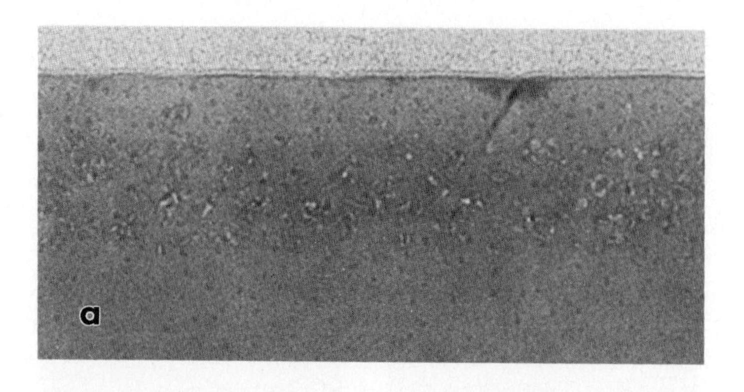

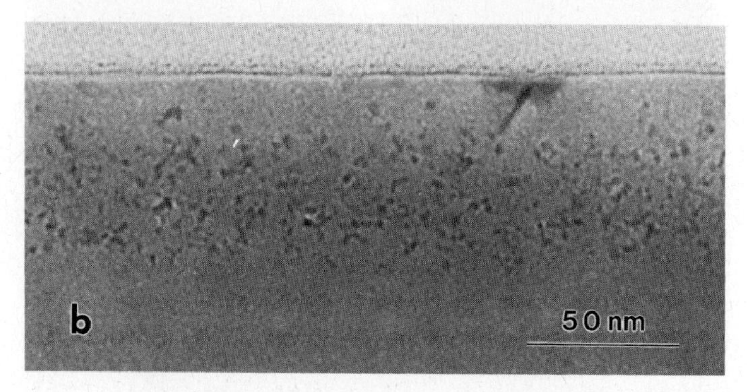

Fig. 4-50. Voids in silicon formed by BF_2 ion implantation and annealing. In analogy to Fresnel fringe contrast, the voids appear bright (with a dark ring) for underfocusing of the image (a) and dark (with a bright ring) for overfocusing (b).

4.6 High-resolution electron microscopy

4.6.1 Image formation

The intensity distribution of the direct beam at the exit surface of the specimen was calculated to describe the diffraction contrast or the scattering contrast in a bright-field image (sections 4.5, 4.1.6). Imaging in the TEM could then be based on the laws of geometrical optics, and the image reproduces this intensity distribution on a magnified scale. Since in the case of phase contrast the image is generated by interference between the wave of the direct beam and the various diffracted and scattered electron waves, wave optics must be used to describe the process of image formation. As already indicated in section 4.1.3, this can be done with the aid of two Fourier transforms.

Let us begin by considering the interaction of the incident wave with the specimen. A plane wave of unit amplitude is modified by the specimen resulting in the wave function $\psi_{sp}(x, y)$ at the exit surface, which is referred to as the transmission function of the specimen. If the specimen changes only the phase of the incident wave (phase object), the transmission function is given by:

$$\psi_{sp}(x, y) = \exp\left[-i\sigma \cdot V_p(x, y)\right] \tag{4-22}$$

where σ is the interaction constant ($\sigma = \pi\lambda/U_0$, U_0 ... acceleration voltage) and $V_p(x, y)$ the projected potential of the crystal integrated along the beam direction. The phase change $\sigma \cdot V_p$ induced by extremely thin specimens comprising light atoms is small *(weak-phase approximation)* and eq. (4-22) becomes

$$\psi_{sp}(x, y) = 1 - i\sigma \cdot V_p(x, y). \tag{4-23}$$

When focused perfectly in an ideal microscope, a pure phase object would generate no contrast because the image intensity $|\psi_i|^2$ is equal to unity: $I = |\psi_i|^2 = |\psi_{sp}|^2 = 1$. As already indicated in section 4.1.6, the phase contrast can be converted into an amplitude contrast by an additional $\pm\pi/2$ phase change of the scattered wave. A phase shift of this type can be obtained by exploiting the spherical aberration and by suitably defocusing the objective lens (see below). In the specific case where only scattered or diffracted beams with a phase shift of $\pi/2$ contribute to the image, an optimum image is obtained with the following intensity distribution [4-43]:

$$I(x, y) = 1 + 2\sigma \cdot V_p(x, y). \tag{4-24}$$

The image intensity is directly proportional to the projected potential of the specimen and the image can therefore be interpreted directly and easily. However, for materials of medium atomic number, the weak-phase approximation applies only for extremely thin specimens of thickness $t \leq 2$ nm [4-44]. If the specimen is thicker, mutual interaction of the various diffracted beams and the direct beam must be taken into account and the contrast is highly dependent on the specimen thickness (section 4.6.2).

The influence of the imaging system must be considered in more detail in the general case of image contrast of phase objects. The waves scattered in a particular direction by the specimen are focused in the back focal plane of the objective lens and form a diffraction spot (Fig. 4-3). In the case of crystalline specimens, the Bragg equation for small angles ($\lambda = 2d \cdot \vartheta$) clearly shows that the scattering angle $\Theta = 2\vartheta$ is inversely proportional to the lattice spacing d: $\Theta = \lambda/d$. The parameter $u = 1/d = \Theta/\lambda$ corresponds to a radial coordinate in the diffraction pattern and represents a spatial frequency. (Since electron microscopes are cylindrically symmetric systems, we will restrict ourselves to spatial frequencies in the radial direction). The diffraction pattern therefore reproduces the spatial frequencies in the specimen in the form of a two-dimensional spectrum. This procedure taking place in an ideal lens – namely obtaining the amplitude distribution in the diffraction pattern from that in the specimen – is described in mathematical terms by a Fourier transform.

The imaging characteristics are influenced firstly by the objective aperture, because it limits the scattering angle to $\Theta < \Theta_0$, when Θ_0 is the aperture cutoff angle. This means that structural details with a spatial frequency $u > u_0 = \Theta_0/\lambda$ are eliminated. Secondly, the aberrations of the objective lens produce a phase shift. This is taken into account by multiplying the amplitude distribution in the diffraction pattern by the *contrast transfer function* (CTF).

For weak-phase objects, this function is expressed by sin $\chi(u)$ [4-43], with the phase factor $\chi(u)$ given by:

$$\chi(u) = \pi \cdot \Delta f \cdot \lambda \cdot u^2 + \frac{\pi}{2} \cdot C_s \cdot \lambda^3 \cdot u^4 . \tag{4-25}$$

$\chi(u)$ consists of two terms. The first defines the phase change due to defocusing the objective lens by an amount Δf and the second that resulting from spherical aberration (aberration constant C_s, section 2.2). C_s and the wavelength λ depend on the microscope used. However, the phase shift χ and hence the CTF can be changed by selecting different values for the defocus Δf. Since C_s is always > 0, the two terms in eq. 4-25 counteract each other when the defocus is negative ($\Delta f < 0$).

For weak-phase objects, optimum agreement between the image and the unknown structure of the specimen is obtained when the largest possible range of spatial frequencies is transferred by the imaging system with a CTF of sin $\chi \approx -1$. In this range of spatial frequencies, $\chi(u) \approx -\pi/2$. This and the condition for an extremum value ($d\chi/du = 0$) result in a special defocus known as the *Scherzer focus*:

$$\Delta f_{sch} = -(C_s \cdot \lambda)^{1/2} . \tag{4-26}$$

Fig. 4-51a shows the CTF at Scherzer focus for a 200 kV microscope. With increasing spatial frequency, a region with sin $\chi \approx -1$ is followed by oscillations in which the sign of the CTF changes with decreasing intervals. The attenuation in amplitude of the oscillation will be discussed later. The zeros in the CTF represent gaps in the image transfer: i.e., these spatial frequencies do not appear in the image. Each oscillation brings about a contrast inversion for a particular range of spatial frequencies. Image structures with spatial frequencies in which the CTF oscillates cannot therefore be directly interpreted. Only spatial frequencies preceding the first zero of the CTF (point A in Fig. 4-51a) are transmitted with uniform contrast. The

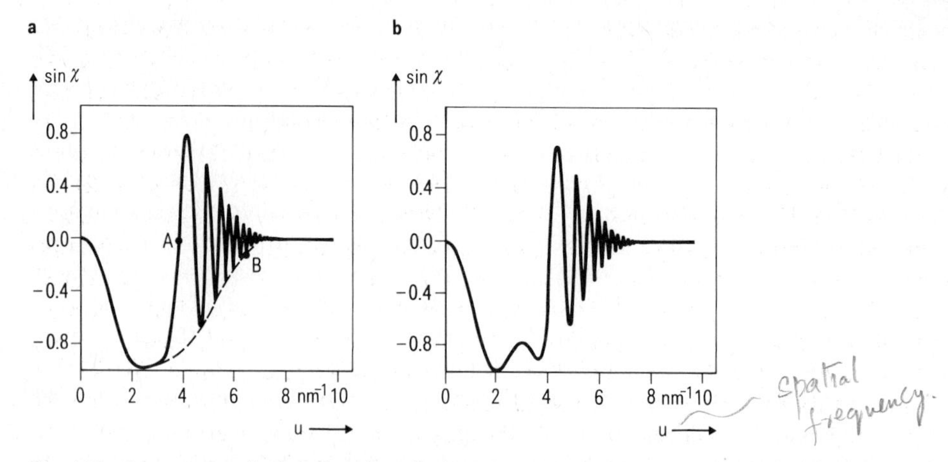

Fig. 4-51. Contrast transfer function sin $\chi(u)$ for Scherzer focus (a) and optimum defocus (b). The first zero of the CTF in the Scherzer focus defines the point resolution (A in Fig. (a)). Point B marks the information resolution limit [4-45].

spacing which corresponds to this first zero is known as *point resolution* or "interpretable resolution". The point resolution represents the resolution limit for non-periodic structures. For the CTF in Fig. 4-51a, the point resolution is 0.26 nm; in other words, object details with spacings down to 0.26 nm are unambiguously imaged. To suppress the confusing contribution of higher spatial frequencies to the image, the size of the objective aperture is optimally chosen such that it cuts off the oscillating part of the CTF.

If a phase change of $\chi(u) = -2\pi/3$ is chosen instead of $-\pi/2$, an *optimum defocus* of

$$\Delta f_{\text{opt}} = -\left(\frac{4}{3} C_s \cdot \lambda\right)^{1/2} \tag{4-27}$$

is obtained, in which the first zero of the CTF is shifted toward slightly larger spatial frequencies than for the Scherzer focus (Fig. 4-51b). The point resolution is therefore smaller (0.24 nm). It follows from eqs. (4-25) and (4-27) that the point resolution at optimum defocus is given by:

$$d_p = 0.66\,(C_s \cdot \lambda^3)^{1/4}\,. \tag{4-28}$$

d_p represents a theoretical limit of resolution which is almost achieved in modern microscopes. As already stated in section 4.2, small C_s values and high voltages (small λ) are required to achieve high resolution. At an acceleration voltage of 200 to 400 kV, a point resolution below 0.2 nm can be obtained.

The CTF sin χ with $\chi(u)$ from eq. (4-25) would produce oscillations with a constant amplitude. In practice, however, the finite source size reduces the interference capability (spatial coherence) and the energy spread of the beam (the result of high voltage and lens current instabilities, cf. section 4.2) produces chromatic aberration. This results in a damping envelope by which the oscillation amplitudes are attenuated (Fig. 4-51). A second resolution limit is defined to characterize the limits to which spatial frequencies are transmitted at all. It is known as the "information resolution limit" and is determined by the spatial frequency at which the envelope drops to $\exp(-2)$ (at B in Fig. 4-51a). The information resolution limit for microscopes with a point resolution of better than 0.2 nm is below 0.15 nm. As indicated above, images cannot be directly interpreted if they contain spatial frequencies in this oscillating part of the CTF, and computer simulation is then required (section 4.6.3).

The influence of the CTF on imaging characteristics is well illustrated by using thin amorphous specimens, since their projected potential exhibits a continuous spectrum of spatial frequencies. The thin, amorphous silicon film in Fig. 4-52a was imaged close to the optimum defocus and shows a fine, granular contrast. Recording an *optical diffractogram* from the micrograph (negative) allows the spatial frequencies contained in the image to be determined. Such a diffractogram appears in the back focal plane of a long-focus optical lens if the negative is illuminated with a broad, parallel laser beam. The optical diffractogram of Fig. 4-52a is shown in Fig. 4-52b. The intensity distribution in the radial direction reproduces the CTF in optimum defocus. Since intensities are displayed in the diffractogram, both positive and negative regions of sin χ appear bright. The innermost, broad, bright ring corresponds to the broad frequency band extending to the first zero of the CTF. Intensities beyond this first zero are barely visible in the diffractogram. Fig. 4-52c shows the silicon film at greater defocus. The first zero of the CTF now lies at considerably smaller spatial frequencies and a number of

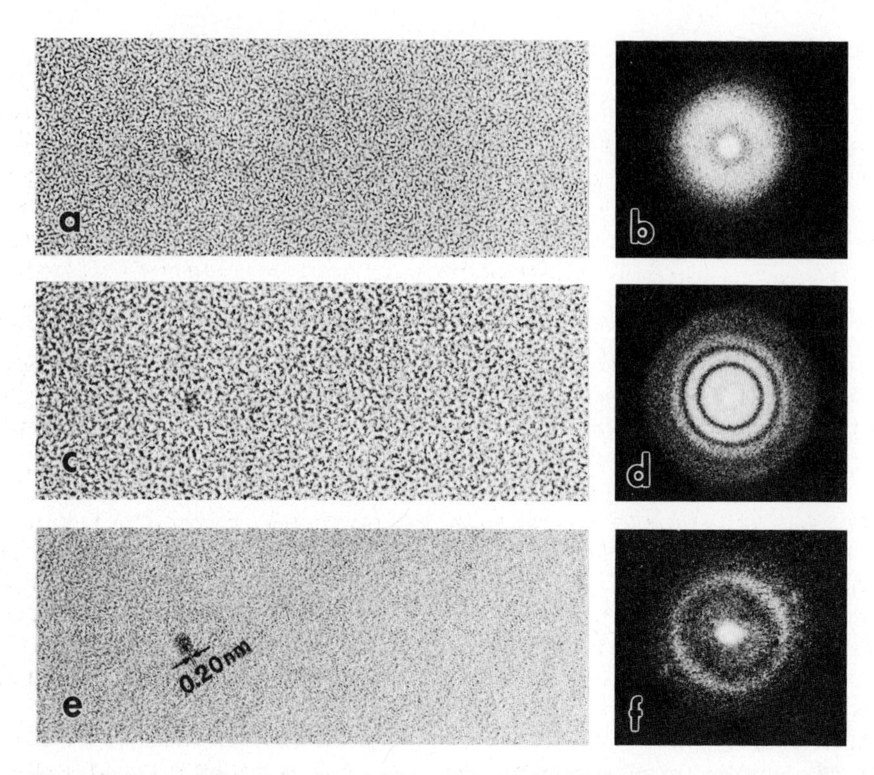

Fig. 4-52. Phase contrast from thin amorphous silicon foil for various defocus: a), c), e) High-resolution micrographs, b), d), f) Corresponding optical diffractograms. a), b) Close to optimum defocus; c), d) Large defocus; d), f) Close to focus of minimum contrast. (Courtesy of JEOL Ltd.)

bright rings can be seen in the optical diffractogram (Fig. 4-52d). They correspond to the oscillations of the CTF. The granular contrast appears much coarser than in the optimum defocus because the several zeros in the CTF at which no image information is transmitted, lie at low spatial frequencies corresponding to large spacings. Minimum contrast is obtained approximately at a defocus of $\Delta f_{\min} = -0.44\,(C_s \cdot \lambda)^{1/4}$ (Fig. 4-52e, f) [4-43]. This value lies between the Gaussian focus ($\Delta f = 0$) and the optimum defocus of eq. (4-27).

The possibility of determining the characteristics of the CTF (such as the point resolution) from high-resolution images of amorphous specimens is used in testing the resolution of microscopes; optical diffractograms are thus used for evaluation. Moreover, the spherical aberration constant C_s and the respective defocus values can also be determined from the diffractograms [4-43]. In addition, optical diffractograms can be used to check the microscope alignment, since any residual astigmatism shows up in a deviation from circular of the diffractogram rings. Optical diffractograms are therefore an important aid for experimental high-resolution microscopy. It is, however, time consuming to expose negatives and record diffractograms on an optical bench. During alignment it is considerably easier and faster to record the high-resolution image (using a TV system with image intensifier) in digital form and then calculate the diffractogram in a computer with the aid of a Fourier transform (section 4.2). Since images consisting of 512×512 pixels require only a computing time of the order of a second, the diffractogram is available on-line.

4.6.2 Lattice imaging of crystals

If a thin crystal is imaged with an electron beam parallel to a low-index zone axis in high-resolution mode, several reflections contribute to image formation. The desired reflections can be selected by choosing the appropriate size and position of the objective aperture. A pattern of bright and dark dots is produced in the image by interference with the direct beam; it reflects the lattice periodicity normal to the direction of projection. (If only one reflection (or $+\boldsymbol{g}$ and $-\boldsymbol{g}$) is used in addition to the direct beam, lattice fringes appear instead of a dot pattern.) If the interplanar spacings are greater than the point resolution, Scherzer focus is used and the crystal is so thin that the weak-phase approximation applies, the lattice image can be interpreted directly: the bright dots correspond to the tunnels in the crystal structure and the dark dots indicate the positions of the atomic columns. Since the projected crystal structure is correctly reproduced, this case is referred to as *structure imaging*. When thicker crystals are imaged, dynamic interaction occurs between the diffracted beams and the direct beam. The dark dots then no longer correspond to the atom positions and a comparison with computer-simulated images is required before the images can be interpreted (section 4.6.3).

If the interplanar spacings have the size of the point resolution or less, pairs of atomic columns may appear as a *single* bright or dark dot (see below). In order to determine the positions of the columns in this case, a focal series of images must be recorded. In this series, the defocus is varied in small steps from image to image. The images in this series are then compared with calculated images. This procedure allows the true crystal structure to be clarified (section 4.6.3). Furthermore, lattice images for which the beam direction is parallel to a zone axis have the following properties [4-43]: at certain defocus and thickness values, dots (or fringes) appear with half the interplanar spacings (half spacings). The contrast of the dots or fringes changes periodically from bright to invisible to dark with defocus and thickness. The periodic change with the defocus is due to the CTF oscillations shifting during defocusing. The spatial frequency of the interplanar spacing in question is thus transmitted once with positive CTF and once with negative CTF. At the zeros in between, this spatial frequency remains invisible.

By way of example, the lattice imaging of *semiconductor crystals* is discussed below. Silicon and germanium crystallize with the diamond cubic structure (Fig. 2-2c) whereas compound semiconductors of the AIII BV type such as GaAs exhibit the sphalerite structure. The latter differs from the former only by the base of its structure having one type A and one type B atom instead of two atoms of the same material (e.g. Si). Figs. 4-53a and b show a lattice model of the diamond structure viewed approximately in the [110] and [100] directions. The exact projections in these directions are shown in Figs. 4-53c and d for the sphalerite structure. The lattice constant a and various interplanar spacings are also indicated.

As already stated, lattice images are easier to interpret when the projected interplanar spacings are larger than the point resolution. To check for which interplanar spacings of silicon this holds, the spatial frequencies of the spacings have to be compared with the CTF at optimum defocus. This is done in Fig. 4-54, which shows as an example the CTFs for a 200 kV TEM with 0.24 nm point resolution (C_s = 1 mm) and a 400 kV TEM with a (theoretical) point resolution of 0.16 nm (C_s = 1 mm). In the [110] projection, only the frequencies of the (111) and (200) spacings lie in front of the first zero of the CTF for the 200 kV microscope, whereas at 400 kV this also holds for the (220) spacings. In the 200 kV TEM, the most closely spaced pairs of atoms in the [110] projection with spacing $a/4$ in Fig. 4-53c ((400) spacing in

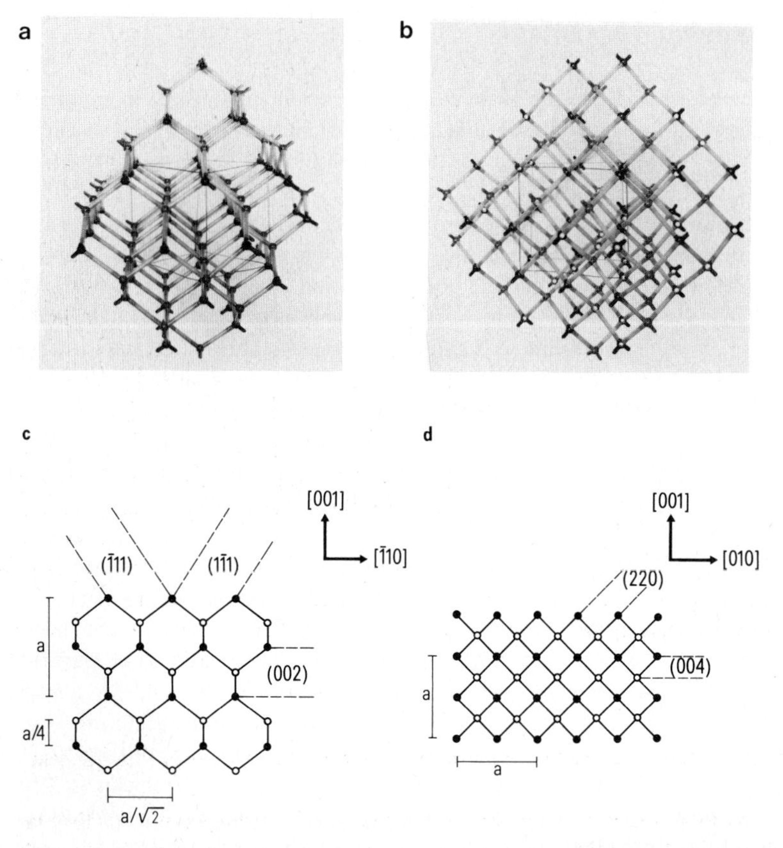

Fig. 4-53. a) b) Model of diamond structure viewed approximately in the [110] (a) and [100] directions (b). c) d) Projections of sphalerite structure in the [110] (c) and [100] directions (d). Full and open circles represent different types of atom, e.g. Ga and As in GaAs. Various lattice planes and spacings are indicated.

Fig. 4-54) are not resolved and appear as a *single* bright or dark dot (Fig. 4-6d). At 400 kV, the (400) spacings are also not in front of the first zero of the CTF but they lie well within the information resolution limit and will be well transmitted if they coincide with an oscillation maximum of the CTF as in Fig. 4-54. In such a case a white double dot pattern will appear which – for a certain specimen thickness – shows the correct spacing of the atom pairs [4-47]. In the [100] projection, the positions of the atoms are correctly reproduced only when viewed in the 400 kV microscope, because the (220) spacing is smaller than the point resolution of the 200 kV microscope. The same applies to the [111] projection. Comparison of the two microscopes demonstrates the advantage of a low point resolution. Different crystal projections – including those with small lattice spacings – can then be used for lattice imaging. True atomic imaging is possible in the ⟨100⟩, ⟨111⟩, and ⟨013⟩ orientations [4-47]. However, to date the [110] projection has principally been used for high-resolution investigations of semiconductor crystals because its {111} spacings are large and hence easy to image. At 200 kV the size of the aperture is generally selected such that the seven innermost beams can pass through. Fig. 4-6d was recorded under these conditions. Because metals have smaller lattice

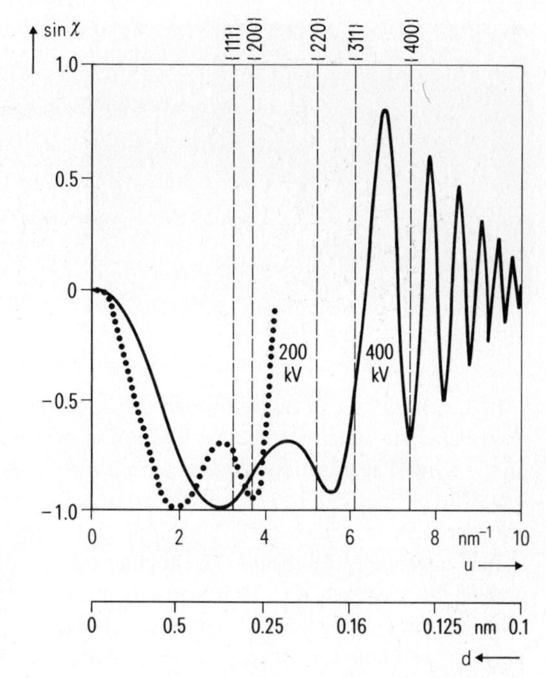

Fig. 4-54.
Contrast transfer function for 400 kV (solid line curve) and 200 kV (dotted curve) microscope at optimum defocus. Both spatial frequencies and spacings are shown on the abscissa. Various lattice spacings for silicon are indicated [4-46].

parameters, a low point resolution is even more important for imaging metals than for semiconductors.

High-resolution investigations may be performed to characterize amorphous-crystalline interfaces, for example. Fig. 4-6d shows that the interface between the thermal oxide and (100) silicon exhibits a roughness of up to two atomic layers. However, since these irregularities also occur along the direction of projection, the atomic positions at the interface cannot be precisely defined. It is apparent that the roughness is best reproduced at specimen regions of minimum thickness. Further examples of amorphous-crystalline interfaces and amorphous interfacial layers are given in Chapter 9.

The positions of the atom columns cannot be determined with precision unless the atomic configuration is constant along the projection direction. With a suitable orientation, this applies to cases such as atomically planar heteroepitaxial interfaces or specific crystal defects: edge dislocations, twin and domain boundaries and tilt grain boundaries, if both the tilt axis and the boundary are parallel to the electron beam. The displacement field of these defects does not depend on the z-direction. By way of example, Fig. 4-55 shows dislocations which compensate the difference in the lattice parameter at the interface between GaAs and InP (misfit dislocations). Since these dislocations run along the [110] direction of projection, they appear as the termination lines of additional, inserted lattice planes.

Lattice images are also used to investigate complex crystal structures with large unit cells. In layer structures, a high density of stacking faults often disrupts the long-range order. Although this proves an insurmountable problem for other methods such as X-ray diffraction, the structure can still be determined with the aid of lattice images. This is shown in section 9.4.4 where a high-temperature superconductor from the BiSrCaCu-oxide system is discussed.

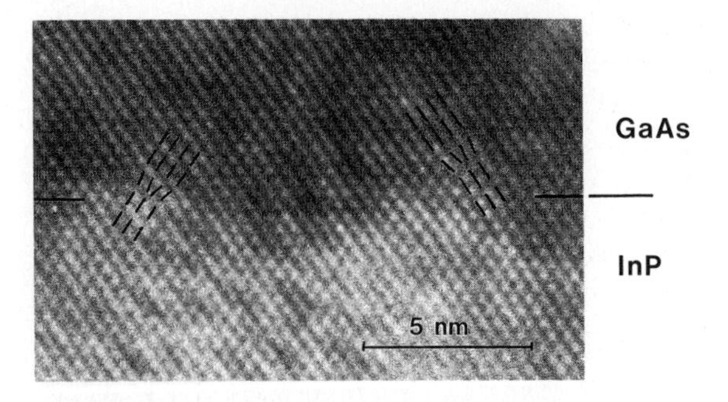

Fig. 4-55.
Lattice image in [110] projection showing misfit dislocations at GaAs-InP interface. The lattice planes around the two dislocations are marked. The GaAs layer was grown by molecular beam epitaxy onto an InP substrate.

Lattice imaging can be performed satisfactorily only when a number of *experimental requirements* are satisfied. These relate to the microscope on the one hand and the specimen on the other. The most important precondition is correct alignment of the microscope (section 4.2). The settings can be checked by means of optical diffraction using amorphous areas of the specimen. In addition, the crystal must be tilted so that the selected zone axis is parallel to the incident electron beam. The angular deviation should be less than 1 mrad. Contamination of the specimen with hydrocarbons must be avoided, because even thin contamination layers may impair the image of specimens as thin as the ones used for high-resolution imaging. Ways of avoiding contamination are explained in section 4.7.3. If LaB_6 cathodes are used, the illumination is so bright that exposure times of 1 – 2 seconds are adequate, even at magnifications as high as $M = 500000:1$. To restrict specimen drift during this time, it is important to ensure high thermal stability in the specimen chamber.

In many instances, it is difficult to thin the specimens to an adequately low thickness of <20 nm. Furthermore, these specimen areas should be uniformly thick and not bent, because the contrast in lattice images is sensitive to changes in both thickness and orientation. If ion-beam milling is chosen for final thinning, an amorphous surface layer is produced on the top and bottom faces of the specimen (section 4.3). One method of keeping this layer thin in proportion to the total specimen thickness is to use very low ion energies (<2 keV) at the final thinning stage.

4.6.3 Computer simulation of lattice images

The fine details of lattice images are sensitive to changes in thickness and defocus and mostly do not permit direct conclusions to be drawn about the projected crystal structure or the atomic positions of defects. As indicated in the preceding section, computer simulation of lattice images is therefore required for image interpretation. The defocus and thickness must be known if experimental images are to be compared with calculated ones. However, both variables can be determined only with limited precision. The defocus can be determined by calibrating the current setting of the objective lens and using the focus of minimum contrast as the reference (starting point for defocusing). This focus value can be adjusted by observing amorphous regions of the specimen. Assuming a wedge-shaped specimen, its thickness can be estimated from the relative position of the specimen area of interest between the first thickness fringe and the edge of the hole.

The calculation of images requires three steps. Initially, a model of the crystal structure or the defect must be set up providing the coordinates for the positions of the atoms. The wave function at the exit surface of the specimen is then calculated for this model, a procedure which must be based on dynamic electron scattering. In the third and final step, the optical transfer characteristic of the imaging system is taken into account. Several methods of calculating the wave function of the specimen are available [4-48]. The multi-slice method allows the most efficient calculations and is used in the majority of cases [4-49, 50]. In this procedure, the crystal is split into numerous thin slices normal to the beam. Since the slices are assumed to be only a few 0.1 nm in thickness, the transmission function for each slice can be expressed by eq. (4-23) (weak-phase object). The wave function at the exit surface of the specimen is calculated in an iterative process involving coupling the wave function of one slice to that of the next one.

In image calculations, both parameters, defocus Δf and thickness t, are always varied. A set of through-focal and through-thickness images will show the combinations of Δf and t at which the actual positions of the atoms are best reproduced by certain image details. When recording experimental images, the goal is to set or use approximately these values of Δf and t. Alternatively, the model of the crystal structure or defect is verified by the attempt to find agreement with experimental images recorded at various Δf and t values. If a critical difference exists between the experimental and theoretical images, the model must be modified and a new set of images calculated. This procedure is repeated until optimum agreement is achieved. The configuration of atomic columns on which the model is based then corresponds to the crystal or defect structure in the specimen.

4.7 Scanning-transmission and analytical electron microscopy

4.7.1 Image contrast in the scanning transmission electron microscope (STEM)

Principle of reciprocity

The signal from the transmitted electrons is generally used in STEM imaging. The requirements imposed on the illumination lens system and the detector configuration were discussed in section 4.2. In order to compare the contrast of STEM images with those from the TEM (designated CTEM or conventional TEM below), we exploit the fact that the ray paths in the STEM and CTEM are related by the principle of reciprocity [4-51]. This principle applies only to elastically scattered electrons and states that in a system with a source at A, the wave amplitude at a point B is the same as that at A when the source is at point B. The reciprocal relationship of the ray paths in the CTEM and STEM is shown in Fig. 4-56. The source A of the CTEM (Fig. 4-56a) is the demagnified image of the crossover (cf. Fig. 4-9). The rays leaving an object point within an angle $2\beta_c$ meet at an image point B. In the STEM (Fig. 4-56b), the electrons emitted from the source B are focused onto the specimen with a divergence angle $2\alpha_s$. The detector A collects the electrons within a cone of an angle $2\beta_s$. Whereas in CTEM all individual image points are recorded simultaneously on the film, the STEM probe scans the specimen and generates the image points one after the other. If the rays of the probe are projected back in the STEM, they scan across a virtual source which corresponds to the image plane in the CTEM. For the principle of reciprocity to be satisfied, the apertures must also fulfill this reciprocal relationship: the divergence angle $2\alpha_c$ of the illumination in the CTEM must be the same as the collection angle $2\beta_s$ in the STEM; likewise, the angle of emergence $2\beta_c$ limited in the CTEM by the objective aperture must be equal to the divergence angle $2\alpha_s$ of the STEM probe. The image contrast is then the same in both the STEM and the CTEM.

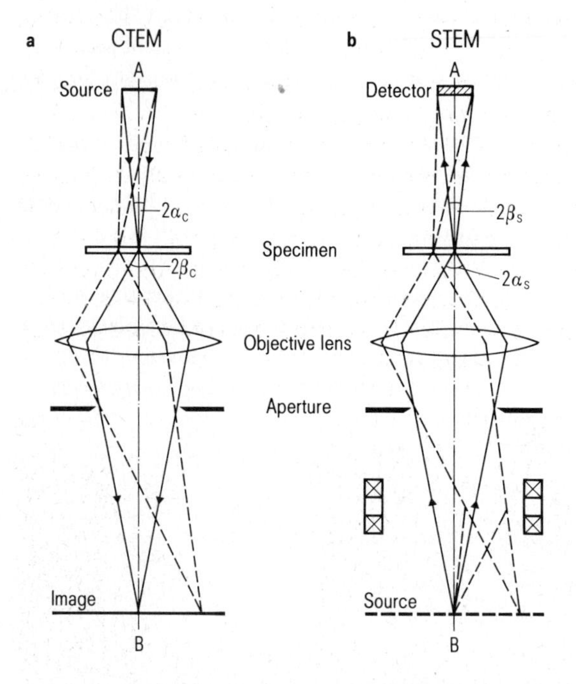

Fig. 4-56. Ray diagrams of CTEM (a) and STEM (b) demonstrating the principle of reciprocity. For the CTEM, the demagnified image of the crossover acts as source. Rays are drawn for the object point at the optical axis (full lines) and for one off-axis point (broken lines).

Generally, however, the above conditions ($\alpha_c = \beta_s$ and $\beta_c = \alpha_s$) are not satisfied in practice. In the CTEM, the condenser is defocused and $2\alpha_c$ has a typical value of 0.5 mrad. For an objective aperture of medium size, the angle of emergence $2\beta_c$ is 5 mrad. In the STEM, a larger divergence angle, for example $2\alpha_s = 10$ mrad, must be used in order to produce a fine probe (cf. eq. (2-23)). An upper limit for α_s follows from the condition that the diffraction disks in the convergent-beam electron diffraction pattern produced with a stationary probe must not overlap: $\alpha_s < \vartheta$. To obtain a high signal-to-noise ratio, the collection angle $2\beta_s$ is made as large as possible. Since the aim is to collect only the direct beam or a single diffracted beam, the Bragg diffraction angle sets an upper limit here too ($\beta_s < \vartheta$), with the result that $2\beta_s$ is typically ≈ 10 mrad. Whereas the divergence angle α_s in the STEM is equal to the emergence angle $2\beta_c$ in the CTEM to within a factor of two, the collection angle $2\beta_s$ in the STEM is more than ten times as large as the beam divergence $2\alpha_c$ in the CTEM. The principle of reciprocity then no longer applies, and the CTEM and STEM images of the same crystalline specimen generally differ in contrast. Moreover, the principle of reciprocity fails for thick specimens where inelastic scattering dominates (see below). However, it is useful for understanding the image contrasts in the STEM.

Image contrast

When the diffraction contrast in a CTEM image is considered, the divergence angle $2\alpha_c$ is generally neglected. According to the principle of reciprocity, however, the large collection angle $2\beta_s$ in the STEM corresponds to a large divergence angle in the CTEM. This can be taken into account by assuming the angle of incidence in the CTEM to vary about a mean angle from $-\beta_s$ to $+\beta_s$. Various excitation errors s result with different angles of incidence. A STEM image with a large angle β_s is then equivalent to an (incoherent) superposition of CTEM images for which s varies through a given range. This range is defined by $s \pm \Delta s$, where $\Delta s = \beta_s |\boldsymbol{g}|$. Eq. (4-19) shows that different effective extinction distances $\xi_{g,\mathrm{eff}}$ derive from the different excitation errors s. In dynamic contrast effects such as thickness fringes, different periodicities are then superposed and their contrast is consequently blurred. Thickness fringes, bend contours and the oscillating contrast of dislocations are thus considerably weakened or disappear entirely (Fig. 4-57). This means that when polycrystalline materials are imaged, the individual grains appear with a more uniform contrast than in the CTEM, because only large differences in their orientation or thickness affect the contrast. Whereas crystal defects in the CTEM generally exhibit contrast details that are darker *and* brighter than the background, defects in the STEM appear only with dark contrast against a uniformly bright background. The invisibility criteria such as $\boldsymbol{g} \cdot \boldsymbol{b} = 0$ used for analysing dislocations and stacking faults in the CTEM can also be used in the STEM [4-53].

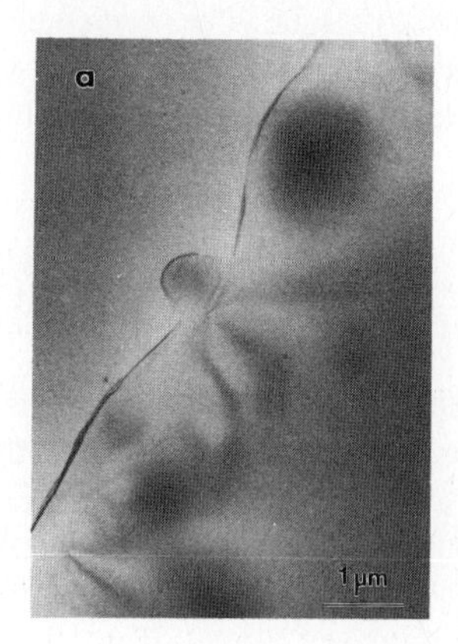

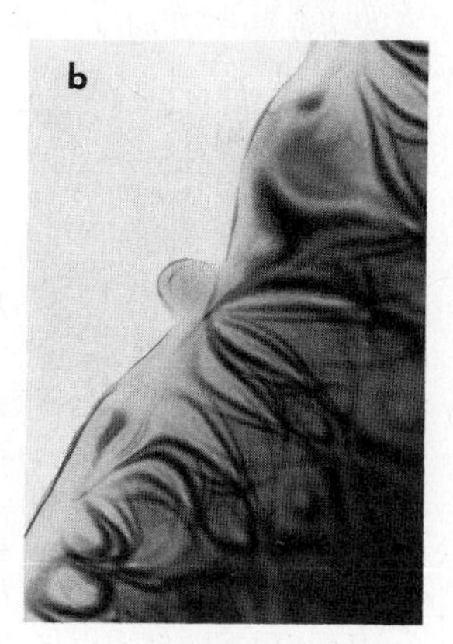

Fig. 4-57.
Comparison of STEM image (a) and CTEM bright-field image (b). Due to the large detector collection angle resulting in a range of various excitation errors s, dynamic contrast effects such as thickness fringes and bend contours are blurred in the STEM image.

Formation of scattering contrast by an amorphous specimen in the STEM is analogous to the process that occurs in the CTEM (section 4.1.6). As with crystalline specimens, a large collection angle is used in order to achieve high image intensity. The scattered electrons can be recorded effectively by an annular dark-field detector. Individual heavy atoms on a thin, light supporting film can then be imaged in a STEM with a field emission gun providing a beam diameter of a few 0.1 nm [4-54]. If a high-angle annular detector with a high inner collection angle is used, only those electrons are detected which are elastically scattered through high angles. In an image formed with this signal, structural features become almost invisible and the contrast is mainly due to differences in atomic number (Z contrast) [4-55].

Inelastic scattering processes become more frequent in thicker specimens. This has the effect of reducing resolution in the CTEM due to the chromatic aberration of the objective lens. This effect cannot occur in the STEM because the lenses located after the specimen have no imaging function. Penetration of amorphous specimens is therefore roughly three times as effective in the STEM as in the CTEM [4-56]. The comparison is more complex for diffraction contrast of defects in crystalline specimens. In this case, the STEM has an advantage only in imaging materials of light elements; for medium and heavy elements the penetration is roughly the same [4-57, 53]. The top-bottom effect limits the resolution of thick specimens in the STEM: elastic and inelastic scattering processes broaden the electron probe as it passes through the specimen. In consequence, details near the top of the specimen appear clearer than those close to the bottom, where the probe is already broadened.

STEM imaging is mainly used during analytical work for selecting the specimen point of interest for microanalysis. Moreover, it is the only imaging mode in dedicated STEMs which are equipped with field emission guns.

4.7.2 Electron energy-loss spectroscopy

Of the methods of analytical electron microscopy illustrated in Fig. 4-7 (section 4.1.7), only electron energy-loss spectroscopy (EELS) will be described here in detail. Electron microdiffraction has already been discussed in section 4.4.4 and X-ray microanalysis of thin foils will be the treated in section 5.6.2. The principle of operation of the magnetic sector-field spectrometer and how it is coupled to the TEM has been described in section 4.2.5. An alternative approach to EELS is to introduce an energy filter between two lenses of the imaging lens system. In addition to energy loss spectroscopy, such an instrument allows direct recording of energy-filtered images.

Energy-loss spectrum

The various inelastic scattering processes in the specimen lead to different energy losses ΔE of the incident electrons. According to the size of ΔE, three regions are distinguished in the energy-loss spectrum (Fig. 4-58). The first of these is the *zero-loss peak*. It includes not only the unscattered electrons, but also electrons which are elastically scattered and those which

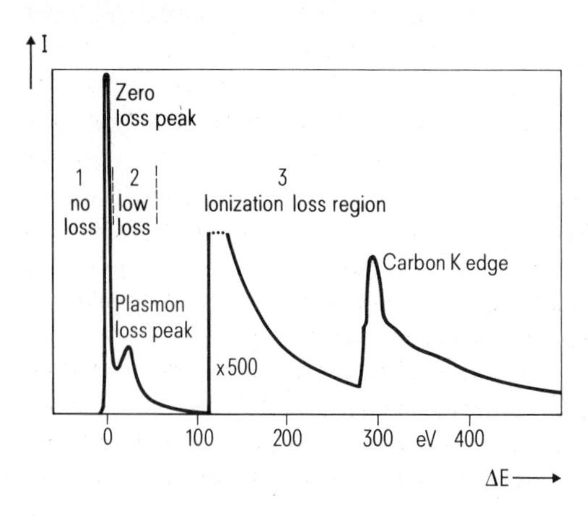

Fig. 4-58.
Electron energy-loss spectrum from thin carbon film. The three different regions of the spectrum are indicated. Since the intensity in region 3 is much smaller than that in regions 1 and 2, region 3 was recorded with 500 times higher gain.

have excited phonons. However, phonon excitation involves the transfer of only a small amount of energy, typically less than 0.1 eV, so that $\Delta E \approx 0$ is satisfied. The width at half maximum of the zero-loss peak is a measure of the energy resolution; it is determined on the one hand by the spectrometer (e.g. by the slit width if alignment is correct), and on the other hand by the energy width of the electron source (approx. 1 eV for a thermionic source). As a rule, the spectrometer is adjusted such that the width of the zero-loss peak is a few eV.

The second part of the spectrum from the zero-loss peak to roughly $\Delta E = 50$ eV is known as the *low-loss region.* Energy losses in this region are due to excitation or ionization of electrons in various binding states (e.g. valence band excitation). In metals and semiconductors, the principal phenomenon is the excitation of plasmons, in other words, collective oscillations of the electron gas. The resulting plasmon peak in the spectrum occurs at energy losses of $\Delta E_{pl} \approx 20$ eV. These losses are proportional to $(n_E)^{1/2}$ where n_E is the density of the free electron gas. This relationship can be used in analysis, for example to determine the local composition of alloys from the peak shift. But this method is applicable in only a limited number of cases. If the thickness of the specimen exceeds the mean free path for plasmon excitation Λ_{pl} (50 nm to 150 nm at 100 kV), the electrons may suffer multiple losses. In this case, a second, weaker peak with twice the energy loss ($2\Delta E_{pl}$) will appear beside the first plasmon peak, and may even be followed by a third. The specimen thickness t can be determined from the intensity ratio of the zero-loss peak I_0 and the first plasmon peak $I_{pl,1}$: $I_{pl,1}/I_0 = t/\Lambda_{pl}$ [4-8]. This is helpful in selecting the optimum specimen thickness (see below).

In the third region of the spectrum (energy-loss greater than 50 eV), the characteristic features are the *edges of the inner-shell ionizations.* These edges are superposed on a high background intensity originating from multiple valence-shell excitations. The ionization losses are produced by electrons which have excited one of the inner shells (e.g. the K shell) of an atom, losing an energy of at least $\Delta E = E_K$, where E_K is the binding energy of the electron emitted from the atom when excited. If kinetic energy is also transferred to this electron, the energy loss is $\Delta E > E_K$. In the energy-loss spectrum, therefore, edges appear with a steep rise in intensity at $\Delta E = E_K$ and a gradual decay towards higher energy losses down to the background level which would have existed in the absence of the ionization edge. The edge energies are characteristic of the different elements and thus permit qualitative analysis in a manner similar to X-ray microanalysis (Chapter 5). Once the intensities of the ionization losses have been determined, quantitative analysis may also be performed (see below). Since the intensity is very low at large energy losses, spectra are usually recorded only up to energy losses of 2000 eV. Depending on the atomic numbers of the elements, the ionization energies of the K, L, M or N shells lie in this range (cf. Fig. 5-3). The mean free paths for inner-shell ionization are of the order of a few μm and are thus roughly two orders of magnitude greater than those for elastic or plasmon scattering. The edge intensities are thus very weak. The mean scattering angles are in the 5 mrad range and, like the angles of plasmon scattering, are therefore considerably smaller than the angles for elastic and phonon scattering, which may be as large as approx. 50 mrad [4-8]. This is of advantage in recording loss electrons.

The fact that the edges have different shapes presents a problem for the *interpretation* of energy-loss spectra. K edges have a characteristic triangular shape with an abrupt onset at the ionization energy E_K. The intensity increase of L and M edges is often more gradual and the maximum is thus delayed to a greater or lesser extent compared with E_L or E_M. Since the edges always lie on top of a rapidly declining background, the change in slope is often the only way of identifying them in the spectrum. If there is uncertainty in determining the element from the onset energy of the edge, the K, L, M and N edges can be distinguished on the basis of their shape. Frequently, a comparison with experimentally determined spectra summarized in a catalogue [4-58] can be of assistance. In the case of heavier elements, it is advisable to record an X-ray spectrum in addition to the energy-loss spectrum.

Where energy resolution is high (≈ 1 eV) and the signal-to-noise ratio is good, fine structures are visible around the edge onset – the "near-edge structures". The shape of these structures depends on the density of the unoccupied bound states and is affected by the bonding conditions and the coordination of the atoms. Beyond the edge, weak periodic modulations occur which extend several 100 eV beyond the edge. This extended energy-loss fine structure (EXELFS) is analogous to the EXAFS effect (extended X-ray absorption fine structure) in X-ray absorption spectra and contains information about the bonds to the nearest neighbor atoms. Thus, both types of fine structure yield information on chemical bonding conditions. They cannot be easily interpreted, but can nevertheless serve as fingerprints for the identification of chemical compounds.

To obtain a good signal-to-noise ratio when recording spectra, a large acceptance angle β is used in the spectrometer. However, the energy resolution drops correspondingly and β must be selected such that the desired energy resolution is still attained. When β is large (e.g. 10 mrad), a large proportion of all ionization processes is recorded. In contrast, the X-ray detector used in X-ray microanalysis (XMA) collects only a small part (approx. 1%) of the characteristic X-rays emitted in all directions. This advantage of EELS is counterbalanced by the drawback of a much poorer peak-to-background ratio than in XMA, because the characteristic edges lie on top of a higher background. The signal-to-noise ratio also improves as specimen thickness increases. However, a thick specimen produces multiple-scattering processes. Electrons which have suffered both an ionization loss and a plasmon loss produce a small replica of the plasmon peak beside the edge. In consequence, the edges produced by thick specimens are not as clear and the changes in their shape make identification more difficult. It is best to use thicknesses between once and twice the mean free path for inelastic scattering [4-8].

Quantitative microanalysis using ionization losses

The intensities in the ionization-loss spectrum can be used for quantitative analysis. The results of X-ray microanalysis (XMA) of thin, unsupported films (section 5.6.2) provide the basis for calculating these intensities. As shown in that section, the ionizations n_A of atoms of an element A per incident electron are:

$$n_A = C_A \cdot \sigma_A \cdot t$$

(C_A is the concentration of element A, i.e. number of atoms A per unit volume, σ_A the ionization cross section, and t the specimen thickness). If the mass fraction w_A is used instead of C_A, with $C_A = w_A \cdot \varrho \cdot N_a / A_A$ (eq. (1-1), ϱ is the density, N_a the Avogadro number, and A the atomic weight), the resulting intensity of the ionization losses I_A for a probe current i is:

$$I_A = \frac{i}{e} \cdot w_A \cdot \left(\frac{N_a}{A_A}\right) \cdot \sigma_A \cdot \varrho \cdot t. \tag{4-29}$$

This equation is obviously similar to eq. (5-36) for the XMA of thin specimens. In contrast to the latter, however, the ionization cross section σ_A is the only physical parameter included. In accordance with [4-59], it is usual to replace the probe current in eq. (4-29) (or i/e) by the intensity of the zero-loss peak I_0. If a specimen is a compound of elements A and B and the intensities of the edges of both elements are measured in a spectrum, the ratio of their mass fractions w_A / w_B is:

$$\frac{w_A}{w_B} = \frac{\sigma_B/A_B}{\sigma_A/A_A} \cdot \frac{I_A}{I_B} . \tag{4-30}$$

As in XMA, parameters such as local specimen thickness and probe current which cannot be determined easily are eliminated by forming the ratio.

To determine the intensities I_A and I_B of the edges, the spectrum must be integrated over a certain energy range Δ starting from the edge (Fig. 4-59). However, this can be done only after the background has been subtracted. It is not possible to describe the background theoretically, but it can be represented by an empirical expression of the form $A \cdot \Delta E^{-r}$, where the constants A and r have to be fitted by comparison with the experimental curve. If this fit is made on the low-loss side of the edge, the background can be extrapolated beyond the edge (Fig. 4-59).

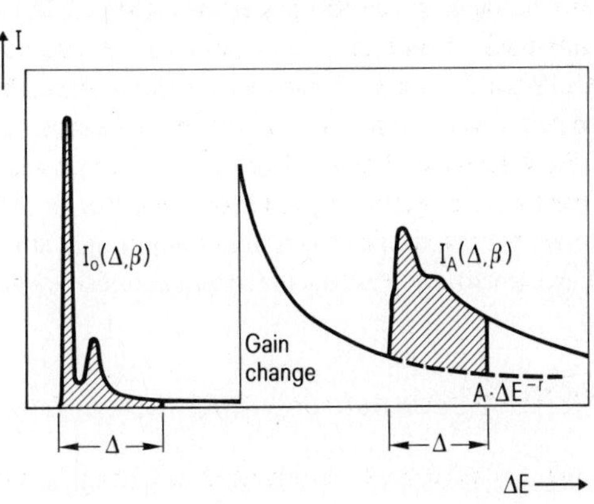

Fig. 4-59.
Schematic energy-loss spectrum showing parameters for quantifying a spectrum. After background extrapolation and subtraction, the intensity of the ionization losses of element A $I_A(\Delta,\beta)$ is obtained by integration over an energy range Δ. Both intensities of ionization losses $I_A(\Delta,\beta)$ and of zero losses $I_0(\Delta,\beta)$ are a function of the energy range Δ and the acceptance angle β of the spectrometer.

Past the edge, the intensity distribution of the ionization losses drops only slowly with increasing energy losses. To determine the total intensity I_A, it would be necessary to integrate the energy-loss spectrum over a large energy range of $\Delta > 200$ eV. If this range is too large, however, errors arise due to inaccuracies in extrapolating the background. An optimum range between 50 and 100 eV is therefore chosen. Apart from Δ, I_A also depends on the angle 2β of the inelastically scattered electrons accepted by the spectrometer. As already indicated, β should be as large as possible to improve the signal-to-noise ratio (e.g. 10 mrad) but in any case larger than the characteristic angle for inelastic scattering $\Theta_E = E_K/2E_0$. Despite the restrictions imposed by Δ and β, 50% of all the energy-loss electrons are recorded under favorable conditions.

The experimentally determined intensity of the ionization losses is thus a function of Δ and β: $I_A(\Delta,\beta)$. To permit the use of eq. (4-29) in practice, the intensity of the zero-loss peak I_0

used instead of i/e and the cross section σ must also be treated as a function of Δ and β. $I_0(\Delta,\beta)$ is obtained by integrating the spectrum beyond the zero-loss peak to the energy loss Δ (Fig. 4-59). The partial ionization cross section $\sigma(\Delta,\beta)$ which is used instead of σ is usually calculated by an approximation method for the K and L edges [4-60]. Eq. (4-30) then allows simple calculation of the composition of a binary alloy. The accuracy of quantitative analyses of this type is typically 20% [4-61]. EELS is less sensitive than XMA in analysing small mass fractions; the detection limit is generally around a few percent. In contrast, EELS is extremely sensitive in detecting small masses [4-8]. Both accuracy and detection limits will improve when parallel detection is employed. The main advantage of EELS when compared to XMA is the possibility of also analysing light elements which are difficult to detect with conventional energy-dispersive spectrometers.

In addition, during quantitative analysis using EELS the following points must be considered:

- As in the XMA of thin single crystals, no reflection should be strongly excited. If electrons are strongly diffracted, the resultant Bloch waves may preferentially ionize a particular type of atom in the crystal lattice.
- In the STEM mode, the divergence angle $2\alpha_0$ of the probe may attain the same order as the acceptance angle 2β of the spectrometer. The collection efficiency is then reduced [4-61].
- Multiple scattering losses which occur in thick specimens change the spectrum and thus impair a quantitative analysis.

Finally, it should be mentioned that the spatial resolution possible when using EELS with a small probe diameter is considerably better than that attainable by XMA. Beam broadening due to elastic scattering (cf. eq. (5-40)) plays no part in EELS, because only those scattered electrons are collected that leave the specimen within the acceptance angle 2β of the spectrometer. The maximum possible broadening is thus $b = 2\beta \cdot t$. When $\beta = 10$ mrad and the specimen thickness t is 100 nm, the value of b is only 2 nm.

4.7.3 Specimen contamination and radiation damage

The mechanism of *specimen contamination* under the electron beam has already been described in section 3.3: hydrocarbon molecules diffusing on the surface of the specimen are cross-linked by the electron beam and thus become fixed. Whereas contamination reduces the SE yield in the SEM, it particularly impairs TEM analysis with small probes [4-62]. The contamination needles or cones which are produced during analysis with a stationary probe degrade spatial resolution in XMA because of additional scattering (cf. section 5.6.2). In EELS, an extra carbon edge is produced and the local specimen thickness increased, leading to undesirable multiple scattering.

Contamination due to residual gas is low in modern microscopes which generally use ion gettering pumps and cryoshields for the specimen chamber. The specimen becomes contaminated during preparation or storage in the laboratory atmosphere. This contamination can be removed or reduced by the following methods: washing in methanol, short ion-beam thinning or slight baking under vacuum. If contamination persists in TEM despite these measures, it can be reduced by high-dose irradiation of a large specimen area (e.g. 100 μm in diameter). The hydrocarbons in this area are then cross-linked and their diffusion reduced. Diffusion of the hydrocarbon molecules on the surface of the specimen and, in turn, the formation of contamination needles can also be prevented by the use of a cold stage (with liquid nitrogen cooling of the specimen).

Radiation damage in inorganic materials is caused by the interaction of the fast electrons with atomic nuclei and electrons. In both cases, permanent atomic displacements result [4-63]. The direct collision of a fast electron with the nucleus may cause *knock-on damage*. The displaced atom represents an interstitial which forms a Frenkel defect together with the remaining vacancy. A certain displacement energy is required if an atom is to be removed from its lattice position through a saddle point to a second stable position. This energy is of the order of 10 to 30 eV and also depends on crystal orientation. The maximum possible energy transfer in an elastic collision between electron and nucleus is defined by the mass of the atom and the electron energy. For knock-on damage to occur, this energy transfer must be greater than the displacement energy. From this a minimum incident electron energy results, known as the threshold energy, for a particular atom or element. This energy is 145 keV for silicon and 400 keV for copper, for example. Although knock-on damage in silicon can occur at energies as low as 200 keV, it becomes particularly significant in high-voltage electron microscopy with voltages above, say 400 kV. The excitation of electrons in metal and semiconductor atoms does not produce displacements. In ion crystals such as alkali halides (e.g. NaCl) in contrast, the excitation energy may be converted into a momentum acting on the nucleus [4-63] and producing point defects. This process is known as *radiolysis*. The point defects which are produced by both types of damage can then migrate via thermal diffusion and may agglomerate to form defect clusters or dislocation loops.

Specimen heating is seldom a cause of damage to the specimen since high-energy electrons transfer little energy to the thin specimens used in the TEM. This follows from Fig. 4-58, inter alia, where the intensity of the energy-loss electrons is low in comparison with the zero-loss intensity. Specimen heating is therefore generally negligible. A significant temperature rise occurs only in thin layers with poor thermal conductivity (e.g. carbon films) subjected to irradiation over a large area [4-2]. As a rule, however, only the observed area of the specimen is irradiated and this area is usually small. Moreover, the thickness of thinned specimens with wedge-shaped hole edges increases with the distance from the edge of the hole, thus ensuring effective heat dissipation.

The imaging of well-insulating specimens such as ceramics in the TEM may be obstructed by electrical *charging*. Although the high-energy electrons pass the specimen, secondary electrons are excited. This results in positive charging which contrasts with the SEM (section 3.4), where the incident electrons remain in the specimen. The charging may reach a level of several 100 V [4-63] and has a detrimental effect on imaging. As in the SEM, a remedy is to coat the specimen with a thin, conductive film. Carbon films 10 to 20 nm in thickness deposited by evaporation (section 3.3) are particularly suitable for the TEM, because they have only a minor effect on imaging. A thin carbon film can also be produced in an ion milling system by placing a piece of carbon (graphite) on the specimen mount and bombarding it with argon ions. This removes carbon by sputtering and within a short time ($<$ 1 minute) the specimen mounted beneath the piece of graphite is coated with a conductive film.

4.8 Other Methods

4.8.1 Overview

The methods used in scanning electron microscopy such as the electron beam induced current (EBIC) and cathodoluminescence (CL) techniques (section 3.6) were also applied to thin specimens in the TEM [4-64]. However, problems are encountered here since surface recombination generally takes place on both surfaces of the TEM specimen. In the case of EBIC, this means that the density of free charge carriers, which could be separated for generating an EBIC signal, is small in thin specimen areas. In CL, the radiative recombination decreases strongly with decreasing specimen thickness. CL of thin specimens in a STEM, however, provides a better spatial resolution than CL of bulk specimens [4-65].

Two further methods or contrast effects will be briefly explained below: the occurrence of Fresnel fringes at the specimen edge and at interfaces, and the imaging of magnetic domains.

4.8.2 Fresnel fringes

In the case of over- or underfocusing of the objective lens, dark and bright fringes are observed at the specimen edge, known as Fresnel fringes. To explain their appearance, the edge of an opaque specimen onto which a parallel beam is incident is shown in Fig. 4-60a. Since we assume a defocused image, we consider the intensity distribution in a plane which lies below the specimen at a distance of the defocus Δf. From the edge of the specimen, scattered waves are emitted which interfere with the incident wave, thus producing the Fresnel fringes. Their intensity distribution is shown in Fig. 4-60 b. It was obtained from wave-optical calculations [4-2]. The positions x_n of the maxima are given by [4-2]

$$x_n = [\lambda \cdot \Delta f \cdot (8n - 5)/4]^{1/2} \quad (4\text{-}31)$$

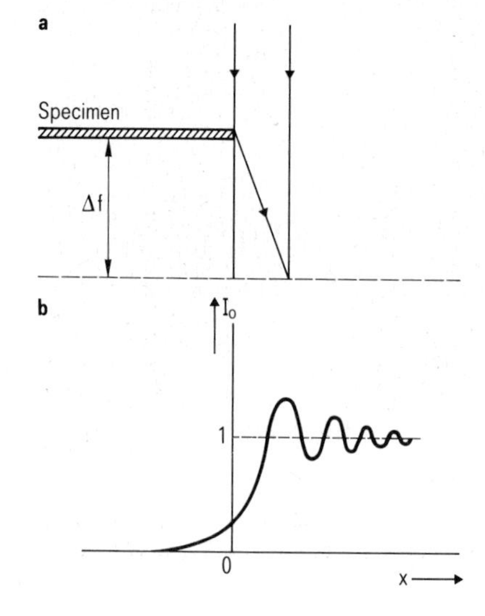

Fig. 4-60.
Formation of Fresnel fringes at specimen edge.
a) Waves which are scattered at the edge interfere with the incoming wave, b) Intensity profile perpendicular to the edge in a plane lying at the defocus Δf below the specimen (overfocusing since the objective lens is overexcited).

(λ is the electron wavelength), and their distances increase with defocus. However, several fringes, i.e. maxima with $n > 1$, are observed only at very low beam divergence (with a very strongly defocused condenser 2). Normally, the beam divergence is so large that the higher-order fringes are blurred and only the first maximum and the first minimum are visible. For overfocusing as in Fig. 4-60, therefore, a bright and a dark fringe appear at the specimen edge with the dark fringe lying outside. For underfocusing the outer fringe is bright. This explains the bright or dark contrast of small voids with under- or overfocusing (Fig. 4-50). Fresnel fringes at small holes in the specimen are used to correct the objective astigmatism. In the case of incorrect adjustment, the defocus Δf differs in different radial directions and the Fresnel fringes therefore show no uniform contrast along the hole edge.

Analogous fringes occur when imaging interfaces lying parallel to the electron beam where the materials on both sides have different internal potentials. Fig. 4-61 shows an example of a cross section through a thin SiO_2 layer lying between the silicon substrate and a polysilicon layer. At both interfaces, bright and dark fringes appear. In contrast to the specimen edge, where the internal potential is always smaller in vacuum than in the specimen, here the internal potential is smaller in the SiO_2 than in silicon. For this reason, in analogy to the Fresnel fringes, the bright fringes point towards the oxide for underfocusing (Fig. 4-61a). The oxide thickness can be most accurately determined at minimum defocus. But the contrast at thin specimen areas is then low. To allow the position of the interface and thus the oxide thickness to be obtained at somewhat greater defocus also, which gives a better contrast, experimental images were compared with calculated contrast profiles [4-66, 67]. This showed that the interfaces can be assumed to be in the middle between the respective bright and dark fringes to an accuracy of about 0.5 nm. Defocused imaging was also used to detect and characterize intergranular films in ceramics [4-68, 69].

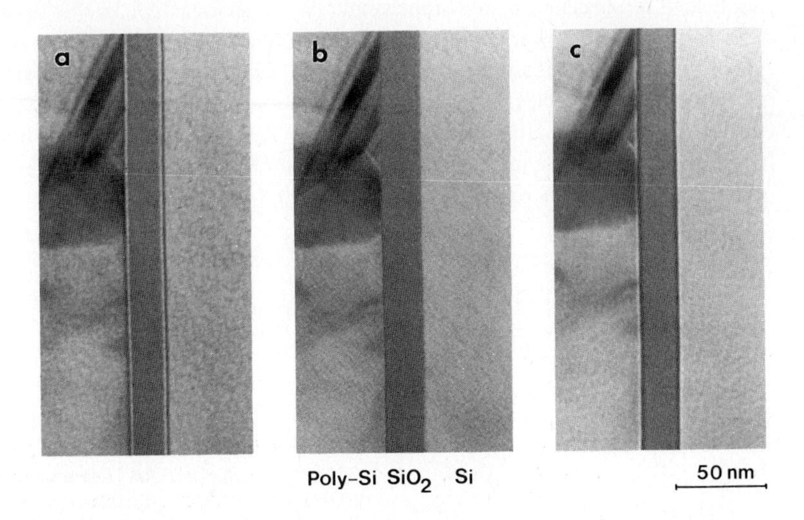

Fig. 4-61. Bright-field images of cross section through SiO_2 layer between the silicon substrate (right) and a polysilicon film (left). a) Underfocus, b) In-focus, c) Overfocus. In the defocused images, Fresnel type fringes occur at the interfaces.

4.8.3 Imaging of magnetic domains

In the case of thin ferromagnetic films, the magnetization generally lies in the film plane. In adjacent domains of the film it is usually rotated by approximately 180°. The contrast effects used to image the domain structure are based on the deflection of the incident electrons by the magnetic induction $\boldsymbol{B}$ in the specimen (Lorentz force). This field of electron microscopy is therefore also known as Lorentz microscopy. The angle of deflection by a magnetic field was specified in eq. (2-11) (section 2.2), the distance l for which the magnetic field acts being given by the specimen thickness t. The deflection angles γ are small due to the small specimen thicknesses of say, 0.1 µm. For 100 keV electrons they have a value of around $\gamma = 10^{-1}$ mrad. To ensure that contrast effects can still be observed at such low deflection, the divergence of the incident beam must be adjusted to about 10^{-2} mrad, i.e. significantly less than γ. This can be done by strongly defocusing condenser lens 2. The domain structure in the specimen can be changed by the high magnetic field of the objective lens. This is most simply avoided by switching off the objective lens and using the first intermediate lens for imaging. An alternative approach is to use specially designed pole pieces which produce low magnetic field at the location of the specimen. Special methods must be used to image the domains or domain walls: *defocused imaging (Fresnel mode) or the displaced aperture method (Foucault mode)* [4-70, 2, 71].

Fresnel mode

Fig. 4-62a shows how the different domains in the specimen deflect the incident electrons in opposite directions. The electron beams then either diverge or partly overlap. By defocusing the objective lens (overfocusing in this case) the intensity distribution is imaged in a plane lying below the specimen by a distance equal to the defocus Δf. In this defocused plane, the domain walls then appear with dark or bright contrast (Fig. 4-62b). Fig. 4-63 shows an example of the domain structure in a nickel-iron film. The contrast of the domain walls is reversed when the type of defocusing is changed (from overfocus to underfocus or vice versa). The contrast width of the domain walls is given by $\Delta x = 2\gamma \cdot \Delta f$ and bears no relation to the real wall width within which the magnetization changes direction. To image the walls with a width of say, 0.2 µm, a defocus of $\Delta f = 1$ mm is required for $\gamma = 10^{-1}$ mrad.

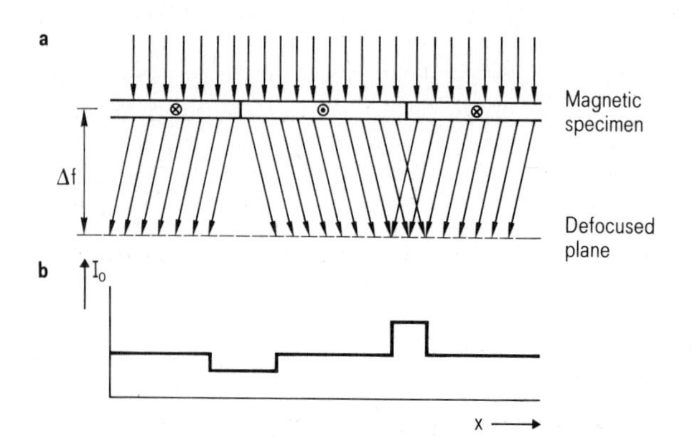

Fig. 4-62. Defocused imaging of magnetic domains. a) Due to the antiparallel magnetization of the domains, the electron trajectories either diverge or converge at the domain walls. b) Image intensity profile across domain structure: the walls appear as dark or bright lines.

Fig. 4-63. Defocused image of magnetic domains in nickel-iron film. The structure of fine dark and bright lines within the domains is due to local fluctuations of the magnetization (ripple). The mean direction of the magnetization is always perpendicular to the ripple structure [4-70].

Foucault mode

If the diffraction pattern is considered, the different deflection of the electrons in the various domains leads to a splitting of the spot of the direct beam into two spots. If the objective aperture is shifted such that only one of the two beams can pass through it, the associated domains appear bright on the whole. In the other domains the electron beam strikes the aperture so that they appear dark. In contrast to the Fresnel mode, the specimen in this case is imaged in focus. This has the advantage of imaging both the domain structure and the microstructure with high resolution.

4.9 References

[4-1] D.E. Newbury in D.C. Joy, A.D. Romig Jr., J.I. Goldstein (Eds.): *Principles of Analytical Electron Microscopy.* Plenum Press, New York 1986, 1

[4-2] L. Reimer, *Transmission Electron Microscopy.* Springer, Berlin 1984

[4-3] H. Bethge, J. Heydenreich (Eds.): *Electron Microscopy in Solid State Physics.* Elsevier, Amsterdam 1987

[4-4] D.J. Joy, A.D. Romig Jr., J.I. Goldstein (Eds.): *Principles of Analytical Electron Microscopy.* Plenum Press, New York 1986

[4-5] R.H. Geiss, A.D. Romig Jr., in [4-4], 29

[4-6] W. Neumann, R. Hillebrand, Th. Krajewski, in [4-3], 265

[4-7] D.B. Williams, J.I. Goldstein, C.E. Fiori, in [4-4], 123

[4-8] D.C. Joy, in [4-4], 249

[4-9] M.H. Loretto, *Electron Beam Analysis of Materials.* Chapman and Hall, London 1984

[4-10] O.L. Krivanek, C.C. Ahn, R.B. Keeney, *Ultramicroscopy 22,* 103 (1987)
[4-11] P.J. Goodhew in A.M. Glauert (Ed.): *Practical Methods in Electron Microscopy, Vol 11: Thin Foil Preparation for Electron Microscopy.* Elsevier, Amsterdam 1985
[4-12] H. Bartsch, in [4-3], 564
[4-13] J.C. Bravman, R.M. Andersen, M.L. McDonald (Eds.), *Specimen Preparation for Transmission Electron Microscopy of Materials.* Mat. Res. Soc. Symp. Proc. Vol. 115, Pittsburgh 1988
[4-14] B.O. Kolbesen, K.R. Mayer, G.E. Schuh, *J. Phys. E 8,* 197 (1975)
[4-15] Gatan Inc., Warrendale, PA 15086 USA, Instruction Manual for Dual Ion Mill
[4-16] N.G. Chew, A.G. Cullis, *Ultramicroscopy 22,* 175 (1987)
[4-17] R.B. Marcus, T.T. Sheng, *Transmission Electron Microscopy of Silicon VLSI Circuits and Structures.* John Wiley & Sons, New York 1983
[4-18] H. Rehme, H. Oppolzer, *Siemens Forsch.- u. Entwickl.-Ber. 14,* 193 (1985)
[4-19] S.J. Klepeis, J.P. Benedict, R.M. Anderson, in [4-13], 179
[4-20] H. Oppolzer in A.G. Cullis, D.B. Holt (Eds.): *Microscopy of Semiconducting Materials, 1985.* Inst. Phys. Conf. Ser. No. 76: Sect. 11. Adam Hilger, Bristol 1985, 461
[4-21] H. Kakibayashi, F. Nagata, *Jap. J. Appl. Phys., 24,* L905 (1985)
[4-22] H. Formanek, *Ultramicroscopy 4,* 227 (1979)
[4-23] M. v. Ardenne, *Tabellen zur angewandten Physik Vol. 1,* VEB Deutscher Verlag der Wiss., Berlin 1962, p. 473
[4-24] W.F. McClune (Ed.): *Powder Diffraction File.* Int. Center for Diffraction Data, Swarthmore 1988
[4-25] P.B. Hirsch, A. Howie, R.B. Nicholson, D.W. Pashley, M.J. Whelan, *Electron Microscopy of Thin Crystals.* Butterworths, London 1965
[4-26] J.W. Edington, *Practical Electron Microscopy in Materials Science.* McMillan, London 1975
[4-27] J.C.H. Spence, R.W. Carpenter, in [4-4], 301
[4-28] P.M. Kelly, A. Jostsons, R.C. Blake, J.G. Napier, *Phys. Stat. Sol. A31,* 771 (1975)
[4-29] J. Tafto, J.C.H. Spence, *J. Appl. Cryst. 15,* 60 (1982)
[4-30] B.F. Buxton, J.A. Eades, J.W. Steeds, G.M. Rackham, *Phil. Trans. R. Soc. 281,* 171 (1976)
[4-31] J. Gjonnes, A.F. Moodie, *Acta Cryst. 19,* 65 (1965)
[4-32] J.W. Steeds, in [4-52], 387
[4-33] D.B. Williams, *Practical Analytical Electron Microscopy in Materials Science.* Verlag Chemie Internat., Weinheim 1984
[4-34] J. Mansfield, *Convergent-Beam Electron Diffraction of Alloy Phases.* Adam Hilger, Bristol 1984
[4-35] M. Tanaka, M. Terauchi, *Convergent-Beam Electron Diffraction.* Jeol, Tokyo 1985
[4-36] Y. Tanishiro, K. Takayanagi, K. Yagi, *J. Microscopy 142,* 211 (1986)
[4-37] G. Thomas, M.J. Goringe, *Transmission Electron Microscopy of Materials.* John Wiley & Sons, New York 1979
[4-38] M. Rühle, M. Wilkens in R.W. Cahn, P. Haasen (Eds.): *Physical Metallurgy.* North-Holland, Amsterdam 1983
[4-39] A.K. Head, P. Humble, L.M. Charebrough, A.J. Morton, C.T. Forwood, *Computed Electron Micrographs and Defect Identification.* North-Holland, Amsterdam 1973
[4-40] S. Amelinckx, R. Gevers, J. Van Landuyt (Eds.): *Diffraction and Imaging Techniques in Material Science.* North-Holland, Amsterdam 1978.
[4-41] A. Howie, M.J. Whelan, *Proc. Roy. Soc. A267,* 206 (1962)
[4-42] M. Wilkens, in [4-40], 185
[4-43] J.C.H. Spence, *Experimental High-Resolution Electron Microscopy.* University Press, Oxford 1981
[4-44] W. Neumann, R. Hillebrand, P. Werner, in [4-3], 97
[4-45] J.C. Jones, *J. Materials Science 19,* 533 (1984)
[4-46] A. Ourmazd, K. Ahlborn, K. Ibeh, T. Honda, *Appl. Phys. Lett. 47,* 685 (1985)
[4-47] A. Bourret, J.L. Rouviere, J. Spendeler, *phys. stat. sol (a) 107,* 481 (1988)
[4-48] P.G. Self, M.A. O'Keefe, P.R. Buseck, A.E.C. Spargo, *Ultramicroscopy 11,* 35 (1983)
[4-49] P. Goodman, A.F. Moodie, *Acta Cryst. A30,* 280 (1974)
[4-50] J.M. Cowley, A.F. Moodie, *Acta Cryst. 10,* 609 (1957)
[4-51] J.M. Cowley, in [4-4], 77
[4-52] J.J. Hren, J.I. Goldstein, D.C. Joy (Eds.): *Introduction to Analytical Electron Microscopy.* Plenum Press, New York 1979
[4-53] C.J. Humphreys, in [4-52], 305

[4-54] M. Isaakson, M. Ohtsuki, M. Utlaut, in [4-52], 343
[4-55] S.J. Pennycook, S.D. Berger, R.J. Culbertson, *J. Microscopy 144,* 229 (1986)
[4-56] T. Groves, *Ultramicroscopy 1,* 15 (1975)
[4-57] H.L. Frazer, I.P. Jones, M.H. Loretto, *Phil. Mag 35,* 159 (1977)
[4-58] C.C. Ahn, O.L. Krivanek, *EELS Atlas,* Gatan Inc., Warrendale, PA 15086, USA
[4-59] R.F. Egerton, C.J. Rossouw, J.J. Whelan in J.A. Venables (Ed.): *Developments in Electron Microscopy and Microanalysis.* Acad. Press, London 1976, 129
[4-60] R.F. Egerton, *Ultramicroscopy 4,* 169 (1979)
[4-61] D.C. Joy, in [4-4], 277
[4-62] J.J. Hren, in [4-4], 353
[4-63] L.W. Hobbs, in [4-52], 437
[4-64] P.M. Petroff, D.V. Lang, J.L. Strudel, R.A. Logan in Om Johari (Ed.): *Scanning Electron Microscopy, vol. I.* SEM Inc., AMF O'Hare 1978, 325
[4-65] J. Cibert, P.M. Petroff, G.J. Dolan, S.J. Pearton, A.C. Gossard, J.H. English, *Appl. Phys. Lett. 49,* 1275 (1986)
[4-66] P. Wurzinger, P. Pongratz, P. Skalicky, *Phil. Mag. A61,* 35 (1990)
[4-67] F.M. Ross, W.M. Stobbs, *Surf. Interface Anal. 12,* 35 (1988)
[4-68] D.R. Clarke, *Ultramicroscopy 4,* 33 (1979)
[4-69] J.N. Ness, W.M. Stobbs, T.F. Page, *Phil. Mag. A54,* 679 (1986)
[4-70] E. Fuchs in G. Schimmel, W. Vogell (Eds.): *Methodensammlung der Elektronenmikroskopie.* Wiss. Verlagsges., Stuttgart 1970, section 4.2.4
[4-71] D. Hesse, K.-P. Meyer in [4-3], 496

5 Electron beam X-ray microanalysis

5.1 Principle

5.1.1 X-ray spectra

X-ray spectrometry is based on the emission of X-rays from the atoms of a solid when struck by particles or waves of sufficiently high energy, followed by the relaxation of the atoms to the ground state after ionization. This effect has been described in section 2.4.3. In X-ray microanalysis, electrons are used to excite the atoms, since they can be focused on very small areas. The X-ray spectrum produced by electron excitation consists of two parts, a continuous spectrum of deceleration radiation, known as bremsstrahlung, and a line spectrum of the characteristic radiation.

The *bremsstrahlung* results from a change in the energy and direction of the incident electron in the positive field of the atomic nucleus. The electron approaches the atom with an energy E_0 and is deflected from it with a reduced energy E_0'. The energy difference $\Delta E = E_0 - E_0'$ can be emitted as an X-ray quantum. According to the energy-frequency relation of quantum mechanics, $E = h \cdot \nu = h \cdot c/\lambda$ (h is Planck's constant, c the speed of light, cf. section 2.2.1), the X-ray quanta correspond to radiation with a wavelength $\lambda = h \cdot c/\Delta E$. From the universal constants $c = 2.998 \cdot 10^8\ \mathrm{ms^{-1}}$ and $h = 4.136 \cdot 10^{-15}$ eV s, the following wavelength dependence on the energy is then obtained

$$\lambda = (12.399/\Delta E) \cdot 10^{-10}\ \mathrm{m}$$

(E in keV).

The many electrons of the incident electron beam are decelerated by the atoms of the material in very different ways, producing a continuous deceleration spectrum. The maximum X-ray energy at the short wavelength end of the continuous spectrum, corresponding to a limiting wavelength λ_{min}, is due to the fact that an X-ray quantum can possess an energy no higher than that of the exciting electrons (Duane-Hunt limit [5-1])

$$\lambda_{min} = h \cdot c/E_0 = h \cdot c/(e \cdot U_0)$$

$$\lambda_{min} = (12.399/U_0) \cdot 10^{-10}\ \mathrm{m} \qquad (5\text{-}1)$$

(U_0 is the acceleration voltage of the incident electrons in V). The fundamental, experimentally determined profiles of a number of deceleration spectra are shown in Fig. 5-1. After a maximum, the intensity drops off sharply with increasing wavelength. This drop is due to the absorption of the long wavelength radiation in the material of the solid.

Although the intensity of the bremsstrahlung is a linear function of the atomic number Z, it contains no additional information on the elemental composition. In contrast, the *charac-*

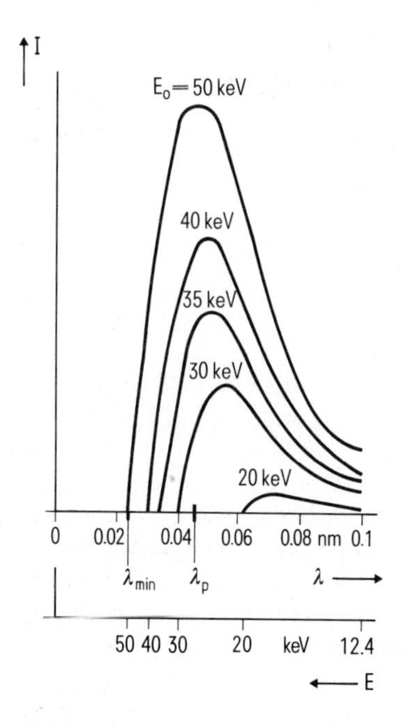

Fig. 5-1.
Basic waveforms of the spectral intensity distribution of the bremsstrahlung of a bulk specimen for various primary energies E_0. With increasing electron energy E_0, the wavelength limit λ_{min} becomes smaller in line with eq. (5-1). The maximum λ_p of the spectral intensity corresponds to deceleration processes in which the electrons emit about 70% of their energy as radiation. λ_{min} and λ_p are specified for $E_0 = 50$ keV. The curves follow the law $I \sim Z \cdot (E_0 - E)/E$ [5-2].

teristic X-ray spectrum exhibits sharp maxima (Fig. 2-42), whose position on the energy scale may be assigned to a specific type of atom according to a simple law. The characteristic spectrum thus plays the decisive role in the material analysis, whereas the continuous spectrum represents a background which hinders the analysis. The characteristic spectra reflect the structure of the electronic shells near the atomic nucleus (cf. section 2.1.2) in the same way as the structures of the outer electron shells of the atoms are reflected by the optical spectra.

Whereas the bremsstrahlung is due to interactive processes between the primary electrons and the atomic nuclei, the characteristic radiation is caused by the interaction of the primary electrons with the atomic shell electrons. High-energy primary electrons can remove atomic electrons from the inner (K,L,M, ...) shells. The precondition for this ionization process is that the energy E_0 is at least as great as the binding energy E_b of the atomic electron in the relevant shell. The ionization probability, which represents a measure of the number of ionization processes occurring, is described by the ionization cross section $\sigma(E)$ [5-3]. $\sigma(E)$ is the effective area of the atom which must be struck by an incident electron in order to eject an atomic electron. The diameter of this area is about 0.001 nm, and is thus very small compared with the atomic diameter, which is of the order of 0.1 nm. Only very few electrons therefore contribute to the generation of X-rays. By far the greatest number of inelastically scattered electrons originate from collisions with the outer atomic electrons. The ionization cross section is a function of the electron energy. It is zero for $E_0 \leq E_b$ and increases rapidly with increasing energy to reach a maximum at about three times the binding energy E_b (Fig. 5-2). Afterwards, σ drops off slowly.

After an atom from an inner shell, e.g. the K shell, has been displaced by a primary electron, the ground state is restored by a sequence of electron transitions (Fig. 2-43). An atomic electron from a shell lying further out, such as the L shell, jumps to the K shell. The energy difference $E_{K\alpha} = E_K - E_L$ can be emitted as an X-ray quantum. The vacancy which was just for-

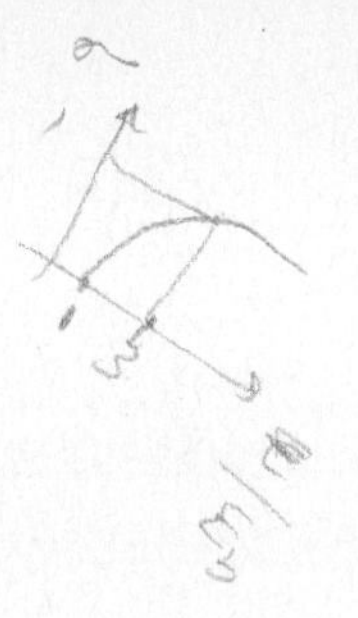

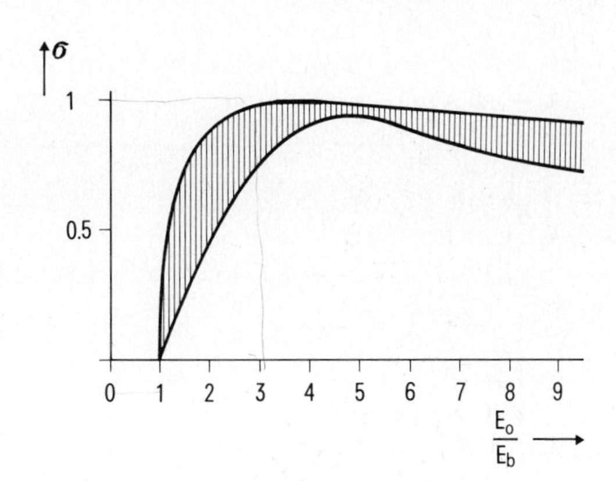

Fig. 5-2.
Relative values of the ionization cross section σ as a function of the ratio of the primary electron energy E_0 to the binding energy E_b. The measured values for various elements are located within the marked region [5-4 to 10]. The mean maximum determined for each element was set equal to one.

med in the L shell is filled by a transition from a shell lying still further out, the M shell for example, in which case an X-ray quantum of energy $E_{L\alpha} = E_L - E_M$ is emitted. Transitions in shells lying still further out occur in the same way. The energies of the characteristic radiation E_c are therefore determined by the spacings between the energy levels of the atom and are thus specific to (characteristic of) the type of atom involved:

$$E_c = \Delta E = E_{n1} - E_{n2} \,. \tag{5-2}$$

In accordance with the step-by-step build-up of the atomic electron shell with increasing atomic number Z in the periodic table of the elements, a simple relation exists between the energy of the X-ray lines and the atomic number. This relation can be calculated from equation (5-2) and the binding energies of the levels involved in the relevant electron transition. Moseley's law then holds as an approximation [5-11]:

$$E_c = K \cdot (Z - \delta)^2 \,. \tag{5-3}$$

In this equation, K is a constant which contains the Rydberg constant (cf. Fig. 2-44). The "screening constant" δ takes into account the fact that the attractive force acting on an atomic electron is not determined by the entire nuclear charge $Z \cdot e$ but is smaller due to screening by the atomic electrons occupying the shells further in. For transitions ending in the K shell $\delta = 1$, for the L spectra $\delta = 7.4$. The energy respectively the wavelength (Fig. 5-3) of the X-ray lines is a simple function of the atomic number because the structure of the inner electron shells is the same for all elements. The shells close to the nucleus of elements with medium and higher atomic numbers are practically unaffected by the valence electrons, which determine the chemical behavior of the elements. This is why the X-ray spectra are insensitive towards the binding state of the element. Only for the light elements can a minor change in the energy of the X-ray lines (chemical shift) be observed between the pure element and its compounds [5-12]. This chemical shift is particularly marked in insulators, where it can provide important additional information for the analysis [5-13 to 15].

Weak lines, which cannot be assigned to the transitions of singly-ionized atoms, are sometimes observed in the proximity of strong lines. They originate from doubly-ionized atoms and are known as satellite lines (see Fig. 5-15).

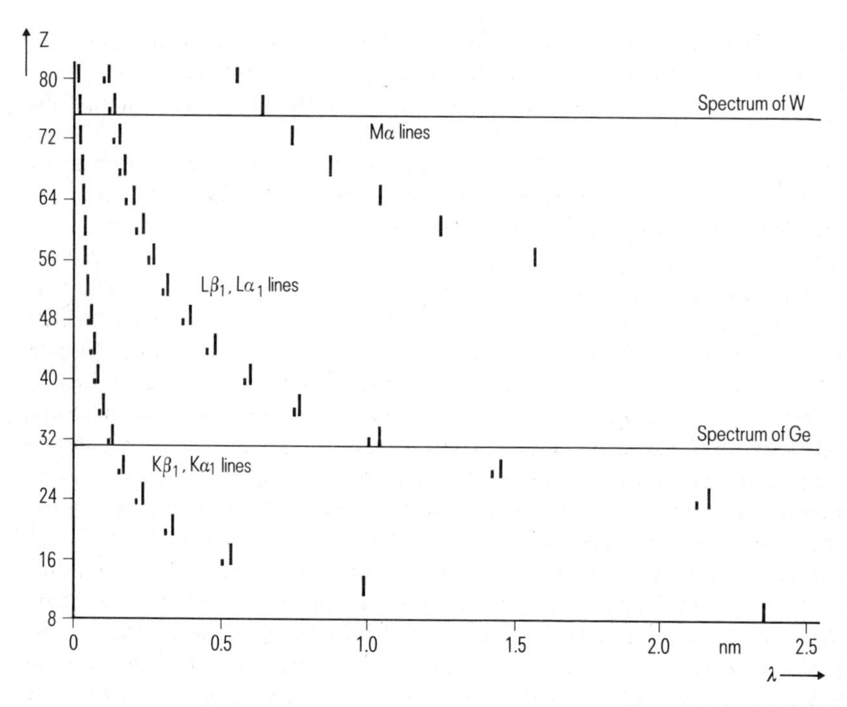

Fig. 5-3. Wavelength spectra of the K, L and M lines arranged in accordance with the atomic number Z in the periodic table of the elements. The lengths of the bars roughly indicate the intensity relationships of the lines within the series (cf. Fig. 2-42b). The spectra of tungsten (W) and germanium (Ge) are marked as examples.

In principle, all elements apart from hydrogen and helium, which possess only a K shell, yield an X-ray spectrum. However, the X-ray lines of the lightest elements are of very low intensity and thus can not easily be exploited by X-ray spectrometry, whose use in microanalytical practice therefore begins at an atomic number of 4.

5.1.2 X-ray microanalysis

In the X-ray microanalyzer, just as in the scanning electron microscope, the electron source is demagnified by the action of electron lenses and directed by a deflection system to defined positions on the specimen [5-16]. The spatial resolution is determined by the size of the volume in which X-rays are excited. With bulk specimens, this volume is pear-shaped and of the order of 1 μm in diameter, thus being somewhat smaller than the scattering volume of the electrons (see section 2.4.1, Fig. 2-32) since the electrons can effect no further ionization after having been decelerated to an energy of $E < E_b$. The spatial resolution cannot be improved, even with a significantly smaller probe, since the electron scatter in the specimen fills out a volume of about 1 μm in diameter. Although the resolution may be somewhat improved by reducing the electron energy, a minimum energy of E_b is nevertheless required in every case. If, however, the specimen is a thin foil with a thickness of the kind required for transmission electron microscopy, small beam spreading occurs due to few scattering processes and analysis with a spatial resolution of the same order of magnitude as the probe diameter is possible. This advantage is exploited in X-ray microanalysis using the transmission electron microscope (section 5.6.2).

Depending on the choice of primary electron energy, the X-rays excited by the incident electrons cover a smaller or greater wavelength (or energy) range. For the elemental analysis, the spectral lines contained in this region must be separated and sorted according to their wavelength or energy. This is accomplished by X-ray spectrometers, which are used to measure the intensity of the radiation as a function of the wavelength (wavelength-dispersive spectrometer, WDS) or the energy (energy-dispersive spectrometer, EDS). The elements contained in the specimen can be identified by comparing the measured spectra with standard spectra (qualitative analysis). Because a complicated but unambiguous relationship exists between the intensity of the characteristic radiation and the concentration of the element which emits this radiation, micro-regions can also be analyzed quantitatively. This is done by relating the intensities measured at the specimen to measurements on standards consisting of pure elements or having a composition similar to that of the specimen. If the specimen surface is ground smooth and polished, and other effects on which the intensity depends (atomic number, absorption, secondary excitation, fluorescence) are taken into account, then quantitative analyses with errors of below 1% can be attained (section 5.4). Insulating specimens can also be analyzed. Charging phenomena due to the exciting electron beam can be avoided by vapor-depositing a thin conducting film of weakly absorbent material (e.g. carbon) onto the surface (section 3.3). Even quantitative analyses are hardly affected by this conductive coating since its thickness is usually small in comparison with the diameter of the scattering volume.

Since the X-ray microprobe is equipped with a deflection system just like the scanning-electron microscope, concentration profiles along a selected spatial coordinate as well as element distribution images of selected specimen areas can be recorded. In the latter case, an image of the specimen surface is obtained on the screen of a CRT (storage monitor) in the "light" of the selected element.

These properties make X-ray microanalysis a universally applicable method for qualitative and quantitative analysis of micro-regions on solid surfaces. Since X-ray spectra are generally not affected essentially by the binding state of the atoms, this method is eminently suited to identify the elements contained in specimens of unknown composition (multi-element analysis) in a single analysis step.

A recent, detailed presentation of the fundamentals of X-ray microanalysis can be found in the book by Scott and Love [5-17], which particularly treats the methods of quantitative analysis, as well as in the books by Heinrich [5-18] and Goldstein et. al. [5-19]. In addition, specialist and review articles on all sectors of X-ray microanalysis can be found in the volumes of the meetings on "X-ray Optics and Microanalysis" and in the "Proceedings of the Electron Microscopy and Analysis Group (EMAG), Institute of Physics".

5.2 Instrumentation

5.2.1 Overall design

The instruments used for electron-beam X-ray microanalysis largely comprise the same electron-optical devices as those required for scanning electron microscopy. X-ray microanalysis can therefore also be performed by fitting X-ray spectrometer attachments to scanning electron microscopes [5-20], transmission electron microscopes [5-21] or scanning transmission electron microscopes [5-22,23]. On the other hand, the instruments specifically developed for X-ray microanalysis always contain equipment for scanning electron microscopy. Dedicated e-beam X-ray microprobes, which naturally offer the highest performance as X-ray microanalysis instruments, are often also known as electron beam micro-analyzers, or microprobes for short (Fig. 5-4). In this book they will be called X-ray microprobes in order to avoid confusion with Auger microprobes (Chapter 6).

The basic design of an X-ray microprobe is shown in Fig. 5-5. The electron-optical part with electron gun, condenser lens, objective lens and double deflection system as well as the equipment for detecting and processing the secondary and backscattered electron signals are fully analogous to the corresponding components of the scanning electron microscope (see Fig. 3-1).

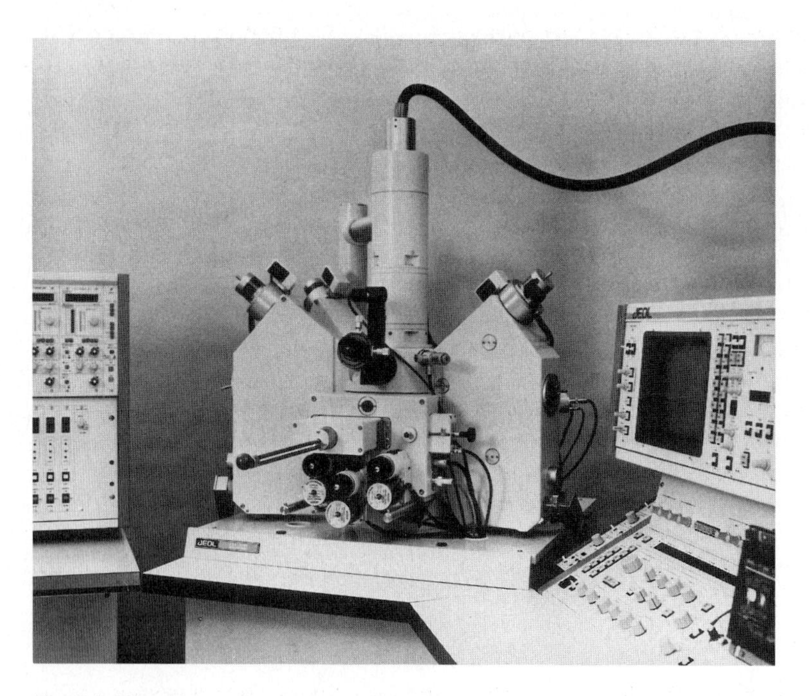

Fig. 5-4. View of a modern X-ray microprobe, of type JXA 8600 by JEOL Ltd., Tokyo, Japan; courtesy of KONTRON PHYSTECH GmbH, D 8057 Eching, Germany.

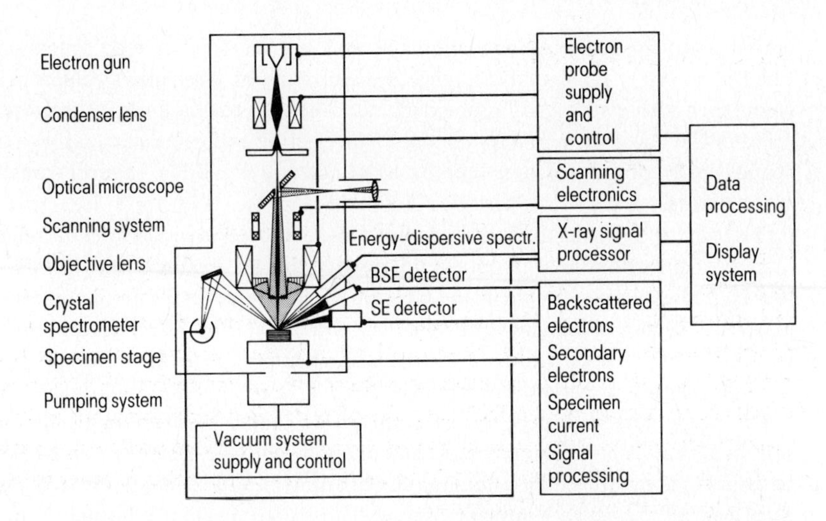

Fig. 5-5. Schematic diagram of the overall setup of an X-ray microanalyzer. As in the scanning electron microscope, the electron probe is usually generated with two condenser lenses (only one is shown here) and an objective (or probe-forming) lens and then deflected to a selected point on the specimen by deflection coils (scanning system). Crystal spectrometers and a semiconductor detector are used to analyze the X-radiation. A reflecting light microscope and scanning electron microscopy equipment are used to image the object.

The equipment used for reflecting light microscopy is typical of that for the X-ray microprobe. The reflecting light microscope can be used to select the site to be analyzed and also to adjust precisely the vertical position z of the specimen. This requires a minimum depth of field, i.e. a maximum numerical aperture of the optical objective lens. This is best realized by using a mirror optical system (Cassegrain) of the kind employed in reflecting telescopes.

To allow the specimen points of interest to be selected repeatably in the optical microscope, a high-precision specimen stage is required. The stage can be used to pinpoint predefined coordinates x, y in the plane of the specimen surface. A vertical drive allows an exact height adjustment z to be made under microscope control. The specimen stage can be accessed usually via an airlock for loading with several specimens and standards.

For measuring the X-ray spectra, X-ray microprobes contain several crystal spectrometers (WDS) and a semiconductor detector (EDS). Because of the fixed geometry of specimen and spectrometer arrangement the spectrometer crystal selects from the radiation emitted by the specimen only that part that is emitted at a specific angle, the take-off angle ψ (angle between the specimen surface and the direction towards the spectrometer).

The same vacuum systems are used as those normally employed for scanning electron microscopy.

5.2.2 Electron-optical column

In commercial instruments, the electron probe is generated by directly heated thermionic tungsten or lanthanum hexaboride cathodes (see section 2.3.1). The higher brightness and spatial resolution of the LaB_6 cathode compared with the tungsten cathode brings only minor advantages to X-ray microanalysis, since the scattering or dissipation volume is always about 1 μm^3 even with a very small probe diameter. The tungsten cathode is therefore preferred because of the lower vacuum requirements. The electron probe is generated by a multistage demagnification of the crossover onto the specimen (see Fig. 2-27). In most cases, two short-focus condenser lenses are provided for this purpose. The third lens, the objective or probe-forming lens, has a relatively long focus to allow a long working distance, which is required for incorporating the optical microscope and to maximize the take-off angle for the X-rays. Lenses and apertures in the electron-optical beam path are dimensioned so that the principal imaging errors, above all spherical aberration (Fig. 2-23), are minimized. Any astigmatism is corrected by means of a stigmator. At a probe energy of 10 keV, probe diameters of about 0.5 µm are attained (see Fig. 2-28).

The probe current can be varied within wide limits, between about 10^{-5} and 10^{-12} A. A Faraday cup is provided for measuring the probe current. Beam currents below 10^{-9} A (cf. section 3.1) are required for operating the instrument as a scanning electron microscope and of at least 10^{-8} A as an X-ray microprobe. The probe current is adjusted to the desired value by modifying the Wehnelt voltage, the condenser lens currents and the diameter of the objective aperture. The energy of the electron probe is also variable within broad limits, and can therefore be adapted to the desired excitation conditions. Usually, energies in the range of between 3 and 50 keV are applied.

The current and energy of the electron probe must be maintained at a very constant value over a long period of time to obtain reliable quantitative results. A deviation of less than 10^{-3} h^{-1} is required for the probe current and of less than 10^{-2} h^{-1} for the acceleration voltage.

Element distribution images (elemental maps) or scanning electron images are formed by equipping the electron-optical column with two pairs of deflection coils (only one is shown in Fig. 5-5) for each direction. These allow the electron probe to be scanned over the specimen line by line, as in the scanning electron microscope (SEM). The scanned area, and thus the magnification, can, as in the SEM, be varied within wide limits. Electron detectors of the same design as those used in the SEM are provided for recording the signals from secondary or backscattered electrons (see section 3.2.4).

5.2.3 Wavelength-dispersive spectrometer (WDS)

The phenomenon of diffraction in single crystals is utilized here for displaying the spectra (see section 2.1.1). In accordance with Bragg's law

$$n \cdot \lambda = 2d \cdot \sin \vartheta \tag{5-4}$$

a single crystal with lattice planes parallel to its surface reflects radiation of a specific wavelength λ at certain angles of incidence in a direction with the same exit angle ϑ ($n = 1, 2, 3, \ldots$; d, lattice plane spacing in the single crystal; ϑ, angle of incidence and reflection of the beam; see Fig. 2-4). Waves of other wavelengths are transmitted through or absorbed by the crystal. If the angle ϑ is changed by rotating the single crystal, radiation of different wavelengths is successively reflected in different directions. The X-ray intensity can then be measured with a detector (gas proportional counter) as a function of the angle and the spectrum recorded in this way.

In X-ray microanalysis, the X-ray source – which is the excited volume in the specimen – has a diameter of the order of 1 μm. With a flat crystal such as that shown in Fig. 2-4 therefore, the Bragg condition is fulfilled only for a very small angular range around the take-off angle ψ (angle between the specimen surface and the direction of radiation towards the crystal) and the reflected intensity is correspondingly small. Bent crystals are therefore used. The radius of curvature is dimensioned so that all rays from the almost punctiform source strike the analyzer crystal with the same angle of incidence ϑ and are thus reflected towards the detector as a focused beam (Fig. 5-6). If source and detector are arranged symmetrically to the crystal along the focusing circle, beams with different wavelengths are successively detected in line with the different Bragg angles.

In the X-ray microanalyzer, however, the position of the source (i.e. the specimen) and the take-off angle ψ are fixed. Since it is nevertheless required that the focusing criteria always be fulfilled when a spectrum is recorded, the crystal and detector must be moved: the crystal along a straight line in the direction ψ (path of the crystal) and the detector on a curved path (detector path) (Fig. 5-7). The wavelength of the beam which is reflected and recorded by the detector is then proportional to the distance between the source and the crystal (linear spectrometer).

However, only a limited part of the spectrum can be analyzed with a single analyzer crystal. An upper limit is set by the fact that sin ϑ (in Bragg's law) cannot exceed a value of 1. A lower limit is set by technical factors: the crystal cannot be placed as close as desired to the specimen. Commercial crystal spectrometers cover an angular range from $\vartheta \approx 15°$ to $\vartheta \approx 65°$. For analyzing all elements accessible to X-ray microanalysis therefore, several (at least four) crystals with different lattice spacings are required for different wavelength regions. They can be

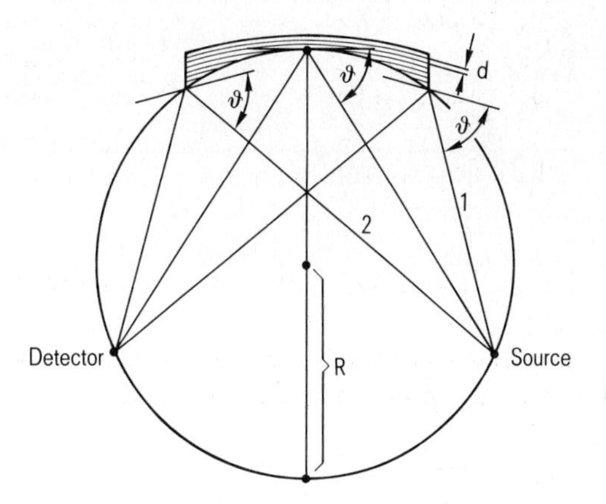

Fig. 5-6. Focusing X-ray spectrometer: the analyzer crystal is cylindrically bent with a radius of $2R$. The source and detector lie on the focusing circle (Rowland circle), and are arranged symmetrically to the crystal (focusing condition). The Bragg condition is fulfilled for all rays between edge rays 1 and 2. Exact focusing of the radiation reflected over the entire crystal surface is obtained when – as shown here – the crystal surface is ground to a cylindrical shape of radius R in addition to its curvature with radius $2R$ (Johannson configuration [5-24]). Without this grinding step, the setup permits only partial focusing (semi-focusing spectrometer according to Johann [5-25]).

interchanged during operation as required. Modern X-ray microprobes are equipped with several WDSs each containing several crystals, so that several elements can be analyzed simultaneously.

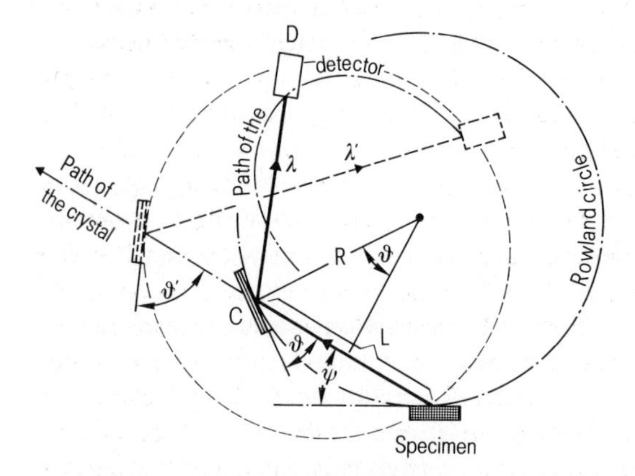

Fig. 5-7. Focusing linear spectrometer with fixed X-ray source and fixed take-off angle ψ: whereas the crystal C is guided along a straight line in a direction ψ with simultaneous rotation, the detector must follow a curved path to stay on the Rowland circle. The spacing $L = 2 \cdot R \cdot \sin\vartheta$ results from the Bragg angle ϑ and the radius R of the Rowland circle. Together with the Bragg equation (5-4), this yields $\lambda = (d/R) \cdot L$ for reflections of the first order ($n = 1$).

Table 5-1. Data of standard analyzer crystals. The values of energy resolution depend on the energy. The higher the energy or shorter the wavelength, the greater is $\Delta E/E$.

Crystal material	Double distance of the reflecting lattice planes $2d$/nm	Resolution $\frac{\Delta E}{E}/10^{-2}$	Wavelength range λ/nm	Element ranges for K- (L-) [M-] radiation, atomic number Z
Lithium fluoride (LiF)	0.403	0.1–2	0.1–0.383	19–36 (52–88)
α-Quartz (SiO_2)	0.669	0.2–1	0.173–0.635	15–32 (40–80)
Pentaerythritol PET ($C_5H_{12}O_4$)	0.874	0.2–0.7	0.218–0.83	13–25 (37–60) [70–92]
Ammonium dihydrogen phosphate ADP ($(NH_4)H_2PO_4$)	1.064	0.3–0.7	0.266–1.01	13–22 (33–58) [66–92]
Thallium acid phthalate, TAP ($C_8H_5O_4Tl$)	2.59	0.2–0.8	0.644–2.56	8–14 (24–39) [57–71]
Lead stearate ($Pb(C_{18}H_{35}O_2)_2$)	10.04	1–2	2.20–9.52	5–8 (16–23)

In order to obtain spectra of maximum intensity with easily separable peaks, the analyzer crystals should have a high reflectivity and a good angular resolution. Their lattices should therefore be as perfect as possible (low density of crystal defects) and the reflecting lattice planes should be densely packed with atoms (cf. section 2.1.1). Good properties are exhibited by artificially grown crystals, which can now be manufactured with a high degree of perfection. Pseudocrystals are used to analyze long wavelength radiation. In this case, the reflecting planes are manufactured by a step-by-step deposition of thin alternating layers with low and high scattering power. The most common analyzer crystals in use are listed in Table 5-1.

Gas-filled counters are used for recording the X-ray quanta in the WDS. Their basic design is shown in Fig. 5-8. X-ray quanta entering the counter tube ionize atoms of the inert gas (e.g. argon), leading to the generation of electron-ion pairs. An X-ray quantum of energy $E = h \cdot \nu$ can generate $h \cdot \nu/\overline{E}_i$ such pairs on average, where $\overline{E}_i$ is the mean ionization energy of the atoms of the counter tube gas ($\overline{E}_i = 20 \ldots 30$ eV). The electrons (photoelectrons) are accelerated by the high electric field in the environment of the counter tube wire and generate further electrons by impact ionization (section 2.5.2). This leads to an avalanche-like rise of the current, a short discharge takes place and a current pulse (duration about 10^{-6} s) can be measured with the usual amplifier setups. The mean number of electrons contained in a current pulse is

$$\overline{N} = (h \cdot \nu/\overline{E}_i) \cdot A \tag{5-5}$$

the gas amplification factor A specifying the degree to which the impact cascade increases the primarily-generated charge carriers. In the case of gas proportional counters, A has a value of between 10^2 and 10^4. The quantum energy can be inferred from the pulse height in accordance with eq. (5-5). However, the energy resolution has a value of about 1 keV, which is not sufficient to allow separation of individual elements. In the WDS, the counter is merely used to measure the intensity, the elements being determined by the analyzer crystal.

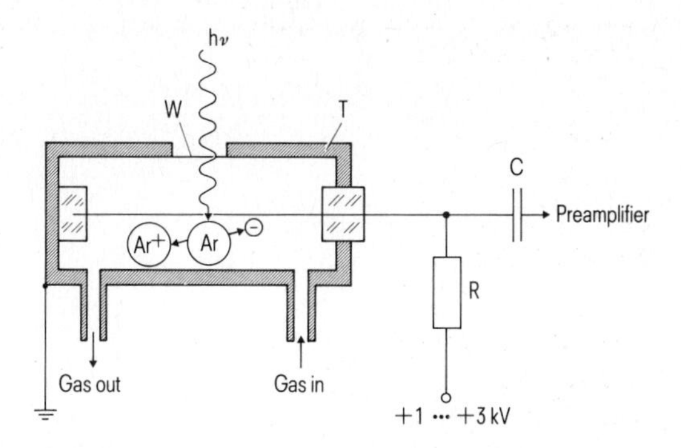

Fig. 5-8. The gas proportional counter (a flow counter tube of the kind used for soft X-rays is shown here in longitudinal section) consists of a cylindrical, metallic tube T which is at ground potential. A thin wire running along its axis (approx. 50 μm diameter, usually of tungsten) is at positive potential. The X-rays enter the gas-filled chamber (90% argon, 10% methane) through a narrow window W made of a thin organic foil (such as Mylar). The X-ray quanta $h \cdot \nu$ ionize the gas (photoelectric effect), thus initiating a short gas discharge.

The counters must be filled with a gas (e.g. argon, or for very hard X-rays, xenon) allowing a sufficiently strong absorption of the X-rays to ensure high efficiency. A window made of a minimum-absorption material (e.g. beryllium) is provided for the X-rays to penetrate from the vacuum to the gas chamber. For detecting very soft X-rays, extremely thin plastic (Mylar) films supported on grids are used as window material. However, due to unavoidable leaks in such thin films, the counter gas must be continually renewed, so that the counter is actually operated as a flow counter tube (Fig. 5-8).

5.2.4 Energy-dispersive spectrometer (EDS)

In contrast to the WDS with its elaborate, high-precision mechanical parts, the EDS is of relatively simple construction. Its principal component is a semiconductor crystal which acts simultaneously as a detector and analyzer [5-26]. In a similar way as in a proportional counter tube, X-ray quanta which penetrate the semiconductor detector generate pairs of charge carriers whose number is proportional to the energy of the incident quanta. However, the energy resolution of the semiconductor detector is between 140 and 170 eV and is thus significantly superior to that of the counter tube. This resolution is sufficient to separate the X-ray lines of adjacent elements from each other. An additional facility for sorting the energy – like the WDS analyzer crystal – is therefore not required. The spectrum is obtained by evaluating the pulses generated and amplified in the detector system according to their height and sorting them. This takes place in a multichannel analyzer. Each channel corresponds to a specific pulse height and, after suitable calibration, to a specific energy. In contrast to the WDS, in which the spectrum must be recorded sequentially, the EDS detects the entire spectrum simultaneously.

The semiconductor detector is a special silicon diode, which is accommodated in a vacuum chamber and kept at a temperature of about 100 K by means of a cold finger (Fig. 5-9). A beryllium window (thickness about 10 μm) closes the evacuated chamber containing the detec-

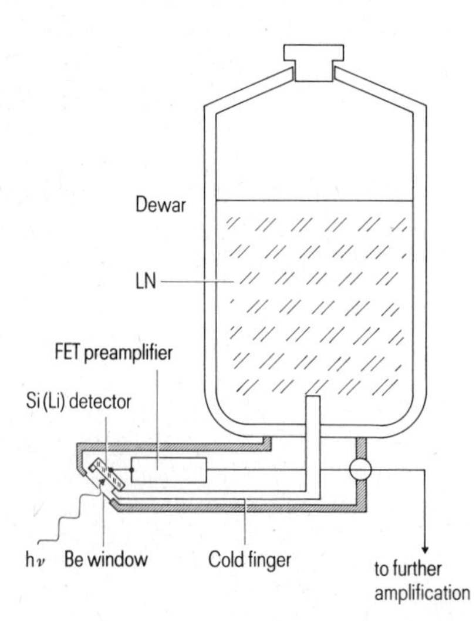

Fig. 5-9.
Basic setup of the energy-dispersive X-ray spectrometer. The Si(Li) detector crystal and the first preamplifier (a field-effect transistor FET) are located in a vacuum chamber. With the aid of a cooling finger, the detector and FET are continually maintained at the temperature of liquid nitrogen (LN, about 100 K). This cooling is required to minimize the signal background caused by the thermal noise and to prevent the Li atoms from drifting under the influence of the applied bias voltage and thus destroying the detector.

tor and preamplifier. The window protects the detector from condensation and contamination as well as from scattered electrons and light. The entire system can be adjusted both horizontally and vertically.

A silicon diode (see section 2.1.3) into which a broad region of intrinsic conductivity is incorporated between the p- and n-conducting layers (pin structure, Fig. 5-10) is used as the detector. In this region, the extrinsic conduction, which is caused by unavoidable impurity atoms, is compensated for by the addition of lithium. The lithium atoms are diffused into the crystal and made to "drift" under the action of an electric field at about 100 °C so that an intrinsic zone several millimeters in width is created. Such devices are therefore called lithium-drifted detectors, or Si(Li) detectors for short.

The intrinsic conduction is caused solely by electrons of the silicon atoms which are supplied with so much energy by thermal or other effects that they are raised from the valence to the conduction band (see Fig. 2-12a). In the case of Si(Li) detectors, this energy is supplied by the penetrating X-ray quanta. Unlike an electron, an X-ray quantum releases its entire energy $E = h \cdot \nu$ during a single excitation process (if the very rare Compton effect is neglected). Every photoelectron generated in this process obtains a kinetic energy $E_{kin} = h \cdot \nu - E_i \approx h \cdot \nu$, in which the ionization energy $E_i \ll h \cdot \nu$ (E_i has a value of several eV, $h \cdot \nu$ several 100 eV to several 10 keV). The energy supplied to the electron is therefore sufficiently high to generate further electron-hole pairs in a manner fully analogous to the generation of charge carriers by electron bombardment (section 3.6.1). The total number of electrons generated on average per X-ray quantum is

$$\overline{N} = h \cdot \nu / \overline{E}_i \tag{5-6}$$

where $\overline{E}_i$ is the mean excitation energy per electron-hole pair (Table 2-2); for silicon $\overline{E}_i$ = 3.8 eV.

If a reverse bias is applied between the p and n layers of the diode, the electrons and holes are accelerated in opposite directions: the electrons migrate to the n layer and the holes to the p layer (Fig. 5-10). In a manner fully analogous to the gas filled counter, a current pulse arises whose size is, according to eq. (5-6), proportional to the quantum energy $h \cdot \nu$.

To attain a high detector efficiency, as many X-ray quanta as possible must be absorbed and all the charge carriers they generate must be collected. For this reason, the layer of intrinsic conductivity (i layer) must be several mm in thickness and the detector area must be made as large as possible (about 30 mm^2).

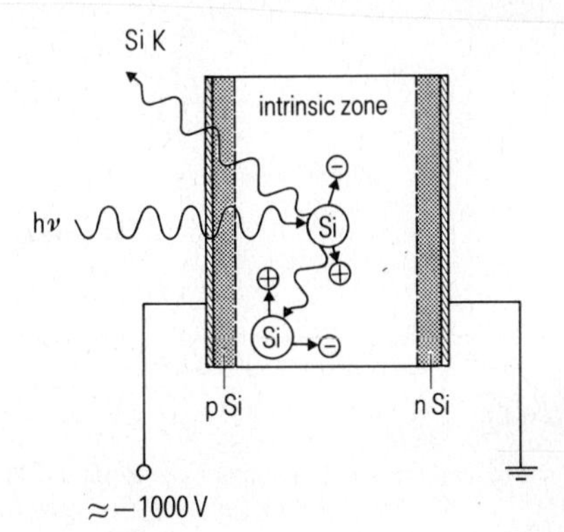

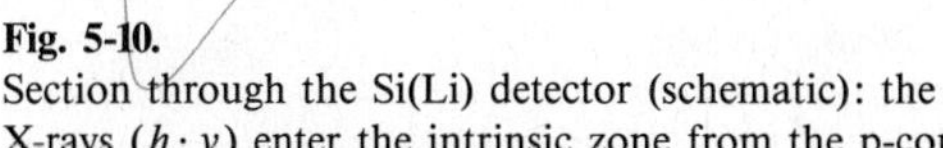

Fig. 5-10.
Section through the Si(Li) detector (schematic): the X-rays ($h \cdot \nu$) enter the intrinsic zone from the p-conducting side. It generates negative and positive charge carriers (electrons and holes) which move towards the electrodes under the action of the electric field. In addition, Si-K fluorescent radiation is emitted, only a small part of which leaves the detector.

As a consequence of the ionization of the Si K shell, not only are photoelectrons generated, but Auger electrons and Si X-ray quanta as well. Most of these are absorbed in the detector, where they again generate electron-hole pairs. The energy E_c of the X-ray quanta then remains completely within the crystal.

A small part of this Si K radiation can, however, leave the detector. The incident quanta of energy E_c then lack this part of the quantum energy after the energy transfer. The energy deposited in the detector is therefore smaller than E_c by the energy of the Si K radiation (essentially Si Kα radiation of 1.74 keV). As a result, the energy $E_{esc} = E_c - E_{SiK\alpha}$ gives rise to a separate pulse, and a small peak, known as the *escape peak*, appears in the spectrum at this point (Fig. 5-11). The intensity of the escape peak is below 2% of the intensity of the parent peak. An escape peak can naturally occur only when the energy of the incoming quanta is greater than that of the Si K absorption edge (1.84 keV). As the energy of the incident radiation increases, the escape peak gets smaller.

Before the X-ray quanta reach the active intrinsic zone in the detector, they must traverse the beryllium window, the gold contacting layer and the p-Si layer (known as the dead layer). The greatest absorption losses occur in passing through the Be window, where the intensity of radiation with an energy of 1 keV is reduced by as much as 40%. The sensitivity of the detector for the analysis of light elements is thereby dramatically reduced. Normally therefore, only elements with $Z \geq 11$ (i.e. from sodium onwards) can be measured. "Windowless" detectors were developed to get around this drawback [5-27 to 30]. As in the gas flow counter, the Be window is replaced by a thin organic film onto which a thin aluminum layer has been vapor-deposited to prevent the incidence of light on the detector. Recently, also ultrathin windows of boron nitride or consisting of a diamond-like film are used for light element analysis. They withstand atmospheric pressure and thus allow venting of the specimen chamber without the need of protecting the window by a special airlock. In addition, a magnetic trap must be set up in front of the detector to screen off scattered electrons. In this low-energy region an overlap of peaks can, however, hardly be avoided due to the low energy resolution.

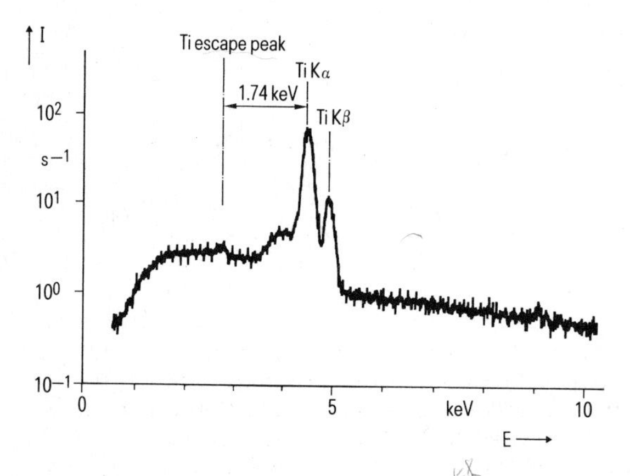

Fig. 5-11. Energy-dispersive spectrum of titanium. The titanium escape peak can just be seen above the background (intensity in a logarithmic scale).

5.2.5 Comparison between WDS and EDS

The properties of these two types of spectrometer differ very significantly. Either may therefore be the more suitable, depending upon the analysis problem at hand. Since the WDS and the EDS mutually complement each other, it is often a good idea to use both spectrometers for a specific problem.

The EDS provides a quick overview of the elements present in the specimen, since it allows the entire detectable spectral range to be recorded simultaneously. This spectrometer can be brought very close to the specimen, so that the collection angle (or the geometrical collection efficiency) of the radiation emitted by the specimen and recorded by the detector is relatively large. Also a sufficient number of current pulses are then obtained in the detector even with a relatively low probe current. This also allows specimens sensitive to radiation to be investigated without significant damage.

The Si(Li) detector has a quantum efficiency of above 90% in the energy range from 2 to 20 keV. This means that almost every incident X-ray quantum generates a measurable current pulse. Since the spectral sensitivity of the detector can be calculated to a good approximation, quantitative analyses can also be performed without standard specimens if the effects of different absorption and atomic number are taken into account (see section 5.4.6). Since, unlike WDS, no exact focusing conditions are required for the EDS, specimens with a rough surface or small particles can be quantitatively analyzed with acceptable accuracy. Furthermore, element distribution images of large areas can also be recorded. Its high quantum efficiency, large geometrical collection efficiency and insensitivity towards the position of the specimen make the Si(Li) detector particularly suitable for analyzing thin specimens in a transmission electron microscope (see section 5.6.2).

A disadvantage of EDS analysis is its poor energy resolution of about 150 eV, which results in relatively wide peaks. The pulses, which may be assigned to a characteristic line, are consequently "smeared" over a relatively large energy range and the peak maximum is correspondingly lower (Fig. 5-12). In order to identify a peak reliably, pulses must be collected over a rela-

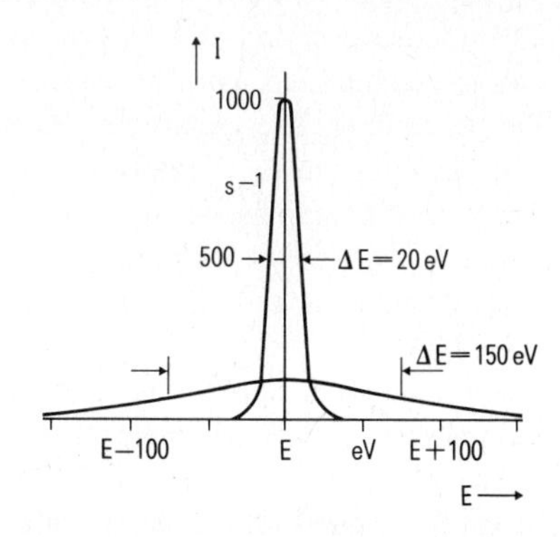

Fig. 5-12. Dependence of peak height on energy resolution ΔE. The energy resolution ΔE of the spectrometer determines the measured peak height as long as ΔE is greater than the natural line width (about 2 eV). A peak which has a height of 1000 s^{-1} at a resolution of ΔE = 20 eV, attains a height of only 130 s^{-1} at ΔE = 150 eV. The natural line width of 2 eV would correspond to a peak height of 10^4 s^{-1}. The energy resolution and line width are specified as the full width of the peak at half the maximum height (FWHM).

tively large energy window. This also leads to a relatively large part of the background radiation being recorded. The peak-to-background ratio is consequently smaller than for narrow peaks. The wide peaks may also cause peak overlap, thus making both the qualitative and quantitative analyses more difficult. Although superimposed lines can be separated with the aid of deconvolution calculations, which are included in the software packages of commercial instruments, weak peaks due to elements of low concentrations can nevertheless be overlooked.

The great advantage of the WDS is its high energy resolution. With good crystals, it is of the same order as the natural line width of the characteristic radiation. The background correction is thus significantly simpler than for the EDS. The WDS therefore allows a higher sensitivity to be obtained in detecting trace elements and a greater accuracy in quantitative analysis than the EDS. However, the precondition for this is that the specimens possess a well-defined, planar surface and that the focusing condition is adhered to very exactly.

Whereas the analysis of elements with atomic numbers below 11 presents usually problems for the EDS, the WDS can in any case be used to analyze light elements down to beryllium, both qualitatively and quantitatively.

5.3 Measurement technique

5.3.1 Signal processing for the WDS

The signal generated in the detector by the ionization process is extraordinarily weak. It can be estimated from equation (5-5) that for a counter filled with argon, for instance, about $2 \cdot 10^5$ electrons are generated for each incident quantum of Cr Kα radiation ($E_{\mathrm{CrK}\alpha}$ = 5.4 keV) when the gas amplification factor $A = 10^3$. This corresponds to a current of 0.03 μA over a period of 1 μs. Low-noise configurations are required to allow such small signals to be amplified. For this reason, the preamplifier is located in the immediate vicinity of the gas counter. The signals emitted from the preamplifier have a size of several 10 mV. The main amplifier raises the pulses to a level of several volts for further processing (Fig. 5-13). The amplified signals are then fed to a pulse height analyzer which evaluates and sorts the pulses on the basis of their amplitude, which corresponds to the quantum energy. A discriminator circuit is then used to sort out undesired pulses (higher-order reflections at the analyzer crystal) above or below a specific energy region. The position and width of this electronic window are automatically adjusted when the spectrum is traversed so that the signal-to-background ratio is optimized over the spectral range covered.

The output pulses from the pulse height analyzer and discriminator are recorded by the rate meter or the scaler and reflect the intensity of the X-rays within the wavelength region between λ and $\lambda + \Delta\lambda$ given by the analyzer crystal of the WDS. Whereas the rate meter displays the mean number of pulses per unit time (pulse rate), the scaler shows all the pulses incident during a preselected measuring time.

The counting system cannot process any desired number of pulses per unit time. While a pulse is being processed, the electronics are unable to detect another pulse. The time interval

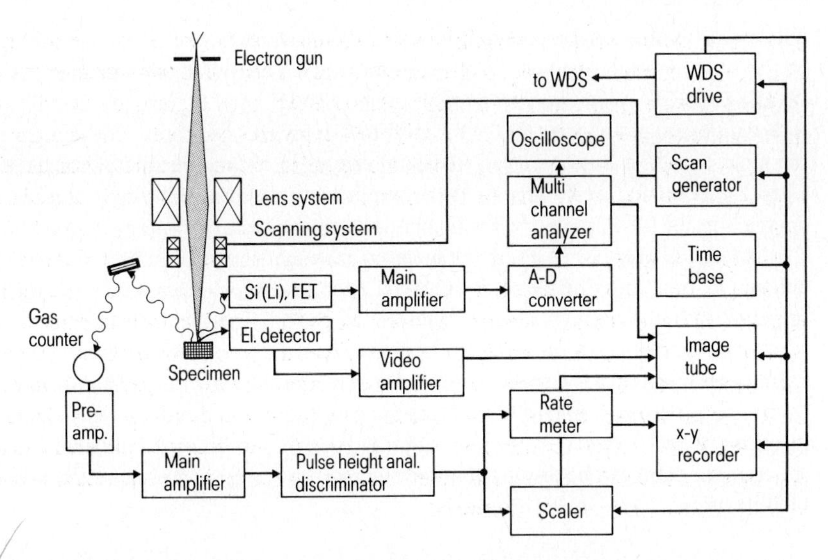

Fig. 5-13. Basic setup of the signal processing unit for an X-ray microprobe. To record the spectrum by means of the WDS, the spectrometer is driven by the WDS drive and scans the selected angular region. The x axis of the x-y plotter is continuously driven synchronously with the spectrometer, whereas the y coordinate is controlled by the amplified detector signal. The entire spectral range picked up by the Si(Li) detector is recorded simultaneously by means of the EDS. The amplified pulses are stored in specific channels of the multichannel analyzer depending on their height (energy). During analysis the X-ray lines can be directly observed on the screen of an oscilloscope to rise up above the background.

between a pulse arriving and readiness for processing the next pulse is the dead time τ. It is essentially determined by the electronic amplification and has a typical value of between 1 and 2 μs. This means that at count rates of 10^5 s^{-1}, about 10% of the pulses are lost. At pulse rates of $\leq 10^4$ s^{-1}, this effect is negligibly small. Dead time losses must be taken into account in quantitative analyses. Modern instruments are usually equipped with computers, by which a dead time correction is carried out automatically.

5.3.2 *Signal processing for the EDS*

In a Si(Li) detector, the pulse is smaller by about two orders of magnitude than in a gas counter. For this reason, not only the detector but also the preamplifier must be cooled in order to keep the thermal noise to a minimum. At the preamplifier output, the pulses have an amplitude of several millivolts.

The Si(Li) detector measures not only the intensities but also the energies of the X-ray spectrum. The main amplifier must therefore process the pulses coming from the preamplifier in a way which avoids impairment of the energy resolution. The signal-to-noise ratio of the amplifier configuration must therefore be optimized, a process which requires a relatively large processing time per pulse (dead time). From this the danger results that two pulses with a short time interval may overlap and thus lead to "false" lines in the spectrum. If a later pulse enters the amplifier when an earlier pulse is just starting to be processed, the amplitudes of the two pulses will be superposed. The amplitude of the measured pulse then corresponds to the sum of the two quantum energies involved (sum peak). This pulse pile-up effect can be reduced by using special pulse pile-up rejection circuits [5-31].

A multichannel analyzer (MCA) is used to display the spectrum. The signals coming from the main amplifier are converted to digital signals in an analog/digital converter (Fig. 5-13). They are evaluated according to their amplitude and correspondingly assigned to specific channels in the MCA memory. Each channel corresponds to an energy interval between E and $E + \Delta E$. Each pulse received by the memory increases the puls number (counts) in a specific channel by one. In principle, every channel operates like a scaler. The energy-selected pulses are summed in the memory during a preselected period. The spectrum appears as a histogram, which can be displayed on an oscilloscope screen or plotted on a printer.

In modern systems, the spectra are stored in a computer memory. The operator can use the computer to perform various procedures for qualitative and quantitative evaluation. Thus the positions of the strongest X-ray lines and the relative intensities for all elements are stored in the computer and can be displayed in the recorded spectrum as line markers. This allows rapid identification of the various X-ray lines.

5.3.3 Specimen preparation and alignment

The aim of an analysis with the X-ray microprobe is to determine the qualitative and quantitative chemical composition of the specimen with maximum accuracy. In contrast, the purpose of the X-ray spectrometer attachments fitted to electron microscopes is to provide analytical information in addition to the electron image of the specimen. The main objective of electron microscopy is to investigate the morphology of the specimen. The scanning electron microscope, for instance, can cope with a specimen having a surface topography which varies considerably in depth. In contrast, the precondition for a reliable analysis with the X-ray microprobe is a surface which is as flat as possible, having the surface quality of a metallographic section produced by means of grinding and polishing. This ensures that the take-off angle ψ is constant and well defined over the entire specimen surface. In addition, shadowing of the X-rays due to surface roughness is avoided. The etching agents often used in metallography should not be used, since they may change the surface composition. Polishing is unnecessary for microtechnology components if the surface to be investigated is already flat, as is the case in planar technology, for instance.

To avoid charging, the specimen must be electrically connected to the specimen holder. Nonconductive surfaces must be coated with a thin conductive film consisting of a material of low atomic number (carbon or aluminum). Such coatings, which are only a few tens of nm thick, can be produced by vapor deposition or sputtering (section 3.3). If standards are required for quantitative analysis (section 5.4), they must also be coated at the same time as the specimen.

The specimen site to be investigated is positioned under the optical microscope. The z-position must be set with particular care, since it is critical for maintaining the focusing condition for the WDS. It is convenient to use high magnification with a small depth of field for this adjustment. An optimum focus with a calibrated z-setting of the optical microscope acts as a criterion for the correct z-position. In the case of the vertical WDS configuration normally used (plane of Rowland circle is parallel to the z-axis), deviations in z by ± 10 µm at an take-off angle of 40° result in a reduction of the peak intensity by around 10%. The sensitivity with respect to a misalignment in the z-axis can be reduced by an oblique arrangement of the WDS (Rowland circle plane tilted with respect to the z-axis). This does, however, lead to a reduction of the take-off angle and an increase of the X-ray absorption path in the specimen.

Selection of the acceleration voltage depends on the elements which are to be analyzed (cf. Fig. 5-3). Voltages between 10 and 25 kV are recommended for exciting the K radiation of elements with atomic numbers between 11 and 30. At higher voltages, which would be required for exciting the K radiation from elements of higher atomic number, the penetration depth is greater (cf. Fig. 2-32 and 33). This can lead to greater errors in the quantitative analysis due to a stronger absorption (see section 5.4) and impairs the spatial resolution. For analyzing heavier elements, the lower-energy L or M spectra should therefore be used.

The probe current must also be matched to the specimen material. Conductive materials tolerate higher current loads than non-conductive ones, which can be heated up by the electron beam to the extent of decomposing. In the case of energy-dispersive analysis, the higher count rates associated with higher probe currents lead to longer dead times and increased pulse pile-ups. EDS analysis is therefore generally carried out with considerably lower probe currents (order of magnitude 1 nA) than WDS analysis (order of magnitude 0.1 μA).

5.3.4 Qualitative analysis

The aim of qualitative analysis is to identify the chemical elements present in the specimen. This is done by comparing the measured spectrum of the characteristic lines with known spectra of the pure elements. Comparative data is available in the form of tables [5-32] or graphs [5-31]. However, agreement of only one line of the measured spectrum with one line of the comparison spectrum does not suffice for an unambiguous identification. For instance, the Ti Kα and Ba $L\alpha_1$ lines have almost the same energy of about 4.5 keV. An element can be determined conclusively only when there is agreement not only with the energy position of several lines of the K or L series or possibly in several series, but also with the relative intensities within a series. The following approximate values of relative intensities of the K series apply to elements of medium mass ($30 \leq Z \leq 50$):

$$I_{K\alpha 1} : I_{K\alpha 2} : I_{K\beta 1+3} : I_{K\beta 2} = 1 : 0.5 : 0.24 : 0.13 \, .$$

The intensity (or the number of generated photons) of a line is designated, with reference to the sum of the intensities of all lines belonging to a series as the "weight" p of a line. Thus

$$p_{K\alpha} = \frac{\text{No. of K}\alpha\text{ photons}}{\text{No. of all photons generated in the K series}} \, ,$$

and for the K series $p_{K\alpha} + p_{K\beta} = 1$.

The intensity ratio $I_{K\alpha 1} : I_{K\alpha 2}$ has the same value, namely 1 : 0.5, for all elements because the shells involved in these transitions have the same electron configuration throughout the entire periodic table. However, the relative intensity of the β lines of the K spectrum changes with atomic number, because the electron occupation of the M and N shells involved depends on the element in question. Going from heavy to light elements, the β lines gradually get weaker. The following holds approximately for the relative intensities in the L spectrum

$$I_{L\alpha} : I_{L\beta1} : I_{L\beta2} = 1 : 0.2 : 0.08$$

and for the M spectra

$$I_{M\alpha} : I_{M\beta} = 1 : 0.6 .$$

Energy-dispersive spectrometer

A quick overview of a large region of the spectrum is obtained by the EDS, since all spectral lines in this region are detected simultaneously (Fig. 5-14). All elements with atomic number ≥11 can be detected by Si(Li) detectors with Be windows.

It must, however, be borne in mind that the limited resolution of the Si(Li) detector means that the EDS reproduces the lines with a full width at half maximum (FWHM) of about 150 eV. The danger of lines overlapping therefore exists, and can lead to components of low concentration in the specimen not being detected. The energy position of a line can, at best, be specified with an uncertainty of one channel width (about 10 eV). For the strong lines in particular, care must be taken to watch escape peaks and sum peaks in order to avoid erroneous interpretations (Fig. 5-11).

A line can be identified as such only when it rises significantly above the background. This is conventionally the case when the count N_P contained in the peak is at least three times as large as the standard deviation $\sigma = \sqrt{N_b}$ of the background at the position of the peak ($N_P \geq 3 \cdot \sqrt{N_b}$, where N_b is the mean pulse number of the background, see section 1.4). The value of σ can be roughly estimated from the variation of the background signal. The peak height should be at least three times this variation.

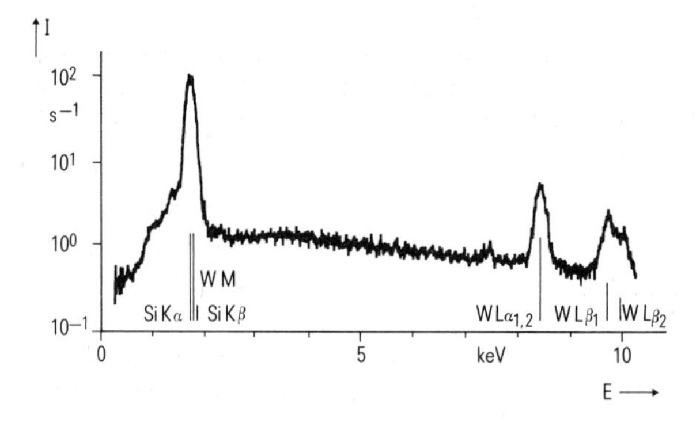

Fig. 5-14. Energy-dispersive spectrum of a tungsten-silicon metallization layer for a microelectronics device. The identification is facilitated by the added line markers, which provide a spectrum of the pure element for comparison. Due to the poor energetic resolution the lines from tungsten (W M), and silicon (Si Kα, Si Kβ) are recorded as one common line (electron energy 20 keV, probe current 4 nA, counting time 200 s).

In order to make the peak rise clearly above the background, more pulses can be collected over a longer time interval. The peak can also be increased by raising the probe current and thus the count rate. However, limits exist to this procedure. For as the count rate increases, so does the dead time and the danger of pulse pile-ups. Pulse rates of 3000 s^{-1} in the overall spectrum should therefore not be exceeded.

Wavelength-dispersive spectrometer

In WDS analysis, the danger of erroneous interpretation of the spectra is significantly lower since the energy resolution is one order of magnitude better than for EDS analysis (Fig. 5-15). This advantage must, however, be offset against the considerably longer time required for recording the spectra. Whereas the EDS already allows to determine the elements present in the specimen within several minutes, the WDS requires at least half an hour for the same result.

Erroneous interpretations are possible because reflections of higher orders ($n = 2, 3, 4, \ldots$ in Bragg's law) are also detected by the analyzer crystal. Interpretation of the spectrum should therefore start with the lines having the shortest wavelengths, since these have the greatest probability of being of first order.

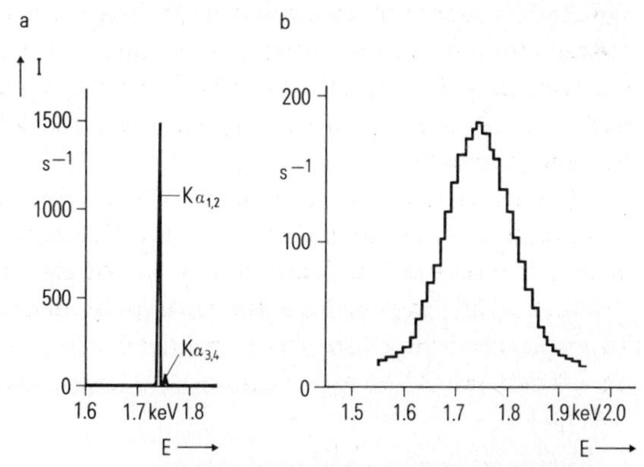

Fig. 5-15. Comparison of spectra, recorded with the WDS (a) and the EDS (b). The full with at half maximum of the Si Kα peak is 70 times greater for the EDS than for the WDS. Consequently, the $K\alpha_{1,2}$ line and the $K\alpha_{3,4}$ satellite line appear as a single line in the energy-dispersive spectrum.

5.3.5 Point, line and area analysis

The combination of optical microscope and X-ray microanalysis allows microregions inspected by microscopy to be pinpointed and analyzed. However, localization with the optical microscope is possible only with an uncertainty of several μm. A more exact assignment can be attained when the specimen exhibits cathodoluminescence (cf. section 3.6.2), i.e. emits visible light under electron bombardement. This is the case for some insulators and semiconductors. The point on the specimen which is struck by the electron beam can then be observed using the optical microscope.

For identifying different phases, it is useful to combine X-ray analysis with a backscattered electron image of the relevant specimen region. Since the backscattering coefficient depends on the atomic number of the specimen (Fig. 2-35), backscattered electron images display regions having different element compositions with different brightness (see section 3.5).

Point analyses are mostly used for the quantitative analysis of the selected microregions, but also for other purposes, such as investigating small particles lying isolated on a surface or being embedded in bulk material in the form of inclusions.

Line scans are representations of the element concentration along a selected line on the surface plane of the specimen. Forward advance along this line is effected by moving the specimen on its stage with the aid of the stepper motor control or by keeping the specimen stationary and deflecting the electron probe. The element profile is recorded on the screen of an image tube or with a plotter. In this recording, the y coordinate is controlled by the X-ray intensity, which is measured point-by-point, whereas the x coordinate is coupled to the stepper motor control or to the deflection voltage (Fig. 5-13). While the mechanical movement allows a relatively long distance to be scanned (up to several mm), scanning by beam deflection is limited to a distance of about 500 μm. During WDS measurements, larger deflections produce such great deviations from the focusing conditions that the intensity drops noticeably.

Area analysis means recording element distribution images (elemental maps). For this purpose, the electron probe is scanned across the specimen with the aid of the deflection system and the scan generator (Fig. 5-13). The spectrometer is permanently set to a selected characteristic line. The X-ray intensity controls the brightness of the oscilloscope CRT. Images are obtained "in the light" of the selected element: regions with higher element concentration appear bright, those with low concentration dark [5-33]. When using the WDS, the limited acceptance region of the spectrometer means that the imaged area is restricted to about $500 \cdot 500\ \mu m^2$. The EDS allows to display larger areas. The magnification of the elemental maps is set, as in the scanning electron microscope, with the aid of the deflection voltages.

Line scans and elemental maps are in most cases used to obtain a qualitative or semiquantitative overview of the distribution of an element (Fig. 5-16). Quantitative concentration profiles can be obtained when the mass fractions of the relevant elements are determined for every

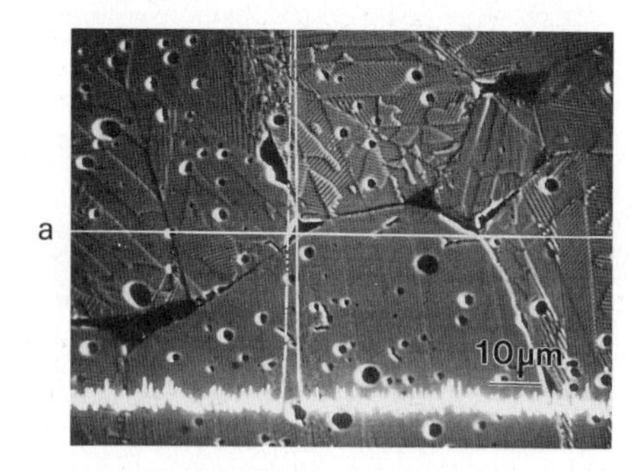

Fig. 5-16. a) Line scan: backscattered electron image of barium titanate ceramic with silicate inclusion. The inclusion was localized and identified by a Si line scan along the straight line shown in the middle of the micrograph.
b) Element mapping: microstructure of a barium titanate ceramic with a low yttrium content. The yttrium X-ray distribution image is superimposed on backscattered electron image and displays the Y-rich second phase.

measuring point by the correction calculations used in quantitative analysis (section 5.4). This procedure requires a time ranging from several tenths of a second to several seconds for each measuring point [5-34].

5.4 *Quantitative analysis*

5.4.1 *Background correction*

Let the intensity measured (as pulses per unit time) at the site in the spectrum of a specific characteristic X-ray line be designated I_{mA}. It is composed of the background intensity I_b and the characteristic X-ray intensity I_A (Fig. 5-17). The first aim of quantitative analysis consists of determining the background intensity and subtracting it from the measured signal. In measurements obtained with the WDS, which supplies very sharp spectral lines, the background at the site of the spectral line is generally calculated from the mean value of the background intensities on both sides of this line (Fig. 5-17a).

In measurements obtained with the EDS, which in contrast yields very broad spectral lines, the linear interpolation of the background intensity can be used only in the high-energy region and with isolated lines. In other cases, more elaborate mathematical correction methods must be used. For example the profile of the bremsstrahlung spectrum can be approximated by analytical functions (Fig. 5-17b) [5-35]. Also the peak area, which contains the sum of all useful pulses, provides a better measure for the peak intensity than the peak height at the maximum because of the low peak-to-background ratio in the case of broad lines (Fig. 5-17c). Since the intensity of the background changes only slowly with energy, whereas the intensity profile of the spectral lines changes quickly, the background can also be isolated from the spectrum with the aid of filter techniques and Fourier transforms [5-36,37]. An experimental determination of the bremsstrahlung spectrum profile can also be used for a correction procedure. The spectrum of carbon can then be used as the "standard spectrum" which is adapted to the real spectrum by scaling factors [5-38].

5.4.2 *Analysis with matched standards*

The intensity I_A of the characteristic radiation of an element A resulting after subtraction of the background is proportional, as a first approximation, to the number of atoms of this element per unit volume, i.e. $I_A \sim C_A$. The proportionality factor depends in a complex way on the interactive effects between the primary electrons and the X-ray quanta excited by them in the specimen. These "matrix effects" are almost identical when the specimen and the standard have similar composition (matched standard). But the measurement result is also influenced by instrumental factors (probe current, primary electron energy, take-off angle, spectrometer setting etc.). The instrument effects can be eliminated by measuring the intensities I_A and I_A^s from the specimen and standard under identical conditions and forming the ratio of these two values, thus obtaining $I_A/I_A^s = C_A/C_A^s$. The concentration C (number of atoms per unit

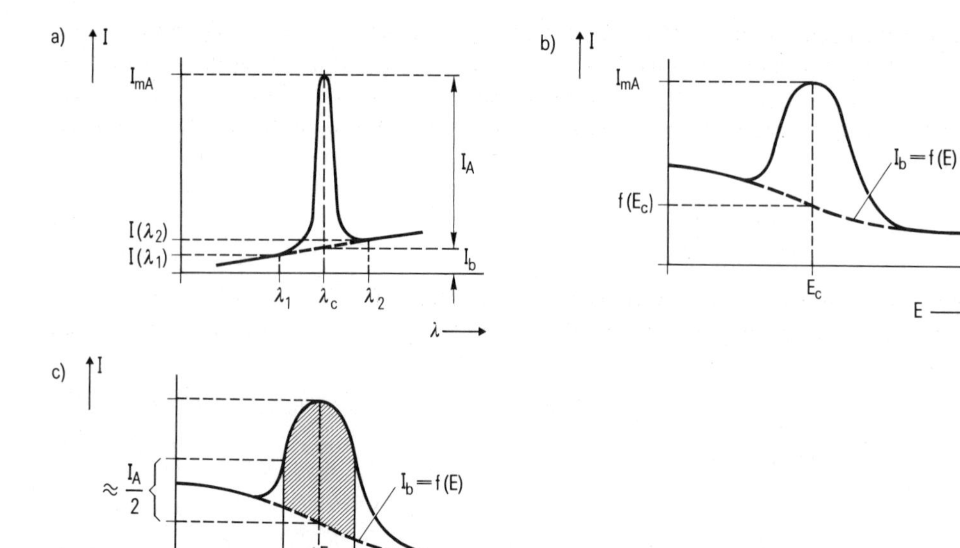

Fig. 5-17. Determination of the intensity I_A of the characteristic peak.
a) the intensity I_A is the difference between the intensity I_{mA} measured at the peak maximum and the mean intensity of the background radiation on both sides of the peak:
$I_A = I_{mA} - 0.5 \cdot [I(\lambda_1) + I(\lambda_2)] = I_{mA} - I_b$.
b) For the wide spectral lines obtained with EDS, the background is approximated by a polynomial $f(E)$ from which it follows that the intensity
$I_A = I_{mA} - I_b(E_c)$.
c) For wide spectral lines, the integral peak intensity (hatched area) provides a better satisfied precision. It is the sum of all pulses measured within the energy interval ΔE substracted by the background within this interval. The background can again be calculated from the approximation function $f(E)$. For ΔE usually the full width at half maximum (FHWM, cf. Fig. 5-12) of the characteristic peak is chosen. For a Gaussian-shaped peak, integration over ΔE includes 74% of the total peak intensity.

volume) can be converted to mass fraction w with the aid of the relation (eq. (1-3)) $C = w \cdot \varrho \cdot N_a / A$, where A is the mass of N_a atoms and N_a the Avogadro number. For approximately equal densities of both specimen and standard ($\varrho \approx \varrho^s$) the following is obtained [5-39, 40]:

$$\frac{I_A}{I_A^s} = k_A = \frac{w_A}{w_A^s}. \tag{5-7}$$

From this, together with the measured X-ray intensities and the known value of w_A^s, the mass fraction w_A in the specimen may be directly determined.

A whole series of "matched standards" is required to analyze all possible compositions in a system comprising several elements. Careful analyses performed on specimens with two [5-41] and even more [5-42] elements have shown that the relationship between the mass fraction

w_A and the intensity ratio $I_A/I_A^s = k_A$ can be approximated by a hyperbolic function. Two points of $w_A(k_A)$ can be specified immediately: at $k_A = 0$ (no element A present) w_A must be 0 and at $k_A = 1$ (only the pure element A present) w_A must be 1. Within a specific multi-component system, experimentally measured values are reproduced extremely well by the function

$$w_A = \frac{\alpha \cdot k_A}{k_A \cdot (\alpha - 1) + 1} \tag{5-8}$$

(Fig. 5-18). The constant α is determined from the values w_A^c and k_A^c of a specimen of known composition (calibration standard) from the same system. Using the relationship in equation (5-8), we obtain:

$$\alpha = \frac{w_A^c \cdot (1 - k_A^c)}{k_A^c \cdot (1 - w_A^c)} . \tag{5-9}$$

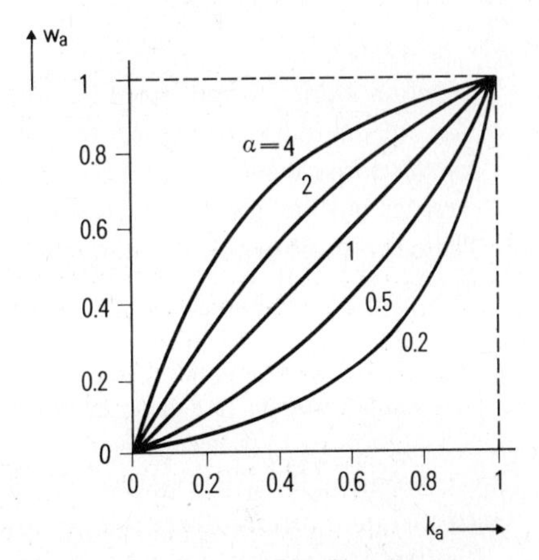

Fig. 5-18.
In multi-element specimens, the dependence of the mass fraction w_A of an element A on the measured intensity ratio $k_A = I_A/I_A^s$ can be approximated very well by the hyperbolic function of eq. (5-8) with empirically determined values of α.

After measuring k_A on specimens of any given composition, equation (5-8) can be used to calculate the mass fractions. But the precondition for this procedure is that the specimens contain the same elements as are present in the calibration standard. Accordingly, every specific system requires its own calibration standards. This procedure therefore does not make a great deal of sense for analyses of specimens with frequently changing element composition. In this case, other iterative correction methods must be used. These methods, which are described in the following sections, do not obviate reference measurements but use standards from pure elements. The assumptions made for eq. (5-7) are then usually not satisfied. Despite this, eq. (5-7) can be used as the first approximation for the iteration, as suggested by Castaing [5-39, 40]. For pure element standards, $w_A^s = w_A^0 = 1$, and the above equation assumes the form:

$$I_A/I_A^0 = k_A = w_A . \tag{5-10}$$

5.4.3 Matrix effects on the intensity

In general, matched standards are not available. The environment of the atoms of the element to be analyzed, the matrix, then differs significantly in the specimen and the standard. For determining the quantitative relationship between intensity and concentration therefore, the interactive processes of electrons and X-ray quanta listed below with the material must be taken into account.

a) The probe electrons are increasingly slowed down along their path in the specimen, some of them being backscattered before they can excite X-rays (cf. section 2.4.2). Due to the electron energy decreasing with depth, the primary X-ray intensity I_g generated in the material has a depth-dependent distribution which is a function of the density or the mean *atomic number* of the specimen material. The distribution of the generated X-ray intensity with depth is described by the function $\phi_0(\varrho z)$:

$$dI_g/d(\varrho z) = \phi_0(\varrho z) \,. \tag{5-11}$$

b) The X-ray intensity I_g generated in the specimen by primary electrons is attenuated by *absorption* along its path from its point of origin to the specimen surface.

c) Apart from the primary excitation of characteristic X-rays of the element A to be analyzed by the primary electrons, secondary excitation *(fluorescence)* can take place due to the X-rays of another element B present in the specimen also being excited.

These three influences of the matrix on the X-ray intensity are treated in more detail in the following.

Depth distribution of the generated X-ray intensity (atomic number effect)

Two fundamentally different ways exist of calculating the intensity I_g of the generated X-rays. One consists of determining the relationship between the physical interactive processes (backscattering, deceleration, ionization) and the generated X-ray intensity. The precondition for the other method is that the depth distribution $\phi_0(\varrho z)$ (eq. 5-11) for the generated X-rays be known. Both approaches are described in the following treatment.

To derive a relationship between primary X-ray intensity, ionization, deceleration and backscattering, let us initially consider the number of ionizations caused in the material by a probe electron penetrating the specimen. Along a path length dz, this electron generates

$$dn_A = C_A \cdot \sigma(E) \cdot dz \tag{5-12}$$

ionizations, where C_A is the number of atoms of an element A per unit volume and $\sigma(E)$ the ionization cross section (see section 5.1) of the material. As the total path in the specimen gets longer, the electron is increasingly slowed down. In the energy range in question here, this deceleration takes place monotonically, so that the differential quotient dE/dz can be used to describe the energy change per path increment. If $-dE/dz$ (the minus sign refers to the energy loss with increasing path length) is substituted into equation (5-12), then

$$dn_A = -C_A \cdot \sigma(E) \cdot dE/(dE/dz) \tag{5-13}$$

is obtained. The electron can ionize atoms only as long as its energy exceeds the critical ionization energy E_b of the atom. The total number n_A of the ionizations generated by an incident electron is obtained by integrating eq. (5-13) within the limits E_b and E_0. For the mass fraction $w_A = C_A \cdot A_A/(\varrho \cdot N_a)$ the following is obtained

$$n_A = -\frac{w_A \cdot N_a}{A_A} \int_{E_b}^{E_0} \frac{\sigma(E)}{dE/d(\varrho z)} dE . \quad (5\text{-}14)$$

The parameter $-dE/d(\varrho z) = S$ characterizes the decelerating force which affects the electron in a material of density ϱ and is called the stopping power. S was originally calculated by Bethe [5-43] for the slowing down of electrons in hydrogen. Different theoretical and empirical formulae were suggested for the heavier elements in the subsequent years [5-44 to 52].

Equation (5-14) specifies the number of ionizations generated by an incident electron in the specimen. To determine the X-ray intensity (pulse rate), the number of ionizations per unit time is required. If the specimen is struck by a constant probe current i_0, the number of ionizations per unit time is proportional to $i_0 \cdot n_A$ if all electrons were to transfer an energy up to $E \geq E_b$ to the specimen. However, some electrons leave the specimen as a result of backscattering (cf. section 2.4) and so are not available for ionization. This means that only the quantity $R \cdot i_0 = (1 - r) \cdot i_0$ of the incident electron current contributes to the ionization, where r is the ratio of the number of ionizations lost by backscattering to the number for which backscattering would not occur. r depends on the backscattering coefficient η (Fig. 2-35) and on the energy distribution of the backscattered electrons (Fig. 2-37). Taking backscattering into account, it follows from equation (5-14) for the intensity of the generated X-rays with $S_A = dE/d(\varrho z)$ that

$$I_{gA} = c \cdot w_A \cdot R_A \cdot \int_{E_b}^{E_0} \frac{\sigma(E)}{S_A(E)} dE \quad (5\text{-}15)$$

where the constant c essentially contains the probe current, the fluorescence yield (cf. section 2.4.3), the atomic mass and Avogadro's number.

For the second case, where the depth distribution $\phi_0(\varrho z)$ is known, the intensity of the generated X-rays is obtained directly by integrating equation (5-11):

$$I_g = \int_0^\infty \phi_0(\varrho z)\, d(\varrho z) . \quad (5\text{-}16)$$

$\phi_0(\varrho z)$ can either be calculated by a Monte Carlo simulation [5-53 to 55], taking into account the interaction processes, or determined experimentally [5-56 to 59]. Most experimental data were obtained by the tracer method, in which the X-ray intensity of a thin film ("mass thickness" $\Delta\varrho z$) of an element B embedded in a material of an element A is measured as a function of the film depth below the surface (Fig. 5-19a). The depth z multiplied by the density ϱ is used in place of the depth z, since this "mass depth" ϱz is independent of the physical and chemical state of the material. The atomic numbers of the elements A and B should be as close to each other as possible; for convenience, the element B (the tracer) should be one

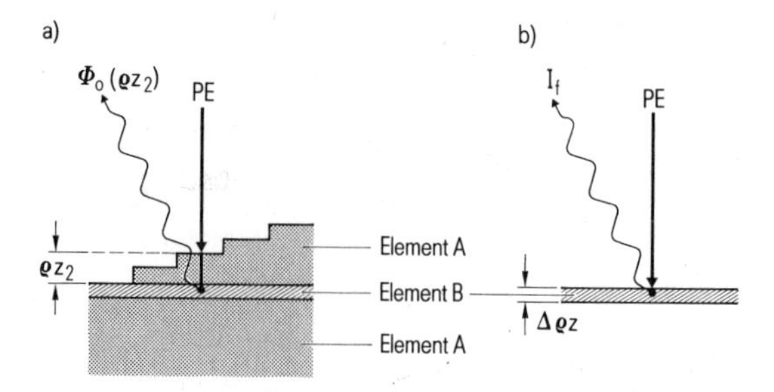

Fig. 5-19. Tracer method for the experimental determination of the depth distribution $\phi(\varrho z)$ of the generated X-ray intensity I_g.
a) The layers of element A are deposited in a stepped fashion onto the thin tracer layer of element B. The electron probe (PE) is directed on one step after the other and the X-ray intensity emitted from the tracer element B is measured at every step and corrected with respect to the absorption along its path through the A layers.
b) The intensity I_f emitted from the unsupported layer from the element B with the same thickness $\Delta \varrho z$ under identical excitation conditions acts as the reference parameter. For approximately identical physical properties of the elements A and B, the values determined in this way yield to the normalized depth distribution $\phi(\varrho z) = \phi_0(\varrho z)/I_f$ for the radiation from element A.

position higher in the periodic table than element A. This ensures that the characteristic radiation of A cannot excite any characteristic fluorescent radiation in B. In addition, the properties of both elements are then almost identical with respect to electron scattering, and the measured X-ray intensity can be assigned to the substrate material A. In order to determine the depth distribution of Cu radiation for example, Zn is used as the tracer and the Zn radiation measured.

After correction for the absorption caused by the embedding material (see below), the X-ray intensity measured in this way as a function of the mass depth ϱz is proportional to the depth distribution $\phi_0(\varrho z)$. The X-ray intensity of an unsupported thin film, which is identical to the tracer layer in terms of material and layer thickness, is then additionally measured under exactly identical conditions (Fig. 5-19b). The measured intensity of the unsupported layer is proportional to the X-ray intensity I_f generated in it by the primary electrons. If the values of the embedded tracer layer are referred to those of the unsupported layer, the normalized depth distribution which now is independent of the measuring conditions is obtained:

$$\phi(\varrho z) = \phi_0(\varrho z)/I_f \, . \tag{5-17}$$

The normalized $\phi(\varrho z)$ curves (Fig. 5-20) have a characteristic profile which is similar for all elements. The X-ray intensity is greater in the top surface layer of the bulk material than in the unsupported layer ($\phi(0) > 1$), since in the former the ionizations caused by backscattered electrons from deeper layers also contribute to the X-ray intensity. With increasing depth, the X-ray intensity initially increases: as a result of the increase in scattering with depth which also leads to trajectories in lateral direction the overall electron paths become greater in each successive layer reached. This increases the probability of generating X-ray quanta. In addition, the ionization cross section σ increases with decreasing electron energy, assuming that the probe energy E_0 is significantly greater than the ionization energy E_b (cf. Fig. 5-2). Since

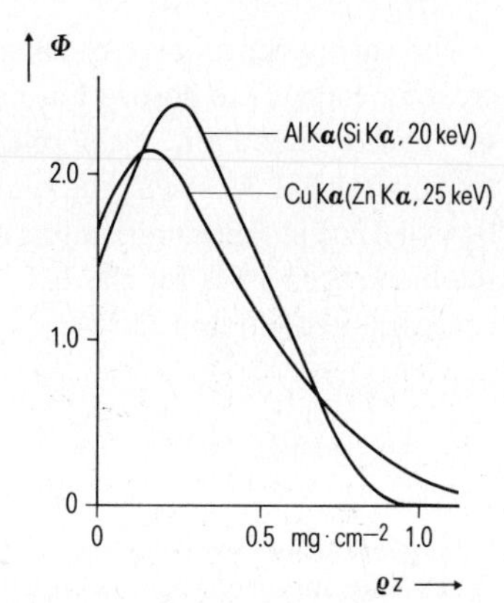

Fig. 5-20.
Depth distribution ϕ (ϱz) for aluminum and copper determined by the tracer method after normalization. The tracer elements and the electron energies are in each case specified in parentheses. In accordance with the lower penetration depth of the primary electrons, the maximum moves closer to the surface with increasing atomic number (as per [5-55, 57, 58]).

only fewer electrons reach greater depths and because the ionization cross section becomes smaller again as the electron energy decreases further, the intensity of generated X-ray quanta is, after a maximum, reduced again with increasing depth.

Using the normalized depth distribution, the total generated X-ray intensity (see eqs. (5-16) and (5-17)) is:

$$I_g = I_f \int_0^\infty \phi(\varrho z) \, d(\varrho z) \, . \tag{5-18}$$

Attenuation of the generated X-ray intensity by absorption

The X-rays generated inside the specimen by the primary electrons is attenuated by absorption along its path from the point of excitation to the specimen surface. The absorption takes place because the X-rays ionize outer shells of the same atom (intrinsic absorption) and other atoms contained in the specimen in a secondary process (photoelectric effect). The degree of absorption is determined by the linear absorption coefficient μ which depends on the energy of the generated X-rays, on the atomic number of the absorbing atoms and on the density of the specimen material. After passing through material of thickness x the X-ray intensity I_0 is reduced to $I = I_0 \cdot \exp(-\mu x)$. In place of μ, it is convenient to use the mass absorption coefficient μ/ϱ referred to the density, since this, like the mass thickness, is essentially independent of the physical and chemical state of the atom.

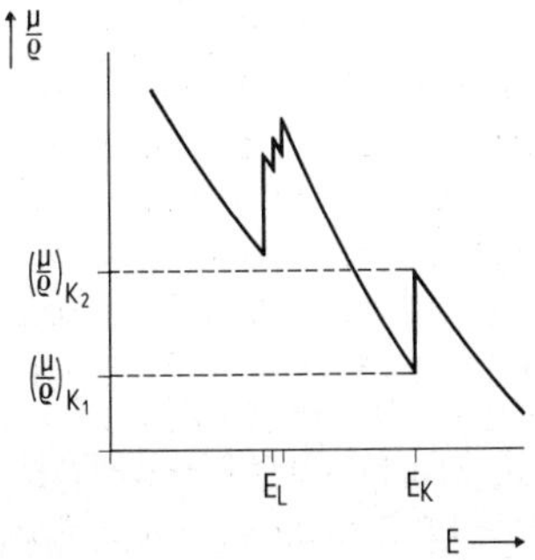

Fig. 5-21.
Mass absorption coefficient μ/ϱ as a function of the energy E of the X-ray photons. The absorption edges (jumps of μ/ϱ) lie at the critical excitation energies for the K, L, ... radiation. The L edge is threefold due to its three sublevels.

The energy dependence of the mass absorption coefficient is shown in Fig. 5-21. With increasing energy, the absorption initially decreases, the material being more easily penetrated by "harder" radiation. If the photon energy attains the critical excitation energy for the electrons in the K,L,M shells, the probability of ionization increases abruptly. The mass absorption coefficient therefore exhibits discontinuities (steps), known as absorption edges, at exactly the binding energies for the K,L,M electrons. The size of the steps can be described by the absorption edge jump ratios

$$\nu = (\mu/\varrho)_2/(\mu/\varrho)_1$$

(Fig. 5-21) which decrease with increasing atomic number Z [5-60]. The mass absorption coefficients for many elements and for various X-ray energies were determined experimentally and represented by empirical formulae as a function of energy, wavelength and atomic number [5-61 to 64]. Since the mass absorption coefficients for a specific element change continually between the absorption edges and also show a systematic dependence on the atomic number for a given X-ray energy, missing experimental values can be calculated by interpolation.

In Fig. 5-22, the absorption of the generated X-ray intensity I_g is shown along its path through the material. Using for simplification

$$\chi = (\mu/\varrho)/\sin\psi$$

the intensity emitted from the specimen then (without taking into account the fluorescence excitation) has the value:

$$I = \int_0^\infty \exp(-\chi\varrho z)\,\mathrm{d}I_g\,.$$

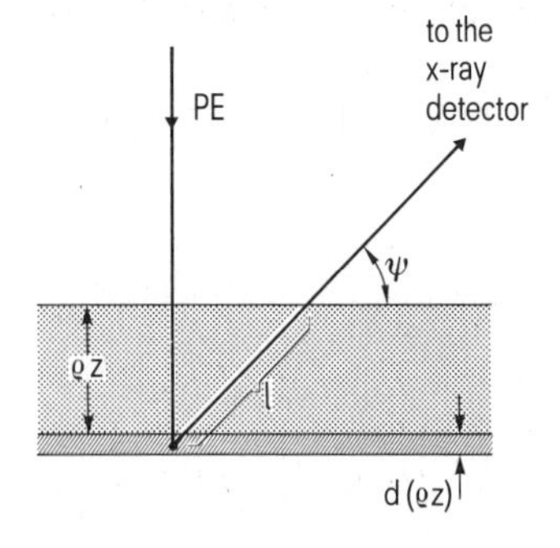

Fig. 5-22. The X-ray quanta generated by the primary electron PE in the layer dz must traverse a distance $l = z/\sin\psi$ in the specimen along the path towards the spectrometer. The intensity dI_g of the X-rays at a depth z is consequently reduced as a result of absorption. The intensity emitted from the specimen therefore has the value
$\mathrm{d}I = \mathrm{d}I_g \cdot \exp(-\mu l) = \mathrm{d}I_g \cdot \exp[(-\mu/\varrho)\cdot\varrho\cdot z/\sin\psi]$
where μ/ϱ is the mass absorption coefficient.

With equations (5-11) and (5-17), one then obtains:

$$I = I_f \cdot \int_0^\infty \phi(\varrho z) \cdot \exp(-\chi \varrho z) \cdot d(\varrho z) \,. \tag{5-19}$$

If this expression is referred to the total intensity generated in the specimen in line with eq. (5-18), the following correction factor is obtained for the absorption:

$$f(\chi) = \frac{I}{I_g} = \frac{\int_0^\infty \phi(\varrho z) \cdot \exp(-\chi \varrho z) \cdot d(\varrho z)}{\int_0^\infty \phi(\varrho z) \cdot d(\varrho z)} \,. \tag{5-20}$$

Increase in the X-ray intensity due to fluorescence excitation

Apart from the characteristic X-rays of interest (of an element A), the X-rays of other elements present in the specimen are also excited by the primary electron bombardment. Where the excited X-rays from one of the other elements (e.g. of an element B) are more energetic than the excitation energy for the X-rays generated from A, fluorescence excitation of A takes place. Both the characteristic radiation of B and the continuous high-energy bremsstrahlung can contribute to the fluorescence. The fluorescence radiation leads to an increase in the primary intensity I_g generated by the incident electrons by the factor

$$1 + \gamma + \delta \,, \tag{5-21}$$

where γ specifies the contribution of the fluorescence radiation due to the characteristic radiation and δ the contribution due to the bremsstrahlung. Detailed derivations for the fluorescence correction are found e.g. in [5-18].

Intensity of the emitted radiation

To take into account all the matrix influences mentioned so far, eq. (5-19), which describes the attenuation of the intensity by absorption, must be supplemented by the contribution of the fluorescence excitation (factor (5-21)). Finally, the following is obtained for the intensity of the radiation emitted from the specimen:

$$I = I_f \cdot \int_0^\infty \phi(\varrho z) \cdot \exp(-\chi \varrho z) \cdot d(\varrho z) \cdot (1 + \gamma + \delta) \tag{5-22}$$

or with the absorption factor $f(\chi)$ from equation (5-20)

$$I = I_f \cdot \int_0^\infty \phi\,(\varrho z)\,\mathrm{d}\,(\varrho z) \cdot f\,(\chi) \cdot (1 + \gamma + \delta)$$

$$= I_g \cdot f\,(\chi) \cdot (1 + \gamma + \delta)\,. \tag{5-23}$$

For the generated intensity I_g, the expression from eq. (5-18) was used.

The intensity formula derived here applies to the case where the specimen normal points in the direction of the exciting electron beam. If the normal deviates from this direction, the specific geometrical relationships must be taken into account.

5.4.4 The ZAF correction factors

Equation (5-23) contains the matrix influences on the *emitted* X-ray intensity, but not the instrument parameters (probe current, primary electron energy, spectrometer setting, etc.), which in addition influence the *measured* intensity significantly. To eliminate these influences, reference measurements arc made on standards and relative X-ray intensities are used to derive the final result. No matched standards are required in this approach, pure element standards being sufficient as a rule.

If the intensities I_A of an element A in a compound specimen and I_A^0 of a standard consisting of the pure element A are expressed by I of equation (5-23) and the ratio of the two then formed, the result is

$$\frac{I_A}{I_A^0} = \frac{I_{fA}}{I_{fA}^0} \cdot \frac{\int_0^\infty [\phi\,(\varrho z)\,\mathrm{d}\,(\varrho z)]_A}{\int_0^\infty [\phi\,(\varrho z)\,\mathrm{d}\,(\varrho z)]_A^0} \cdot \frac{f\,(\chi_A)}{f\,(\chi_A^0)} \cdot (1 + \gamma_A + \delta_A)$$

$$= w_A \cdot F_Z \cdot F_A \cdot F_F\,. \tag{5-24}$$

This takes into account that in the pure element standard no fluorescence excitation can occur due to other characteristic radiation ($\gamma_A^0 = 0$) and that secondary excitation can be neglected due to harder bremsstrahlung radiation ($\delta_A^0 \approx 0$). F_Z, F_A and F_F are the correction factors which take into account the atomic number effect, the absorption and the fluorescence excitation respectively. I_{fA} and I_{fA}^0 are the intensities which are generated by primary electrons in a compound layer with mass fraction w_A and in the pure element layer. Since these layers, which are needed for determining the normalized distributions $\phi_A\,(\varrho z)$ and $\phi_A^0\,(\varrho z)$, are very thin (see Fig. 5-19), matrix influences can be neglected. Further, both layers possess the same mass thickness $\Delta \varrho z$. As a consequence, equation (5-10) holds correctly for the ratio of these two intensities: $I_{fA}/I_{fA}^0 = w_A$.

The correction factors F_Z, F_A and F_F contain parameters (density, absorption coefficient, excitation energy, backscattering coefficient, atomic mass, atomic number) which differ in the specimen and standard in line with their different compositions. They are therefore themselves dependent on the mass fractions w_i of all elements contained in the specimen (i = A, B, C, ...). An iterative procedure must therefore be adopted to determine the mass fraction w_A. In the first step, the w_i values from the Castaing approximation $w_i = k_i$ (eq. (5-10)) are determined. Together with equation (5-24), this yields a second approximation which must as a rule be followed by further approximations. The iteration method is described in more detail in the next section.

Many formulae for the individual correction factors, which were found either empirically or theoretically, have been described in the literature. Since it is difficult to specify an integratable function for the depth distribution $\phi(\varrho z)$ with the available parameters, particular methods were developed in which the calculation of the three correction factors is based on simplified physical models of the interactive processes of excitation and scattering [5-65 to 78]. In calculating the atomic number correction factor, eq. (5-15) is used for the generated X-ray intensity. Although some of these methods are over 25 years old, they have proved their value for the analysis of elements with atomic numbers $Z > 10$ and are still largely used today.

In recent years, methods have increasingly been developed in which the depth distributions, $\phi(\varrho z)$, determined either experimentally or with the aid of Monte Carlo calculations, were approximated by analytical functions [5-55, 58, 59; 5-79 to 86]. By this means, the two correction factors F_Z and F_A can be combined into a factor F_{ZA} (cf. eqs. (5-24) and (5-20)):

$$\frac{I_A}{I_A^0} = w_A \cdot \frac{\int_0^\infty [\phi(\varrho z) \cdot \exp(-\chi \varrho z) \cdot \mathrm{d}(\varrho z)]_A}{\int_0^\infty [\phi(\varrho z) \cdot \exp(-\chi \varrho z) \cdot \mathrm{d}(\varrho z)]_A^0} \cdot (1 + \gamma_A + \delta_A)$$

$$= w_A \cdot F_{ZA} \cdot F_F . \tag{5-25}$$

The correction methods based on this approach also supply very precise results for the analysis of lighter elements. They are now also being offered in commercial systems [5-83].

5.4.5 ZAF Iteration procedure

Initially, the intensities I_i of the selected (intense) spectral lines of all elements i contained in the specimen (i = A, B, C, ...), as well as the intensities I_i^0 of the associated standards, must be determined. They are used as a first approximation in the iteration procedure in line with equation (5-10)

$$w_{i1} = k_i = \frac{I_i}{I_i^0} . \tag{5-26}$$

The sum of the (true) mass fractions in the specimen must total 1. Since this is not generally the case for the measured values in eq. (5-26), the w_{i1} values must be normalized so that their sum equals 1:

$$w_{i1}^* = w_{i1} / \sum_i w_{i1} . \tag{5-27}$$

The values of w_{i1}^* are then used to calculate the factors for the atomic number correction F_Z, the absorption correction F_A and the fluorescence correction F_F. The 2nd approximation is obtained by substituting them into eq. (5-24)

$$w_{i2} = k_i / F_{ZAF}(w_{i1}^*) \tag{5-28}$$

($F_{ZAF} = F_Z \cdot F_A \cdot F_F$). After normalizing w_{i2} to w_{i2}^*, w_{i2}^* can again be used to calculate a correction factor $F_{ZAF}(w_{i2}^*)$ and thus to obtain a third approximation w_{i3}. The iteration should then be continued until the atomic number fractions no longer change significantly. However, this procedure converges extremely slowly and can sometimes even diverge [5-87].

An additional step is therefore introduced into the iteration procedure, which allows the relationship as per eq. (5-8) between the mass fractions w_i and the measured intensity ratios k_i to be taken into account:

$$w_i = \frac{\alpha_i \cdot k_i}{k_i \cdot (\alpha_i - 1) + 1} . \tag{5-29}$$

To calculate the mass fraction w_i from the measured intensity ratios k_i with eq. (5-29), the constant α_i is required. In the first approximation α_{i1} is calculated from equation (5-29) solved for α (see eq. (5-9)) by substituting k_{i1} values which are obtained with the aid of the correction factors $F_{ZAF}(w_{i1}^*)$:

$$k_{i1} = w_{i1}^* \cdot F_{ZAF}(w_{i1}^*) . \tag{5-30}$$

For the normalized mass fractions w_{i1}^*, the values in equation (5-27) are used, thus obtaining the following for α_{i1}

$$\alpha_{i1} = \frac{w_{i1}^* \cdot (1 - k_{i1})}{k_{i1} \cdot (1 - w_{i1}^*)} . \tag{5-31}$$

With the α_{i1} values calculated in this way, the second approximation is obtained with the aid of an equation analogous to equation (5-29):

$$w_{i2} = \frac{k_i \cdot w_{i1}^* \cdot (1 - k_{i1})}{k_i \cdot (w_{i1}^* - k_{i1}) + k_{i1} \cdot (1 - w_{i1}^*)} . \tag{5-32}$$

For the next step, w_{i2} must be normalized (w_{i2}^*) in turn, k_{i2} calculated following equation (5-30) with the correction factors $F_{ZAF}(w_{i2}^*)$ and thus α_{i2} calculated following equation (5-31). The third approximation is then obtained by substitution into an equation analogous to eq. (5-32):

$$w_{i3} = \frac{k_i \cdot w_{i2}^* \cdot (1 - k_{i2})}{k_i \cdot (w_{i2}^* - k_{i2}) + k_{i2} \cdot (1 - w_{i2}^*)} .$$

The iteration is completed when two successively calculated k_{in} values no longer differ significantly. In general, the procedure converges quickly; rarely are more than three iteration steps required. It is initially recommended not to normalize the result of the last iteration step. Conclusions about erroneous measurements (e.g. unidentified element) or incorrect data input can be drawn from deviations from $\Sigma w_i = 1$. Further, it may be the case that the ZAF program used is not adapted to the analysis problem at hand (extreme differences in atomic numbers, thin layers, boundary layers, small particles, light elements).

These correction procedures are today largely performed by computers which are already included into the X-ray microprobes. The commercial computer programs used are essentially based on the physical principles described in the previous sections. However, different analytical expressions are used for the correction formulae, depending on the simplifications which were made in matching the analytical formulae to the experimentally determined values. Comprehensive representations of the various methods can be found in the work of Love and Scott [5-78] as well as in the books by Heinrich [5-88] and Scott and Love [5-17]. Various studies have shown that the programs most frequently used supply comparable good results [5-89, 90, 91].

The methods of Duncumb and Reed [5-44] and Philibert and Tixier [5-45] are the most frequently used for the atomic number correction. The uncertainty caused by the Z correction is in every case smaller than the measurement uncertainty in determining the intensity ratio k_A (eq. (5-10)). Of the three correction factors F_Z, F_A and F_F, the absorption correction F_A is associated with the greatest error because not enough experimental data are available for the depth distribution of the generated X-rays $\phi(\varrho z)$ and for the mass absorption coefficients.

With newly developed empirical formulae for the depth distribution $\phi(\varrho z)$, very precise correction factors can be obtained in those cases in which reliable data are available [5-59, 84, 86]. In the analysis of light elements ($Z < 10$) also, the mean errors remain within acceptable limits at values of around 4%. It is to be expected that those correction methods which use matched $\phi(\varrho z)$ curves will become more widely accepted as reliable data, determined experimentally or with the aid of Monte Carlo calculations [5-53, 92], become available.

With the conventional ZAF correction method and the use of pure element standards for elements with $Z > 11$ and wavelengths below about 0.3 nm, analysis errors of between $\pm 1\%$ and $\pm 3\%$ must be expected as long as the overall correction factor F_{ZAF} does not deviate from 1 by more than $\pm 30\%$. With larger wavelengths and greater deviations, the analysis error is about $\pm 5\%$. Significantly greater errors can occur with light elements because the absorption coefficients for these elements are either unknown or not sufficiently precise (section 5.5).

5.4.6 *Theoretical standards*

With the aid of the formulae for the generated X-ray intensity I_{gA}, it is in principle possible to calculate the intensity I_A for any pure element A measured with the X-ray detector [5-93, 94, 95]. These values can then be used as intensities I_A^{th} of "theoretical standards" and thus to determine $k_A = I_A / I_A^{th}$ (cf. eq. (5-10)), where I_A is the measured intensity of the specimen (composed of several elements). The precondition here is that the instrument parameters are known precisely and can be set repeatably during the period of the measurement. This precondition is satisfied in the energy-dispersive spectrometer systems which can be used to collect and record the spectral lines of most elements simultaneously. The Si(Li) detector has the further advantage that its detection efficiency for X-ray quanta is constant over a large spectral region and has a value of almost 100% (cf. section 5.2.3).

However, a series of drastic simplifications must be made for calculating the intensities of the theoretical standards. The analysis results obtained with these are therefore burdened with greater uncertainties than those obtained by measurements on real pure element standards. Relative errors of between 5 and 20% must be expected for mole fractions greater than 1% and considerably greater errors for fractions below 1%. The advantage of this method in energy-dispersive analysis, however, lies especially in the fact that semi-quantitative results can be obtained rapidly from only one single measurement.

5.5 Analysis with low-energy radiation

Quantitative analysis with the aid of low-energy radiation ($E \leq 1$ keV, $\lambda \leq 1.2$ nm) is more difficult to perform than analysis using radiation of higher energy for the following reasons:

- With decreasing energy (increasing wavelength) of the X-rays, the fluorescence yield is reduced (cf. Fig. 2-47). As a result, the detection sensitivity is lower with the use of "soft" X-rays than with harder radiation. Apart from this, the background correction is then burdened with larger errors (lower signal-to-background ratio).
- With decreasing energy, the absorption increases strongly (cf. Fig. 5-21). Since errors in the absorption correction factor have a stronger influence than errors in the other correction factors, the results of correction calculations can contain large errors. Only inaccurate data, some of it with a large scatter, is available for the absorption coefficients of long wavelength radiation. Greatest use is made of the values published by Heinrich [5-96]. The high absorption also means that contamination or conductive coatings have a much greater effect with low-energy than with high-energy radiation.
- The low-energy radiation originates from transitions between the outer electron shells of the atoms (K spectrum of the light elements ($Z < 11$), L,M,N spectra of the heavier elements). These transitions are strongly influenced by the surroundings (bonding) of the atom, leading to changes in the shape, intensity and energy (chemical shift) of the characteristic lines [5-97,98].
- In the light elements, only the K spectrum is available for analysis. This means that when lines of other elements overlap, recourse cannot be made to other spectral regions.
- For the light elements, standards which are defined with sufficient accuracy are often not available.
- Low-energy radiation is more difficult to detect than high-energy radiation (cf. sections 5.2.3 and 5.2.4).

The errors due to these influences can be reduced by the following measures:

- Well-matched standards are used whose physico-chemical properties are similar to those of the specimen. The correction factors then do not deviate much from unity.
- The surface of insulating specimens must be coated with conductive layers made up of elements with low atomic number (carbon or aluminum) and not containing the element to be analyzed.
- To reduce the absorption path length, the instrument is operated at a low excitation energy E_0 and a large take-off angle ψ. However, E_0 should not be selected to be too small, since

the excitation probability also becomes smaller with decreasing excitation energy (Fig. 5-2).

- To avoid contamination layers on the surface, the specimens must be carefully cleaned and stored in high vacuum. During the analysis, anti-contamination devices (cold traps) must be used.

In practice, however, the measures suggested above cannot always be applied. Thus a reduction of the excitation energy E_0 is no longer useful when short wavelength radiation must be measured in addition to long wavelength radiation (analysis of light and heavy elements in a specimen). In commercial instruments, the take-off angle ψ is generally fixed in a manner allowing no free choice. Further, matched standards are not always available, so that pure element standards must be used. In this case, satisfactory results can be obtained with correction methods developed on the basis of depth distributions $\phi(\varrho z)$ obtained experimentally or by Monte Carlo calculations (section 5.4).

5.6 Thin films and particles

5.6.1 Thin films on substrates

When the thickness of surface films on bulk substrates is greater than the excitation depth in which the primary X-rays are generated, the composition of the films can be determined in the same way as in bulk specimens (section 5.4).

For the maximum excitation depth z_x, an analogous dependence on the primary electron energy E_0 applies as for the penetration depth: $z_p \sim E_0^n$ (cf. Fig. 2-33). However, z_x is smaller than z_p since X-ray quanta can be generated only as long as the electron energy is greater than the critical excitation energy ($E_0 \geq E_b$). z_x may therefore be represented by formulae of the form

$$z_x = a \cdot \frac{A}{Z \cdot \varrho} \cdot (E_0^n - E_b^n)$$

(A, mass of N_a atoms; Z, atomic number; ϱ, density). For the factor a and the exponents n, numerous value pairs were specified. These values can be determined experimentally (e.g. [5-99]) or by Monte Carlo calculations (e.g. [5-100]). According to [5-100] the relationship

$$z_x = 0.021 \cdot \frac{A}{Z \cdot \varrho} \cdot (E_0^{1.75} - E_b^{1.75}) \tag{5-33}$$

(z_x in µm, A in g, ϱ in g/cm^3 and E_0, E_b in keV) yields good results for electron energies $E_0 > 5$ keV and atomic numbers 13 to 82. For copper ($E_K = 8$ keV), the maximum excitation depth is 1.7 µm when a primary energy of $E_0 = 30$ keV is used. If a copper film of 0.5 µm thickness is still to be treated as bulk material, operation must be with a primary energy lower than 30 keV, i.e. about 17 keV. At still thinner films, even lower primary electron energies are required. Due to the lower ionization cross section, however, the X-ray intensity is then also lower (see Fig. 5-2). If low-energy X-ray lines are available for the analysis, such as the copper Lα line with $E_L \approx 1$ keV, then the primary electron energy can again be considerably reduced, say to 5 keV. The excitation depth z_x then decreases to somewhat below 0.1 μm. When using such low energy X-ray lines, however, the absorption effects are high and the results of the correction calculations are correspondingly inacurrate (cf. section 5.5).

For thin films, whose thickness is less than the maximum excitation depth, special correction methods must be used, which above all take into account the depth dependence of the X-ray generation. They can be used for the quantitative determination of both the composition and the film thickness. The general precondition for analyzing the elements in coating layers is that the element to be determined in the coating film is not present in the substrate.

In the first instance, let the case of a pure element film of an element A on the substrate of an element B be considered. To determine the film thickness t, the X-ray intensity I_A^F emitted from the layer is compared with that of the pure element standard I_A^0. With the depth distributions $\phi_A^F(\varrho z)$ for the film-substrate combination and $\phi_A^0(\varrho z)$ for the bulk standard, the following relationship applies (cf. eq. (5-16))

$$\frac{I_A^F}{I_A^0} = \frac{\int_0^{\varrho t} \phi_A^F(\varrho z)\, d(\varrho z)}{\int_0^{\infty} \phi_A^0(\varrho z)\, d(\varrho z)} = k_A \tag{5-34}$$

when absorption is neglected. The depth distributions are dependent on the primary electron energy, the atomic number and other parameters. ϕ_A^F for the film-substrate combination also depends on the substrate and the film thickness. The influence of the substrate results from the fact that electrons backscattered from the substrate generate different amounts of secondary X-ray quanta in the film than if the film were to lie on a substrate of the same element. If the atomic number of the substrate is, say, higher than that of the film, then the higher backscattering coefficient gives rise to a higher proportion of ionizations due to the electrons backscattered at the substrate. The values of ϕ_A^F are then greater than those of ϕ_A^0.

If the scattering properties of the layer and substrate differ only slightly, then $\phi_A^F \approx \phi_A^0$ can be set as an approximation. The intensity ratio k_A can then be calculated from eq. (5-34) by integrating $\phi_A^0(\varrho z)$ up to a mass thickness ϱt and dividing it by the integral in the limits from zero to infinity. Fig. 5-23 contrasts ϕ_{Cu} and k_{Cu} for copper, both as functions of ϱz. The $k_{Cu}(\varrho z)$ curve can be used as a calibration curve from which the mass thickness can be read off at a measured k_{Cu} value. For very small film thicknesses, it can be assumed that $\phi_A^F(0)$ = constant and the relationship between k_A and ϱz is linear. Up to layer thicknesses where $\phi_A^F(\varrho z) > \phi_A^F(0)$, the $k_A(\varrho z)$ curve deviates only slightly from a straight line (cf. Fig. 5-23b). The layer thickness to which this approximately linear relationship applies increases with the primary electron energy. At greater film thicknesses, which come into the region of the excitation depth, k_A approaches the value 1.

If X-ray microanalysis is to be used for measuring film thicknesses for only one type of a film-substrate combination, the simplest approach is to determine the calibration curve $k_A(\varrho z)$ for a suitable primary electron energy by experimental means. If, however, very different film-substrate combinations have to be measured frequently, this procedure is too elabo-

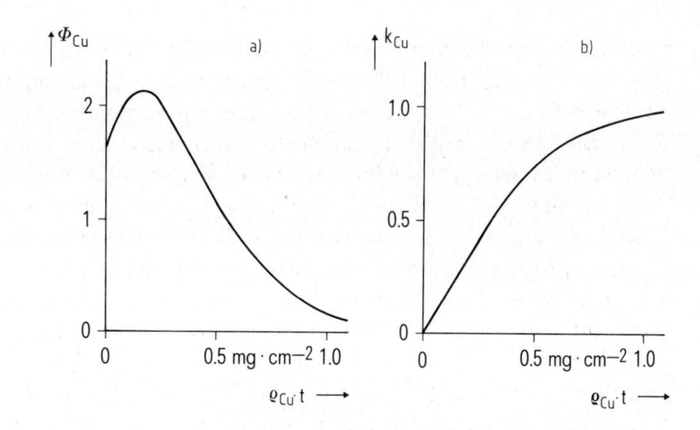

Fig. 5-23. Depth distribution ϕ_{Cu} for copper at an electron energy of 25 keV (a) as per [5-101] and the function k_{Cu} determined from it by graphical integration (b). The mass thickness $\varrho_{Cu} \cdot t$ of a copper layer on a substrate with similar scattering properties (e.g. zinc) can be determined from the measured intensity ratio I^F_{Cu}/I^0_{Cu} with the aid of the curve b).

rate and the calibration curves must be calculated theoretically. For the general case in which the atomic numbers of the layer and substrate differ considerably, the approximation $\phi^F_A \approx \phi^0_A$ mentioned above does not apply. To set up a calibration curve, it is then not sufficient to know only a single depth distribution. $\phi^F_A(\varrho z)$ must be specifically determined for every film-substrate combination as well as for the different film thicknesses and k_A then calculated according to eq. (5-34). Whereas the depth distributions $\phi^0_A(\varrho z)$ were determined from experimental data for numerous bulk pure element standards, no $\phi^F_A(\varrho z)$ curves are available due to the numerous possibilities of combining the layer and substrate. For the depth distributions $\phi^F_A(\varrho z)$ therefore, theoretically calculated expressions must be used in each case. These can be obtained from analytical expressions whose parameters are normalized to experimental data [5-102,103,104] or obtained by means of Monte Carlo calculations [5-105,106].

Since the absorption in the thin film is small compared with that in the substrate, it is sufficient, especially for very thin films (e.g. for copper with film thicknesses below 0.1 μm) merely to correct the X-ray intensity of the standard in eq. (5-34) by means of the absorption factor (eq. (5-20)). For somewhat greater film thicknesses (for copper up to about 0.5 μm) the absorption correction for the film can be approximated by a factor based on a punctiform X-ray source in the middle of the film:

$$f'(\chi) = \exp\left[-\chi(\varrho t/2)\right] . \tag{5-35}$$

In general, a fluorescence correction can be neglected. In special cases, however, (e.g. an atomic number difference of $\Delta Z = 2$ for medium atomic number) the substrate radiation can strongly excite X-ray emission in the film, leading to errors up to 20% if no fluorescence correction is performed [5-107].

The analysis of multi-element films involves measuring the intensity ratio k_i for every element. If the films are very thin (e.g. $t \ll z_p$), then matrix effects in the film can be neglected and calibration curves for pure element films can be used to determine the mass thicknesses

$(\varrho t)_i$ for the various elements from the k_i values. The entire film thickness then results from the sum of the individual thicknesses:

$$(\varrho t)_{\text{tot}} = \sum_i (\varrho t)_i$$

and the following applies for the mass fractions:

$$w_i = (\varrho t)_i / (\varrho t)_{\text{tot}} \ .$$

When calculating the depth distribution $\phi_i^{\mathrm{F}}(\varrho z)$ for somewhat larger film thicknesses, the composition of the film must also be taken into account. The composition then forms an additional parameter for the calibration curve $k_i(\varrho z)$ [5-108].

5.6.2 Unsupported thin films

The term unsupported films refers to thinned foils and thin layers deposited directly on "electron-transparent" supporting films or removed from the substrate after deposition. The supporting films which in addition are supported by a grid should consist of light elements, such as collodion or carbon, so that excitations in them can be neglected in comparison with the film to be analyzed. Unsupported thin films are the kind of specimens used in the transmission electron microscope (TEM) (section 4.3). X-ray microanalysis in the TEM or the scanning transmission electron microscope (STEM) was already mentioned in section 4.1 as the most important method in analytical electron microscopy. Because of the limited space in the specimen surroundings, only energy-dispersive spectrometers are used in the TEM or STEM. The associated experimental problems are described in section 4.2 (Fig. 4-13).

The main advantage of X-ray microanalysis of thin films is the improved spatial resolution which can be achieved in the thin specimen compared with bulk specimens due to the lower scattering of the high energy primary electrons (100 to 200 keV) (Fig. 2-33).

Quantitative analysis

The quantitative determination of the composition is significantly simpler for thin specimens than for bulk specimens. For very thin films, the absorption and fluorescence excitation is negligible and the ionization cross section σ is constant due to the negligible energy losses. Equation (5-12) for the ionizations n_{A} generated per primary electron can therefore be simply integrated and yields

$$n_{\mathrm{A}} = C_{\mathrm{A}} \cdot \sigma_{\mathrm{A}} \cdot t$$

where t is the film or specimen thickness. In calculating the intensity of the generated X-rays I_{gA}, which is equal to the emitted radiation in the case of thin specimens, the fluorescence yield ω_A and the fraction p_A (weight of an X-ray line, cf. section 5.3.4) of the recorded line in the entire K, L or M spectrum must also be taken into account (with $C_A = w_A \cdot \varrho \cdot N_a/A_A$)

$$I_{gA} = \frac{i_0}{e} \cdot w_A \cdot \omega_A \cdot \frac{N_a}{A_A} \cdot \sigma_A \cdot p_A \cdot \varrho \cdot t \tag{5-36}$$

(i_0, probe current; e, electronic charge; N_a, Avogadro's number; A_A, mass of N_a atoms of element A). The intensity I_A recorded by the detector also depends on the collection angle Ω and the detector efficiency ε_A:

$$I_A = I_{gA} \cdot \frac{\Omega}{4\pi} \cdot \varepsilon_A \,.$$

Since the entire X-ray spectrum is recorded simultaneously in energy-dispersive analysis, the simplest approach is to form intensity ratios from the various elements present in the specimen. Parameters which are not easy to measure, such as probe current, collection angle and local specimen thickness, are thereby eliminated. For a homogeneous binary alloy of elements A and B, the ratio of the mass fractions can be determined from the intensity ratio in this way with equation (5-36):

$$\frac{w_A}{w_B} = \frac{(\omega \cdot p \cdot \sigma/A)_B \cdot \varepsilon_B}{(\omega \cdot p \cdot \sigma/A)_A \cdot \varepsilon_A} \cdot \frac{I_A}{I_B} \,. \tag{5-37}$$

This equation is usually written in the following form:

$$\frac{w_A}{w_B} = k_{AB} \cdot \frac{I_A}{I_B} \tag{5-38}$$

and represents the basis for the X-ray analysis of thin specimens. This relation was first used by Cliff and Lorimer [5-109]. The parameter k_{AB} is constant for a specific incident electron energy and is known as the k_{AB} or Cliff-Lorimer factor. It must not be confused with the intensity ratio of the specimen and standard (cf. equation (5-7), for example). In a binary alloy, w_A and w_B can be determined independently since in addition to equation (5-38) $w_A + w_B = 1$ holds true.

The k_{AB} factors can be determined experimentally by taking measurements from specimens of known composition or calculated theoretically from eq. (5-37). They are always referred to a specific element (e.g. to silicon in [5-109]) and then tabulated as k_{ASi}, k_{BSi}, ... A specific k_{AB} factor is obtained from

$$k_{AB} = k_{ASi}/k_{BSi} \,.$$

Fig. 5-24 shows experimentally determined k_{AFe} factors for Kα lines, which are referred to iron (Fe) [5-110]. In contrast to this, the curves specify the range of theoretically determined k_{AFe} factors. The uncertainty of the theoretical values is due primarily to an inaccurate knowledge of the ionization cross sections.

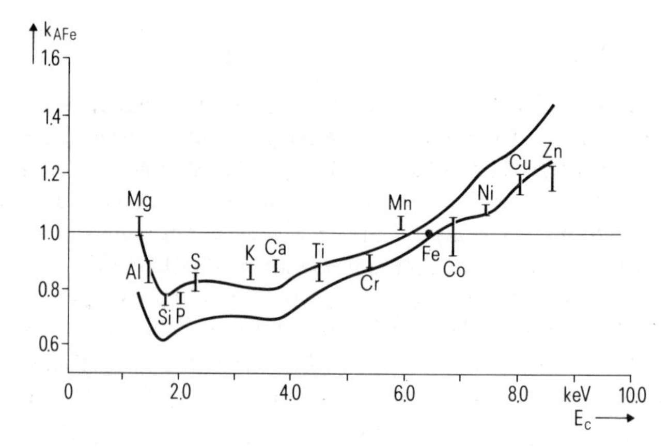

Fig. 5-24. k_{AFe}-factors for Kα lines of different elements. The limits of the calculated values are marked for comparison by the two curves (as per Wood et al. [5-110]).

With somewhat thicker specimens, absorption must be taken into account. For this purpose, the term exp $(-\chi \varrho z)$ (cf. Fig. 5-22; $\chi = (\mu/\varrho)/\sin \psi$) is integrated over the specimen thickness t. The k_{AB} factor without absorption is thereby modified in the following way:

$$k_{AB}^{abs} = k_{AB} \cdot \frac{\chi_A \cdot [1 - \exp(\chi_B \varrho t)]}{\chi_B \cdot [1 - \exp(\chi_A \varrho t)]} . \tag{5-39}$$

The difference between the mass absorption coefficients of the two elements A and B is important in deciding whether absorption should be taken into account or not. According to [5-111], absorption can be neglected when

$$(\chi_A - \chi_B) \cdot \varrho \cdot t/2 < 0.1 .$$

To be able to apply this criterion, the local specimen thickness must be known. It may be determined or estimated in various ways:

- The number of thickness fringes (see section 4.5) is a measure of the specimen thickness.
- The thickness can also be determined from the contamination spots, which in general grow on both the top and bottom surface of the specimen during the X-ray measurement [5-112]. For this purpose, the specimen is tilted by a specific angle after the X-ray analysis. The local thickness can then be calculated from the spacing between top and bottom spot which now appear separated.
- Comparing the X-ray intensity of the specimen with that of a calibration layer of known thickness provides another method.
- Finally, the fringe spacing in the diffraction disks of convergent beam diffraction patterns can be used to determine the thickness (section 4.4.4).

Beam broadening

The spatial resolution of X-ray microanalysis of thin specimens is determined partly by the probe diameter and partly by broadening of the beam as a result of scattering processes in the specimen. The beam broadening is especially important when very small probe diameters (<2 nm), of the kind possible with field emission cathodes, are available. The elastic scattering of the electrons at the atomic nuclei is responsible for the beam broadening, since larger scattering angles occur only in this case (cf. sections 2.4 and 4.1). An analytical expression for the beam broadening was derived by Goldstein et al. [5-111]. As outlined in Fig. 5-25, it was assumed that every electron is elastically scattered once at the centre of the specimen. The diameter of the scattering cone at the exit surface is designated as the beam broadening parameter b which contains 90% of the scattered electrons (and thus also 90% of the X-ray generation):

$$b = 625 \cdot \frac{Z}{E_0} \cdot \sqrt{\frac{\varrho}{A} \cdot t^3} \tag{5-40}$$

(b in cm, primary electron energy E_0 in keV, density ϱ in g/cm^3, mass A of N_a atoms in g and specimen thickness t in cm). Table 5-2 lists values of b calculated for some elements by eq. (5-40) for various specimen thicknesses. A beam broadening below 10 nm can be attained only with very low specimen thicknesses (t < 100 nm). For thicker specimens, (t > 200 nm) the beam broadening is so great that there is no point in using very finely focussed electron probes (d_0 < 10 nm). Monte Carlo calculations are also particularly suitable for calculating the beam broadening. In [5-113], various such calculations were compared with experimental data as well as with the values obtained with eq. (5-40). It was apparent that this equation also supplies reasonable quantitative results for larger specimen thicknesses in which the assumption of single scattering no longer applies.

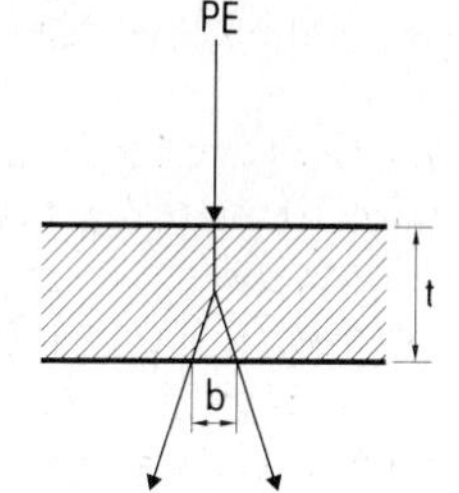

Fig. 5-25.
Model for calculating the broadening b of the primary electron beam (PE) in a thin specimen of thickness t. It is assumed that only single scattering takes place in the centre of the layer.

Table 5-2. Beam broadening parameter b from equation (5-40) for various elements and specimen thicknesses.

Specimen thickness	Beam broadening parameter b in nm for			
in nm	C	Si	Cu	Au
50	1.8	2.8	8	17
100	5	8	21	49
200	14	22	61	140
500	58	90	240	–

In the choice of parameters for the electron probe (diameter, current), it should be noted that high spatial resolution and high detection sensitivity (see section 5.7) are mutually exclusive. In order to detect small mass fractions of another element in the presence of a predominant element, good counting statistics and thus a high pulse rate with a reasonably long measurement time are required. The pulse rate increases linearly with the probe current and the specimen thickness. However, a higher probe current can be attained only with a larger probe diameter (cf. eq. (2-23) and Fig. 2-28), and the beam broadening increases with increasing specimen thickness. Even at small probe diameters (<2 nm), field emission cathodes supply such high currents (approx. 1 nA) that measurements can be made at thin specimen sites and a good spatial resolution (<10 nm) attained together with a high detection sensitivity (0.1%) [5-114].

5.6.3 Particles

When analyzing particles or small phases which are *larger* than the excitation volume, erroneous results may be obtained due to fluorescence effects. Since the absorption of X-rays is significantly lower than that of electrons, the range of the generated X-rays in the specimen is significantly greater than the diameter of the dissipation volume. As a result, these X-rays can still excite fluorescence radiation at a relatively large distance from their source (up to about 100 μm) (Fig. 5-26). This fluorescence excitation is relatively large when elements with adjacent atomic numbers are present. In the spectrum, therefore, in addition to the lines of an element A which is present in the phase to be investigated, lines of an element B from an adjacent second phase can appear. Where the investigated element is present in both phases with different concentrations, an erroneous concentration can be measured due to this fluorescence effect.

In practice, this effect is only small. Even in the extreme case of the analysis of a pure nickel particle which is so large that it exactly fills the excitation volume (Fig. 5-26b) and is embedded in a pure iron matrix, only a mass fraction of 13% iron is excited by fluorescence [5-115].

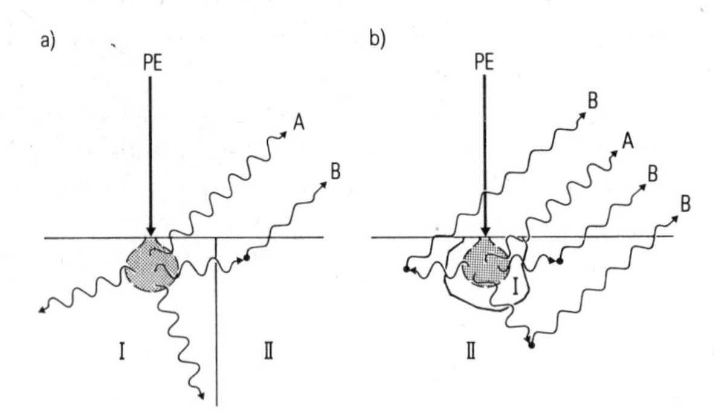

Fig. 5-26. Fluorescence excitation in the vicinity of a phase boundary (a) and in the environment of an embedded particle (b). The radiation from element A excited in phase I by the electron probe can excite element B in the adjacent phase II to produce fluorescence radiation as long as the atomic number of A is greater than that of B. Since the X-rays have a significantly greater range than the primary electrons PE (their range is indicated by the "excitation volume"), the radiation from B can also be emitted at a relatively large distance from the incidence point of the primary electrons.

This effect must also be taken into account when concentration profiles are to be determined across a phase boundary. With the aid of correction procedures [5-116,117] based on the fluorescence correction and taking into account the special geometrical conditions, quantitative determinations may be made with an uncertainty of a few percent.

In the analysis of particles *smaller* than the excitation volume (e.g. inclusions, precipitates) the elements of the matrix are generally also excited by the primary electrons. An unambiguous identification of the elements in the particles is therefore not possible. In these cases it is recommended that the particles be extracted from the matrix, e. g. by preparing an extraction replica (Fig. 4-21) and analyzed in a TEM or STEM. For an analysis in a SEM or microprobe, it is recommended that the particles are prepared on carbon mounts.

The peak-to-background method [5-118] is suitable for the quantitative analysis of such isolated particles. It is based on the assumption that the bremsstrahlung is influenced by matrix effects and the specimen geometry in a similar way as the characteristic radiation. Consequently, the ratio of peak intensity to background intensity is approximately independent of the specimen shape as well as of absorption and backscattering effects. As a measure of the relative mass concentration of two elements A and B in the particle, the ratio

$$\frac{w_A}{w_B} = g_{AB} \cdot \frac{(I_P/I_b)_A}{(I_P/I_b)_B}$$

can therefore be used. In this equation, I_P is the integral peak intensity after subtracting the background and I_b the intensity of the background at the energy of the characteristic peak. g_{AB} is a factor which depends on the elements A and B and is determined by taking measurements on calibration standards. The measured results have an uncertainty of between 10 and 20% [5-119].

More exact results can be attained when the effects of absorption and atomic number are taken into account and the analysis is performed with the aid of the depth distribution $\phi\,(\varrho z)$ [5-120]. For this purpose, $\phi\,(\varrho z)$ is determined by Monte Carlo calculations or approximated by a Gaussian curve and integrated for special particle shapes (e.g. spherical, cylindrical or plate-like). The calculated intensity ratios lie on parabolic curves when plotted as a function of the particle size. The composition of the particles can be determined with an uncertainty of about 5% when the k values (I_A/I_A^0) are referred to the particle sizes and shapes, which must be separately determined. To minimize the influence of the beam profile and to attain a uniform excitation, the electron probe should be focussed as finely as possible and the scanned area should be somewhat larger than the particle itself.

5.7 *Detection limits*

An element A in a given specimen can be said to be detectable when the intensity of its characteristic radiation can just be distinguished from the background with sufficient certainty. The number N_A of the pulses of the peak must clearly exceed the number N_b of the background pulses. According to statistical laws, an element A exists with a probability of more than 99%

if the number N_A of peak pulses is equal to at least three times the standard deviation of the background pulses N_b (see section 1.4, eq. (1-14)):

$$N_{A\min} = 3 \cdot \sqrt{N_b} \ .$$

From this, with the intensities $I_A = N_A/t$ and $I_b = N_b/t$ (t measuring time), the contribution of the statistical error to the limit of detection is:

$$I_{A\min} \cdot t = 3 \cdot \sqrt{I_b} \cdot t \tag{5-41}$$

[5-121]. The following then applies for the minimum detectable mass fraction $w_{A\min}$ of an element A

$$w_{A\min} \approx \frac{I_{A\min}}{I_A^0} = \frac{3}{I_A^0} \cdot \sqrt{\frac{I_b}{t}} \tag{5-42}$$

when the mass fraction is set approximately equal to the measured intensity ratio of the specimen and standard. Whereas $w_{A\min}$ is inversely proportional to I_A^0, $w_{A\min}$ increases only with the root of the background intensity I_b. The sensitivity consequently increases with increasing intensity I_A^0 of the characteristic radiation.

Therefore, sensitivity becomes higher with increasing probe current. The probe current cannot be choosen too high, however, since the increasing specimen heating which also increases with the probe current, can lead to specimen damage, especially in the case of poorly conducting or insulating specimens [5-122, 123]. In addition, the probe diameter also gets larger with increasing current (see Fig. 2-27).

Further, with increasing probe energy the intensity of the characteristic radiation increases more strongly than that of the bremsstrahlung. But the absorption losses also increase at the same time due to the greater penetration depth. In the case of bulk specimens therefore, a value of $E_0 \approx 5 \cdot E_b$ should not be significantly exceeded.

Eq. (5-42) shows that the sensitivity also increases with the measuring time. However, measurements cannot be made over infinite periods. The measuring time cannot even be extended significantly above the order of 100 seconds because the stability of the instruments is limited in terms of time. For a specified total measuring time, the statistical error can be minimized by suitably selecting the measurement times for the peak (t_{mA}) and background (t_b). For this, the following relationship holds:

$$t_{mA}/t_b = \sqrt{I_{mA}/I_b}$$

(I_{mA} as in Fig. 5-17a). At a typical peak intensity of $I_{mA} = 10^5 \text{ s}^{-1}$ for an element of medium atomic number with a background intensity of $I_b = 100 \text{ s}^{-1}$ and a counting time of 400 s, a minimum detectable mass fraction of $w_{A\min} = 1.5 \cdot 10^{-5} = 0.0015\%$ can be calculated from eq. (5-42) for bulk specimens. Due to the poor signal-to-background ratio, the limit of detection is significantly worse in a specimen consisting of a matrix with high mean atomic number in which an element of low atomic number is to be measured.

The detection limit, however, does not depend only on the statistical nature of X-ray emission. It is also a function of the spectrometer properties, the specimen composition, and the instrument alignment. The detection limit calculated purely from the statistics can therefore be regarded only as a theoretical limit in the favorable case of a specimen containing elements of similar physical and chemical properties. For WDS analyses, the limit for the minimum detectable mass fraction w_{Amin} attainable in analytical practice for bulk specimens is somewhat below 0.01% for elements of atomic number $Z \geq 11$ and at around 0.1% for elements of $Z \leq 10$. Since an analysis volume of about 1 μm^3 contains a mass of about 10^{-11} to 10^{-10} g, the minimum detectable mass is of the order of 10^{-15} g. The detection limits are about an order of magnitude higher for EDS analyses than for WDS analyses.

Conditions are more favorable for the analysis of thin electron-transparent specimens in the TEM or STEM. Absorption and scattering effects play only a minor role, and the advantages of higher probe energies (better signal-to-background ratio, higher brightness, better lateral resolution) can be fully utilized. As indicated by theoretical estimates [5-23], the minimum detectable mass has a magnitude of 10^{-21} g (corresponding to a mass of about 20 copper atoms) when a field emission cathode and a probe energy of 100 keV are used. The minimum detectable mass fraction for thin specimens is of the order of 0.1%.

5.8 References

[5-1] W. Duane, F.L. Hunt, *Phys. Rev. 6,* 166 (1915)
[5-2] M.A. Kramers, *Phil. Mag. 46,* 863 (1923)
[5-3] G.J. Powell in K.F.J. Heinrich, D.E. Newbury, H. Yakowitz (Eds.): *Quantitative Electron-Probe Microanalysis.* Nat. Bur. Stand. Spec. Publ. 460, US Dept. Commerce, Washington 1976, 97
[5-4] D.L. Webster, W.W. Hansen, F.B. Duveneck. *Phys. Rev. 43,* 839 (1933)
[5-5] L.T. Pockman, D.L. Webster, P. Kirkpatrick, K. Harworth, *Phys. Rev. 71,* 330 (1947)
[5-6] M. Green, *Ph. D. Thesis,* Cambridge University (1962)
[5-7] H.E. Bishop, J.C. Rivière, *J. Appl. Phys. 40,* 1740 (1969)
[5-8] W. Hink, A. Ziegler, *Z. Phys. 226,* 222 (1969)
[5-9] D.E. Smith. T.E. Gallon, *J. Phys. D 7,* 191 (1974)
[5-10] K. Goto, K. Ishikawa, T. Koshikawa, R. Shimizu, *Surf. Sci. 47,* 477 (1975)
[5-11] G.G.J. Moseley, *Phil. Mag. 27,* 703 (1914)
[5-12] D.F. Fabian, L.M. Watson, C.A.W. Marshall, *Rep. Prog. Phys. 34,* 601 (1971)
[5-13] D.W. Fischer in B.C. Henke, J.B. Newkirk, G.R. Mallet (Eds.): *Advances in X-ray Analysis.* Plenum Press, New York 1970, 159
[5-14] G. Love, M.G.C. Cox, V.D. Scott, *J. Phys. D: Appl. Phys. 7,* 2131 (1974)
[5-15] H. Oppolzer, E. Wolfgang, *Microchimica Acta Suppl. 6,* 311 (1975)
[5-16] R. Castaing, *Advances in Electronics and Electron Physics 13,* 317 (1960)
[5-17] V.D. Scott, G. Love (Eds.): *Quantitative electron-probe microanalysis.* Ellis Horwood Ltd., Chichester 1983
[5-18] K.F.J. Heinrich, *Electron Beam X-ray Microanalysis.* Van Nostrand Reinhold Comp., New York 1981
[5-20] L. Reimer, *Scanning Electron Microscopy.* Springer, Berlin 1985
[5-21] E. Fuchs, *Rev. Sci. Instr. 37,* 623 (1966)
[5-22] H. Oppolzer, U. Knauer, *Mikrochimica Acta, Suppl. 8,* 243 (1979)
[5-23] H. Oppolzer, U. Knauer in: *Scanning Electron Microscopy 1979(I).* SEM Inc., AMF O'Hare, Illinois 1979, 111
[5-24] T. Johannson, *Zeit. Phys. 82,* 507 (1933)
[5-25] H.H. Johann, *Zeit. Phys. 69,* 185 (1933)
[5-26] R. Fitzgerald, K.Keil, K.F.J.Heinrich, *Science 159,* 528 (1968)

[5-27] J.C.Russ, A.O. Sandborg, *Nat. Bureau of Standards, Spec. Publ. 604,* 71 (1981)
[5-28] P.J.Staham, *J. Physique, Coll. C2,* 175 (1984)
[5-29] P.J. Goodhew in G.J. Tatlock (Ed.): *Electron Microscopy and Analysis 1985.* Inst. Phys. Conf. Ser. 78, A. Hilger Ltd., Bristol 1985, 183
[5-30] D.J. Bloomfield, G. Love, V.D. Scott in G.J. Tatlock (Ed.): *Electron Microscopy and Analysis 1985.* Inst. Phys. Conf. Ser. 78, A. Hilger Ltd. Bristol 1985, 193
[5-31] C.E. Fiori, D.E. Newbury, *SEM 1978/I.* SEM Inc. AMF O'Hare, Illinois 1978, 401
[5-32] E.W. White, G.G. Johnson Jr., *X-ray Emission and Absorption Wavelengths and Two-Theta Tables.* American Society for Teasting and Materials, Data Series DS 37A, Philadelphia 1970
[5-33] K.F.J. Heinrich, *National Bureau of Standards, Techn. Note 278.* US Dept. Commerce, Washington 1967
[5-34] R.L. Myklebust, R.B. Marienko, D.E. Newbury, D.S. Bright, in G. Tatlock (Eds.): *Electron Microscopy and Analysis 1985.* Inst. of Phys. Conf. Ser. No. 78, A. Hilger, Bristol 1986, 219
[5-35] C.E. Fiori, R. Myklebust, K.F.J. Heinrich, H. Yakowitz, *Anal. Chem. 48,* 172 (1976)
[5-36] J.C. Russ, *Proc. 7th Nat. Conf. Electron Probe Analysis Soc.,* San Francisco, 30 (1972)
[5-37] P.J. Statham, *Anal. Chem. 42,* 2149 (1977)
[5-38] D.G.W. Smith, S.B.J. Reed, *X-ray Spectr. 10,* 198 (1981)
[5-39] R. Castaing, *Ph. D. Thesis,* University of Paris, ONERA Publ. No. 55, 1951
[5-40] R. Castaing in L. Marton (Ed.): *Adv. Electronics and Electron Phys. 13.* Academic Press, New York, 1960, 317
[5-41] T.O. Ziebold, R.E. Ogilvie, *Anal. Chem. 36,* 322 (1964)
[5-42] A.E. Bence, A. Albee, *J. Geol. 76,* 382 (1968)
[5-43] H.A. Bethe, *Ann. Phys. Leipzig 5,* 325 (1930)
[5-44] P. Duncumb, S.J.B. Reed in K.F.J. Heinrich (Ed.): *Quantitative Electron-Probe Microanalysis.* Nat. Bur. Stand. Spec. Publ. 298, US Dept. Commerce, Washington, 1968, 133
[5-45] J. Philibert, R. Tixier in K.F.J. Heinrich (Ed.): *Quantitative Electron-Probe Microanalysis.* Nat. Bur. Stand. Spec. Publ. 298, US Dept. Commerce, Washington, 1968, 13
[5-46] G. Love, M.G.C. Cox, V.D. Scott, *J. Phys. D: Appl. Phys. 11,* 7 (1968)
[5-47] H.A. Bethe, J. Ashkin, *Experimental Nuclear Physics 1,* 252 (1953)
[5-48] A.T. Nelms, *NBS Circular 577.* US Dept. of Commerce, Washington 1956
[5-49] A.T. Nelms, *NBS Suppl. Circular 577.* US Dept. Commerce, Washington 1958
[5-50] F. Bloch, *Zeit. Phys. 81,* 363 (1933)
[5-51] M.J. Berger, S.M. Seltzer in: *Studies of Penetration of Charged Particles in Matter.* Nat. Res. Council Pub. 1133, National Academy of Sciences, Washington 1964, 205
[5-52] P. Duncumb, P.K. Shields-Mason, C. Da Casa in G. Möllenstedt, K.H. Gaukler (Eds.): *X-ray Optics and Microanalysis.* Springer, Berlin 1969, 146
[5-53] G. Love, M.G.C. Cox, V.D. Scott, *J. Phys. D: Appl. Phys. 10,* 7 (1977)
[5-54] M. Gaber, H.J. Fitting, *phys. stat. sol.(a) 90,* 669 (1985)
[5-55] D.A. Sewell, G. Love, V.D. Scott, *J. Phys. D: Appl. Phys. 18,* 1233 (1985)
[5-56] R. Castaing, J. Descamps, *J. Radium 16,* 304 (1955)
[5-57] J.D. Brown, *Adv. Electronics Suppl. 6,* 45 (1969)
[5-58] R.H. Packwood, J.D. Brown, *X-ray Spectrom. 10,* 138 (1981)
[5-59] D.A. Sewell, G. Love, V.D. Scott, *J. Phys. D: Appl. Phys. 18,* 1245 (1985)
[5-60] H. Rindfleisch, *Ann. Phys. 28,* 409 (1937)
[5-61] K.F.J. Heinrich in T.D. Mc Kinley, K.F.J. Heinrich, D.B. Wittry (Eds.): *The Electron Microprobe.* J. Wiley, New York 1966, 159
[5-62] G. Springer, B. Nolan, *Can. J. Spectrosc. 21,* 134 (1976)
[5-63] R. Theisen, D. Vollath, *Tabellen der Massenschwächungskoeffizienten.* Verlag Stahleisen, Düsseldorf 1967
[5-64] T.P. Thinh, J. Leroux, *X-ray Spectrom. 8,* 85 (1979)
[5-65] H.E. Bishop in R. Castaing, P. Deschamps, J. Philibert (Eds.): *Optique des Rayons X et Microanalyse.* Hermann, Paris 1966, 153
[5-66] G. Springer, *Microchimica Acta 3,* 587 (1966)
[5-67] H. Yakowitz, R.L. Myklebust, K.F.J. Heinrich, *Nat. Bur. Stand., Techn. Note 796.* US Dept. of Commerce, Washington 1973

[5-68] J. Philibert in H.H. Pattee, V.E. Cosslett, A. Engström (Eds.): *Proc. 3rd Int. Conf. on X-ray Optics and Microanalysis.* Academic Press, New York 1963, 379

[5-69] P. Duncumb, P.K. Shields in T.D. Mc Kinley, K.F.J. Heinrich, D.B. Wittry (Eds.): *The Electron Microprobe.* J. Wiley, New York 1966, 284

[5-70] K.F.J. Heinrich, *National Bureau of Standards, Technical Note 521.* US Dept. of Commerce, Washington 1969

[5-71] M. Green, V.E. Cosslett, *Proc. Phys. Soc. 78,* 1206 (1961)

[5-72] S.J.B. Reed in V.D. Scott, G. Love (Eds.): *Quantitative Electron Probe Microanalysis.* J. Wiley, New York 1983, 191

[5-73] S.J.B. Reed, *Brit. J. Appl. Phys. 16,* 913 (1965)

[5-74] J. Henoc in K.F.J. Heinrich (Ed.): *Quantitative Electron Probe Microanalysis.* Nat. Bureau of Stand. Spec. Publ. 298, US Dept. of Commerce, Washington 1968, 197

[5-75] J. Henoc, K.F.J. Heinrich, R.L. Myklebust, *Nat. Bureau of Stand., Technical Note 769.* US Dept. Commerce, Washington 1973, 129

[5-76] G. Springer in E. Preuss (Ed.): *Quantitative Analysis with Electron Microprobes and Secondary Ion Mass Spectrometry.* Kernforschungsanlage Jülich, West-Germany 1973

[5-77] G. Springer, B. Rosner in G. Möllenstedt, K.H. Gaukler (Eds.): *V. Int. Congr. on X-ray Optics and Microanalysis,* Tübingen 1968. Springer, Berlin 1969, 170

[5-78] G. Love, V.D. Scott, *Scanning 4,* 111 (1981)

[5-79] A.R. Buchner, W. Pitsch, *Z. Metallkunde 62,* 392 (1971)

[5-80] L. Parobek, J.D. Brown, *X-ray Spectrom. 7,* 26 (1978)

[5-81] H.E. Bishop, *J. Phys. D: Appl. Phys. 7,* 2009 (1974)

[5-82] G. Love, V.D. Scott, *J. Phys. D: Appl. Phys. 11,* 1369 (1978)

[5-83] J.P. Pouchou, F. Pichoir, *Rech. Aérosp. 3,* 13 (1984)

[5-84] S. Tanua, K. Nagashima, *Microchimica Acta 1,* 299 (1983)

[5-85] G.F. Bastin, F.J.J. van Loo, H.J.M. Heijligers, *X-ray Spectrom. 13,* 91 (1984)

[5-86] D.A. Sewell, G. Love, V.D. Scott, *J. Phys. D: Appl. Phys 18,* 1269 (1985)

[5-87] K.F.J. Heinrich, *Microchimica Acta, Suppl. IV,* 252 (1970)

[5-88] K.F.J. Heinrich (Ed.): *Quantitative Electron-Probe Microanalysis,* Nat. Bur. Stand., Spec. Publ. 298. US Dept. Commerce, Washington 1968

[5-89] D.R. Beaman, J.A. Isasi, *Anal. Chem. 42,* 1540 (1970)

[5-90] G. Love, M.G.C. Cox, V.D. Scott, *J. Phys. D: Appl. Phys. 8,* 1686 (1975)

[5-91] G. Love, M.G.C. Cox, V.D. Scott, *J. Phys. D: Appl. Phys. 9,* 7 (1976)

[5-92] M. Gaber, H.-J. Fitting, *phys. stat. sol. (a) 90,* 669 (1985)

[5-93] J.C. Russ, *Proc. 9th Nat. Conf. Electron Probe Analysis,* Ottawa, 22a (1974)

[5-94] M.J. Nasir, *J. Microsc. 108,* 79 (1976)

[5-95] J. Heckel, P. Jugelt, *X-ray Spectrom. 13,* 159 (1984)

[5-96] K.F.J. Heinrich in T.D. McKinley, K.F.J. Heinrich, D.B. Wittry (Eds.): *The Electron Microprobe.* J. Wiley, New York 1966, 296

[5-97] W.L. Baun in A.J. Tousimis, L. Marton (Eds.): *Advances in Electronics and Electron Physics, Suppl. 6.* Academic Press New York, 1969, 155

[5-98] D.F. Fabian, L.M. Watson, C.A.W. Marshall, *Rep. Prog. Phys. 34,* 601 (1971)

[5-99] R. Castaing in L. Marton (Ed.): Advances in electronics and electron physics, XIII. Academic Press, New York 1960, 317

[5-100] M. Gaber, H.-J. Fitting, *phys. stat. sol. (a) 90,* 669 (1985)

[5-101] J.D. Brown in J.D. Brown, D.F. Kyser, H. Niedrig (Eds.): *Electron Beam Interactions with Solids for Microscopy, Microanalysis and Microlithography.* Proceedings of the 1st Pfefferkorn Conference, Monterey, CA, USA. SEM Inc., AMF O'Hare, Chicago 1984, 137

[5-102] W. Reuter in G. Shinoda, K. Kohra, T.Ichinokawa (Eds.): *Proc. Sixth Int. Conf. on X-ray optics and microanalysis (Osaka 1971).* University of Tokyo Press, Tokyo 1972, 121

[5-103] H. Yakowitz, D.E. Newbury in O. Johari (Ed.): *Scanning Electron Microscopy 1976/I.* AMF O'Hare (Chigaco) 1976, 151

[5-104] J.D. Brown, R.H. Packwood, *Appl. Surf. Sci. 26,* 294 (1986)

[5-105] R.B. Bolon, E. Lifshin in O. Johari (Ed.): *Scanning Electron Microscopy 1973.* AMF O'Hare (Chigaco) 1973, 285

[5-106] H.E. Bishop, D.M. Poole, *J. Phys. D: Appl. Phys. 6,* 1142 (1973)

[5-107] M.G.C. Cox, G. Love, V.D. Scott, *J. Phys. D 12,* 1441 (1979)
[5-108] D.F. Kyser, K. Murata, *IBM J. Res. and Develop. 18,* 352 (1974)
[5-109] G. Cliff, G.W. Lorimer, *J. Microscopy 103,* 203 (1975)
[5-110] J.E. Wood, D.B. Williams, J.I. Goldstein, *J. Microscopy 133,* 255 (1984)
[5-111] J.I. Goldstein, J.L. Costley, G.W. Lorimer, S.J.B. Reed in O. Johari (Ed.): *Scanning Electron Microscopy 1977/I.* IITRI, Chigaco 1977, 315
[5-112] G.W. Lorimer, G. Cliff, J.N. Clark in J.A. Venables (Ed.): *Developments in Electron Microscopy and Analysis.* Academic Press, London 1976, 153
[5-113] S.M. Allen, *Phil. Mag. A 43,* 324 (1981)
[5-114] H. Oppolzer, W. Eckers, H.C. Schaber, *J. de Physique 46, suppl. No. 4,* C4 – 523 (1985)
[5-115] S.B.J. Reed in V.D. Scott, G.Love (Eds.): *Quantitative Electron-Probe Microanalysis.* Ellis Horwood Ltd., Chichester 1983, 215
[5-116] G.F. Bastin, F.J.J. van Loo, P.C.J. Vosters, J.W.G.A. Vrolijk, *Scanning 5,* 172 (1983)
[5-117] G.F. Bastin, F.J.J. van Loo, P.J.C. Vosters, J.W.G.A. Vrolijk, *Spectrochimica Acta 39 B,* 1517 (1984)
[5-118] P.J. Statham, *Microchimica Acta Suppl. 8,* 229 (1979)
[5-119] P.J. Statham, J.B. Pawley, *Scanning Electron Microscopy 1978/I.* SEM Inc. AMF O'Hare (Chicago) 1978, 469
[5-120] J.T. Armstrong in K.F.J. Heinrich (Ed.): *Microbeam Analysis 1982.* San Francisco Press, San Francisco 1982, 175
[5-121] H. Kaiser, *Z. Anal. Chemie 209,* 1 (1965)
[5-122] G.S. Amalsi, J. Blair, R.E. Ogilvie, R.J. Schwartz, *J. Appl. Phys. 36,* 1848 (1965)
[5-123] J.L. Lineweaver, *J. Appl. Phys. 34,* 1786 (1936)

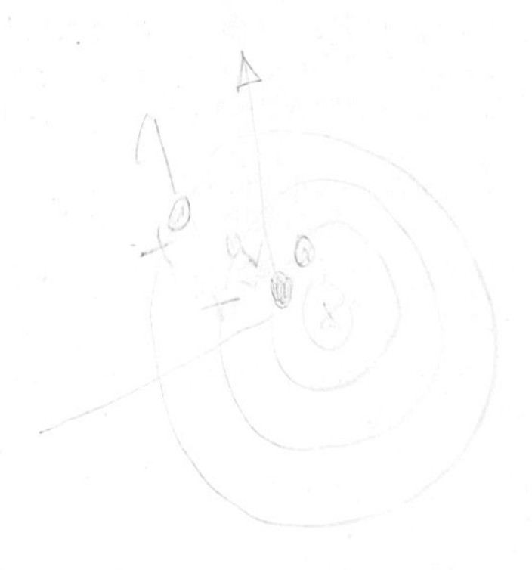

6 Auger electron microanalysis

6.1 Principle

Auger electron microanalysis, also called Auger microanalysis for short, is the analysis of the Auger electrons emitted by the relaxation of atoms with vacancies in their inner electron shells. This process is described in detail in section 2.4.3. Due to their origin, Auger electrons possess certain element-specific energies which are basically independent of the primary electron energy. The type of atom from which the emitted Auger electrons originate can therefore be deduced by determining the Auger electron energy by a spectrometer. The number or current of Auger electrons of a specific energy is a measure (although merely a very approximate one) of the quantity of the corresponding element present in the volume under analysis. Like X-ray spectrometry, therefore, Auger electron spectrometry is a method for the qualitative and quantitative analysis of elements. It is a suitable way of identifying the elements present in a specimen of unknown composition by a single spectrum (multielement analysis).

As is described in section 2.4.3, the Auger process in general involves three electron shells. After emission of an electron from a shell W due to ionization by a primary electron of sufficient energy, the atom's ground state is restored by a sequence of electron transitions (Fig. 2-43). An electron from an outer shell X is captured by the shell W and the difference between the binding energies is transferred to another electron from a shell Y which leaves the atom as an Auger electron. This electron is characterized in terms of the three electron shells involved (W, X and Y) as a "WXY Auger electron". The kinetic energy E_{WXY} of the Auger electron is the difference between the binding energies involved in the Auger process [6-1]:

$$E_{WXY} = E_w - E_X - E'_Y \, .$$

It should be noted that during the emission of the Auger electron from the shell Y an electron is missing in the shell X so that the nuclear charge is less shielded than in a neutral atom. As a result, $E'_Y > E_Y$, where E_Y is the binding energy of the neutral atom. In an atom with atomic number Z, a missing electron in an inner shell has, to a first approximation, the same effect as an increase of the nuclear charge number by one. The following approximation therefore applies for the kinetic Energy of the Auger electron in terms of the binding energies in the neutral atom E_W, E_X and E_Y:

$$E_{WXY}(Z) = E_W(Z) - E_X(Z) - E_Y(Z+1) \, .$$

Since it is, in principle, not possible to distinguish whether the Auger electron originates from the X shell and the electron which is captured by the W shell originates from the Y shell or vice versa, the arithmetic mean of both cases is taken [6-2]. The following is then obtained as a rough approximation for the energy of the Auger electron WXY obtained from an element Z:

$$E_{WXY}(Z) \approx E_W(Z) - \tfrac{1}{2}\,[E_X(Z) + E_Y(Z+1)] - \tfrac{1}{2}\,[E_X(Z+1) + E_Y(Z)] \, . \quad (6\text{-}1)$$

More precise values are obtained from theoretical calculations based on quantum mechanical considerations of the inter-electronic interactions.

The preceding considerations are strictly valid for free atoms only. In a solid, all three stages of the Auger process

- generation of the ionized state
- refill of the core hole and emission of the Auger electron
- emission of the Auger electron from the solid

are affected by vicinal atoms.

These influences, however, do not fundamentally alter the spectra, because the binding energy of an atom in a solid is small compared to typical kinetic energies of Auger electrons. In some favorable cases, these energy shifts, which are due to the chemical environment and are therefore known as chemical shifts, can provide clues to the chemical state of the atom. But since the Auger effect involves three electron shells, the interpretation of chemical shifts is complicated. Photoelectron spectroscopy (ESCA) is therefore a more popular way of obtaining the chemical information, since the single electron shell involved renders the interpretation much easier. Auger electron analysis is consequently used preferentially for the analysis of the elemental composition, especially in microanalysis applications. If the chemical environment of the atom in question is additionally taken into account, the energies of the Auger electrons can be calculated with reasonable accuracy [6-3].

Since Auger emission is based on the energy differences between (at least) three electron shells and, unlike X-ray emission, is not subject to any selection rules, an enormous number of possible Auger transitions exist. Their probability or the detectable intensity, however, depends crucially on the mutual coupling of the two electrons involved in the recombination and on the number of similar electrons present in the atom. Due to the variety of contributing transitions of widely differing intensities, the spectra are most often interpreted by comparison with standard spectra from handbooks [6-4,5,6]. The most intensive transitions which are consequently best suited for microanalysis involve adjacent shells, such as the KLL, LMM or MNN series (Fig. 2-45). With the exception of hydrogen and helium, all elements possess prominent lines of these series in the energy range of up to 3 keV. The probability for the emission of X-rays, which competes with the emission of Auger electrons, is small in this range. Auger spectrometry can be used to detect all elements except hydrogen and helium. The two lightest elements are inaccessible to this analysis because the Auger process requires at least two occupied energy states and three electrons (cf. section 2.4.3). An isolated lithium atom cannot emit Auger electrons either, since it has only one electron in its outer shell. In contrast, Auger emission can take place from lithium atoms which are bound in a solid, since in this case the valence band provides additional electrons.

The Auger electrons are generated within a dissipation volume whose extension depends on the primary electron energy (see section 2.4.1). However, only Auger electrons from the outermost atomic layer are emitted without energy loss. Electrons from deeper layers lose their specific Auger energy due to inelastic scattering and contribute to the large smooth background in the spectra or are recaptured within the material of the specimen. The inelastic escape depths of Auger electrons range from about 0.5 nm up to 5 nm, depending on their energy. Auger electron spectroscopy is thus a method for the analysis of surface layers of about this thickness range. The escape depth, also called "the information depth λ_i", of the Auger electrons varies little with atomic number, since the predominant energy loss process is caused by interactions with the valence electrons depending only slightly on the material (Fig. 6-1).

Auger spectrometry differs from X-ray spectrometry not only by its surface specificity but also by its high sensitivity in analyzing light elements (cf. Fig. 2-47). Since Auger electrons of element-specific energy originate only from a thin surface layer, the scatter volume, which gets broader with increasing depth, has only a minimal effect on the spatial resolution. Local analysis by an Auger microprobe on smooth surfaces may be performed in areas down to about 0.1 μm diameter.

Auger microprobes are often equipped with ion etching equipment which is used to abrade the specimen by means of ion bombardment. If the intensity of the Auger peak of a selected element is traced during abrasion, it is possible to measure the concentration profile of this element with depth (depth profiling, Fig. 6-2).

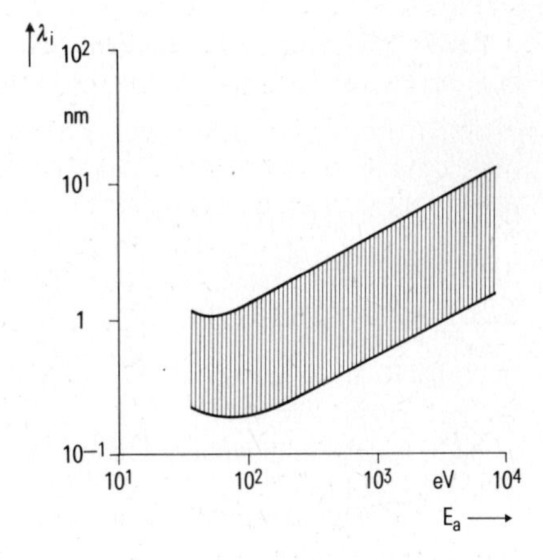

Fig. 6-1.
Information depth λ_i of Auger electrons as a function of their energy E_a. The shaded area represents results from measurements obtained by various authors and for various materials [6-7,8,9].

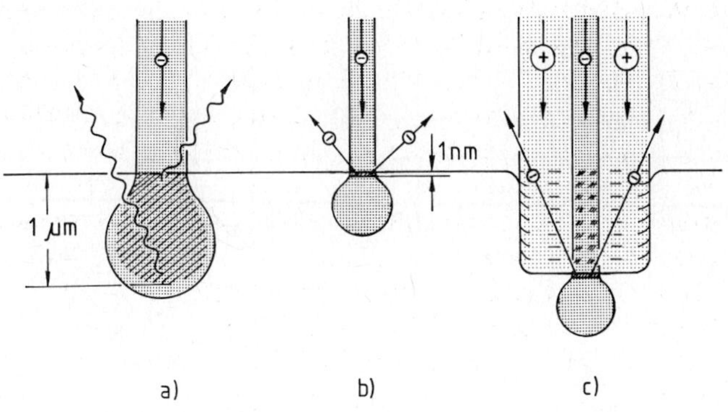

Fig. 6-2. Comparison of information depth and lateral resolution in X-ray microanalysis and Auger microanalysis. In the former (a), lateral resolution and information depth are approximately equal to the diameter of the energy dissipation volume, which has a typical value of about 1 μm^3. In Auger microanalysis (b), the lateral resolution is determined by the diameter of the thin surface layer with a typical thickness of 1 nm, from which the Auger electrons are emitted without energy loss. Although this diameter is greater than the probe diameter, because Auger electrons in the surface layer are also excited by backscattered electrons originating from the scatter volume, the great majority of the element-specific Auger electrons originate from the region within the diameter of the excited surface area. By means of step-by-step abrasion of the surface with a beam of positive ions, the Auger microprobe can be used to determine the dependence of elemental concentration on depth (c depth profiling).

The Auger microprobe can be used to determine not only elemental concentrations in selected small regions but also to map the elemental distribution over larger areas. In this process, the electron probe scans the specimen in a raster pattern with the spectrometer set to a selected range of the spectrum. An image of the specimen surface then appears on the screen of a CRT (storage monitor) with the signal related to the selected element (Auger map). The recording electron beam of the monitor runs synchronously in the same raster scan as the electron probe and its brightness is controlled by the intensity of the selected Auger electron signal. This is the same principle as that used in X-ray microprobes (see chapter 5).

Auger microanalysis is thus used for characterization of thin surface layers as well as for depth profiling, exploiting its high sensitivity for light elements (e.g. contaminants such as O, C, N, Cl, F, S).

Recent comprehensive treatments of the basis and applications of Auger electron microanalysis can be found in the books by Briggs et al. [6-10] and Thompson et al. [6-11].

6.2 Instrumentation

6.2.1 Overall design

In principle, an Auger microprobe consists of an electron source, a specimen stage, an electron energy separator, an electron detector and (optionally) an ion gun. All of these are accommodated in a vacuum chamber. Also necessary are power supplies and electronic equipment for collection and processing of the measured data (Figs. 6-3 and 6-4).

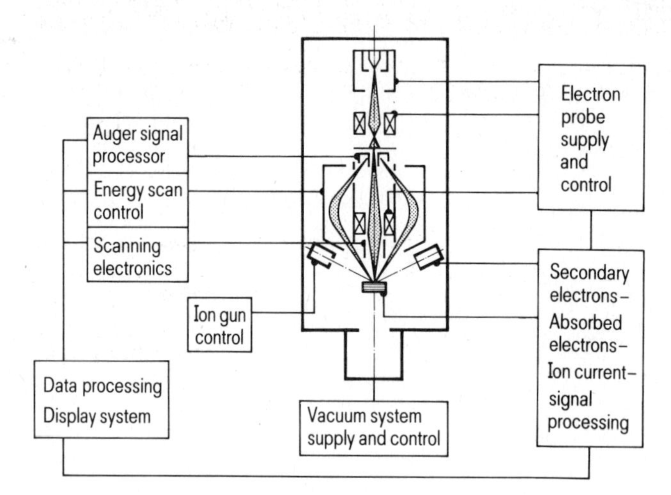

Fig. 6-3. Schema of the overall design of an Auger microanalyzer. The electron probe (primary electron beam) is shaped by two electron lenses (condenser lens and objective lens). A cylindrical electrostatic mirror is used as an energy-dispersive analyzer for the electrons emitted from the specimen. Within this analyzer, electrons of selected energy are guided on parabola-like paths to the detector. The deflection plates are used to guide the electron probe in a raster scan over the specimen. The secondary electron detector is used to record scanning electron images (as in the scanning electron microscope). The specimen surface can be successively abraded by an ion beam which strikes the specimen from the side.

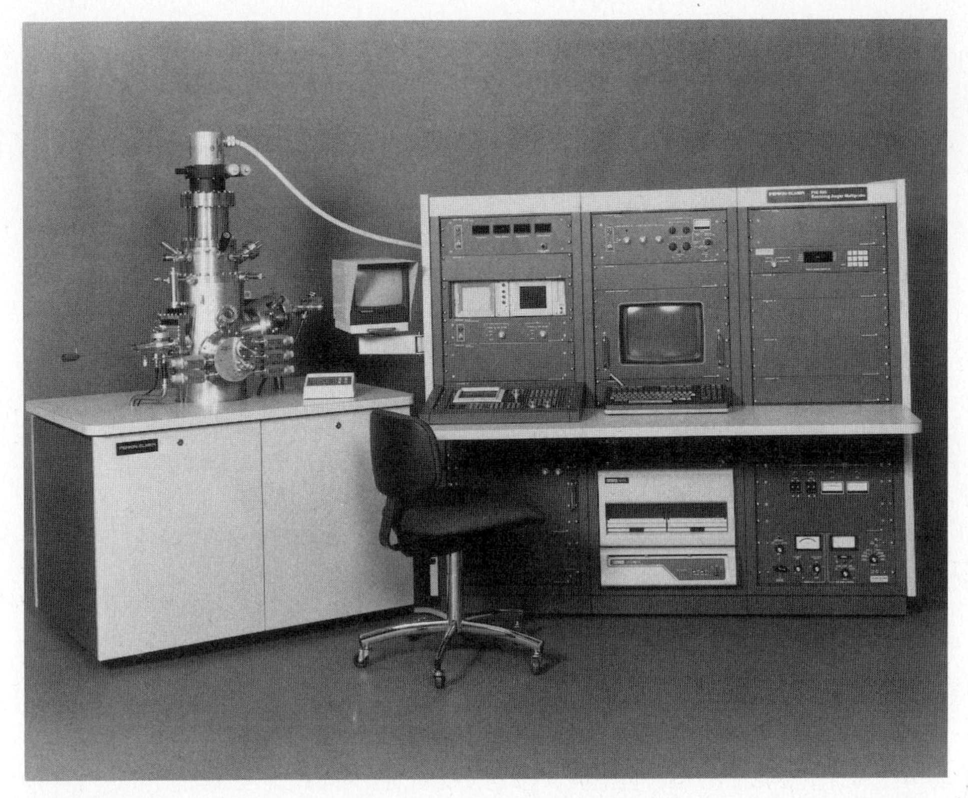

Fig. 6-4. View of a modern Auger electron microanalyzer, PHI® Multiprobe 600 by Perkin-Elmer (courtesy of Perkin-Elmer, Physical Electronics Division).

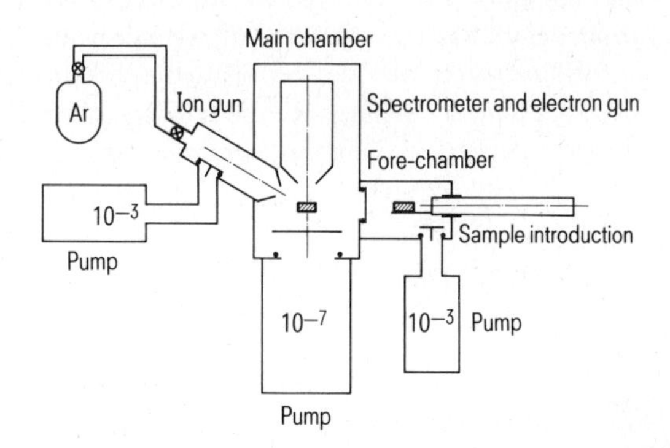

Fig. 6-5.
Schematic diagram of the vacuum system of an Auger microprobe. The diagram shows the conditions during analysis. The fore-chamber (airlock) is used to introduce samples without disturbing the vacuum in the main chamber. The given numbers specify the respective pressures in Pa.

Due to its high surface sensitivity, unadulterated results can be obtained from Auger spectrometry only under extremely clean vacuum conditions (ultra-high vacuum) (Fig. 6-5). The vacuum system should be designed to allow all types of gas to be pumped away with sufficient speed. Combinations of mechanical fore-pumps, turbomolecular pumps, sorption pumps, ion pumps and titanium sublimation pumps have proved effective for this purpose. Stainless steel is used for the vacuum chamber, and wherever possible the seals are of copper or gold so that the chamber can be baked, to desorb physisorbed gaseous species from the walls after pump-down. However, some valves make use of Viton® seals, thus limiting the maximum heating temperature. The analysis chamber is usually kept permanently at a pressure of about 10^{-7} Pa by a combination of an ion pump and a titanium sublimation pump. The exchange of specimens is performed without breaking the vacuum via a separate introduction stage. For this purpose this stage must be evacuated independently. Within the analysis chamber the specimen is attached to a manipulator stage, allowing 3-dimensional alignment. Rotation around up to two axes and equipment for heating or cooling the specimen may also be provided.

6.2.2 Electron probe

Auger microprobes of high spatial resolution require an electron probe which is as fine as possible. At the same time rather high electron currents are necessary to ensure reasonable sensitivity, since the Auger signals are weak (see Fig. 2-40). Use is therefore made of electron emitters with high brightness, namely LaB_6 or field emission cathodes (cf. section 2.3.1). Focusing takes place in a number of stages by means of electrostatic or electromagnetic lenses (Fig. 2-27). A stigmator is provided to correct for the astigmatism. The accelerating voltage can be controlled typically from about 1 kV to several 10 kV, in order to adapt the electron probe optimally to the analysis problem.

Beam diameters in the range of 0.1 μm can be attained with LaB_6 cathodes and electrostatic lenses, and in the range of 20 nm with magnetic lenses. The probe current and probe diameter can be influenced by the Wehnelt voltage and by adjusting the focusing lenses. As the probe diameter is decreased, the probe current is reduced and consequently so is the detection sensitivity (Fig. 6-6, cf. section 2.3.2). As a result, a compromise must frequently be made between spatial resolution and detection sensitivity. As in scanning electron microscopes, the electron gun is also equipped with an electrostatic beam deflection system, which can be used to image the surface by means of secondary electrons, and to define the area of analysis.

Field-emission cathodes can be used to obtain smaller probe diameters with higher current densities than is possible with thermionic cathodes (see section 2.3.1) [6-12 to 15]. In general, however, Auger microanalysis requires electron currents of several tens of nA at electron ener-

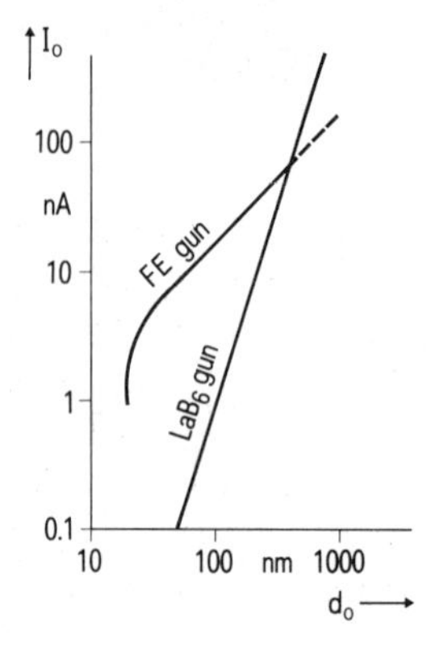

Fig. 6-6.
Experimentally measured dependence of the electron probe diameter d_0 on the electron current I_0 for a thermionic gun with LaB_6 cathode and magnetic lenses (LaB_6 gun, Multiprobe PHI® by Perkin-Elmer) and for a field-emission gun with electrostatic lenses (FE gun [6-13]); the electron energy is 10 keV in each case.

gies above 2 keV in order to obtain sufficiently intense Auger signals. The probe diameter is then of the order of 100 nm, and thus not significantly smaller than for LaB_6 cathodes (Fig. 6-6). This is why Auger microprobes with field-emission cathodes, which require more control and produce more noise in the spectra than LaB_6 cathodes, have hitherto not found wide use.

6.2.3 *Ion gun*

In order to extend the analyzed volume in depth, to eliminate surface contamination, to investigate thin film stoichiometry or contamination, or to examine surface-parallel interfaces, many Auger microprobes are equipped with an ion gun for in situ sputter erosion of the specimen.

The ion gun is operated with an inert gas in order to avoid chemical reaction with the specimen material. The normal choice is argon, which is fed via a leak valve to the discharge chamber in the ion gun. The ions are generated by impact ionization processes in a gas discharge (cf. section 2.5.3). The discharge is maintained by an electric field supported by thermionic electrons and/or by a magnetic field, depending on the type of ion gun. In the discharge chamber, ion currents of several tens of μA are generated. The positive ions are extracted from the discharge chamber by means of an emission aperture which is at negative potential and then focused with electrostatic lenses.

The emission aperture also acts as a differential pressure stage, so that the pressure in the analysis chamber does not rise above 10^{-5} Pa with the ion gun activated. This pressure can be tolerated during analysis since it is merely due to the inert sputter gas. In order to provide homogeneous erosion over the area of analysis, a deflection system can be used to scan the ion beam over a field of several mm^2. The ion current density on the specimen may attain values up to several 100 $\mu A/cm^2$, and together with the ion energy and the angle of incidence controls the erosion rate. The erosion rate can be influenced, with the beam in a fixed position, by regulating the ion current, or, with the ion current at a constant value, by varying the scan field. The sputter crater has sloping walls, corresponding to the Gaussian profile of the ion beam. The eroded zone has to be aligned thoroughly with the incidence area of the primary electron beam which defines the field of analysis. In order to avoid misleading contributions from the slopes of the erosion crater and to compensate for residual misalignment, the eroded area is generally chosen to be much larger than the analyzed area.

6.2.4 *Spectrometer*

A cylindrical mirror analyzer (CMA) has proved to be an effective energy dispersion system for Auger microanalysis [6-16 to 19]. It consists of two concentric hollow cylinders between which the electrons are guided to the detector through a radial electric field (Fig. 6-7). The outer cylinder is at variable negative potential, while the inner one is kept at ground potential. Electrons which are emitted from the specimen on the cylinder axis can pass through both apertures of the inner cylinder only when their energy is in a certain ratio to the potential applied to the outer cylinder (Fig. 6-8). The CMA is an extension of the plate capacitor [6-20]. An electron which enters the capacitor's homogeneous field behaves quite analogously to a pro-

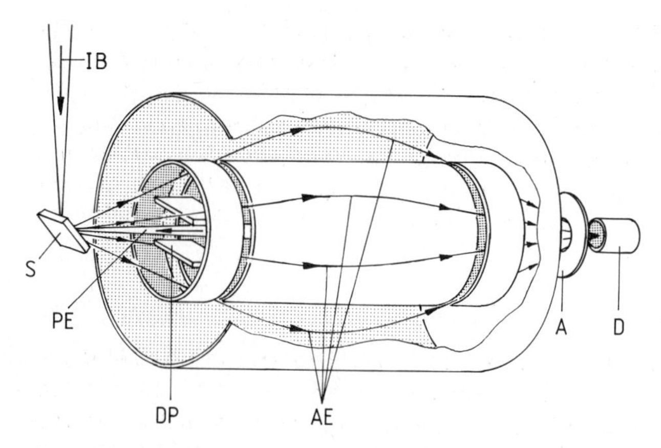

Fig. 6-7. Cylindrical mirror analyzer (CMA) with concentric electron gun for energy analysis of Auger electrons (AE). The Auger electrons excited by the primary electrons PE in the specimen S enter the intermediate space between inner and outer cylinders through an annular slit. The electrons are guided to the detector D along parabola-like paths by an electric field generated by the potential U_{CMA} between the two cylinders. The electron source can be arranged axially within the internal cylinder as shown here, or outside the spectrometer (Fig. 6-3), or obliquely to the spectrometer axis (Fig. 6-12). The energy resolution can be regulated by the emission aperture A. The deflection plates DP allow the primary electron beam to be scanned over the specimen. The ion beam IB may strike the specimen obliquely.

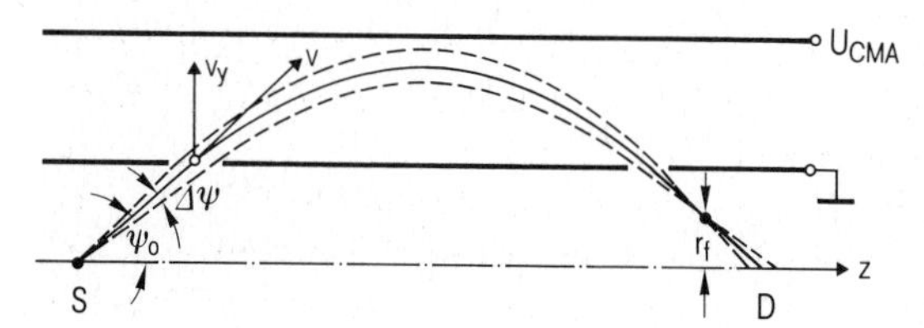

Fig. 6-8. Reflection of the electrons in the cylinder mirror. The electric field generated by the voltage U_{CMA} between the two cylinders has a retarding effect on the incident electrons. Reflection occurs when the velocity component $v_y = v \cdot \sin \psi_0$ is retarded to zero. This is the case when the corresponding component of the kinetic energy of the electrons $E_y = (m/2) \cdot v_y^2$ is at least equal to or smaller than the potential energy in the analyzer field $e \cdot U_{CMA}$. From this it follows with $v_y = v \cdot \sin \psi_0$ that:

$$\frac{e \cdot U_{CMA}}{E} = \frac{U_{CMA}}{U} \geq \sin^2 \psi_0 .$$

In the vicinity of the focus D, the electrons emitted within the aperture angle $\Delta\psi$ are focused onto a ring of radius r_f.

jectile thrown into the earth's gravitational field at an angle ψ. This projectile traverses a parabolic curve if air resistance is neglected. The distance on the horizontal z axis (the "range of projection") is in both cases

$$z \sim \sin 2\psi$$

and attains its maximum at $\psi_0 = 45°$. Small deviations $\Delta\psi$ around ψ_0 change z_{max} only in the second order. Within $\pm\Delta\psi$, the plate capacitor therefore focuses monoenergetic electrons.

The field in the CMA is not homogeneous as in the plate capacitor, but is inversely proportional to the radius due to radial symmetry. The optimum angle of incidence is thus not 45° but $\psi_0 = 42.3°$. At this angle, electrons of the same energy (more exactly: electrons with energies in the range from E to $E + \Delta E$) are focused within $\pm\Delta\psi$ to the first and second orders. Since the CMA also covers the complete azimuthal angle, its entrance aperture is rather large. It covers about 10% of the hemisphere, so that a comparable fraction of all emitted electrons is collected. The energy of the transmitted electrons is proportional to the potential of the outer cylinder U_{CMA}

$$E = U \cdot e = K \cdot U_{CMA} \tag{6-2}$$

where K is the spectrometer constant. The electron spectrum is obtained by recording the electron current picked up by the detector as a function of the cylinder potential, and the energy scale is calibrated according to equation (6-2). The potential applied to the outer cylinder is smaller than the potential U corresponding to the electron energy (Fig. 6-8). Due to $\psi_0 = 42.3°$ it follows from $U_{CMA}/U \geq \sin^2 \psi$ that $U_{CMA}/U > 0.45$. In commercial instruments, U_{CMA}/U is between 0.6 and 0.7.

An important measure for the energy separation of two adjacent Auger peaks is the relative energy resolution. It is defined as

$$R = \Delta E/E \tag{6-3}$$

and becomes worse as ΔE increases. The relative energy resolution increases with the aperture angle $\Delta\psi$. For a CMA it is a constant at fixed $\Delta\psi$. This means that the energy window ΔE is narrow for electrons of low energy and wide for electrons of high energy. At a typical resolution $R = 0.5\%$, ΔE has, according to equation (6-3), a value of 0.5 eV at an energy of 100 eV. Since the natural line width of the Auger peaks ranges from below 1 eV to about 10 eV, the energy window can be smaller, comparable to or broader than the line width. This has to be borne in mind when experimental intensities are compared with calculated ones. However, for practical element detection and comparision with standard spectra, it has proved sufficient to work with a fixed energy resolution of about 0.5%.

Microanalysis often requires high sensitivity for the detection of contaminants. The analyzer should therefore transmit as many electrons as possible from the selected energy window. The number of transmitted electrons increases as the aperture angle is enlarged. However, ΔE also increases in this case, so that not only the element-specific Auger electrons are detected but also electrons from the background region outside the Auger peak of interest. The optimum signal-to-background ratio, which can be set by means of suitable spectrometer parameters, is reached when ΔE is equal to the natural line width of the Auger peak. However, it is impractical to adapt the mechanically defined spectral resolution of a CMA to different natural line widths between about 1 eV and 10 eV. As a compromise, a fixed resolution of about 0.5% is often chosen, which is considerably smaller in the low energy range and essentially broader in the high energy range than most of the Auger peaks. For a fixed resolution the transmission of the spectrometer is given by a simple dependence on the energy. The current I received by the detector is proportional to the width of the energy window and to the number of electrons $N(E)$ in the region from E to $E + \Delta E$

$$I(E) = c \cdot \int_{E}^{E+\Delta E} N(\varepsilon)\,\mathrm{d}\varepsilon \approx c \cdot N(E) \cdot \Delta E$$

(c is a constant), or with introduction of the energy resolution eq. (6-3)

$$I(E) = c \cdot R \cdot E \cdot N(E) \, . \qquad (6\text{-}4)$$

Since R is constant, the current transmitted through the analyzer increases with the energy, and in recording a spectrum the product $E \cdot N(E)$ is measured.

The properties of the CMA are, strictly speaking, valid only for an electron point source on the cylinder axis and positioned exactly at the focal point in front of the entry slit. If the exciting electron beam moves radially away from the focal point, the energy resolution and the signal intensity deteriorate. The area around the focal point which can still be analyzed without a major loss in intensity is known as the "source definition" of the analyzer. It depends on a number of factors, including the energy resolution of the analyzer, and may attain between a few tenths of a millimeter and 1 mm in diameter. A scanning Auger micrograph cannot therefore image more than such a region (scanning Auger microscopy, see section 6.6).

6.3 Measurement technique

6.3.1 Spectra representation

The Auger spectrum is a representation of the number of emitted electrons as a function of their energy, which covers the energy range where distinct Auger transitions are anticipated, i. e. some range between about 50 eV and 3 keV. At high intensities, the Auger current is determined by analog measurements, while small intensities are recorded by measuring the number N of electrons per time interval Δt. Since only relative intensities are required for the analysis, it is sufficient to measure a parameter which is proportional to the intensity. It is usual to specify the number N of electrons as a measure of the intensity, and this for all intensity ranges. The Auger spectra can be recorded in the *direct* mode $N(E)$ or – in accordance with the operating mode of the CMA – in the form $E \cdot N(E)$, or in the *differential* mode $\mathrm{d}[N(E)]/\mathrm{d}E$ or $\mathrm{d}[E \cdot N(E)]/\mathrm{d}E$.

Differential spectrum

In practice, the differential spectrum is the preferred choice, since it allows very effective suppression of the intense background which dominates the direct spectrum (see Fig. 2-40). Peaks which are very weak in the direct spectrum are clearly discernible in the differential spectrum (Fig. 6-9). The differentiation can be performed numerically after measuring the direct spectrum or in analog form during spectrum acquisition by modulating the pass energy of the spectrometer [6-21]. The numerical differentiation is often performed by polynomial fit routines in order to reduce the noise in the measured data. In differentiating by the modulation method, an energy modulation $E_{\sim} = E_s \cdot \sin \omega t$ is superimposed on the spectrometer pass energy. The current $I(E + E_{\sim})$ is then collected at the detector. When developed in a Taylor series, it has the following form

$$I(E + E_{\sim}) = I(E) + E_s \frac{dI}{dE} \sin \omega t + \frac{E_s^2}{2} \cdot \frac{d^2 I}{dE^2} \sin^2 \omega t + \ldots$$

A phase-sensitive lock-in amplifier, connected to the detector and tuned to the first harmonic ω then returns the signal (cf. eq. (6-4)):

$$I(\omega) = E_s \cdot \frac{dI}{dE} = E_s \cdot c \cdot R \cdot \left[N(E) + E \cdot \frac{d\,[N(E)]}{dE} \right].$$

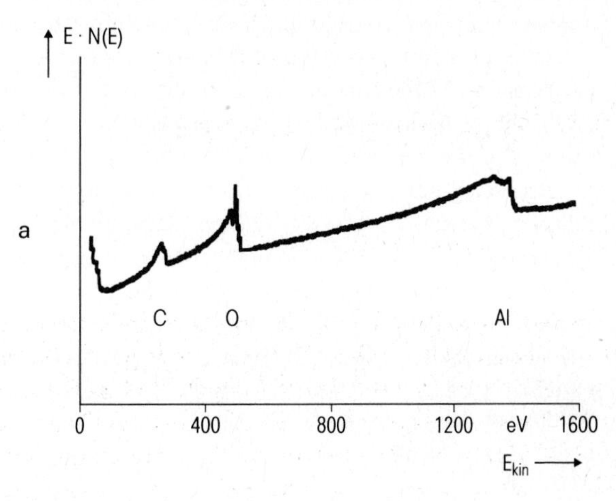

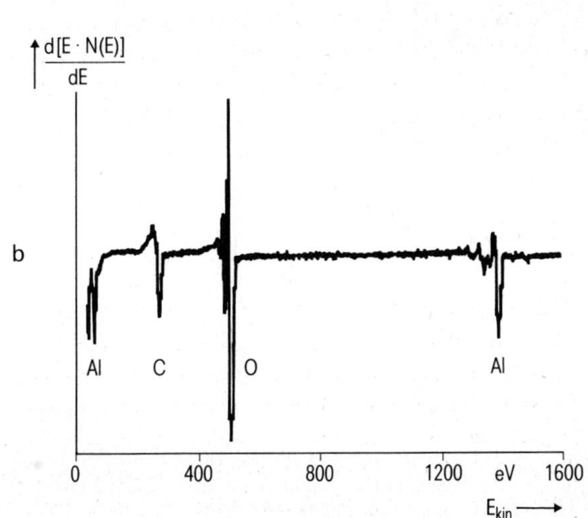

Fig. 6-9.
Auger spectrum of an oxidized aluminum surface in direct mode $E \cdot N(E) = f(E)$ (a) and in differential mode $d\,[E \cdot N(E)]/dE = f'(E)$ (b). An element is identified by assigning the energy to the element-specific transitions (cf. Fig. 2-45). The position of the sharp negative excursion in the differential spectrum is defined as the Auger energy.

Above about 50 eV where the contribution of true secondary electrons becomes small, the first term in $I(\omega)$ can be neglected. The signal from the amplifier output therefore essentially represents the differentiated spectrum:

$$I \sim E_s \cdot R \cdot E \cdot \frac{d\,[N(E)]}{dE} . \tag{6-5}$$

The differential signal increases with the modulation amplitude E_s, or for the numerical differentiation, with the differentiation increment. The signal-to-background ratio is thus improved. On the other hand, the energy resolution gets worse (as for a deteriorated spectrometer resolution R). The modulation amplitude E_s or the differentiation increment should therefore be matched to the line width of the Auger peak, but should not exceed a few eV. For Auger peaks which have a shape similar to that of a Gaussian function (singlet peaks), the dependence of the peak intensity on the modulation amplitude shown in Fig. 6-10 is obtained [6-22]. The curve drops again after a maximum at $E_s/w = 2$, since for large modulation amplitudes a large region is included, which does not, however, contribute to the signal.

In the differential spectrum, the energy position of the Auger electron current maximum, i.e. the energy of the relevant Auger transition, corresponds to the zero crossover between minimum and maximum. But this "zero passage" cannot be determined accurately. Therefore the position of the minimum of the differential spectrum, which is most often more pronounced then the maximum, is generally termed the energy of the Auger peak. The energy defined in this way depends on the modulation amplitude E_s or on the differentiating increment E_D. The negative peak is shifted to higher energies, by increasing the modulation amplitude or the differentiating increment. Its position is also a function of the energy resolution of the spectrometer and the deviation can reach several eV (Fig. 6-11 [6-22]). This must be taken into consideration, especially when peak energies measured with different instruments are to be compared.

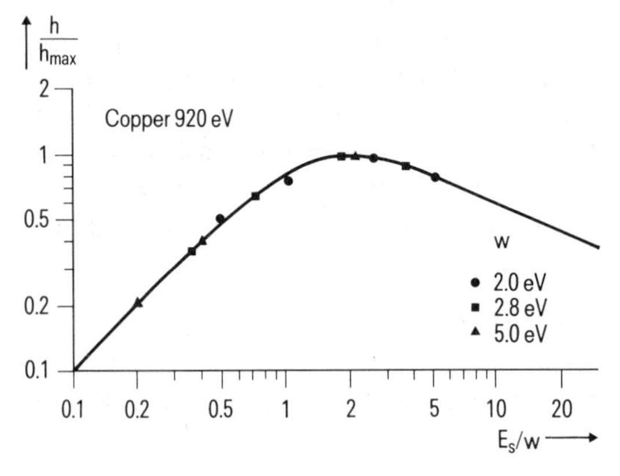

Fig. 6-10. Intensity of the differential LMM peak of Cu (in relative units) at 920 eV as a function of the modulation amplitude E_s. E_s is referred here to different peak widths w, which in this case are given by the spectrometer resolution. The negative excursion of the peak was used as a measure of intensity (see Fig. 6-18d). The values lie on a "universal curve" in good agreement with theoretical data (bold line), based on a Gaussian profile as an approximation to the peak shape (as per Anthony and Seah [6-22]).

Considerable differences were found between instruments when measured peak energies and intensities were compared in a round robin study in which 28 laboratories participated [6-23]. These differences are thought to be due to inaccurately calibrated instruments. For analyses requiring highly accurate energy and intensity values, the modulation amplitude should therefore be calibrated with calibration standards. The Ag MNN-Auger doublet at 350 eV is often used for this purpose [6-24,25].

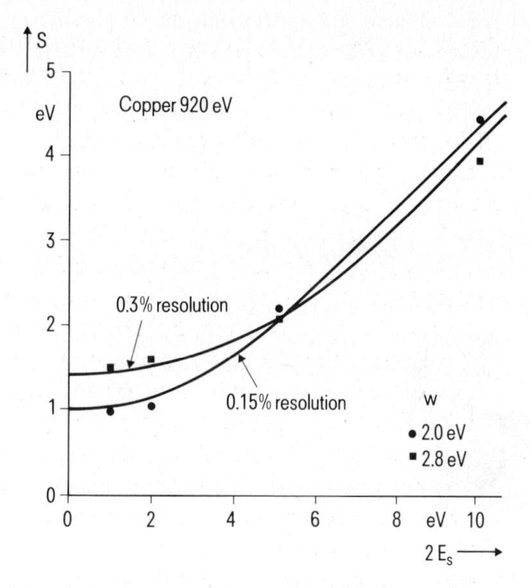

Fig. 6-11.
Energetic shift s of the negative maximum of the Cu peak at 920 eV as a function of the modulation amplitude E_s for two different spectrometer resolutions w. The continuous curves represent calculated data with an assumed Gaussian peak shape, as in Fig. 6-10 (as per Anthony and Seah [6-22]).

Direct spectrum

The direct spectrum can be obtained as a function of the spectrometer pass energy by recording the detector signal either amplified by a dc amplifier (analog amplification) or in case of small intensities by counting the number of electrons per unit time (pulse count mode). Direct spectra can also be recorded with the aid of the phase-sensitive lock-in amplifier, which is tuned to a periodic blanking of the primary electron beam [6-26,27]. This is achieved by using a beam blanking system to modulate the primary electron beam with the same ac voltage that controls the lock-in amplifier. A signal

$$I(E) \sim E \cdot \mathrm{N}(E) \cdot \sin \omega t$$

is obtained from the electron detector which can be fed into the lock-in amplifier which returns the signal

$$I(\omega) \sim E \cdot N(E) \qquad (6\text{-}6)$$

which is proportional to the direct spectrum.

The direct spectrum has a number of advantages over the differential spectrum. The signal losses inevitably attending the energy modulation do not occur here. The signal-to-noise ratio is therefore better, and there are no modulation effects on the position of the peak maximum. According to a theoretical and experimental study from Seah and Hunt [6-28], the signal-to-noise ratio of the Cu peak at around 915 eV measured by pulse counting is 3.3 times higher than a peak-to-peak height measurement in the differential spectrum with a 5 eV spectrometer modulation. In the case of numerical computer differentiation of the direct spectrum, however, the signal-to-noise ratio is only slightly deteriorated. When the Cu peak is measured by analog amplification, the signal-to-noise ratio is smaller by a factor of $\sqrt{2}$, and when measured by the beam blanking technique with an equal on/off ratio, it is smaller by a factor 2, compared with the measurement by the pulse count mode.

The advantage of recording the direct spectrum by analog amplification is mainly due to the fact that this technique supplies only one kind of spectrum for both high and low currents.

6.3.2 Signal processing

The electron current transmitted by the energy analyzer varies between 10^{-18} and 10^{-7} A, depending on excitation cross section, primary current, and transmission of the spectrometer. Secondary electron multipliers must be used to detect the very small currents ($1.6 \cdot 10^{-18}$ A

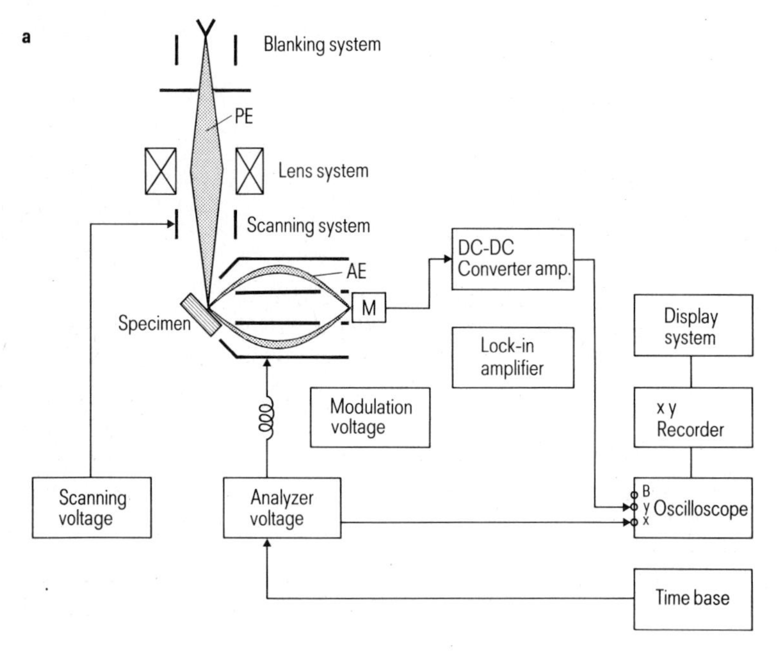

Fig. 6-12. a) Recording of the direct spectrum $E \cdot N(E)$: the Auger signal of the output of multiplier M is proportional to $E \cdot N(E)$. In the dc-dc converter amplifier, it is converted to an ac signal, which is amplified and then rectified again. It is then fed to the vertical deflection (y) of the oscilloscope, the horizontal deflection (x) being controlled by the analyzer voltage U_{CMA}. The oscilloscope shows the momentary value $E \cdot N(E)$, while the x-y plotter records the Auger spectrum from E_a to E_b during continuous tuning of the analyzer voltage from $U_{CMA,a} = E_a/K$ to $U_{CMA,b} = E_b/K$ (cf. eq. (6-2)).

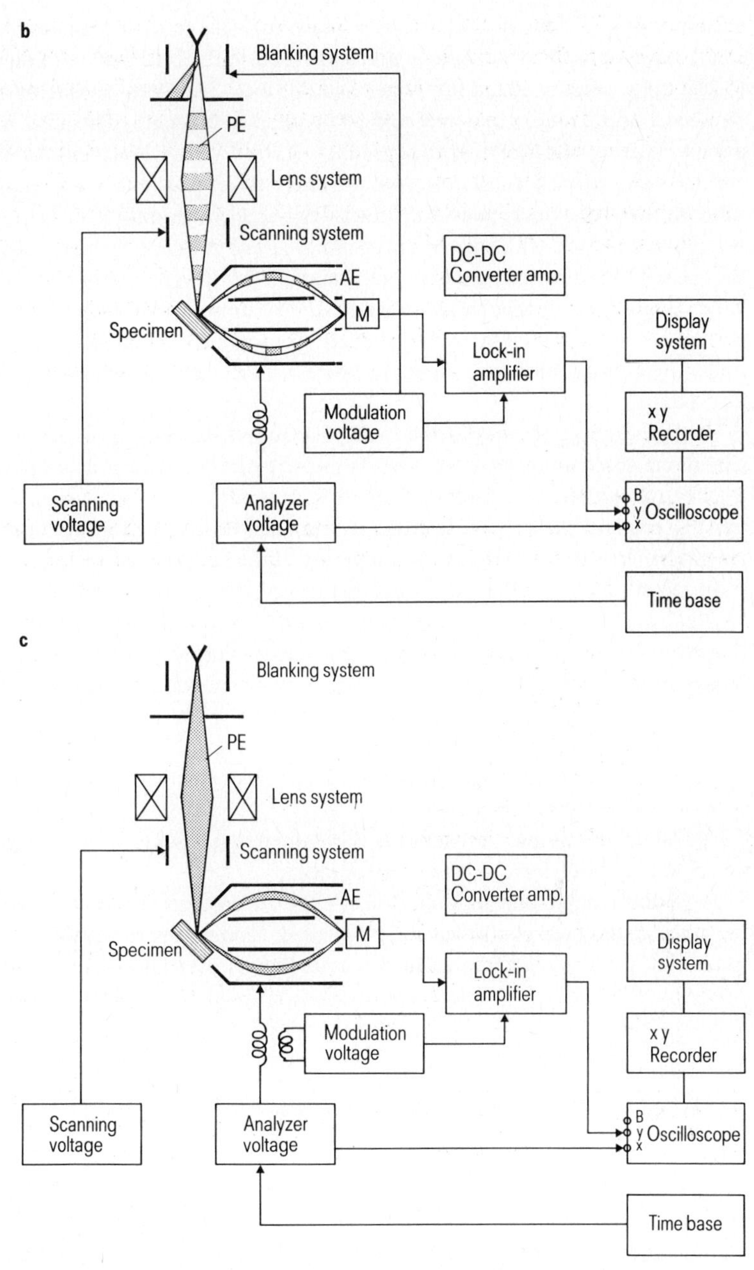

Fig. 6-12. b) Recording of the direct spectrum $E \cdot N(E)$ with a lock-in amplifier: primary electron current (PE) and lock-in amplifier are controlled with the same frequency and phase by the modulation voltage generator. A signal which is proportional to $E \cdot N(E)$ is then obtained from the amplifier.
c) Recording of the differential spectrum $\mathrm{d}[E \cdot N(E)]/\mathrm{d}E$: an ac voltage is superimposed on the analyzer voltage. The lock-in amplifier, which is tuned to this ac frequency and phase, supplies the oscilloscope with a signal proportional to $\mathrm{d}[E \cdot N(E)]/\mathrm{d}E$.

corresponds to a flow of 10 electrons per second). Due to their high amplification factor and small dimensions, discrete dynode multipliers or channel electron multipliers (Channeltrons® [6-29]) are especially suited for Auger microprobes. The latter consist of a curved glass tube between 1 and 2 mm in diameter and 50 to 100 mm in length. The inner wall is covered with a semi-isolating material of high secondary electron yield (cf. section 2.4.3). A voltage of several kV drops linearly across the approximately 10^9 Ω resistance between the two ends of the tube. An electron which enters the tube is thus accelerated until it strikes the wall and triggers several secondaries, which again are accelerated and produce further secondaries. In this way the incident electron can generate an avalanche of 10^8 secondary electrons, a current pulse which is used for further signal processing. The noise, which may be due to such factors as cosmic radiation or radioactivity, is very low, having a value of less than one pulse in 10 seconds. Electron multipliers are able to process more than 10^6 pulses per second until saturation occurs.

Further signal processing at the multiplier output depends upon the operating mode used. The direct spectrum can be recorded by pulse counting, with a dc amplifier (Fig. 6-12a), or a lock-in amplifier (Fig. 6-12b). If the lock-in amplifier is used, however, the electron probe must be modulated. The differential spectrum can be obtained by modulating the spectrometer pass energy (Fig. 6-12c), or by numerical differentiation of the direct spectrum.

To record depth profiles, the x-coordinates of the recording devices are controlled by the sputter time and the y-coordinates by the intensity of one or several selected Auger peaks. The quasi-simultaneous analysis of several elements is enabled by switching the spectrometer pass energy in regular cycles to preselected ranges which cover the Auger peaks of interest. To record the intensities I_{min} and I_{max} of the differential spectrum or the peak maximum and the background value of the direct spectrum (see section 6.4, Fig. 6-18) in every measurement cycle as accurately as possible, the analyzer voltage is continuously tuned within a small window in which these values are sought.

The lateral distribution of elements is obtained in line scans or maps according to the same principles as those for the X-ray microprobe.

To obtain reliable and reproducible results, all parameters which affect the measurement must be carefully set, controlled and monitored. Modern instruments are controlled by a computer which enables preprogrammed measurements over an extended period of time.

6.3.3 Specimen preparation and alignment

The specimens to be investigated by Auger microanalysis must be sampled, stored and transported such that their surface composition remains unchanged. They must not be touched without gloves or stored in containers which release additives, such as many plastic materials. The specimens should be prepared in a dust-free environment, e.g. in a laminar box. In insulating or poorly conducting specimens, it is sometimes necessary to coat the surface with a conductive layer to drain off excess charge caused by the primary electron beam.

Prior to analytical measurement, the specimen, the electron gun and the ion gun must be aligned. When a CMA is used, the position of the specimen on the z axis (spectrometer axis) is of crucial importance. To obtain accurate and reproducible energy values, the specimen must be positioned as exactly as possible at the focal point of the spectrometer (Fig. 6-13), where the properties of the analyzer are optimized (focusing, maximum transmission). The adjustment is performed by exposing the specimen to primary electrons of an energy E_0 which is still accessible to the spectrometer (e. g. $E_0 = 2$ keV). If the instrument is properly calibrated, the optimum specimen position is given when the elastically scattered primaries appear at an energy E_0 in the spectrum. To ensure reproducibility of the energy positions of the Auger peaks of 0.1 eV, the alignment must be performed to within about 6 μm [6-30]. In practical microanalysis, however, where minor shifts of peak positions can be tolerated, an alignment error of ±50 μm is still permissible due to the relatively large energy width of the Auger peak. For this method of specimen alignment, the calibrated and constant energy of the primary electrons is a basic precondition (at 2 keV at least 0.1%; [6-31]).

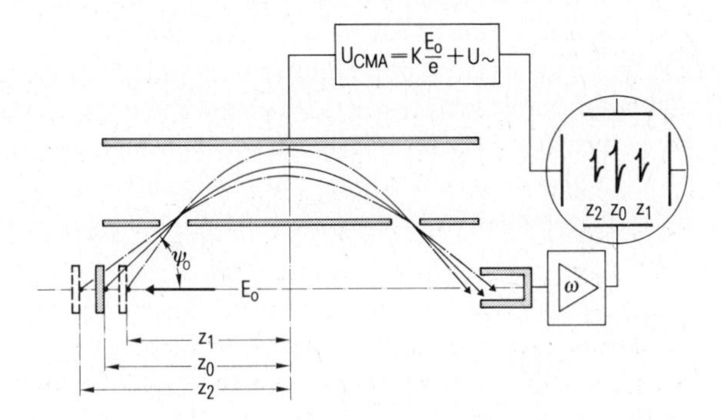

Fig. 6-13. Adjustment of the specimen in front of the CMA with the aid of the elastically backscattered electrons. With the specimen correctly adjusted at the focal point z_0, the CMA receives the electrons at the angle of incidence of $\psi_0 = 42.3°$, where its transmission is a maximum. On the oscilloscope screen, the signal of the elastic peak has a maximum amplitude. For values which deviate from z_0, smaller peaks are recorded. Because either larger or smaller analyzer voltages are required for electron transmission under inappropriate adjustments z_1, z_2, in such cases the peak of elastically scattered electrons of energy E_0 is detected at apparently lower (z_1) or higher (z_2) values, respectively.

6.3.4 *Characteristic Auger spectra, qualitative analysis*

An overview of the qualitative chemical composition of the investigated specimen is obtained by recording the entire spectrum with standard settings of the electron probe and spectrometer. The energy E_0 of the primary electrons must in every case be higher than the binding energy E_b of the electrons in the relevant electron shell in order to excite Auger transitions. The cross section for exciting Auger transitions is proportional to the ionization cross section (cf. section 5.1, Fig. 5-2), increasing with the E_0/E_b ratio and attaining a maximum at $E_0/E_b = 3...5$. For larger values of E_0/E_b the ionization cross section changes only insignificantly.

In the overview spectra, the most intense Auger peaks are given for elements with atomic numbers up to about $Z = 14$ by the KLL spectra, then up to about $Z = 40$ by the LMM spectra, and above $Z = 40$ by the MNN spectra. These are also the peaks most commonly used for analysis in practice (cf. Fig. 2-44). Spectrum libraries are available to aid identification of the measured peaks. They list the predominant Auger peaks of the elements in the differential

[6-4,5] or in the direct [6-6] mode. The spectra of the lightest elements Li, Be, B and C, which can be analyzed by Auger spectrometry, are generally characterized by only one peak which is accompanied by weaker satellites on the low-energy side. With increasing atomic number, adjacent peaks appear, their intensity increases and the spectra become more complex (Fig. 6-14). Apart from the peaks associated with core level transitions, many peaks are assigned to transitions which comprise the valence band (these transitions are termed LVV, MVV, ... for example).

The discernibility of peaks in the differential spectrum depends on the modulation amplitude as well as on the energy resolution of the spectrometer. In cases when a high sensitivity is required, this behavior can be exploited. The signal increases when using a greater modulation amplitude or an adapted energy resolution (Fig. 6-15).

The shape and energies of the Auger peaks depend on the chemical environment of the atom. Chemically-caused energy shifts can have values of up to several eV, and are well observable in Auger spectroscopy. Significant chemical shifts are shown by many oxides (Fig. 6-16) and non-metallic surface layers which are chemisorbed onto metals [6-32 to 38]. These shifts are particularly large for transitions which include the valence band, but core level transitions of solids also show such effects. Chemically influenced changes of peak shape are also known for many compounds. However, due to the limited energy resolution of the CMA, these chang-

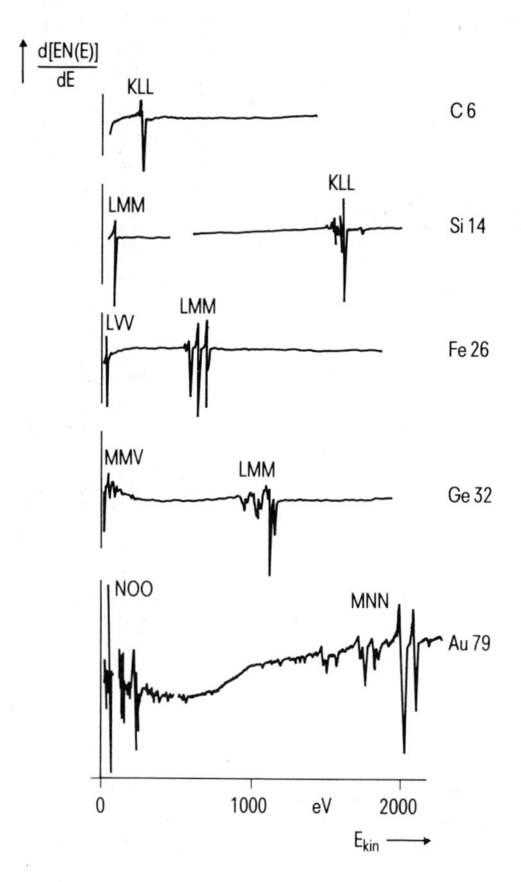

Fig. 6-14.
Characteristic Auger spectra of some selected elements. Whereas the lighter elements essentially show a single peak, the spectra of heavy elements show multiplets of intense peaks. Typical triplets are shown by the transition metals (e.g. Fe). With increasing atomic number, the spectra become more complex (from [6-4]).

es are not exploited in analyses as routinely as in photo-electron spectroscopy. One well-known example of a prominent change of peak slope is the carbon LVV peak [6-39].

Incorrect alignment of the specimen or decalibration of the spectrometer may also lead to deviations of the measured values from the specifications in the spectrum libraries. If the values given in the various libraries are compared, deviations will be noticed in the kinetic energy for some peaks. To obtain an unambiguous identification of the elements, it is therefore insufficient merely to compare the energy positions of single characteristic peaks; rather, the entire spectrum must be compared.

When analyzing an initially unknown specimen, it must be kept in mind that every surface, which in this context is to be understood as an interface between a solid and a gas or a solid and a vacuum, is contaminated whenever it had been exposed to etching agents, solvents, rinsers, or at least to the normal laboratory atmosphere. This layer, which in general contains some of the most common contaminant elements such as C, N, O, Na, Si, Cl or S, can be thicker than the escape depth of the Auger electrons so that the Auger analysis records only the contaminant elements.

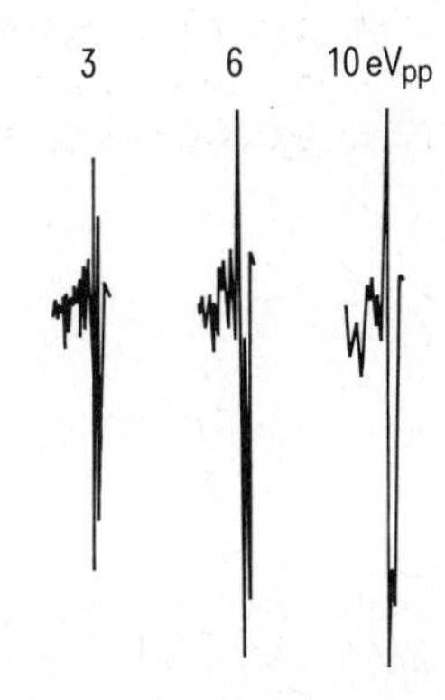

Fig. 6-15.
With increasing modulation amplitude (measured from peak-to-peak: eV_{pp}) the sensitivity increases but the resolution of the MNN doublet of iodine (511 eV) decreases (from [6-5]).

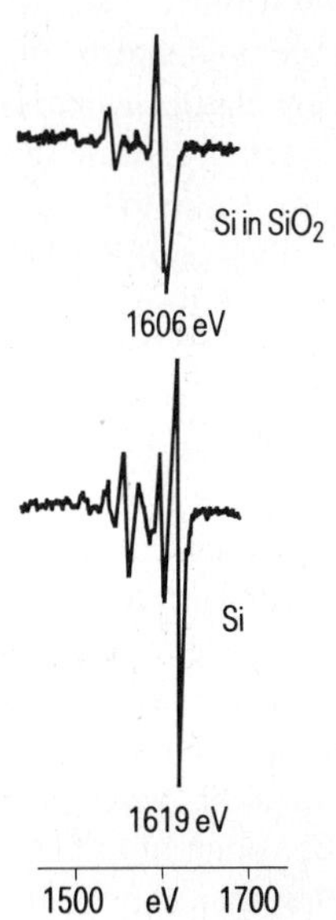

Fig. 6-16.
Chemical effect of oxidation on the KLL peak of silicon. The peak of the oxide is shifted by 13 eV towards lower energies, compared to that of pure silicon.

In many cases, however, an analysis of the adsorbate-covered surface is nevertheless useful or necessary. For surfaces where even small amounts of contamination are detrimental, such as substrate material in semiconductor manufacturing or the electric contacts of miniaturized switching relays, comparative surface analysis can provide a clue to the source of contamination [6-40]. Contamination layers often contain constituents which are very volatile when exposed to the electron beam. If these constituents, such as Cl and F, are of interest, the electron dose must be kept as small as possible. This can be done by scanning the electron probe over an enlarged area (provided it still lies within the source definition of the analyzer). If the region below the contamination layer is of interest, the contamination must be removed by sputtering. By monitoring the signals from contaminants during sputtering, it must be decided whether only the surface or the underlying layer is contaminated. Especially when the surface is rough, or when it is covered by granular contaminations so that removal of the surface layer is inhomogeneous, it is difficult to obtain unambiguous information about the subsurface state.

6.4 Quantitive analysis

6.4.1 Matrix effects

In determining the quantitative relationship between Auger intensity and element concentration, all factors which affect the detected intensity must be taken into account. The number of Auger transitions $N_A(z)$ within a unit volume of the specimen is proportional to the ionization cross section $\sigma_A(E_0)$ (see Fig. 5-2), the Auger yield a_A (see section 2.4.3, Fig. 2-47) and to the number of atoms A per unit volume ($C_A(z) = N_A(z)/V$; atomic concentration) at depth z. Moreover the intensity is affected by the atomic environment of the element A in the solid matrix, in which the primary as well as the secondary electrons and Auger electrons are elastically and inelastically scattered. The contribution of backscattering of the primary electrons is described by the backscatter factor R. The influence of the inelastic scattering of the Auger electrons is described in terms of the "inelastic mean free path (IMFP)", and the influence of the inelastic *and* elastic scattering in terms of the "attenuation length" of the Auger electrons.

The inelastic mean free path

Only those Auger electrons which were emitted from the specimen without energy losses can be unequivocally attributed to a transition of a specific element and thus are useful for analysis. The probability of emission from the material without inelastic scattering decreases rapidly with increasing depth. The inelastic mean free path (IMFP) is defined as the path length λ_p which the Auger electrons can traverse with a probability of e^{-1} without inelastic scattering. The Auger current therefore decreases as $\exp(-l/\lambda_p)$, depending on the distance l from the point of origin. Due to elastic scattering, the distance from the point of origin is generally less than the corresponding path length. Thus the emission depth of an electron from a solid is given by the "attenuation length" λ [6-41] which is smaller than the IMFP λ_p, depending on the cross section of elastic scattering. If the angle of emission ψ corresponds to the angle between surface normal and spectrometer entrance, the term $\lambda_i = \lambda \cdot \cos \psi$ is the "information depth", also called the "escape depth" (see Fig. 6-1), for the respective arrangement. Auger electrons which are generated at a depth z leave the specimen with an intensity decreased by a factor exp $[-z(\lambda \cdot \cos \psi)]$ (Fig. 6-17), where ψ is the angle of emission. The Auger current emitted within a small cone in the direction ψ exhibits the proportionality

$$I_g \sim \int_0^\infty C_A(z) \cdot \exp\left[-z/(\lambda \cdot \cos \psi)\right] dz . \tag{6-7}$$

λ depends on the sample material and the energy of the considered Auger transition.

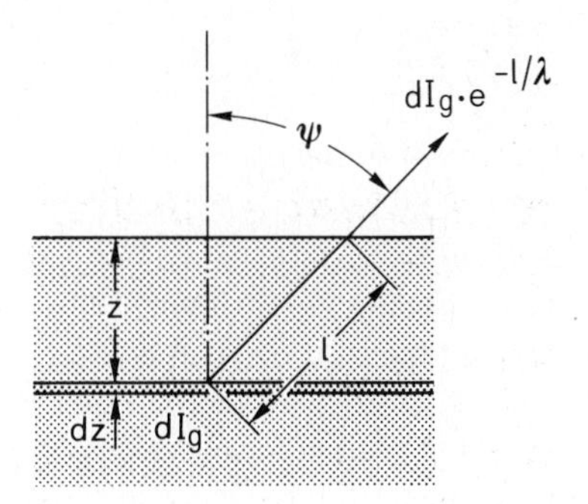

Fig. 6-17. Dependence of the detected Auger intensity on the attenuation length of low energy electrons. An Auger current of intensity $dI_g \sim C_A(z)\, dz$ is generated in a thin layer (thickness dz) at a distance z from the surface. This current is reduced by

$$\exp(-l/\lambda) = \exp\left[-z/(\lambda \cdot \cos \psi)\right]$$

along its path l within the material in the direction ψ to the spectrometer entrance slit (ψ is the angle between the surface normal and the spectrometer slit).

The backscatter factor R

Backscattered electrons with sufficiently high energies add further excitation to that produced by direct primary electrons. This further excitation is described by the backscattering coefficient η (see section 2.4) which depends on the material and the energies of the primary electrons and the excited Auger transition. Thus the total Auger yield is increased due to backscattering by a factor $R = 1 + \eta$.

The detected Auger intensity

The detected Auger intensity is further affected by surface roughness, sputter effects, chemical shift and crystallographic effects [6-42]. In an amorphous flat specimen consisting of elements with similar chemical properties, these effects can be neglected. The Auger intensity I_A of the element A is then

$$I_A = k_A \cdot a_A \cdot \sigma_A(E_0) \cdot R_M(E_A) \cdot \int_0^\infty C_A(z) \cdot \exp(-z/\lambda_M \cos \psi)\, dz \tag{6-8}$$

where k_A is related to effects of the instrument, namely primary electron current, spectrometer transmission, detector efficiency and – in the event of the differential signal being recorded – the modulation amplitude. a_A is the Auger transition probability and $\sigma_A(E_0)$ the ionization cross section. The index M indicates the dependence of R_M and λ_M on the matrix material. For a specimen with a homogeneous element distribution, the integral becomes $C_A \cdot \lambda_M(E_A) \cdot \cos \psi$ and equation (6-8) reduces to:

$$I_A = k_A \cdot a_A \cdot \sigma_A(E_0) \cdot R_M(E_A) \cdot C_A \cdot \lambda_M(E_A) \cdot \cos \psi\,. \tag{6-9}$$

Definition of the characteristic Auger intensity

Because of the impossibility of exactly separating the characteristic Auger signal from the intense background of inelastically scattered primary electrons and secondary electrons, it is very difficult to apply equation (6-9) in practical analytical problems. The Auger intensity emitted from a specific transition of an element in the specimen consists of both a zero-loss peak and a tail of inelastically scattered Auger electrons whose energies range from below the peak energy to zero. The total Auger intensity would therefore be obtained if all background and all other Auger contributions were subtracted from the direct spectrum and the intensity of the specific Auger transition were integrated from $E = 0$ up to the respective energy. However, this procedure would mean the loss of the surface sensitivity of the method, since all electrons which had suffered energy losses are not excluded. Not only do the many different scattering processes contribute to the background but also those electrons which originate from higher energy transitions and have suffered energy losses. These electrons with energy losses come from depths which may be many times greater than the escape depth of the Auger electrons discernible in the Auger spectrum [6-43]. Thus, the shape and intensity of the background is affected by the composition of layers which are far from the region which is accessible to Auger analysis. An exact background correction would have to eliminate all nonspecific contributions, which has hitherto proved impossible.

For routine analysis, comparative procedures are used and the intensity measurements from the investigated specimen are compared with measurements of standards with exactly known composition at equivalent instrument settings (similar to quantitative X-ray microanalysis). For common instrument settings, the spectra as well as the sensitivity factors of many monoelement and single composed standards are listed in handbooks [6-4,5,6]. By comparing the measured spectra either with the analyst's own standard spectra, with tabulated standard spectra or with sensitivity factors, influences which are independent of the matrix, such as the instrumental setting are avoided or at least reduced. In every case, however, the element-specific signal must be distinguished from the background.

Since exact background correction procedures are not available, several empirical approaches are in use to extract Auger intensities from electron spectra [6-44 to 47]. In the direct spectrum, the ratio of the peak height to the height of the background is sometimes used for the intensity (Fig. 6-18a,b) [6-48]. Most widely used for quantification, however, are the differential spectra. Here the peak-to-peak height is commonly taken as a measure of the Auger intensity (Fig. 6-18c). This peak-to-peak height of the differential spectrum emphasizes all sharp

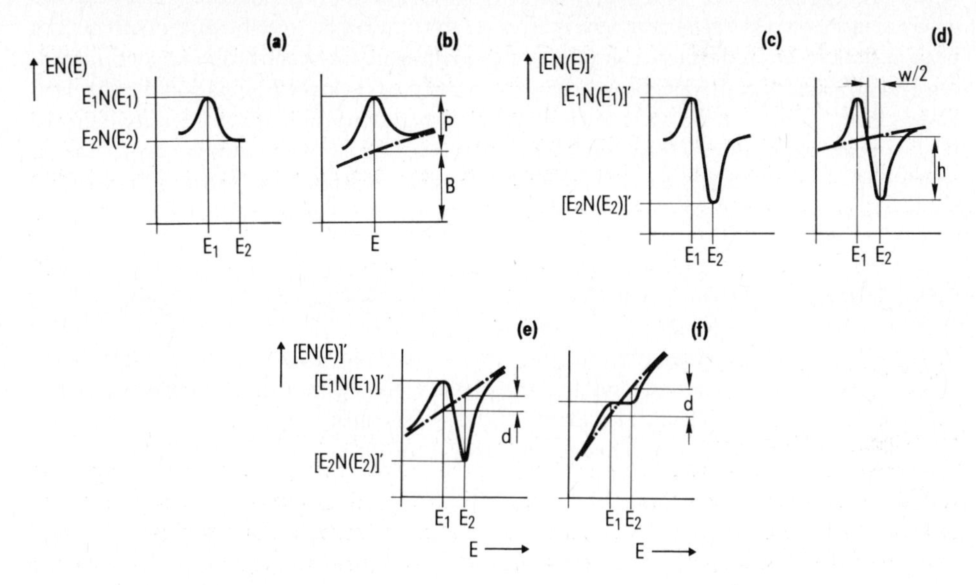

Fig. 6-18. Various definitions of the "Auger intensity" are in use:
1. *Direct spectrum* $E \cdot N(E)$: From the measured pulse counts $E \cdot N(E_1)$ and $E \cdot N(E_2)$ at energy E_1 of the peak maximum and energy E_2 of the background a peak intensity may be defined as (a):

$$I = E_1 \cdot N(E_1) - E_2 \cdot N(E_2) .$$

It is often useful to use an intensity which is normalized to the background [6-48]:

$$I = \frac{E_1 \cdot N(E_1) - E_2 \cdot N(E_2)}{E_2 \cdot N(E_2)}$$

or, with the designations in (b):

$$I = P/B .$$

2. *Differential spectrum* $[E \cdot N(E)]'$: In general, the height difference between the extrema of the positive and the negative excursions ("peak-to-peak height") (c) is used as the measure of intensity

$$I = [E_1 \cdot N(E_1)]' - [E_2 \cdot N(E_2)]' .$$

According to [6-22,43], the amplitude of the negative peak (d) represents a measure which is less dependent on the background and on chemical effects:

$$I = h$$

(w designates the peak width). When the background is steeply inclined, uncritical use of the peak-to-peak height can give rise to large errors. In the extreme case, this value can even become zero (or negative), although the peak intensity remains finite (f). This effect of the background slope can be approximately corrected by [6-52] (e):

$$I = [E_1 \cdot N(E_1)]' - [E_2 \cdot N(E_2)]' + d .$$

details of the $N(E)$ or $E \cdot N(E)$ spectra, so that it is especially sensitive to the no-loss Auger features. The slowly sloping background of inelastic and secondary electrons is widely suppressed by differentiation, however, and has to be considered, when low-intensity Auger peaks have to be evaluated (Fig. 6-18e) [6-49]. The peak-to-peak height is approximately proportional to the $N(E)$ peak area. This proportionality holds exactly true when the $N(E)$ peak has a Gaussian shape. The $N(E)$ peak area is assumed to represent the specific Auger current. The peak-to-peak height is therefore regarded as a good measure for quantification. Since the negative excursion of the differential spectrum is less affected by loss electrons and the chemical environment, it may, according to Seah [6-39], be a better measure of the intensity than the peak-to-peak height (Fig. 6-18d). But here problems may arise from defining the zero level, which is somewhat arbitrary and in lock-in based systems is sensitive to instrumental settings.

6.4.2 Analysis with standards

As mentioned above, quantitative analysis with standards involves measuring not only the Auger signal I_A of the specimen but also the Auger signal I_A^0 of a standard which contains only the element A (pure element standard). If the measurements are performed under identical instrument settings, then for homogeneous specimens the ratio of the two intensities is obtained in accordance with eq. (6-9):

$$\frac{I_A}{I_A^0} = \frac{C_A \cdot R_M \cdot \lambda_M}{C_A^0 \cdot R_A \cdot \lambda_A} . \tag{6-10}$$

If the mole fraction $X_A = N_A/N_M$ (N_A is the number of A atoms, N_M the number of all atoms in the specimen) is introduced, the following is obtained with atomic volume a_A^3 for the pure-element standard and a_M^3 as the corresponding mean value in the specimen:

$$C_A^0 = \frac{N_A^0}{V} = \frac{N_A^0}{N_A^0 \cdot a_A^3} = \frac{1}{a_A^3} ;$$

$$C_A = \frac{N_A}{V} = \frac{N_A}{N_M \cdot a_M^3} = \frac{X_A}{a_M^3} . \tag{6-11}$$

From equations (6-10) and (6-11) the mole fraction X_A is given by:

$$X_A = \frac{I_A}{I_A^0} \cdot \left[\frac{a_M}{a_A}\right]^3 \cdot \frac{R_A}{R_M} \cdot \frac{\lambda_A}{\lambda_M} . \tag{6-12}$$

If a standard with exactly known composition X_A^s similar to that of the specimen is used in place of a pure-element standard (matched standard), the effects of atomic diameter, backscattered electrons and attenuation length can be neglected, so that eq. (6-12) is reduced to:

$$X_A = X_A^s \cdot \frac{I_A}{I_A^s} . \tag{6-13}$$

If well-matched standards are used, quantification with a precision of 1% or better is possible by eq. (6-13).

The use of equation (6-12) presupposes that atomic volumes, backscatter factors and attenuation lengths are sufficiently well known. However, quantitative analyses which are often of sufficient accuracy for many cases can also be performed with pure-element standards without these data. For this purpose, the Auger intensity of the corresponding pure-element standards must be measured for every element contained in the specimen. To a first approximation, when the effects of different atomic densities, backscattering factors and attenuation lengths are neglected, X_A is given by:

$$X_A = \frac{I_A/I_A^0}{\sum_i I_i/I_i^0} \,. \tag{6-14}$$

The results so obtained may, however, suffer from serious errors.

In the case of binary alloys, good quantitative results may be obtained with pure element standards if the matrix effects are taken into account by correction factors [6-50,51]. In this case, the relations between the Auger signals of the two alloy components A and B and the associated pure-element standards are measured and these relative intensities then expressed by a ratio. Using eq. (6-12) the following ratio is obtained:

$$\frac{I_A/I_A^0}{I_B/I_B^0} = \frac{X_A}{X_B} \cdot \left[\frac{a_A}{a_B}\right]^3 \cdot \frac{R_B(E_B) \cdot R_M(E_A)}{R_M(E_B) \cdot R_A(E_A)} \cdot \frac{\lambda_B(E_B) \cdot \lambda_M(E_A)}{\lambda_M(E_B) \cdot \lambda_A(E_A)} = \frac{X_A}{X_B} \cdot \frac{1}{F_{AB}} \,, \tag{6-15}$$

with

$$F_{AB} = \left[\frac{a_B}{a_A}\right]^3 \cdot \frac{R_M(E_B) \cdot R_A(E_A)}{R_B(E_B) \cdot R_M(E_A)} \cdot \frac{\lambda_M(E_B) \cdot \lambda_A(E_A)}{\lambda_B(E_B) \cdot \lambda_M(E_A)}$$

as the Auger matrix factor. Hall and Morabito [6-50] calculated the matrix factor for the two extremes of nearly 100% of the fraction of element A and nearly 100% of the fraction of element B:

$$F_{AB}(X_A \to 0) \,; \qquad F_{BA}(X_A \to 1) \,. \tag{6-16}$$

For the calculations of the backscatter factor and of the attenuation length they used empirical formulae [6-8,52,53,54]. They showed that for the calculated 4860 binary systems the two limiting values in eq. (6-16) are practically equal, even if both elements exhibit great matrix differences. The values of the factors F_{AB} and F_{BA} yielded scatter between 0.2 and 5. Although most of these factors had a value of 1 with a standard deviation of about 0.5, an error of up to a factor of 5 may result when the matrix factor is neglected and eq. (6-14) alone is employed for quantification. The method described here for binary alloys can be easily extended to compounds with more than two ingredients.

An iterative correction of the matrix effects – analogous to the ZAF correction in electron beam X-ray microanalysis (cf. section 5.4) – is suggested by Sekine et al. [6-55]. For this purpose, equation (6-10) is used in the following form:

$$\frac{I_A}{I_A^0} = X_A \cdot \frac{C(X_1, X_2, \ldots X_n)}{C_A^0} \cdot \frac{R_M(X_1, X_2, \ldots X_n)}{R_A} \cdot \frac{\lambda_M(X_1, X_2, \ldots X_n)}{\lambda_A} \,. \tag{6-17}$$

In equation (6-17), C represents the number of all atoms per unit volume in the specimen and C_A^0 the number of atoms per unit volume in the pure-element standard A. The matrix dependence is taken into account in that C, R_M and λ_M are a function of the mole fractions X_1, X_2, ... X_n. The matrix factor may then be represented as a product of three correction factors:

$$X_A = \frac{I_A}{I_A^0} \cdot \frac{C_A^0}{C(X_1, X_2, \ldots X_n)} \cdot \frac{R_A}{R_M(X_1, X_2, \ldots X_n)} \cdot \frac{\lambda_A}{\lambda_M(X_1, X_2, \ldots X_n)}$$

$$= \frac{I_A}{I_A^0} \cdot \beta . \quad (6\text{-}18)$$

The following procedure is used to calculate these correction factors: the number of all atoms per unit volume, the atomic concentration, is the sum of all individual atomic concentrations C_i of the elements present in the specimen, $C = \Sigma C_i$. For the mole fraction $X_i = N_i/N_M$, it is assumed as an approximation that the mean atomic diameter in the specimen is equal to that in the standard. Then $X_i = N_i/N_i^0 = C_i/C_i^0$
and the value C in the correction factor is:

$$C = \sum_i X_i C_i^0 . \quad (6\text{-}19)$$

The correction factors in equation (6-18) can be calculated by using empirical formulae for the backscatter factor and the inelastic mean free path and by setting $\beta_0 = 1$ as the first approximation. The values X_{A1}, X_{B1}, ... resulting from this yield a second approximation X_{A2}, X_{B2}, ... when corrected in the manner described. This iterative procedure is continued until X_{An} no longer changes. Relative errors of below 10% were obtained by this method for palladium-silver alloys.

A simple method of quantitative Auger microanalysis using the $E \cdot N(E)$ spectrum was described by Janssen et al. [6-56] and Langeron et al. [6-48]. The peak-to-background ratio P/B of the $E \cdot N(E)$ spectrum (Fig. 6-18b) is used as a measure of the intensity. The mole fraction X_A of the element A in a specimen composed of several elements is obtained from the first approximation of equation (6-14) with $\beta = 1$:

$$X_A = \frac{P_A/B_A}{P_A^0/B_A^0} \quad (6\text{-}20)$$

where P_A^0/B_A^0 is the peak-to-background ratio of the pure-element standard (Fig. 6-18b). It could be shown that for the majority of pure specimens the ratio P/B for a given specimen is independent of various equipment parameters (current, focusing and deflection of primary electron beam, detector amplification) within an uncertainty of about 5%. P/B is also relatively insensitive to topography and to deviations of the specimen from the ideal z_0 position in front of the spectrometer. Tolerances of ± 2 mm still yield acceptable results, as was shown [6-48].

6.4.3 Analysis with sensitivity factors

In routine analysis, especially in cases with frequently changing element compositions, the analysis with standards is seldom used, because it is difficult to maintain exactly identical instrument settings when measuring the specimen and standard. In addition, an adequate number of appropriate standards is not always available. Most often, therefore, the less precise results which are obtained with tabulated "sensitivity factors" have to be accepted in microanalytical practice. This method can be easily applied and yields results with an accuracy sufficient for most tasks in routine analysis.

The sensitivity factors S_A, which are listed in handbooks [6-4,5,6] are defined as follows:

$$S_A = \frac{I_A^0}{I_{Ag}} . \tag{6-21}$$

I_{Ag} and I_A^0 refer to Auger intensities determined under identical equipment conditions for silver (Ag) and a pure element A, whose mole fraction in the investigated specimen is to be determined. The Auger intensities can be obtained from the differential spectrum (Fig. 6-18c) or from the direct spectrum (Fig. 6-18a), in general leading to two different sets of sensitivity factors. For historical and practical reasons, however, the method of sensitivity factors is almost exclusively applied to the differential spectra. In cases where no pure-element standard is available, such as for oxygen, the sensitivity factor is determined from a defined compound $A_n B_m$ (e.g. Fe_2O_3):

$$S_A = \frac{m+n}{n} \cdot \frac{I_A^0}{I_{Ag}} . \tag{6-22}$$

Different equipment settings during the measurements of standard A and silver can be corrected by applying a factor

$$S_A = \frac{I_A^0}{I_{Ag} \cdot D_A} \tag{6-23}$$

with

$$D_A = \frac{L_A^0 \cdot E_{mA}^0 \cdot i_{0A}^0}{L_{Ag} \cdot E_{mAg} \cdot i_{0Ag}} = \frac{d_A^0}{d_{Ag}} ,$$

where L_A and L_{Ag} are the lock-in amplifier sensitivities, E_{mA}^0 and E_{mAg} the modulation amplitudes, and i_{0A}^0 and i_{0Ag} the primary electron currents. The method using the silver standard is mainly applied to the collection of data which are listed in handbooks [6-4,5,6] (Fig. 6-19 and 6-20).

In most cases of routine analysis the mole fraction X_A is determined without pure element standards. For this purpose, the Auger signals of all elements in the specimen must be measured. The mole fraction then follows from the relation (cf. eqs. (6-14) and (6-21)):

$$X_A = \frac{I_A}{S_A \cdot d_A} \Big/ \sum_i \frac{I_i}{S_i \cdot d_i} . \tag{6-24}$$

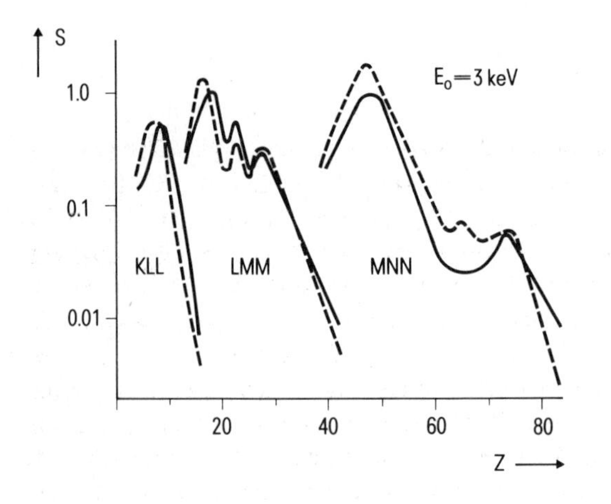

Fig. 6-19.
Experimental relative sensitivity factors S_A for quantitative Auger analysis from the handbook [6-4] for a primary electron energy E_0 of 3 keV. These values apply to the peak-to-peak heights in the differential spectrum. The broken curves correspond to the profile of calculated data [6-58].

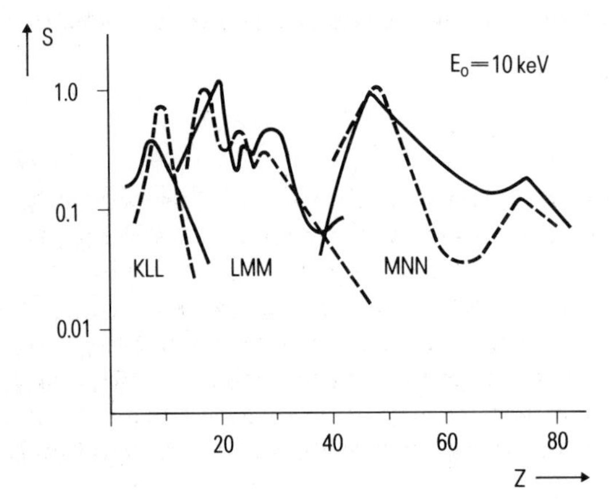

Fig. 6-20.
Experimental relative sensitivity factors S_A valid for intensities from the direct spectrum at a primary electron energy of 10 keV [6-26]. The experimental factors used for the differential spectrum are plotted for comparison (broken curves). They are taken from the handbook [6-4], and the vertical scale is normalized to a value of 1 for the Ag-MNN peak.

The intensities I_i (i = A, B, C ...) have to be measured by only a single measurement run. In this way, the influences of instrument settings are eliminated or at least reduced. However, it may be necessary to measure particular spectral ranges with different instrument settings (e. g. lock-in sensitivity, modulation amplitude or differential increment) in order to obtain all peaks with optimum sensitivity and resolution. Such changed settings must be corrected by the factors d_i. In modern instruments these correction procedures are performed automatically. The results obtained by this approximation may show errors of up to several times 10%. Commercial Auger microprobes are equipped with programs including these sensitivity factors and also allow flexible adaption of the sensitivity for preselected elements to the analysis problem.

The sensitivity factors can also be calculated theoretically. Eq. (6-9) has been taken as a basis for evolving approaches to semi-empirical formulae. Such an approach must consider all factors which significantly affect the measured Auger intensity [6-57,58,59]. The calculated values agree with the experimentally determined curves (Fig. 6-19) within a factor of about 2 to 3 [6-58] for most elements. These semi-empirical approaches conveniently complement the measured sensitivity factors in cases where the latter are unknown or uncertain (e.g. at primary energies above 10 keV).

6.5 Depth profiling

Auger spectrometry permits the concentration of an element to be measured as a function of the distance z from the surface (depth) in various ways [6-60]. By varying the effective emission depth, which depends on the emission angle and the Auger electron energy, the depth profile can be measured without destroying the specimen [6-61]. This allows a depth up to about 5 nm to be analyzed. Owing to its wide entrance aperture, a normal CMA does not allow an effective variation of the emission angle. Therefore this method cannot be applied in CMA-based instruments without modification. For depth profiling of thicker layers, in the range from several nanometers up to the micrometer region, use is predominantly made of succesive abrasion of the surface by sputtering (Fig. 6-2c) [6-62,63]. Sputtering with noble gas ions allows depths down to about 1 μm to be analyzed within acceptable periods of time. If depth profiles are to be measured beyond this range, other methods which allow either a more rapid or a more controllable abrasion are employed, e. g. oblique grinding or selective etching.

Sputter depth profiling

In this method the Auger signal I_A is measured as a function of the sputter time t (Fig. 6-21). To obtain the dependence of the mole fraction on intensity and the dependence of the sputter depth on the time, both must be known (see section 2.6.5). To determine the atomic concentration or the mole fraction as a function of the intensity, the methods described in section 6.4 are used. For routine analyses, use is normally made of the sensitivity factors (Fig. 6-19,20,21). It must, however, be noted that the surface at any time is modified by interaction with the ions (section 2.6) and may therefore deviate from its bulk composition of interest (see equation (2-47) and Fig. 2-70). The basic effects which occur during sputtering also lead to a degradation

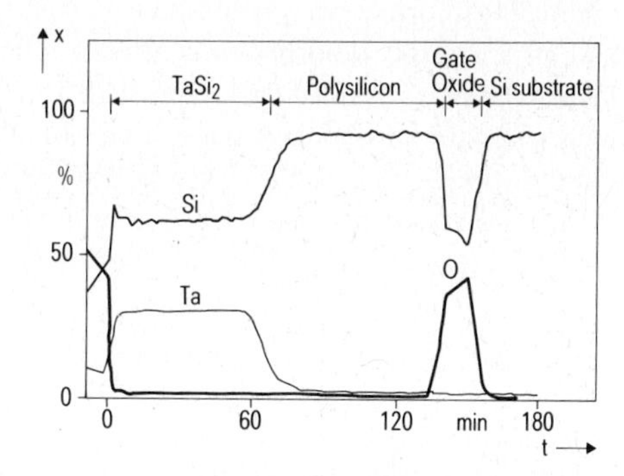

Fig. 6-21. Depth profile analysis of a sequence of layers on a silicon substrate. The diagram shows the chemical composition (calculated from the peak-to-peak height with sensitivity factors) in dependence on the sputter time. The uppermost layer (surface at $t = 0$) consists of tantalum silicide ($TaSi_2$) with a polycrystalline silicon layer underneath. The substrate is (mono-crystalline) silicon with an oxidized surface. To convert the time scale into a depth scale the (depth dependent) sputter rate $\dot{z}$ is required. For this depth profile the sputter rate was about 3 nm/min. Accordingly, the thickness of the $TaSi_2$ layer is about 200 nm. This example is discussed in detail in section 9.2.

of the depth resolution. The fundamental dependence of the depth z attained by sputtering on the sputter time t is described by equations (2-44) and (2-46). When the abrasion rate $\dot{z}$ is known and is constant, the depth can be determined by equation (2-44). By selecting the current density j_0 of the ions, the abrasion rate can be varied and defined within broad limits. In the case of an unknown but constant abrasion rate, the depth scale can be calibrated by a subsequent measurement of the sputter crater.

In general, however, the sputter rate and thus the abrasion rate $\dot{z}$ are not constant. The depth scale must then be calibrated against matched standards or determined by simultaneous measurement of the specimen during sputtering. Such in-situ measurements can be performed

- by means of laser interferometry on the crater base [6-64] or, where the specimens consist of thin layers,
- by determining the product of thickness and density or the layer thickness of the residual layer from the characteristic X-ray emission [6-65].

In addition the depth scale can be determined very precisely if cross sections for transmission electron microscopy (TEM) can be prepared from the same specimen [6-66].

However, even if the depth scale has been determined very precisely, there are certain factors which may give rise to significant deviations of the measured profile from the actual depth distribution of elements. This deviation is described by the depth resolution. It represents the uncertainty Δz defined in Fig. 2-68 and is a measure of the degree to which the original composition of the specimen is "blurred" in a depth profile by the effect of the abrasion [6-63]. The depth resolution generally gets worse with increasing depth and depends partly on parameters which are directly under the analyst's control and partly on specimen-specific factors [6-67 to 71].

The instrument-specific parameters include not only instabilities of the electron and ion sources, but also the geometry and energy of the electron and ion beams. To avoid crater edge effects, the electron beam diameter should be significantly smaller than the crater diameter. A relatively large crater area can be obtained by scanning the ion beam (Fig. 6-22). In addition, the electron beam should be directed accurately onto the crater center [6-72]. Furthermore, the information which can be derived from an Auger depth profile is impaired if surface atoms are locally desorbed or sputtered under the influence of the electron irradiation [6-73]. Since this effect is most noticeable with a finely focused electron beam, great care is necessary when volatile components have to be traced at high spatial resolution. Further effects which impair the depth resolution are heating of the specimen by high electron current densities, which gives rise to diffusion processes [6-74] and to residual gases in the specimen chamber which may adsorb on the surface during analysis and thus may cause misleading results. On the other hand, diffusion processes at high temperatures can level the surface and thus reduce the roughness if very low sputter rates are used [6-75]. The roughness of the sputter crater can increase during oblique ion bombardment because the erosion rate may depend strongly on the orientation of micro-facets on imperfectly plane surfaces [6-76]. On the other hand, grazing incidence also may provide optimal depth resolution for sequences of plane layers [6-77]. A slow rotation (about once per minute) of the specimen during sputtering also improves the depth resolution [6-78].

Other mechanisms which impair the depth resolution originate from the various interactions between the ions and the material (section 2.6). The best depth resolutions, Δz, are attained with extremely flat specimens of amorphous material, as was experimentally determined with SiO_2 on Si [6-78] and Ta_2O_5 on Ta [6-80]. Values of Δz below 2 nm were obtained for sputter depths of several 100 nm with argon ions of 1 keV. Ta_2O_5 films are prepared by anodic oxidation of polished Ta and are well suited to test the analyst's own equipment [6-80,81]. The depth resolution Δz which can be obtained for polycrystalline metals is greater [6-82] and increases with depth z roughly in accordance with the relation $\Delta z \sim z^{1/2}$, as was measured on vapor-deposited layers [6-70]. The depth resolution for monocrystalline material ranges between the values for amorphous and polycrystalline material (Fig. 6-23). Generally the attainable depth resolution depends strongly on the material, especially for polycrystalline specimens.

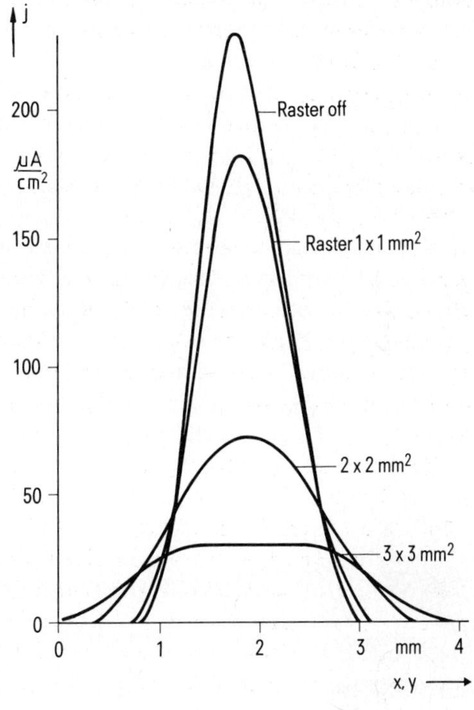

Fig. 6-22.
Ion current density distribution of Ar^+ ions of 4.5 keV energy along a local coordinate at fixed ion beam (raster off) and at scan amplitudes which guide the ion beam on the specimen over areas of $x \cdot y = 1 \cdot 1$ mm², $2 \cdot 2$ mm² and $3 \cdot 3$ mm². For an amplitude which corresponds to a deflection of more than x, y = 2 mm, the center of the crater base becomes flat in accordance with the ion current distribution during sputtering. (Differentially pumped ion gun of type PHI® Model 04-303).

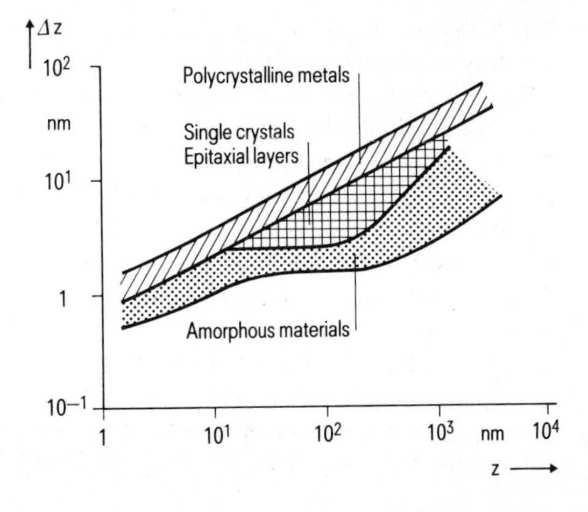

Fig. 6-23.
Depth resolution Δz as a function of the sputter depth z (Argon ions, 1 keV), measured at amorphous Ta_2O_5 and SiO_2 layers and at various polycrystalline metallic vapor-deposited layers (as per Seah and Hunt [6-69]).

Crater edge profiling

Since the crater edge exposes the layer sequence from the topmost surface down to the crater bottom, the crater edge can be used to obtain a depth profile analysis [6-83]. This has the advantage that the depth scale is given directly by the crater geometry, which can be measured with relatively high precision by optical interferometers or with stylus profilometers. In addition, a complete energy spectrum can be recorded from a number of selected points along the crater edge in a relatively short time. Since sputtering and measurement are decoupled, instrumental settings and the measuring process can be optimized for the specific problem. Furthermore, only a small area must be abraded because the ion crater need not necessarily coincide with the electron beam during sputtering [6-84]. The attainable depth resolution, however, is mostly inferior to that of conventional depth profiling.

Angle lapping

If depth profiles are to be recorded over larger depths ($z \gg 1$ μm), a microsection of the specimen must be produced [6-85]. A method which has proved effective involves grinding a very flat spherical crater into the surface using a rotating steel ball covered with diamond paste ("ball cratering" [6-86 to 88]). In this case, the depth resolution is determined essentially by the surface roughness produced by the grinding process; it is of the order of $\Delta z \approx 100$ nm and independent of depth. An investigation by Lea et al. [6-89] has revealed that for depths z greater than 2.5 μm, ball cratering is superior to sputtering as far as depth resolution is concerned.

6.6 Scanning Auger microscopy

Like the X-ray microprobe, the Auger microprobe can be used to image the element distribution. This is done by selecting a particular element-specific Auger peak with the spectrometer. While the electron probe is scanned over the specimen, the brightness of a synchronized video display is controlled by the Auger intensity detected in each case. In this way, images are obtained "in the light" of the selected elements (Fig. 6-24). Due to the low emission depth of the Auger electrons, the spatial resolution is determined essentially by the diameter of the electron probe (see Fig. 6-2) [6-90 to 92]. Scanning Auger microscopy therefore produces element distribution images with fundamentally higher lateral resolution than those obtained with X-ray microanalysis. Due to the low signal intensities, however, the smallest obtainable probe diameters and correspondingly low probe currents are not suited for Auger microanalysis. To obtain useful Auger images within acceptable recording times, probe currents of at least 5 to 10 nA are required [6-93]. The probe diameter then has a value of at least about 100 nm, provided that LaB_6 cathodes with acceleration voltages above 10 kV are used (Fig. 2-28). With field emission cathodes, the resolution can be improved to about 50 nm [6-13,14] and in special cases down to 8 nm [6-94]. Since a dwell time of at least 10 ms is necessary for recording one pixel, images with e. g. $128 \cdot 128$ pixels require recording times of several 100 seconds [6-95].

Fig. 6-24.
Scanning Auger images from a mask on an integrated circuit: element mapping for Ga, As and Au and corresponding secondary electron micrograph (SEM). 200 · 200 pixels, primary electron energy 10 keV, probe current 16 nA, recording time 27 min.

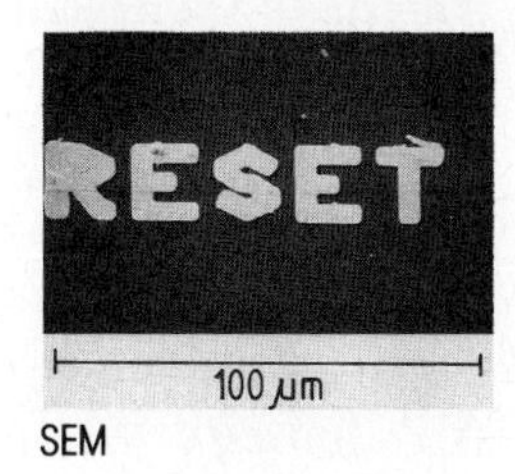

Local inhomogeneities of the same order of magnitude as the probe diameter, such as roughness [6-96] and variations in the crystal orientation and the backscatter factor, can drastically effect the Auger signal and thus lead to incorrect element distributions in the images [6-97]. Since the background depends on these inhomogeneities in much the same way as the peak intensity does, this effect may be approximately corrected by using the peak-to-background ratio as a measure of the Auger intensity during image recording. Normalization of the Auger intensity $E_1 \cdot N(E_1) - E_2 \cdot N(E_2)$ to the background $E_2 \cdot N(E_2)$

$$I = \frac{E_1 \cdot N(E_1) - E_2 \cdot N(E_2)}{E_2 \cdot N(E_2)} \tag{6-25}$$

(Fig. 6-17a) [6-61] or

$$I = \frac{E_1 \cdot N(E_1) - E_2 \cdot N(E_2)}{E_1 \cdot N(E_1) + E_2 \cdot N(E_2)} \tag{6-26}$$

[6-98, 99] are recommended for measurements using the direct spectrum. However, by use of these peak-to-background ratios the backscattering contribution from buried inhomogeneities may also be overcompensated [6-100], so that a careful interpretation of the images is always necessary.

To keep the image recording time within economical limits, only few spectral data can be measured for determining the peak and background intensities [6-101], so that only very crude corrections are possible. In general, therefore, an uncertainty of at least ±50% has to be accepted when element distribution images are evaluated quantitatively.

If the distribution of several elements or the occurrence of different phases in a specimen is to be determined, there is no point in recording individual element maps in sequence. The unavoidable drift due to the long recording times greatly impairs local composition assignment. It is therefore suggested [6-102] that at every point scanned by electrons, measurements be taken of the peak and background intensities of all elements to be represented. Different phases from, say, two elements A and B may thus be determined in the following way: the values of the peaks $E_{1A} \cdot N(E_{1A})$ and $E_{1B} \cdot N(E_{1B})$ as well as their associated background values $E_{2A} \cdot N(E_{2A})$ and $E_{2B} \cdot N(E_{2B})$ are measured at every point, and the ratio

$$R = \frac{E_{1A} \cdot N(E_{1A}) - E_{2A} \cdot N(E_{2A})}{E_{1B} \cdot N(E_{1B}) - E_{2B} \cdot N(E_{2B})} \tag{6-27}$$

is obtained. The frequency of occurrence of R is plotted in a diagram (histogram) as a function of R. If this histogram is calibrated with the aid of standards, the phases occurring in the specimen can be determined from the cluster points. This procedure can also be extended to three or more elements.

6.7 Limits of detection

The smallest detectable mass or the smallest detectable concentration of an element is limited by the shot noise and the signal background. In Auger microanalysis, the signal background, which originates from inelastically scattered and secondary electrons, is the crucial factor when considering the signal-to-noise ratio. The shot noise of the strong background may far exceed the signal of the specific Auger transition alone, if not enough events have been counted. To attain a high sensitivity, the equipment parameters must be selected so that the signal-to-noise ratio is as large as possible. Thus the counting time should be chosen as large as possible and/or a maximum number of events should be measured by setting a high beam current and/or a high spectrometer transmission. As is shown in section 6.3.4, the Auger signal increases with the excitation energy E_0 and slowly decays again after a maximum at $E_0/E_b \approx 4$. The strong background from inelastic backscattered electrons and secondary electrons (Fig. 2-40) drops more markedly than the Auger signal after this maximum. The signal-to-background ratio therefore becomes greater with increasing excitation energy, i.e. the sensitivity increases [6-103]. For this reason, the excitation energy should be higher than $4 \cdot E_b$ (e.g. 10 keV).

The signal-to-noise ratio is additionally dependent on the mode of signal acquisition [6-104 to 107]. In single-pulse counting for recording the normal spectrum $E \cdot N(E)$, the signal-to-noise ratio is better than when recording the differential spectrum $\mathrm{d}[E \cdot N(E)]/\mathrm{d}E$ with a lock-in amplifier [6-104]. Depending on the spatial resolution, the smallest detectable mole fractions lie between 10^{-4} and 10^{-2} [6-108]. As shown by measurements on pure-element specimens, the smallest analyzable mass is of the order of 10^{-19} to 10^{-17} g in the most favorable case. This corresponds to an accumulation of typically about 10^5 atoms in the volume under analysis.

To obtain high spatial resolution, a small probe diameter is required. As the probe energy increases, the diameter decreases (Fig. 2-28). However, the energy dissipation volume also increases (Fig. 2-32). Within this volume, backscattered electrons or X-rays may additionally excite Auger electrons, which are then emitted from a specimen area greater than that struck by the primary electron beam [6-91]. The majority of element-specific Auger electrons, however, originate from an area whose diameter is approximately equal to the probe diameter [6-109]. Thus, in most cases the spatial resolution is of the order of the probe diameter. A spatial resolution which is exactly the same as the diameter of the electron probe can only be attained with thin-film specimens [6-110].

The spatial resolution attainable in practice depends greatly on the investigated material. The smaller the probe diameter, the greater the current densities required to obtain a sufficiently high Auger signal. But excessively high current densities may damage the specimen. The electron radiation may also break bonds, remove surface atoms, generate defects, heat the specimen and induce diffusion processes [6-73,111,112]. For many oxides, insulators and organic compounds, the critical doses leading to verifiable radiation damage are, at radiation energies of several keV, of the order of 10^{-3} to 10^{-1} C/cm^2 [6-113]. For example, SiO_2 is dissociated by a dose of 0.6 C/cm^2 (E_0 = 2 keV) and KCl by a dose of $3 \cdot 10^{-2}$ C/cm^2 (E_0 = 1.5 keV). For an electron probe of 100 nm diameter and a current of 1 μA, this would correspond to irradiation times of 50 and 2 seconds, respectively. Therefore the attainable spatial resolution limits, somewhere between 0.05 μm and 1μm, are in many cases not compatible with an unaffected surface state. The most favorable case is when the investigated material is available in the form of a metallic conducting specimen. In contrast, specimens of organic materials require an analysis area with a diameter of at least several 100 μm [6-114]. To avoid irradiation damage reliably, the probe current would have to be reduced drastically, which necessarily leads to an impairment of the detection limits.

For insulators and semiconductors, the analysis is impaired by electrical charges [6-115,116], which cannot be avoided by coating with gold or carbon as in scanning electron microscopy or X-ray microanalysis. If such steps were taken, only the conductive covering layer would make a significant contribution to the Auger signal. The charges can be diminished only by reducing the probe current, reducing the probe energy, increasing the angle of incidence and shortening the leakage path. In most cases, these measures will allow spectra of insulators to be obtained, but the need to select special excitation conditions will make comparison of the specimens with each other, and thus quantitative evaluation, more difficult [6-117].

6.8 References

[6-1] E.H.S. Burhop, W.H. Asaad, *Adv. At. Mol. Phys. 8,* 163 (1972)

[6-2] M.F. Chung, L.H. Jenkins, *Surf. Sci. 21,* 253 (1970)

[6-3] F.P. Larkins, *Appl. Surf. Sci. 13,* 4 (1982)

[6-4] L.E. Davis, N.C. McDonald, P.W. Palmberg, G.A. Riach, R.E. Weber, *Handbook of Auger Electron Spectroscopy.* Physical Electronics Division, Perkin-Elmer Corp., Eden Prairie, Minnesota 1978

[6-5] G. McGuire, *Auger Electron Spectroscopy Reference Manual.* Plenum Press, New York 1979

[6-6] T. Sekine, Y. Nagasawa, M. Kudoh, Y. Sakai, A.S. Parkes, J.D. Geller, A. Mogami, K. Hirata, *Handbook of Auger Electron Spectroscopy.* JEOL Ltd., Tokyo 1982

[6-7] C.R. Brundle, *J. Vac. Sci. Technol. 11,* 212 (1974)

[6-8] M.P. Seah, W.A. Dench, *Surf. Interface Anal. 1,* 2 (1979)

[6-9] M.A. Burke, J.J. Schreurs, *Surf. Interface Anal. 4,* 42 (1982)

[6-10] D. Briggs, M.P. Seah (Ed.): *Practical Surface Analysis by Auger and X-ray Photoelectron Spectroscopy.* J. Wiley & Sons Ltd., Chichester 1983
[6-11] M. Thompson, M.D. Baker, A. Christie, J.F. Tyson, Auger Electron Spectroscopy (P.J: Elving, J.D.Winefordner, I.M. Kolthoff (Ed.): *Chemical Analysis Vol.74).* Wiley & Sons, New York 1985
[6-12] J.A. Venables, A.P. Janssen, C.J. Harland, B.A. Joyce, *Phil. Mag. 34,* 459 (1976)
[6-13] H. Todokore, Y. Sakitani, S. Fukuhara, Y. Okojiama, *J. Electron Microsc. 30,* 107 (1981)
[6-14] M. Prutton, R. Browning, M.M. El Gomati, D. Peacock, *Vacuum 32,* 351 (1982)
[6-15] P. Morin, F. Simoudet, *J. Physique, Coll. C2 45,* 307 (1984)
[6-16] E. Blauth, *Z. Phys. 147,* 228 (1957)
[6-17] H.Z. Sar-El, *Rev. Sci. Instr. 38,* 1210 (1967)
[6-18] P.W. Palmberg, G.K. Bohn, J.C. Tracy, *Phys. Lett. 15,* 254 (1969)
[6-19] H.E. Bishop, J.P. Coad, J.C. Rivière, *J. Electron Spectrosc. 1,* 389 (1972/73)
[6-20] G.A. Harrower, *Rev. Sci. Instr. 26,* 850 (1955)
[6-21] L.A. Harris, *J. Appl. Phys. 39,* 1419 (1969)
[6-22] M.T. Anthony, M.P. Seah, *J. Electron Spectrosc. Relat. Phenom. 32,* 73 (1983)
[6-23] C.J. Powell, N.E. Erickson, T.E. Madey, *J. Electron Spectrosc. Relat. Phenom. 25,* 87 (1982)
[6-24] M.P. Seah, M.T. Anthony, W.A. Dench, *J. Phys. E 16,* 848 (1983)
[6-25] M.P. Seah, M.T. Anthony, *J. Electron Spectrosc. Relat. Phenom. 32,* 87 (1983)
[6-26] T. Sekine, A. Mogami, M. Kudoh, K. Hirata, *Vacuum 34,* 631 (1984)
[6-27] A. Mogami, *Surf. Sci. Interface Anal. 7,* 241 (1985)
[6-28] M.P. Seah, C.P. Hunt, *Rev. Sci. Instrum. 59,* 217 (1988)
[6-29] G. Eschard, B.W. Manley, *Acta Electronica 4,* 19 (1971)
[6-30] E.N. Sickafus, D.M. Holloway, *Surf. Sci. 51,* 131 (1957)
[6-31] R. Payling, *Appl. Surf. Sci. 22/23,* 215 (1985)
[6-32] A.P. Janssen, R. Schoonmaker, J.A.D. Matthew, A. Chambers, *Solid State Commun. 14,* 1263 (1974)
[6-33] A. Barrie, F.J. Street, *J. Electron Spectrosc. Relat. Phenom. 7,* 1 (1975)
[6-34] C. Wijers, M.R. Adriaens, B. Feuerbacher, *Surf. Sci. 80,* 317 (1977)
[6-35] H.H. Madden, J.E. Houston, *J. Vac. Sci. Technol. 14,* 1 (1977)
[6-36] R.E. Clausing, D.S. Easton, G.L. Powell, *Surf. Sci. 36,* 377 (1973)
[6-37] E. Kny, *J. Vac. Sci. Technol. 17,* 658 (1980)
[6-38] J. Kleefeld, J.J. Levenson, *Thin Solid Films 64,* 389 (1979)
[6-39] P. Dolizy, F Grolière, *Surf. Interface Anal. 5,* 4 (1983)
[6-40] K.L. Mittal (Ed.): *Surface Contamination: Detection and Control, Vol.1 and 2.* Plenum Press, New York 1979
[6-41] A. Jablonsky, B. Lesiak, H. Ebel, M.F. Ebel, *Surf. Interface Anal. 12,* 87 (1988)
[6-42] P. Morin, *Surf. Sci. 164,* 127 (1985)
[6-43] M.P. Seah, *Surf. Interface Anal. 1,* 86 (1979)
[6-44] P. Staib, J. Kirschner, *J. Appl. Phys. 3,* 421 (1974)
[6-45] E.N. Sickafus, *Phys. Rev. B 16,* 1448 (1977)
[6-46] K. Ishikawa, Y. Tomida, *J. Vac Sci. Technol. 15,* 1123 (1978)
[6-47] J.T. Grant, *J. Vac. Sci. Technol. A 2,* 1135 (1984)
[6-48] J.P. Langeron, L. Minel, J.L.Vignes, S. Bouquet, F. Pellerin, G. Lorang, A. Ailloud, J. Lettéricy, *Solid State Communic. 49,* 405 (1984)
[6-49] F. Labohm, *Surf. Interface Anal. 4,* 194 (1982)
[6-50] P.M. Hall, J.M. Morabito, *Surf. Sci. 83,* 391 (1979)
[6-51] M.P. Seah in D. Briggs, M.P. Seah (Ed.): *Practical Surface Analysis by Auger and X-ray Photoelectron Spectroscopy.* J. Wiley & Sons Ltd., Chichester 1983, 181
[6-52] S. Ichimura, R. Shimizu, *Surf. Sci. 112,* 386 (1981)
[6-53] W. Reuter in G. Shinoda, K. Kohara, T. Ichinokawa (Ed.): *Proc. 6th Intern. Conf. on X-ray Optics and Microanalysis.* University of Tokyo Press, Tokyo 1971, 121
[6-54] D.R. Penn, *J. Electron Spectrosc. Related Phenom. 9,* 26 (1976)
[6-55] T. Sekine, K. Hirata, A. Mogami, *Surf. Sci. 125,* 565 (1983)
[6-56] A.P. Janssen, C.J. Harland, J.A. Venables, *Surf. Sci. 62,* 277 (1977)
[6-57] S. Mroczkowski, D. Lichtman, *Surf. Sci. 131,* 159 (1983)
[6-58] R. Payling, *J. Electron Spectrosc. Related Phenom. 37,* 225 (1985)

[6-59] R. Zhang, S. Chu, *Chinese Phys. Lett. 2,* 333 (1985)
[6-60] S. Hofmann, *Surf. Interface Anal. 2,* 148 (1980)
[6-61] S. Hofmann, *Analysis 9,* 181 (1981)
[6-62] H.J. Mathieu in H. Oechsner (Ed.): *Thin Film and Depth Profile Analysis.* Springer, Berlin, Heidelberg 1984, 39
[6-63] S. Hofmann, J.M. Sanz in H. Oechsner (Ed.): *Thin Film and Depth Profile Analysis.* Springer, Berlin, Heidelberg 1984, 141
[6-64] J. Kempf in A. Benninghoven, C.A. Evans, R.A. Powell, R. Shimizu, H.A. Storms (Ed.): *SIMS II,* 2nd Int. Conf. Stanford, Springer, New York 1979, 97
[6-65] J. Kirschner, H.W. Etzkorn, *Appl. Surf. Sci. 14,* 221 (1982-83)
[6-66] R.v. Criegern, T. Hillmer, V. Huber, H. Oppolzer, I. Weitzel, *Fresenius Z. Anal. Chem. 319,* 861 (1984)
[6-67] H.W. Werner, *Surf. Interface Anal. 4,* 1 (1982)
[6-68] E. Zinner, *J. Electrochem. Soc. 130,* 199 C (1983)
[6-69] M.P. Seah, C.P. Hunt, *Surf. Interface Anal. 5,* 33 (1983)
[6-70] A. Zalar, S. Hofmann, Z. Zabkar, *Thin Solid Films 131,* 149 (1985)
[6-71] M.P. Seah, C. Lea, *Thin Solid Films 81,* 257 (1981)
[6-72] S. Duncan, R. Smith, D.E. Sykes, J. M. Walls, *Surf. Interface Anal. 5,* 71 (1983)
[6-73] ASTM Publ. E 983-84, *Surf. Interface Anal. 10,* 173 (1987)
[6-74] S. Hofmann, A. Zalar, *Thin Solid Films 56,* 337 (1979)
[6-75] M.P. Seah, M. Kuhlein, *Surf. Sci. 150,* 273 (1985)
[6-76] M. Keenlyside, F.H. Scott, G.C. Wood, *Surf. Interface Anal. 5,* 64 (1983)
[6-77] D.E. Sykes, D.D. Hall, R.E. Thurstans, J.M. Walls, *Appl. Surf. Sci. 5,* 103 (1980)
[6-78] A. Zalar, *Thin Solid Films 124,* 223 (1985)
[6-79] C.F. Cook, C.R. Helms, D.C. Fox, *J. Vac. Sci. Technol. 17,* 44 (1980)
[6-80] C.P. Hunt, M.T. Anthony. M.P. Seah, *Surf. Interface Anal. 6,* 92 (1984)
[6-81] M.P. Seah, H.J. Mathiew, C.P. Hunt, *Surf. Sci. 139,* 549 (1984)
[6-82] M.P. Seah, M.E. Jones, *Thin Solid Films 115,* 203 (1984)
[6-83] N.J. Taylor, J.S. Johannessen, W.E. Spicer, *Appl. Phys. Lett. 29,* 497 (1976)
[6-84] H. Poppa, L. Cota-Araiza, *Thin Solid Films 115,* 217 (1984)
[6-85] L.L.Levenson in O. Johari (Ed.): *Scanning Electron Microscopy 1984 III.* SEM Inc., AMF O'Hare, Chicago 1984, 1211
[6-86] J.M. Walls, D.D. Hall, D.E. Sykes, *Surf. Interface Anal. 1,* 204 (1979)
[6-87] L.A. Larson, M. Prutton, H. Poppa, *J. Vac. Sci. Technol. 20,* 1403 (1982)
[6-88] V. Thompson, H.E. Hintermann, L. Chollet, *Surf. Technol. 8,* 421 (1979)
[6-89] C. Lea, M.P. Seah, *Thin Solid Films 81,* 67 (1981)
[6-90] M.M. El Gomati, M. Prutton, *Surf. Sci. 72,* 485 (1978)
[6-91] J. Cazeau, *Surf. Sci. 125,* 335 (1983)
[6-92] M.M. El Gomati, A.P. Janssen, M. Prutton, J.A. Venables, *Surf. Sci. 85,* 309 (1979)
[6-93] J.A. Venables, A.P. Janssen in J.M. Sturgess (Ed.): *Electron Microscopy 1987 III,* 9th. Int. Congr. Electron Microsc. Toronto 1978. The Imperial Press, Ontario 1987, 280
[6-94] J. Cazaux, J. Chazelas, M.N. Charasse, J.P. Hirtz, *Ultramicroscopy 25,* 31 (1988)
[6-95] M. Prutton in O. Johari (Ed.): *Scanning Electron Microscopy 1982 II.* SEM Inc., AMF O'Hare, Chicago 1982, 83
[6-96] O.K.T. Wu, E.M. Butler, *J. Vac. Sci. Technol. 20,* 453 (1982)
[6-97] M.M. El Gomati, J.A.D. Matthew, M. Prutton, *Appl. Surf. Sci. 24,* 147 (1985)
[6-98] R. Browning. D.C. Peacock, M. Prutton, *Appl. Surf. Sci. 22/23,* 145 (1985)
[6-99] M. Prutton, L.A. Larson, H. Poppa, *J. Appl. Phys. 54,* 374 (1983)
[6-100] M.M. El Gomati, *Inst. Phys. Conf. Series No. 78,* Chapter 7, EMAG 1985, 235 (1985)
[6-101] G. Todd, H. Poppa, *J. Vac. Sci. Technol. 15,* 672 (1987)
[6-102] R. Browning, *Inst. Phys. Conf. Series 78,* Chapter 7, EMAG 85, 231 (1985)
[6-103] H.E. Bishop, Inst. Phys. Conf. Ser. 61, Chapter 9, EMAG 1981, 435 (1981)
[6-104] R.v. Criegern, T. Hillmer, I. Weitzel, *Fresenius Z. Anal. Chem. 314,* 293 (1983)
[6-105] A. Mogami, T. Sekine in D.G. Brandou (Ed.): *Electron Microscopy 1976.* Tal International Publishing Comp., Jerusalem 1976, 422
[6-106] J.E. Houston, *Appl. Phys. Lett. 24,* 42 (1974)

[6-107] J.T. Grant, T.W. Haas, J.E. Houston, *Surf. Sci. 42,* 1 (1974)
[6-108] H.W. Werner, R.P.H. Garten, *Rep. Prog. Phys. 47,* 221 (1984)
[6-109] J. Cazaux, *J. Microscopy 145,* 257 (1987)
[6-110] D.B. Wittry in P. Brederoo, V.E. Cosslett (Eds.): *European Congr. Electron Microsc., The Haag, 3.* Seventh European Congress on Electron Microscopy Foundation, Leiden 1980, 14
[6-111] T.E. Madey, J.T. Yates, *J. Vac. Sci. Technol. 8,* 525 (1971)
[6-112] H.H. Madden, *J. Vac. Sci. Technol. 13,* 228 (1976)
[6-113] C.G. Pantano, T.E. Madey, *Appl. Surf. Sci. 7,* 115 (1981)
[6-114] L.L. Levenson in O. Johari (Ed.): *Scanning Electron Microscopy 1982.* SEM Inc., AMF O'Hare, Chicago 1982, 925
[6-115] K. Röll, *Appl. Surf. Sci. 5,* 388 (1980)
[6-116] M.G. Vasilyev, D.V. Klyachko, V.G. Krigel, I.V. Razumoskaya, *Phys. Chem. Mech. Surf. 2 (5),* 1525 (1984)
[6-117] J.L. Hock, D. Snider, J. Kovacich, D. Lichtman, *Appl. Surf. Sci. 10,* 405 (1982)

7 Secondary Ion Mass Spectrometry

7.1 Principle

Secondary ion mass spectrometry (SIMS) takes advantage of the fact that when a solid surface is bombarded with ions (primary ions PI), components of this surface are abraded by sputtering, and some of the sputtered particles are emitted as ions (secondary ions SI; see sections 2.6.4 and 2.6.5). Analytical information is then obtained by separating the SIs according to their masses with the aid of a mass spectrometer and subsequently recording the individual species with a detector.

As was seen in section 2.6, sputtering is only one of many interactive effects which occur when a solid is bombarded with ions: the PIs are implanted in the specimen and change the structure and composition of a layer close to the surface. When the bombardment is normal to the surface, the thickness of this altered layer corresponds approximately to the projected range (R_p, Fig. 2-54) of the ions. When a layer with a thickness of about twice the projected range has been sputtered ($d \approx 2\ R_p$), an equilibrium is, as a rule, set up between implantation and sputtering in which the composition of the sputtered material corresponds to that of the bulk specimen (section 2.6.4). The proportion of ions in the total number of sputtered particles is in most cases low ($<5\%$). It depends greatly on matrix effects and can be considerably increased by reactive sputtering, for instance with oxygen or cesium (section 2.6.5).

The following conclusions can be drawn from the foregoing considerations:

a) As in all mass spectrometric methods, *all elements and their isotopes* can be analyzed.
b) SIMS is *destructive*, because the particles to be analyzed must be abraded from the specimen.
c) SIMS is *surface-specific*, since the exit depth of the sputter particles is only a few atomic layers (approx. 1 nm, section 2.6.4).
d) By sputtering of successive atomic layers, SIMS can be used for *depth profiling* (section 2.6.6).
e) Since ion beams can be focused (section 2.5.4), SIMS is in principle also suitable for *micro-field analysis*.

Summarizing reviews of SIMS can be found in [7-1] and in the Proceedings of the International Conferences on Secondary Ion Mass Spectrometry [7-2 to 7-6 and 7-59]. Extensive work covering basic concepts, instrumental aspects, applications and trends of SIMS has been published quite recently [7-7].

Secondary ion intensity

The secondary ion intensity recorded by the electronics of a SIMS apparatus is given by:

$$I_i^{\pm} = \beta_i \cdot P_i^{\pm} \cdot \dot{N}_i \qquad (7\text{-}1)$$

where $I_i^{\pm}$ is the number of positive or negative SIs of an element i recorded per time interval. The instrument constant β_i represents the overall transmission, and comprises the transfer of the SIs of this element to the mass spectrometer, the transmission of the mass spectrometer and the detector efficiency. $P_i^{\pm}$ is the ionization probability and $\dot{N}_i$ the sputter rate for atoms of element i.

From equations (2-36), (2-37) and (7-1), the following applies for sputtering in *equilibrium* (and *only* then!)

$$I_i^{\pm} = \beta_i \cdot P_i^{\pm} \cdot Y \cdot X_i \cdot \dot{N}_0 \tag{7-2}$$

where Y is the total sputter yield of the multielement specimen; X_i is the mole fraction of the element i in the specimen and $\dot{N}_0 = i_0/q$ the PI flux. In equilibrium sputtering therefore, the secondary ion intensity is proportional to the mole fraction ($I_i^{\pm} \sim X_i$). The index i refers to a specific element. A mass spectrometer does not, however, separate according to elements but to their isotopes. Equation (7-2) therefore applies only to monoisotopic elements. In the case of multi-isotopic elements, the secondary ion intensity of an isotope k of an element i is given by $I_{ik}^{\pm} = \nu_{ik} I_i^{\pm}$, where ν_{ik} is the isotopic abundance (proportion of the isotope k in the total number of i atoms).

Useful yield

The product of overall transmission and ionization probability is often combined as the useful yield [7-8]:

$$\tau_i^{\pm} = \beta_i \cdot P_i^{\pm} \, . \tag{7-3}$$

This parameter characterizes the number of *ions recorded* per time interval of an element i in relation to the number of *sputtered atoms* of the same element during the same time interval. $\tau_i^{\pm}$ therefore determines the amount of material used up during the measurement for a given number of recorded SIs.

Since SIMS is a destructive method of analysis, every atom can be sputtered once only and every generated ion can be registered once only (if at all). To obtain a specific measurement signal, i.e. to record a specific number $n_i^{\pm}$ of ions of a given element (or isotope) i with the detector, a specific volume must be sputtered. The magnitude V of this volume depends on the concentration of that element in the specimen and on the useful yield. The following relationship follows from eq. (7-1) with eqs. (7-3), (2-36) and (2-44) for sputtering in equilibrium:

$$I_i^{\pm} = \tau_i^{\pm} \cdot X_i \cdot \dot{N} = \tau_i^{\pm} \cdot X_i \cdot C \cdot \dot{V} \tag{7-4}$$

where $A\dot{z}$ was substituted by $\dot{V}$. By integrating $I_i^{\pm}$ over the measuring time, $n_i^{\pm}$ is obtained, and thus for the required volume [7-8, 9]

$$V = n_i^{\pm} / (\tau_i^{\pm} \cdot X_i \cdot C) \, . \tag{7-5}$$

This relation will be used in section 7.4.5 to compute the detection limit for a given sputter volume. It tells us nothing about the shape of the sputter volume, which can be modified in line with the requirements of analysis: large-area analysis of thin films or microanalysis of small regions.

X_i and C are parameters defined by the specimen; $n_i^{\pm}$ depends on the required measurement accuracy: the relative standard deviation of a measured value is $\varepsilon = \sqrt{n}/n = 1/\sqrt{n}$. It follows from this that at least 100 pulses must be counted in order to ensure a standard deviation not exceeding 10%.

The required sputter volume for a given statistical accuracy of the measurement can thus be influenced only by way of the useful yield. Eq. (7-3) shows us that $\tau_i^{\pm}$ may be improved by reactive sputtering (increasing the ionization probability $P_i^{\pm}$; see section 2.6.5) or by optimizing the equipment (increasing the overall transmission β_i; see section 7.2.5).

7.2 Instrumentation

7.2.1 Fundamental setup

SIMS is often combined with other analysis techniques, such as Auger electron spectroscopy (chapter 6) in a single UHV instrument. In the present treatment we will restrict ourselves to describing dedicated instruments with high spatial resolution of the kind suitable for microanalysis. Two types of equipment, which differ significantly in their ion-optical configurations, will be considered [2-43]: *Ion microprobes* (Fig. 7-1a) scan the specimen with a focused PI beam. As in X-ray and Auger microprobes (chapters 5 and 6), the specimen can be analyzed point by point with a stationary probe or a scan image can be generated via beam scanning and simultaneous recording of a selected signal. In *ion microscopes* (Fig. 7-1b), a larger area of the specimen surface may be simultaneously irradiated with PIs. A stationary ion image is produced by ion-optical imaging with the aid of ions of selected mass. Point analyses are possible by using an aperture in the image plane (or in an intermediate image plane, if several lenses are employed). Common to *both* types of equipment are the device components shown schematically in Fig. 7-2.

Vacuum system and specimen chamber

Residual gas components such as hydrogen, carbon, nitrogen and oxygen are adsorbed onto the specimen surface and can be incorporated in the specimen by recoil implantation (section 2.6.3). They then appear in the SI spectrum, mostly as molecular ions in combination with constituents of the specimen itself (e.g. SiO^+, SiH^+, SiC^+ etc. with silicon specimens). They can also influence the SI yield significantly (section 2.6.5). The investigation must therefore proceed under ultra-high vacuum conditions, as in Auger microanalysis (section 6.2.1). To meet this requirement, the entire apparatus from ion source to detector is accommodated in a bakable UHV chamber. It is common practice to evacuate the PI side separately in order

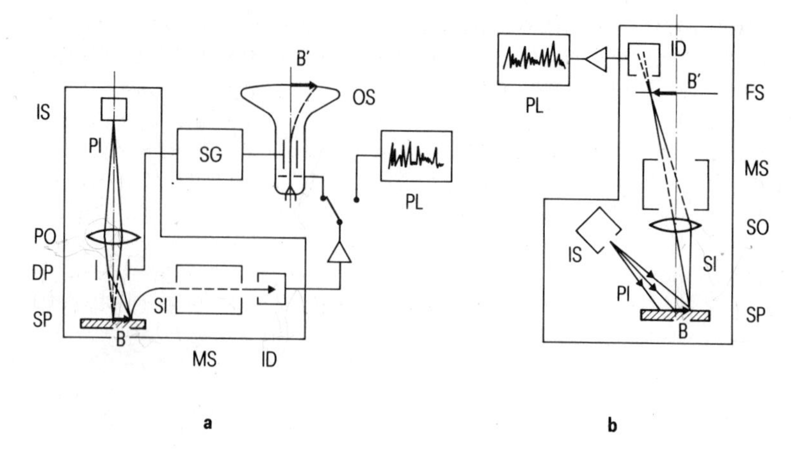

Fig. 7-1. Equipment for SIMS analysis, according to [2-43] (ion lenses shown schematically as glass lenses). a) *Ion microprobe:* the primary ion optics PO (one or more lenses) focuses the primary ion beam PI onto the specimen SP. The beam is positioned or scanned with the aid of the deflection plates DP. The secondary ion species (SI) are separated with the mass spectrometer MS and recorded by the ion detector ID. A scan image B′ of the specimen area B can be produced on the screen of the oscilloscope OS using a selected SI signal, or (with stationary PI beam) a mass spectrum of a selected specimen point can be recorded on plotter PL. IS ion source; SG scanning generator. b) *Ion microscope:* a large area of the specimen is irradiated with PIs. The secondary ion optics SO (one or more lenses) projects an SI image B′ of the specimen area B onto a fluorescent screen FS using an SI species which was selected with the aid of the mass spectrometer. If FS is replaced by an aperture (when several lenses are used, this is located in an intermediate image plane), the SI signal from a selected specimen point can be made to pass into the detector and, by continuously tuning the mass spectrometer, a spectrum of this point can be recorded.

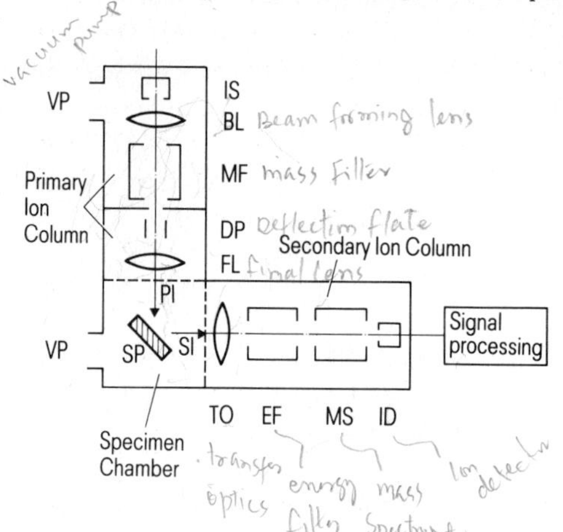

Fig. 7-2.
Basic SIMS setup, according to [2-43]. *Primary ion column:* IS ion source, BL beam forming lens, MF mass filter, DP deflection plates, FL final lens. *Secondary ion column:* TO transfer optics, EF energy filter, MS mass spectrometer, ID secondary ion detector, PI primary ions, SP specimen, SI secondary ions, VP vacuum pumps.

to keep the specimen chamber free from gas molecules that originate from the ion source but do not contribute to the PI beam. A narrow aperture allows differential pumping of the two parts of the vacuum system. A vacuum of between 10^{-7} and 10^{-6} Pa is typically maintained in the specimen chamber, even when the ion source is in operation, with the aid of a titanium sublimation pump or a cryopump in addition to ion getter pumps. Turbomolecular pumps are used to evacuate the primary ion column and, if required, the specimen airlock.

The specimen (or a set of several specimens) is mounted in a holder and can be introduced, with the aid of an airlock and a transfer rod, into the specimen chamber and subsequently attached to a specimen stage without breaking the vacuum. The stage may be driven by stepping motors to position the specimen and tilt it in order to set the required PI angle of incidence. The specimen stage is electrically isolated with respect to earth. The specimen voltage can therefore be varied, for instance to compensate for charging of insulating or poorly conducting specimens. For the same purpose, modern SIMS instruments are equipped with an electron source: positive charging due to ion bombardment can be largely compensated for by simultaneous electron bombardment (section 7.3.2). An optical microscope mounted at a window in the specimen chamber is the usual means for observing the specimen, but secondary electron imaging may also be used.

Primary ion column

The ion optics on the primary side of a SIMS device are designed to focus the ions of specific species extracted from the ion source onto the specimen. The devices used as *ion sources* are dealt with in section 2.5.3. Those which are the most important for depth profiling are plasma ion sources for noble gases and oxygen, and surface ionization sources for cesium. Due to their poor total beam current, liquid metal ion sources are used only for imaging SIMS and surface analysis [2-50].

The PIs emerge from the ion source as a divergent beam which (when extended backwards) forms a virtual crossover. This must be imaged in reduced form onto the specimen (*beam forming,* section 2.5.4). This is effected by means of one or several electrostatic tube or aperture lenses, the last of which is known as the final lens. A system with more than one lens offers the advantage of allowing both the probe current and the probe diameter (or demagnification) to be varied within wide limits; values used in practice are listed in Table 2-3.

Like the residual gas components, the contaminant element ions contained in the PI beam may also disturb the result of the analysis. They can be implanted into the specimen and show up again in the mass spectrum as apparent trace elements. On top of this, reactive contaminant ions affect the SI yield in an uncontrolled way (section 2.6.5). A PI *beam purification* step is therefore necessary. This is implemented by mass filters, which are usually magnetic sector fields or Wien filters.

The effect of a magnetic sector field [7-10] is based on the phenomenon that particles of different masses traverse paths of different curvatures in the magnetic field (see section 2.2.1) and are thus separated. By appropriately selecting the magnetic field strength, it can be ensured that only PIs of a specific species can pass the exit aperture of the magnetic field sector. Magnetic sector fields may also be applied in the mass spectrometers on the secondary side of the instruments (section 7.2.2).

In a Wien filter [7-11], homogeneous electric and magnetic fields are arranged normal to each other (Fig. 7-3); such filters are therefore also known as ExB filters. Both fields exert forces on an ion moving in the axial direction. The field strengths are adjusted in such a way that the deflecting force of one field is exactly compensated by that of the other. The compensation condition is, however, fulfilled only for a single ion velocity v_0 in each case. Since v_0 depends on the ion energy and the ion mass, velocity filtering acts as a selection in both energy

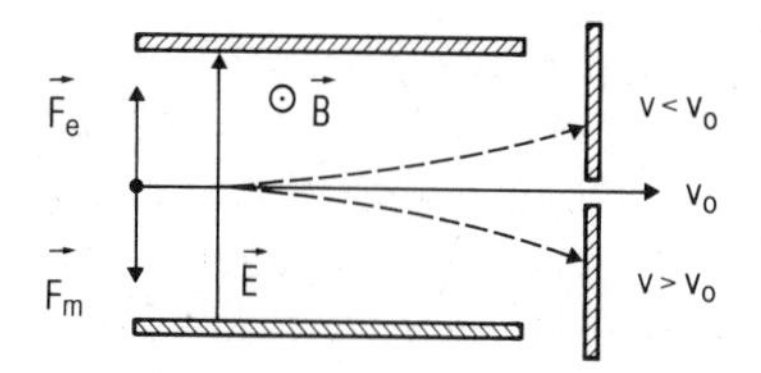

Fig. 7-3. Mode of operation of the Wien filter, according to [2-38]: a homogeneous electric field $\boldsymbol{E}$ and a homogeneous magnetic field $\boldsymbol{B}$ are superimposed ($\boldsymbol{B} \perp \boldsymbol{E}$; $\boldsymbol{B}$ and $\boldsymbol{E} \perp \boldsymbol{v}_0$) so that the accelerating forces $\boldsymbol{F}_e$ and $\boldsymbol{F}_m$ acting on a (positive) ion moving with velocity $\boldsymbol{v}_0$ have opposite directions. It follows from eqs. (2-6) and (2-10) that the forces exactly compensate each other when $E = v_0 B$. Consequently, only particles of a specific speed v_0, which can be set by selecting the appropriate field strengths, can emerge from the exit aperture of the filter (velocity filtering).

(required due to the energy spread of the ion sources; section 2.5.3) and mass (to eliminate the contaminant ions).

Since a Wien filter is a "direct-vision" device, it is possible for high-energy neutral particles, which may be produced by charge exchange during or after the acceleration, to pass through the filter and strike the specimen in an uncontrolled way. When using a Wien filter therefore, the PI beam must be deflected at some point by a few degrees to remove the neutral particles from the beam path [7-12].

All equipment is provided with a beam deflection system for positioning the beam and scanning a given specimen surface. Even devices which are not designed for scan imaging make use of beam scanning in order to generate, on a larger surface, a uniform mean current density, and thus create an even crater bottom during depth profiling (section 7.3.4). This is done with the aid of electrostatic deflection plates, one pair of which acts in the x direction and the other in the y direction (only one of these is shown in Fig. 7-2). They can be arranged in front of or behind the final lens [7-13].

Secondary ion column

The ion-optical facilities on the secondary side of a SIMS apparatus have to ensure that the SIs emitted from the specimen surface are guided as effectively as possible to the detector. This involves collecting the emitted SIs, transferring them to the mass spectrometer and transmitting them through it.

Magnetic sector fields or dynamic quadrupole mass filters are the devices used as *mass spectrometers* for depth profiling SIMS. In either case, the SIs have previously to pass an electrostatic energy analyzer, since the natural width of the SI energy distribution (see Fig. 2-67) is too great to ensure an adequate mass resolution [2-41]. Because of their fundamental significance for SIMS, mass spectrometers will be treated separately in the next section. Time-of-flight (TOF) mass spectrometers, preferentially used for static SIMS conditions, are treated briefly in section 7.3.3.

The SIs are emitted from the specimen surface into the half-space 2π. The purpose of the *transfer optics* is to collect SIs from the largest possible surface area and solid angle and to transfer them to the detection area and acceptance angle of the mass spectrometer. Such *emittance-acceptance matching* [7-14] allows optimization of the overall transmission β_i and thus, as evident from eq. (7-1), of the SI intensity.

The most effective way to collect SIs is to accelerate them as they leave the specimen surface [2-43]. In the simplest case, this is done with the aid of an extraction voltage applied, in the case of ion probes with quadrupole mass spectrometers, to the entrance aperture of the energy analyzer [7-15]. In the ion microscope, the specimen surface directly adjoins the electric field of the first lens (objective lens), in which the SIs are accelerated and focused to produce an ion image. Additional lenses (not shown in Fig. 7-2) transfer the ions to the energy and mass analyzer.

Signal processing

The most important electronic components and their interaction for the various modes of operation are shown schematically in Fig. 7-4.

The *SI detectors* (ID) are usually open electron multipliers: channeltrons (section 6.3.2) or Cu/Be dynode multipliers (for operating principle see Fig. 3-5). They are operated in the pulse-counting mode, i.e. each electron avalanche generated by an ion is individually counted. In conjunction with the electronics connected to them, these detectors must be able to process pulse rates of about 10^6 s^{-1} without significant dead time losses. On the other hand, their background level should be lower than 1 s^{-1}. They must, for instance, be able to measure concentrations in a dopant depth profile from the percentage region down to the ppb region. Should the counting rate exceed the upper limit value, which occurs in the analysis of major constituents, the multiplier can, in some equipments, be replaced by a Faraday cup and pulse counting by dc measurements.

The *signal processing unit* includes, inter alia, a preamplifier, a pulse shaper, a pulse discriminator (for suppressing the low- amplitude noise pulses) and a pulse counter. After being converted into matched dc signals, the pulse rate can be used for the y deflection of a plotter or monitor or for controlling the brightness of an oscilloscope.

The deflection voltages for the deflection plates (DP) are supplied by the *scan generator.* The PI beam is either scanned over a given specimen surface (Fig. 7-4 b,c) or positioned onto a fixed point on the specimen (Fig. 7-4a). During scan imaging (Fig. 7-4b), the scan generator additionally ensures that the electron beam in the oscilloscope is synchronized with the PI beam.

The mass detected is set by the *spectrometer control unit,* i.e. by setting specific supply voltages or currents in line with the type of mass spectrometer used. The energy filter control and specimen voltage control must be matched to the prevailing experimental conditions. The *specimen voltage* may be varied in order to compensate for charging in poorly conducting specimens or to measure the SI energy distribution (section 7.3.2). Time-dependent settings are synchronized by the *time base unit.* A computer, which is not shown in the figures, controls the interaction of all components and permits a largely automatic analysis sequence.

7.2.2 Mass spectrometer

In a mass spectrometer [7-10, 16], ion beams are treated so that only ions of uniform mass (more precisely: of uniform mass to charge ratio; see later) can reach the detector through the exit aperture. By continuously altering one of the operating parameters (e.g. the electric

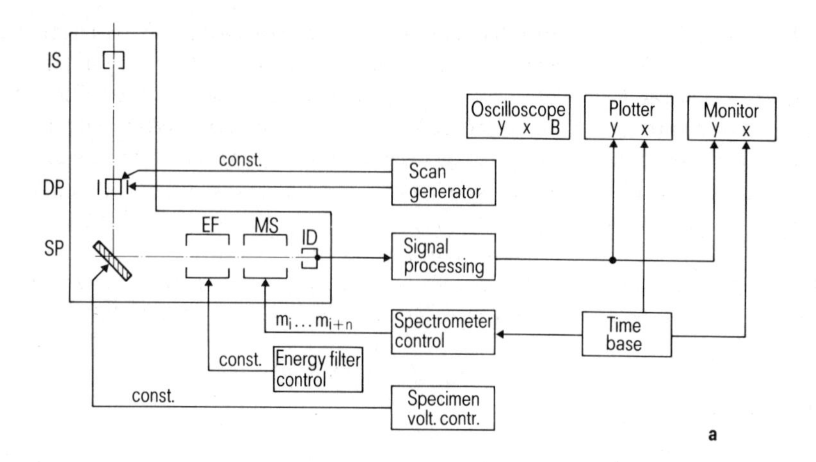
IS
DP
SP
Oscilloscope
y x B
Plotter
y x
Monitor
y x
const.
Scan generator
EF
MS
ID
Signal processing
$m_i \dots m_{i+n}$
Spectrometer control
Time base
const.
Energy filter control
const.
Specimen volt. contr.
a

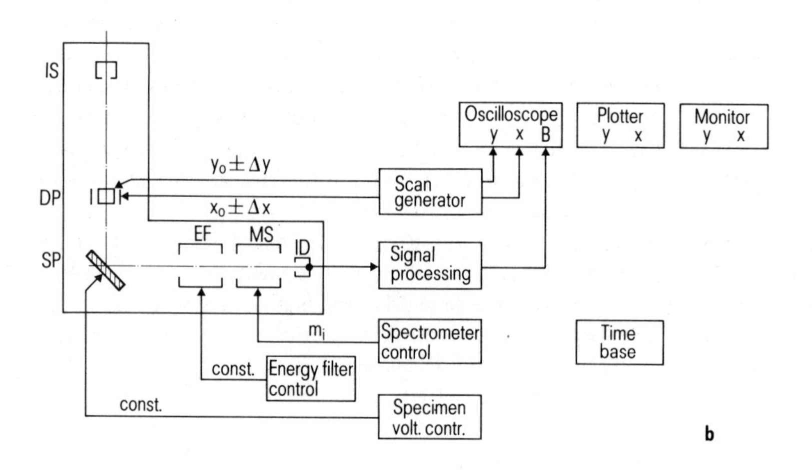
IS
DP
SP
Oscilloscope
y x B
Plotter
y x
Monitor
y x
$y_0 \pm \Delta y$
$x_0 \pm \Delta x$
Scan generator
EF
MS
ID
Signal processing
m_i
Spectrometer control
Time base
const.
Energy filter control
const.
Specimen volt. contr.
b

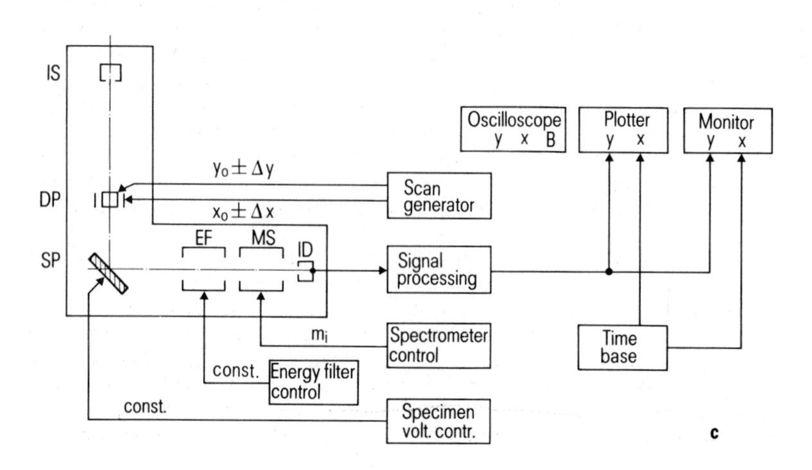
IS
DP
SP
Oscilloscope
y x B
Plotter
y x
Monitor
y x
$y_0 \pm \Delta y$
$x_0 \pm \Delta x$
Scan generator
EF
MS
ID
Signal processing
m_i
Spectrometer control
Time base
const.
Energy filter control
const.
Specimen volt. contr.
c

or magnetic field strength), it can be ensured that ions of different mass can pass the exit aperture in succession and be recorded. The plot of the ion intensities as a function of the relevant operating parameter (which can be assigned a mass scale) is known as the *mass spectrum*.

The quality of a mass spectrometer is characterized by two parameters: the *mass resolution* $m/\Delta m$ is the reciprocal value of the relative mass difference between two ions which can still be distinguished as ions of different masses. The FWHM (full width at half the maximum) value of a specific line in the mass spectrum is frequently measured for Δm. The *transmission* is understood to be the "permeability" of the equipment to ions of a certain species, i.e. the ratio of the intensity passing through the exit aperture to the intensity entering the entrance aperture (this parameter being different from the overall transmission discussed in section 7.1).

Sector field mass spectrometer

The mode of operation of sector field mass spectrometers is based on the different effects of electric and magnetic fields on moving ions. Let us initially consider an electric sector field, i.e. a capacitor with two cylindrical deflection plates (Fig. 7-5). The deflection voltage is so adjusted that the equipotential area mid-way between the electrodes has zero potential. The entrance aperture is located at O (slit normal to the drawing plane). An ion coming from this point and striking this middle region tangentially at point P is guided along it with uniform velocity. The deflecting force $F_e = qE$ acts on the ion according to eq. (2-6), where E is the field strength in the central area of the capacitor.

The deflecting force is counteracted by the equivalent centrifugal force $F_{ze} = m \cdot v^2/r_e$, where v is the ion velocity. It can, according to eq. (2-7), be described by $v = \sqrt{2(q/m)U}$ (U is the acceleration voltage). It follows from these relations that the path radius

$$r_e = m \cdot v^2/q \cdot E = 2U/E\,. \tag{7-6}$$

Fig. 7-4. Signal processing with different operating modes.
a) *Recording of a mass spectrum:* the measuring signal is amplified and processed by the signal processing unit and supplied to the *y*-deflectors of the plotter and monitor. Their *x*-deflectors are synchronized via the time base with the spectrometer control, which scans the mass spectrometer (MS) over a selected mass range (from m_i to m_{i+n}). The specimen voltage and energy filter (EF) are set to fixed optimum values; the same applies to the scan generator (unless an average is to be taken over a specific specimen surface during the measurement).
b) *Recording of a scanning image:* the scan generator guides the PI beam in a raster scan over the specimen surface to be imaged and synchronously scans the electron beam in the oscilloscope over the screen. The mass spectrometer is set to a specific mass m_i and the measurement signal is used to control the brightness B of the oscilloscope.
c) *Recording of a depth profile:* mass spectrometer, energy filter and specimen voltage are set to fixed values, as in case b). The scan generator scans the PI beam over a specific specimen surface to produce a flat crater bottom during sputtering (section 7.3.4). The signal is averaged over the crater area (or a section of it) and used to control the *y*-deflection of the plotter and monitor, whose *x*-deflections are controlled by the time base unit. The result is a representation of the intensity of ions of mass m_i as a function of the sputter time. The intensity profile of several ionic masses can be measured quasi-simultaneously by switching the setting of the spectrometer during the sputter process alternately back and forth between the relevant masses.

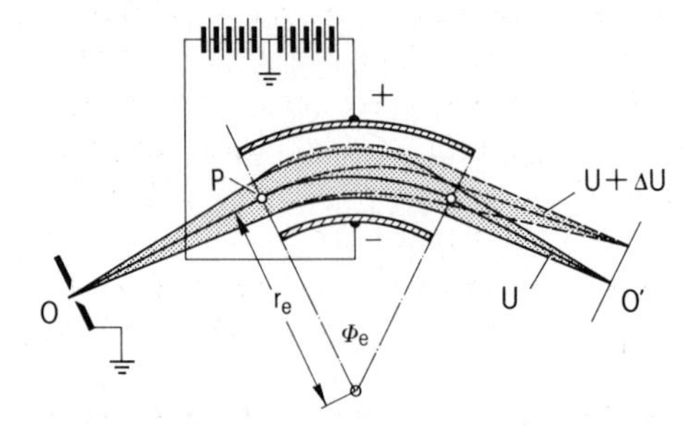

Fig. 7-5. Focusing and dispersing effect of an electric sector field [7-16]: positive ions of energy qU, which are emitted from point O as a narrow, divergent beam, are focused at point O′. The trajectory in the middle of the beam is guided along the "earthed" equipotential surface central between the electrodes. Ions with higher energy $q(U + \Delta U)$ are more weakly deflected and focused at a different point.

This applies to both positive and negative ions. Let us initially consider a beam of positive ions of energy qU coming from O. Those ions in the beam which enter the capacitor more towards the positive electrode must travel against the field. They therefore become slower and their path becomes more curved the closer they approach the positive electrode. In contrast, the ions entering the field more towards the negative electrode traverse a less strongly curved path. The result is a directional focusing: the ion beams emerging from the entry slit at O meet again on a line normal to the drawing plane (focal line) at O′. If negative ions are to be focused, the field polarity must be reversed.

For small energy differences $q \cdot \Delta U$, the following is obtained by differentiating eq. (7-6):

$$\Delta r_e / r_e = \Delta U / U \,. \tag{7-7}$$

Ions of deviating energy, such as $q(U + \Delta U)$, consequently traverse paths with a different curvature and are focused at a different point (hatched bundle in Fig. 7-5) from ions of energy qU. An electric sector field therefore acts as an energy filter.

Fully analogous considerations can be employed for a magnetic sector field, in which the field lines are normal to the direction of ion motion (let the space between the electrodes in Fig. 7-5 be thought of as permeated by lines of magnetic force protruding perpendicular to the drawing plane). The deflecting force is in this case described in eq. (2-10) by $F_m = q \cdot vB$. In terms of the centrifugal force $F_{zm} = m \cdot v^2/r_m$, the following are obtained in place of eqs. (7-6) and (7-7):

$$r_m = m \cdot v/q \cdot B = (1/B)\sqrt{2m \cdot U/q} \tag{7-8}$$

$$\frac{\Delta r_m}{r_m} = \frac{1}{2} \cdot \frac{\Delta m}{m} + \frac{1}{2} \cdot \frac{\Delta U}{U} \,. \tag{7-9}$$

According to these relations, a magnetic sector field disperses ions both by mass and by energy. This means that the lines in the mass spectrum are smeared due to the broad energy distribution of the ions and the mass resolution is consequently reduced to zero. An electric sector field must therefore be combined with the magnetic sector field to obtain an energy filter. It is convenient to arrange the setup such that the energy-dispersing action of the electric field exactly compensates that of the magnetic field. The ions separated on the basis of their energy by the electric sector field are then reunited by the magnetic sector field (Fig. 7-6). An arrangement of this kind is a double-focusing spectrometer, which *focuses on the basis of direction and energy* and *disperses only on the basis of mass.* But the precondition for this is that all ions must have the same charge q. For ions of different charge, the ratio m/q in equations (7-6) and (7-8) must be regarded as a variable. This means that the dispersion takes place in accordance with the "m/q value" in this case.

Combinations of electric and magnetic sector fields are double-focusing as long as the radii r_e, r_m and the angles and lengths marked in Fig. 7-6 are related to each other in a specific way (double focusing condition [7-16]). The acceptance of the spectrometer is designated as the (4-dimensional) product of beam cross-sectional area and solid angle which can be tolerated at a required mass resolution [7-14]. The beam cross section and angle can be narrowed by the entrance aperture (EN) and the angular aperture (A) respectively. The energy region ("energy window") which is to contribute to the mass spectrum may be varied by the energy slit (ES). All apertures and slits must be selected to be very narrrow if a high mass resolution $m/\Delta m$ is to be achieved. An increase in resolution is therefore obtained at the expense of the transmission and vice versa.

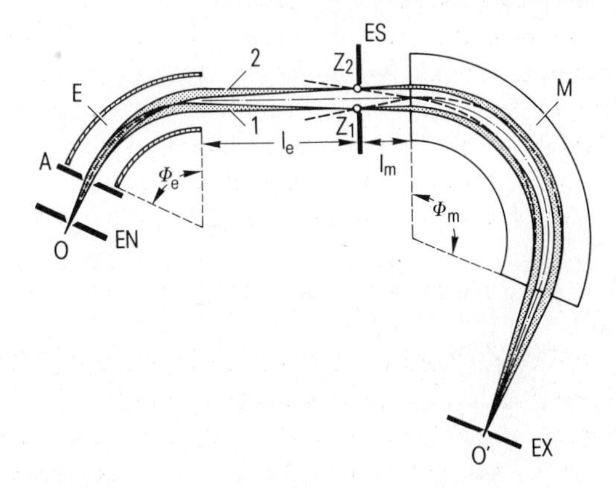

Fig. 7-6. Double-focusing mass spectrometer [7-16]: ion beams of different energies emitted from point O (shown are beams 1 and 2 with $q(U - \Delta U)$ and $q(U + \Delta U)$ respectively) are focused by the electric sector field E at different points Z_1 and Z_2. The dispersion $\overline{Z_1Z_2}$ is cancelled by the dispersing effect of the magnetic sector field M, i.e. beams of different energy (but equal mass m) are focused at a single point O′. Due to the mass dependence of the dispersion in the magnetic sector field, ions of deviating mass ($m \pm \Delta m$) are focused at other points. Consequently, this configuration disperses only by mass, but it focuses in a dual way, namely by energy and direction. EN entrance slit; ES energy slit; EX exit slit.

Quadrupole mass spectrometer

The operation parameters of *static* mass spectrometers (such as those described hitherto) remain constant with time. They are slowly changed only for recording a mass spectrum.

Dynamic mass spectrometers [7-17] are characterized by the fact that temporal variation of one of the operating parameters (e.g. the electric field) is essential for the dispersion. In a quadrupole mass spectrometer [7-18, 19], mass separation takes place by the excitation of ion oscillations in an electric quadrupole field, which consists of the sum of a static and a high-frequency component.

The field is generated by four rod electrodes with a circular (in the ideal case: hyperbolic) cross section. They are arranged and wired as shown in Fig. 7-7. A direct voltage U_a and a high-frequency alternating voltage $U_b \cos \omega t$ are applied to the electrodes. The electric field depends on x, y and t, but (if the influences of the stray fields at the end of the rods are neglected) not on z. Ions injected in the longitudinal z direction therefore travel with a constant velocity component $\dot{z}$ through the field but oscillate along their path in the x and y directions under the influence of the high-frequency field.

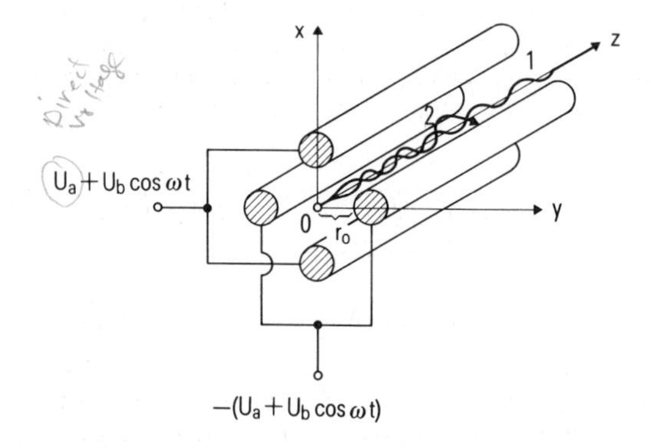

Fig. 7-7. Principle of the quadrupole mass spectrometer: the voltages $U_a + U_b \cos \omega t$ and $-(U_a + U_b \cos \omega t)$ are each applied to two opposite rod electrodes. Ions injected at point O in the z direction perform lateral oscillations in the high-frequency quadrupole field. For specified field parameters and ions of uniform charge, the shape of these oscillations depends on the mass. Ions of specific mass move along "stable" paths (1) and can leave the spectrometer in the direction of the detector. Ions of deviating mass perform instable oscillations with increasing amplitude (path 2), strike the electrodes and are thus eliminated.

Due to the polarity of the electrodes in the quadrupole, the oscillation components in the xz and yz planes have different shapes. Whether an ion can pass through the quadrupole field or not depends on these shapes. Only those ions reach the detector whose amplitudes of oscillation are limited in both directions. Their paths are said to be "stable" (1 in Fig. 7-7). The other ions move along "unstable" paths (2): their oscillation amplitudes increase quickly with time and these ions strike the electrodes or housing walls and are eliminated.

With the field parameters (U_a, U_b, ω, r_0) specified, the stability problem is determined only by the specific charge q/m of the ions or (if only ions of uniform charge are considered) by the ion mass. To understand this, let us initially assume that the ac amplitude $U_b = 0$. Positive ions are then accelerated towards the axis in the xz plane due to the repelling action of the positive electrodes, i.e. they are focused. In the yz plane, they are attracted by the negative voltage at the electrodes, i.e. they are defocused.

If an alternating voltage is superimposed onto the direct voltage, nothing is significantly changed as long as $U_b < U_a$ (curve 2 in Fig. 7-8). However, as soon as $U_b > U_a$, there are

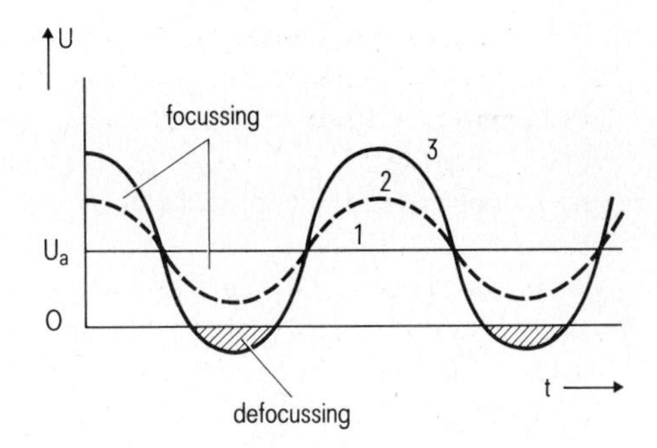

Fig. 7-8. Effect of dc and ac voltage components on positive ions in the *xz* plane of the quadrupole field. The diagram shows the time slope of the electric potential $U = U_a + U_b \cos \omega t$ at the two electrodes located opposite each other in the *xz* plane and electrically short-circuited. As long as these are at positive potential, they have a focusing effect on positive ions. This is the case for $0 \leqq U_b < U_a$ (curves 1 and 2). For $U_b > U_a$ (curve 3) the potential at the electrodes is negative at times (hatched regions). In these phases, the electrodes have a defocusing effect on positive ions.

phases in which the action of the steady field becomes "overcompensated" by the ac component, i.e. the *xz* plane contains "defocusing phases" (hatched areas in Fig. 7-8) and the *yz* plane contains "focusing phases".

The effect of all this on the ion paths depends on the ion mass: because of their smaller inertia, light ions can follow the alternating field more easily than heavy ions. They are therefore preferentially defocused in the *xz* plane, whereas the steady field gains the upper hand in focusing the heavy ions. The *xz* component of the quadrupole field therefore acts as a high-pass mass filter. The situation in the *yz* plane is exactly reversed: whereas the heavy ions are affected by the defocusing action of the steady field, the path of the light ions is stabilized by the alternating-field component. The action of the *yz* component of the quadrupole field therefore corresponds to a low-pass mass filter. Both directions together yield a filter with a specific bandwidth for ions whose masses allow them to traverse stable paths in *both* planes and therefore to pass the filter. The band-width depends on the ratio U_a/U_b. This may be represented by the "stability diagram" (Fig. 7-9), in which the following parameters are plotted

$$a = 8q \cdot U_a/(m \cdot r_0^2 \cdot \omega^2) \tag{7-10a}$$

$$b = 4q \cdot U_b/(m \cdot r_0^2 \cdot \omega^2)\,. \tag{7-10b}$$

These parameters play a role in the solution of the equations of motion for the ions in the quadrupole field [7-18] and contain all the relevant operating parameters and ion properties. Every point in the *a, b* diagram is therefore an "operating point". Those operating points which correspond to stable paths in both the *xz* and *yz* planes lie within the trilateral region marked in Fig. 7-9.

Let us consider a field specified by U_a, U_b, ω, r_0 and ions of uniform charge q. Then all ions of the same mass have the same operating point (a, b). Since according to eq. (7-10a, b) the ratio $a/b = 2U_a/U_b$ is independent of the mass, ions of different masses are located on a straight line $a = 2(U_a/U_b)\,b$ passing through the origin of the stability diagram. The sec-

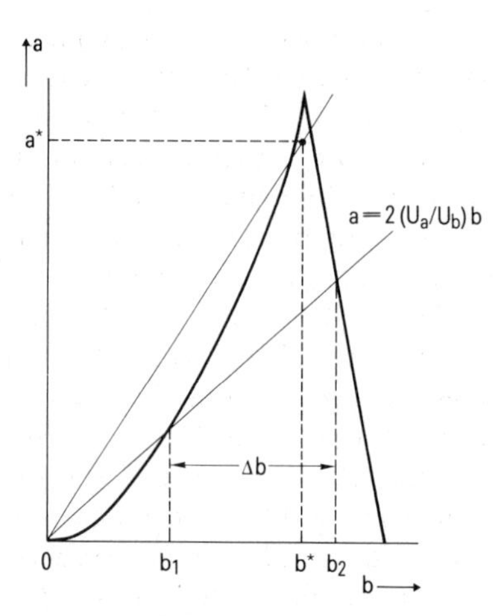

Fig. 7-9. Stability diagram for a quadrupole mass spectrometer (schematic) [7-18]: the parameters $a = 8q \cdot U_a/(m \cdot r_0^2 \cdot \omega^2)$ and $b = 4q \cdot U_b/(m \cdot r_0^2 \cdot \omega^2)$ are plotted. The operating points (a,b) which lead to stable paths lie within the framed, trilateral region. For given field parameters and ions of uniform charge, the operating points for ions of different mass lie on a straight line $a = 2\,(U_a/U_b)\,b$. This determines the stability interval $\Delta b = b_2 - b_1$ (and thus a mass interval Δm) for ions which can pass the quadrupole on stable paths. The larger U_a/U_b is set, the steeper is the "mass transit line" and the narrower the stability interval. At the operating point (a^*, b^*) only ions of a specific mass m^* are let through. By proportionally changing U_a and U_b (i.e. U_a/U_b = const.), a mass spectrum can be recorded; by changing the ratio U_a/U_b, the resolution $(m/\Delta m)$ and the transmission can be varied.

tion of this straight line lying within the stability region characterizes the stability interval $\Delta b = b_2 - b_1$, from which the bandwidth, i.e. the stable mass region $|\Delta m| = m_2 - m_1$ can be determined in line with equation (7-10b). A steeper slope is obtained by increasing the voltage ratio U_a/U_b. The stability interval is thus narrowed until finally, at the operating point a^*, b^*, only ions of a specific mass (more exactly: ions from a very narrow mass window) can traverse the field in a stable manner.

The relationship between a selected ion mass m^* and the field parameters follows from equation (7-10a, b):

$$m^* = \frac{8q \cdot U_a}{a^* \cdot r_0^2 \cdot \omega^2} = \frac{4q \cdot U_b}{b^* \cdot r_0^2 \cdot \omega^2}\,. \tag{7-11}$$

Accordingly, a mass spectrum can be recorded when the voltages U_a and U_b are changed simultaneously and proportionally. In principle the mass setting can also be varied by changing the frequency ω. But this method is not used in practice, because it would be too elaborate.

The narrowing of the stable mass region Δm signifies a high mass resolution $m/\Delta m$, but also a low transmission. Both parameters may be changed within wide limits by changing the ratio U_a/U_b (i.e. the slope of the straight line in Fig. 7-9). In the case of low mass resolution ($m/\Delta m < 80$), a spectrometer transmission of 100% may be attained; the best value for the mass resolution of a quadrupole mass spectrometer (of 1m length!) published in the literature is around 1500 [7-18].

As has been shown, the stability character of the ion paths depends only on the position of the operating point in the *a, b* plane, but not on the initial conditions. However, care must be taken that the oscillation amplitudes of the stable paths remain smaller than the distance r_0 of the electrodes from the spectrometer axis. Otherwise, ions from the stable mass region reach the electrodes and the transmission is reduced. The entrance points x_0, y_0 and the transverse components $\dot{x}_0$, $\dot{y}_0$ of the entrance velocity must not therefore exceed certain limit values (the "acceptance" region [7-14]). In addition, the ions in the quadrupole must experience a sufficiently large number of oscillations so that the unstable paths are sufficiently "shaken up" and the relevant ions reliably eliminated. For a specified spectrometer length l and frequency ν, this condition sets an upper limit to the entrance energy. Typical values are: $l = 25$ cm; $r_0 = 8$ mm; $\nu = \omega/2\pi = 1.5$ MHz; maximum entrance energy 30 eV [7-18, 20].

For the reasons mentioned above, a quadrupole mass spectrometer must have an energy filter preconnected, although its mass resolution is, in principle, not (as in the magnetic sector field) dependent on the entrance energy. The SIs emitted from the specimen are, as explained in the previous section, accelerated by an extraction voltage of several hundred Volts to the entrance aperture of the energy filter. A capacitor with plane [7-21] or cylindrical [7-22] electrodes, which guides ions from a sufficiently narrow energy range (energy window) towards the mass spectrometer, may be used as an energy filter. Before entering the quadrupole field, these ions are retarded again and focused electrostatically, so that they fulfil the initial conditions for the quadrupole mass spectrometer. A configuration of this kind, in which the ions are initially accelerated and then decelerated again, is also called an accel-decel system [7-23].

For the sake of a simple presentation, we have considered ions of uniform charge and therefore assumed q in the equations (7-10a, b) to be constant. But these equations can be so rewritten that (m/q) appears in the denominator. It is then clear that the considerations with respect to path stability, resolution and transmission apply just as much to ions of different charge as to ions of uniform charge, provided that the mass m is in all cases replaced by (m/q).

Comparison of mass spectrometers

Of the two mass spectrometers described, the quadrupole is simpler in design and operation. When switching over from one mass line to another, only electrical parameters are changed. Since no hysteresis effects occur, as in the magnetic sector field, rapid switching back and forth between widely spaced mass lines is possible. This is of importance in the quasi-simultaneous measurement of concentration profiles of different elements, for example (section 7.3.4).

The most significant difference between the two types of equipment relates to their mass resolution and their transmission. Because of its double-focusing property, the system consisting of an electric and a magnetic sector field is in this respect superior to the quadrupole: a higher mass resolution can be attained for equal transmission and a greater transmission for equal mass resolution. This also applies to the overall transmission (referring not to the intensity of ions entering the spectrometer but to the intensity of the relevant ionic species emitted from the specimen), which is important for SIMS. To achieve this, however, optimal emittance-acceptance matching by the SI optics (section 7.2.1) must be ensured [7-8, 14].

The quadrupole mass spectrometers of practical use for SIMS are usually designed for a resolution of about 500. This is perfectly sufficient for separating all elements and isotopes, since even the main isotope of the element uranium with mass number $M = 238$ only requires a resolution of $M/\Delta M = m/\Delta m = 238$ to separate it from the isotope with mass number 237.

However, problems arise due to molecular ions, whose mass lines may superimpose on those of atomic ions with nominally equal mass numbers. They differ from atomic ions merely in the mass defect, which is due to differences in the binding energy (see section 7.3.3) but represents only a very small fraction of the mass difference of adjacent isotopes. If it is necessary to separate such superimposed mass lines by mass spectrometry, a very much higher resolution of several 1000 to 10,000 is required, and this can be attained with a double-focusing mass spectrometer. Quadrupole devices can only be used in such cases in conjunction with the method of energy filtering described in section 7.3.2.

7.2.3 Ion microprobe

Ion microprobes according to the principle presented in Fig. 7-1a were first implemented by Liebl with a double-focusing mass spectrometer [7-24] and by Wittmaack with a quadrupole mass spectrometer [7-25]. The first commercial equipments were IMMA (Ion Microprobe Mass Analyser) from Applied Research Laboratories ARL [7-26] and a-DIDA (Dynamic In-Depth Analyser) from Atomika GmbH [7-23]. The best established ion microprobes on the market are at present those produced by Perkin Elmer/Atomika GmbH, ISA/Riber S.A., Vacuum Generators Scientific Ltd. and more recently the Cameca IMS 4F (see section 7.2.5). Besides these, a series of laboratory instruments has also been developed. Review papers on ion probes can be found in [2-41, 7-27,7].

As an example of this type of instrument, the ATOMIKA 6500 ion microprobe [7-28] which is in use in the authors' laboratories will be described below. The arrangement of the main components is shown in Fig. 7-10. Worth noting are the two PI columns, one of them equipped with a plasma ion source for gas ions (Ar^+, O_2^+, O^-), the other with a surface ionization source for generating Cs^+ ions. Both ion columns can be kept simultaneously in stand-by mode, so that a rapid changeover from one type of ion to another is possible. Their axes are positioned normal to each other and intersect on the object plane; the ion-optical configuration of both PI columns is identical, one of them is shown in Fig. 7-11.

Primary ion column

The ion sources are described in section 2.5.3. Depending on the polarity of the ions to be emitted, a positive or negative high voltage ($U_0 = 0.5 \ldots 15$ kV) is applied to the sources. The (virtual) crossover in the ion source (1 in Fig. 7-11) is demagnified in two stages and imaged onto the specimen surface. An electrostatic lens with cylindrical electrodes is used as the beam forming lens (2), its first electrode acting as the extraction electrode and its second electrode accelerating the ions to mass potential. The intermediate image of the crossover (at 4) generated by this arrangement is further demagnified by an aperture lens (final lens 7) and imaged onto the specimen (8). The diameter of the irradiated spot is the probe diameter (Table 2-3) and determines the spatial resolution attainable in the analysis.

Beam "purification" is performed by a Wien filter (3 and Fig. 7-3), and the neutral particles are suppressed by deflecting the PI beam with the deflector (5) by 2° (this deflection is not

a

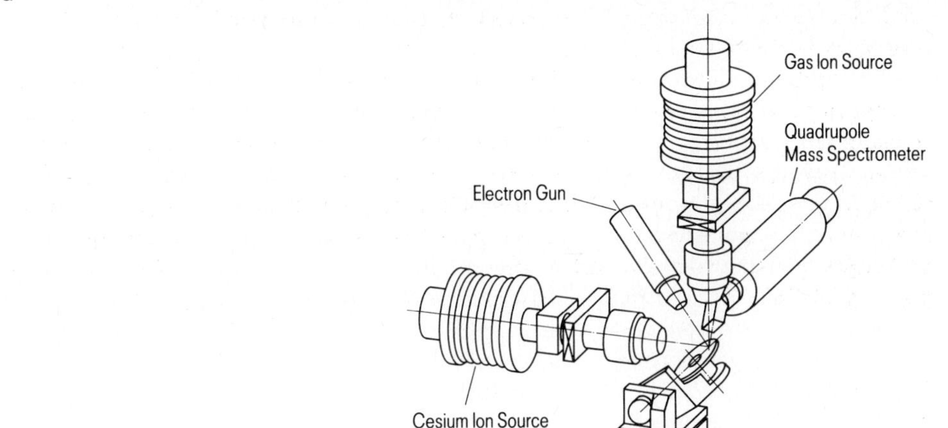

b

Fig. 7-10. Secondary ion mass spectrometer ATOMIKA 6500 (ion microprobe): a) view; b) arrangement of the main components. (Courtesy of Perkin Elmer/Atomika GmbH.)

shown in Fig. 7-11). Peripheral rays, which would increase the probe diameter due to spherical aberration, can be eliminated with the interchangeable primary beam aperture (4) in the intermediate image plane of the crossover. An aperture of 10 μm diameter permits a probe diameter of 1 μm to be implemented (acceleration voltage $U_0 = 15$ kV, probe current $I_0 < 10^{-9}$ A). The beam can be positioned onto the specimen with the aid of the deflector (6) and scanned over an area of up to 500 μm × 500 μm.

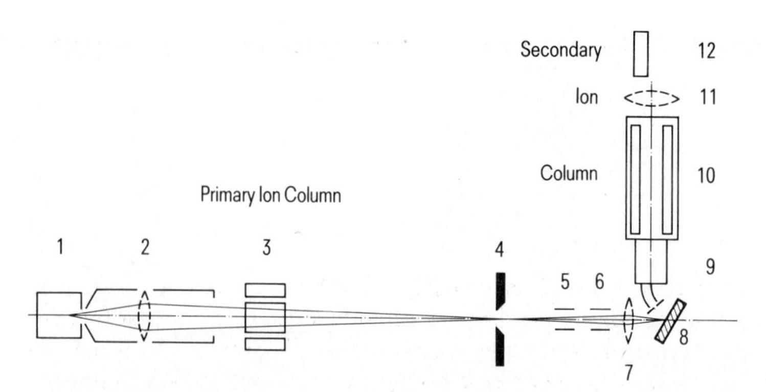

Fig. 7-11. Secondary ion mass spectrometer ATOMIKA 6500; ion-optical configuration: 1 ion source (plasma or cesium source); 2 beam forming lens; 3 primary ion mass filter (Wien filter); 4 primary beam aperture; 5 deflection plates for elimination of neutrals; 6 primary beam deflector (scanning unit); 7 f.inal lens; 8 specimen; 9 acceleration-deceleration system with energy filter; 10 quadrupole mass spectrometer; 11 detector optics; 12 channeltron SI detector. (Courtesy of Perkin Elmer/Atomika GmbH.)

Secondary ion detection system

The ATOMIKA 6500 Ion Microprobe is, just like its predecessors, equipped with a quadrupole mass spectrometer. An acceleration-deceleration system [7-21, 22] with a cylinder capacitor (9) as an energy filter is used for transferring the SIs to the quadrupole. An extraction voltage of about 100 V applied between the specimen and the capacitor entrance accelerates the SIs to the energy filter. The SIs which pass through the sector field are retarded again and focused before entering the quadrupole spectrometer (section 7.2.2).

A channeltron multiplier (12) is used as a detector. The ions emerging from the quadrupole are transferred to the channeltron by the detector optics (11) so as to obtain maximum amplification. Simultaneously, electrons as well as ions scattered or sputtered at the quadrupole rods are largely eliminated from the SI beam in order to improve the signal-to-noise ratio. The signal processing components (Fig. 7-4) as well as the electronics for the PI column and the mass spectrometer are controlled by a DEC PDP 11/73 computer. Special software (SIMS Data System SDS 800 [7-29]) ensures high operator convenience and allows rapid changeover between the various operating modes of the device.

7.2.4 Ion microscope

Ion microscopes operating on the principle shown in Fig. 7-1b are based on work by Castaing and Slodzian [7-30, 8]. The first commercial instrument of this type was the IMS 300 ion microscope by Cameca S.A. [7-31], followed by the CAMECA IMS 3F [7-32] and 4F [7-33] ion microanalyzers. The latter is an upgraded version of the IMS 3F and contains elements of a microprobe. These elements will be treated in section 7.2.5. The ion microscopic arrangements described below are the same for both instruments, the main components being those shown in Fig. 7-2.

Primary ion column

On the primary side, the apparatus (Fig. 7-12b) is equipped with two *ion sources*: a duoplasmatron (1) is used to generate positive or negative gas ions such as Ar^+, O_2^+ and O^-, and a surface ionization source (2) supplies Cs^+ ions (section 2.5.3). By reversing the polarity of a magnetic prism (3), these sources can be alternately selected. The prism is a magnetic sector field and acts simultaneously as a PI mass filter (section 7.2.1).

Three electrostatic lenses, whose optical axis forms an angle of 30° with that of the SI column, are provided for *beam forming*. The first two lenses (4) act as condenser lenses; the third lens (final lens 7) focuses the PI beam onto the specimen (8). The diameter of the irradiated spot may be varied between 1 μm (even less with the IMS 4F) and 200 μm and depends on the PI species, the excitation of the lenses and the selection of the primary beam aperture (5). Several electrostatic deflectors and a stigmator, not shown in Fig. 7-12b, are used for aligning and optimizing the beam. The double deflector (6) is used to position the beam onto the specimen and to scan it, if required, over an object field of up to 500 μm × 500 μm or to deflect it into a Faraday cup for measuring the probe current.

Primary ion impact energy

Since the specimen in the ion microscope is not at ground potential, the energy E_0 with which the PIs strike the specimen is not identical to that with which they are accelerated from the ion source.

The acceleration voltage is applied between the ion source and the extraction electrode. Since the latter is always at ground potential, the potential U_{is} at the ion source must be positive when the charge of the PIs is positive, and vice-versa. Exactly the same applies to the specimen potential U_s with respect to the polarity of the SIs. These are accelerated right from the specimen surface with the aid of an electrode at ground potential. The spacing between the specimen and the acceleration electrode is only a few millimeters. This means that the PI beam must pass through a side aperture in this electrode in order to reach the specimen surface (Fig. 7-13). The PIs must therefore traverse the SI acceleration field and are retarded or accelerated by it depending on the sign of their own charge and the field polarity. Also, they are deflected from their straight-line path and strike the specimen at an angle ψ, which deviates from 30° (the angle between the optical axes of the primary and the secondary columns). Care must therefore be taken, with the aid of the deflection system (6 in Fig. 7-12b), to ensure that the PI beam is nevertheless correctly positioned.

The acceleration voltage may be regulated in steps between $|U_{is}| = 5$ kV and 20 kV. The specimen voltage has a value of $|U_s| = 4.5$ kV, which may be varied only within narrow limits (e.g. for voltage adjustment to counteract the charging of insulating specimens). Values for the case of $|U_{is}| = 10$ kV and $|U_s| = 4.5$ kV are shown in Table 7-1 for clarification.

CAMECA

a

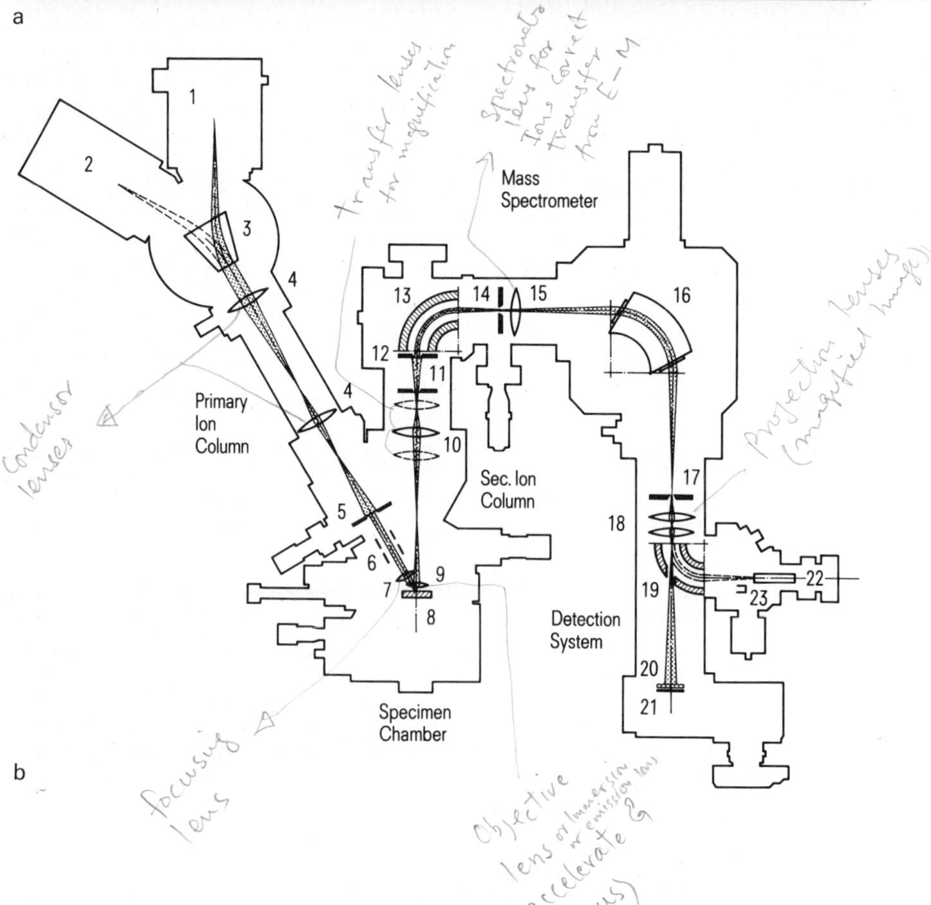
1
2
3
4
Primary
Ion
Column
5
6
7
8
9
Specimen
Chamber
10
11
12
13
14
15
Sec. Ion
Column
Mass
Spectrometer
16
17
18
19
20
21
22
23
Detection
System

b

Secondary ion column

The *objective lens* (9 in Fig. 7-12b) is designed to both accelerate and focus the emitted SIs so that a real, magnified image of the specimen surface is produced [7-8]. Such a lens is called an *emission lens* (or, by a somewhat inadequate analogy to optics, an *immersion lens*). The acceleration and magnification fields of the lens can be considered separately. The acceleration takes place in a homogeneous electric field between the object (which simultaneously acts as an electrode of the emission lens) and the acceleration electrode (Fig. 7-13). Directly adjacent to it is a magnifying field which is set up in the apertures of the lens electrodes.

The image formed by the emission lens is further magnified by one of three *transfer lenses* (10 in Fig. 7-12b) and passed through the entrance slit (11) into the mass analyzer. The functions of the transfer lenses and the apertures located in the SI beam path are explained later in connection with the optimization of the instrument parameters.

The *analyzer* is a double-focusing mass spectrometer of the kind described in section 7.2.2 and shown in Fig. 7-6. Located between the electric and magnetic sector fields are the energy slit (14 in Fig. 7-12b) and (unlike Fig. 7-6) an electrostatic spectrometer lens (15). This lens transfers the ions correctly from the electrical to the magnetic part of the spectrometer across a space which is inherent in the design of the apparatus. The resulting spectrum of mass lines is imaged in the plane of the exit slit (17).

Detection system

Two projection lenses (18 in Fig. 7-12b) can be used to produce a magnified image of the exit slit in the final image plane; an open slit allows a section of the mass spectrum to be observed as a series of lines. The projection lenses may also be adjusted in such a manner that they project, with the aid of the beams which have passed the spectrometer exit slit, a real image of the object in the final image plane. Suitable setting of the spectrometer and of the slit will thus produce an image of the specimen using only ions of a specific mass.

An image converter is fitted into the final image plane to record the image. It consists of a channel plate (20) and a fluorescent screen (21). The ions generate avalanches of secondary electrons in the microscopically fine plate channels, each of which acts as an electron multiplier. These electrons are accelerated onto the fluorescent screen and excite it to emit light. The fluorescent screen may be observed by visual inspection or by using a TV camera, or the image may be photographed.

With the aid of an electrostatic sector field (19) located between the projection lenses and the image converter, the SI beam can be deflected by 90°. The ions then strike a horizontally

Fig. 7-12. Secondary ion mass spectrometer of Cameca (ion microscope). a) view of the CAMECA IMS 4F; b) ion-optical configuration of both the IMS 3F and 4F version:1 duoplasmatron ion source; 2 cesium ion source; 3 magnetic prism (also primary ion mass filter); 4 beam forming lenses; 5 primary beam aperture; 6 primary beam deflector (scanning unit); 7 final lens; 8 specimen; 9 objective lens (emission lens); 10 transfer lenses; 11 spectrometer entrance slit; 12 field aperture; 13 electrostatic sector-field analyzer; 14 energy slit; 15 spectrometer lens; 16 magnetic sector-field analyzer; 17 spectrometer exit slit; 18 projection lenses; 19 electric deflection field; 20 channel plate; 21 fluorescence screen; 22 electron multiplier; 23 Faraday cup. (Courtesy of Cameca S.A.)

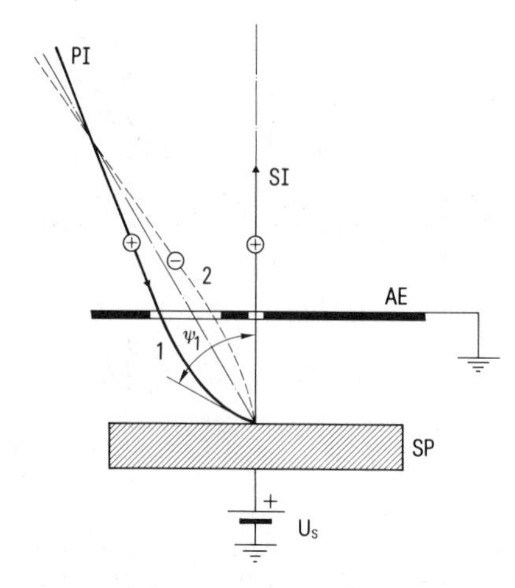

Fig. 7-13.
Influence of the secondary ion (SI) acceleration field in the emission lens between specimen (SP) and acceleration electrode (AE) on the paths of the primary ions (PI). The diagram shows cases 1: positive PIs and positive SIs (continuous lines) and 2: negative PIs (broken line) and positive SIs. The impact angles ψ are different in the two cases.

Table 7-1. Impact energy $E_0 = q \cdot |U_{is} - U_s|$ and impact angle ψ of singly-charged primary ions as a function of the primary ion and secondary ion polarity.

PI polarity	+	+	−	−
SI polarity	+	−	+	−
ion source voltage U_{is}/kV	+10	+10	−10	−10
specimen voltage U_s/kV	+ 4.5	− 4.5	+ 4.5	− 4.5
PI impact energy E_0/keV	5.5	14.5	14.5	5.5
PI impact angle ψ	>30°	<30°	<30°	>30°

arranged electron multiplier (22) or, alternatively, a retractable Faraday cup (23). When the sector field is deactivated, the ions pass through a hole in the external sector field electrode to the channel plate and images may be recorded. The signal electronics, which amplify and process the received signal, fulfil the requirements outlined in section 7.2.1. A computer controls the interaction of the components used for signal processing as well as the various equipment functions.

Optimization of instrument parameters

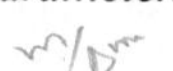

The spatial resolution, mass resolution and sensitivity must be matched to very different requirements. The beam path between the specimen (SP) and the field aperture (FA) is shown in Fig. 7-14 to explain the steps which may be taken to achieve this. The object section of radius $\overline{A_0B_0}$ is magnified in two steps by the emission lens and a transfer lens (images I′ and I″) and imaged onto the plane of the field aperture. At the same time, an image CO″ of the crossover CO′ is produced in the plane of the spectrometer entrance slit (EN). This ensures that at a given aperture (slit width) a beam of maximum intensity enters the spectrometer.

The size of the object field imaged in the opening of the FA (diameter 1.8 mm) depends on the magnification of the transfer optics. The three possible lens combinations (emission

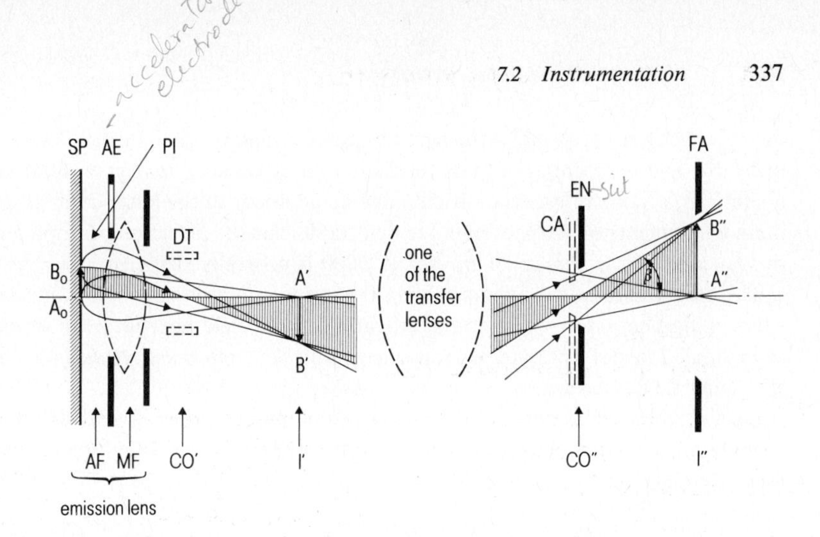

Fig. 7-14. Path of rays between specimen (SP) and field aperture (FA). The emission lens produces a real, magnified image I' with radius $\overline{A'B'}$ of a specimen section with radius $\overline{A_0B_0}$. This image is then further magnified by one of the three transfer lenses. An image I'' of radius $\overline{A''B''}$ is formed in the plane of the field aperture. An image CO'' of the crossover CO' is formed in the plane of the spectrometer entrance slit EN. AE, AF and MF represent the acceleration electrode, acceleration field and magnification field of the emission lens respectively; CA is the contrast aperture (exchangeable); DT are the dynamic transfer system deflection plates (only for Cameca IMS 4F; see section 7.2.5).

lens and one of three transfer lenses) correspond to object field diameters ($2\overline{A_0B_0}$) of 400 μm, 150 μm and 25 μm. At constant overall transmission, the signal intensity would decrease with the object field area, but this loss of intensity is compensated for by the increase in collection angle as the object field area decreases. This means that the transfer optics can be used to optimize the collection efficiency, and thus (for a given mass resolution) the overall transmission, for three analyzed object regions of different sizes.

The *spatial resolution* of the image is determined by the spherical and chromatic aberrations of the objective lens (see section 2.2.2). These two aberrations are large because the SIs are emitted into a large solid angle and with widely differing energies between 0 and several hundred eV. Peripheral beams can be eliminated by using a contrast aperture (CA) close to the crossover plane CO'' (where the spectrometer entry slit is located). This relates to ions emitted at large angles and with high initial energy (i.e. with a large lateral component of their initial velocity). The spatial resolution may be further improved by narrowing the energy slit (see Figs. 7-6 and 7-12) so that ions which leave the device at a small angle to the optical axis but with high energy are also extracted. Both options must be used to optimize the spatial resolution; values down to ≤ 1 μm are then attained.

For *microfield analysis*, a section of the image (I'') produced by the transfer optics can be selected. This is done by using an aperture changing device with field apertures (FA) of different size. Under favorable conditions, this technique permits analyses of object fields down to about 1.5 μm diameter (smallest field aperture at highest magnification). But the analysis of larger fields (such as for depth profiling) may likewise make it necessary to narrow the imaged object field (usually from 150 μm to 60 μm diameter) with the aid of a field aperture to eliminate peripheral beams from the sputter crater (see section 7.3.4). The size of the selected field aperture simultaneously defines the angle of entry into the spectrometer (2β) and thus, like the spectrometer entry slit, influences the *mass resolution* (see section 7.2.2). In the case of open slits and apertures, the mass resolution has a value of $m/\Delta m = 600$; under optimum conditions, a value in excess of 10000 may be attained. In the latter case, of course, only the main components of the specimen are detectable. For the detection of doping concentrations in semiconductors, a value of $m/\Delta m \leqq 5000$ is generally sufficient.

The measures required for optimizing the ion microscope may be summarized as follows:

For *microfield analysis:* high magnification, small field aperture; for *high spatial resolution:* small contrast aperture, narrow energy slit; for *high mass resolution:* narrow entrance slit, small field aperture.

Each of these measures (and particularly the interaction of several of them) reduces the overall transmission and thus the *sensitivity.* A compromise between these features must therefore be sought for every analysis.

7.2.5 Comparison of probe and microscope arrangements

It is difficult to compare different versions of these types of instrument because they differ in many of their components and properties. They will have differing advantages and disadvantages depending on the desired function and mode of operation in any given case. Only some of the aspects which are important for microanalysis will be treated below. Others, which are related to the different types of mass spectrometers, have already been discussed in section 7.2.2.

Overall transmission

It follows from the discussions in section 7.1 that the overall transmission β_i is the critical instrumental parameter for effective SI collection. If we assume modern equipment to have detectors with the same efficiency, their overall transmission is determined by the mass spectrometer and the transfer optics. The instruments described in the two previous sections differ quite significantly with respect to these components.

The double-focusing mass spectrometer of the CAMECA ion microscope has, for a given mass resolution, the higher transmission (section 7.2.2). This instrument also shows the greater efficiency of SI transfer from specimen to mass spectrometer. This is due firstly to the relatively high field strength ($U_s/a = 4.5$ kV/5 mm $\approx 10^4$ V/cm) in the acceleration space of the emission lens (Figs. 7-13 and 14), in which the SIs emitted from an object point (such as A_0) into the half-space 2π are formed into a narrow beam. Secondly, its SI optics are flexible enough to match the object field to be analyzed (with radius $\overline{A_0B_0}$) to the acceptance of the mass spectrometer (solid angle with aperture 2β), which depends on the required mass resolution.

The quadrupole-based ion microprobe (Fig. 7-11) does not have this feature. The extraction field strength at the specimen is low (about 50 V/cm) and the SI transfer from the specimen to the spectrometer is therefore less effective. On the whole, a significantly smaller overall transmission and thus (with the same PI species and impact angle used) a lower useful yield is obtained than with the ion microscope. In a comparative study, useful yields of boron ($^{11}B^+$) in silicon were measured with O_2 ion bombardment in different instruments [7-34]:

with quadrupole-based ion microprobes, the best value was $3 \cdot 10^{-4}$ (Atomika a-DIDA), whereas ion microscopes attained a useful yield of $6 \cdot 10^{-3}$ (Cameca IMS 3F).

Spatial resolution

The spatial resolution of an ion microscope is given by the aberrations of the SI optics (section 7.2.4). The dominant contributions to this come from the spherical and chromatic aberrations of the acceleration field of the emission lens (field strength U_s/a). They limit the spatial resolution to

$$\delta = \frac{U_{ex}}{U_s/a} \tag{7-12}$$

where U_{ex} is the volt equivalent of the exit energy qU_{ex}. Due to its effect on the PI beam (Fig. 7-13), the electric field strength cannot be made significantly greater than $U_s/a \approx 10^4$ V/cm. For this reason, the energy bandwidth ($0 \ldots qU_{ex}$) for the ions passed through to the detector must be narrowed down, i.e. the transmission must be reduced if the value of δ is to be improved. On the other hand, the transmission required for an analysis determines the attainable spatial resolution. In line with eq. (7-12), with $qU_{ex} \leqq 1$ eV, it is thus possible to realize a resolution of $\delta < 1$ µm at the expense of the detection sensitivity [7-9].

In the latter case, a better approach is to collect as many SIs as possible independently of their energy, i.e. to work under high transmission conditions, and to implement the required spatial resolution by focusing the PI beam to $d_0 < 1$ µm [7-9, 24]. Here, the ion probe is superior to the ion microscope, but only if the overall transmission is equally good for both. This is the case for the Cameca IMS 4 F secondary ion mass spectrometer [7-33], which can be operated both in microscope and microprobe modes. On the secondary side, it is equipped with the same lenses as its predecessor type IMS 3 F, but contains a novel cesium microbeam ion source (section 2.5.3) and a dynamic SI transfer system [7-8, 35].

The microbeam source allows a probe diameter down to $d_0 = 0.1$ µm to be attained (Table 2-3). The dynamic transfer system has the following function: in probe operation, the image (e.g. I″ in Fig. 7-14) is not generated simultaneously but sequentially, i.e. the image point scans across the image plane in line with the PI beam scanning process. It follows from Fig. 7-14 that only ions from a limited object field of radius $\overline{A_0B_0}$ are optimally transferred to the acceptance angle (aperture 2β) of the mass spectrometer. The transmission therefore decreases dramatically when the PI beam strikes points outside this region.

A deflection system (DT in Fig. 7-14) located in the crossover plane CO′ and synchronized with the PI scanning system deflects all beams so that their subsequent path corresponds to that of ions emitted from an axial point, i.e. they strike the image planes I′ and I″ at A′ and A″ respectively (instead of at B′ or B″, for instance). In this mode of operation, the spectrometer receives even the SIs from off-axis object points in the same way as near-axis beams. The result is thus a high transmission which is constant over the entire scanned object field and permits the spatial resolution given by the PI probe diameter ($\leqq 0.5$ µm with the Cs microbeam source) to be exploited.

In the ion microprobe with a quadrupole mass spectrometer, the transmission does not depend on the beam position to the same degree; a dynamic readjustment of the kind necessary for transfer optics with an emission lens is therefore not required. But the transmission is (for a given mass resolution, see above) in principle lower. This places limits on exploiting spatial resolutions in the submicron range. Hitherto a spatial resolution of 1.0 μm, given by the beam diameter, has been attained with the ATOMIKA 6500 microprobe with oxygen bombardment [7-28].

Specimen manipulation, crater edge and memory effects

In microanalysis, the size of the analyzed area must be distinguished from the spatial resolution: as a rule, only a section of the scanned area is utilized for the analysis (by means of gating techniques; see section 7.3.4). This is necessary in depth profiling due to the unavoidable crater edge effects. The size of the analyzed (gated) area is related to the spatial resolution only to the extent that the latter must be significantly smaller than the former.

The advantage of the quadrupole-based ion probe is its simpler ion optical arrangement and thus simpler operation. The specimen does not form an integral part of the ion optics, therefore it is less restricted with regard to its shape, is easier to manipulate, and the PI beam is not influenced by a high electric field at the specimen surface. Impact energy and impact angle may be selected independently of each other. In contrast, the experimental possibilities of an ion microscope are more restricted in this respect as can be seen from Fig. 7-13 and Table 7-1.

On the other hand, using PIs and SIs of the same polarity (first and last column of Table 7-1) a high impact angle ψ may be realized in ion microscopes together with very low impact energies, e.g. $E_0 \leqq 1$ keV, using ion source voltages $|U_{is}| \leqq 5.5$ kV. This is a prerequisite for obtaining an optimum depth resolution (section 2.6.6). In the case of low PI energy, ion microprobes suffer from a corresponding increase in beam diameter. This results in very poor lateral resolution and increased crater edge effects in depth profiling. In contrast, for the ion microscope the resolution is given by the SI optics and the results are therefore mainly independent of the PI beam quality. Consequently, depth profiles taken at low PI energy may show less background in the case of ion microscopes than in the case of ion microprobes (see different gating techniques, section 7.3.4).

The more open arrangement of components inside the specimen chamber of quadrupole-based ion probes is more compatible with the requirement of very high vacuum. Also contamination of the components by sputter products is reduced and therefore there is an accompanying reduction in memory effects (which could considerably impair the detection limit).

7.3 Measurement technique

7.3.1 Specimen treatment

Preparation

As a rule, specimens for SIMS analysis require no special preparation (for an exception see Fig 9-8). Repeated cleaning in an ultrasonic bath with ultra-clean solvents is recommended if the surface adsorbates or contaminants are not to be analyzed. The surfaces must be as smooth and even as possible. This is particularly important when the ion microscope is used for analysis, since the specimen constitutes an integral part of the ion optics in this instrument and an uneven surface would disturb the homogeneity of the SI acceleration field (section 7.2.4). Roughness and contaminants always produce shadow effects and non-uniform sputtering, and lead to deterioration of the depth resolution in depth profiling (section 2.6.6).

A range of specimen holders is available for specimens of different shapes and sizes; in special cases suitable holders can be readily constructed by the operators. If there is no alternative, the specimen must be broken up or sawn into smaller pieces. As in all other manipulations, extreme care must be taken in such cases to ensure that the surface to be analyzed is not damaged or contaminated; the same rules apply, with respect to care, as for Auger electron spectrometry (section 6.3.3).

Handling of insulators

Whereas metallic specimens and doped semiconductors possess a sufficiently high conductivity, insulators (such as glass, thick oxide layers or semi-insulating gallium arsenide) charge up during analysis. The magnitude and sign of the charge depends on the charge balance on the specimen, i.e. on the polarity and current intensity of the striking and emitted charge carriers (PI, SI, SE). An equilibrium potential which can deviate by up to several hundred volts from the voltage applied to the specimen holder is established on the specimen surface.

The yield of SIs is, in general, very low ($Y^{\pm} \ll 1$) and therefore has a negligible effect on the charge balance on the specimen. In contrast, the SE yield is $\delta > 1$, i.e. several SEs are emitted per PI. The direction of the SI extraction field on the specimen surface also plays a part in this process [7-36]: where negative SIs are to be accelerated (cases b and d in Table 7-2), the field polarity is such that the SEs are also extracted (effective SE yield $\delta_{eff} > 1$). In the case of positive SIs (cases a and c), the SEs are retarded in the extraction field and fall back onto the specimen ($\delta_{eff} \approx 0$). Due to the higher strength of the extraction field required for ion microscopes, this effect is more marked in these than in quadrupole-based ion probes. Considered in isolation, the SE emission leads to positive charging of the specimen (cases b and d). This is increased still further by bombardment with positive PIs (case b) but reduced by negative ion bombardment (case d).

Table 7-2. Charging of insulating specimens (qualitative) as a result of primary ion (PI) bombardment and secondary electron (SE) emission for ions of different polarity (cases a–d, see text).

Case	PI polarity	Charging due to PI bombardment (1)	SI polarity	Effective SE yield δ_{eff}	Charging due to SE emission (2)	Charge resulting from (1) and (2)
a	+	pos.	+	≈ 0	≈ 0	pos.
b	+	pos.	−	> 1	pos.	strongly pos.
c	−	neg.	+	≈ 0	≈ 0	neg.
d	−	neg.	−	> 1	pos.	compensated

It can be seen from the table that only in case d it is possible to (approximately) compensate for the charging due to the PIs and SEs. A compensation of this kind can also be effected in case c by means of an auxiliary electrode (tantalum aperture on the specimen surface [7-36, 37]), since in such a case most of the SEs which fall back strike the aperture rather than the insulating specimen and so are drained off; a similar situation then occurs as in case d. By varying the experimental conditions, especially the polarity of the PIs and their energy (which influences the yield of the secondary particles) in this way, the charging of the specimen can be stabilized at a low value, but it cannot actually be reduced to zero.

A metal aperture or a conducting film (such as 50 nm of gold [7-38]) on the insulator surface will favor charge draining. In the latter case, contaminants can get onto the specimen during coating. They are transported by atomic mixing into layers close to the surface and can falsify the analysis of trace elements [7-36, 37].

Charges disturb the ion-optical imaging and the SI transfer into the mass analyzer. This latter process may be associated with an intensity loss of several orders of magnitude. Ultimately, specimen charging can also lead to electromigration of highly mobile ions (section 2.6.3) and can falsify depth profiles, e.g. of alkali metals in glass materials. Where charging cannot be avoided by the steps mentioned above, its results must be corrected or compensated. Ways of doing this will be treated in the following section in connection with the SI energy distribution.

7.3.2 Energy distribution and energy filtering

Measuring the energy distribution

The energy of an SI is composed of a kinetic part (exit energy E_{ex}), and a potential part given by the electric potential of the specimen surface (specimen voltage U_s):

$$E = E_{ex} + q \cdot U_s \, . \tag{7-13}$$

The SI energy distribution (number of recorded SIs as a function of E_{ex}) can therefore be measured by setting the energy filter to a fixed value $E = E_w$ (optimized with respect to the acceptance of the mass filter) and varying the specimen voltage U_s. According to eq. (7-13), $E_{ex} = E_w - q\,U_s$, and the specimen voltage U_s must consequently be reduced in the case of positive ions ($q > 0$) if SIs of increasing kinetic energy are to be collected (see Fig,. 2-67).

The energy window, set to $E = E_w$ with the energy filter, has a finite width ΔE_w which must be matched to the analytical requirements (mass resolution). In order to obtain a high signal, the specimen voltage should be selected such that the intensity maximum of the relevant ionic species coincides with this energy window (Fig. 7-15).

Compensating charges on insulating specimens

One consequence of charged insulating specimens can be seen from Fig. 7-15: charging is equivalent to a change ΔU_s in the specimen voltage and causes a shift of the energy distribution in the energy scale. If the charging is small, it may suffice to widen the energy window (i.e. the energy slit in the case of double-focusing spectrometers). Although this impairs the

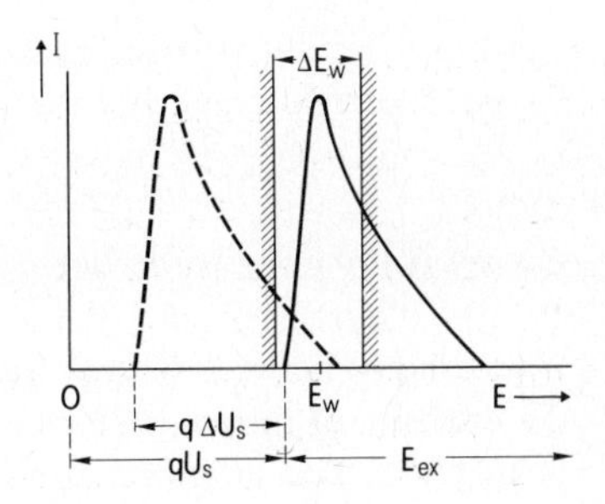

Fig. 7-15. Energy distribution of secondary ions: signal intensity I in normalized, logarithmic representation as a function of the overall energy E. The energy of the SIs is composed of the potential energy $q\,U_s$ (U_s specimen voltage) and the kinetic energy E_{ex} (exit energy). E_w characterizes the middle of the "energy window" and ΔE_w its width. Charging of the specimen is equivalent to a change ΔU_s of the specimen voltage; it causes a shift of the energy distribution by $q\Delta U_s$.

mass resolution, it has the advantage of a small relative intensity change due to the charging. With stronger charging, however, the shift of the energy distribution can be so great (broken curve) that only a fraction of the original intensity passes through the energy window. In order to correct this intensity loss, modern instruments provide the following facilities [7-9, 38]:

a) Readjustment of the energy window: by correcting E_w, the energy window can be so shifted as to allow the original intensity to be captured again. This adjustment, however, is simple only in equipment containing a double-focusing spectrometer. In that case it can be performed by simple mechanical shifting of the energy slit (section 7.2.2).

b) Readjustment of the specimen voltage: an additional voltage $(-\Delta U_s)$ is applied to the specimen to cancel the energy shift $q\,\Delta U_s$ due to charging. The disadvantage of this method is that whereas the charging takes place locally in the sputter crater the additional voltage affects the potential on the entire specimen surface and thus the potential distribution in the space in front of the specimen as well as the SI transfer. The optimum value for the specimen voltage must therefore be determined experimentally.

   In the case of depth profiling through regions of different conductivity, specimen charging can change during the measurement, e.g. at a pn-junction. Techniques of automatic potential adjustment have been developed for such cases. To monitor the charging, a reference ion must be included in the measurement in a quasi-simultaneous mode, i.e. it must be measured alternately with the ions of interest. This reference ion may, for instance, be an ion of the substrate material emitted with a suitable intensity over the entire profile depth. The specimen voltage is varied by a computer for every individual measurement until the maximum intensity of the reference ion is found. The actual measurement, e.g. of the dopant ion of interest, is then performed at this setting if the intensity maxima of the reference ion and of the dopant ion are located on (more or less) the same point of the energy scale. Otherwise, the measurement of the dopant must be performed with a constant offset voltage on the specimen, which must be defined before the measurement.

c) Simultaneous electron bombardment: with the use of positive ions (cases a,b in Table 7-2), very high positive charges may occur which can no longer be compensated for by adjusting the specimen voltage. In such cases it is helpful to irradiate the specimen simultaneously with low-energy electrons [7-39, 40]. Modern SIMS instruments are therefore equipped with a controllable auxiliary electron source. Since automatic adjustment of the electron current has not yet been realized, it is recommended that this method be combined with that described under b) to effect an accurate charge compensation.

Energy filtering for separating atomic and molecular ions

It was shown in Fig. 2-67 that the energy distributions of atomic and molecular ions differ: the intensity of molecular ions drops much more steeply after the maximum than that of atomic ions. This difference can be exploited for separating these two kinds of ions [7-41] if they have almost the same mass and the equipment or the analysis problem do not permit a mass resolution sufficient for their separation (line superposition, see section 7.3.3).

In such a case, the specimen voltage and the position of the energy window will be so selected as to ensure that essentially only atomic ions pass through the energy window (voltage offset: Fig. 7-16). This step must, however, be traded off against a lower intensity of atomic ions. In order to ensure sufficiently high detection despite this, the energy window must be opened as wide as the other analytical requirements permit. In every case, an optimum parameter setting requires careful measurement of the energy distribution of all the ionic species involved.

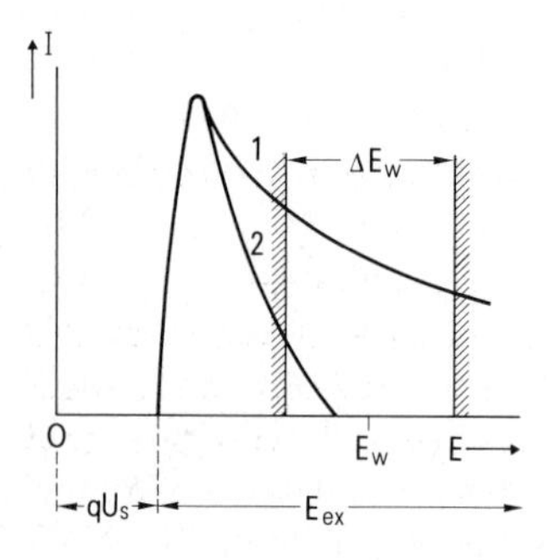

Fig. 7-16. Energy distribution of atomic ions (curve 1) and molecular ions (curve 2) of nominally equal mass. Same representation as in Fig. 7-15. The specimen voltage U_s is so selected that, essentially, only atomic ions and hardly any molecular ions pass through the energy window ΔE_w at E_w ("voltage offset").

7.3.3 Mass spectra and surface analysis

Mass number

Up to this point, the usual dimension of mass, designated by m and measured in grams, has been used for the masses of the atoms and molecules. It is, however, of advantage for practical work in mass spectroscopy to specify the particle masses as multiples of an atomic mass parameter for which the designation m_u has been introduced. The unit of this parameter is

$$1\ [m_u] = 1\,\mathrm{u} \tag{7-14}$$

(u is the abbreviation for "atomic mass unit", often designated as amu in other publications). It is defined as the twelfth part of the mass of the carbon isotope ^{12}C; its value is $1\,u \approx 1.66 \cdot 10^{-24}$ g.

This definition has the advantage that the numerical value of the mass of an atom (or molecule) measured in the unit u is, within an accuracy sufficient for many cases, an integer known as the mass number M. It corresponds in physical terms to the number of nucleons (protons plus neutrons) in the relevant atom. The following relation holds for the mass of an atom (isotope)

$$m \approx M \cdot u \,. \tag{7-15}$$

Accordingly, the carbon isotope containing 12 nucleons (6 protons and 6 neutrons) has a mass number $M = 12$. To characterize a particle more precisely, it is common to write the relevant mass number as a superscript preceding the chemical symbol, for example ^{12}C. For molecules, the mass number is equal to the sum of the mass numbers of the atoms involved, e.g. $^{72}(^{28}Si_2\ ^{16}O)$. According to eq. (7-15), the mass of this molecule is then $m_{Si_2O} \approx 72$ u.

Mass defect

Eq. (7-15) applies only as an approximation, i.e. a difference (even though a small one) does exist between the nuclear mass and the sum of the masses of the nucleons. This is due to the release of energy which occurs when nucleons combine to form an atomic nucleus in a nuclear reaction. According to the equivalence principle of the theory of relativity, an energy loss ΔE is matched by a mass loss $\Delta m = \Delta E/c^2$ (c velocity of light). This "mass defect" depends on the type and number of the nucleons concerned and is in all cases $\Delta m < 0.1$ u or, expressed as a fraction of u, $\Delta M < 0.1$.

Atomic and molecular ions can have the same mass number, e.g. ^{31}P and $^{31}(^{30}Si\,^{1}H)$. They do, however, differ in their mass defects and can therefore be separated by mass spectrometry. But the mass difference is extremely small. Thus the mass line of ^{31}P is located at $31 - \Delta M_P = 30.97376$ and that of $^{31}(SiH)$ at $31 - \Delta M_{SiH} = 30.98158$ [7-42]. In order to separate this difference of $\Delta M = \Delta M_P - \Delta M_{SiH} = 0.00782$, a mass resolution of $M/\Delta M \geqq 31/0.00782 \approx 4000$ is required, which can be attained only with a double-focusing mass spectrometer (section 7.2.2). For instruments with quadrupole spectrometers or in cases where problems of measurement (e.g. drift of the magnetic field in long-term measurements) preclude setting such a high mass resolution, the only alternative is to try the method of energy filtering described in the previous section for separating atomic and molecular ions. But in many cases (e.g. P and SiH mentioned above), the difference in the energy distributions may be too small for an effective separation.

Mass spectra

It follows from the properties of a mass spectrometer (section 7.2.2) that the SIs are influenced on the basis of their ratio of mass to charge. If the mass spectrometer is continuously tuned (see Fig. 7-4a), a mass spectrum of the kind shown in Fig. 7-17 is obtained. This figure shows

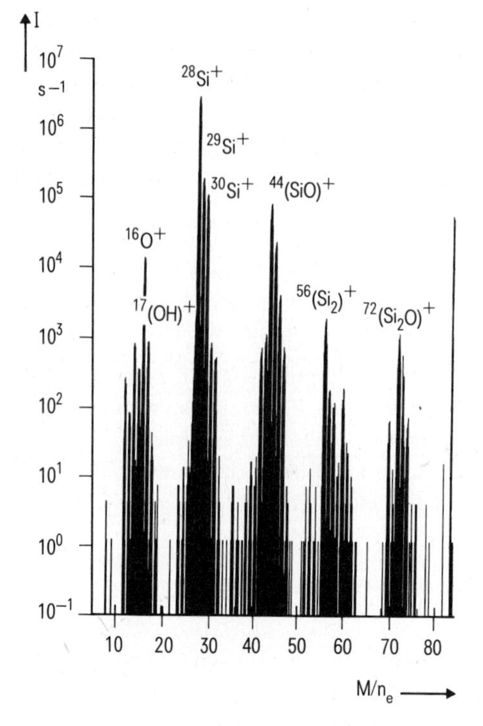

Fig. 7-17.
Mass spectrum of positive ions of ultra-pure silicon (I pulse rate, M mass number, n_e number of elemental charges). Normal incidence bombardment with scanned O_2^+ ion beam, E_0 = 12 keV, $I_0 = 10^{-8}$ A, A = 0.5 mm², t = 1600 s.

the intensities of positive SIs of ultra-pure silicon generated by oxygen bombardment. The mass-to-charge ratio is plotted in units of M/n_e, where n_e specifies the number of elemental charges. It is, however, common practice (and will also be adopted in the following) to represent mass spectra only as a function of the mass number M. Strictly speaking, the abscissa scale then applies only to singly charged ions (n_e = 1). Doubly charged ions appear at half the mass number, e.g. $^{28}Si^{2+}$ at M = 14. The same principle applies to ions of higher charge state.

Although the mass spectrum in Fig. 7-17 was recorded from very pure, undoped silicon, the spectrum obtained is extremely rich in lines. The most intense lines (peaks) are those of the three isotopes of silicon with mass numbers 28, 29 and 30; their relative heights correspond to their isotopic abundances. In addition, lines of oxygen PIs as well as those of silicon-oxygen molecules, such as $^{44}(SiO)^+$ and $^{72}(Si_2O)^+$ are found. But multiply charged ions as well as clusters comprising the constituents of the residual gas from the vacuum apparatus, e.g. $^{17}(OH)^+$, also contribute to the line spectrum.

Overlap of mass lines

The signal intensities (pulse rate in counts per second) in the mass spectrum extend over many decades from nearly 10^7 s^{-1} to less than 1 s^{-1}. A detection range from main constituents down to trace concentrations of the ppm order, in the most favorable cases even below this, can therefore be expected.

The detection limit for ions of a specific species in a given matrix is defined by the signal background, i.e. by the noise of the detector and signal electronics, by line superposition and, in some cases, by other effects (section 7.3.4). In the case of dopants in silicon, the mass lines shown in Fig. 7-17 form the signal background which is specific to this specimen under oxygen bombardment. At a mass number 11, which corresponds to the main isotope of the dopant boron, the spectrum is free of lines. A particularly low detection limit can therefore be expected for boron.

The situation is quite different for phosphorus: the line $^{31}(^{30}Si\,^{1}H)$ is already found in the silicon spectrum at mass number 31, which corresponds to ^{31}P (as already described above). A phosphorus concentration in silicon can therefore be measured only down to values at which the ^{31}P signal lies above the signal background formed by the $^{31}(SiH)$ (the hydrogen coming from the residual gas). The same applies to the dopant element arsenic, which has only one isotope with mass number 75 and is superimposed by the mass line $^{75}(^{29}Si\,^{30}Si\,^{16}O)$. Line superpositions of this kind are a serious problem for SIMS, and greatly impair unequivocal assignment of the mass lines as well as quantitative statements. The measuring techniques used to separate atomic and molecular ions (high mass resolution and energy discrimination) have already been treated.

In the literature, line superpositions are usually designated as *mass interferences.* We wish to avoid this designation since what is involved here has nothing in common with interference effects in the sense defined in physics.

Static SIMS

From the measurement data relating to Fig. 7-17 it follows that the mean current density during the analysis was $j_0 = 2 \cdot 10^{-6}$ A/cm^2. The sputter rate has a value of $\dot{z} = 6.25 \cdot 10^{-3}$ nm/s under these conditions, so that a layer of 10 nm has been sputtered *during* the measurement. This mode of operation is suitable for qualitative overview analyses of layers and for the overall analysis of homogeneous bulk material, but not for analyzing the uppermost atomic layers.

In order to obtain "quasi-static" experimental conditions, the most important thing is to reduce the PI current density by several orders of magnitude. At a value of 10^{-9} A/cm^2, the sputter rate is reduced to a mere 10^{-4} monolayers per second, giving the uppermost atom layer a "lifetime" of several hours [7-43]. This step is, however, associated with a major drop in signal intensity. To compensate for this, the irradiated area was first of all increased from its normal size of 0.5 mm^2 to about 10 mm^2 using constant current density.

The methods of "static SIMS" have been introduced and extensively studied by Benninghoven [7-43]. Under optimized conditions, it is possible to detect constituents which correspond to only 1 ppm of a monolayer. In addition, the chemical composition of the uppermost atomic layer (e.g. of organic adsorbates on metal surfaces) can be inferred from the molecular ions recorded in the mass spectrum. This is because the extremely low ion dose density required for one specimen means a high probability of every SI being emitted from a point on the specimen which has not yet been damaged by previous PIs. This method has acquired great importance in the study of surface reactions and adsorption phenomena [7-43], and in the detection and identification of organic molecules ("organic SIMS" [7-44]).

More recently, time-of-flight (TOF) mass spectrometers have been successfully used for static SIMS [7-45]. As an example, the static SIMS mass spectrum of a contaminated GaAs

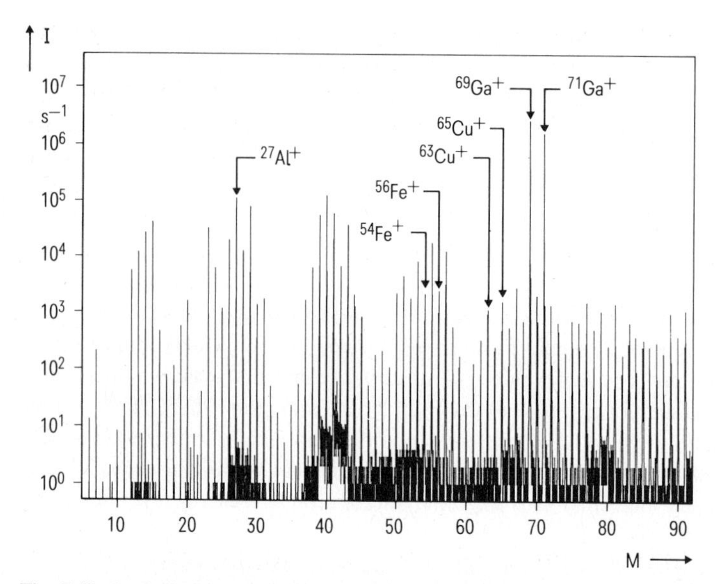

Fig. 7-18. Static SIMS mass spectrum of positive ions of a contaminated GaAs surface (low mass range) [7-46]. Primary ion bombardment: 10 keV Ar^+, dose $4 \cdot 10^8$ ions $\triangleq$ $6.4 \cdot 10^{-11}$ As; $A = 10^{-2}$ mm^2. In addition to the ^{69}Ga and ^{71}Ga lines, a large number of other lines can be seen. These belong to metallic impurities and to hydrocarbon ions, originating from organic surface contaminants.

surface is shown in Fig. 7-18 [7-46]. The main advantage of the TOF mass spectrometer is its ability to transmit ions of all masses (having the same sign of charge) quasi-simultaneously. The material consumption for the measurement of a mass spectrum is therefore less, by several orders of magnitude, than in the case of the other types of spectrometer (section 7.2.2). However, TOF mass spectrometers operate in pulse mode and the PI beam must be synchronously pulsed. As a consequence of this, the sputter removal rates are extremely low, thus making this spectrometer desirable for monolayer analysis but not for the depth analysis techniques described in the following sections.

7.3.4 Depth profiling

Dynamic SIMS

In surface analysis, especially under conditions of static SIMS, care must be taken to ensure that as little as possible of the surface to be analyzed is sputtered away during the measurement process. For depth profiling, in contrast, a sufficiently large depth must be recorded within a practicable measuring time. This mode of operation, in which a comparatively high PI current density must be applied, is known as "dynamic SIMS".

The principle of the measurement technique for depth profiling is shown in Fig. 7-4c. The result obtained is the intensity of one SI signal (or several SI signals) as a function of the sputter time; an example is shown in Fig. 7-19a. In some cases, such as when comparing similar

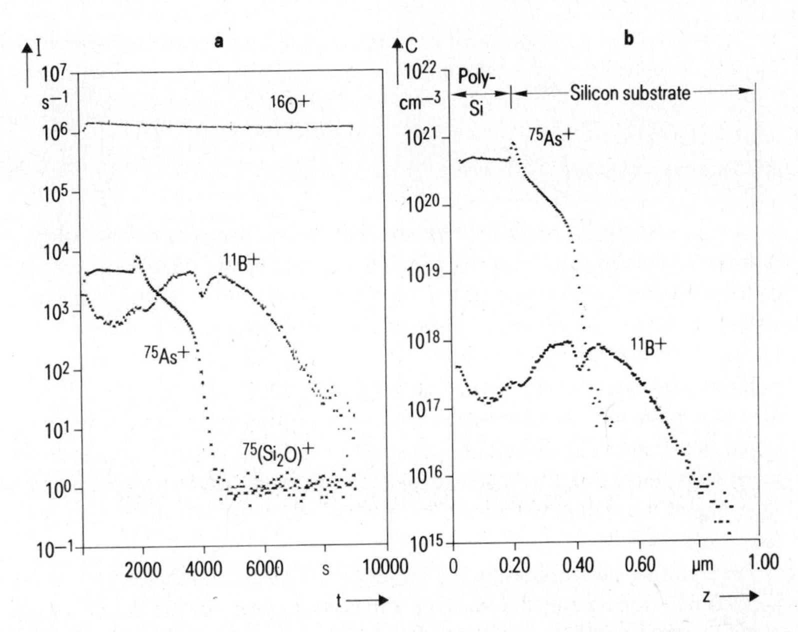

Fig. 7-19. SIMS depth profile of boron and arsenic in silicon (bipolar transistor structure with polysilicon emitter diffusion source), normal bombardment with scanned O_2^+ ion beam, E_0 = 12 keV, $I_0 = 10^{-6}$ A, scanned area A = 0.8 × 0.8 mm², gate 32% (linear), cycle time 50 s. a) pulse rate as a function of the sputter time (reference signal: oxygen; background for arsenic: Si_2O); b) same profiles after quantification: concentration as a function of the depth.

specimens, such qualitative information may be sufficient. But in most cases the result must be quantified, i.e the signal intensity must be assigned a concentration scale and the sputter time a depth scale (Fig. 7-19b). The methods of quantification are treated in section 7.4. At this point, we will only deal with the technical preconditions for depth profiling. The quality of a profile is characterized by its depth resolution and its dynamic range.

Depth resolution

The depth resolution attainable in sputter depth profiling depends on the properties of the specimen (surface roughness, inhomogeneities, structure) and on the interaction effects between the PI beam and the specimen (implantation, atomic mixing, preferential sputtering). These influences have already been discussed in section 2.6.6. A further requirement, in terms of equipment, for ensuring good depth resolution is a homogeneous PI current density across the entire surface to be analyzed. The current density in the PI beam is, however, not constant over the beam diameter, but follows a Gaussian distribution with a maximum in the center and tails of variable width at the edges. In order to generate an even crater bottom despite this shape, the PI beam is usually scanned over the specimen [7-47]. With a sufficiently large deflection (several times greater than the beam diameter), the temporal average of the current density within a central part of the sputter crater is equal for all object points. All modern units are therefore equipped with a system for scanning the PI beam, even if they are not specifically designed for scan imaging (section 7.2.1).

Raster scan technique

As a rule, scanning is performed line by line, so that a square or quadrilateral crater contour results. One sweep (e.g. from the left upper corner to the right lower corner of the crater) is called a "frame". The length of the edges, the aspect ratio and the frame time can be selected within limits; typical values are given in Fig. 7-19. After having traversed a frame, several different operating options are possible. Most usually, the PI beam "jumps" back to the starting point and starts at the next frame, i.e. begins sputtering the next layer. If several ionic species are to be measured, the mass spectrometer may be switched over to the next ionic mass in between two frames (or during one or several sacrificed frames). If the depth distribution of n elements (including the reference ion) is to be measured quasi-simultaneously, a minimum of n frames is required before one "cycle" is completed and the system switches back to the first ionic mass.

The counting electronics sum the pulses per frame and computes a temporal average (counts per second) from these; the result is a measuring point in the depth profile. Depending on the expected signal intensity, different counting times per frame can be set: it is therefore appropriate to select a significantly shorter counting time for an ion species of high pulse rate (e.g. a matrix reference ion) than for a dopant species which is present only in traces. The entire measurement process (control of the scan generator, the mass spectrometer and, if required, further components) is controlled by the computer and must be defined before starting the depth profile.

The cycle time results from the sum of the frame times and the time required for parameter switchover and for processing the measurement data. The number of measurement cycles is frequently indicated in place of the sputter time. A profile over a depth of say, 1 μm, necessitates between 100 and 300 cycles, depending on the requirements.

Dynamic range

The signal range (difference between highest and lowest measured value) which can be covered in a depth profile is designated as the dynamic range. Its upper limit is given by the nonlinearity of the amplifier arrangement at pulse rates above $10^6\ s^{-1}$. The experimental conditions (e.g. PI current, apertures) must be selected to ensure that the strongest signal (in Fig. 7-19a the reference signal, but in most cases the profile maximum) remains just within the linear recording range. The lower limit of the dynamic range is given by the detection limit (section 7.4.5), i.e. the signal background at the relevant point on the mass scale.

In the case of the ion ^{75}As, the background is formed by the molecule ion $^{75}(Si_2O)$, so that the depth profile in Fig. 7-19 could be traced only as far as to an arsenic concentration of about $5 \cdot 10^{16}\ cm^{-3}$ (Fig. 7-19b). The lowest value for the boron profile obtained in this measurement was $10^{15}\ cm^{-3}$. The discussion in the previous section showed that a very low detection limit can be expected for boron. This is, however, sometimes impaired in depth profiling, particularly by *crater edge effects* [7-47].

The crater edge is a more or less steep ramp, whose shape and slope are given by the current density distribution in the PI beam and its impact angle. Whenever the scanning PI beam approaches the edges, parts of the ramp are also sputtered away. As a result, constituents which originate from higher layers of the specimen (quite often with high dopant concentration) contribute to the measured signal and may by far exceed the signal intensity generated in the central part of the crater area.

Gating technique

To suppress these crater edge contributions, gating techniques are commonly applied. This can be done in the ion microscope with the field aperture (section 7.2.4); typical values are 250 μm × 250 μm scanned area, 60 μm diameter of the analyzed region. This method is also called *optical gating*. In ion probes, an *electronic gating* technique is commonly used instead [7-47]. It utilizes electronic methods to ensure that the counting channel in the measuring setup is opened only when the ion beam is present in an interior region (the gated area) of the scan field. The size of the gate can be selected. It is specified as the ratio of the length of the gate edge to that of the sputter crater edge.

The smaller the gate selected for a given scanning area, the greater the suppression of the edge effects and the greater the dynamic range, but the smaller the number of recorded SIs. An optimum ratio will therefore have to be sought between the crater and the gate area. This may require repeated measurements with different gate settings, which is extremely time-consuming.

Topographical data acquisition

The time required can be considerably reduced when the measured data of a depth profile are stored as a sequence of ion images and the optimum gate is afterwards determined (retrospective depth profiling). This requires a procedure for topographical data acquisition and storage as described by Rüdenauer [7-48] for *three-dimensional element analysis*. An enormous quantity of data must be handled for this purpose: thus, the recording of say, three ion species, 256 × 256 measuring points (pixels) per frame, 256 pulses per pixel and 200 measuring cycles results in a data volume of about 40 Mbyte per depth profile. A simple local correlation to the analyzed volume naturally requires a flat initial surface and a spatially constant sputter rate. This precondition is fulfilled, for example, in depth profiling of dopants in semiconductor crystals, but no longer in structured devices.

More extensive options are offered by the topographical data acquisition technique. It can be used to eliminate the signal contributions of any desired sub-areas from the depth profile when inhomogeneities (such as contaminants) are observed within the analyzed surface; this is known as *contamination blanking* [7-49]. Such inhomogeneities may be determined by observing the simultaneously recorded ion scan field or by making a computer printout of the signal contributions of individual sub-areas after the measurement. The topographical data acquisition technique may also be helpful for depth profiling of laterally structured specimens. It permits the size, shape and position of the gate to be subsequently matched to the structure to be analyzed. The requirements concerning specimen positioning are less stringent in this case, and drift effects occurring during the measurement may be subsequently corrected.

Checkerboard and spiral scan techniques

A considerable data reduction, as compared to the total data acquisition technique, is possible if the measured values are not stored separately for each of the, say, 256 × 256 pixels per frame, but summed over sub-areas. The number or size of the sub-areas must be so dimensioned that a sufficient spatial resolution is ensured. A "checkerboard" technique is described in [7-49] in which the analyzed region (the central 50% of the crater area) is subdivided into a matrix of 8 × 8 sub-areas (Fig. 7-20a). This arrangement can be used to select four concentric gates of different size between 19% and 75% of the crater edge length. Fig. 7-20b demonstrates on the basis of an example how the measurement result can be optimized in this way without repeated measurements: the optimum selection in this case is the 38% gate; it leads to the same dynamic range as the 19% gate, but yields four times as many counts, i.e. superior counting statistics. An example of contamination blanking using the checkerboard technique is given in Fig. 7-21.

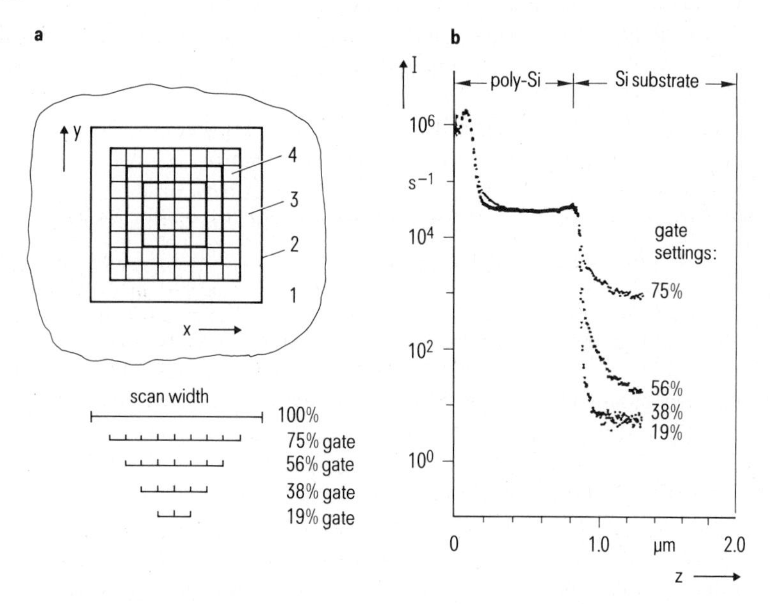

Fig. 7-20. Variable gate setting. a) topographical data acquisition by checkerboard technique [7-49]. 1 specimen surface; 2 crater edge; 3 marginal zone (no data acquisition); 4 checkerboard area (acquisition of 8 × 8 = 64 data elements per frame). b) depth profile of boron in silicon with four different gate settings obtained from a single depth profiling run. Optimum value: 38% gate.

In [7-29], a 16 × 16 checkerboard (matrix gate) as well as a spiral scan technique are presented. The latter technique (Fig. 7-22) allows an even greater data reduction: thus, if gate sizes of 10%, 20% ...80% are specified, then only eight values need to be stored per frame instead of 256 (for 16 × 16) or 64 (for 8 × 8 sub-areas). However, this method is of advantage only when concentric gates are to be selected for the scan area.

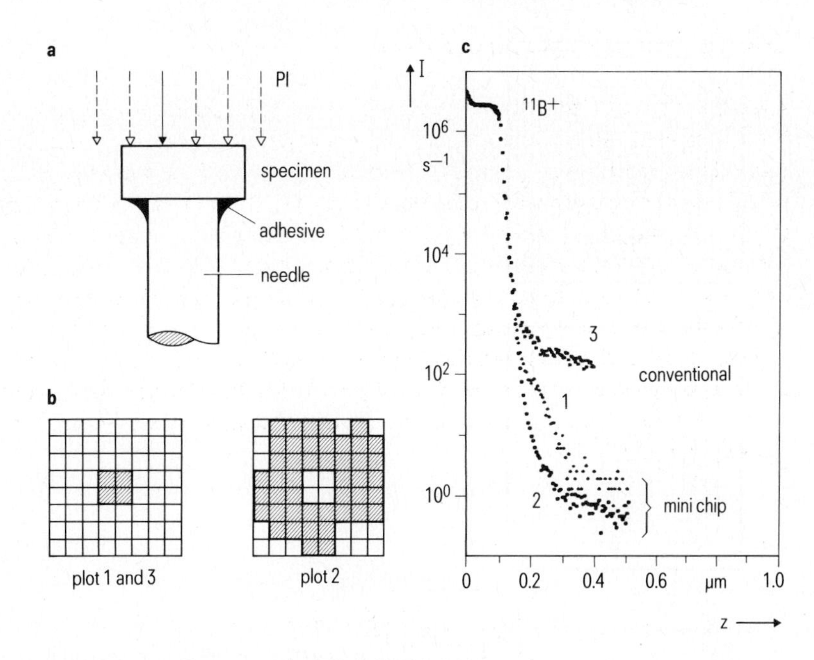

Fig. 7-21. Improvement of the dynamic range by environment-free specimen preparation (mini-chip) and contamination blanking [7-49]. a) Configuration of the mini-chip specimen on a needle acting as a specimen holder; b) gate settings (hatched sub-areas) for the plots in Fig. c; c) depth profiles of boron in silicon: curves 1 and 2 of the mini-chip (only one depth profiling run), curve 3 of a conventional specimen for comparison; curves 1 and 3 are plots with a centric 19% gate, curve 2 a plot which excludes some contamination spots. Note that the evaluated sample area is 11 times larger in case 2 than in cases 1 and 3 (hence better statistics).

Mini-chip technique

Even when selecting very small gates, contributions of the crater edge to the signal cannot be completely excluded. They could, for instance, be caused by long tails on the PI beam or by scattered PIs or impurity ions which strike the crater edge independently of the position of the PI beam and generate SIs there in an uncontrolled way. In addition, particles sputtered at the crater edge can be redeposited on the crater bottom and be sputtered again from there. The most consistent method for avoiding crater edge effects is the "mini-chip" technique [7-49]: in this, the specimen is cut so small (e.g. 0.7 mm × 0.7 mm), and mounted in such a way in the specimen chamber, that it lies completely within the area scanned by the PI beam. The entire specimen surface is then sputtered away, no crater can be formed and thus no crater edge effects can occur.

Diamond saws of the kind which are used for cutting up wafers in semiconductor production ("dicing saws") are suitable tools for cutting out the mini-chips. Specimen holders may be made from long steel needles of, say, 0.4 mm diameter, on one end of which the mini-chips are mounted, using a conducting adhesive, with the aid of a special device (Fig. 7-21a). In this way, any influences from the specimen holder can be avoided; the specimen is practically "environment-free". A result of such a technique is shown in Fig. 7-21c. It also shows the method

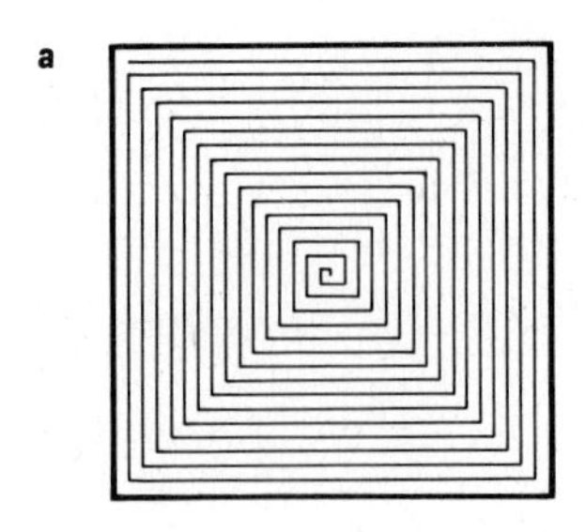

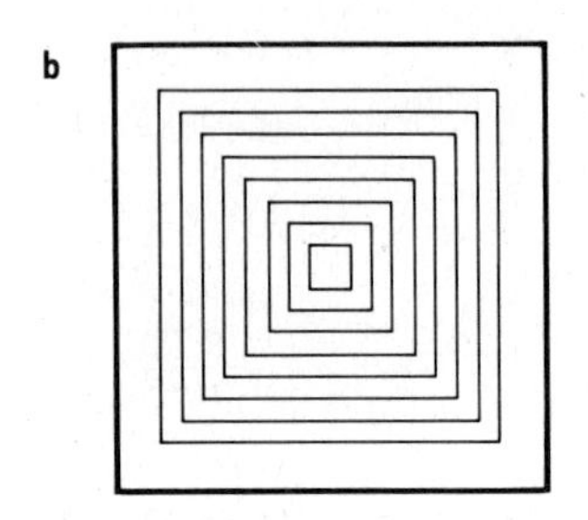

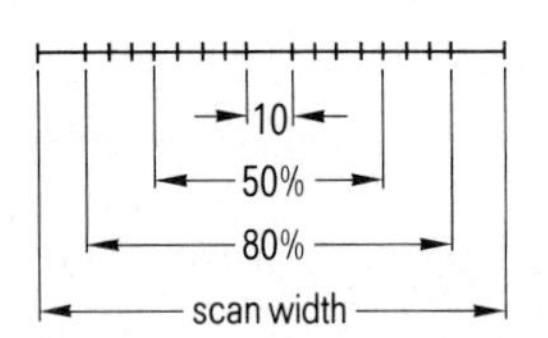

Fig. 7-22.
Variable gate setting by spiral scan technique [7-29]: a) trace of the PI beam per frame, starting in the middle of the crater; b) acquisition of eight data elements per frame at 10%, 20%, ... 80% gate width, i.e. eight different gates may be subsequently selected.

of contamination blanking described above, employed with the aim of eliminating the effects of residues on the specimen (e.g. adhering sawdust particles): for curve 2 the contents of all quadrants have been eliminated which in the low-concentration part of the profile exhibited pulse rates which were clearly in excess of the average rate. In this case, it was possible to obtain a dynamic range of 7 orders of magnitude [7-49].

7.4 Quantitative analysis

7.4.1 Theoretical models

In quantitative analysis, the problem arises of determining the mole fraction X_i or the concentration C_i of the relevant element in a multielement specimen from a measured SI intensity $I_i^{\pm}$. For monoisotopic elements, the relationship between these parameters is given by eq. (7-2), which we may write in the form

$$I_i^{\pm} = (\beta_i \cdot P_i^{\pm} \cdot Y \cdot i_0/q)\, X_i \,. \tag{7-16}$$

To obtain a quantitative evaluation of X_i, the parameters contained in parentheses in eq. (7-16) must be determined individually or as a product. Theoretical and empirical techniques have been developed for this purpose; comparative overviews can be found in [7-50, 51].

The theoretical models relate to the ionization process during sputtering. Examples are the quantum-mechanical tunnel model and the bond breaking model (section 2.6.5), which lead to eqs. (2-42a,b; 2-43) for the ionization probabilities. But none of these models is sufficiently advanced to be useful for the quantitative evaluation of measured SI intensities.

The most successful model to date is a thermodynamic model for SI emission in reactive sputtering applied by Andersen and Hinthorne [2-99]. The authors postulate a plasma-like state in the specimen volume struck by the PIs (oxygen) and assume a local thermal equilibrium (LTE) there. Under these conditions, the ratio of the sputter rates of ions and uncharged atoms (e.g. $\dot{N}_i^+/\dot{N}_i^0$) can be calculated according to the laws of thermodynamics and a semi-quantitative analysis can be performed on this basis. However, the determination of unknown parameters presupposes a knowledge of the concentrations of at least two "internal standards". The CARISMA program [2-99] was developed for the very elaborate computing procedure involved. The results of numerous analyses of metals and insulators deviate from reference values by up to a factor of 2 [2-99].

Morgan and Werner [7-52] simplified this computing procedure. In their approximation,

$$\dot{N}_i^+/\dot{N}_i^0 = K_i \cdot \exp(-E_i/kT)\,. \tag{7-17}$$

The factor K_i contains matrix parameters as well as electronic partition functions of the positive ions and the neutral atoms of the element *i*, known from the literature. E_i is the dissociation energy and k the Boltzmann constant. The temperature T is the only fitting parameter used here. It is determined by reference measurements on an element r of known mole fraction X_r.

Since the great majority of the emitted sputter particles are generally neutral atoms, the total number of particles of an element i sputtered per time interval can be approximated by $\dot{N}_i \approx \dot{N}_i^0$. In this approximation, eq. (7-17) supplies the ionization probability $P_i^+ = \dot{N}_i^+/\dot{N}_i$. Thus, with $\beta_i \approx \beta_r$ and X_i/X_r from eq. (7-16), the following is obtained

$$X_i = \frac{I_i^{\pm}/P_i^{\pm}}{I_r^{\pm}/P_r^{\pm}} \cdot X_r\,, \tag{7-18}$$

where $I_i^{\pm}$, $I_r^{\pm}$ are measured values and X_r, $P_r^{\pm}$ reference values. The QUASIMS program was developed for this evaluation, and it is shown in [7-53] that, as compared to the LTE model [2-99], this approach involves no significant loss in accuracy.

The successes of the thermodynamic model, which produces quite good results in some cases, are amazing since many experimental findings contradict a thermal character of the ionization process [7-54]. In particular, the physical significance of the temperature contained in eq. (7-17), for which values between 5000 and 15000 K are obtained, is disputed [7-55]. To this extent, the LTE model makes no contribution towards clarifying the physical nature of the ionization process. The relatively good results of thermodynamic approaches may be due to the fact that the "temperature" in eq. (7-17) is matched to the experimental results by fitting, and that the other models also show an exponential dependence of the ionization probabilities on physical parameters (cf. eqs. 2-42a, b; 2-43).

Like every other theoretical model, the thermodynamic model describes only the ionization process during sputtering. Effects of the apparatus are neglected by the approximation $\beta_i \approx \beta_r$. The overall transmission does not, however, vary only from instrument to instrument, but also, to a significant extent, from element to element. Even in analyses using the same instrument with the same settings, the theoretically determined concentrations, if not corrected for instrument effects, can contain a considerable error which is not due to the model.

7.4.2 Quantification with sensitivity factors

Empirical methods permit quantitative analyses without the use of questionable models and little-known physical parameters. Instead, calibration must be performed with the aid of standards. The success of this approach depends on a number of important conditions:

a) The composition of the standards must be sufficiently well known, either by specific preparation or by determination through other methods of analysis.
b) A sufficiently large number of standards is required; they must contain all the elements to be analyzed and cover the concentration range of these elements in the specimen (e.g. for setting up calibration curves).
c) Due to the low specimen volume which is captured in the SIMS analysis, the standards must be homogeneous down to microregions (in glass melts and monophase alloys, for example).
d) The composition of the specimen and the standards must be similar, so that matrix effects can be largely eliminated.
e) All measurements on the specimen and on the standards must be performed with the same instrument using standardized settings to exclude effects of the apparatus as far as possible.

Absolute sensitivity factors

In the simplest case, the product in the parentheses of eq. (7-16) can be regarded as an absolute sensitivity factor S_i and be determined by a reference measurement [7-50]. The mole fraction is then obtained from the simple relation

$$I_i^{\pm} = S_i \cdot X_i \, . \tag{7-19}$$

In this case, the sensitivity factors must be determined for all the elements. The calibration curves are linear only in the region of low element concentrations (e.g. < 10% in the analysis of steels), and values of 5% are attained for the relative analysis error [7-50].

Relative sensitivity factors

A somewhat greater degree of flexibility is obtained when relative sensitivity factors S_{ir} are used [7-1]. Some of the factors in the Parentheses of eq. (7-16) can be eliminated by measuring relative intensities and selecting an element contained in the specimen (conveniently a main constituent) as the reference element (r). The following intensity relationships are then obtained from eq. (7-16)

$$I_i^{\pm}/I_r^{\pm} = S_{ir} \cdot X_i^{\pm}/X_r^{\pm} \tag{7-20}$$

where $S_{ir} = \beta_i \cdot P_i^{\pm}/\beta_r \cdot P_r^{\pm}$.

If the specimen is made up of n elements (including element r), then $(n-1)$ equations of the form (7-20) are obtained. Together with the standardization condition $\Sigma X_i = 1$, n defining equations were obtained for the n unknown mole fractions X_1 to X_n.

The overall transmission and ionization probabilities are included in the relative sensitivity factors as ratios. It can therefore be expected that in each case apparatus- and matrix-dependent effects can at least be partly compensated for. Calibration curves were determined for glasses and oxides (in which the oxygen content of the matrix has a similar amplification effect on the SI yields of all elements) which were linear over several orders of magnitude. Unknown concentrations may be determined with the aid of these calibration curves with a relative error of about 10% [7-50, 7-1].

With the aid of the relation $S_{ir'} = S_{ir} \cdot S_{rr'}$, sensitivity factors can be converted from one reference element (r) to another (r′), but a higher analysis error must then be accepted [7-50].

7.4.3 Quantification of depth profiles

The preparation of standards which meet the high requirements mentioned in the previous section is costly and in many cases impossible. An extremely useful alternative is ion implantation in homogeneous matrices (e.g. substrate crystals). The dose of implanted ions can be measured, its lateral distribution is homogeneous and the depth distribution known (section 2.6.2).

This – likewise empirical – method is used above all for quantifying dopant depth profiles in semiconductors [7-56] (Fig. 7-19). In many cases, even the dose implanted in the specimen is known, which then has the character of an internal standard. In other cases, a reference specimen of the same substrate material implanted with the same dopant can be easily produced. Differences in the matrix effects are then not to be expected; but the specimen and standard must be measured at identical apparatus settings.

The dopant contents are usually specified in concentrations C_i (number of atoms per unit volume). Consequently, a relation of the form

$$I_i^{\pm} = k_i \cdot C_i \tag{7-21}$$

is sought for the quantification. A relationship between C_i and the implantation dose ϕ_i (neglecting small backscatter contributions) is given by eq. (2-32), and the following is then obtained with eq. (7-21)

$$\phi_i = \int_0^{\bar{z}} C_i(z) \cdot \mathrm{d}z = (1/k_i) \int_0^{\bar{z}} I_i^{\pm}(z) \cdot \mathrm{d}z = (1/k_i) \cdot \bar{I}_i^{\pm} \cdot \bar{z} \, . \tag{7-22}$$

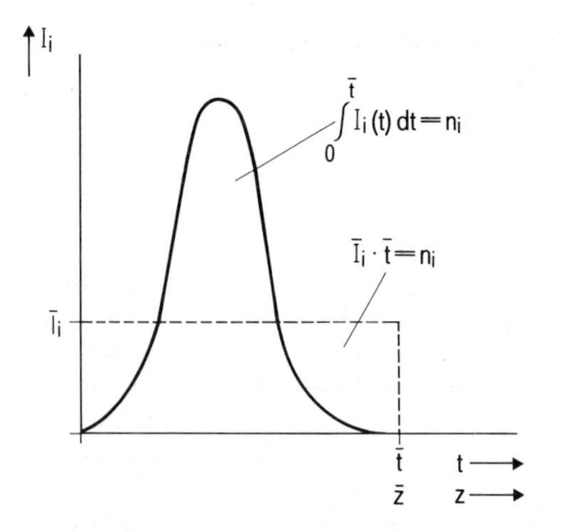

Fig. 7-23.
Quantification of depth profiles. The area under the measured curve corresponds to the total number of measured pulses and is proportional to the implantation dose. $\bar{I}_i$ is the mean value of the pulse rate over the measuring time $\bar{t}$ ($\bar{z}$ sputter crater depth).

In this relation, the depth $\bar{z}$ of the sputter crater was set instead of ∞ for the integration limit. It must, however, be greater than the depth of the entire implantation profile (Fig. 7-23), so that all particles implanted per surface element are reliably recorded. $\bar{I}_i^{\pm}$ is the intensity averaged over the crater depth, so that

$$\int_0^{\bar{t}} I_i^{\pm}(t) \cdot \mathrm{d}t = \bar{I}_i^{\pm} \cdot \bar{t} = \bar{n}_i \tag{7-23}$$

is the total number of pulses measured during the sputter time $\bar{t}$ (i.e. down to depth $\bar{z}$) (area under the curve of Fig. 7-23). It then follows from eqs. (7-22) and (7-23) that the calibration factor

$$k_i = \bar{n}_i \cdot \dot{z}/\phi_i\,, \tag{7-24}$$

where $\dot{z} = \bar{z}/\bar{t}$ represents the sputter rate, which is assumed to be constant over the crater depth (section 2.6.6).

The sputter crater depth can be subsequently measured to an accuracy of about 1% with an interference microscope or a mechanical profilometer (e.g. Talystep). Together with the uncertainty in determining the implantation dose, this produces a relative error of between 15 and 20% for the quantification.

The following relation may be obtained from eqs. (7-21) and (7-24)

$$C_{i\max} = (\phi_i/\bar{n}_i \cdot \dot{z})\, I_{i\max}^{\pm} \tag{7-25}$$

for the frequently required concentration in the distribution maximum; it is proportional to the implantation dose.

If the depth profile extends over layers of different composition, the proportions in the individual layers must be quantified separately. Only in the case of oxide layers, e.g. SiO_2 on silicon, may the same calibration factor be approximately valid for both layer and substrate. This is due to the fact that under normal bombardment with oxygen, an SiO_2 surface layer is formed on silicon. Therefore, even when profiling through SiO_2 into silicon, the SI-emitting surface is always SiO_2 and no matrix effect is encountered.

7.4.4 Comparison of quantification methods

The strong dependence of the physical parameters (sputter yield and ionization probability) on the matrix composition represents the greatest uncertainty factor in the quantitative evaluation based on equation (7-16). In contrast to this, the instrument parameters can be determined with sufficient accuracy and set repeatably (at least on the same apparatus) if sufficient effort is expended.

The theoretical models have the advantage of being independent of external standards. Their disadvantage is that the SI emission is based on different processes (section 2.6.5) so that no unified theoretical description exists. In particular, there is no experimental confirmation for the thermodynamic LTE model. Also, instrumental effects are not included, so that they require an additional correction.

The empirical methods place high demands on both the reproducibility of the instrument parameters and the quality of the standards. The latter must either be produced in a specified way, or existing specimens must be characterized by other, sufficiently sensitive methods of analysis. Both procedures are very elaborate.

The final results are, however, worth the effort: comparisons of the results of analyses obtained by theoretical and empirical methods clearly show the superiority of the latter [7-50, 51]. Empirical methods should consequently be used as long as matched standards, with the properties outlined in section 7.4.2, can be obtained. For microelectronics applications, where it is mainly a matter of analyzing dopant depth profiles, calibration with the aid of ion implantation of known dose (section 7.4.3) is the simplest and best-matched method.

7.4.5 Detection limits

Equation (7-16) enables a value for the detection limit to be estimated. By integrating over the measuring time $\bar{t}$, the number n_i of the collected ions or measured pulses is obtained on the left hand side. If we assume that all the parameters on the right hand side are constant during the measuring time, it follows from eq. (7-16), with $\tau_i^{\pm} = \beta_i \cdot P_i^{\pm}$ (useful yield), that

$$X_i = \frac{q \cdot n_i^{\pm}}{\tau_i^{\pm} \cdot Y \cdot i_0 \cdot \bar{t}} \,. \tag{7-26}$$

It was estimated in section 7.1 that at least $n_i = 100$ pulses must be counted in order to ensure that the relative standard deviation does not exceed 10%. The following values are assumed for the other parameters:

$\tau_i^{\pm} \leqq 10^{-2}$; $Y \leqq 5$; $I_0 \leqq 10^{-6}$ A; $\bar{t} \approx 30$ s. It then follows for singly charged ions that $X_{i\,\mathrm{min}} = 10^{-11}$, which is 0.01 ppb and corresponds to a concentration $C_{i\,\mathrm{min}} = 5 \cdot 10^{11}$ cm^{-3} for a dopant in silicon.

A large amount of material would be sputtered off the specimen during such an experiment. Thus, this estimation is relevant only for *bulk analysis* in the case of enough material being available for one measuring point.

For a more realistic estimation of the detection limit in the *depth profiling mode*, we prefer eq. (7-5), substituting $V = (\pi d^2/4)\Delta z$ for the analyzed volume:

$$X_i = \frac{4 n_i^{\pm}}{\tau_i^{\pm} \cdot C \cdot \pi \cdot d^2 \cdot \Delta z} \, . \tag{7-27}$$

Furthermore we assume $d = 100$ μm for the diameter of the analyzed area and $\Delta z = 10$ nm (same order of magnitude as the depth resolution) for the thickness of the layer sputtered off during one measurement. With $n_i^{\pm} = 10$ (giving a standard deviation $\varepsilon = 1/\sqrt{n} \approx 30\%$), $\tau_i^{\pm} \leqq 10^{-2}$ and $C \approx 5 \cdot 10^{22}$ cm^{-3} for silicon we obtain $X_{i\,\min} = 2 \cdot 10^{-10} = 0.2$ ppb corresponding to $C_{i\,\min} = 10^{13}$ cm^{-3}. Almost the same limits are true for GaAs.

For comparison, Table 7-3 lists a number of experimentally determined detection limits published hitherto. Some of them are much worse than the theoretically estimated values. They may suffer from line superpositions (section 7.3.3), crater edge effects (section 7.3.4), memory effects (section 7.2.5) or a constant background count rate of other origin (e.g. from detector noise). Further optimization of the instruments and methods may allow even better detection limits to be achieved.

Table 7-3. Typical SIMS detection limits in depth profiling mode.

Matrix	Element	Detected ion	Primary ion	$C_{\min}$ cm^{-3}	Ref.
Si	B	$^{11}B^+$	O_2^+	$5 \cdot 10^{13}$	[7-57]
	Al	$^{27}Al^+$	O_2^+	$2 \cdot 10^{14}$	[7-57]
	P	$^{31}P^+$	O_2^+	$5 \cdot 10^{15}$	[7-38]
	P	$^{31}P^-$	Cs^+	$5 \cdot 10^{15}$	[7-57]
	As	$^{75}As^+$	O_2^+	$2 \cdot 10^{16}$	[7-38]
	As	$^{103}AsSi^-$	Cs^+	$3 \cdot 10^{13}$	[7-57]
	Sb	$^{121}Sb^+$	O_2^+	$5 \cdot 10^{16}$	[7-38]
GaAs	Be	$^{9}Be^+$	O_2^+	$8 \cdot 10^{13}$	[7-58]
	Si	$^{28}Si^+$	O_2^+	$4 \cdot 10^{14}$	[7-58]
	Cr	$^{52}Cr^+$	O_2^+	$4 \cdot 10^{13}$	[7-58]
	Mn	$^{55}Mn^+$	O_2^+	$4 \cdot 10^{13}$	[7-58]
	Fe	$^{56}Fe^+$	O_2^+	$7 \cdot 10^{13}$	[7-58]

7.5 References

[7-1] J.A. McHugh in A.W. Czanderna (Ed.): *Methods of Surface Analysis.* Elsevier, Amsterdam, Oxford, New York 1975, 273

[7-2] A. Benninghoven et al., *Secondary Ion Mass Spectrometry, SIMS II.* Springer, Berlin, Heidelberg, New York 1979

[7-3] A. Benninghoven et al., *Secondary Ion Mass Spectrometry, SIMS III.* Springer, Berlin, Heidelberg, New York 1982
[7-4] A. Benninghoven et al., *Secondary Ion Mass Spectrometry, SIMS IV.* Springer, Berlin, Heidelberg, New York, Tokyo 1984
[7-5] A. Benninghoven et al., *Secondary Ion Mass Spectrometry, SIMS V.* Springer, Berlin, Heidelberg, New York, Tokyo 1986
[7-6] A. Benninghoven et al., *Secondary Ion Mass Spectrometry, SIMS VI.* John Wiley & Sons, Chichester, New York, Brisbane, Toronto, Singapore 1988
[7-7] A. Benninghoven, F.G. Rüdenauer, H.W. Werner, *Secondary Ion Mass Spectrometry, Basic Concepts, Instrumental Aspects, Applications and Trends.* Wiley, New York 1987
[7-8] G. Slodzian in K.F.J. Heinrich, D.E.Newbury (Eds.): *Secondary Ion Mass Spectrometry.* Nat. Bur. Stand. (U.S.), Spec. Publ. 427, 1975, 33
[7-9] G. Slodzian in A. Septier (Ed.): *Applied Charged Particle Optics, Part B.* Academic Press, New York, London, Toronto, Sydney, San Francisco 1980, 1
[7-10] R. Jayaram, *Mass Spectrometry.* Plenum Press, New York 1966
[7-11] R.L. Seliger, *J. Appl. Phys. 43,* 2352 (1972)
[7-12] K. Wittmaack, J.B. Clegg, *Appl. Phys. Lett. 37,* 285 (1980)
[7-13] H. Liebl, *Vacuum 33,* 525 (1983)
[7-14] F.G.Rüdenauer, *Int. J. Mass Spectrom. Ion Phys. 6,* 309 (1971)
[7-15] K. Wittmaack, *Rev. Sci. Instr. 47,* 157 (1976)
[7-16] C. Brunnée, H. Voshage, *Massenspektrometrie.* Karl Thiemig KG, München 1964
[7-17] E. W. Blauth, *Dynamische Massenspektrometer.* Friedr. Vieweg & Sohn, Braunschweig 1965
[7-18] W. Paul, H.P. Reinhard, U. v.Zahn, *Z. Phys. 152,* 143 (1958)
[7-19] P.H. Dawson, *Quadrupole Mass Spectrometry and its Applications.* Elsevier, Amsterdam, Oxford, New York 1976
[7-20] H. Liebl, *Nuclear Instrum. Meth. 191,* 183 (1981)
[7-21] K. Wittmaack, J. Maul, F. Schulz in R. Bakish (Ed.): *Proc. Sixth Int. Conf. Electron and Ion Beam Science and Technology.* The Electrochemical Soc., Princeton 1974, 164
[7-22] F.G. Rüdenauer, W. Steiger, V. Kraus, *Beitr. elektronenmikr. Direktabb. Oberfl. 11,* 17 (1978)
[7-23] K. Wittmaack in O. Meyer et al. (Eds.): *Ion Beam Surface Layer Analysis, Vol. 2.* Plenum Publ. Corp., New York 1976, 649
[7-24] H. Liebl, *J. Appl. Phys. 38,* 5277 (1967)
[7-25] K. Wittmaack, J. Maul, F. Schulz, *Int. J. Mass. Spectrom. Ion Phys. 11,* 23 (1973)
[7-26] T.A. Whatley, C.B. Slack, E.Davidson, *Proc. 6th Int. Conf. X-ray Optics and Microanalysis.* University of Tokyo Press 1972, 417
[7-27] K. Wittmaack, *Vacuum 32,* 65 (1982)
[7-28] H. Frenzel, J.L. Maul, H. Mertens, R.Raab, Ch. Scholze in [7-6], 219
[7-29] Ch. Scholze, H.Frenzel, J.L.Maul, *J. Vac. Sci. Technol. A5,* 1247 (1987)
[7-30] R. Castaing, G. Slodzian, *J. Microscopie 1,* 395 (1962)
[7-31] M. Lepareur, *Revue Technique Thomson-CSF 12,* 225 (1980)
[7-32] J.M. Rouberol, M. Lepareur, B. Autier, J.M. Gourgout in D.R. Beaman et al. (Eds.): *X-ray Optics and Microanalysis.* Pendell Publ. Comp., Midland, Mich., USA (1980), 322
[7-33] H.N. Migeon, C. Le Pipec, J.J. Le Goux in [7-5], 155
[7-34] J.B. Clegg et al., *Surface and Interface Analysis 6,* 162 (1984)
[7-35] H. Liebl, *Adv. Mass Spectrom. 7A,* 751 (1978)
[7-36] H.W. Werner, A.E. Morgan, *J. Appl. Phys. 47,* 1232 (1976)
[7-37] B.M.J. Smets, R.G. Gossink, *Fresenius Z. Anal. Chemie 314,* 285 (1983)
[7-38] G. Stingeder, *Dissertation,* Technische Universität Wien, 1983
[7-39] K. Wittmaack, *J. Appl. Phys. 50,* 493 (1979)
[7-40] W. Reuter, M.L. Yu, M.A. Frisch, M.B. Small, J. Appl. Phys. *51,* 850 (1980)
[7-41] K. Wittmaack, *Appl. Phys. Lett. 29,* 552 (1976)
[7-42] R.A. Burdo, G.H. Morrison, *Table of Atomic and Molecular Lines for Spark Source Mass Spectrometry of Complex Sample-Graphite Mixes.* Department of Chemistry, Cornell University, Ithaka, N.Y., USA
[7-43] A. Benninghoven, *Surface Sci. 35,* 427 (1973)
[7-44] A. Benninghoven, *J. Vac. Sci. Technol. 3,* 451 (1985)

[7-45] P. Steffens, E. Niehuis, T. Friese, D. Greifendorf, A. Benninghoven, *J. Vac. Sci. Technol. A3,* 1322 (1985)
[7-46] E. Niehuis, T. Heller, U. Jürgens, A. Benninghoven, *J. Vac. Sci. Technol., B 7,* 512 (1989)
[7-47] K. Wittmaack, *Appl. Phys. 12,* 149 (1977)
[7-48] F .G. Rüdenauer, W. Steiger, *Microchimica Acta (Wien) 1981 II,* 375
[7-49] R. v.Criegern, I. Weitzel, J. Fottner in [7-4], 308
[7-50] D.E. Newbury, *Scanning 3,* 110 (1980)
[7-51] G.H. Morrison in [7-3], 244
[7-52] A.E. Morgan, H.W. Werner, *Anal. Chemistry 48,* 699 (1976)
[7-53] A.E. Morgan, H.W. Werner, *Microchimica Acta II,* 31 (1978)
[7-54] K. Wittmaack, *Nuclear Instrum. Methods 168,* 343 (1980)
[7-55] H.W. Werner, *Vacuum 24,* 493 (1974)
[7-56] H.W. Werner, *Acta Electronica 19,* 53 (1976)
[7-57] H. Zeiniger, private communication
[7-58] J.B. Clegg in [7-3], 308
[7-59] A. Benninghoven et. al., *Secondary Ion Mass Spectrometry,* SIMS VII. John Wiley & Sons, Chichester, New York, Brisbane, Toronto, Singapore 1990

8 Electron beam testing

8.1 Principle

The particle beam methods treated in the previous chapters are used exclusively for materials analysis. They involve measuring signals which originate in the interactions between the particle beams and the specimen material. This chapter describes particle beam methods which have a completely different objective, namely to probe electrical circuits.

In the past, measurements of electrical circuits could be performed with the aid of extremely fine mechanical probes. For probing of very-large-scale integrated (VLSI) circuits, however, mechanical probes can no longer be used: interconnections with a width of 1 μm or less are destroyed by the mechanical contact, the measured results are falsified by the capacitance of the probes and circuit operation is modified.

Use of secondary electrons

A suitable alternative to mechanical probes are electron probes of the kind used in the scanning electron microscope (SEM) as well as in X-ray and Auger electron microanalysers (Chap. 3,5 and 6). The acceleration voltage and probe current can be chosen so that the device under test is neither mechanically nor electrically damaged and the measured result in no way impaired. Electron beam testing thus takes place without contact and is largely nonloading.

The measured signal must provide information about the voltages on the specimen. Particles suitable for this purpose are the secondary electrons (SE) emitted from the point of impact of the primary electrons, since the SE energy E referred to ground potential depends on the specimen voltage:

$$E = E_{ex} - eU_P . \qquad (8\text{-}1)$$

Here E_{ex} is the kinetic exit energy and $-eU_P$ describes the potential energy component given by the voltage U_P at the exit point of the SEs. It should be noted that a positive voltage corresponds to a negative potential energy due to the negative electron charge ($q = -e$).

As was shown in Fig. 2-40 (range III), SEs do not possess a uniform exit energy, their energy distribution extending over several tens of electron volts. According to eq. (8-1), a measuring point voltage differing from zero produces a shift of the entire energy distribution curve on the energy scale (Fig. 8-1). Because of the linear relationship between E and U_P expressed by eq. (8-1), this energy shift gives a direct measure of the voltage at the measuring point. However, SE detectors are not energy-dispersive systems and are therefore unable to measure the shift of the energy distribution. In order to obtain useful results in spite of this drawback, a number of procedures and devices have been developed. They will be discussed in the following sections. Review papers on electron beam testing can also be found in [8-1 to 8-7].

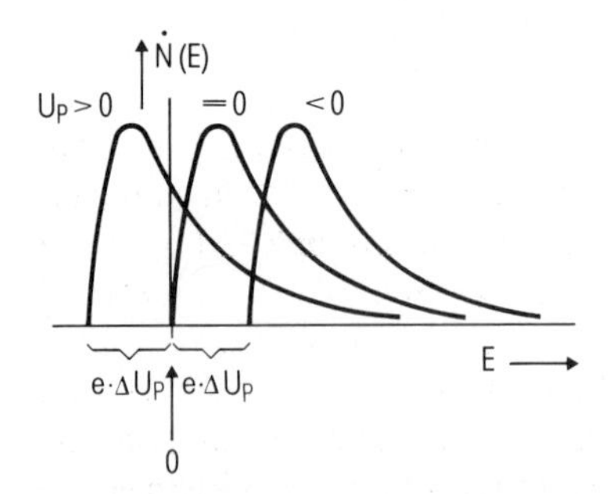

Fig. 8-1. Energy distribution of secondary electrons: $\dot{N}(E)$ is the number of SEs emitted per time and energy interval and E the energy referred to ground potential. The middle curve applies to the case of SEs starting from a grounded measuring point ($U_P = 0$). Voltages deviating from zero lead to a shift in the energy distribution. Due to the negative electron charge, the shift is to lower energies for $U_P > 0$ and higher energies for $U_P < 0$.

8.2 Modes of operation

8.2.1 Qualitative modes

As described in section 3.4, the contrast in secondary electron imaging depends essentially on the topography and material of the specimen. "Voltage contrast" is observed when regions of differing voltage are present next to each other on the surface of a specimen. Fig. 8-2 shows an example of an integrated circuit in the grounded state and after application of a dc voltage. A new kind of contrast can be clearly seen on the second micrograph. It is due to the different voltage states of the aluminium interconnections.

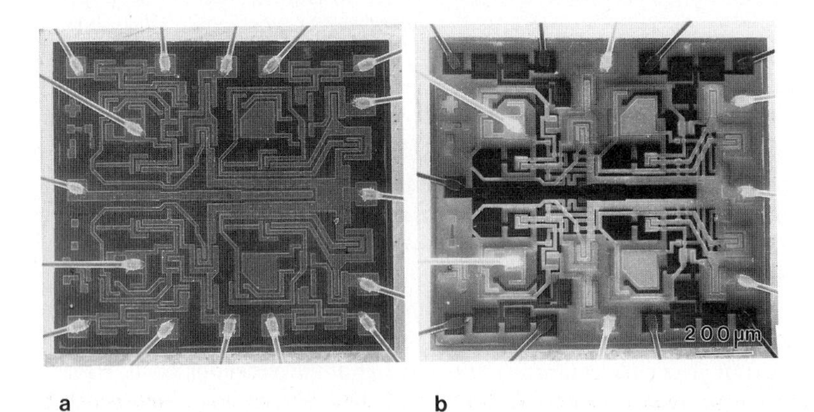

Fig. 8-2. Scanning electron microscope micrographs of an integrated logic circuit [8-6]. a) Normal SE image, all interconnections grounded. b) SE image with voltage contrast; dark interconnections: +12 V; bright interconnections: 0 V.

Voltage contrast

The phenomenon of voltage contrast will be described on the basis of Fig. 8-3, which is a schematic diagram of the specimen environment. The specimen is a silicon circuit with two interconnections, one at 0 V and the other at 10 V. An electric field, which is indicated by broken equipotential lines, is set up between the specimen and the SE detector. The boundary conditions for this field distribution are given by the grounded surroundings (specimen chamber, specimen holder, pole piece of the last lens), by a voltage of, say, +300 V at the detector and the voltage distribution on the specimen itself. The different voltages of adjacent interconnections give rise to "microfields" which influence the paths of the SE in a sensitive way.

SEs which start from a grounded interconnection (Fig. 8-3a) are accelerated immediately. The total emitted SE current (corresponding to the total area under the energy distribution curve in Fig. 8-3c) reaches the detector and controls the brightness of the CRT in the SEM. The electron probe (PE beam) in Fig. 8-3b strikes the interconnection with $U_P = +10$ V. There, the SEs encounter a local retarding field and are retarded. In this example, the retarding

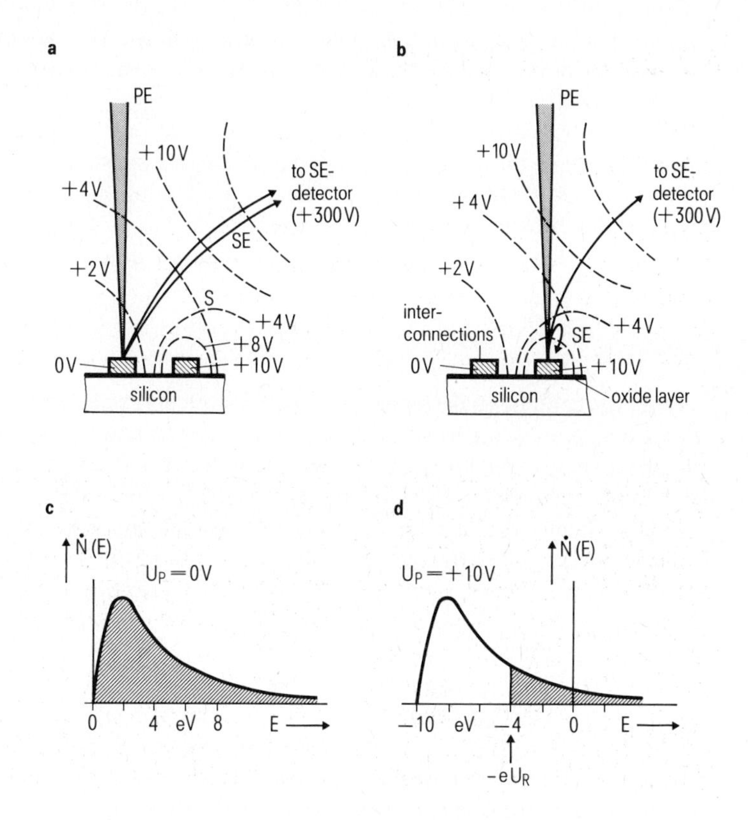

Fig. 8-3. Principle of voltage contrast imaging [8-6]. a) The PE beam strikes an interconnection with $U_P = 0$ V: all generated SEs (total area under the energy distribution curve in Fig. c) reach the detector. b) The PEs strike an interconnection with $U_P = +10$ V: the SEs are retarded at a potential threshold, only the higher-energy SEs (hatched area under the curve in Fig. d) reach the detector. U_R is the retarding voltage at the saddle point S, other designations as for Fig. 8-1.

voltage at the saddle point S (indicated in Fig. 8-3a) is $U_R = 4$ V. Only a small part of the SEs emitted with exit energies $E_{ex} \geqq -e(U_R - U_P) = 6$ eV reaches the detector, the rest falling back onto the specimen. In this case, the size of the image signal corresponds to the hatched area in Fig. 8-3d and is described by

$$I_{SE} = e \int_{-eU_R}^{50\,\mathrm{eV}} \dot{N}(E)\,\mathrm{d}E \qquad (8\text{-}2)$$

where E is obtained from eq. (8-1). The upper integration limit is selected somewhat arbitrarily: the SE energy distribution curve (Fig. 2-40) passes continuously into the energy distribution of the higher-energy backscattered electrons. Since it is impossible to distinguish between electrons of these two types, it is usual to assume a boundary at about 50 eV, at which the energy distribution curve shows a minimum.

In the voltage contrast method, therefore, interconnections with high voltage appear dark on the screen whereas those with low voltage are shown bright. This applies independently of the selection of the zero point on the energy scale, i.e. including the region of negative voltages. The microfields play a determining role in voltage contrast. They are formed by local deviations from the mean potential on the specimen surface – irrespective of whether this is positive or negative.

Voltage contrast imaging is eminently suited to investigating digital circuits, in which (for positive logic) the "high" and low" switching states are shown dark and bright respectively (section 8.2.3). But voltage contrast can also be used to investigate analog circuits. Different grey levels can be assigned to various voltage levels and as a rule potential differences of 2 V (in favorable cases down to 50 mV) can be qualitatively distinguished.

Stroboscopic voltage contrast

Every voltage contrast image represents a specific switching state. The simplest way of observing switching processes is by TV imaging. This requires the switching frequency of the device to be small compared with the TV image frequency. It is therefore limited to those circuits which can be operated at a sufficiently low frequency of a few Hz.

Dynamic circuits, such as modern memory chips or microprocessors, operate in the frequency range between a few kHz and 100 MHz. The inertia of the measuring instruments means that processes operating at such high speeds cannot be directly recorded. However, the periodicity of these processes allows individual switching states to be recorded with the aid of stroboscopic techniques [8-8].

The most commonly used method is to "illuminate" the specimen intermittently as in optical stroboscopy. This is done by periodically interrupting the PE beam and splitting it up into a sequence of short PE pulses (beam blanking, section 8.3.2). The pulse frequency (repetition rate) must be synchronized with the switching frequency of the specimen (device signal). The phase relationship between the two is fixed and remains constant, the PE beam being scanned across the specimen as in normal SEM imaging. This ensures that the SE pulses contributing to the image all originate from the same phase of the device signal.

The principle of stroboscopic voltage contrast imaging is shown in Fig. 8-4: the two micrographs in Fig. c and d show a section of an integrated circuit with an external voltage connec-

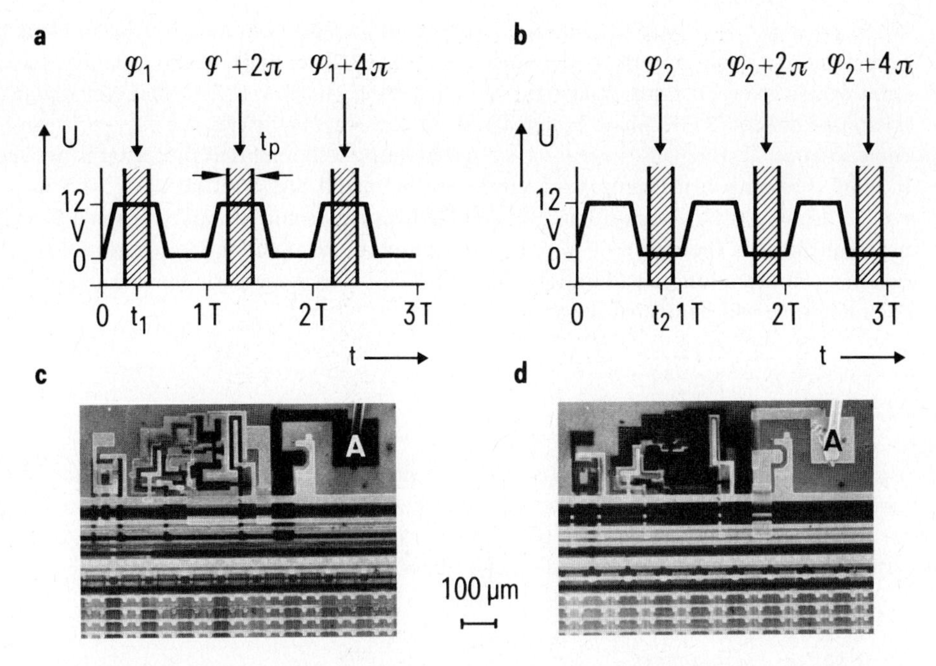

Fig. 8-4. Principle of stroboscopic voltage contrast imaging [8-6]. a, b) Voltage $U(t)$ at the voltage feed A and PE pulses (hatched bars) in the phase position φ_1 or φ_2; c, d) Associated voltage contrast patterns. T period of the device signal, t_p pulse width.

tion A. The voltage $U(t)$ at this pad is shown in Figs. a and b as a trapezoidal curve. The width and position of the hatched bars specify the duration t_p and the phase position φ of the PE pulses. The PE beam shown in Fig. 8-4a is always unblanked exactly at the same time as a voltage of +12 V is applied at A. Fig. 8-4c shows the associated switching state $U(x, y, \varphi_1)$. If the phase position of the PE pulses is shifted by half a period, then the voltage at A is $U = 0\,\text{V}$ (Fig. 8-4b) and a different brightness distribution is set up in the voltage contrast pattern according to $U(x, y, \varphi_2)$ (Fig. 8-4d).

The two stroboscopic voltage contrast images are equivalent to instantaneous micrographs which could be obtained at times $t_1 = \varphi_1(T/2\pi)$ and $t_2 = \varphi_2(T/2\pi)$ respectively, if suitably rapid and sensitive recording devices were available. Here T refers to the period of the device signal, which in the present example corresponds to the pulse interval T_p (T_p can also be an integral multiple of T, see section 8.3.1).

Circuit designers are interested in the switching states prevailing at specific times within a period of the device signal. In practice, therefore, stroboscopic voltage contrast recordings are defined in terms of time instead of phase (which would be more correct), i.e. $U(x, y, t_1)$ in place of $U(x, y, \varphi_1)$, for example. A corresponding situation exists for the methods of waveform measurement (section 8.2.2), and the modes of logic state presentation (section 8.2.3). The ratio

$$c = T_p/t_p = 1/(f_p \cdot t_p) \tag{8-3}$$

is designated the duty cycle (f_p pulse frequency or pulse repetition rate). The beam blanking system and the signal processing electronics (section 8.3.2) set limits to the repetition rate in stroboscopic voltage contrast imaging. For VLSI-testing values up to 200 MHz (for testing of microwave devices: 3 GHz) have been realized. At very low frequencies, the duty cycle can become so large, i.e. the pauses between two pulses become so long that the signal is drowned in noise (section 8.4.1). Examples of values which are still just practical are $t_p = 5$ ns and $c = 10^4$, from which it follows that $f_p = 20$ kHz. The same limits are valid for the device signal frequency f in the case of $f = f_p$ (for synchronization condition see section 8.3.1). The question of integration time (duration of the PE beam at each measuring point) and thus the recording time will be treated in section 8.4.2.

8.2.2 Quantitative modes

The brightness in the static or stroboscopic voltage contrast image provides *qualitative* information about the voltage distribution on the imaged circuit. This information is insufficient for testing integrated circuits. Such testing requires the electric potential $U(x_0, y_0)$ to be determined *quantitatively* at individual circuit nodes P (x_0, y_0).

Static voltage measurement

Voltage contrast imaging makes use of the SE current described by eq. (8-2) as the signal. However, due to the microfields at the circuit surface, this is not unambiguously dependent on the measuring point voltage U_P. In Fig. 8-3, the retarding voltage U_R at the saddle point (lower limit of integration in eq. (8-2)) is influenced by the surroundings. This means that points of equal potential but with different surroundings (e.g. width of the relevant interconnection, spacing and voltage of adjacent interconnections) give rise to different SE signals [8-5,9] and an exact voltage measurement is therefore not possible.

The quantitative modes of operation therefore require not only the energy of the SEs to be measured but also the local microfields at the circuit surface to be suppressed. The principle is shown in Fig. 8-5 on the basis of a retarding field spectrometer: a grid electrode G1 above the circuit generates a strong extraction field of 600 V/mm. This largely compensates the local retarding fields and the SEs generated at the measuring point P are accelerated to the extraction electrode G1. They are then retarded again in the retarding field RF between grids G1 and G2. The effective retarding voltage is given by the voltage U_R of grid G2 (the extraction electrode potential U_E does not enter into the energy balance). This gives a well-defined value (namely eU_R) for the lower integration limit of eq. (8-2) and together with eq. (8-1) the I_{SE} signal is unambiguously dependent on the measuring point voltage U_P. (A certain "smearing" of the spectrometer cut-off characteristic due to the inhomogeneous field distribution in the grid mesh [8-5] is neglected in this treatment.)

In Fig. 8-5 b and c, the energy distribution curves and the measuring signal curves (resulting from integrating the energy distribution curves in accordance with eq. 8-2) are shown for two cases: at $U'_P = 0$ V and $U'_R = -6$ V all SEs with exit energies $E_{ex} \geqq 6$ eV contribute to the measuring signal $I_{SE} = I'_{SE}$. If the voltage at the measuring point changes, say by $\Delta U_P = +2$ V (or the probe is transferred to another measuring point with $U''_P = +2$ V), this corre-

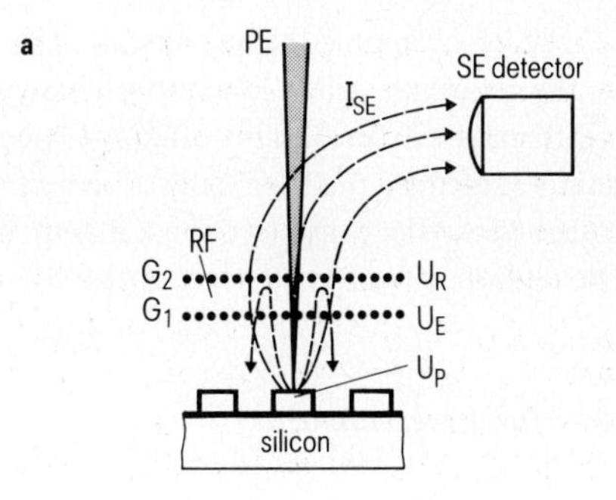

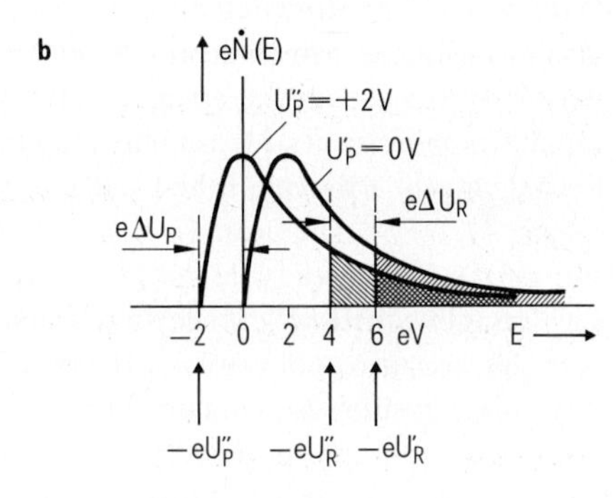

Fig. 8-5.
Principle of voltage measurement [8-6]. a) Configuration with retarding field spectrometer RF: grid G_1 extraction electrode with voltage U_E, grid G_2 retarding electrode with voltage U_R; measuring point voltage U_P. b) SE energy distribution curves (designations as Fig. 8-1) and c) integral curves I_{SE} (U_R) according to eq. (8-2). Shown are the cases $U_P' = 0$ V; $U_R' = -6$ V and $U_P'' = +2$ V; $U_R'' = -4$ V. In both cases, the detector picks up the signal $I_{SE} = I_{SE}'$ (hatched areas in b).

sponds to a shift of the energy distribution curve of about -2 eV. With an unchanged setting of the spectrometer ($U_R' = -6$ V), fewer SEs would reach the detector (checkered area in b). Because of the shape of the energy distribution curve, the signal change would not be linear. A measurement parameter which depends linearly on the measuring point voltage is obtained when the voltage at the spectrometer is subsequently controlled so that the recorded current has the same I_{SE}' value as before. This is the case when $U_R'' = -4$ V is selected. The control voltage $\Delta U_R = U_R'' - U_R' = +2$ V is equal to the parameter ΔU_P to be measured. In practice, the subsequent linearization is performed continuously with the aid of an electronic feedback loop (see Fig. 8-12) [8-5, 10].

All this, however, assumes that the change ΔU_P of the measuring point voltage occurs so that the two curves are shifted without changing the curve profile, i.e. that voltage contrast is caused by the potential barriers alone (LFE 1: type 1 local field effect). In reality, the microfields do not only produce local potential barriers but also influence the trajectories of those

SEs which overcome the barriers [8-5,11]. The angular distribution with which these SEs arrive at the retarding grid is therefore changed, and consequently also the number of electrons which pass this grid. This effect of the microfields (LFE2: type 2 local field effect) is not compensated by the high extraction voltage. It can, however, be reduced by a suitable choice of the operating point (section 8.4.1) or with the aid of a suitable configuration and shape of the retarding field grids (section 8.3.3).

Waveform measurement

With the configuration described in the previous section, voltages which change with time can also be measured. The switching frequency is, however, limited by the bandwidth of the feedback loop and usually has a maximum value of 300 kHz [8-10]. This is insufficient for testing rapid dynamic circuits. Measurement must therefore be phase-selective, similar to stroboscopic imaging: the PE beam is blanked and unblanked synchronously with the device signal frequency ($f_p = f$). The beam is, however, fixed at a measuring point and only the phase position of the PE pulses is shifted [8-2,7]. This procedure, known in electrical measurement as the "sampling method", is shown schematically in Fig. 8-6. The PE beam is unblanked exactly as the measuring point voltage traverses a specific value, e.g. 2 V (hatched bars in Fig. 8-6a). This phase position is maintained until the measured value has been determined with a sufficient signal-to-noise ratio. Only then is the phase position changed so that a new measured value is recorded, say 3 V (broken bars). If the time increments are selected to be sufficiently small, the entire voltage profile $U(\varphi)$ can gradually be measured (Fig. 8-6b). In practice, this is done continuously and the result is output as a curve on an oscilloscope or plotter. Measuring time and real time are not identical: for a period of $T = 1$ µs, the measuring time may be 30 s. For practical reasons, the time within the device signal period is mostly specified instead of the phase as the independent parameter, exactly as in stroboscopic voltage contrast imaging (section 8.2.1), i.e. $U(t)$ in place of $U(\varphi)$.

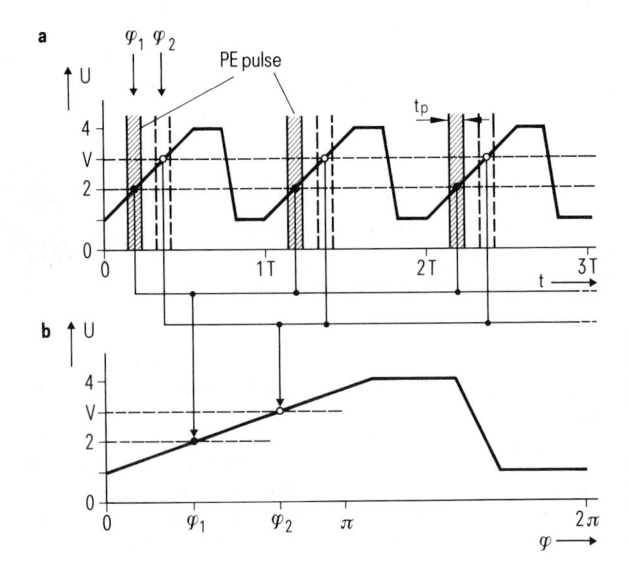

Fig. 8-6.
Principle of waveform measurement (sampling mode) [8-6]. a) Waveform $U(t)$ at the measuring point; hatched bars: phase position of the PE pulses for 2 V measurement; striped bars: for 3 V. T period, t_p pulse width, b) Measured curve.

The same limits apply to the device signal frequency as in the case of stroboscopic voltage contrast imaging, namely 20 kHz $\leqq f \leqq$ 200 MHz for VLSI testing (section 8.2.1). The attainable time resolution depends on the temporal width of the PE pulses (section 8.4.2). In conjunction with a phase modulation technique [8-10], which eliminates drift effects as well as material and topographical influences on the SE signal, potential differences down to 10 mV can still be measured using the sampling mode.

8.2.3 Modes of logic state presentation

In section 8.2.1 it was shown that the "low" state (or "0") of a circuit results in a high SE signal, and the "high" state (or "1") in a low SE signal (Fig. 8-3). In this section, methods are described which are specially suited to monitoring such "logic states" and are frequently used to test integrated circuits. They do not require an exact measurement of the SE signals. However, the measurement electronics must distinguish reliably between the two states "0" and "1" and evaluate them correctly.

Voltage coding

In voltage contrast imaging in TV mode, disordered bar patterns appear on the oscilloscope when the circuit is operated at such high frequencies that the voltage changes several times during a line scan. This effect can be exploited to analyze the circuit by synchronizing its frequency f with the TV line frequency f_l, i.e. by selecting $f = nf_l$ (n = 1, 2, 3). This method

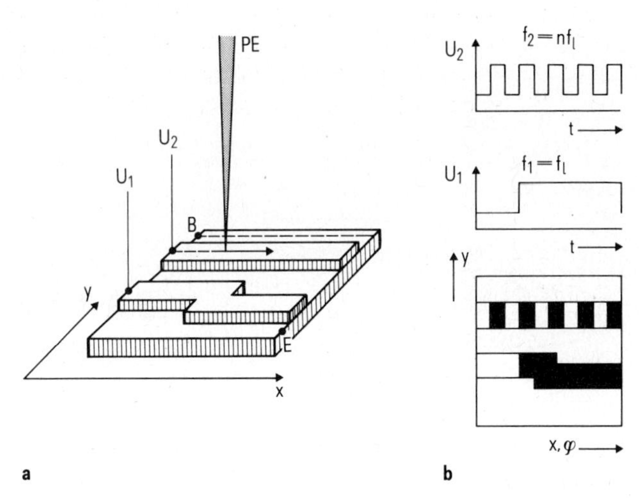

Fig. 8-7. Principle of voltage coding according to [8-7]: a) Section from an integrated circuit with two interconnections at which the pulsed voltages $U_1(t)$ with a frequency f_1 and $U_2(t)$ with a frequency $f_2 = nf_1$ are applied. The (non-pulsed) PE beam scans the circuit line-by-line in TV mode, starting at B and ending at E. The frequencies f_1 and f_2 are synchronized with the TV line frequency f_l. b) Due to the voltage contrast, a stationary bar pattern appears on the oscilloscope corresponding to the synchronization $f = nf_l$ (upper interconnection n = 5, lower one n = 1).

is known as voltage coding [8-2, 7]; its principle is shown in Fig. 8-7a. The voltage coding pattern (Fig. 8-7b) is a projection of the logic states onto the interconnections, i.e. information regarding the phase dependence of the voltage $U(\varphi)$ is superimposed onto that regarding the spatial voltage distribution $U(x, y)$. Due to the synchronization of signal and line frequency, x and φ are linked by $x/l = \varphi/(2n\pi)$ (l line length). When changing the phase relationship, the bars shift in the x direction on the oscilloscope.

The lower limit for the switching frequency is given by $n = 1$, i.e. one contrast change per line (lower interconnection in Fig. 8-7b). In this case $f = f_l \approx 15$ kHz in accordance with the prevailing TV standard (e.g. 625 lines/image; 25 images/s). Since little more than 50 bar pairs can be distinguished on each line on the screen, voltage coding is practicable only for frequencies up to $f = 50 \cdot f_l \approx 750$ kHz.

Logic state mapping

Logic state mapping is a procedure which is independent of the TV frequency and is also suitable for representing logic states of higher frequencies [8-2,7]. In this mode, the PE beam scans back and forth along only one line. It is pulsed as in stroboscopic voltage contrast imaging but the phase position of the PE pulses is shifted after every line scan. This requires a phase control which is synchronized with the line frequency. The principle is shown in Fig. 8-8. The result is a voltage contrast recording which is similar to the voltage coding image. However, the phase information is no longer superimposed by the spatial information. Fig. 8-9 shows an example of logic state mapping as well as waveform measurements on the same interconnections for comparison [8-12].

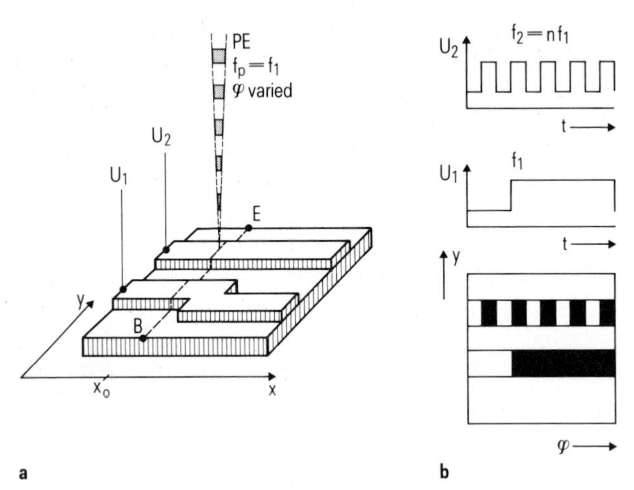

Fig. 8-8. Principle of logic-state mapping according to [8-7]. a) Circuit as in Fig. 8-7. The PE beam is pulsed and scanned back and forth along a straight line $x = x_0$ normal to the interconnections between B and E. The pulse frequency corresponds to the smallest circuit frequency to be examined ($f_p = f_l$). After every line scan, the phase φ of the PE pulse is shifted by a specific amount. b) At the oscilloscope the y shift is synchronized with the y shift of the PE beam and the x shift with the phase shift of the PE pulse. A bar pattern results which describes the variation of the voltage for every point y on the straight line BE as a function of the phase $U(\varphi)$.

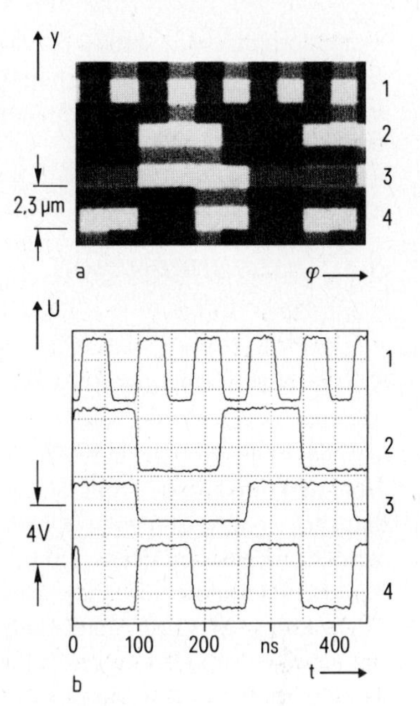

Fig. 8-9.
Electron beam voltage measurements on 1.1 µm wide interconnections of the 4 Mbit memory chip [8-12]: a) Logic state mapping (U_0 = 1 kV; $I_0 = 10^{-9}$ A; f_p = 1 MHz; t_p = 1 ns; recording time: 20 s) b) Waveform measurements (sampling mode) on one point of each of the four interconnections (U_0 = 1 kV; $I_0 = 10^{-9}$ A; f_p = 1 MHz; t_p = 1 ns; measuring time per curve: 30 s).

Timing diagram

The information obtainable through logic state mapping can also be displayed as a phase or timing diagram [8-2, 13]. In this case, the PE beam does not need to scan the entire line BE at every phase position but merely to jump from one interconnection to another. The measuring points need not all lie on a single line either. The dwell time per measuring point must be selected so that the logic states "0" or "1" can be reliably evaluated. When all measuring points have been recorded, the phase position of the PE pulses is shifted by an adjustable value and the PE beam jumps back to its starting point. This process is repeated until the entire program has been run. The result is displayed on the screen of a logic analyzer; an example is shown in Fig. 9-90b. Due to the much smaller number of measuring points, the measuring time is shorter than for recording a logic image. It can be further reduced by selecting large phase increments. This is permitted when a high time resolution is not required and only the temporal sequence of logic states on the interconnections is of interest. This procedure leads to a considerable data reduction and less loading of the circuit.

Real-time logic state analysis

The methods described up to now allow the detection only of periodically repeating states. An attempt has also been made to record unique or nonperiodic switching processes [8-14].

For this purpose the PE beam is kept focussed on a single measuring point. The time-dependent SE signal is recorded by a very fast detector-amplifier configuration and fed to a logic analyzer. To obtain a sufficiently high signal-to-noise ratio, such as is required for a reliable evaluation of logic states, operation must be with a very high probe current of about $I_0 = 10^{-7}$ A (permanent current!). This leads to problems regarding spatial resolution and specimen damage (section 8.3.1) so that this method is not suitable for testing VLSI circuits.

8.3 Instrumentation

8.3.1 Requirements on electron probes

The aim of electron beam testing is to measure the voltage on an electrical circuit at a specific point and time (phase). Spatial, time and voltage resolution (section 8.4) depend on the probe diameter, acceleration voltage, probe current and electron beam pulse width. For designing or selecting an electron beam testing system as well as optimizing the measuring conditions it is important to know how these parameters should be chosen.

The testing of VLSI circuits and microwave devices is subject to different requirements. At present most applications are in the field of VLSI testing. The following description therefore emphasizes the instrumentation required for this purpose. An electron beam test system for microwave devices has been described in [8-15].

Probe diameter

As a rule, voltage measurements with the electron probe are performed on circuit nodes in the uppermost interconnection level of integrated circuits. In this case it is absolutely necessary to avoid PEs striking the adjacent passivation layer, as charging would then lead to incorrect measurements.

Experience shows that the PEs strike a spot on the specimen surface whose diameter is approximately five times as large as the probe diameter d_0 described by eq. (2-23). This is due to the lateral tails of the current density distribution (Gaussian distribution) as well as to the influence of aberrations and instrument instabilities (section 2.3.2). In addition, probe movements resulting from the beam blanking process can contribute to an increase in the effective probe diameter [8-1]. To test highly-integrated circuits with interconnection widths of $w = 1$ µm and less therefore, probe diameters $d_0 \leqq 0.2$ µm are required, i.e. $d_0 = w/5$ [8-7].

Acceleration voltage

Because of the PE radiation, the specimens are loaded with electrical charge and energy. Both of these can disturb the operation of an electrical circuit and lead to incorrect measurements. Selection of a suitable PE energy E_0 can ensure that the coefficient of total electron yield is $\sigma = \delta + \eta = 1$ (Fig. 3-16 and Tab. 3-1). Under these conditions (e.g. $E_0 = 2.8$ keV for aluminium) the *charge* supplied is drawn off again during the emission of SEs and BSEs and no current then flows through the circuit to ground.

However, only a small part of the *energy* supplied is emitted again, the main portion being absorbed by the specimen. In this way, electron-hole pairs are generated in the semiconductor by ionization (section 3.6.1). The excess of minority charge carriers in each case can lead to serious disturbances of the charge equilibrium in a circuit (section 2.4.3). In the dielectric layers also, such as silicon oxide, electron-hole pairs are generated by electron bombardment. The electrons are highly mobile and flow away from the oxide layer, so that the less mobile holes produce a positive space charge in the oxide. In addition, the charge carriers generated are captured at the oxide-substrate interface and can change the population of the interface states [8-4, 16]. The consequence of all these effects depends greatly on the type of device in question. Most sensitive are MOS circuits, which respond to electron bombardment with changes in cutoff voltage and dielectric strength and with increased leakage currents [8-17]. In practice, acceleration voltages of 1 kV or less are aimed for, so that current load and radiation damage are largely avoided [8-18].

The choice of the acceleration voltage is also of importance where electrical voltages are to be measured through a passivation layer. There are two ways of doing this [8-4]:

a) Operation can be at such a *high acceleration voltage* that the PE beam penetrates the passivation layer down to the interconnection lying below it (e.g. $U_0 \geqq 10$ kV at 1 μm SiO_2 layer). The charge carriers generated by ionization (section 3.6.3) then form a conductive channel in the dielectric and the layer surface acquires the voltage of the interconnection lying below it. Due to possible radiation damage, however, this procedure is particularly unsuitable for MOS circuits.

b) Operation is at *low acceleration voltages* so that charge equilibrium prevails at the layer surface ($U_0 \approx 1$ kV; $\sigma \approx 1$). Under these conditions, voltage contrast can also be observed as a result of capacitive coupling between the irradiated layer surface and the interconnection lying below it [8-18]. However, in this process only ac components of the signal are visible. This is sufficient in most cases, since these involve checking of alternating signals.

Electron beam pulse width and repetition rate

Stroboscopic voltage contrast imaging, waveform measurement and logic state mapping are phase-selective methods. The PE beam may therefore be briefly unblanked only once per period (or once every 2nd or 3rd period) of the device signal to be measured. The synchronization condition is therefore $f_p = f/n$ ($n = 1, 2, 3, \ldots$), where f_p is the pulse frequency (repetition rate) and f the device signal frequency. Variable repetition rates of some kHz to some 100 MHz are required for testing integrated circuits and up to several GHz for microwave devices.

To measure the rising edge of a device signal by means of the sampling mode, the electron pulse width t_p must be small compared with the rise time t_r. For the case of a linear rising edge and a square-wave pulse, the condition $t_p = t_r/5$ can be calculated; the measured result is then exact within the time interval in which the device signal increases from 10% to 90% of its maximum [8-19]. To measure typical rising edges in integrated circuits, pulse widths down to values of $t_p < 1$ ns are required, in microwave devices down to $t_p < 50$ ps [8-20].

Probe current

The intermittent PE radiation used in stroboscopic methods and the limitation of the SE current when using a spectrometer produce a considerably reduced SE signal compared with a normal SE image in the SEM. If, for instance, we look at the case of stroboscopic voltage contrast imaging, in which the objective is to image a switching state of short duration (required pulse width $t_p = 10$ ns) repeated at a frequency $f = 100$ kHz, then according to eq. (8-3) with $f_p = f$, the duty cycle is $c = 10^3$. This means that every object point is excited to SE emission in only 1/1000 of the integration time (the dwell time of the PE beam at each point during which the SE signal is integrated). The effective probe current is $I_{0\text{eff}} = I_0/1000$ and the signal yield has a corresponding value (under otherwise identical conditions) which is a thousand times smaller than in the unpulsed operating mode. To balance this signal loss, the probe current must be higher by three orders of magnitude than otherwise customary in SEM operation, i.e. $I_0 \geqq 10^{-9}$ A.

Equally high demands must be made on the probe current when measuring the waveform via the sampling mode. In this case, the spectrometer properties, its setting and the required voltage resolution are included in the estimate (sections 8.4.1 and 8.4.2).

8.3.2 Fundamental setup

Today, instruments are available for electron beam testing which are adapted to the special requirements of device testing (section 8.3.3). In principle, they are scanning electron microcopes (SEM) as shown in Fig. 3-1, and most electron beam testers in operation are still based on a commercial SEM. Various additional components are required, depending on the measurement techniques to be used. They can be assembled by the user or procured as complete units [8-21].

Fig. 8-10 is a schematic representation of the most important components of an electron beam testing system (without the electron lenses) for two different applications.

Electron source

Extreme demands are made on the electron gun. The descriptions in the previous section have shown the requirement for a very small probe diameter d_0 with a simultaneously high probe current I_0 and low acceleration voltage U_0. According to eqs. (2-23) and (2-21), these requirements are not mutually compatible. An optimum parameter setting must therefore be sought for every measurement job. Modern instruments make use of electron guns with LaB_6 cathodes whose high brightness allows a smaller diameter than conventional tungsten cathodes under otherwise identical conditions (Fig. 2-28). Electron beam testing systems with field emission guns have also been implemented [8-22,23].

Beam blanking system

High demands are also made on the beam blanking system. It must attain pulse frequencies (repetition rates) up to several 100 MHz and pulse widths down to 0.1 ns to meet VLSI requirements. The testing of microwave devices needs repetition rates in the GHz range and pulse widths below 10 ps [8-15]. Another requirement is that the probe diameter and position on the specimen must not be degraded significantly by the blanking process. These requirements can largely be satisfied with relative ease by deflecting the PE beam via an aperture (A in Fig. 8-10a) with the aid of a deflection capacitor. This capacitor is supplied with sinusoidal ac voltage or pulsed dc voltage of variable frequency. It is of advantage here to superimpose a dc voltage onto the deflection voltage so that the PE beam is normally off and is unblanked only by the deflection pulse. The attainable values of repetition rate and pulse width depend on the size of the capacitor, on its location within the electron-optical column and on the performance of the pulse generator. The various options and attainable values have been discussed in detail in the literature [8-1, 24]. Recently, pulses below 10 ps at a repetition rate up to 3 GHz have been demonstrated with this type of blanking system [8-15, 25]. This makes it applicable to VLSI circuits as well as microwave devices.

Secondary electron spectrometer

For quantitative testing, an electron spectrometer must be used. Except for newer developments (section 8.3.4), which will initially be excluded, the spectrometer is inserted between the pole piece of the probe forming lens and the specimen. Fig. 8-10b outlines a retarding field spectrometer (ES) consisting of two grid electrodes. Only those SEs which can overcome the retarding field reach the SE detector SD and contribute to the measuring signal. The requirements which an SE spectrometer must satisfy, and two different versions, are treated in section 8.3.3.

Specimen chamber

The specimen chamber is designed on the basis of the type of specimen and the method of investigation used. For static voltage contrast imaging of discrete devices, the only requirements are electrical feed-throughs in the housing and a socket in the specimen stage (SS in Fig. 8-10).

For investigating circuits on wafers, a large-area specimen chamber must be used. Such investigations are required if errors are to be analyzed and error sources eliminated at the earliest possible production stage. Electron beam testers with *large-area specimen chambers* are among those described in [8-3, 7]. It should be possible with such a configuration to bring every chip on the wafer into the field of view and to make suitable contacts for supplying the driving signals. This latter requirement is achieved with the aid of a probe system mounted on an adapter card. The probes are needles, pre-adjusted so that they fit onto the pads of the relevant circuit type. With the aid of motor-controlled drives, the circuits can be aligned with respect to the probe system, the probes set down and the contacted circuit examined.

Sensitive voltage measurements can be performed only on discrete devices. It is then difficult to transfer the signals generated by the drive unit without disturbance and losses as far as the circuit connections. With modern microprocessors, this transfer requires more than 200 shielded cables and vacuum feed-throughs. Because of the power losses in the cables and resistors, drivers are additionally required to set the voltage level needed to operate the circuit.

The drive problem can be solved elegantly with the aid of a *mini specimen chamber.* In this design, the circuit to be examined is arranged so that its connections are located outside the chamber. For this purpose, the circuit socket must be designed to seal the vacuum of the specimen chamber. Maximum benefit is obtained with this solution in conjunction with an inverted SEM column: the circuit connections are then particularly simple to access, the feed lines are short and the driver circuit can be fitted outside the vacuum and be easily interchanged [8-7, 26].

Instrument control, signal processing and automation

The most important electronic components required to operate an electron beam tester are shown schematically in Fig. 8-10. Their interaction is described there for the cases of stroboscopic voltage contrast imaging (Fig. 10a) and for waveform measurement (Fig. 10b). Com-

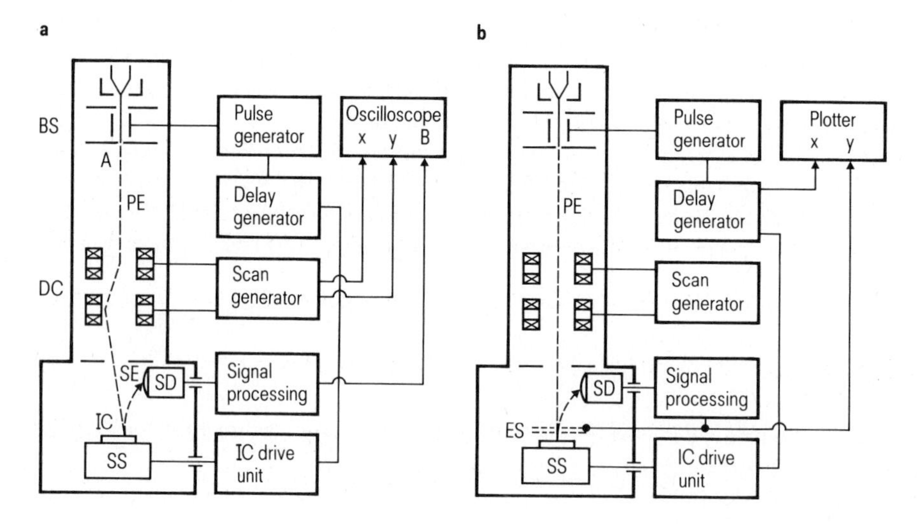

Fig. 8-10. Electron beam testing system for different modes of operation (the computer used for controlling the system is not shown in this schematic representation). a) Detection of *stroboscopic voltage contrast* pattern: as in the scanning electron microscope, the PE beam is scanned by means of a scan generator and deflection coils (DC) over the specimen (in this case an integrated circuit IC). In addition, the beam is interrupted to produce a sequence of pulses by the beam blanking system BS through periodic deflection out of the aperture A. The deflection voltage at the BS is generated by a pulse generator which is synchronized with the IC drive unit. The phase position of the PE pulses relative to the IC signal under test is set with the aid of the delay generator and remains constant during the recording. The signal received by the SE detector SD (determined by the voltages on the IC) is amplified by the signal processing unit and used to control the brightness (B) of the oscilloscope. SS specimen stage.
b) *Waveform measurement* using the sampling mode: the pulsed PE beam (beam blanking as in a) is set to a specific measuring point (IC node). The SE reach the SE spectrometer ES (in this case a retarding field spectrometer with two planar grid electrodes). The voltage at the retarding field electrode is controlled by a feedback loop (see Fig. 8-12) contained in the signal processing unit. The control voltage is a measure of the measuring point voltage (section 8.2.2) and simultaneously controls the y shift of the plotter (or monitor). The x shift (time or phase axis) is controlled by the delay generator, which simultaneously shifts the phase position of the PE pulses during the measurement relative to the IC signal. The waveform is then "scanned" and recorded as a function of the phase (time).

puters are usually used for supervising the instrument parameters and controlling the imaging or measurement procedure, as well as for recording and processing the results. They also permit linkup with a CAD (computer aided design) workstation to create an integrated electron beam measurement system [8-27, 28]. A system of this kind allows the following time-intensive procedural steps in electron beam testing to be speeded up dramatically:

a) Locating the internal test point of interest and positioning the electron beam probe exactly onto this point.
b) Setting suitable electrical test stimuli for the device under test.
c) Comparing the measured values with simulation values, and if required, performing a resimulation for fault analysis.

A configuration allowing fully-automatic control of the measuring instrument with simultaneous availability of the CAD data is extremely user-friendly. It allows the chip designer to test even complex circuits quickly without having to concern himself about the operating details of the electron-optical configuration.

8.3.3 Secondary electron spectrometer

The following demands are made on an SE spectrometer (or SE analyzer):

a) Minimal influence on the PE beam to ensure precise beam positioning.
b) High sensitivity, so that small SE signal changes (i.e. potential differences $\ll 0.1$ eV; section 8.4.1) can be detected.
c) High transmission, so that the very low signal yield for some modes of operation (section 8.3.1) is not further reduced.
d) High extraction field between the specimen and the spectrometer to suppress the effect of local retarding fields.
e) Low sensitivity with respect to changes in the SE angular distribution due to local microfields (section 8.2.2).

There is an additional requirement for an SE spectrometer to be used in an SEM with a conventional electron-optical column, i.e.:

f) Low height to ensure a short working distance between the probe forming lens and the specimen and thus a small probe diameter (section 3.4.3).

An overview of numerous SE analyzers and their performance data is given in [8-5]. Most of the configurations are retarding field spectrometers with planar or hemispherical grids or retarding field lenses. Later developments are published, inter alia, in [8-29,30,31]. Two configurations will now be described whose low height means that they optimally fulfil requirement f) and so are suited for testing VLSI circuits.

Fig. 8-11a shows a retarding field spectrometer with planar grids [8-10]. A high extraction field strength of 600 V/mm is generated with the grid electrode G_1. The SEs accelerated by it are then retarded again and selected in the retarding field between grids G_1 and G_2. The high-energy SEs which can overcome the retarding potential at G_2 are finally accelerated laterally from grid electrode G_3 towards the detector. For this purpose, a voltage of only about 100 V, whose influence on the PE beam is sufficiently small, must be applied to G_3.

A modified configuration is shown in Fig. 8-11b [8-31]. Here the SEs are, just as in a), initially accelerated to electrode G_1. They then traverse the field-free space in a straight line between grids G_1 and G_2 and finally reach the hemispherical retarding field between grids G_2

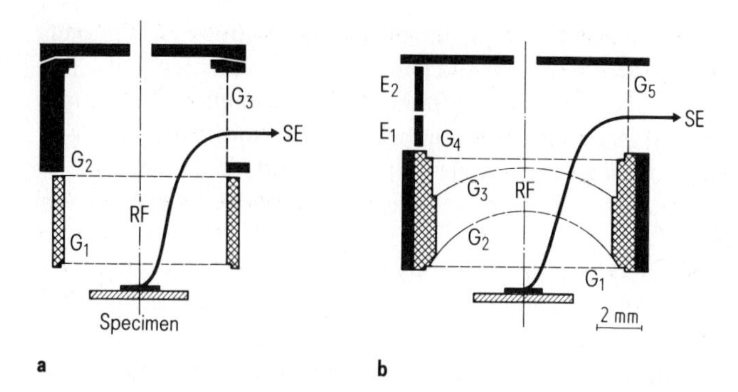

Fig. 8-11. Retarding field spectrometers. a) Configuration with planar grids [8-10]: the SEs are accelerated to the extraction grid G_1 and retarded in the retarding field (RF) between grids G_1 and G_2. SEs which can overcome the retarding voltage at G_2 are extracted from the deflection grid G_3 (+100 V) and reach the detector from there. b) Configuration with planar and hemisperical grids [8-31]: the retarding field (RF) lies between the spherical grids G_2 (same voltage as the planar extraction grid G_1) and G_3 (same voltage as the shielding grid G_4). All SE trajectories enter the RF normally. The deflection unit comprises grid G_5 (+180 V) and the electrodes E_1 (0 V) and E_2 (−20 V).

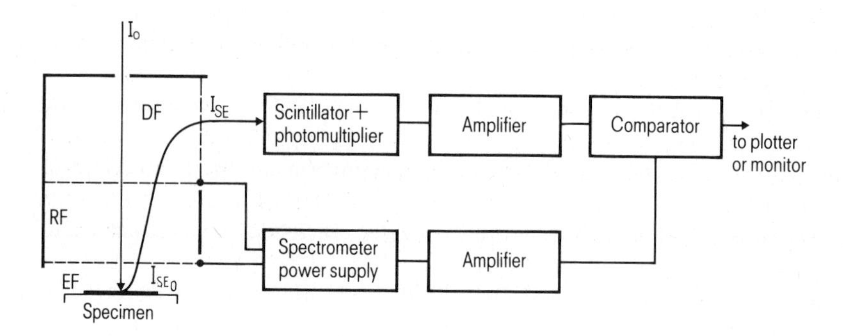

Fig. 8-12. Schematic diagram of a feedback loop for waveform measurement by means of the sampling mode: the SE signal I_{SE} leaving the spectrometer is detected by the scintillator-photomultipier setup, integrated in the amplifier and compared in the comparator with the set nominal value. After amplification, the deviation from the nominal value controls the voltage at the spectrometer so that the SE signal remains constant when the specimen voltage is changed. The bandwidth of this configuration is 300 kHz. I_{SE0} is the SE current leaving the specimen; EF extraction field; RF retarding field; DF deflection field.

and G_3. The spherical grids are arranged so that their common mid-point coincides with the virtual SE source (virtual image of the measuring point emitting the SEs) [8-29]. All SE trajectories are therefore incident upon the grids G_2 and G_3 in a normal direction, independent of their original direction of exit. The edge regions of these electrodes are solid in order to reduce the number of backscattered electrons passing the analyzer. The shielding grid G_4 is at the same voltage as grid G_3 and amplifies its effect. This is required to reduce the smearing of the spectrometer characteristic resulting from the inhomogeneous field distribution in the meshes of the retarding grid (see section 8.2.2). The geometry of the deflection unit is cylindrically symmetric. The deflection field is generated by the grounded housing, the deflection grid G_5 (+180 V) and the electrodes E_1 (0 V) and E_2 (−20 V).

In both configurations, all voltages are shifted together with the retarding voltage; this is possible because only the retarding voltage enters into the energy balance (sect. 8.2.2). The grids have a high transmission (250 μm mesh size, 10 or 25 μm bar width), and they are arranged so that in a spot measurement the PE beam can reach the measuring point, which is always kept at the center of the image by driving the specimen stage, unhindered. Both spectrometers are operated in conjunction with a feedback loop to ensure a linear SE signal (section 8.2.2). This principle is presented in Fig. 8-12.

8.3.4 Dedicated electron beam tester

Even when operation is with miniaturized SE spectromers as shown in Figs. 8-11, the working distance cannot be made smaller than 10 mm. Under these circumstances, the requirements derived in section 8.3.1 for testing VLSI circuits (d_0 = 0.1 ... 0.2 μm; $U_0 \leqq 1$ kV; $I_0 \geqq 10^{-9}$ A) cannot be realized. Completely new electron-optical configurations were recently suggested and developed for this purpose [8-23, 32]. They contain:

a) New types of beam forming systems with a high-brightness electron source optimized for low acceleration voltages.

b) Probe forming lenses (objective lenses) with integrated SE spectrometers which allow a reduction of the working distance down to values between 1 and 3 mm and are largely insensitive with respect to local field effects.

An electron beam tester with a field emission gun is presented in [8-23]. The objective lens is split into two: a hemispherical retarding grid analyzer is accommodated in the intermediate space between the "sub-objective lens" and the "main objective lens" [Fig. 8-13]. The SEs are extracted from the specimen with the aid of a planar grid and accelerated through the bore of the main objective lens to the retarding field analyzer. At U_0 = 1 kV and $I_0 = 5 \cdot 10^{-9}$ A, a probe diameter of 0.1 μm was realized.

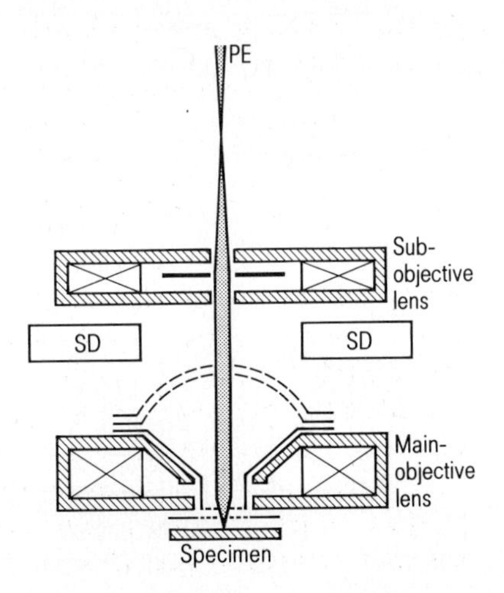

Fig. 8-13.
Schematic diagram of an improved voltage analyzer: a hemispherical retarding field spectrometer is placed between the sub- and main objective lenses. Since the PE beam runs parallel between these lenses, sufficient space can be made available for the analyzer. SD secondary electron detectors. According to [8-23].

Another dedicated electron-beam test system is described in [8-32, 33]. In this case, the electron-optical column is optimized not only with respect to brightness, lens aberrations and spectrometer efficiency, but also with respect to electron-electron interactions, which at low PE energies lead to a widening of the probe diameter (Boersch effect, section 2.3). The optimization was realized by reducing the column length to 275 mm and guiding the PE beam through long sections of the column (especially where the current density is high) with a relatively high energy. The electron gun system contains an immersion condenser lens in addition to a triode gun with LaB_6 cathode. It combines a magnetic pole piece lens and an electrostatic tube lens. The electrons are accelerated in the triode gun (where the current density is high) by a strong extraction field and retarded in the condenser lens to their final energy (0.7 to 2.5 keV). The beam blanking system is integrated into the anode of the triode gun.

The objective lens of this dedicated test system is also designed as a combination of electrostatic and electromagnetic fields [8-32,33]. Deflection coils, the stigmator and a hemispherical retarding field spectrometer are integrated into it. The action of this configuration on PEs and SEs is shown separately in Fig. 8-14. The SE generated from the PE beam are extracted with the aid of the planar grid electrode G_1 (extraction field strength 1 kV/mm) and finally focussed by the objective lens. The focus coincides with the common midpoint of the two hemispherical grids. This has the advantage that all SE pass normally through the spectrometer grids, independently of their exit angle. This not only increases the accuracy of measurement but also reduces the effect of the local microfields.

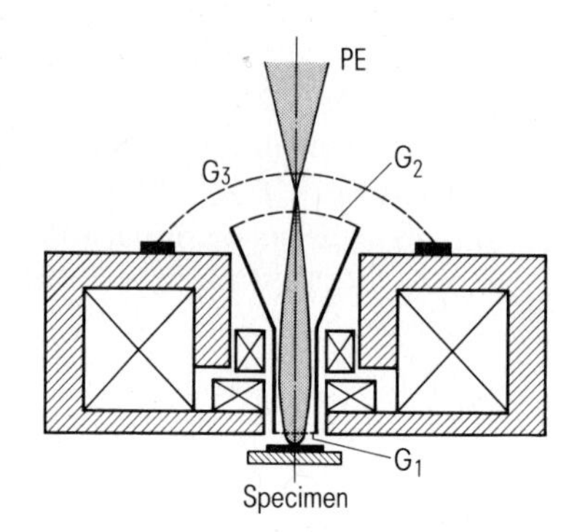

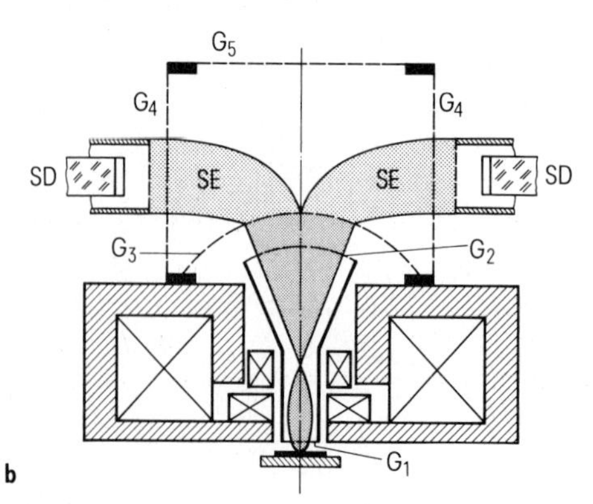

Fig. 8-14.
Schematic diagram of a compound spectrometer objective lens with stigmator and deflection coils optimized for low PE energies (0.7 to 2.5 keV) [8-12,32]: a) The PE beam is focussed on the specimen by superimposed electric and magnetic fields. The PEs are accelerated between grid electrodes G_3 and G_2 and decelerated again between G_1 and the IC specimen.
b) The emitted SEs are simultaneously extracted by G_1 and focussed by the magnetic lens field onto the center of the hemispherical retarding field analyzer (retarding field between G_2 and G_3). G_4 deflection grids, G_5 suppressor grid, SD secondary electron detectors (not shown in Fig. a for clearness).

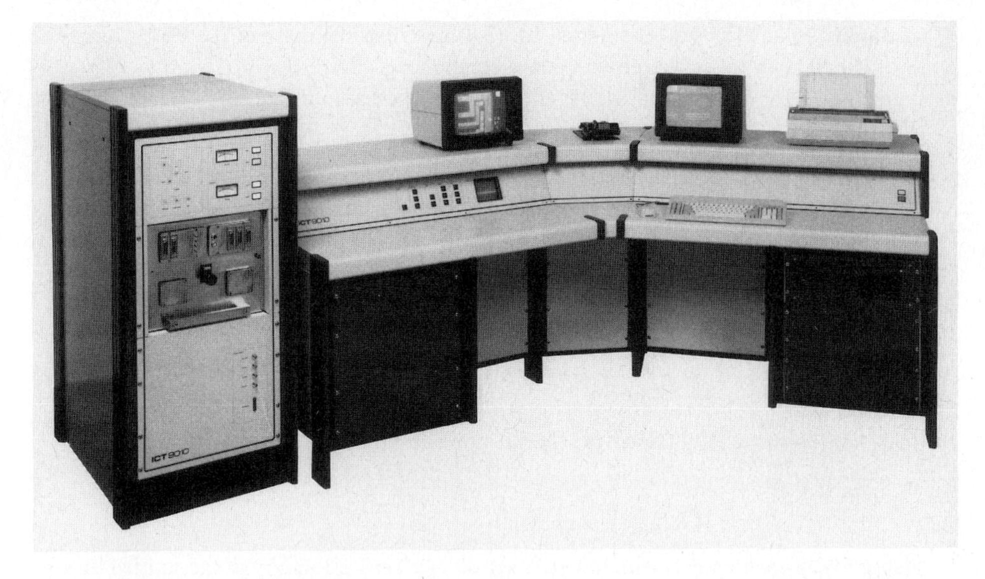

Fig. 8-15. View of a dedicated electron beam test system: E-beam tester 9010 by ICT (Courtesy of ICT Integrated Circuit Testing GmbH, Klausnerring 1a, D-8011 Heimstetten near Munich).

Fig. 8-15 is a view of the dedicated electron beam test system just described. Using this equipment, a probe diameter of 0.12 μm was measured at $E_0 = 1$ keV, $I_0 = 2.5 \cdot 10^{-9}$ A and a working distance of 2 mm [8-12].

8.4 Measurement technique

8.4.1 Voltage resolution

The voltage resolution is specified as the smallest voltage difference which can still be detected at a given noise.

Noise of secondary electron current

The emission of electrons is a statistical process which results in fluctuations of the emission current, known as "shot noise". The emission current is thus composed of the dc current component I_0 and a noise component. It is usual to describe the latter by its "mean square fluctuations". In the case of a high-vacuum diode (operated in the saturation region), the mean square fluctuations of the noise current measured in a frequency interval Δf (bandwidth) are given by the well known Schottky formula

$$\overline{i_0^2} = 2 \cdot e \cdot I_0 \cdot \Delta f \,. \qquad (8\text{-}4)$$

When applied to an electron beam tester, eq. (8-4) describes the noise of the PE current. The noise of the SE current is composed of two components [8-4]:

a) In the first place, the noise of the PE current is transferred to the SE current in accordance with the ratio of the SE yield δ, so that the following component is obtained for the mean square fluctuations

$$\overline{i_{SE}^2}' = \delta^2 \cdot \overline{i_0^2} . \tag{8-5}$$

b) The emission of the SEs is itself a statistical process and gives rise to fluctuations of the SE current I_{SE0} emitted from the specimen. By analogy with eq. (8-4) we obtain

$$\overline{i_{SE}^2}'' = 2 \cdot e \cdot I_{SE0} \cdot \Delta f = 2 \cdot e \cdot \delta \cdot I_0 \cdot \Delta f . \tag{8-6}$$

In total, the following is obtained for the SE noise from eqs. (8-4) to (8-6)

$$\overline{i_{SE}^2}' + \overline{i_{SE}^2}'' = \overline{i_{SE}^2} = 2 \cdot e \cdot \delta \cdot (1 + \delta) \cdot I_0 \cdot \Delta f . \tag{8-7}$$

In quantitative measurements with the SE spectrometer, only a fraction of the emitted SE current is detected, namely

$$I_{SE} = \gamma \cdot \beta \cdot I_{SE0} = \gamma \cdot \beta \cdot \delta \cdot I_0 . \tag{8-8}$$

Here β is the spectrometer transmission and γ the fraction of SEs which can overcome the retarding voltage U_R set at the retarding field spectrometer. Both parameters have an effect on the noise component and the following is finally obtained in place of eq. (8-7)

$$\overline{i_{SE}^2} = 2 \cdot e \cdot \gamma \cdot \beta \cdot \delta \cdot (1 + \delta) \cdot I_0 \cdot \Delta f . \tag{8-9}$$

The noise of the detection system is of minor importance compared with the shot noise. Noise contributions from backscattered electrons and tertiary electrons can be suppressed by suitable shielding in the spectrometer. Under these preconditions, eq. (8-9) describes the noise of the measured SE signal.

Minimum detectable voltage

Fig. 8-16 shows a spectrometer characteristic in which the fraction γ of the SEs passing the spectrometer is a function of the retarding voltage U_R. By selecting U_R^*, the operating point is defined and thus the signal $I_{SE}^* = \gamma^* \cdot \beta \cdot \delta \cdot I_0$ set at the detector according to eq. (8-8).

A low voltage change ΔU_P at the measuring point shifts the characteristic along the abscissa (as in Fig. 8-5c) and produces a signal change of ΔI_{SE}.

The relationship between ΔU_P and the measured value ΔI_{SE} is given by the slope $a^* = \Delta\gamma/\Delta U_R$ of the characteristic at the operating point. The following is then obtained with $\Delta\gamma = \Delta I_{SE}/(\beta \cdot \delta \cdot I_0)$ according to eq. (8-8) and $\Delta U_R = \Delta U_P$

$$\Delta I_{SE} = a^* \cdot \beta \cdot \delta \cdot I_0 \cdot \Delta U_P . \tag{8-10}$$

According to the laws of statistics (section 1.4), the ΔI_{SE} signal is regarded as detected with a certainty of more than 99% when it is three times higher than the noise level, i.e. when the condition $\Delta I_{SE} \geqq 3\sqrt{i_{SE}^2}$ is satisfied. The following is obtained from eqs. (8-9) and (8-10) when solving for ΔU_P and designating the lower limit of ΔU_P by ΔU_{min}

$$\Delta U_{min} = \frac{3\sqrt{2e \cdot \gamma^* \cdot \beta \cdot \delta(1+\delta)}}{a^* \cdot \beta \cdot \delta} \cdot \sqrt{\frac{\Delta f}{I_0}} . \qquad (8\text{-}11)$$

If we additionally assume that at a duty cycle c only a current I_0/c reaches the specimen as an average over time, we obtain a relation of the form

$$\Delta U_{min} = C^* \sqrt{c \cdot \Delta f / I_0} , \qquad (8\text{-}12)$$

known as the Gopinath formula [8-35].

According to this, the minimum detectable voltage difference gets smaller as the factor C^* decreases. C^* contains only the SE yield δ in addition to the spectrometer parameters. High values of δ favour the attainment of a high voltage resolution (small ΔU_{min}). For this reason (in agreement with the requirements stated in section 8.3.1) operation should be at low acceleration voltages at which a high SE yield can be expected. At a defined acceleration voltage, C^* is a spectrometer constant which depends only on the selection of the operating point (i.e. on U_R^*). C^* and ΔU_{min} can be numerically calculated with the aid of an approximate solution for the slope a^* [8-35,36]. In practice, the procedure adopted is to calibrate the spectrometer at its operating point.

Spectrometer operating point

The optimum position of the operating point is given by the point of inflexion of the spectrometer characteristic at which the slope is a maximum (Fig. 8-16): a low voltage change there results in a maximum signal change. The following numerical values were determined for the spectrometer constants at the optimum operating point:

Post-lens spectrometer with planar grids (Fig. 8-11a) [8-31]: $C^* = 5 \cdot 10^{-8}\ \mathrm{V} \cdot \sqrt{\mathrm{As}}$.
Post-lens spectrometer with hemispherical grids (Fig. 8-11b) [8-31]: $C^* = 1.7 \cdot 10^{-8}\ \mathrm{V} \cdot \sqrt{\mathrm{As}}$.
In-lens spectrometer with hemispherical grids (Fig. 8-13) [8-32]: $C^* = 5 \cdot 10^{-9}\ \mathrm{V} \cdot \sqrt{\mathrm{As}}$.

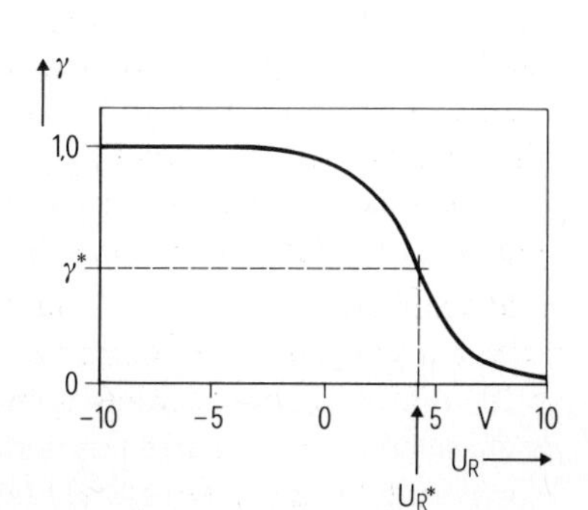

Fig. 8-16.
Characteristic of a retarding field spectrometer as shown in Fig. 8-11a. Fraction γ of the SE current which passes the spectrometer at a retarding voltage U_R. The operating point is specified by γ^*, U_R^*.

With respect to the local microfields, it may be necessary to select an operating point at a higher retarding voltage. More of the low-energy SEs, whose trajectories are still too strongly affected by the local microfields, are eliminated by this means (section 8.2.2). This is particularly necessary for retarding field spectrometers with planar grids. In hemispherical analyzers, the measurement is less sensitive to the angular distribution of the electrons to be analyzed. This allows a more advantageous selection of the operating point and thus a higher SE signal.

Minimum charge per measuring point

It follows from eq. (8-12) that a minimum PE charge per measuring point (phase point in the waveform measurement; object point in quantitative voltage contrast imaging) is required to detect a specific voltage difference.

The bandwidth Δf of the detection system must be large enough to detect the frequency range of the signals to be measured. The measuring signal changes with every change from one measuring point (object or phase point) to another. The alternating frequency f_s of the signal therefore depends on the dwell time t_s of the PE beam (integration time) per measuring point. As a first approximation, $f_s = 1/t_s$ applies. With $\Delta f = f_s$, the following is obtained

$$\Delta U_{\text{min}} = \frac{C^*}{\sqrt{(I_0/c)\, t_s}} \,. \tag{8-13}$$

The product under the root is the electrical charge q_0 supplied to each measuring point. To obtain a voltage resolution of, say $\Delta U_{\text{min}} = 100$ mV, a charge of

$$q_0 = C^{*2}/\Delta U_{\text{min}}^2 = 2.5 \cdot 10^{-13}\ \text{As}$$

is required when the spectrometer constant $C^* = 5 \cdot 10^{-8}\ \text{V} \cdot \sqrt{\text{As}}$. This corresponds to about 10^6 electrons.

8.4.2 Relationship between measurement parameters

Spatial, time and voltage resolution

The quality of a voltage measurement depends on the values of the spatial, time and voltage resolution which can be obtained under the boundary conditions mentioned in section 8.3.1.

The spatial resolution is given by the *probe diameter* d_0. Dedicated test systems can be used to obtain values of $d_0 \geqq 0.1$ µm [8-33] at the required low PE energies. These allow voltage measurements to be made on interconnections with a width $w \geqq 0.5$ µm.

The time resolution depends primarily on the attainable *pulse width* t_p presuming appropriate stability of the phase position. By means of beam blanking with the aid of capacitor plates, values of $t_p = 10$ ps could be attained at PE energies between 2 and 30 keV [8-1,24].

This allows the measurement of signal rising edges of 50 ps (or 20 ps if a reduced accuracy is accepted [8-20]). Very short pulse widths were also attained with travelling-wave structures (t_p = 5 ps) [8-37,38] and with re-entrant cavity structures (t_p = 0.2 ps) [8-39]. However, these configurations presuppose a specified frequency or PE energy and are therefore not universally applicable.

A value of $\Delta U_{min} \approx 1$ mV has been published for the *voltage resolution* using a retarding field spectrometer with planar grids (Fig. 8-11a) and a special circuit for compensating drift effects [8-10].

The values listed above are not independent of each other. To set up the relationship between them, let eq. (8-3) be substituted in eq. (8-13) for the duty cycle:

$$\Delta U_{min} = \frac{C^*}{\sqrt{f_p \cdot t_p \cdot t_s \cdot I_0}} . \tag{8-14}$$

The pulse repetition rate f_p is given by the synchronization condition (section 8.3.1) and the spectrometer constant C^* by the spectrometer and the selection of the operating point. According to eq. (2-23), the probe current is linked with the probe diameter as follows:

$$d_0 \sim \sqrt{I_0}$$

where the proportionality factor depends on the brightness of the electron gun.

Eq. (8-14) was derived using simplifying assumptions. In conjunction with eq. (2-23), however, it clearly shows how the individual parameters mutually affect each other. It is very helpful when selecting the measuring conditions.

Numerical example

The measuring time per phase point t_s which is required to detect a specific potential difference can be estimated from eqs. (2-23) and (8-14). It is calculated in the following example:

A device signal of specified duration, e.g. a signal rising edge of duration t_r = 5 ns, is to be measured on an interconnection of width w = 1 μm. According to section 8.3.1 this measurement requires a probe diameter $d_0 = w/5$ = 0.2 μm and a pulse width $t_p = t_r/5$ = 1 ns.

Instead of calculating I_0 according to eq. (2-23), we simply take the value used in Fig. 2-28: according to this, for d_0 = 0.2 μm the probe current is $I_0 = 5 \cdot 10^{-9}$ A when operating with an LaB_6 cathode and a PE energy of E_0 = 2.5 keV.

For the repetition rate of the device signal we assume f = 2 MHz and stipulate a value of $f_p = f$ = 2 MHz for the pulse repetition rate to ensure that the PE beam is unblanked once in every period. In this case, the duty cycle has a value of $c = 1/(f_p \cdot t_p)$ = 500. For the spectrometer constant, let us assume $C^* = 5 \cdot 10^{-8}\ \mathrm{V} \cdot \sqrt{\mathrm{As}}$ and stipulate ΔU_{min} = 10 mV.

Using these highlighted values, eq. (8-14) supplies a value of t_s = 2.5 s for the measuring time per phase point. Since five measurements are required for a signal rise time of at least 5 ns and a pulse width of 1 ns, the total measuring time must exceed 12.5 s.

8.4.3 Survey of electron beam test methods

The most suitable method must be selected for every measuring assignment and every type of circuit. To provide a basis for this selection, all the methods treated here and the information obtainable from them are listed in Table 8-1. Further methods, which are less frequently used, such as frequency tracing and frequency mapping, are described in sources including [8-7,40].

All the methods listed in Table 8-1 are used for testing semiconductor devices, especially integrated circuits. They can, however, also be used to examine other materials and devices, e.g. the domain structure of *ferroelectric* crystals (such as triglycine sulphate) [8-41] and surface acoustic waves in *piezoelectric* crystals (e.g. quarz, lithium niobate) [8-42,43]. Examples will be given in section 9.5.

Table 8-1. Main modes of electron beam testing and the information they provide. *U* voltage; x, y, φ, t coordinates of space, phase and time.

Method	Section	Information	
Static voltage contrast	8.2.1	$U(x, y)$	qualitative
TV voltage contrast	8.2.1	$U(x, y, t)$	"
Stroboscopic voltage contrast	8.2.1	$U(x, y, t_0)$	"
Static voltage measurement	8.2.2	$U(x_0, y_0)$	quantitative
Waveform measurement	8.2.2	$U(x_0, y_0)$	"
Voltage coding	8.2.3	$U(x, y, \varphi)$	qualitative
Logic state mapping	8.2.3	$U(x_0, y, \varphi)$	"
Timing diagram	8.2.3	$U(x_i, y_i, \varphi)$ $(i = 1, 2, \dots n)$	"
Real time logic state analysis	8.2.3	$U(x_0, y_0, t)$	qualitative

8.5 References

[8-1] H. Fujioka, K. Ura, *Scanning 5,* 3 (1983)
[8-2] H.P. Feuerbaum, ibid. 14
[8-3] E. Wolfgang, ibid. 71
[8-4] E. Menzel, E. Kubalek, ibid. 103
[8-5] E. Menzel, E. Kubalek, ibid. 151
[8-6] H. Rehme, in M. Zerbst (Ed.): *Meß- und Prüftechnik (Halbleiterelektronik, Band 20).* Springer, Berlin, Heidelberg, New York, Tokyo 1986
[8-7] E. Wolfgang, *Microelectronic Engineering 4,* 77 (1986)
[8-8] A. Gopinath, M.S. Hill, *Scanning Electron Microscopy 1974/I,* IITRI, Chicago 1974, 235
[8-9] K. Ura, H. Fujioka, T. Yokobayaski, *Proc. 7th Europ. Congr. Electron Microscopy,* Leiden 1980, 330
[8-10] H.P. Feuerbaum, *Scanning Electron Microscopy 1979/I,* SEM Inc., AMF O'Hare 1979, 285
[8-11] K. Nakamae, H. Fujioka, K. Ura, *J. Phys. D: Appl. Phys. 14,* 1939 (1981)
[8-12] J. Kölzer, F. Fox, D. Sommer, Electrochem. Soc., Symposium "Beam Testing of Circuits and Devices", Extended Abstracts Vol. 88-2, Chicago, Ill. 1988, p. 564

[8-13] P. Fazekas, H.P. Feuerbaum, E. Wolfgang, *Electronics 14,* 105 (1981)
[8-14] M. Ostrow, E. Menzel, E. Postulka, S. Görlich, E. Kubalek, *Scanning Electron Microscopy 1982/II,* SEM Inc., AMF O'Hare 1982, 563
[8-15] M. Brunner, D. Winkler, R. Schmitt, B. Lischke, *Scanning 9,* 201 (1987)
[8-16] T.P. Ma, G. Scoggan, R. Leone, *Appl. Phys. Letters 27,* 61 (1975)
[8-17] W.J. Kerry, K.O. Leedy, K.F. Galloway, *Scanning Electron Microscopy 1976/IV,* IITRI, Chicago 1976, 507
[8-18] L. Kotorman, *Scanning Electron Microscopy 1980/IV,* SEM Inc., AMF O'Hare 1980, 77
[8-19] E. Plies, J. Otto, *Scanning Electron Microscopy 1985/IV,* SEM Inc., AMF O'Hare 1985, 1491
[8-20] B. Lischke, D. Winkler, R. Schmitt, *Microelectronics Engineering, 7,* 21 (1987)
[8-21] D. Ranasinghe, G. Proctor, N. Richardson, *Microcircuit Engineering,* Lausanne 1981, 522
[8-22] L.W. Swanson, D. Tuggle, Jia-Zheng Li, *Thin Solid Films 106,* 241 (1983)
[8-23] H. Todokoro, S. Yoneda, S. Seitou, S. Hosoki, *Proc. XIth Int. Congr. Electron Microscopy,* Kyoto 1986, Vol. I, 621
[8-24] E. Menzel, E. Kubalek, *Scanning Electron Microscopy 1979/I,* SEM Inc., AMF O'Hare 1979, 305
[8-25] D. Winkler, R. Schmitt, M. Brunner, B. Lischke, *Optik, 78,* 165 (1988)
[8-26] P. Köllensberger, A. Krupp, M. Sturm, R. Weyl, F. Widulla, E. Wolfgang, Intern. Test Conf., IEEE 1984, 550
[8-27] S. Concina, N. Richardson, Intern. Test Conf., IEEE 1987, 554
[8-28] S. Görlich, H. Harbeck, P. Keßler, E. Wolfgang, K. Zibert, Intern. Test Conf., IEEE 1987, 566
[8-29] K. Nakamae, H. Fujioka, K. Ura, *J. Phys. E: Sci. Instrum. 18,* 437 (1985)
[8-30] K. Nakamae, H. Fujioka, K. Ura, T. Takagi, S. Tagashima, *J. Phys. E: Sci. Instrum. 19,* 847 (1986)
[8-31] S. Görlich, P. Keßler, E. Plies, Microelectronic Engineering 7, 147 (1987)
[8-32] J. Frosien, E. Kehrberg, M. Sturm, H.P. Feuerbaum, *J. Electrochem. Soc., Solid State Science and Technology 135,* 2038 (1988)
[8-33] J. Frosien, E. Plies, *Microelectronic Engineering 7,* 163 (1987)
[8-34] E. Plies, J. Kölzer, *Proc. XIth Int. Congr. Electron Microscopy,* Kyoto 1986, Vol. I, 625
[8-35] A. Gopinath, *J. Phys. E.: Sci. Instr. 10,* 911 (1977)
[8-36] Y.C. Lin, T.E. Everhart, *J. Vac. Sci. Technol. 16,* 1856 (1979)
[8-37] G.Y. Robinson, *Rev. Sci. Instr. 42,* 251 (1971)
[8-38] A. Gopinath, M.S. Hill, *Scanning Electron Microscopy 1973/I,* IITRI, Chicago 1973, 197
[8-39] T. Hosokawa, H. Fujioka, K. Ura, *Rev. Sci. Instr. 49,* 624 (1978)
[8-40] H.D. Brust, F. Fox, E. Wolfgang, *Proceedings Microcircuit Engineering 1984,* Academic Press, London 1984, 411
[8-41] Y. Uchikawa, S. Ikeda, *Scanning Electron Microscopy 1981/I,* SEM Inc., AMF O'Hare 1981, 209
[8-42] H. Bahadur, R. Parshad, *Scanning Electron Microscopy 1980/I,* SEM Inc., AMF O'Hare 1980, 509
[8-43] H.P. Feuerbaum, H.P. Grassl, U. Knauer R. Veith, *Scanning Electron Microscopy 1983,* SEM Inc., AMF O'Hare 1983, 55

9 Applications

9.1 Analysis strategy

Modern analytical techniques cannot be used successfully without a well-founded knowledge of the fundamentals of analysis, such as that set out in this book for the methods of microanalysis. Experience in and routine use of the often very complex equipment are also essential factors. Accordingly, the scientist or engineer who is faced with analytical problems will not as a rule be in a position to carry out the analysis himself. Instead, he will assign the job to an analyst. Successful analytical work then calls for close cooperation between analyst and client. The analyst should make sure that he does not play the role of an assistant. He can be a partner among equals if he does not see himself merely as a specialist for analysis, but familiarizes himself with his client's problems. In other words, he must study the relevant technological processes and the associated material problems. He will have to be both highly competent and flexible to respond to frequently changing types of problems.

Besides close contact with the client, it is important for analysts to maintain good cooperation among themselves. Each analyst should be aware of the limits of his particular technique and should not hesitate to take advantage of other methods that may offer additional information. Not infrequently, more than one analytical technique will have to be applied to solve a particular problem.

Special care is called for when preparing and carrying out a microanalysis. As a rule, only very small quantities of specimens are available, an analysis cannot be repeated any number of times and very minor contamination can falsify the results.

Before the measurement, the analyst must obtain as accurate a picture as possible of the history of the specimens (e.g. of relevant process steps). He will have to hold in-depth discussions with the client and take note of the results that the latter suspects or anticipates, but not let his work be influenced by this knowledge during the analysis.

Special precautions must be taken to protect the specimen from contamination. Staff handling the specimens should wear gloves and work in a dust-free, chemically inert environment. This applies to preparatory work carried out on the client's site as well as transport to and handling in the analyst's laboratories.

The nature of the analytical problem (qualification, quantification, localization) and the required accuracy determine the analytical techniques to be used. For example, secondary ion mass spectrometry will be needed to determine the depth distribution of the dopant in a semiconductor wafer. Where the problem is less well defined, a start should be made with techniques that furnish a rapid overview of the specimen without much effort. In many instances, an inspection in an optical microscope or with the scanning electron microscope will be enough to identify the problem or indicate which techniques should be applied next. The use of more sophisticated techniques is justified only when simpler methods prove insufficient. Sometimes it may even be necessary to adapt the methods in order to solve a specific analytical problem. Thus, the requirements of an industrial analytical laboratory differ from those of a university laboratory, for example, where selected specimens can be analyzed to demonstrate the results achievable by a specific method or a specific instrument.

Whenever possible, a specimen should be analyzed several times or a number of identical specimens should be examined. The more comparable results are available, the more reliable is the mean value and the smaller the analysis error (standard deviation from equation (1-2)). If they are not recorded automatically, the instrument settings for the measurements should be entered in a log. While the analytical work is in progress, it is advisable to remain in close contact with the client and it may even be worthwhile to discuss preliminary results with him. Once the work has been completed, the results, their interpretation and possible conclusions should be described in an analysis report. The report should be comprehensible even without detailed knowledge of the analytical technique used. The client will be interested solely in the results and the consequences for his future work and not in any difficulties that may have occurred during the analysis. The results and consequences should be described succinctly and supported by a few (not too many) graphs or figures, where this is relevant. Moreover, they should be documented to allow ready access even at a later date. If similar problems of analysis turn up subsequently, access to the results of earlier work will simplify and speed up future work.

Finally, it should be borne in mind that the time available for carrying out an analysis will often vary considerably. When problems that reduce the yield of a production line are examined, a result must be obtained within the minimum possible time. Initial results may often be of assistance. Frequently, analyses during technological research and development demand more comprehensive investigations, for example to obtain an insight into relevant material effects. Also, for this kind of analysis the results should be available soon, i.e. within one or a few weeks. This is because the conclusions drawn from an analysis should be utilized for planning the next experiments.

Some selected applications taken from the fields in which the authors work are discussed below. The associated analyses were carried out in the course of the development phase of new materials and processes primarily for semiconductor technology or are failure analyses of defective devices. Problems of a very similar kind are also dealt with in many other laboratories that pursue research and development in advanced microtechnologies.

9.2 Silicon technology

9.2.1 Introductory remarks

Today's technological advances are largely induced by developments in the field of microelectronics. Silicon technology plays a major role in these developments: it was the key to very large scale integration (VLSI), i.e., the integration of more than 10^6 electrically active elements on a single chip measuring approx. 1 cm^2. Today, silicon is one of the most investigated and best understood chemical elements. It is produced on a large scale and to a very high degree of perfection and purity (impurity mole fraction $\leq 10^{-10}$).

Manufacturing of integrated silicon circuits involves various recurrent process steps: film deposition, patterning, doping, cleaning. Almost 500 individual steps of this kind are needed to produce a 4 Mbit DRAM (dynamic random access memory). The associated geometric dimensions (line widths, channel lengths, junction depths, layer thicknesses) vary from about 1 μm down to 10 nm.

Analysis has played a crucial role in these remarkable advances in silicon technology: highly sensitive methods of trace-element analysis (e.g., optical emission spectrometry, infrared spectrometry, neutron activation analysis, atomic absorption spectrometry, ion chromatography) are needed to check the extremely low impurity level of the substrate crystals and their surfaces as well as the trace impurities of the chemicals used in the numerous process steps. X-ray techniques and electron microscopy allow the perfection of the substrate crystal lattice to be assessed and reveal lattice defects induced during certain process steps. Together with optical inspection and electrical testing methods, the microanalytical techniques described here are indispensable as a means of characterizing and monitoring the geometry and material properties of device structures in micron and submicron regions.

9.2.2 Shallow doping profiles and residual implantation damage

The increasing integration density of VLSI circuits requires a reduction not only in lateral structural dimensions (line width, channel length, etc.), but also of depths, such as the thicknesses of the dielectric layers and the depths of the doped regions. Junction depths of less than 0.1 μm are required nowadays.

The most commonly used method of doping is ion implantation with boron for p-doping and arsenic or phosphorus for n-doping (section 2.1.2). Ion implantation allows the doping depth and concentration to be varied within wide limits by selecting the acceleration voltage and the dose. However, the nuclear collisions that occur during implantation, damage the crystal lattice (section 2.6.3: structural modifications). This in turn has an effect on those semiconductor properties that depend on the lattice perfection. Hence, the lattice defects produced by implantation damage have to be annealed by heat treatment. This process simultaneously activates the dopant atoms electrically, in other words incorporates them into the crystal lattice as ions. Nonetheless, residual damage often remains even after annealing.

Extremely shallow doping profiles are difficult to obtain for the following reasons:

a) The dopant atoms diffuse to excessive depths during the heat treatment needed to anneal the damage and to activate the dopants (Fig. 9-1a).
b) The channeling effect which occurs during implantation in single crystalline specimens (section 2.6.2) can lead to an excessive range. In the implantation process, the substrate crystals are slightly tilted to prevent the ion beam from impinging in a direction from any of the "open channels" in the lattice: nonetheless, the implantation profile often exhibits a hump caused by ions that have been scattered into these open channels and so have penetrated deeper than they would in amorphous material (compare curves a and b in Fig. 9-2).

New processes must therefore be developed to produce shallow doping profiles. This is particularly the case for boron, an element whose diffusion coefficient in silicon is higher than that of other dopants.

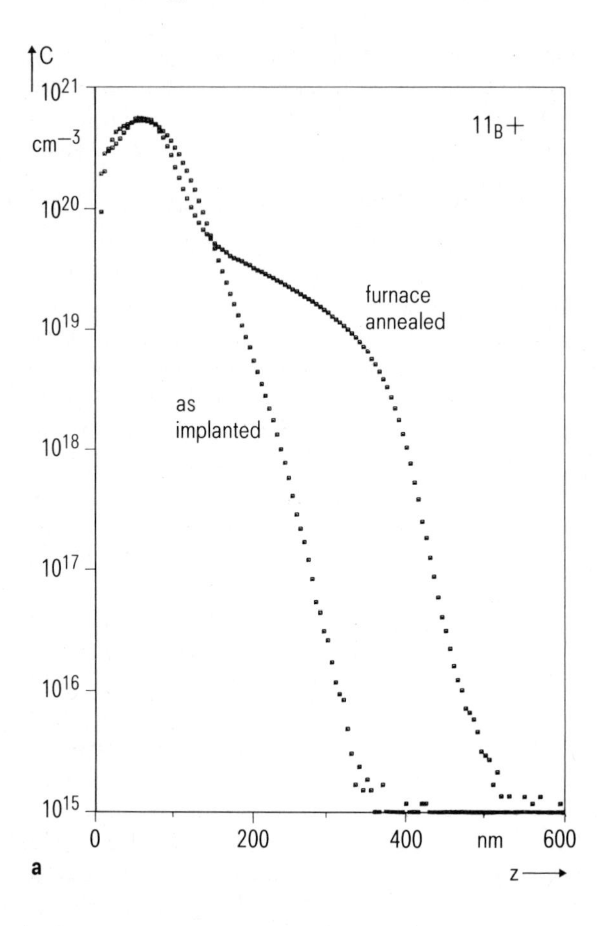

Rapid thermal annealing

Temperatures of at least 900 °C are initially employed to anneal the damage and electrically activate the implanted atoms. The distance covered by the dopant atoms during diffusion depends on the temperature and the annealing time, to which the heating and cooling times also contribute.

The effect of different annealing processes on the depth distribution of the dopant atoms is best investigated by SIMS. Taking boron implantation as an example, Fig. 9-1a shows the result of conventional furnace annealing. 30 minutes annealing at 900 °C allows diffusion to increase the penetration depth (measured at a concentration of $C = 10^{17}$ cm^{-3}) by a factor of 1.5. During "rapid thermal annealing" (RTA) higher temperatures of $\geq$1000 °C are applied for a short period of time of several seconds. Such annealing conditions can be produced by exposing the substrate to an intensive light source (e.g. an iodine quartz lamp) [9-1]. As Fig. 9-1b shows, diffusion is slight under such conditions: the penetration depth is only increased by a factor of 1.17 in this case. Moreover, spreading-resistance measurements show that such short annealing times are sufficient to activate the dopant atoms. Rapid annealing processes are consequently increasingly used to produce shallow doping profiles.

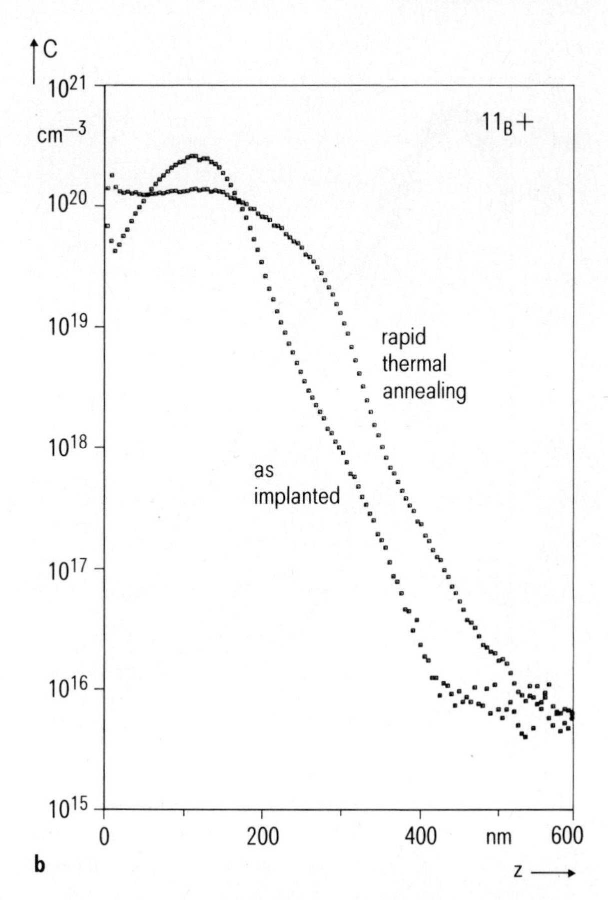

Fig. 9-1. SIMS depth profiles of boron in (100) silicon. a) After ion implantation (25 keV; $5 \cdot 10^{15}$ cm^{-2}) and furnace annealing (900 °C; 30 min). b) After ion implantation (35 keV; $3 \cdot 10^{15}$ cm^{-2}) and rapid thermal annealing (1000 °C; 10 s). SIMS measuring conditions: a-DIDA; normal bombardment ($\psi = 0°$) with O_2^+; $E_0 = 12$ keV; $I_0 = 10^{-6}$ A; crater 750 × 750 μm^2; electronic gate 38% (as shown in Fig. 7-20). Quantification (also of the following depth profiles) was performed as described in section 7.4.3.

Effect of pre-amorphization on doping profiles

The channeling effect can be suppressed considerably by pre-amorphization of the substrate crystal. This requires an additional implantation of the substrate with electrically inactive, high-energy ions (e.g. silicon or germanium) prior to the doping implantation. The resulting amorphous zone at the surface of the substrate (section 2.6.3) prevents single-crystal effects on the implantation profile.

Fig. 9-2 shows depth profiles of boron-implanted silicon with and without pre-amorphization (curves a and b). In this case, the amorphization involved two subsequent implantations of silicon at different energies and doses. Also after annealing (curves c and d), the pre-amorphized specimen exhibits a clearly shallower doping depth and a steeper gradient.

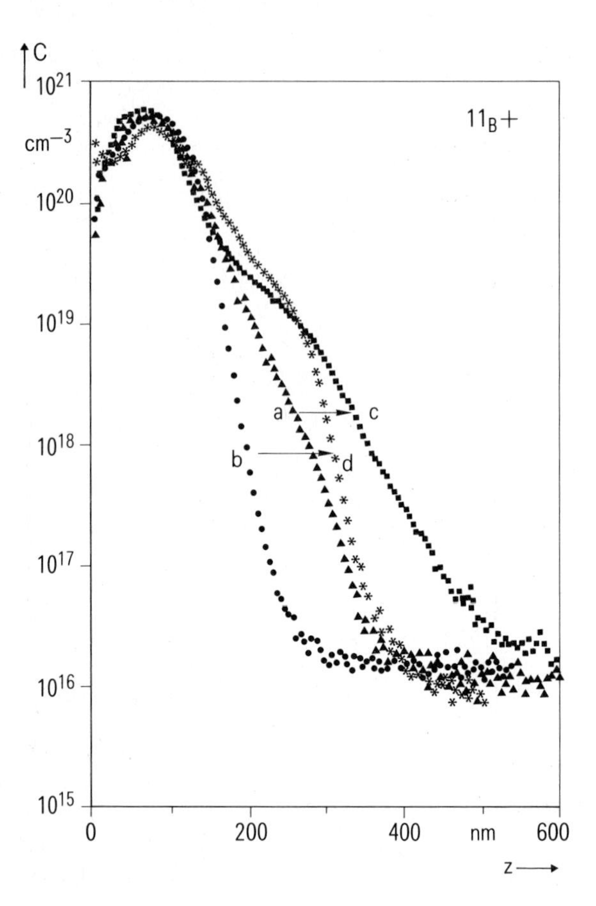

Fig. 9-2.
SIMS depth profiles of boron in silicon. Curve a: as-implanted (25 keV; $5 \cdot 10^{15}$ cm^{-2}). Curve b: as (a), but pre-amorphized by silicon implantations (100 keV; $1 \cdot 10^{15}$ cm^{-2} and 60 keV; $6 \cdot 10^{14}$ cm^{-2}). Curve c: as (a), but annealed at 900 °C for 30 min. Curve d: as (b), but annealed at 900 °C for 30 min. SIMS measuring conditions: as for Fig. 9-1.

Particularly shallow doping profiles can be obtained by combining pre-amorphization with a rapid annealing process. Fig. 9-3 shows the boron depth profiles of specimens after pre-amorphization with silicon, boron implantation and rapid annealing for 10 s at temperatures between 800 °C and 1100 °C. After annealing at 800 °C for 10 s, the doping depth measured at a concentration of $C = 10^{18}$ cm^{-3} is no more than 100 nm. At 1100 °C, it increases to 250 nm. In order to improve the depth resolution for SIMS profiling a lower primary ion energy ($E_0 = 6$ keV) and an increased incidence angle with respect to the surface normal ($\psi = 60°$) were employed than those used for the boron profiles in Figs. 9-1 and 9-2. Both measures result in a lower signal yield and a higher background: For reasons associated with ion optics, the lower energy causes a lower probe current. Due to the increased incidence angle of oxygen bombardment, the silicon surface is not saturated with oxygen (section 2.6.4: chemically influenced sputtering) and the ionization probability is therefore lower than for oxygen bombardment under normal incidence (section 2.6.5).

Fig. 9-4 shows the effect of pre-amorphization on phosphorus and arsenic profiles for a source-drain implantation. In the case of phosphorus, pre-amorphization yields a clearly shallower doping profile. This is not the case for arsenic because the arsenic implantation (with a dose ten times higher than that used for pre-amorphization) itself amorphizes the substrate

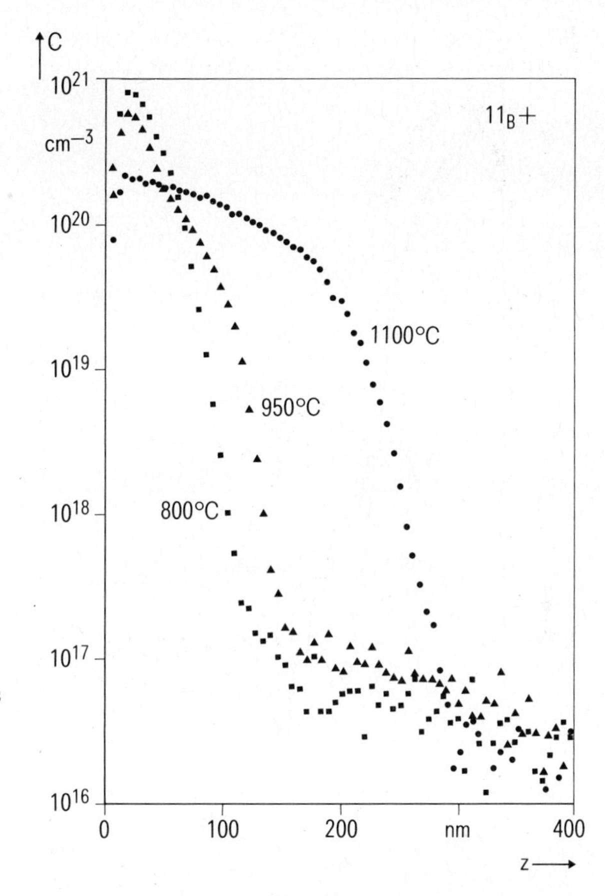

Fig. 9-3.
SIMS depth profiles of boron in silicon. Pre-amorphization (Si; 100 keV; $1 \cdot 10^{15}$ cm^{-2}), doping implantation (B; 15 keV; $4 \cdot 10^{15}$ cm^{-2}), rapid thermal annealing at various temperatures for 10 s. SIMS measuring conditions: a-DIDA; inclined bombardment (ψ = 60°) with O_2^+; E_0 = 6 keV; I_0 = $2 \cdot 10^{-7}$ A; crater 325 × 325 μm^2; electronic gate 19% (as shown in Fig. 7-20).

surface. In contrast, the phosphorus implantation involves a comparatively small dose which is below the limit for amorphization. Since phosphorus is implanted before arsenic, its implantation depth depends on whether or not the specimen was pre-amorphized. This is not true for the arsenic implantation.

The SIMS analysis of phosphorus and arsenic calls for special measures to eliminate those molecular ions the mass lines of which would overlap the mass lines of ^{31}P and ^{75}As (section 7.3.3: overlap of mass lines). High mass resolution and a voltage offset (section 7.3.2: energy filtering for separating atomic and molecular ions) were used to measure the depth profiles in Fig. 9-4.

Residual damage after pre-amorphization

As already indicated, ion implantation results in damage to the crystal lattice. As shown in section 2.6.3 (structural modifications), the damage distribution exhibits a maximum like the implantation profile, but at a shallower depth (Fig. 2-59). In the vicinity of this damage maximum, boron implantation produces a zone of small defect clusters that are difficult to anneal

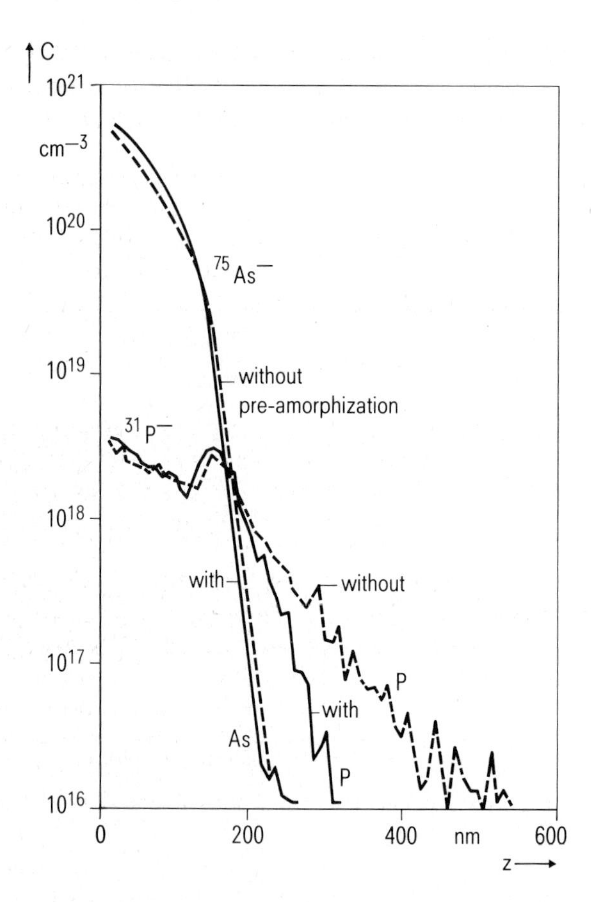

Fig. 9-4. SIMS depth profiles: effect of pre-amorphization (Si; 60 keV; $5 \cdot 10^{14}$ cm^{-2}) on doping profiles of phosphorus (60 keV; $4 \cdot 10^{13}$ cm^{-2}) and arsenic (50 keV; $5 \cdot 10^{15}$ cm^{-2}) after annealing (900 °C; 40 min). SIMS measuring conditions: CAMECA IMS 4F; microscope mode; Cs^+ bombardment; $E_0 = 14.5$ keV; $I_0 = 1.5 \cdot 10^{-7}$ A; crater 250 × 250 μm^2; optical gate diameter 60 μm, specimen voltage offset 40 V; mass resolution $M/\Delta M = 3000$.

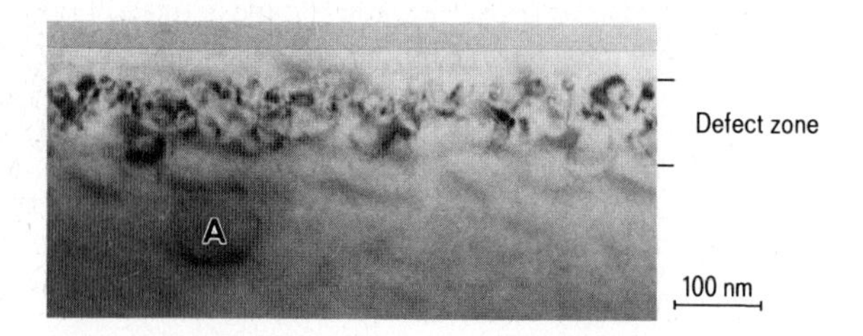

Fig. 9-5. Residual implantation damage after boron implantation (40 keV; $4 \cdot 10^{15}$ cm^{-2}) through 105 nm thick scattering oxide, after annealing at 900 °C for 40 min in N_2 atmosphere (TEM bright-field image of thin cross section). In addition to the numerous small dislocation loops in the marked defect zone, individual loops extend to greater depth (at A).

and form the nuclei of numerous small dislocation loops. The depth distribution of defects is best examined by TEM imaging of cross-sectional specimens: a defect zone of this type containing small loops can be seen clearly in Fig. 9-5. The image also reveals individual dislocation loops (at A) that extend even deeper.

Boron implantation at room temperature will not amorphize the substrate, no matter how high the doses are. This makes pre-amorphization by silicon implantation indispensable to avoid the channeling effect. An amorphized zone produced in this way can clearly be seen in Fig. 9-6a to extend up to the scattering oxide on the surface. The interface between the amorphous zone and the crystalline substrate (a/c interface) is slightly rough. Beneath the interface, a zone with small point-defect clusters is visible. During annealing, the amorphous zone recrystallizes without defects. A defect zone consisting of isolated dislocation loops forms only at the depth of the a/c interface and the point-defect clusters (Fig. 9-6b). (Since the scattering oxide thicknesses and the implantation energies differ, Figs. 9-6 a and b are not directly comparable.) The loops lie on the various {111} planes and appear as rings or lines in projection, depending on their habit planes. Like those in Fig. 4-42, these loops are of the vacancy type. The defect density is much lower than that of the specimen without pre-amorphization (Fig. 9-5), but the individual loops are larger.

The amorphous zone becomes thicker as the Si implantation energy rises. This also produces a rougher a/c interface, since the gradient of the damage profile becomes less steep. The rougher interface leads to an increase of the defect density after annealing and also explains why the defects known as "hairpin dislocations" (at A in Fig. 9-7a) occur in addition to the dislocation loops at high implantation energies (>150 keV). Fig. 9-7b shows that close to a rough a/c interface, crystalline islands remain in the amorphous zone which are slightly tilted with respect to the substrate. If the recrystallization front advances to an island like this, a dislocation segment whose ends propagate to the surface is formed to compensate for the misalignment [9-2].

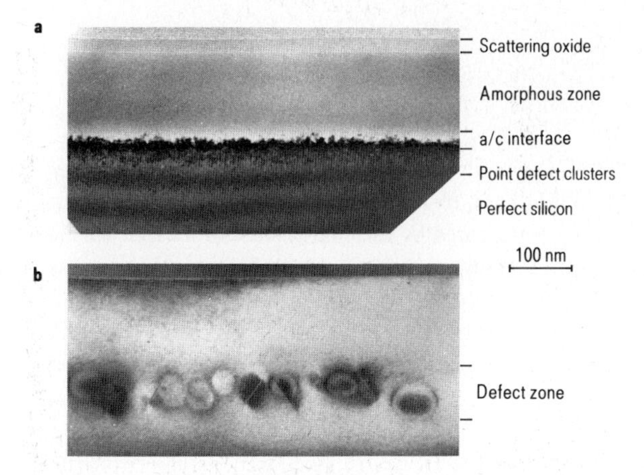

Fig. 9-6. a) Amorphous zone after pre-amorphization by Si implantation (100 keV; $5 \cdot 10^{14}$ cm^{-2}) and B implantation (25 keV; $4 \cdot 10^{15}$ cm^{-2}) with 20 nm scattering oxide. The Si substrate appears dark and exhibits thickness fringes since a reflection was strongly excited. b) Residual defects after pre-amorphization by two Si implantations (100 keV and 150 keV; $5 \cdot 10^{14}$ cm^{-2} each) and annealing at 900 °C for 40 min in N_2 atmosphere (same conditions as for the specimen in Fig. 9-5).

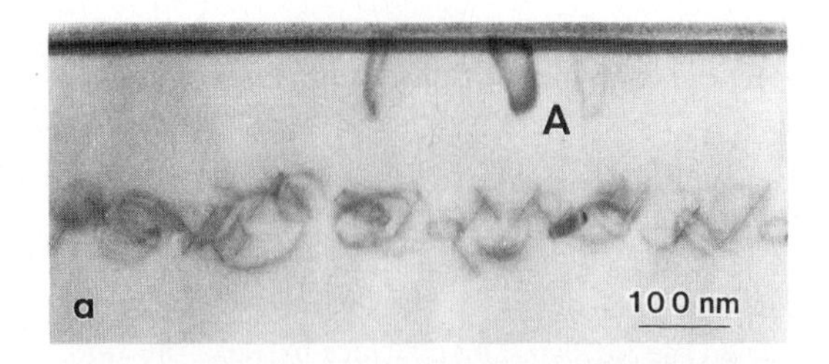

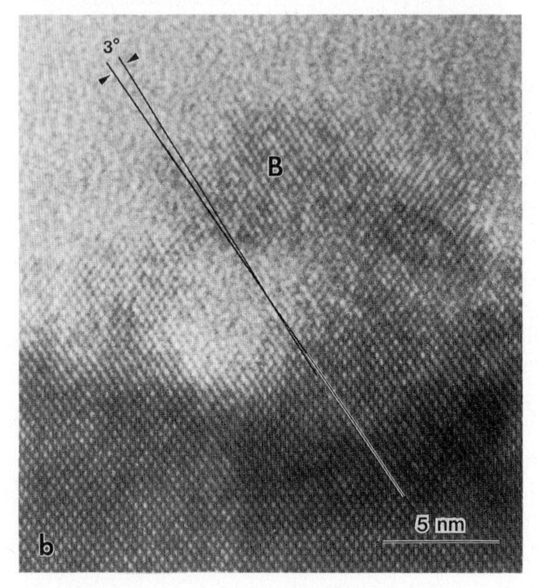

Fig. 9-7.
a) Residual defects after Si implantation (170 keV; $1.3 \cdot 10^{15}$ cm^{-2}), B implantation (30 keV; $4 \cdot 10^{15}$ cm^{-2}), and annealing (900 °C, 30 min, N_2). In addition to the zone of dislocation loops, hairpin dislocations are formed (at A).
b) HREM image of amorphous-crystalline interface after Si implantation (170 keV, $1.3 \cdot 10^{15}$ cm^{-2}). Misoriented crystalline islands (B) in the amorphous zone give rise to the formation of hairpin dislocations.

Doping of trenches

Shallow profiles are also required for doping the capacitors in the memory cells of the 4-Mbit DRAM. As in former generations of dynamic memory devices the information stored in each cell is represented by the charge state of a capacitor. However, the high integration level of the 4-Mbit memory makes it impossible to use the conventional planar arrangement of the memory capacitors parallel to the substrate surface. Instead of this, the vertical dimension is utilized in addition in order to accommodate a memory capacity as high as possible on a smallest possible substrate area. This is realized by etching trenches (typically 1 μm × 1 μm wide and 4 μm deep) in the substrate and using the trench walls as one of the capacitor electrodes (Fig. 3-9).

The first electrode of such trench capacitors is produced by doping the trench wall (Fig. 9-8a). For this purpose, chemical vapor deposition (CVD) is used to coat the trenches with an SiO_2 layer containing arsenic. In a subsequent annealing process (1000 °C; 60 min), arsenic is diffused out of this SiO_2 layer into the trench wall. Removal of the SiO_2 layer is followed by several process steps. These include a thermal oxidation process to form the capacitor dielectric and the deposition and doping of a polysilicon layer as the second electrode.

a

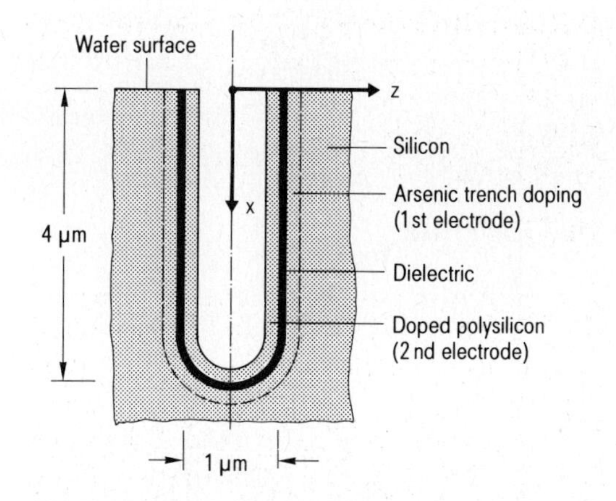

b

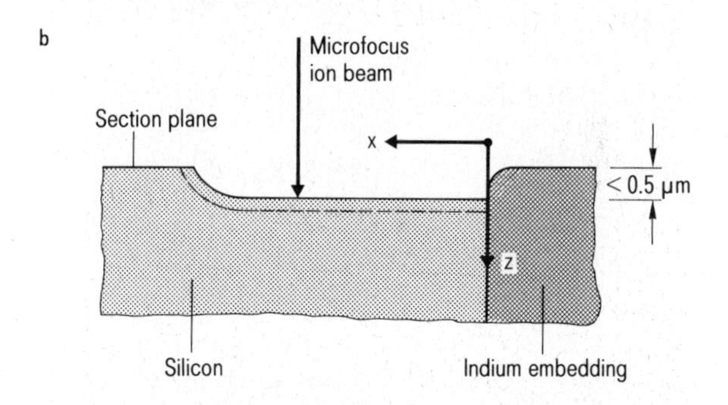

Fig. 9-8.
Trench capacitor of memory cell in 4 Mbit DRAM (schematic view). a) cross section of trench; size normal to the drawing plane: 1 μm, for test trenches: 200 μm. b) Cross section of specimen prepared for measurement of arsenic depth distribution $C(z)$ normal to the trench wall by means of micro-area SIMS (see text for preparation technique).

In principle, the arsenic profile after diffusion and the effects of the subsequent processes can be examined by conventional SIMS on planar, unpatterned test specimens (curve a in Fig. 9-9). However, systematic differences have been observed between the capacity of the trench cell and the analogous planar structure, indicating geometrical influences on the CVD process. It is therefore important to allow direct determination of the arsenic dopant distribution in the sidewalls of the trenches.

These sidewalls have surface areas of a few μm^2. Measuring depth profiles on such small areas means operating under extreme conditions: a lateral resolution of <1 μm with simultaneously high sensitivity (micro-area mode) [9-3,4]. Curve b in Fig. 9-9 shows the depth profile of an area measuring 3 μm^2 of a planar specimen (scanning field 6×6 μm^2; electronic gate 1.8×1.8 μm^2). The curve correctly reproduces the profile down to approx. $5 \cdot 10^{17}$ cm^{-3}. At values lower than this curve (b) deviates from curve (a) measured under conventional conditions on a large surface. A better result was not to be expected because of the very small number of dopant atoms within the analyzed volume. Computing eq. (7-27) with $d = 1$ μm (diameter of the analyzed area) results in a detection limit of $X_{i\,\min} = 2 \cdot 10^{-6}$ corresponding to $C_{i\,\min} = 10^{17}$ cm^{-3}. Nevertheless it follows from Fig. 9-9 that depth profiles can be measured correctly over a large concentration range even in micro-areas.

To measure the trench doping directly, test trenches were etched and a special preparation technique developed. The test trenches are 200 μm long, 1 μm wide and 4 μm deep. After diffusion of the arsenic into the trench walls, these specimens have been subjected only to those

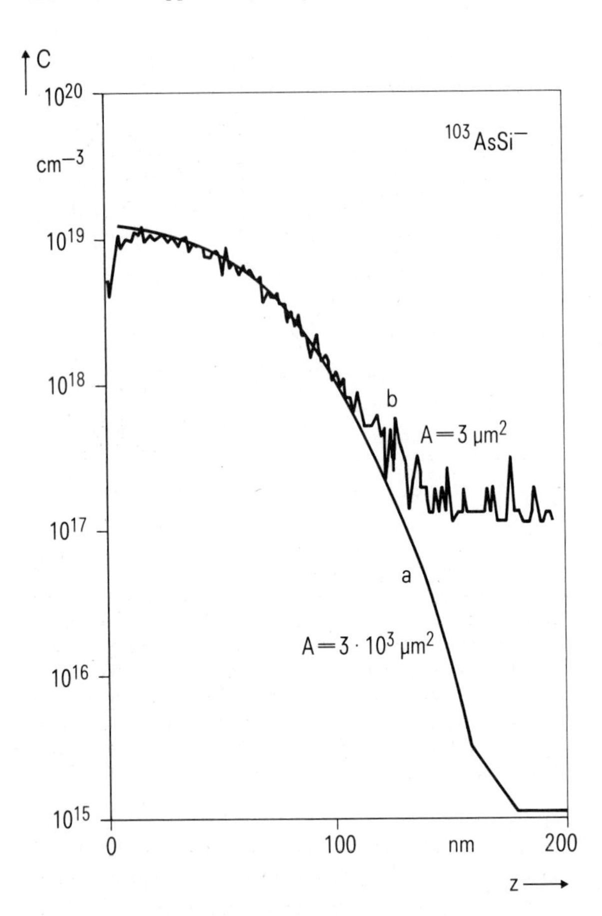

Fig. 9-9.
SIMS depth profiles of arsenic diffused into the silicon substrate from an SiO_2 layer containing arsenic (layer removed after diffusion). CAMECA IMS 4F with microfocus cesium gun. Curve a: microscope mode; $E_0 = 14.5$ keV; $I_0 = 10^{-7}$ A; crater 250 × 250 μm²; optical gate 60 μm diameter. Curve b: micro-area mode; $E_0 = 14.5$ keV; $I_0 = 10^{-11}$ A; $d_0 =$ 0.6 μm; crater 6 × 6 μm²; electronic gate 30% (as shown in Fig. 7-20). Due to an improved signal yield, the molecule ion ^{103}AsSi was recorded instead of ^{75}As. Quantification with ion-implanted reference specimen.

annealing processes that affect the arsenic distribution; the oxide layers produced by these processes were etched off. Two further steps of specimen preparation are necessary [9-3]:

a) Access to the trench walls must be provided. This is achieved by mechanically grinding and polishing the specimen normal to the wafer surface and parallel to the long dimension of a row of test trenches until one of the sidewalls of a trench is removed (Fig. 9-8b). This is done using the arrangement shown in Fig. 3-8. The other wall (measuring 4 μm × 200 μm) is exposed after this preparation step.
b) The Cameca IMS systems require a planar sample surface around the measured area in order to ensure the proper electrical field shape for extracting the secondary ions. A second preparation step was therefore necessary in order to generate a planar conductive surrounding for the trench wall. This was achieved by embedding the sample in a low-melting metal (indium).

Fig. 9-10 shows arsenic depth profiles of 3 μm² areas of a planar specimen and a trench wall after the same process steps. The difference in the total areas beneath the curves indicates that the trench wall has been doped with a lower dose than the planar specimen. The maximum concentration is $C = 1.5 \cdot 10^{19}$ cm^{-3} for the planar specimen and $C = 1 \cdot 10^{19}$ cm^{-3} for the trench wall. The difference is thought to be due to the oxide acting as a diffusion source

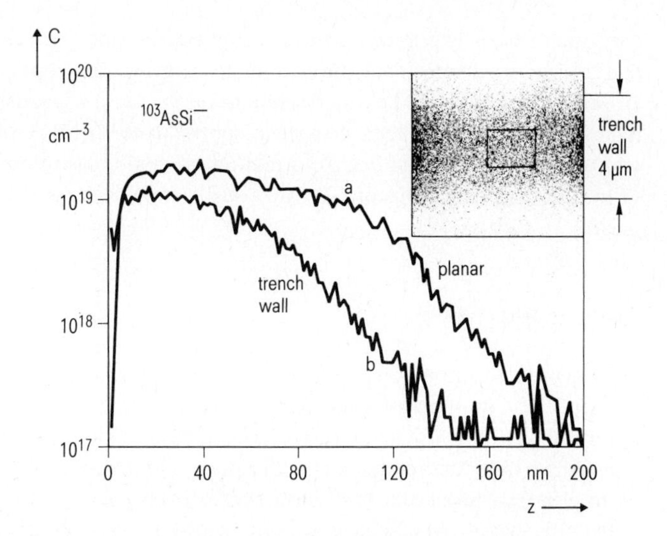

Fig. 9-10. Arsenic depth profiles of trench wall and planar region (Same specimen processing conditions as Fig. 9-9); CAMECA IMS 4F with microfocus cesium gun; micro-area mode; retrospective profiling with the aid of digital image storage system. $E_0 = 14.5$ keV; $I_0 = 10^{-11}$ A; $d_0 = 0.6$ µm; crater 6×6 µm^2, gate 3 µm^2. Curve a: measured on planar specimen; curve b: measured on trench wall (z-axis, see Fig. 9-8). Quantification with ion-implanted reference specimens (see text). Insert: lateral arsenic distribution $C(x, y)$ for curve b. The image consists of 256×256 pixels. Those pixels appear for which the pulse rate (summed over the depth range z = 20 nm to 80 nm) $I > 1$ s^{-1}. The rectangle drawn in on the picture corresponds to an area measuring 1.5×2.0 µm^2 and indicates the size and position of the gate.

being thicker on the planar surface than in the trench. This effect is under further investigation. Profiles measured from 1 µm^2 areas of original trenches have been published in [9-104].

The depth profiles in Fig. 9-10 were obtained with the aid of a digital image storage system. This system permits the area of interest to be selected retrospectively from the stored measurement data from a large crater area extending beyond the trench wall (retrospective depth profiling, section 7.3.4). The area that contributes to the profile can then be optimally matched to the size of the trench wall and an adjustment made to compensate for any drift effects that may have occurred during measurement. The insert in Fig. 9-10 shows the lateral distribution of arsenic in the crater surface and the gate selected for curve b.

Arsenic-implanted reference specimens were used to quantify the depth profiles. As a reference for the trench wall analysis (curve b) a sample with the same topography was used: undoped test trenches were prepared as described in sections a and b above and were then implanted with a defined arsenic dose. This procedure excludes any topographical effects on the quantification.

9.2.3 Localizing pn-junctions

A precise knowledge of both the depth and the lateral extension of the doped areas is essential for the development of VLSI circuits. Deviations from the specified depth or lateral diffusion beneath the mask edges can have considerable effects on the electrical properties. It is standard practice to determine depth distributions by one-dimensional simulation calculations and to use two-dimensional simulations to calculate the dopant distribution at mask edges. However, the results are heavily dependent on the parameters used and require verification or modification by experiment. The same applies to electrical measurements (e.g. of $C(U)$ curves) that provide only indirect information on depth. Therefore, analytical methods capable of localizing pn-junctions directly with sufficiently high spatial resolution (≤ 0.1 µm) are required.

Depth of pn-junctions

With depth resolution down to a few nm, SIMS represents an excellent means of determining the depth of a pn-junction beneath a surface or interface. However, SIMS will not distinguish electrically active and inactive atoms. The dopant atoms must therefore be 100% activated by annealing before measurement can take place. Otherwise the SIMS results cannot be correlated to electrical properties (e.g. sheet resistance).

Fig. 9-11 shows the arsenic and boron profiles in a silicon wafer coated with a polysilicon layer [9-5]; their quantification was shown in Fig. 7-19. Boron was implanted into the substrate before polysilicon deposition. Arsenic was then implanted in the polysilicon layer and subsequently driven into the substrate by annealing (polysilicon emitters for bipolar transistors [9-6]). It can be seen that the arsenic concentration is nearly constant in the polysilicon but shows a peak at the interface, and that boron also diffused from the substrate into the polysilicon layer.

When the finite depth resolution is neglected, the pn-junction is at the point where the two curves intersect: arsenic-doping predominates to the left of the intersection, this area is n-conducting and forms the emitter. Boron doping predominates on the other side of the intersection, this area is p-conducting and acts as the base of the transistor. The conclusion to be drawn from the SIMS profiles is that the emitter-base junction is located at a depth of 240 nm

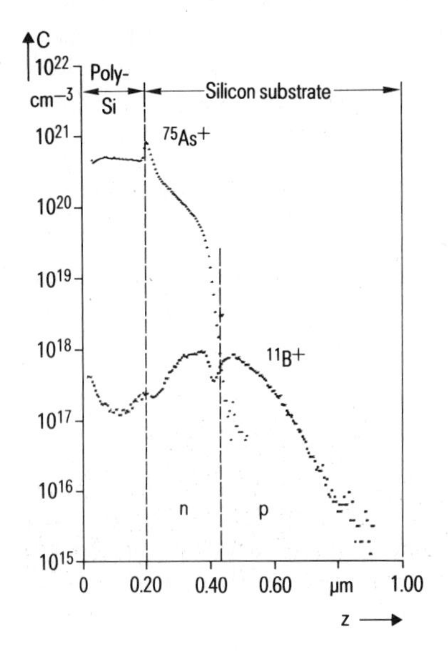

Fig. 9-11.
SIMS depth profiles of boron and arsenic for bipolar transistor with polysilicon emitter [9-5]. The emitter-base junction is at the depth where the profiles intersect, i.e. at a distance of 240 nm from the polysilicon-substrate interface. For SIMS measuring conditions and quantification see Fig. 7-19.

beneath the interface between the polysilicon and the substrate. Polysilicon diffusion sources will be discussed in more detail in section 9.2.4.

The minimum of the boron concentration close to the pn-junction is striking. It is due to a local electric field produced by the steep arsenic concentration gradient and displacing part of the boron ions in the crystal lattice. The boron profile in Fig. 9-11 is typical of this effect.

Lateral effects

A change in the channel length of MOS transistors is one effect that results when dopant atoms diffuse under the mask edges formed by the gate. This effect is particularly serious for modern short-channel transistors.

Fig. 9-12a is a schematic view of a section through an MOS transistor, showing the geometrical parameters of interest. The spatial resolution of SIMS is not sufficient to measure these parameters. However, the doped regions can be imaged in the TEM if the cross sectional specimens are subjected to selective etching [9-7]. But this approach for obtaining quantitative results works well only with arsenic-doped n^+-regions in p-silicon (such as those in question here). Contours of sufficient sharpness are formed only with the steep concentration gradient of arsenic profiles.

Initial specimen preparation is as described in section 4.3.5 (Fig. 4-18). Prior to etching, the specimen must be exposed to buffered hydrofluoric acid for a short period to remove the native oxide layer always present on silicon. Immediately thereafter, the specimen together with its carrier ring is immersed in a mixture of 0.5% hydrofluoric acid in nitric acid for preferential etching of the n^+-silicon (see section 3.3.4). Since the specimen is extremely thin, a very short etching time of no more than a few seconds is used. Relatively sharp contours are obtained due to the steep arsenic profiles and after careful removal of etching residues. Under optimum imaging conditions the etched contour can be localized to an accuracy of 20 nm in the TEM.

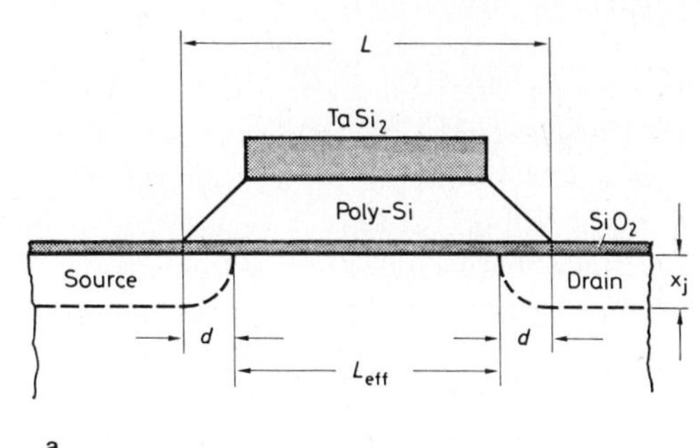

Fig. 9-12.
Cross sections through MOS transistor: a) schematic view (L width of the $TaSi_2$ polysilicon gate electrode, d underdiffusion, $L_{eff} = L-2d$ effective channel length, x_j depth of the pn-junction); b) TEM cross section with selectively etched n^+-doped regions of source and drain. [9-8]

The example [9-8] illustrated in Fig. 9-12b shows a TEM cross section after selective etching from which the geometrical parameters were measured. They can also be determined indirectly by measuring the capacitances of multi-transistor structures [9-9]. Mean values derived from a number of TEM images of one test structure are compared with the results of electrical measurements in Table 9-1. When comparing the results, it must be borne in mind that the etched contours in the TEM images do not correspond exactly to the electrical pn-junctions. A comparison with SIMS measurements of specimens with large doped areas shows that the etched contour corresponds to an arsenic concentration of $C \approx 10^{19}$ cm^{-3}. The exact position of the pn-junction depends on the substrate doping and the gradient of the arsenic doping: in the case under consideration, doping in the channel is only $C = 10^{16}$ cm^{-3}. Due to the steep gradient of the arsenic concentration profile, the electrical pn-junction is only approx. 50 nm deeper than the etched contour. Within the range of measuring accuracy, this is also revealed by a comparison between the TEM results and the electrically measured values: underdiffusion d at the mask edge obtained from the TEM images is slightly smaller than the value obtained by measuring capacitances, whereas the opposite is true for the effective channel length. This good agreement shows that for arsenic doping the less elaborate capacitance measurements are well suited for monitoring of underdiffusion.

Table 9-1. Geometrical parameters of an MOS transistor test structure determined from TEM cross sections (mean values) and capacitance measurements: L_{eff}, d, x_j as in Fig. 9-12 [9-8].

	L_{eff}	d	x_j
TEM imaging	2.04 µm	0.28 µm	0.28 µm
Capacitance measurement	1.91 µm	0.30 µm	–

Contour of pn-junctions

In some instances, the entire two-dimensional contour of the pn-junction is of interest, and not just the depth and lateral extension of a doped area.

In a CMOS inverter on an n-doped substrate, the n-channel transistor is located in a p-doped well (Fig. 9-13a). A sequence of p^+-n-p-n^+ regions forms a parasitic thyristor between neighboring transistors. If this thyristor turns on, a phenomenon known as "latch-up", the device can be destroyed. Charges or potentials on the field oxide (parasitic gate) can promote such turning on. The n-well of the p-channel transistor is designed to reduce this effect.

In a conventional process, the n- and p-wells overlap. The weakly doped overlap region does not contribute to reducing latch-up sensitivity, but it takes up space. A further factor is the complex boundary of the p-well in this area. Fig. 9-13b showing the contour of this junction was produced by means of a computer-aided evaluation of an EBIC micrograph (section 3.6.3, Fig. 3-30). Two critical spots can be seen (arrowed), at which the pn-junction is very severely bent. At these spots latch up is favored [9-10] and the intended effect of the n-well is at least partially destroyed. By varying a number of process steps, it was possible to prevent the p- and n-wells from overlapping. This shortened the critical p^+-n^+ spacing in the inverter, without increasing the latch-up sensitivity [9-10]. Figs. 9-13c to e show the results of several process variants [9-11].

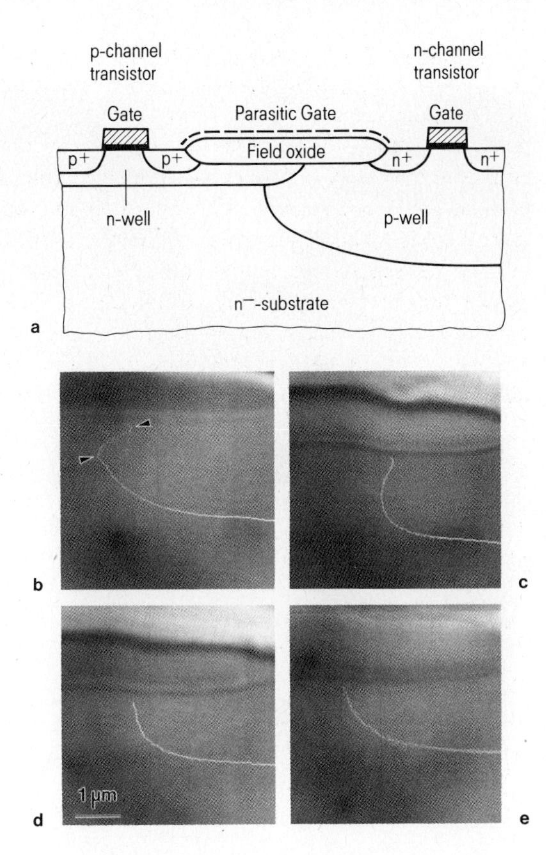

Fig. 9-13.
Cross sections of CMOS inverter. a) schematic view after conventional processing; b) EBIC micrograph of the boundary of the p-well (by computer-processing only the EBIC maximum is mapped); c) to e) images as in (b), but for different process variants reducing the overlap between n- and p-wells. Primary electron energy $E_0 = 5$ keV. [9-11]

9.2.4 Polysilicon diffusion sources

Bipolar transistor with self-aligned polysilicon emitter

If polysilicon layers are deposited directly onto the silicon substrate and doped with a shallow implantation, these layers can then be used as diffusion sources for doping the substrate. This method is used in modern bipolar technology. Fig. 9-14a is a cross-sectional TEM micrograph of a bipolar transistor with a polysilicon emitter produced in this way. Two polysilicon layers are used, acting both as diffusion sources and as contacting layers. From the p^+-polysilicon in the base contact, boron is diffused to obtain the base connections. The emitter is produced by diffusing arsenic out of the n^+-polysilicon of the emitter contact. Besides a polysilicon emitter with its advantage of high current gain, the bipolar transistor in Fig. 9-14 also has a self-aligned emitter-base structure, which results in a drastic reduction of base contact resistance and parasitic capacitances. Self-alignment is realized by using oxide sidewall spacers providing lateral isolation between the base contact and the emitter region. The etching processes used to pattern the p^+-polysilicon and to open the emitter window must be controlled carefully if excessive etching of the silicon substrate is to be avoided. The heights of the steps actually etched by this process can be determined precisely from TEM cross sections – as illustrat-

a

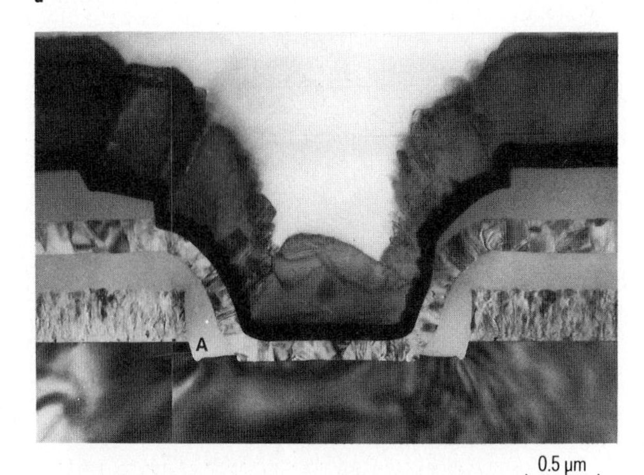

b

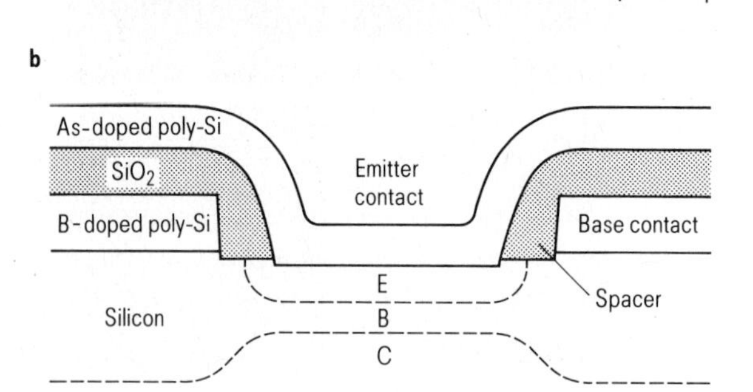

Fig. 9-14.
Cross section of a bipolar transistor with polysilicon emitter and self-aligned emitter-base structure.
a) TEM micrograph, b) schematic drawing. In Fig. (a), the dark contrast in the silicon substrate is due to the crystal orientation being close to the ⟨110⟩ pole. In the emitter contact, the n^+-polysilicon is covered by the metallization consisting of a PtSi and an aluminum layer. In Fig (b), the pn-junctions between emitter E, base B, and collector C are indicated.

ed at A in Fig. 9-14a. The boron-doped p^+-polysilicon exhibits the fine-grained columnar microstructure typical of polysilicon after deposition, because boron doping does not change the microstructure. In contrast, the high arsenic doping in the n^+-polysilicon results in strong grain growth (Fig. 9-14a).

The pn-junctions between emitter, base and collector are outlined in Fig. 9-14b. During technology development, the junction depths are first determined by numerical process modelling. Specific modelling parameters that describe diffusion have to be fitted by comparison with experimental dopant profiles. Good agreement between simulated profiles and experimental SIMS profiles can then be obtained [9-6]. SIMS measurements are required to determine specific dopant depth profiles. Section 9.2.3 includes an example of the depth profiles of arsenic and boron as in an emitter window (Fig. 9-11). The arsenic and boron profiles of a modern 1 µm bipolar process are shown in Fig. 9-15a [9-12]. With a value of about 65 nm, the emitter junction depth in this latter case is much shallower than that in Fig. 9-11, since the arsenic was driven in by rapid thermal annealing (see below). The base extends to the depth at which the boron profile reaches the constant doping level of the epi-layer. In Fig. 9-15a, the base width is no more than approx. 170 nm. It must be as small as possible to obtain bipolar circuits with short switching times. However, narrow, controllable base widths can only be obtained together with very shallow emitter junctions. Fig. 9-15b shows the dopant depth profiles of a bipolar transistor down to greater depths. The collector below the shallow arsenic

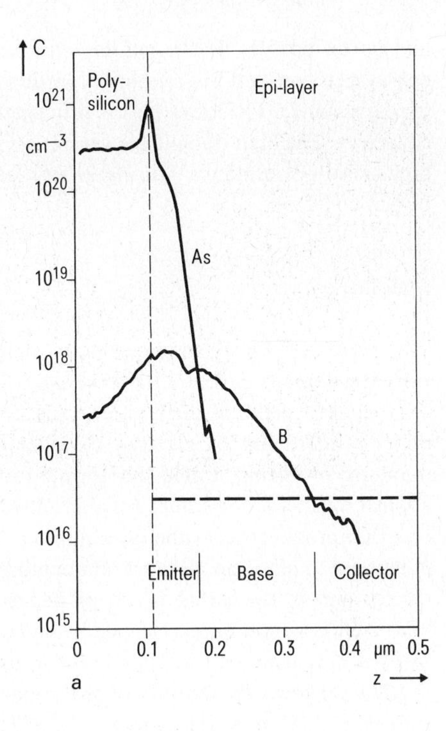

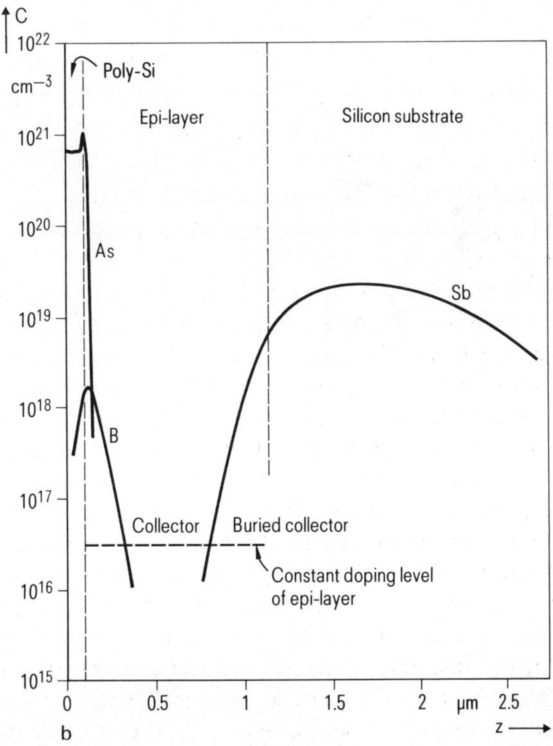

Fig. 9-15.
SIMS profiles of a bipolar transistor. a) Arsenic and boron profiles of shallow emitter and base [9-12]. To measure the shallow profiles with sufficient depth resolution, O_2 primary ions with a low energy of 3 keV and an incidence angle of 60° to the surface normal were used.
b) Dopant profiles of total transistor structure including antimony profile of buried collector.

and boron profiles of the emitter and base consists of two parts. For the upper part, an epi-layer of constant but relatively low dopant concentration is used to reduce the base collector capacitance. The lower part is produced by antimony implantation before epitaxy (buried collector). The high dopant concentration reduces the collector resistance and thus improves high-speed performance. During growth of the epi-layer, antimony diffuses into it from the buried collector.

Interface effects

The interface between the polysilicon and the silicon substrate is of particular significance, since the dopants have to diffuse across it. Moreover, this interface has a major effect on the electrical properties of the bipolar transistors [9-13]. The properties of the interface are determined primarily by the substrate cleaning process prior to polysilicon deposition. Dip etching in diluted hydrofluoric acid (HF dip) to remove the native oxide from the silicon is frequently applied for this process step. An alternative approach is to use standard RCA cleaning, a process that produces an oxide layer approx. 1–2 nm in thickness. However, the silicon surface is exposed to air at an elevated temperature during insertion of the wafer into the deposition reactor and in the initial phase of deposition. This produces a thin layer of native oxide (0.5–1 nm) even on wafers cleaned by an HF dip, with the result that a thin layer of interfacial oxide is always present after polysilicon deposition.

After shallow implantation of the dopant in the polysilicon, the dopant is driven into the substrate by an annealing process. The interfacial oxides described above do not represent major diffusion barriers [9-14]. As already indicated in section 9.2.3, annealing at low temperatures and for short times is needed to produce shallow junctions. When arsenic is diffused by furnace annealing, emitter junction depths of 100 to 150 nm are obtained at drive-in tempera-

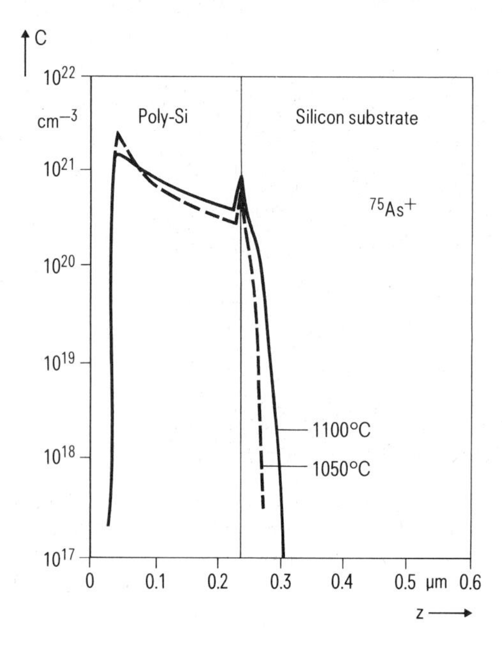

Fig. 9-16.
SIMS depth profiles of arsenic in polysilicon emitter structure after rapid thermal annealing. Arsenic ion implantation: 80 keV, $2 \cdot 10^{16}$ cm^{-2}; annealing: 1050 °C or 1100 °C for 5 s. For annealing at 1050 °C/5 s the junction depth (defined as the depth where the dopant concentration has dropped to 10^{18} cm^{-3}) is only 35 nm. [9-16]

tures of between 900 °C and 950 °C [9-15]. The junction depths obtainable with rapid thermal annealing (section 9.2.2) are even shallower (Fig. 9-16). If annealing is carried out at 1050 °C for 5 s, the emitter junction depth is only 35 nm [9-16]. A low primary-ion energy of about 2 keV is needed if profiles as shallow as this are to be measured with sufficient depth resolution by SIMS. Equally shallow junctions can also be produced for boron under these annealing conditions [9-16].

The morphology of the interfacial oxide and the microstructure of the polysilicon layer may be altered during the annealing process. Both these effects can be examined in detail by observing cross-sectional TEM specimens. Due to their small thickness, it is advisable to use the high-resolution mode to image interfacial oxides. Fig. 9-17 shows HREM micrographs of interfacial oxides after RCA cleaning. The specimen in Fig. 9-17a was furnace annealed at 900 °C after boron implantation. Since the reactive ion-etching process used to open the emitter window roughened the silicon surface to some extent, the two interfaces of the oxide appear blurred. The interfacial oxide is clearly visible by the granular contrast typical of amorphous materials (cf. Fig. 4-52). A continuous layer approx. 1.5 nm thick is formed. At the higher annealing temperatures (≥1000 °C) used for rapid thermal annealing, the interfacial oxide breaks up and agglomerates to form particles (A in Fig. 9-17b). Between the particles, the polysilicon starts to regrow epitaxially onto the silicon substrate (B in Fig. 9-17b). The epitaxial realignment at RCA-cleaned interfaces after annealing at 1100 °C for 5 s, however, remains restricted to small islands like this [9-16].

Specimens cleaned by an HF dip before the polysilicon is deposited have a thinner interfacial oxide layer that breaks up more readily. Epitaxial regrowth then begins at lower temperatures. Fig. 9-18a shows a boron-implanted specimen after furnace annealing at 950 °C. On average, some 0.1 μm of the polysilicon layer has regrown epitaxially. The oxide particles at the interface (arrowed) show up as dark spots due to scattering contrast. As already indicated, high arsenic doping induces strong grain growth at temperatures above approximately 900 °C. Hence, epitaxial realignment is much faster than in the case of boron doping. As a result, epi-

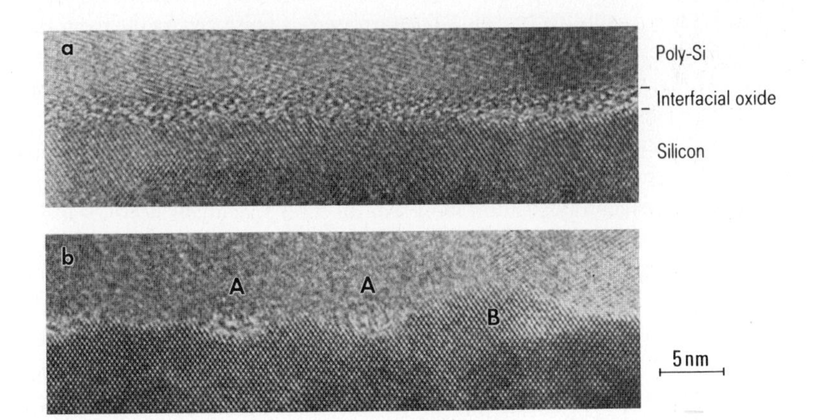

Fig. 9-17. Interfacial oxides between polysilicon and silicon substrate from specimens with RCA cleaning prior to polysilicon deposition; HREM micrographs with ⟨110⟩ projection of the silicon lattice. a) Continuous oxide film after furnace annealing at 900 °C, b) Breaking-up of oxide film into particles A after rapid thermal annealing at 1100 °C for 5 s. An epitaxial island grows at B. [9-35]

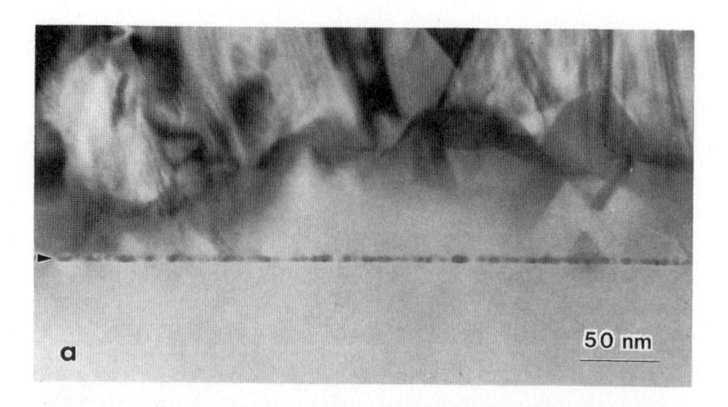

Fig. 9-18. Epitaxial realignment of polysilicon after breaking up of the interfacial oxide (after HF dip) into particles (arrow). Bright-field images of TEM cross sections. a) Partial realignment of boron-doped polysilicon after 950 °C annealing, b) Complete realignment of arsenic-doped polysilicon after rapid thermal annealing at 1050 °C for 5 s. The epitaxial layer contains numerous microtwins. [9-6,16]

taxial regrowth of the polysilicon layer is generally complete in the case of arsenic doping (Fig. 9-18b). Again, the fine dark line (arrowed) is produced by the oxide particles at the original interface. The dark areas are twin lamellae. In this bright-field image they appear dark, because in contrast to the matrix their orientation is such that the electrons are strongly diffracted. They lie on {111} planes and nucleate at the interface.

Fig. 9-19 illustrates the effects produced by different interfacial oxides on dopant diffusion. The figure shows three boron profiles of specimens subjected to different surface treatments prior to polysilicon deposition [9-6]. For all three specimens boron was driven-in for 30 min at 950 °C. The RCA-cleaned specimen exhibits a uniform boron distribution in the polysilicon, because differences in concentration are easily compensated by the strong diffusion along the grain boundaries. The numerous grain boundaries in polysilicon result in the diffusivity being roughly 100 times higher than in single crystalline silicon. The grain boundaries also have the effect that boron can diffuse in polysilicon when the concentration is above the solubility limit. In contrast, at the interface in the single crystalline silicon only the saturation concentration can be achieved which corresponds to the solubility limit (i.e. the concentration of the electrically active boron). As a result, a discontinuity in the boron concentration forms at the interface. The concentration peak at the interface is due to dopant segregation. A segregation peak also occurs at the interface for arsenic doping (Fig. 9-15, 16). However, the concentration is continuous across the interface in this case [9-14].

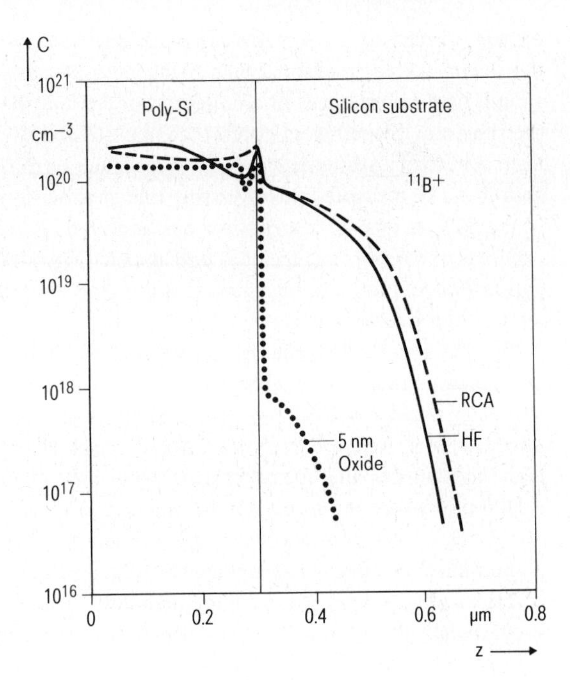

Fig. 9-19.
Influence of interface treatment on boron diffusion; SIMS depth profiles after diffusion at 950 °C for 30 min from samples with various interface treatments: dip etch in diluted HF (full line), RCA cleaning (dashed line), and thermal oxidation (5 nm) at 700 °C (dotted line). [9-6]

The boron concentration of the specimen with HF dip decreases from about the middle of the polysilicon layer and, with the exception of the segregation peak, exhibits no discontinuity at the interface. This is due to the epitaxial realignment of the lower part of the layer (cf. Fig. 9-18a). The discontinuity of the concentration is now located at the interface between the epitaxially regrown polysilicon and the remainder. However, since this interface is not smooth it appears as a continuous transition in the SIMS profile. Epitaxial realignment causes a retarded supply of dopant atoms toward the interface. As a result, the profile of the specimen with HF dip is slightly shallower than that of the RCA specimen. The third profile in Fig. 9-19 is from a specimen at whose interface a thermal oxide approx. 5 nm in thickness was grown at 700 °C. The steep drop in the boron concentration beneath the interface shows that this oxide represents a diffusion barrier. The shoulder at a concentration of approx. $\leq 10^{18}$ cm^{-3} is probably due to the tail of boron implantation into the polysilicon.

As well as affecting the dopant profiles, epitaxial realignment also influences the electrical properties of the polysilicon layers: The sheet resistance of boron-doped specimens falls almost to the value for single crystalline silicon [9-16], and arsenic-doped polysilicon exhibits very low emitter contact resistances after realignment.

Arsenic segregation

The SIMS profiles of both boron and arsenic exhibit concentration peaks at the interfaces indicating dopant segregation. Since there is a thin oxide film present at the interface, a peak of this kind may be caused by chemically enhanced ion emission due to oxygen (section 2.6.5). However, this would be in contradiction to the following observation: vertical sputtering with oxygen primary ions (as is the case with the SIMS profiles of Fig. 9-19) completely oxidizes the silicon surface (section 2.6.4: chemically influenced sputtering) [2-79]. As a result, there should be no differences in secondary-ion yield at the interface between the polysilicon and the substrate. Nonetheless, to exclude any possibility of such an effect, thin TEM cross sections were also investigated by X-ray microanalysis (XMA) with high spatial resolution. Segregation can be measured directly using this method. Since boron is difficult to detect by XMA in the STEM, arsenic segregation was analyzed. As already indicated, the segregation peak occurs for both boron (Fig. 9-19) *and* arsenic (Fig. 9-15, 21c). In addition, this interface segregation was compared to segregation at the grain boundaries of the polysilicon [9-17]. The latter will be discussed first.

Fig. 9-20a shows the microstructure of a highly arsenic-doped polysilicon layer. A row of point measurements were taken across grain boundaries that were parallel to the electron beam (A in Fig. 9-20a). A STEM with a field emission gun providing a beam 1–2 nm in diameter in conjunction with a current of 1 nA was used to obtain high spatial resolution [9-17]. The high current permits measurements with an adequate signal at thin specimen regions (≤100 nm) where the beam broadening due to elastic scattering is small (cf. Table 5-2). The measuring points were selected to be only 5 nm apart in the vicinity of the grain boundary. To minimize the effect of specimen drift, the position of each measuring point was monitored and the necessary corrections made manually during the 100 s measurement time. Fig. 9-20b shows the result of one such measurement, the arsenic profile across a grain boundary. In the

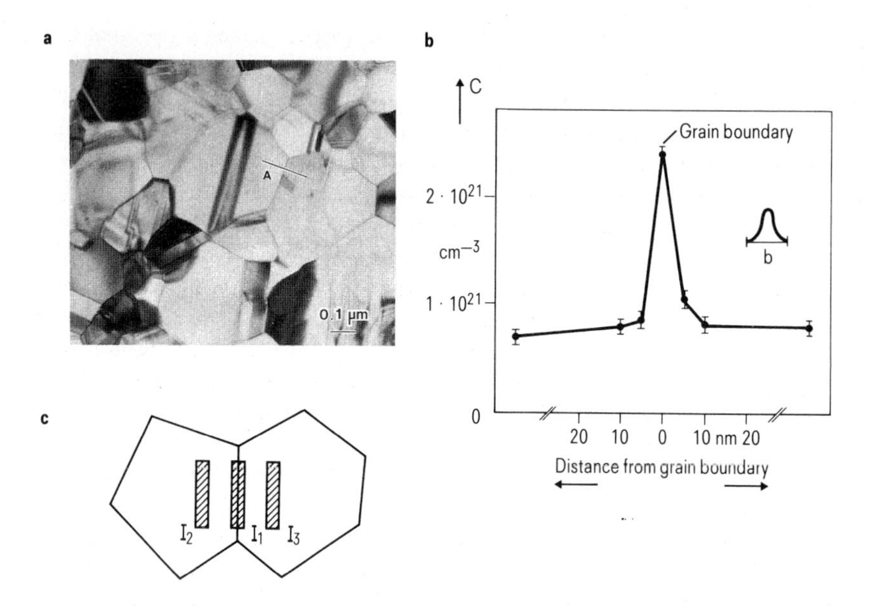

Fig. 9-20. Arsenic segregation at grain boundaries in polysilicon. a) TEM bright-field image of highly arsenic-doped polysilicon layer. The line A indicates a row of XMA measurement points across the grain boundary. b) Arsenic concentration profile across grain boundary measured by XMA in a STEM equipped with a field emission gun. c) Schematic drawing showing analyzed specimen volumina at the grain boundary (intensity I_1) and in the grain interior (intensities I_2 and I_3. [9-17]

interior of the grains, the arsenic concentration is $0.8 \cdot 10^{19}$ cm^{-3}, rising by a factor of 3 at the grain boundary. The error bars are the result of counting statistics. The insert in Fig. 9-20b is a schematic plot of the spatial distribution of X-ray production against beam broadening b from equation (5-40). This distribution shows that the increase in the arsenic concentration at a distance of 5 nm from the grain boundary is due to the tails of the X-ray production distribution and that the width of the zone of high arsenic concentration must be less than 5 nm. This width is probably no more than a few monolayers [9-18].

To obtain a quantitative assessment of the density of the segregated dopant atoms (e.g. expressed as fractions of a monolayer), an elongated specimen volume containing the grain boundary was analyzed (intensity I_1 in Fig. 9-20c). Volumes to the right and left and of the same size but containing no grain boundaries were also analyzed (intensities I_2 and I_3). When the arsenic intensity from the grain interior, i.e. $(I_2 + I_3)/2$, is subtracted from I_1, the remainder is the segregated intensity. Interface densities ranging from 0.7 to $1.5 \cdot 10^{15}$ cm^{-2} arsenic atoms were found at different grain boundaries. The scatter is associated with the varying atomic configuration of the grain boundaries. The average of $1.1 \cdot 10^{15}$ cm^{-2} for the arsenic segregation density corresponds roughly to one monolayer. Segregation also takes place at the interface between polysilicon and thermal oxide, the amount being again roughly one monolayer [9-17]. If segregation at the upper and lower interfaces of the polysilicon is taken into account along with that at the grain boundaries, the fraction of segregated arsenic can be calculated from the average grain size. For a grain size of 0.3 μm and a total arsenic concentration of $8 \cdot 10^{20}$ cm^{-3}, roughly 30% of the arsenic atoms are trapped at the grain boundaries (and interfaces), where they are presumed to be electrically inactive.

Arsenic segregation at the interface between polysilicon and silicon substrate can be measured in a similar way. Polycrystalline regions were observed beside epitaxially realigned regions in a specimen with HF dip after annealing at 950 °C (Fig. 9-21a). The arsenic concentration profiles normal to the interface exhibit clear differences (Fig. 9-21b). In the polycrystalline region, the arsenic concentration at the interface is again higher by a factor of about 3. Comparison with the grain boundaries reveals that here, too, roughly one monolayer is segregated at the interface. Arsenic segregation in the epitaxial region is no more than approximately one third of that in the polycrystalline region and is due to segregation at the interface of the SiO_2 particles . As indicated above, segregation of about one monolayer also takes place at the silicon-SiO_2 interface. The microscopic interface area, however, is reduced by the agglomeration of the interfacial oxide into particles. As a result, the arsenic concentration measured at an interface like this is lower. In the polycrystalline region, arsenic segregation stems primarily from the grain boundary to the silicon substrate, because the interfacial oxide is broken up here as well (cf. Fig. 9-17b).

The SIMS profile of this specimen reveals a clear interface peak (Fig. 9-21c). The integral across the peak yields a total arsenic interface density of $0.8 \cdot 10^{15}$ cm^{-2}. This is an average over a 200×200 μm^2 sampling area containing both aligned and non-aligned regions: it agrees very closely with the average of microscopic segregation densities determined by XMA in the aligned and non-aligned regions (approx. 1.1 and $0.4 \cdot 10^{15}$ cm^{-2}). Since XMA is not dependent on matrix effects when thin specimens are analyzed, this comparison shows that the interface peak in the SIMS profile corresponds to an actual arsenic enrichment. The SIMS profile in Fig. 9-21c also shows that the arsenic concentration in the polysilicon decreases steadily toward the interface. The cause is the epitaxial realignment in parts of the specimen leading to reduced arsenic diffusivity and a retarded supply of dopant atoms.

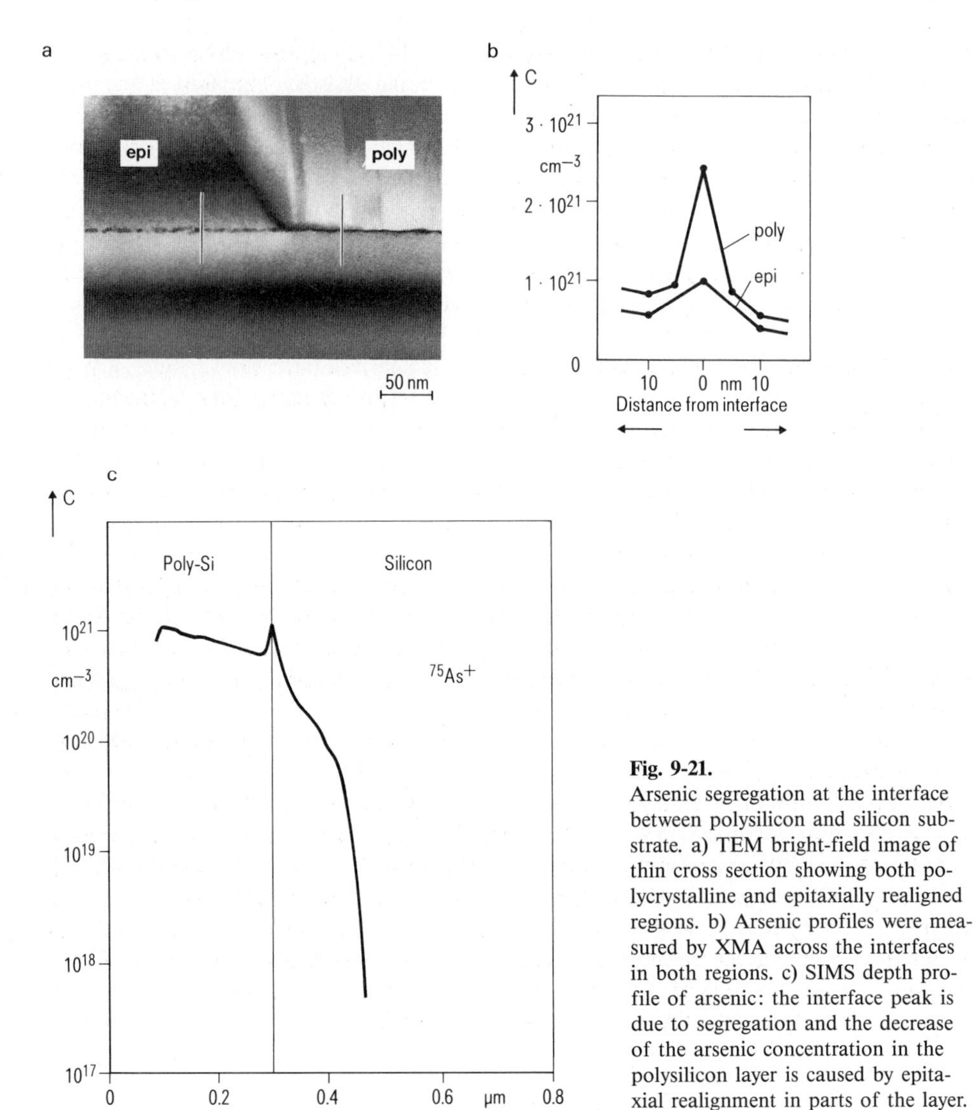

Fig. 9-21.
Arsenic segregation at the interface between polysilicon and silicon substrate. a) TEM bright-field image of thin cross section showing both polycrystalline and epitaxially realigned regions. b) Arsenic profiles were measured by XMA across the interfaces in both regions. c) SIMS depth profile of arsenic: the interface peak is due to segregation and the decrease of the arsenic concentration in the polysilicon layer is caused by epitaxial realignment in parts of the layer. [9-6]

9.2.5 Refractory metal silicide films

$TaSi_2$-polysilicon films (polycides)

Modern CMOS circuits such as memory devices generally require three polysilicon levels for the gate and capacitor electrodes and for the interconnections. The conductivity of polysilicon is limited, however, primarily by the solubility limit of dopants. To overcome this restriction, double layers of a metal silicide and polysilicon – also called polycides for short – are used for the uppermost polysilicon level in particular. The silicides of refractory metals such as titanium, molybdenum and tantalum have the advantage of being compatible with existing process technology: they are stable at temperatures as high as 900 °C and are resistant to chemical wafer cleaning and etching solutions containing HF. Sheet resistances only one tenth that of highly doped polysilicon can be achieved with these silicides. Resistance-capacitance (RC) losses in VLSI circuits with polycide interconnections can thus be reduced in this way, which in turn reduces delay times. The polysilicon is still needed along with the silicide, firstly because the well defined interface with the SiO_2 is retained and secondly because it acts as a silicon source or drain during the annealing process, thus assuring reproducible manufacturing of stoichiometric disilicides. Murarka [9-19] has given an overview on the application of silicides in silicon technology. Here follows a discussion of analyses of $TaSi_2$-polysilicon films [9-20, 21], although it should be mentioned that the results obtained for other silicides such as $MoSi_2$ and $TiSi_2$ are similar.

Sputter deposition is the most commonly used method of producing the silicide films, whose composition should be close to the stoichiometric value with a mole fraction of 33% tantalum. The sputtering processes used differ: when compound targets are employed, their composition determines the composition of the film. Co-sputtering from single-element targets involves depositing alternating thin layers of silicon and metal; the mean composition can be adjusted by varying the thickness of the individual layers. Technologically not as simple as the use of compound targets, this sputtering method is used less frequently in fabrication. However, it has the advantage that the material used for the first layer can be chosen at will. Starting with a tantalum layer helps to reduce and/or break up an interfacial oxide and therefore supports a homogeneous reaction. An uppermost silicon layer provides passivation with its native oxide.

Fig. 9-22a shows a TEM cross section through such a co-sputtered Ta-Si multilayer on polysilicon. The layers are amorphous. Since tantalum has an atomic number much higher than that of silicon, the tantalum layers appear dark in the image. The slightly wavy surface of the highly phosphorus-doped polysilicon layer means that the layers are not smooth and hence, it is only at very thin regions of the TEM specimen close to the top of the film that the individual sputtered layers are clearly distinguishable. Very much thinner layers are observable as separate lines when multilayers are sputtered onto smooth substrates. Thus, the double layers of Ta and Si of the multilayer shown in Fig. 9-22b are only 3 nm thick [9-22].

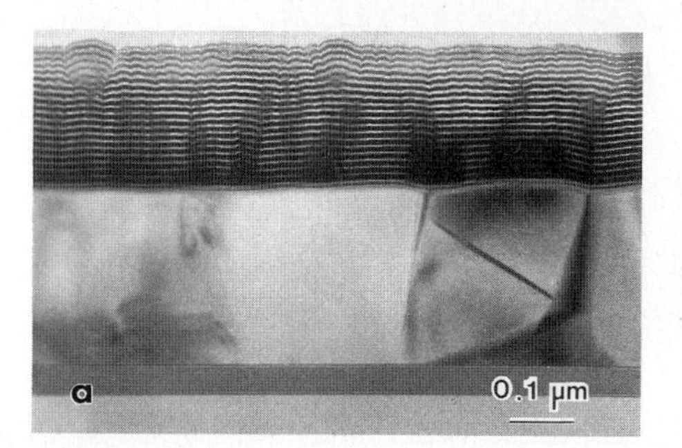

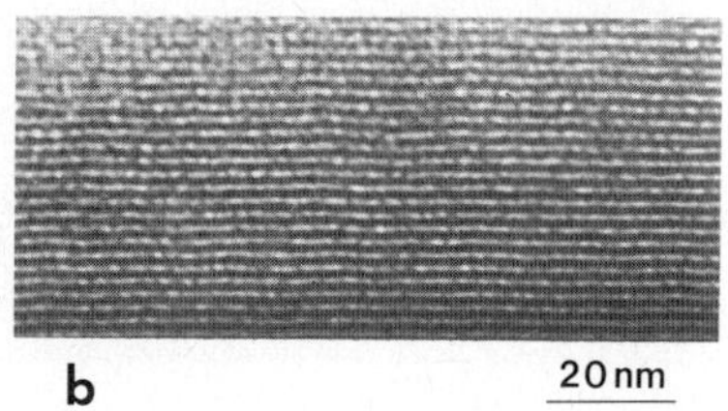

Fig. 9-22. TEM cross sections of tantalum-silicon multilayers produced by co-sputtering from single element targets. a) Multilayer with 12 nm thickness of double layer on top of highly phosphorus-doped polysilicon and gate oxide. (b) Multilayer with 3 nm thick double layers. [9-22]

Measuring conditions allowing good depth resolution were selected in order to resolve the individual layers in Auger depth profiles (section 6.5): a low energy of the argon ions of 1 keV together with a shallow angle of incidence of 5°. Quantification involved applying the standard sensitivity factors to convert the Auger intensities into mole fractions (section 6.4). Fig. 9-23a shows such a depth profile of a multilayer consisting of 20 Ta-Si double layers each 12 nm thick that were sputtered onto a smooth substrate. The individual layers that make up the structure are clearly distinguishable. The uppermost silicon layer and the bottom tantalum layer were thicker than the rest and thus yield a mole fraction close to 100% of their respective elements. The Ta and Si fractions of the remaining, thinner layers vary only between 20% and

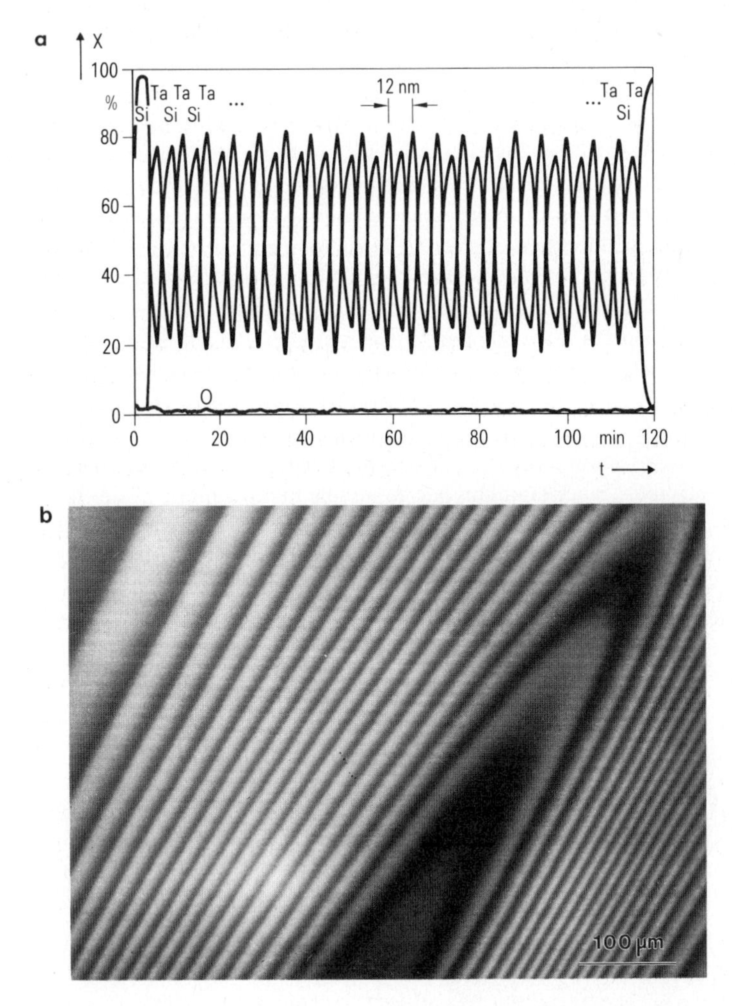

Fig. 9-23. a) Auger depth profiles of tantalum-silicon multilayer with 12 nm thick double layers. Sputtering: 1 keV argon ions at 5° incidence angle (i.e. 85° to surface normal). For the analysis, a stationary electron beam (E_0 = 3 keV, I_0 = 100 nA) was directed normaly onto the specimen and the low-energy Auger peaks of Si (92 eV) and Ta (179 eV) were recorded. b) Secondary electron image of crater edge region after sputtering through the multilayer. [9-21]

80%. While this may indicate interdiffusion, it may also be the result of atomic mixing due to ion bombardment during depth profiling. The amplitude of these oscillations with sputtering time (depth) is not significantly reduced; in other words depth resolution remains virtually constant. This is due to the fact that, as the TEM images show, both types of layer are amorphous and removal under the ion beam is therefore laterally uniform (section 2.6.6). After sputtering through the multilayer, secondary-electron imaging of the crater-edge region provides a means of viewing the individual layers in the Auger microprobe (Fig. 9-23b). The shallow angle of ion bombardment has produced a crater shaped like an elongated ellipse. The shallow section through the multilayer at the crater edge corresponds to a wedge angle of $\leq 0.03°$ that would be virtually unobtainable by conventional angle lapping (section 6.5). At the crater edge, parts of the sequence of layers can be selected at will for further investigation by means of point analyses or line scans.

The sputtered Ta-Si layers have to be *annealed* to produce the low-resistive disilicide. The minimum sheet resistance is already reached after less than 10 min at a temperature of 900 °C [9-20]. Fig. 9-24a shows that annealing produces polycrystalline $TaSi_2$ with a mean grain size of approx. 80 nm. Small pores (arrowed) often appear at the grain boundaries. At the interface between $TaSi_2$ and polysilicon, the individual $TaSi_2$ grains form rounded boundaries with the larger polysilicon grains, thereby minimizing the energy of their grain boundary configuration. This type of interface reaction increases the roughness of the interface. The Auger depth profiles of polycide films before and after annealing are compared in Fig. 9-24b. Since the conditions for profiling were not optimized as regards depth resolution, the individual layers are not resolved in the profile. After sputtering, the mean Ta mole fraction is a nominal 35%. This value was indeed measured by X-ray microanalysis. The value in the Auger profile deviates from 35%, because a relative error of 10% can arise when standard sensitivity factors are used for quantification. Nonetheless, comparison of the two profiles in Fig. 9-24b shows that 2% may be taken as a reliable value for the change in the Ta mole fraction caused by annealing. The 2% difference indicates that stoichiometric disilicide with 33% Ta is formed during annealing. This means a reaction during which some of the polysilicon is consumed. Comparison of the profiles clearly reveals the reduction in polysilicon thickness. Furthermore, the transition of the profiles at the $TaSi_2$-polysilicon interface becomes less abrupt. This phenomenon should not be confused with an interdiffusion effect, but is primarily due to crystallization of the originally amorphous layers and roughening of the interface during annealing. Moreover, sputtering increases the surface roughness of polycrystalline layers and this in turn impairs depth resolution (surface modifications, sections 2.6.3 and 2.6.6).

Sputtering Ta-Si layers with a Ta mole fraction of 35% results in a slight excess of tantalum. This promotes the interfacial reaction, because some silicon is consumed in the silicidation process. The composition of the layers after sputtering has to be monitored carefully if this reaction is to be reproducible. The required measurement error of $\leq 1\%$ mole fraction can be obtained by X-ray microanalysis if the measuring conditions are suitably selected. If the Si K and Ta M X-ray lines are used for the analysis, an excitation energy of 5 keV is sufficient. The excitation depth is approx. 0.1 µm. Consequently, Ta-Si layers as thin as 0.2 µm can still be considered as bulk specimens. The large difference between the atomic numbers of Si and Ta means that errors which cannot be tolerated may arise in quantification using the ZAF correction (section 5.4). Therefore, a matched standard in the form of a polished $TaSi_2$ single crystal must be used here instead of pure-element standards. On account of the comparatively large excitation depth of X-ray microanalysis, the Ta-Si multilayers with 12 nm thick double layers appear as homogeneous material. This is no longer true for thicker double layers (>20 nm). The results of X-ray microanalysis then show slight differences, depending on whether the uppermost layer consists of Ta or Si. Long measurement times of several 100 s are required if the Ta content is to be determined with an error of only 1%. The result showed that the composition does in fact decrease from 35 to 33% mole fraction tantalum when the multilayer is annealed.

If annealing is followed by *thermal oxidation*, an amorphous layer containing no tantalum grows on the surface of the polycide (Fig. 9-25), as can be concluded from the Auger profiles. The oxygen and silicon fractions of this layer are just as large as those of the gate oxide, which means that the layer consists of SiO_2. The chemical etching properties, too, are the same as those of a thermal oxide on (100) silicon. According to Murarka et al. [9-23], all analytical observations can be explained by the following oxidation mechanism: silicon from the $TaSi_2$ oxidizes to SiO_2 on the $TaSi_2$ surface, while at the same time Si diffuses from the polysilicon

a

0.1 µm

b

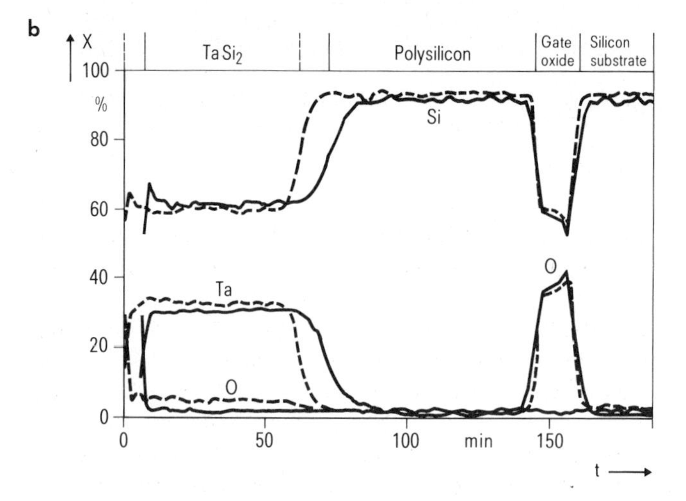

Fig. 9-24. a) TEM cross section of $TaSi_2$-polysilicon layer after annealing at 900 °C showing rough interface and pores (arrows) in the polycrystalline $TaSi_2$. b) Auger depth profiles of tantalum, silicon and oxygen prior (dashed line) and after annealing (full line). The profiles were superimposed such that the location of the gate oxide coincides. Therefore, the origin of the abscissa is only correct for the annealed sample. Standard sputtering conditions were used with 2 keV argon ions incident at an angle of 55° to the surface normal. [9-20,21]

a

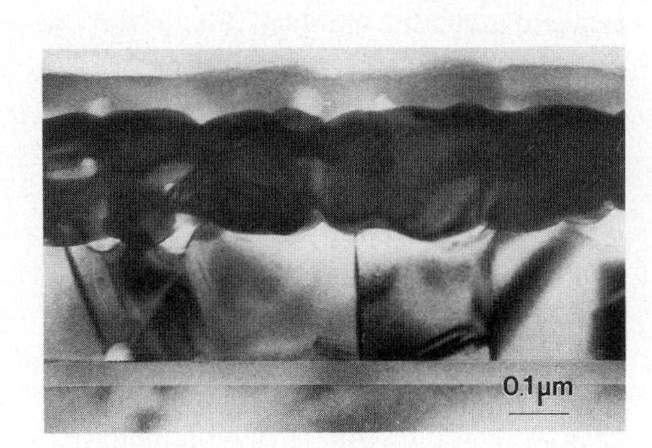

b

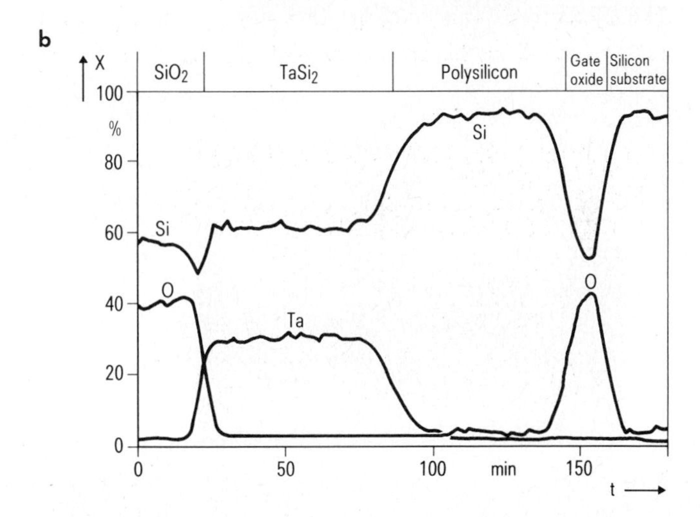

Fig. 9-25. $TaSi_2$-polysilicon layer after annealing and thermal oxidation in dry O_2 (900 °C, 300 min). a) TEM cross section, b) Auger depth profiles. During oxidation a SiO_2 layer grows at the top surface containing large pores. [9-20, 21]

into the $TaSi_2$. Silicon transport through the $TaSi_2$ is so fast that it does not restrict the oxidation rate. Thus, $TaSi_2$ polycide layers can be thermally oxidized in much the same way as polysilicon films, even on sidewalls.

Oxidation of the specimen in Fig. 9-25 was carried out in dry oxygen at 900 °C. An oxidation time of 300 min was needed to grow the approx. 70 nm thick oxide. Over this prolonged period, the average grain size of the $TaSi_2$ increased from 80 nm to 150 nm (Fig. 9-25a). Fig. 9-25a also reveals that the thermal oxide has numerous pores, extending from the interface with the $TaSi_2$ well beyond the middle of the oxide. These pores were assumed to be responsible for the poor breakdown strength of such oxides, which was more than one order of magnitude lower than that of oxides on (100) silicon [9-20].

The removal of the native oxide on the polysilicon prior to Ta-Si deposition represents a critical process step. Etching in diluted hydrofluoric acid and backsputtering are two of the methods used for removing the oxide. If it is not removed, it forms an interfacial oxide between the polysilicon and the $TaSi_2$: this oxide appears as a bright line in the TEM cross section and is about 2 nm thick (Fig. 9-26a). An increase in the oxygen signal is then also observable at the interface in the Auger profile (Fig. 9-26b). A value of 1.5 nm can be estimated for the oxide thickness on the basis of the oxygen peak area, agreeing well with the value measured in the TEM. No interface reaction occurs when a specimen with an interfacial oxide of this kind is silicided. During thermal oxidation the entire $TaSi_2$ layer is oxidized resulting in a mixed Ta-Si oxide (Fig. 9-27), because the interfacial oxide represents a diffusion barrier for the transport of silicon through the $TaSi_2$. The Ta-Si oxide thus produced is of no value as gate material or for interconnections – hence the necessity for removing the interfacial oxide. Removal can be verified both in the TEM and in Auger profiles. Since some silicon is always consumed at a tantalum mole fraction of 35%, the etching process in which the interfacial oxide is removed requires close monitoring to ensure a laterally uniform interface reaction. The effects that can arise by local reactions in contact holes are discussed in section 9.2.6.

a

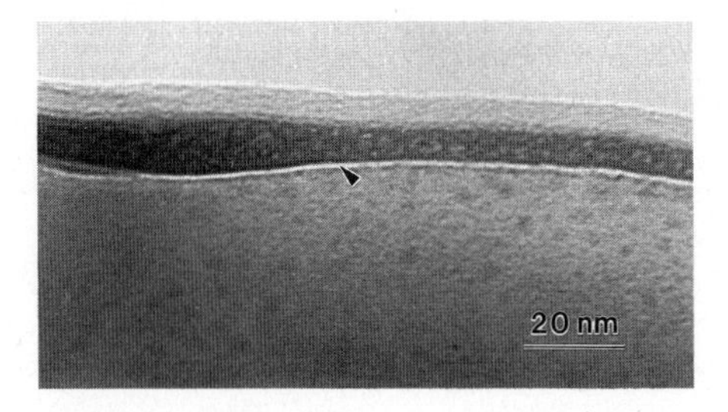

b

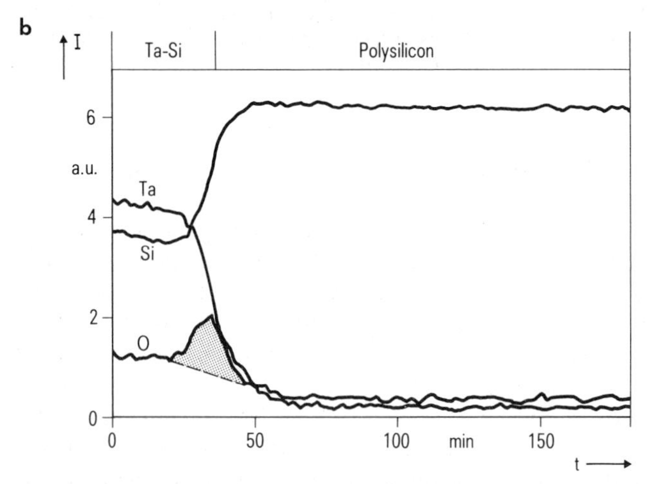

Fig. 9-26. Interfacial oxide remaining between Ta-Si and polysilicon if the native oxide on the polysilicon is not removed prior to Ta-Si deposition. a) TEM cross section, b) Auger depth profiles. Most of the Ta-Si was presputtered prior to recording the profile across the interface with low sputter rate. [9-20,21]

a

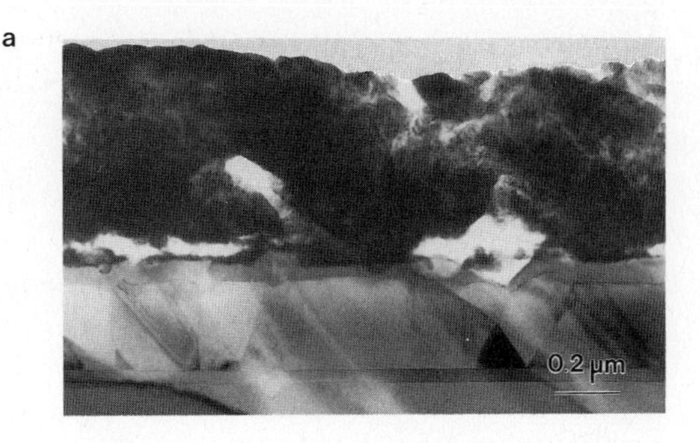

b

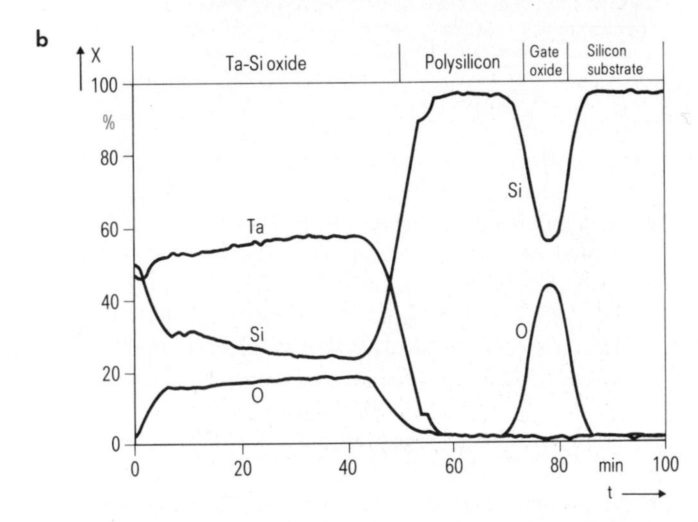

Fig. 9-27. Thermal oxidation of $TaSi_2$-polysilicon layer with interfacial oxide acting as diffusion barrier leads to the growth of a mixed Ta-Si oxide. a) TEM cross section, b) Auger depth profiles. [9-20,21]

Growth of titanium silicide on (100) silicon (salicide process)

With the aid of a self-aligned silicide ("salicide") process, a silicide can be grown in a contact hole, for example, without additional patterning. This process involves depositing a thin metal layer on the silicon wafer with the patterned oxide (Fig. 9-28a). When the wafer is annealed at a moderately high temperature, the metal reacts with the silicon substrate only in the oxide windows (Fig. 9-28b). When the metal which has not reacted is removed from the oxide by selective wet chemical etching a silicide is left that has grown only in the oxide windows (Fig. 9-28c). Since the silicide-silicon interface was produced by a chemical reaction, it is free of contaminants and has a low contact resistance (cf. section 9.2.6). This makes the process interesting for the ULSI (ultra large scale integration) technologies of the future. Suitable metals include platinum and cobalt, but also titanium [9-19]. The silicide of the latter metal exhibits the highest thermal stability. If a salicide process of this kind is applied to an MOS transistor structure in which the sidewalls of the polysilicon gate are masked by oxide spacers, a sili-

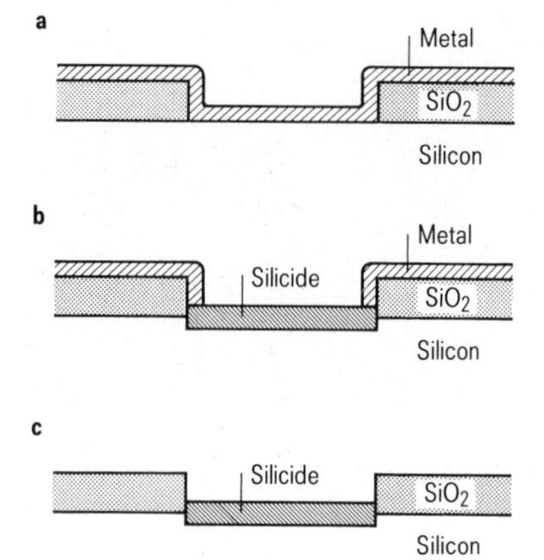

Fig. 9-28.
Principle of self-aligned silicide (salicide) process.
a) Deposition of a metal film on patterned oxide.
b) Annealing to form the silicide in the oxide windows.
c) Selective etching of unreacted metal on SiO_2.

cide is grown on the polysilicon at the same time as growth proceeds in the source/drain contacts. As a result, sheet resistivity is reduced both in the polycide (cf. preceding subsection) and in the source/drain regions [9-24]. Two-stage annealing is called for when titanium is used. In the first step the temperature has to be kept below 700 °C so that the titanium will react only with the silicon and not with the oxide. Following selective etching, the second annealing step is carried out at a higher temperature (e.g. 900 °C) to form the low-resistive $TiSi_2$ of the desired phase (C54). A nitrogen atmosphere is essential for the first annealing process to suppress lateral silicide growth that would cause short-circuiting between adjacent silicide regions [9-25]. During annealing in nitrogen a nitride forms on the titanium surface. Rapid thermal annealing (RTA) is preferable to furnace annealing for the first step, because it more easily avoids oxidation of the titanium which would prevent selective etching later on [9-26]. An overview of the reaction of titanium with silicon, as well as with SiO_2 and Si_3N_4 during RTA is given in [9-27]. In the following, Auger and TEM analyses are used to explain the reaction of sputtered titanium films with (100) silicon during the first annealing step [9-28].

Two factors hinder the Auger analysis of the TiN layer that forms on the surface during RTA in nitrogen. Firstly, the Ti_{LMM} peak overlaps the N_{KLL} peak which is the predominant Auger transition of nitrogen. This has a detrimental effect on both qualitative and quantitative analysis. Secondly, the line shape depends on chemical bonding in the titanium compounds, giving rise to additional errors in quantitative analysis. These difficulties were overcome by taking reference specimens with the compositions of the titanium compounds in question as standards. Fig. 9-29 summarizes the line shapes of the main Auger transitions of titanium in Ti, TiO_x, $TiN_{0.94}$ and TiS_2. The most intense peaks are the LMM and LMV transitions near 385 eV and 420 eV, respectively. With the emission of valence electrons, the LMV line is sensitive to chemical bonding, which results in splitting of the oxide and nitride LMV lines. Changes in line shape also occur for the LMM peak, although valence electrons are not directly involved in this transition. These spectra of reference specimens can serve as "fingerprints" for information on composition and phase.

The quantitative analysis of the composition usually entails converting the peak-to-peak heights of the derivative spectra into mole fractions by applying sensitivity factors (section 6.4). The problems outlined above rendered this method extremely difficult in this case. Alternatively, quantitative information was obtained by comparing the spectra and the signals with

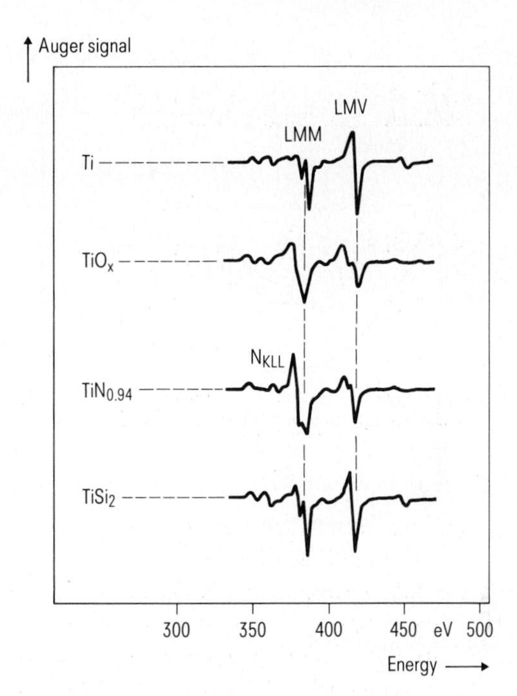

Fig. 9-29.
Auger spectra of Ti, TiO_x, $TiN_{0.94}$ and $TiSi_2$. The line shapes of both LMM (385 eV) and LMV (420 eV) lines of Ti change with chemical bonding. The intensitiy ratio of the major peaks I_{385}/I_{420} is characteristic of the various compounds: 0.67 for Ti, 1.85 for $TiN_{0.94}$, and 0.9 for $TiSi_2$. Since the Ti_{LMM} line is superimposed onto the N_{KLL} line, a value of $I_{385}/I_{420} > 1$ indicates the presence of nitrogen (only in the absence of oxygen). [9-28]

those of the reference specimens. This procedure was restricted to the compositional range close to the standards, greatly reducing the errors due to matrix effects and line shape changes (section 6.4.2). Hence, the peak-to-peak heights (in arbitrary units) in the depth profiles in Fig. 9-30a can be taken as an approximate, relative measure of composition. To estimate the nitrogen content, the profile of the line at 385 eV is shown beside the depth profile of the Ti line at 420 eV. The line at 385 eV represents the sum of Ti and N. The ratios of the two lines I_{385} and I_{420} are characteristic of the different titanium compounds (cf. caption to Fig. 9-29). Suitable data processing yields a more accurate quantitative analysis of TiN_x (section 9.2.6).

During specimen fabrication, the (100) silicon wafers were dip-etched in diluted HF prior to sputter deposition of the 100 nm thick titanium films. Fig. 9-30a shows Auger depth profiles of the Si_{KLL}, Ti_{LMV} and O_{KLL} lines as well as the superimposed Ti_{LMM} and N_{KLL} Auger transitions recorded after annealing in an N_2 atmosphere for 60 s at 700 °C. The line shapes of the Ti_{LMM} peak family are shown in Fig. 9-30b for the various characteristic depths marked in Fig. 9-30a, so that they can be compared with the reference spectra in Fig. 9-29. A high oxygen intensity is found at the top surface: this intensity corresponds to a Ti oxide or oxinitride that is known to form on top of TiN [9-29]. This oxide appears again in spectrum 1 (cf. TiO_x in Fig. 9-29). Once the oxide has been sputtered off, the observable spectrum is typical of TiN (compare spectrum 2 with the spectrum for $TiN_{0.94}$ in Fig. 9-29). The I_{385}/I_{420} intensity ratio also indicates a composition close to TiN. The nitrogen content decreases gradually with depth. After a slow rise, the silicon profile forms a plateau found also in the titanium profile which has the inverse shape. This region corresponds to the silicide formed by solid-state reaction of silicon and titanium. At greater depths, both spectrum 4 and the ratio of the Ti and Si intensities are indicative of the presence of $TiSi_2$. Regions nearer the surface have a higher ratio of Ti to Si, in other words they are richer in titanium. Whereas oxygen

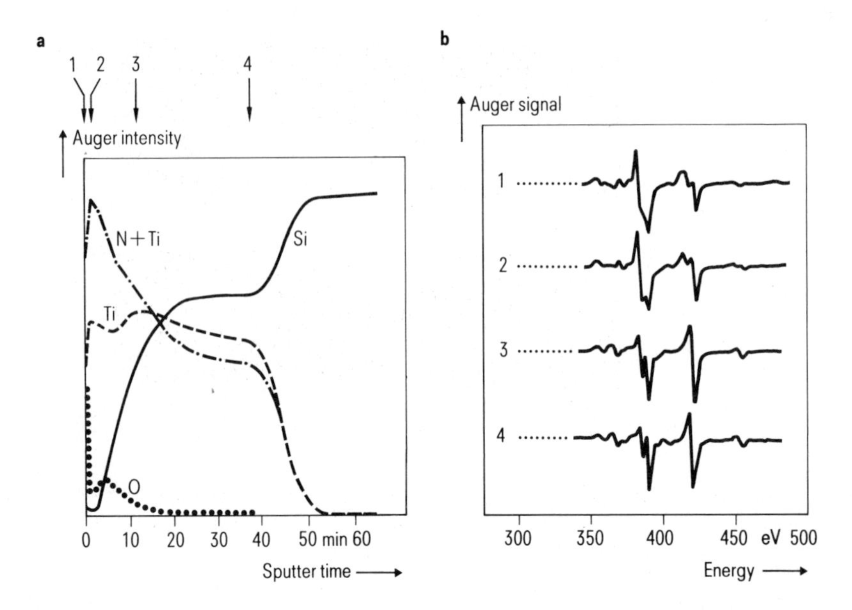

Fig. 9-30. a) Auger depth profiles of the Si_{LVV}, Ti_{LMV}, O_{KLL}, and the overlapping Ti_{LMM} and N_{KLL} transitions for a 100 nm thick Ti film annealed at 700 °C for 60 s in N_2 ambient. For the Auger intensities, peak-to-peak heights are plotted in arbitrary units. b) Auger line shapes of the Ti_{LMM} peak family at characteristic depths marked in Fig. (a) by the arrows 1 to 4. [9-28]

is below the detection limit in as-deposited films, it is clearly detectable as a contaminant in the TiN_x region in annealed specimens and even has a maximum at the silicide interface (Fig. 9-30a).

Fig. 9-31 shows TEM cross sections of 100 nm thick Ti films annealed in nitrogen at 700 °C for 1 s and 60 s, respectively. The approximately 2.5 nm thick oxide or oxinitride layer can be seen on the surface of both specimens. Only part of the Ti layer has reacted with the Si substrate after an annealing time of 1 s (Fig. 9-31a). Lateral inhomogeneities of contamination or a residual oxide at the Ti-Si interface result in a non-uniform reaction in the initial stage. Where a lot of non-reacted Ti is still present, the reaction zone is thin (on the left in Fig. 9-31a). After annealing for 60 s, a TiN_x layer approx. 20 nm thick has formed on the surface. As the Auger profiles in Fig. 9-30a show, the average nitrogen content of this layer is slightly lower than that of TiN. The region of prime interest is the silicide. Two layers are distinguishable in the reaction zones of both specimens: the lower layer contains grains with a lateral extension of approx. 100 nm. The grains in the upper layer are smaller and exhibit numerous crystal defects. The two layers were examined by X-ray microanalysis using EDS in a conventional TEM with a probe diameter of 20 nm. Whereas the lower layer consists of $TiSi_2$ as also indicated by the Auger profiles, analysis showed the composition of the upper layer to be close to TiSi. Since the interface between the $TiSi_2$ and TiSi is very rough, the upper interface of the silicide zone is very broad in the Auger profiles in Fig. 9-30a and the TiSi zone does not appear as a separate layer. The Auger profile also shows that nitrogen is dissolved in the upper region of the TiSi layer, but does not penetrate into the $TiSi_2$.

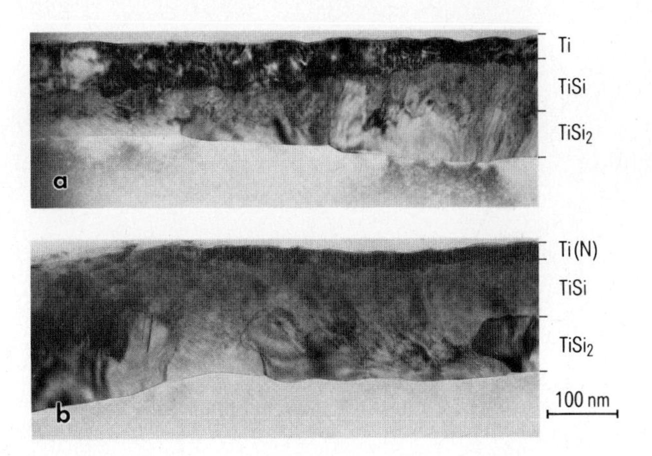

Fig. 9-31. TEM cross sections of 100 nm thick Ti film on Si, annealed at 700 °C in N_2 for 1 s (a) and 60 s (b). The two layers of the reaction zone were analyzed by EDS to consist of TiSi and $TiSi_2$. For Fig. (b) compare the Auger profiles of Fig. 9-30a. [9-28]

Fig. 9-32 is a comparison of the Auger profiles and TEM cross sections of two specimens annealed at 645 °C for 60 seconds, one in a nitrogen atmosphere and the other in argon. The non-reacted titanium layer of the specimen annealed in N_2 has a high nitrogen content at the surface. It is unclear whether TiN has been produced locally or whether a high fraction of nitrogen is dissolved in the Ti. A lateral mixture of Ti and TiN phases may also be present. Obviously, nitride is not detectable in the specimen annealed in argon. A comparison with the TEM cross sections in Fig. 9-32 reveals that the TiSi layer of the specimen annealed in argon is thinner and that all the interfaces are smoother than is the case with N_2 annealing. This also explains why the Auger profiles in Fig. 9-32b show a more uniform silicide composition and sharper interfaces than Fig. 9-32a. Differences in the contaminants may well be the reason why TiSi grows differently in Ar and N_2. In particular, N_2 seems to stabilize the TiSi phase. X-ray diffraction revealed that the $TiSi_2$ consists of the metastable $TiSi_2$ phase (C49). The stable modification (C54) with low electrical resistance grows only at the higher temperatures applied in the second annealing step after the non-reacted Ti and TiN have been removed by etching.

The investigations showed that during annealing in N_2 two competing reactions occur: silicide grows outward from the bottom interface of the Ti film, while a Ti nitride grows inward from the top surface. The TiN and TiSi reaction fronts approach until all the available titanium is consumed. Further reactions are possible only by complex transformation of the phases that have already formed. Further studies of the growth kinetics by numerous Auger profiles showed that the silicidation and nitridation processes are both diffusion-limited [9-28]. Since no nitride is formed during annealing in Ar, silicidation is able to convert all the Ti into silicide. As indicated at the beginning of this section, however, a salicide process requires annealing in N_2 to suppress lateral silicide growth.

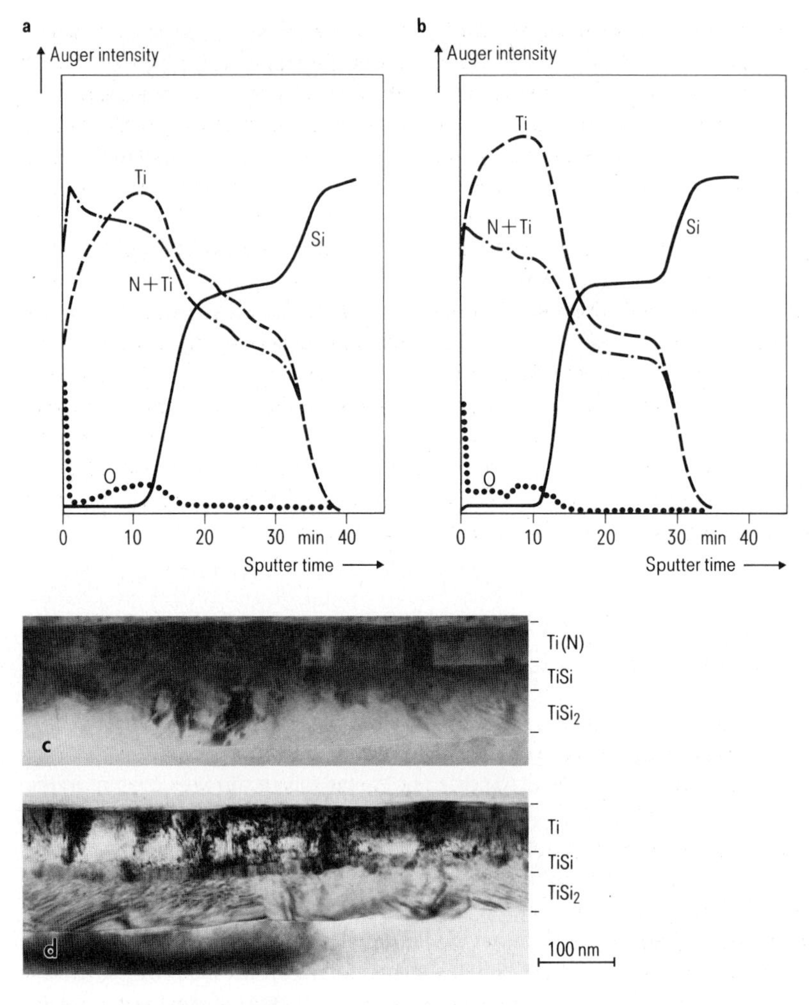

Fig. 9-32. 100 nm thick Ti film on Si annealed at 645 °C for 60 s in N_2 (a, c) and Ar (b, d) ambient. a), b) Auger depth profiles of Si, Ti, O, and overlapping Ti and N lines (peak-to-peak heights in arbitrary units). c), d) corresponding cross-sectional TEM images. [9-28]

9.2.6 Metal Contacts to Silicon

Introduction

Contact holes, too, become smaller as lateral device dimensions decrease – the bit line contact of the 4 Mbit DRAM has a diameter of only 0.8 μm. Although the contact holes become smaller, the thickness of the isolating dielectric layers remains largely constant. Hence, small contact holes have a high aspect ratio, defined as the ratio of depth to diameter. Contact holes like this need steep sidewalls and can consequently be produced only by anisotropic etching processes. Reactive ion etching (RIE) in a CHF_3/O_2 plasma is one such etching process for SiO_2 characterized by high anisotropy and also by high selectivity to the Si substrate.

To ensure reliable contacts, a certain step coverage must be achieved during deposition of the metallization. Step coverage is the ratio of layer thickness on a vertical sidewall to the nominal layer thickness and is a function of the aspect ratio and the deposition process. The sputter deposition processes commonly used, suffer from shadowing effects that reduce film thickness at the bottom and on the sidewalls of the contact hole. Step coverage is usually studied in the SEM by using cleaved cross sections.

Material requirements on both the metallization systems (e.g. sheet resistance) and on the contacts to the silicon substrate and between the individual metallization levels become more stringent as device dimensions decrease. Since higher current densities pass the contact interface, in particular lower contact resistances must be achieved than those acceptable in the past. Undesired interdiffusion between the metallization system and the silicon in the contact can increase contact resistance or cause short circuits. Diffusion barriers (of TiN, for example) can be used to prevent this interdiffusion. A salicide process (section 9.2.5) will also fulfill the same function of metallurgically separating the metallization and Si substrate.

Polymer film and lattice damage induced by reactive ion etching (RIE)

As indicated above, SiO_2 can be etched with high anisotropy and selectivity with RIE in a CHF_3/O_2 plasma in order to produce structures such as contact holes to the Si substrate. High selectivity is obtained by a polymer film which forms on the Si surface as soon as the oxide is removed, protecting it from further attack. During etching of the oxide, however, volatile reaction products are formed because of the additional free oxygen from the SiO_2 and no such protective film is created. Auger spectra revealed the polymer film to contain mainly carbon and fluorine [9-30, 31]. In Fig. 9-33a, the polymer film appears as an amorphous layer approx. 3 nm thick on the silicon [9-32]. The surface of the wafer was sputter-coated with a 10 nm thick Au-Pd film before the TEM specimen was prepared, so that the polymer film would be clearly distinguishable from the epoxy cement used in preparing the cross section.

A bias voltage of several 100 V is applied to make the RIE process rapid and anisotropic. However, bombardment with energetic ions damages the silicon lattice. The defects produced in this way were examined in the TEM [9-32, 33]. The damage is not sufficiently large to amorphize the silicon surface (Fig. 9-33a). Ion bombardment does, however, produce numerous defect clusters extending down to a depth of approx. 40 nm (Fig. 9-33b). Defect density increases with overetch time, i.e. the etching time after the etching front has reached the silicon surface. Besides a pronounced mottled contrast, sharp lines between contrast lobes indicative of platelike defects are also visible in Fig. 9-33b. The defects are located roughly on {111} and {100} planes and exhibit a strain field contrast indicating displacement perpendicular to the habit planes. HREM images of defects on {111} planes showed one or two rows of bright dots, but no inserted lattice planes were observable (Fig. 9-33c). The defects are probably associated with hydrogen and/or carbon.

Since the RIE-induced modifications of the Si surface impair device performance [9-34], the exposed surface is subjected to a post-etch treatment to remove the polymer film and the

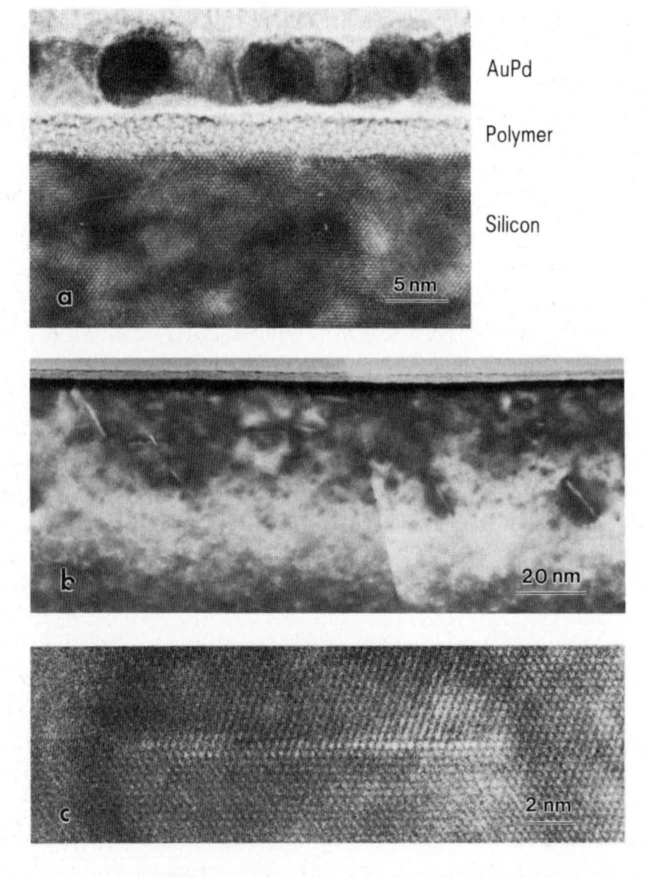

Fig. 9-33. a) HREM micrograph of cross section showing the amorphous polymer film on the silicon surface after RIE of SiO_2 in CHF_3/O_2 plasma. The AuPd film was deposited prior to TEM preparation to separate the polymer film from the epoxy cement. b) RIE-induced defect clusters. Bright-field image of cross section with electron beam parallel to ⟨110⟩ direction (multibeam condition). c) HREM image of platelike defect on {111} plane; ⟨110⟩ projection. [9-32,33]

damage layer close to the surface. Etching in an Ar plasma removes only the polymer film, whereas the damaged surface layer is also removed in a mixture of Ar and NF_3.

Interdiffusion in contact holes

The *Al-Si metallization system* has seen widespread use in silicon technology. The addition of a mass fraction of about 1% Si inhibits the dissolution of silicon from the substrate in the aluminum alloy during annealing, a process that is usually carried out at temperatures around 450 °C. The solid solubility of Si in Al at 450 °C is about 0.8% by mass fraction. If the silicon content is below the solubility limit because of local Si depletion or excessive temperatures, interdiffusion takes place during contact sintering: silicon diffuses from the substrate into the

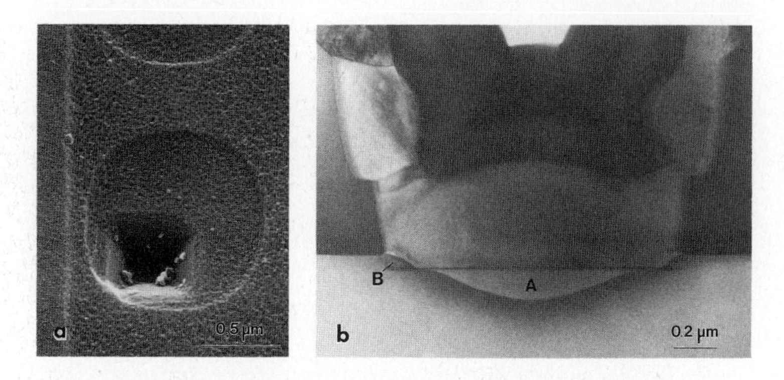

Fig. 9-34. a) The aluminum spike in the contact hole appears as a pit on the Si surface after etching off all device layers; SE micrograph. b) TEM cross section of contact hole showing both Al spike (A) (intermediate stage) and epitaxial Si islands (B) at the corners.

aluminum and vice versa. A feature known as an aluminum spike then is formed in the contact hole. If all the technology layers are removed by etching, the spikes can be seen as pyramid-like pits in the Si substrate (Fig. 9-34a). An Al spike (A) that is not yet fully developed is also present in the contact hole in Fig. 9-34b. The formation of spikes results in junction leakage or shorting in devices with junction depths of a few 0.1 μm.

The excess silicon in the Al(1%Si) metallization precipitates during cooling and forms Si particles that mostly lie at the grain boundaries and at the bottom interface of the Al layer. However, silicon precipitation may also take place in the contact hole, in which case the silicon grows epitaxially on the substrate. Fig. 9-35a shows that small Si islands grow where the interfacial layer is broken up. The interfacial layer is a residual oxide and/or contaminants left behind in the contact hole after the chemical cleaning process that always precedes Al sputtering. In Fig. 9-35b, a large part of the contact hole is covered by epitaxial Si. The small oxide particles make the original interface show up weakly as a dark line. The epitaxial islands contain microtwins. Small epitaxial regions B are also to be seen in the corners of the contact hole shown in Fig. 9-34b. The occurrence of regions like this along with an Al spike in the same contact hole relates to the specific pattern of the metallization and complex temperature cycles [9-35].

Since the epitaxial silicon precipitates from the aluminum, it is p-conductive and produces regions with non-ohmic contact behavior for contacts to n-Si. The contact resistance of contacts to p-Si is increased. Whereas close control of the Si content and annealing conditions is sufficient to cope with the spiking problem, the problem of epitaxial precipitation can only be overcome by barrier layers.

The use and characterization of $TaSi_2$-polysilicon double layers were described in section 9.2.5. In contrast to these polycide layers, the use of $TaSi_2$ layers as interconnections without the polysilicon underlayer would be of advantage in that contacts are possible to both n^+ and p^+ diffusion regions, extending the application to CMOS. However, the process is susceptible to interdiffusion effects in the contacts similar to those that arise with Al(Si) interconnections [9-36]. As described in section 9.2.5, the Ta-Si multilayers are amorphous after co-sputtering from single-element targets and require annealing at high temperatures (900 °C) to form the low-resistive disilicide. The composition of the sputtered layer cannot be controlled to an accu-

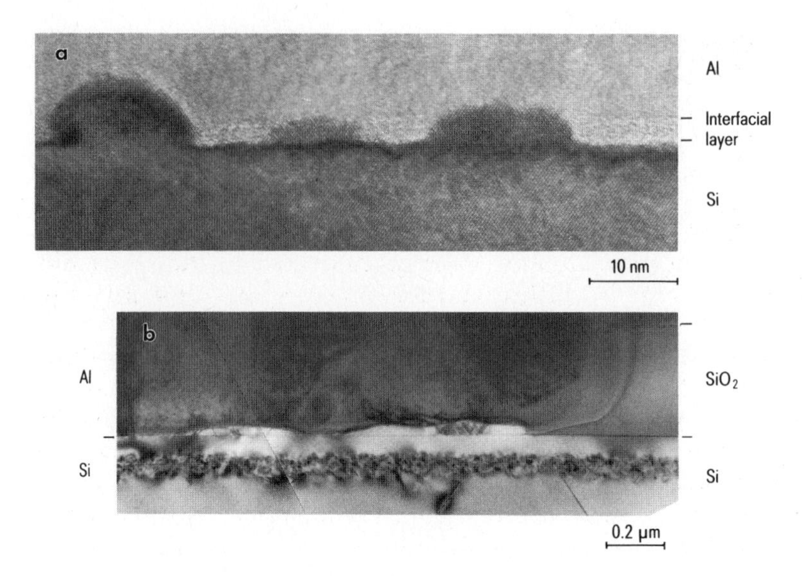

Fig. 9-35. Epitaxial silicon islands in Al (1 %Si) contacts to silicon. a) HREM image of small islands penetrating the interfacial layer. b) Large islands cover most of the contact hole. The defects in the substrate were caused by boron implantation. [9-35]

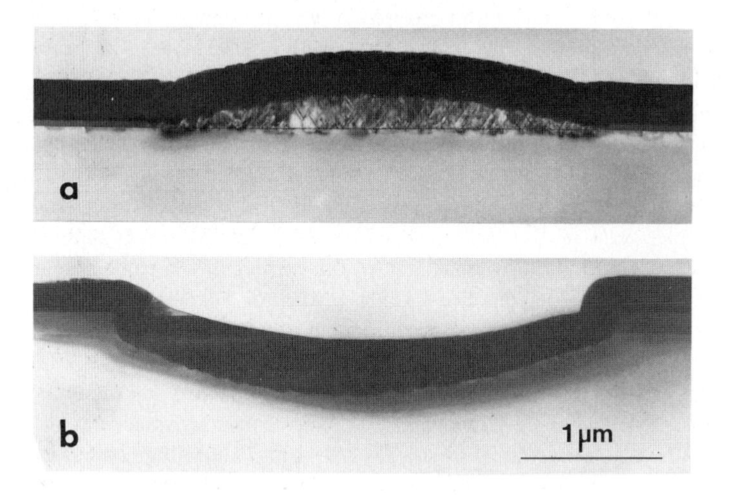

Fig. 9-36. Interdiffusion in $TaSi_2$ contacts to silicon due to small deviations from $TaSi_2$ stoichiometry. a) An epitaxial Si hillock grows for a layer with excess Si. b) A pit is formed in the case of Si deficiency. [9-36]

racy better than 1 % mole fraction. This is also the error for determining composition by X-ray microanalysis (section 9.2.5). To investigate the effect of a deviation from stoichiometry on the high-temperature stability of contacts, layers were produced with deviations of +2% and −2% from the stoichiometric composition (66.7% Si).

In the main, the interconnections lie on oxide and have an interface to the Si substrate only in the small contacts. This means that as with the Al contact holes, TEM cross sections provide the only means of studying such local effects in detail. Fig. 9-36 shows TEM cross sections through contact holes with diameters of 4 μm after annealing at 900 °C. A large hillock of epitaxial silicon containing numerous microtwins has grown in the contact hole of the Ta-Si layer with excess silicon (Fig. 9-36a). Non-linear current-voltage characteristics result, because the epitaxial Si is only lightly doped, if at all. In the layer that is Si-deficient (Fig. 9-36b), silicon from the substrate is consumed and deep pits are produced leading to leakage currents in the contacts. The severe interdiffusion effects are indicative of large-scale lateral mass transport in the Ta-Si layer. This transport extends over a region of several 10 μm and probably takes place in the initial phase of crystallization, during which the numerous grain boundaries permit high diffusivity [9-36]. Taking place even at such slight deviations from stoichiometry, this severe interdiffusion renders sputter-deposited $TaSi_2$ layers useless for the application in question. Deviations from stoichiometry are of no significance where $TaSi_2$-polysilicon double layers are concerned, because they are easily compensated by a reaction between the layers that takes place along the entire interface. Diffusion in the vertical direction is all that is required for this reaction.

TiN-Ti diffusion barriers for Al(Si) contacts

One of the preceding subsections described how interdiffusion (spiking) and epitaxial precipitation of Si can take place in Al(1 %Si) contacts to silicon. These problems, which are critical for small contact holes in particular, can be overcome by a barrier layer. A layer of this nature must act as a diffusion barrier and requires stable interfaces, but it must also have good electrical conductivity [9-37]. On account of its high thermal stability, TiN is one of the most promising materials [9-38]. To improve adhesion on SiO_2 and obtain a low contact resistance on p- and n-silicon, a thin layer of pure titanium is deposited before the TiN. If the TiN layer is produced by reactive sputtering from a Ti target in an Ar/N_2 gas atmosphere, the composition will be primarily dependent on the Ar/N_2 flow ratio. TiN_x layers with $x \approx 1$ exhibit favourable properties.

Since the barrier effect is largely dependent on the composition of the TiN_x layers, quantitative analysis with high accuracy is required. Auger electron spectrometry (AES) is capable of quantitatively analyzing these very thin layers (thicknesses of approximately 100 nm). The overlapping and line shape changes of the Ti and N Auger transitions, however, have a detrimental effect on analysis, as described in section 9.2.5. Nonetheless, accurate results can be obtained if suitable standards are used. Quantitative analyses demand full consideration of the line shape changes with chemical bonding (cf. Fig. 9-29). As usual, the peak-to-peak heights in the derivative spectrum are used to provide a measure of the composition. These heights are proportional to the peak area in the direct spectrum and thus to the number of emitted Auger electrons (section 6.4). The peak-to-peak heights, however, change with the line shape. This effect can be eliminated by reducing the energy resolution to standardize the line shape [9-39]. In this case, energy resolution was reduced by strong numerical smoothing of the experimental data. A series of reference specimens of known composition determined by RBS (Rutherford backscattering spectrometry) and XMA measurements were used for calibration in [9-39]. Fig. 9-37 is a plot of the intensity ratio I_{385}/I_{420} of the overlapping Ti and

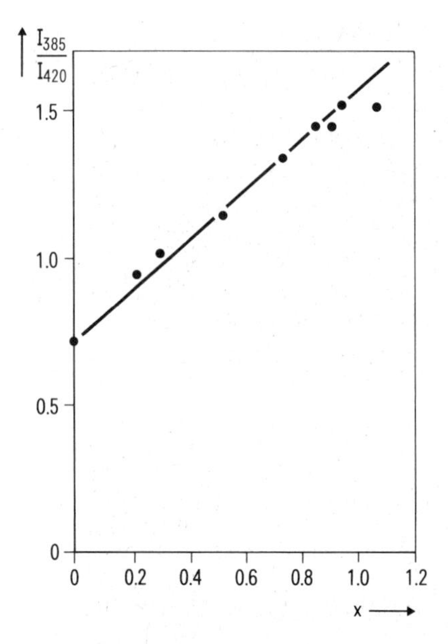

Fig. 9-37.
Intensity ratio I_{385}/I_{420} of the overlapping Ti_{LMM} and N_{KLL} Auger lines (385 eV) and the isolated Ti_{LMV} line (420 eV) versus the N/Ti mole fraction ratio x of TiN_x compounds. The Auger intensities were taken from spectra with reduced resolution and exhibit a linear relationship with x. [9-39]

N lines at 385 eV and the isolated Ti line at 420 eV versus the known N/Ti ratio x. The result is the linear equation

$$x = S \cdot I_{385}/I_{420} - T$$

in which the sensitivity factors S and T can be taken from Fig. 9-37. The factor T takes the overlap of the N and Ti lines into account that becomes apparent in Fig. 9-37 in a shift of the straight line from the origin. When an unknown specimen is analyzed, its composition can be calculated with the aid of this equation. The error in determining x is $\pm 5\%$ [9-39]. However, since the surface is depleted in nitrogen during the removal of TiN_x layers by argon bombardment (cf. section 2.6.3), the specimen to be analyzed must be sputtered under the same conditions as the reference specimens.

For investigation of the thermal stability of the TiN-Ti barrier, specimens annealed at 450 °C and 550 °C were examined by TEM and AES. Fig. 9-38 shows the TEM cross sections of the two specimens. The TiN has a pronounced columnar microstructure. Crystallites with a diameter of 5 – 10 nm extend almost throughout the entire film. The bright lines (for underfocusing) between the grains correspond to Fresnel fringes and would appear dark in an overfocused image (cf. Fig. 4-50). They indicate voids or narrow regions of low density at the grain boundaries which, however, are probably "stuffed" with oxygen due to exposure to air after deposition. Auger analysis showed the TiN to contain indeed several percent oxygen. The Ti layer reacted with the Si substrate in both specimens. Whereas the reaction zone is amorphous in the specimen annealed at 450 °C (Fig. 9-38a), it is crystalline with grains some 100 nm in size after annealing at 550 °C (Fig. 9-38b). The composition of these zones was measured by X-ray microanalysis of the thin cross sections. Since the thickness of the zones is of the order of 10 nm, a STEM equipped with a field emission gun was used for these measurements, providing a spatial resolution <10 nm for XMA (section 5.6.3). The measurements revealed that

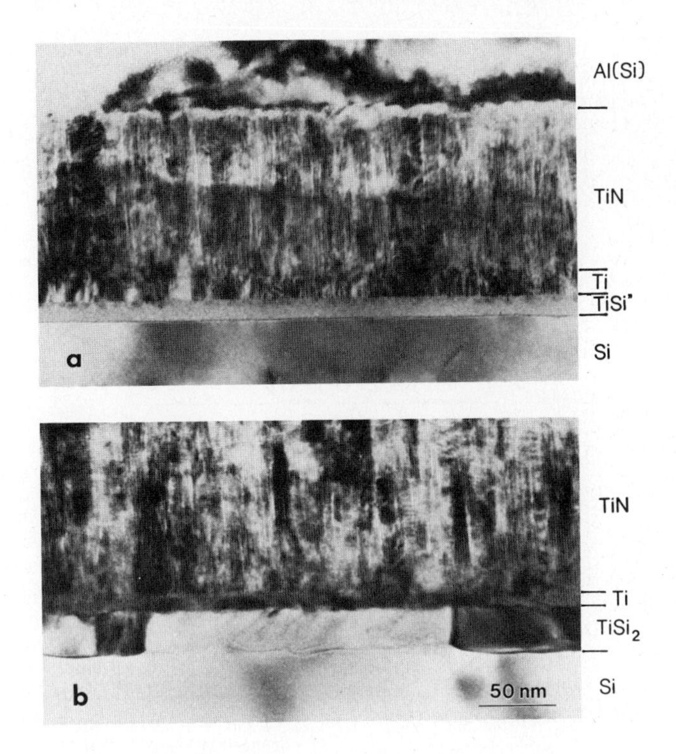

Fig. 9-38. TEM cross sections of TiN-Ti barrier layer between Al(Si) metallization and Si substrate after 30 min annealing at 450 °C (a) and 550 °C (b). EDS analysis of the reaction zones yielded a composition close to TiSi for the 450 °C anneal and $TiSi_2$ for 550 °C. [9-40]

the amorphous layer has a composition close to TiSi, whereas the crystalline layer consists of $TiSi_2$ [9-40].

A crater has to be sputtered through the 800 nm thick Al(Si) film close to the barrier before Auger profiles can be recorded from the barrier. Since the surfaces left by sputtering of aluminum are very rough, the interfaces in the profiles appear to be much wider than they actually are and thin intermediate layers can no longer be analyzed. Angle lapping is used to avoid sputtering through thick films (section 6.5). This entails grinding a shallow spherical dimple in the surface with a ball (ball cratering). If the sputter depth profile is recorded at a point where there is little aluminum covering the TiN, the interfaces are clearly defined (Fig. 9-39) [9-41]. Once again, the isolated Ti_{LMV} and the overlapping Ti_{LMM} and N_{KLL} lines are recorded along with the Al and Si lines. Although the Ti layer can be clearly identified in the Ti profile of the specimen annealed at 450 °C (Fig. 9-39a), the amorphous reaction zone cannot be discerned. (Only after selective chemical etching of the complete aluminum layer and by using a grazing-incidence ion beam for sputtering can the amorphous interfacial silicide be resolved in the Auger profiles as well). In contrast, the $TiSi_2$ layer that grows at 550 °C is clearly visible by the shoulders of the Si and Ti profiles (Fig. 9-39b). A slight reaction with the Al(Si) has also taken place at the upper interface of the barrier. The TiN barrier itself contains a few percent Si and Al, indicative of failure of the barrier at 550 °C. Electric measurements revealed high diode leakage currents.

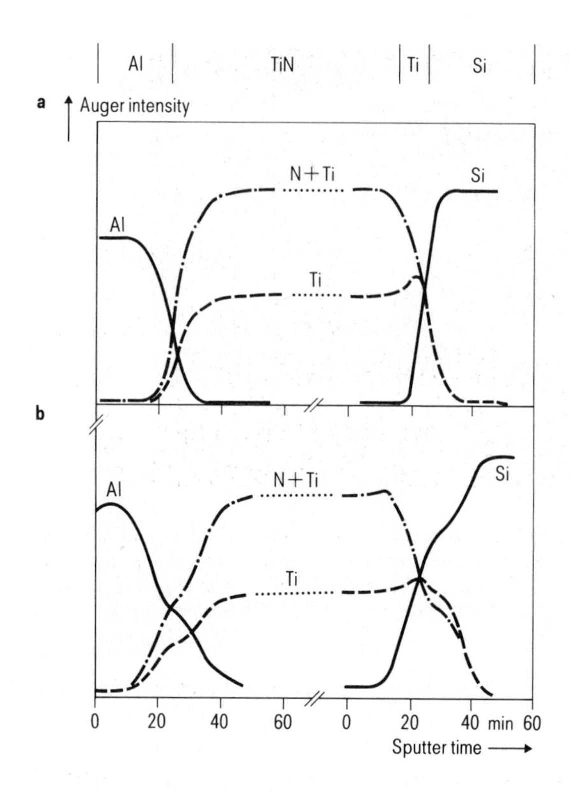

Fig. 9-39.
Auger depth profiles of TiN-Ti barrier layer after annealing at 450 °C (a) and 550 °C (b). To avoid broadening of the interfaces by roughening of the Al surface during sputtering, angle lapping by ball cratering was employed. The depth profile can then be recorded from an area where only a thin part of the Al layer lies above the TiN-Ti barrier of interest.

9.2.7 Submicron device structures

SEM and TEM techniques

Widespread use is made of *scanning electron microscopy* for in-process control in the development and production of integrated circuits. To avoid specimen damage and prevent charging, low-voltage operation is required. However, this limits the resolution to about 10 nm at 1 kV beam voltage in today's field emission SEMs (section 3.4.4). SEM inspection within a process line (in-line SEM) must be nondestructive and precludes any specimen preparation steps, since these would prevent re-introduction of the wafers into the production process. Line widths have to be measured and pattern fidelity checked, which may involve assessing the slope of sidewalls, for example, or inspecting a wafer for residues left after an etching process. Besides in-line SEM, examinations outside the line are required which involve destructive specimen preparation steps. These include the preparation of cross sections for technology assessment and the removal of layers by selective etching for failure analysis (section 9.6).

Cross sections provide detailed information, for example on edge profiles of contact holes or trench capacitors, and can be prepared by cleavage or by polishing (section 3.3.4); they are usually etched to delineate the interfaces, coated with a conductive film and then examined in the SEM at voltages of appoximately 20 kV, high enough to yield resolutions of a few nm. This will generally be sufficient to image the desired details. Fig. 3-9 is a cross section through

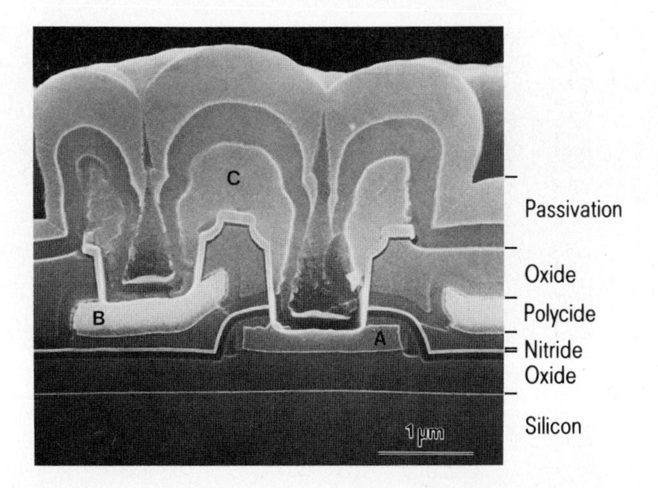

Fig. 9-40. Polished cross section through aluminum contact holes to polysilicon 2(A) and 3(B) in the periphery of a 4 Mbit DRAM (SE micrograph). A Ti-TiN barrier layer is present below the aluminum (C). The polysilicon 3(B) is a $MoSi_2$-polysilicon double layer (polycide). The passivation and the various oxides consist of different layers. (Courtesy E. Demm)

a memory cell with a trench capacitor prepared from a 4 Mbit DRAM chip. Fig. 9-40 is a cross section through the periphery of a 4 Mbit DRAM and shows aluminum contacts to polysilicon 2 and 3.

However, problems arise with the thin dielectric layers that are only some 10 nm thick in today's integrated circuits (ICs). This figure is close to the resolution limit of the SEM, while at the same time the conductive coating covering the sidewalls of the etched steps falsifies the film thickness (Fig. 9-41a). The thickness of the oxide in Fig. 9-41a cannot be determined unless the thickness of the coating is known. If – after etching of the SiO_2 – the silicon substrate and the polysilicon can be etched with different rates such that a stair profile is formed, the effect of the coating is eliminated (Fig. 9-41b). The oxide thickness can then be measured to an accuracy limited only by the resolution of SE imaging. In an SEM with field emission cathode, this limit is approx. 2 nm. In practice, however, this resolution can only be obtained on specimens providing high contrast and when very fine-grained coatings are used.

The best way of imaging the geometrical configuration of thin dielectrics in detail is to prepare thin cross sections (section 4.3.5) and image them in the *transmission electron microscope.* Since the electrons pass through the specimen, the interfaces are only projected as lines in the image when they are parallel to the electron beam. Most IC structures are oriented parallel to the flat of the (100) wafers and are thus oriented along $\langle 011 \rangle$ directions. Hence, the foil normal of the cross sections is close to $\langle 011 \rangle$. Selected area diffraction patterns must be used to tilt the specimen to the correct orientation. The common way of imaging a planar film structure is to tilt the specimen along the $\{400\}$ Kikuchi lines away from the [011] pole and align the (100) plane exactly parallel with the electron beam. Excitation of the (400) and $(\bar{4}00)$ reflection is then equally weak (s is large) and the silicon substrate appears bright in the image (Fig. 9-47a). A structure that is not planar will require imaging in the [011] pole so that all the interfaces are projected as lines. The fact that numerous reflections are excited means that the substrate appears dark (Fig. 9-45). Device structures are examined more easily by using a

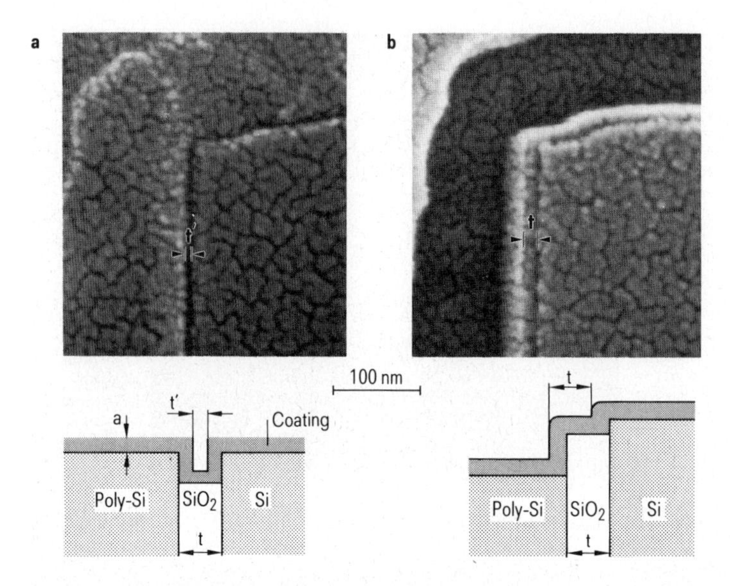

Fig. 9-41. SE micrographs of cross sections through thin thermal oxide between silicon substrate and polysilicon layer. a) The oxide was selectively etched and the sample coated with a conductive film of thickness a. The oxide thickness is not correctly reproduced since the thickness appearing in Fig. (a) is $t' = t - 2a$. b) If the highly phosphorus-doped polysilicon is also etched back (more strongly than the silicon substrate), a stair type profile is produced and the oxide thickness t is displayed correctly. The granular structure stems from the AuPd coating. (Courtesy E. Haudek)

pattern of parallel lines [4-17]. If there are no line patterns, the section plane must pass through regions in which the edges of the patterns are parallel (section 4.3.5). Whereas the interfaces parallel to the surface of the wafer are usually smooth, irregularities may appear at the sidewalls due to the etching processes.

A resolution of approx. 0.5 nm can be achieved when the interfaces in thin specimens are imaged in the bright-field mode. The objective aperture has to be large enough to allow high spatial frequencies to pass through, while at the same time excluding any disturbing reflections from contributing to the image. Slight defocus often proves beneficial in bright-field imaging, because the Fresnel fringes then make the interfaces show up in clearer contrast (Fig. 4-61, section 4.8.2). The true position of the interface is then between the bright and the dark fringe, with a deviation of <0.5 nm [4-66]. High-resolution microscopy provides even better resolution: Fig. 9-42 shows a thin thermal oxide between the silicon substrate and a polysilicon film. The Si lattice is projected in [110] orientation. Lattice fringes can be seen in four individual grains of the polysilicon. The silicon-SiO_2 interface exhibits a slight roughness of between one and two atomic layers (cf. Fig. 4-6d). Since it is present along the direction of projection, too, this roughness means that no quantitative conclusions can be drawn about the atomic configuration at the interface. Nonetheless, with the internal scale of the lattice fringes, HREM images such as the one in Fig. 9-42 are eminently suitable for determining the thickness of thin oxides. During observation and alignment at high magnifications, however, the specimen is irradiated with an intensity high enough to produce modifications. A reduction

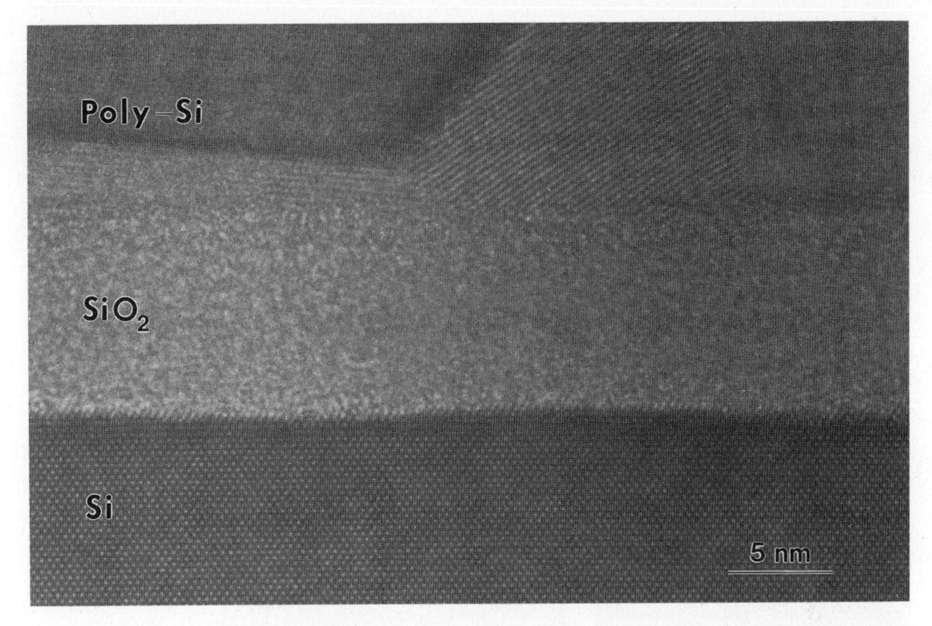

Fig. 9-42. Thin thermal oxide between silicon substrate and polysilicon layer. High-resolution image with silicon lattice in ⟨110⟩ projection.

of up to 20% in oxide thickness has been observed [9-35] after irradiation with a high dose (e.g. 10^4 A s/cm^2). This reduction in oxide thickness is probably caused by oxygen loss when electron irradiation breaks the atomic bonds. As a result, low doses have to be used for HREM investigations of this kind. The same applies when SiO-Si_3N_4-SiO_2 triple layer dielectrics are imaged (Fig. 4-6a). Here, too, electron bombardment produces a change in the film structure [9-42]: free nitrogen produced when the atomic bonds are broken migrates to the Si-SiO_2 interfaces, where it creates artificially nitrogen-enriched layers. The same effect has also been observed when Auger depth profiles are recorded [9-42].

Oxide thinning at the edge of the field oxide

Regions of field oxide separate the active elements in MOS and bipolar circuits. These regions are produced by selective oxidation in which the field oxide grows in the windows of a nitride mask. Lateral oxidation under the mask edges gives the edges of the field oxide the shape of a bird's beak. Fig. 9-43a shows a bird's beak merging into the thin gate oxide. A polysilicon layer was deposited onto the wafer following a number of process steps and growth of the gate oxide. The shape and the length of the oxide beak depend on a number of process parameters and they can be inspected in the scanning electron microscope. The short beaks that are needed for high packing densities can be obtained with thick nitrides and by overetching the field oxide.

The length of the bird's beak is not the only problem that arises in selective oxidation. During gate oxidation the thickness of the oxide at the tip of the beak is reduced (Figs. 9-43b,c).

This oxide thinning depends on the shape of the beak [9-43]. For the sharp beak in Fig. 9-43a with an angle of only 20°, the 11 nm gate oxide is reduced in thickness at the tip of the beak by no more than 22% (Fig. 9-43b). The blunt beak with an angle of 60° in Fig. 9-43c, in contrast, produces a more severe oxide thinning of 35%. This dependence on the angle has also been observed for oxide thinning at etched field oxide edges [9-44]. The degree of oxide thinning cannot be completely explained by modelling the two-dimensional oxidation process at the edge [9-45]. Mechanical stress at the oxide edge is assumed to be responsible for reducing the oxidation rate.

Fig. 9-43. a) Bird's beak configuration at the edge of field oxide. b), c) Thinning of gate oxide at the tip of the bird's beak. Bright-field images of thin cross sections. [9-43]

Dielectrics in trench capacitors

In 4 Mbit DRAMs, the capacitor of a memory cell will usually be arranged along the walls of a trench, so as to yield sufficient capacitance while minimizing the cell area. Fig. 3-9 shows an SEM cross section through a 4 Mbit memory chip. The scanning electron microscope can be used to inspect the shape of the trench after reactive ion etching. However, measuring the thickness of thin dielectrics on the trench walls requires TEM analysis.

Fig. 9-44 is a horizontal section through a trench capacitor with a thermal oxide as dielectric [9-40]. Etching of the trench was followed by a chemical etching process to round the bottom of the trench. By this process, trenches were produced with sidewalls having a preferential orientation along {100} planes. The oxide is not of uniform thickness along the perimeter of the trench, due to differences in the oxidation rate on the various crystallographic planes: oxidation on {110} planes is faster than on {100} planes. Whereas the oxide on the {100} sidewalls is approx. 13.5 nm thick, its thickness increases by approx. 30% to 17.5 nm in the rounded corners where the sidewalls are parallel to {110} planes.

The top and bottom edges of the trench are critical points with respect to thermal oxidation, because at temperatures below 1000 °C, the oxide thickness is reduced at the edges [9-46]. This effect always occurs at the upper edges of the trench, where patterning produces a sharp edge (Figs. 9-45a,b). Besides reducing the thickness of the oxide, this sharp-edged geometry also impairs the oxide's breakdown strength. At the bottom of the trench, this problem can be avoided if well rounded edges are produced by the etching process (Fig. 9-45c). In accordance with their different crystallographic orientations, the (100) trench bottom has a thinner oxide than the {110} trench sidewalls.

Sufficient capacitance is obtained only when thin oxides are employed which are critical for yield and reliability. Triple layer dielectrics with an SiO_2-Si_3N_4-SiO_2 structure benefit from the fact that the nitride exhibits a higher dielectric constant. The thickness of a nitride dielectric yielding the same capacitance as an oxide can therefore be enhanced. The oxide layers are needed to maintain well defined Si-SiO_2 interfaces. As already illustrated with the aid of

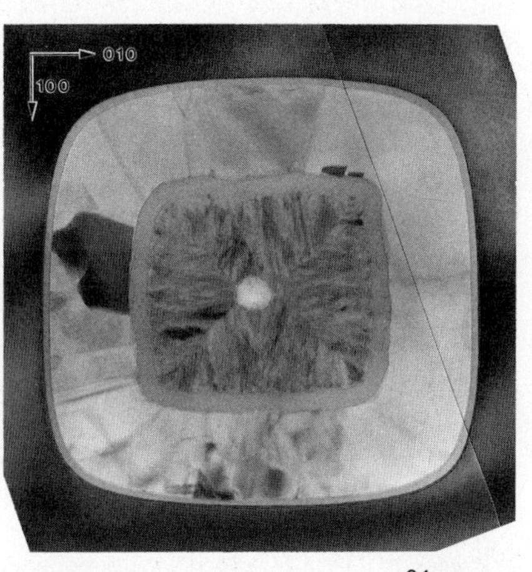

0.1 µm

Fig. 9-44.
Horizontal section through trench capacitor with an oxide dielectric between the Si substrate and the phosphorus-doped (large-grained) polysilicon electrode (bright-field image). The oxide thickness along the perimeter is non-uniform. The inner surface of the polysilicon electrode was thermally oxidized. Then, undoped (fine-grained) polysilicon was deposited to fill the hole in the center. [9-40]

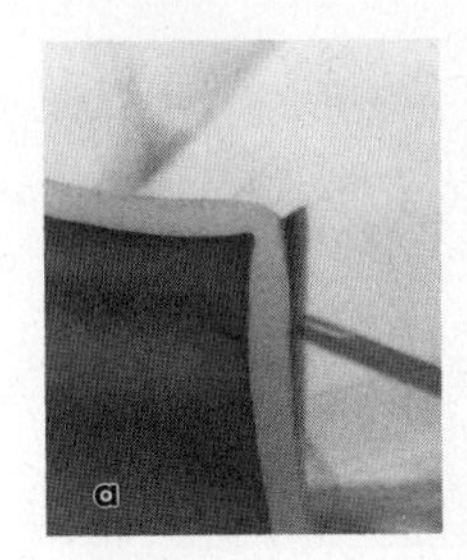

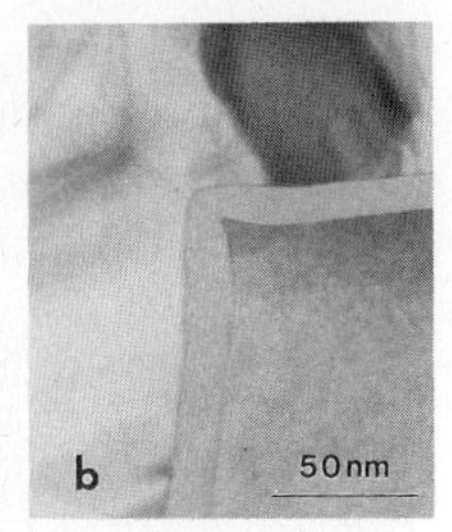

c
0.1µm

Fig. 9-45.
a), b) Reduction of oxide thickness at the sharp upper edges of a trench. c) No reduction at rounded edges of trench bottom. Bright-field images of thin cross sections. [9-35]

Fig. 4-6a, the nitride of the triple layer dielectric appears darker than the oxide layers because of its higher density. The configuration of a triple layer dielectric at the upper edge of a trench is shown in Fig. 9-46a [9-40]. Again, the thermal bottom oxide is thinner at the edge. The nitride, in contrast, coats the edge with uniform thickness, because it was deposited from the vapor phase. The top oxide can be produced by deposition or by thermal oxidation of the nitride. In the latter case, its thickness cannot be determined by ellipsometry and TEM cross sections represent the only means of measurement. Fig. 9-46b shows the configuration at the edge of a polysilicon gate after a reoxidation: small oxide beaks have formed by lateral oxidation above and below the nitride layer. The upper interface of a nitride layer is often slightly rough (Fig. 9-47a). When a very thin nitride is subjected to thermal oxidation to produce the top oxide, the nitride roughness can lead to the formation of pores in the nitride. Since the oxidation rate of silicon is much higher than that of Si_3N_4, the bottom oxide tends to form lens-shaped areas (swellings) at the pores (Fig. 9-47b, c).

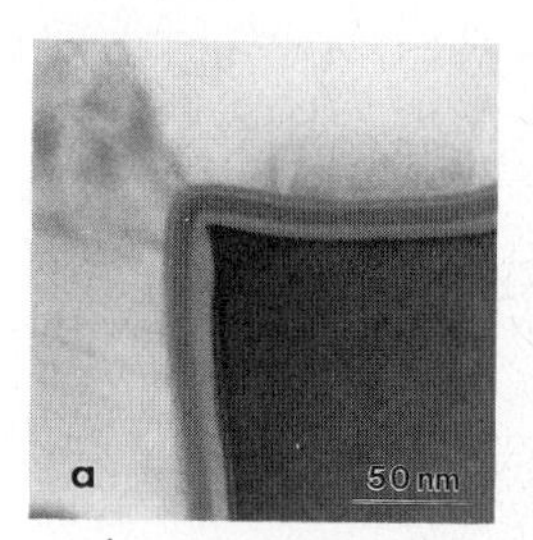

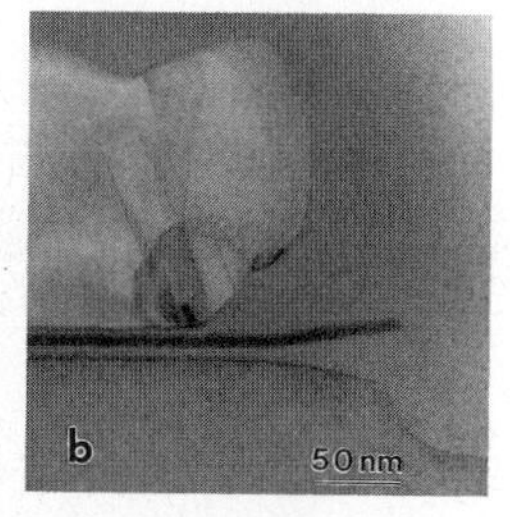

Fig. 9-46. Configuration of triple layer dielectric at upper trench edge (a) and at the edge of a polysilicon gate after reoxidation (b). Bright-field images of thin cross sections. [9-40]

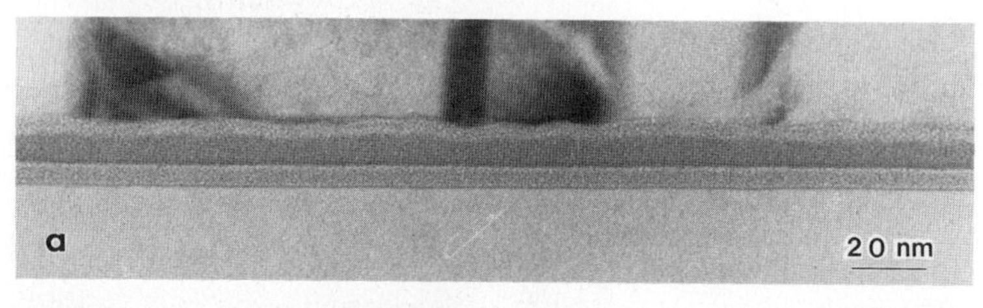

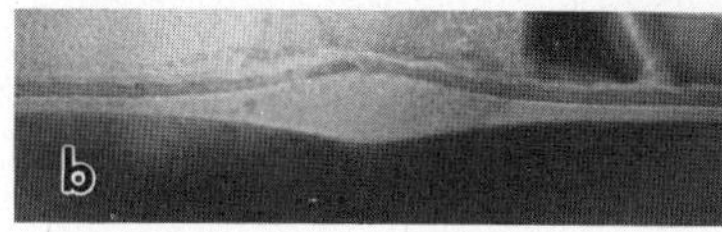

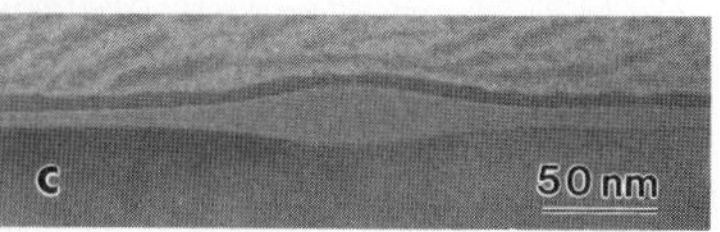

Fig. 9-47. a) Triple layer dielectric exhibiting a slight roughness at the upper nitride interface. b), c) At pores of the nitride, islands of increased oxide thickness grow during thermal oxidation. Bright-field images of thin cross sections.

Interpolysilicon dielectric

The stacked capacitor constitutes another approach to the problem of reducing the size of the memory cell. Since such a capacitor consists of two polysilicon layers, it can be stacked on top of the transistor. As with the trench capacitor, thin dielectrics with high breakdown strength are of crucial importance. The grooves and sharp tips clearly indicate that the thin thermal oxide in Fig. 9-48a has poor breakdown properties. The specimen in Fig. 9-48b provides an explanation for the causes of the grooves in the bottom polysilicon layer [9-35]. This specimen was imaged at an earlier stage of processing, after high-dose arsenic implantation of the bottom polysilicon and annealing. In the initial phase of annealing, a thin oxide is grown to prevent the loss of dopant atoms. This leads to preferential oxidation at the grain boundaries, due to the fact that the arsenic dopant atoms segregate at the grain boundaries and the oxidation rate rises sharply with dopant concentration. When this oxide is etched off and the polysilicon thermally oxidized to produce the dielectric, the result is the morphology shown in Fig. 9-48a. Smooth interfaces and high breakdown strength are obtained if a nitride layer is employed which is deposited immediately after arsenic implantation and before annealing (Fig. 9-48c). The native oxide on the polysilicon is still visible as a bright line at the bottom interface of the nitride.

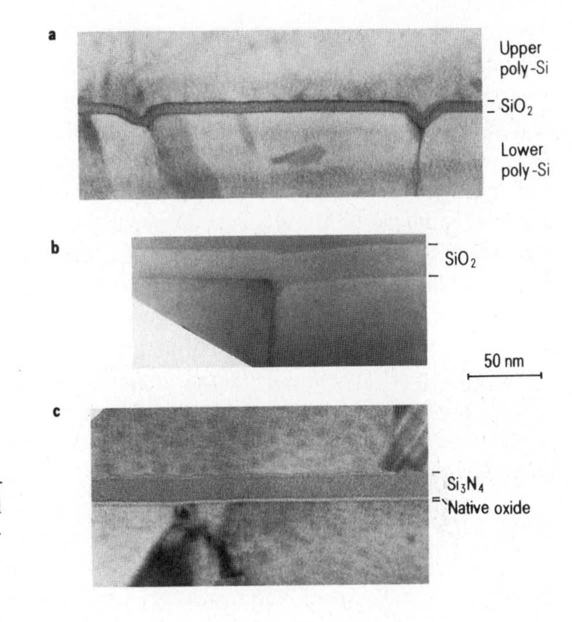

Fig. 9-48.
Interpolysilicon dielectrics for stacked capacitor. a) Thermal oxide; b) Preferential oxidation at grain boundary; c) Si_3N_4 dielectric. Bright-field images of thin cross sections. [9-35]

9.3 Compound semiconductors

9.3.1 Introductory remarks

The most important members of the compound semiconductor family are the III-V compounds of elements of the 3rd and 5th groups in the periodic table such as GaAs and InP. These compounds are direct semiconductors (see section 3.6.2) and differ from silicon by their high probability of radiative recombination processes (Fig. 3-25) and the greater mobility of their charge carriers. Their field of application consequently comprises optoelectronics (light-emitting diodes, laser diodes) and high-frequency electronics (high-frequency transistors, high-speed integrated circuits).

One major advantage offered by compound semiconductors is that the elements are interchangeable within a group (e.g. aluminum can be replaced by gallium, and phosphorus by arsenic), which means that compounds can be produced from three or even four elements. The *lattice constants* and the *band gap* are the major material parameters that can be specifically changed by varying the composition of the crystals (band-gap engineering).

The majority of compound semiconductor devices are based on *heterostructures*. These are structures of thin, single crystalline layers of different compositions grown epitaxially on a substrate crystal. Besides liquid phase epitaxy (LPE), two growth methods are increasingly used: vapor phase epitaxy from metal-organic compounds (MOVPE) and molecular beam epitaxy (MBE). To avoid mechanical stress in the layers that could produce lattice defects, meticulous care is needed in matching the lattice constants a of the various layers to the substrate. High-quality optoelectronic devices call for a lattice mismatch of $\Delta a/a \leq 3 \cdot 10^{-4}$.

Important heterostructure devices include the double heterostructure laser diode [9-47] and the hetero-bipolar transistor [9-48] whose modes of operation cannot be discussed here. In any case, the composition of the layers must be controlled during the growth process: the lattice constants have to be matched to the substrate in question while the band gap must *simultaneously* fulfill the electrical or optical requirements (e.g. the wavelength of a laser diode must match the transmission properties of optical waveguides). In the widely used GaAs-$Ga_{1-x}Al_xAs$ system (written as GaAs-(Ga,Al)As in simplified form) the lattices are naturally well matched, the mismatch being $\Delta a/a = 1.4 \cdot 10^{-4}$ for $x = 1$ and room temperature. However, the band gap of GaAs places an upper wavelength limit of approx. 0.9 μm on lasers in this system. Quaternary systems such as $In_{1-x}Ga_xAs_{1-y}P_y$ (written as (In,Ga)(As,P) in simplified form) on InP substrates allow greater wavelengths (e.g. 1.5 μm). In the case of these latter systems, the two independent compositional parameters x and y must be adjusted accurately to obtain good lattice match as well as the desired band gap.

The growth processes have to be controlled in such a way as to produce layers of the desired thicknesses and with abrupt interfaces, while maintaining the correct compositional parameters. Furthermore, the layers have to be doped in a way that will produce the specified electrical characteristics.

The electrical states of the charge carriers can also be influenced by making the thickness of the layers very small: a very thin layer of less than 10 nm represents a quantum-mechanical potential well for the charge carriers. In a potential well which is always enclosed between two barrier layers, only certain, discrete energy states are possible. This property of very thin layers can be exploited as a further degree of freedom in designing the properties of heterostructure devices, as in the case of the single quantum well (SQW) or multiple quantum well (MQW) laser diodes [9-49].

It is up to the analyst to control the complex heteroepitaxial structures with regard to the composition and thickness of the individual layers, the perfection of the crystal lattice, the smoothness and abruptness of the interfaces and the distribution of the dopant atoms.

9.3.2 Characterization of heteroepitaxial layer structures

X-ray methods represent the best way of determining whether the correct lattice match is obtained when the layers are grown. Multiple crystal X-ray diffractometry will reveal differences of $\Delta a/a \approx 3 \cdot 10^{-5}$ in layers that are not too thin [9-50]. X-ray microanalysis will yield an accuracy of about 1% for the composition of layers approx. 1 μm thick which can be considered as bulk specimens. Nonetheless, photoluminescence is frequently used for this purpose.

Cross sections are needed if these layer structures are to be characterized in detail. As indicated in sections 3.3 and 4.3, III-V crystals are easy to cleave along the {110} planes. When such a cleavage plane is subjected to selective etching, the layer structure can be imaged with good contrast in the scanning electron microscope. Fig. 9-49 shows a GaAs-(Ga,Al)As multilayer in which the (Ga,Al)As layers have been etched back selectively in a solution of KOH and potassium ferricyanide in water. The conductivity of the semiconductor layers is high enough to render sputter coating with a conductive film unnecessary. However, the resolution of the SEM is not high enough to precisely measure the layer thicknesses in the region of 10 nm and less such as those that occur in multiple quantum-well (MQW) structures. The period is all that can be determined when a superlattice is examined. The superior resolution of the TEM is needed to measure the thickness of the individual layers with sufficient accuracy (cf. section 9.2.7). There are several methods of imaging heteroepitaxial layer structures in the TEM; the examples below will serve as outline descriptions.

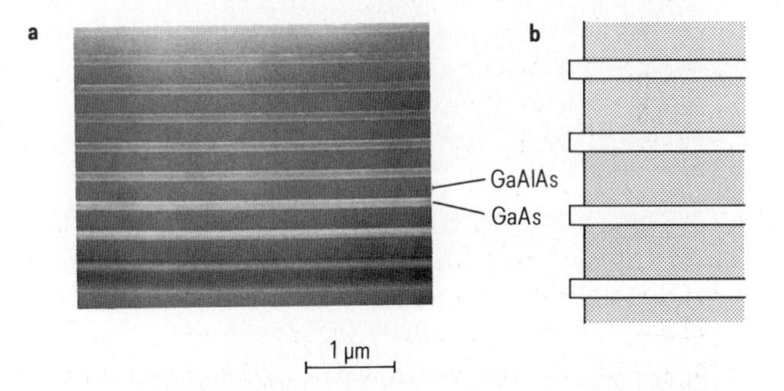

Fig. 9-49. a) Secondary electron micrograph of cleaved cross section through GaAs-(Ga,Al)As multilayer. b) Schematic drawing showing the selective etching of the (Ga,Al)As layers. In Fig. (a) the edges of the GaAs layers appear bright due to diffusion contrast (section 3.4).

(200) dark-field imaging

Dark-field imaging with the intensity of the (200) reflection is most commonly used for thin cross sections [9-51]. The contrast derives from differences in the structure factor F_{200} which for $Ga_{1-x}Al_xAs$ is as follows:

$$F_{200} = 4\,[(1-x)\cdot f_{Ga} + x\cdot f_{Al} - f_{As}]\ .$$

F_{200} is proportional to the difference between the atomic scattering factors f_{Ga} and f_{Al} of the group III atoms and f_{As} of the group V atom. The structure factor $F_{200} = 4\,(f_{Ga} - f_{As})$ for GaAs ($x = 0$) is small, since the difference between f_{Ga} and f_{As} is slight ($F_{200} = 0$ for the diamond lattice (Si or Ge)). Hence, it follows that in the case of $Ga_{1-x}Al_xAs$, the intensity of the (200) reflection is:

$$I_{200} = F_{200}^2 \approx 16x^2\,(f_{Al} - f_{Ga})^2\ .$$

The intensity of the (200) reflection is proportional to the square of x. Therefore, the GaAs layers in the (200) dark-field (DF) images in Figs 9-50 to 52 always appear dark, whereas the (Ga,Al)As layers show bright contrast. Changes in the Al content of $x \geq 0.05$ from one (Ga,Al)As layer to the next show up as variations in contrast. Quantitative assessments of the composition can be derived from these variations [9-52]. However, this is not straightforward because the intensity of the (200) reflection also depends on specimen thickness and deviation from Bragg orientation. On account of the small objective aperture needed to select the (200) reflection, the spatial resolution is limited to 0.5 nm, although this value is sufficient in most cases.

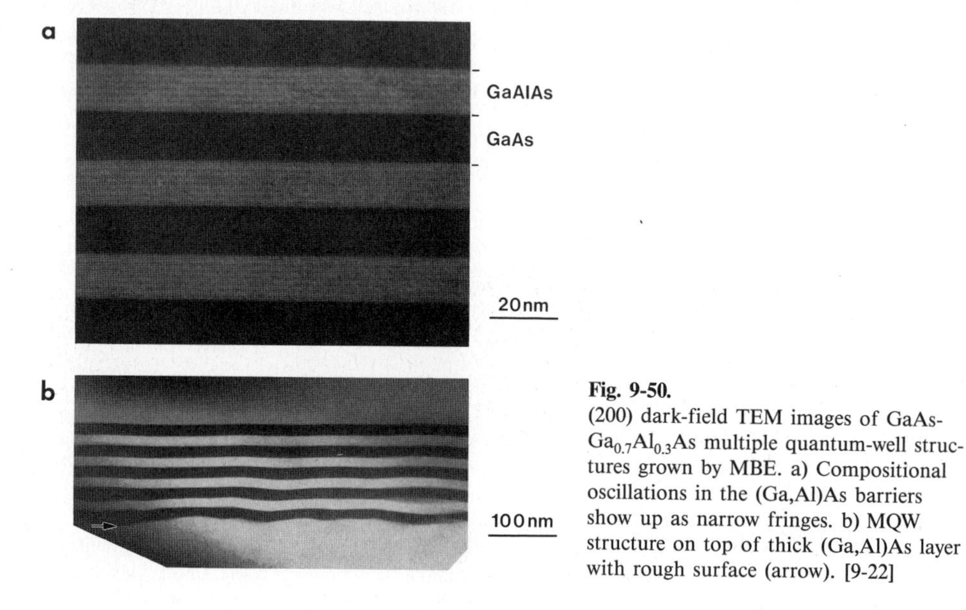

Fig. 9-50.
(200) dark-field TEM images of GaAs-$Ga_{0.7}Al_{0.3}As$ multiple quantum-well structures grown by MBE. a) Compositional oscillations in the (Ga,Al)As barriers show up as narrow fringes. b) MQW structure on top of thick (Ga,Al)As layer with rough surface (arrow). [9-22]

Fig. 9-50a shows part of a multiple quantum-well (MQW) structure grown by MBE. Both the GaAs wells and the (Ga,Al)As barriers are 14 nm thick. The abruptness of the interfaces is at or below the 0.5 nm limit of resolution for (200) DF images. The (Ga,Al)As barriers show striations normal to the direction of growth with a period of about 2 nm. The striations correspond to compositional oscillations induced by substrate rotation during MBE growth and may be attributed to small variations in the flux profile of the group III elements over the surface of the substrate [9-53]. MBE growth of thicker (Ga,Al)As layers may lead to the formation of rough surfaces. The interface arrowed in Fig. 9-50b is the surface of a $Ga_{0.7}Al_{0.3}As$ layer of this kind measuring 0.35 μm in thickness and showing undulations with differences of up to 15 nm in height [9-22]. A higher Al content produces larger undulations. The GaAs-(Ga,Al)As MQW structure on top of the thick (Ga,Al)As layer levels the undulations almost completely (Fig. 9-50b). Caused primarily by the GaAs layers, most of the leveling is accomplished after the first few layers.

Given proper control of the gas flows, MOVPE will grow multilayers with interfaces nearly as abrupt as those obtainable by MBE. The individual layers of the MQW structure in Fig. 9-51a grown by MOVPE exhibit sharp interfaces. However, the contrast of the (Ga,Al)As barrier layers is not uniform in this particular case. The dark band in the lower part of the barrier layers indicates a region where the Al content is reduced by more than 50% [9-22]. The multilayer in Fig. 9-51b was grown by MOVPE at a low growth rate with growth interruptions at the interfaces. Due to the low growth rate, multilayers grown without interruptions also exhibited similarly sharp interfaces of an abruptness within the 0.5 nm resolution limit of (200) DF images [9-22].

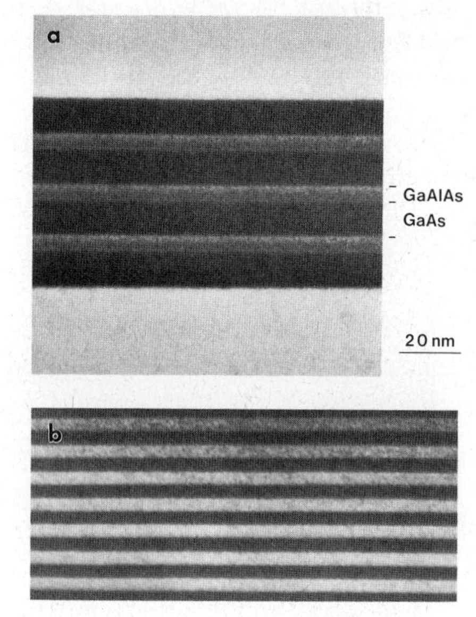

Fig. 9-51.
(200) dark-field images of GaAs-$Ga_{0.7}Al_{0.3}As$ multilayers grown by MOVPE. a) MQW structure with non-uniform Al content in the barriers. b) Multilayers grown with low growth rate. The granular contrast is due to defect clusters near the specimen surface induced by ion milling. [9-22]

Zinc diffusion is known to induce disorder in GaAs-(Ga,Al)As MQW structures [9-54]. An Si_3N_4 mask was used to diffuse Zn locally into a GaAs-(Ga,Al)As superlattice grown by MBE [9-22]. Fig. 9-52 shows the transition region between the disordered zone and the intact superlattice at the edge of the Si_3N_4 mask. Due to underdiffusion, the disordered zone extends 0.5 μm beneath the mask edge A (Fig. 9-52a). The width of the transition region in lateral direction is only 0.2 μm (Fig. 9-52b). Values as low as this can be explained by the steep Zn concentration gradient of the diffusion front [9-55].

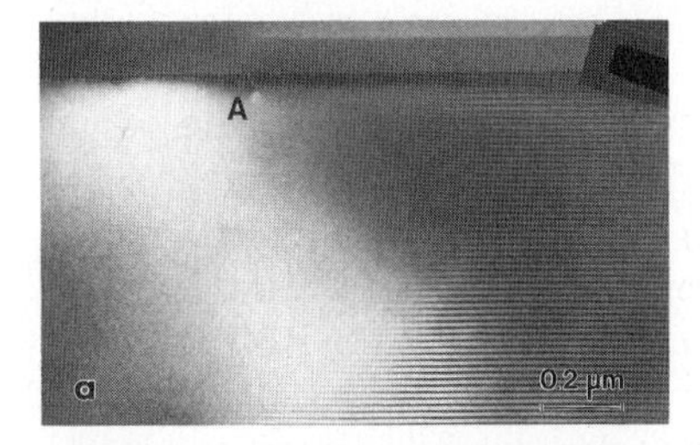

Fig. 9-52. Zinc-diffusion-induced disordering of GaAs-$Ga_{0.7}Al_{0.3}As$ superlattice. a) Region around Si_3N_4 mask edge. b) Disordered-ordered transition region. (200) DF images [9-22].

As explained above, systems other than GaAs-(Ga,Al)As are required for optoelectronic applications in the 1.3 to 1.6 μm wavelength range, e.g. (In,Ga)(As,P). However, problems with the phosphorus sources make compounds containing phosphorus difficult to grow by both MOVPE and MBE. Therefore, the (In,Ga)As-(In,Al)As system is of great interest since it avoids these problems of phosphorus. Precise control of the composition is called for if layers lattice-matched to the InP substrate are to be grown: $In_{0.53}Ga_{0.47}As$ and $In_{0.52}Al_{0.48}As$. Fig. 9-53a shows an MQW structure grown by MBE with a well thickness of 2.5 nm. Growth and luminescence properties were described in [9-56]. The (In,Al)As barrier layers appear dark since the mean value of the atomic scattering factors of Al and In is close to that of Ga. In contrast, the (In,Ga)As wells appear bright because the large atomic scattering factor of In increases the (200) structure factor. The interfaces are abrupt within the 0.5 nm resolution of (200) DF images. The MQW structure in Fig. 9-53b was grown by MOVPE and also exhibits sharp interfaces [9-57]. By carefully optimizing the flow rates, a lattice mismatch of $\Delta a/a \leq 4 \cdot 10^{-4}$ could be obtained.

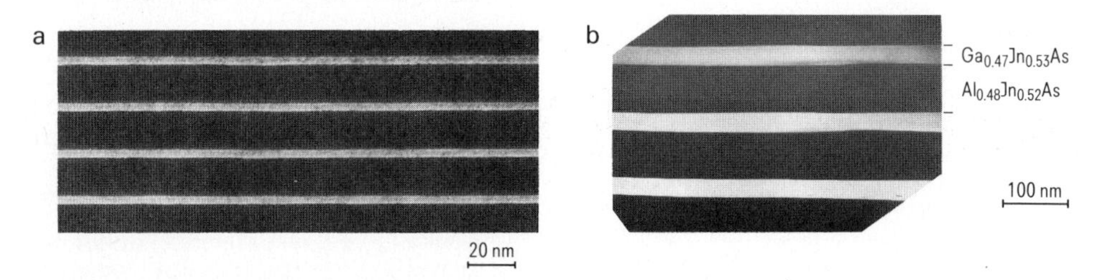

Fig. 9-53. (In,Ga)As-(In,Al)As MQW structures grown by MBE (a) and MOVPE (b). (200) DF images [9-22].

[100] bright-field imaging of cleaved wedge specimens

Conventional TEM cross sections are particularly useful for studying laterally non-uniform effects such as non-planar interfaces (Fig. 9-50b) or the lateral extent and transition width of disordered zones (Fig. 9-52). In most instances, however, depth information is all that is required. As described in section 4.3.5, this information can also be obtained by imaging the edge of a cleaved wedge specimen (Fig. 4-20) [9-58]. Since specimen preparation by cleavage is much faster than the conventional procedure (Fig. 4-18), this technique is becoming more popular.

Just as with thin cross sections, the (200) dark-field method can be used to image the wedge-shaped specimens (Fig. 4-20b, 9-54a). In Fig. 9-54a, the cleaved edge is on the left of the image (arrowed) and is not directly visible. Thickness increases toward the right, in accordance with the geometry of the specimen (Fig. 9-54c). In the thin regions immediately beside the cleaved edge, intensity in the (Ga,Al)As layers increases with thickness (cf. Fig. 4-35). However, thickness fringes do not appear in thicker regions, because the extinction distance of the (200) reflection is very large. As the result of increased electron scattering and "absorption" by the objective aperture, the intensity drops again at distances larger than approx. 50 nm from the cleaved edge.

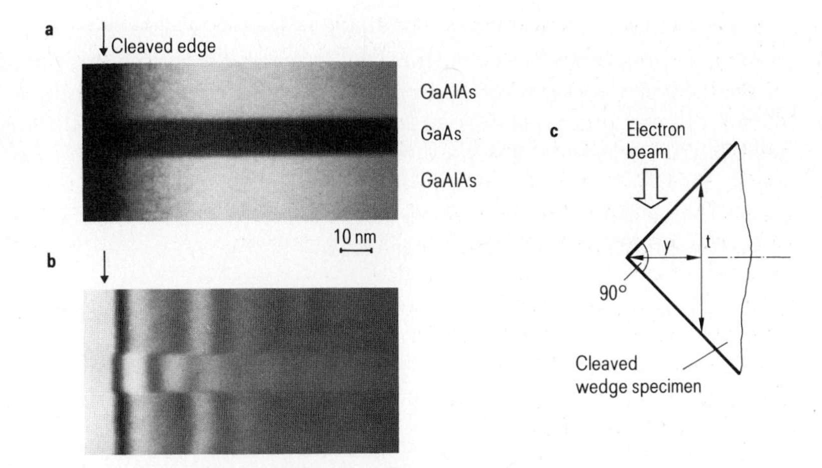

Fig. 9-54. a), b) Cleaved wedge specimen of GaAs-(Ga,Al)As layer structure. By cleavage along two perpendicular {110} planes a 90° wedge is produced (for (100) wafers) which can be directly examined in transmission when it is oriented normal to the electron beam. In the (200) dark-field image (a) both (Ga,Al)As layers appear bright. In the [100] bright-field image (b) the thickness fringes are shifted in layers with different Al content. c) Specimen geometry showing relation between specimen thickness t and distance y from cleaved edge ($y = t/2$).

Besides displaying the layer-structure, imaging of cleaved wedges also allows compositional analysis. This entails tilting the wedge specimen exactly into the [100] zone axis and recording a bright-field image. The precise orientation is controlled with the aid of the diffraction pattern. Multiple-beam conditions result, because numerous reflections are excited. Fig. 9-54b shows such a bright-field image of the GaAs-(Ga,Al)As layer structure of Fig. 9-54a. Thickness fringes parallel to the cleaved edge can now be seen in the layers (cf. Fig. 4-33). The defined geometry of the 90° wedge means that the position of the thickness fringes (distance y from the cleaved edge) correlates unambiguously with specimen thickness t: $y = t/2$ (Fig. 9-54c). The periodicity of the depth oscillations and hence the position of the thickness fringes depends on the structure factor which, in turn, varies with the Al content of the layer. The fringes in layers of different Al content are thus displaced with respect to each other. When the position of the thickness fringes is calculated with the aid of the dynamical theory of electron diffraction, the Al concentration can be quantitatively determined [9-58] (see below). As with the (200) dark-field images, the abruptness of the interfaces can be assessed with a resolution of about 0.5 nm. Bending of the thickness fringes within a layer or at an interface is indicative of a change in the Al content or a transition region at the interface.

In a two-beam case with $s = 0$, the spacing of the thickness fringes is determined by the extinction distance as expressed by equation (4-18). However, under the multiple-beam conditions that prevail here, the fringes are not periodic and it becomes necessary to calculate the intensity profile theoretically, as indicated above. In Figs. 9-55a and c, the bright-field intensities are shown as functions of specimen thickness t for Al contents $x = 0.25$ and $x = 0$. The distance y from the cleaved edge rather than the specimen thickness t is plotted along the abscissa ($y = t/2$).

The magnification in Fig. 9-55b was chosen to match the scale of *y*. When the calculated profiles are compared with the experimental image, it can be seen that the positions of the maxima and minima agree closely with those of the bright and dark fringes. The unknown Al content *x* can be determined to an accuracy of 5 to 10% by matching the experimental and theoretical intensity profiles of the layers. The discrepancies are probably due to the lack of precise knowledge of various parameters used to describe inelastic electron scattering in the theoretical calculation [9-59]. Accuracy can be improved with the aid of calibration measurements using specimens of known composition.

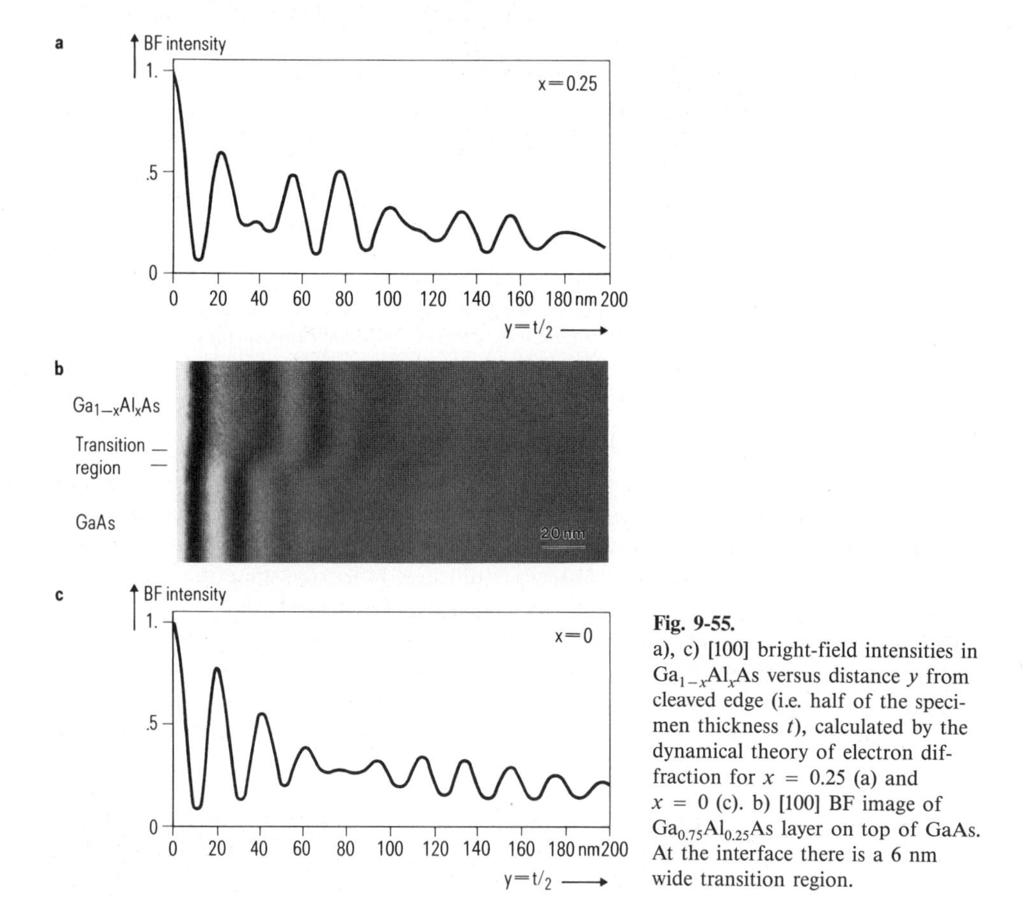

Fig. 9-55.
a), c) [100] bright-field intensities in $Ga_{1-x}Al_xAs$ versus distance *y* from cleaved edge (i.e. half of the specimen thickness *t*), calculated by the dynamical theory of electron diffraction for $x = 0.25$ (a) and $x = 0$ (c). b) [100] BF image of $Ga_{0.75}Al_{0.25}As$ layer on top of GaAs. At the interface there is a 6 nm wide transition region.

Fig. 9-56 shows a comparison of the BF image of a wedge specimen with the DF image of a thin cross section. The layer structure consists of GaAs quantum wells of various thicknesses. There is a modulation in the Al content of the (Ga,Al)As barriers that shows up as zigzag displacements of the thickness fringes in the BF image of the wedge specimen, whereas dark fringes appear in the (200) DF image. Moreover, points at which the thickness of the GaAs layers drops to zero are frequently seen in the DF image of the cross-sectional specimen (arrowed in Fig. 9-56b). In the BF image of the wedge, the lateral region that can be imaged with

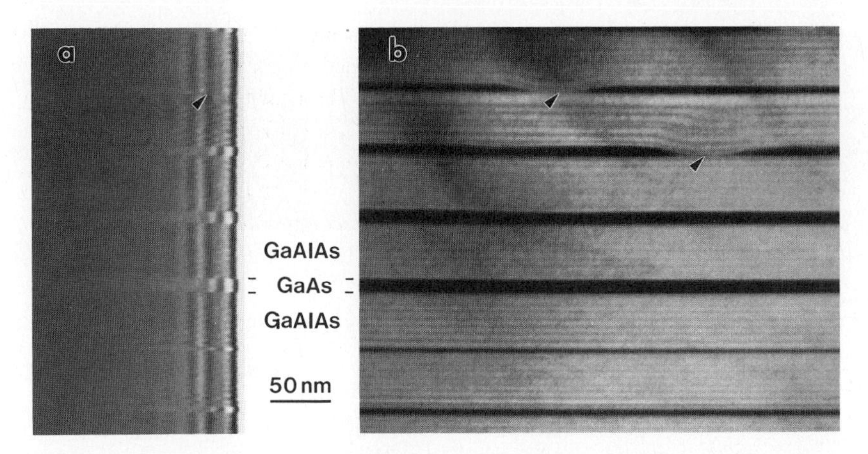

Fig. 9-56. GaAs quantum-wells of varying thickness with (Ga,Al)As barriers. a) [100] BF image of wedge specimen, b) (200) DF image of thin cross section. The barriers show a modulation in Al content. The arrows indicate regions with zero thickness of the GaAs wells.

acceptable resolution at a beam voltage of 200 kV is limited to approx. 0.15 μm. A reduction in thickness can be found at one point only (arrowed in Fig. 9-56a). If laterally inhomogeneous layer growth is anticipated, it is thus better to examine conventional cross-sectional specimens.

The two GaAs-(Ga,Al)As multilayers in Fig. 9-57 were grown by MOVPE within the same specimen so that they could both be imaged at the same time and under the same conditions in the TEM. The multilayer in Fig. 9-57a was deposited with growth interruptions at the interfaces and exhibits abrupt interfaces. In contrast, the multilayer in Fig. 9-57b exhibits transition zones approx. 1.5 nm in width at the interfaces, because growth was not interrupted and the growth rate was higher than that of the specimen in Fig. 9-51b.

Fig. 9-58 shows GaAs single quantum wells of decreasing thickness. These wells were grown by MOVPE with barriers of (Ga,Al)As. An arrow indicates the position of the second intensity minimum in each GaAs layer. In the thicker layers (Figs. 9-58a,b,c), these minima are at the position characteristic for GaAs. In the two thin layers, one of which is just over 1 nm

Fig. 9-57. GaAs-$Ga_{0.6}Al_{0.4}Al$ superlattice grown by MOVPE with (a) and without (b) growth interruptions at the interfaces. [100] BF image of cleaved wedge specimens.

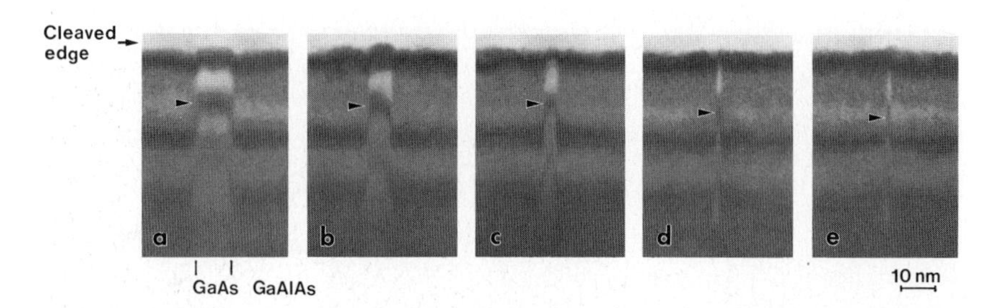

Fig. 9-58. MOVPE grown GaAs single quantum wells of decreasing thickness between (Ga,Al)As barriers; [100] BF images of wedge specimen. The shift of the second dark thickness fringe in the two thinnest GaAs wells (Figs. d, e) compared to the others indicates incorporation of Al during MOVPE growth.

and the other just under 1 nm thick (Figs. 9-58d,e), the minimum is shifted toward the thicker regions, indicating the presence of aluminum. The interfaces of the thick layers are not quite abrupt, exhibiting instead a narrow transition zone measuring approx. 0.7 nm in width. This explains the Al content in the extremely thin layers.

High-resolution imaging

The resolution attainable by the techniques described above is not sufficient to characterize superlattices with layer thicknesses of a few atomic layers or to assess the interfaces on the atomic scale, and resort must be made to high-resolution electron microscopy (HREM). However, a problem is encountered in obtaining sufficient contrast between layers of different compositions, as this demands very precisely defined imaging conditions. Normally, thin cross sectional specimens are irradiated in the [110] direction, with seven beams contributing to the image (cf Fig. 4-6c): four {111} and two {200} diffracted beams along with the (000) direct beam. The image is dominated by the {111} beams that are relatively insensitive to changes in composition and as a result, differences in the mean intensities of the layers are small. In Scherzer focus (section 4.6.1), even very different layers such as GaAs and AlAs are scarcely distinguishable [9-60]. Nonetheless, clear differences in the fine structure of the dot patterns do arise when a different defocus is selected. With the aid of computer-simulated images, defocus values can be found that will yield a similar dot pattern for a given thickness range and thus a more uniform image. The HREM image shown in Fig. 9-59 was recorded at a beam voltage of 400 kV using a microscope with a point resolution of 0.18 nm. Whereas the mean intensities of the GaAs and AlAs layers are virtually the same, there is a big difference in the fine structure of the dots: in the GaAs, the dots are slightly elongated normal to the plane of the layer, whereas double dots parallel to the layer can be seen in the AlAs. The interfaces of the layers which where grown by MOVPE appear to have an abruptness of roughly one atomic layer.

Δx of the specimen in Fig. 9-59 has a value of 1. The differences in the fine structure of layers with smaller differences in their Al content is slighter and at $\Delta x = 0.3$, for example, the layers can only be distinguished with difficulty. In this case, it is better to prepare a cross section with a {100} specimen plane and examine it in the $\langle 100 \rangle$ direction [9-61]. In the diffrac-

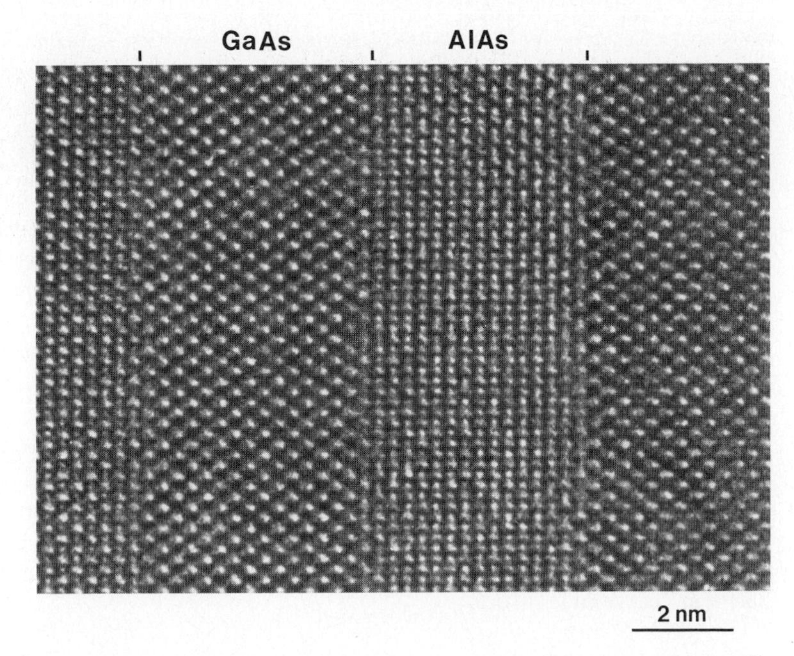

Fig. 9-59. High-resolution image of GaAs-AlAs superlattice in [110] projection. The layers exhibit a different fine structure of the dot pattern. The abruptness of the interfaces is about one atomic layer.

tion pattern, the four reflections of the {200} type are closest to the direct beam. With an objective aperture chosen to allow only these four {200} beams to contribute to image formation along with the (000) beam, the contrast between different layers is very high [9-62]. Since the {200} beams are extremely sensitive to composition, differences in mean intensities are large. Cleaved wedge specimens, too, can be imaged in [100] projection using HREM.

9.3.3 Depth profiling of doped heterostructures

The performance of heterostructure devices critically depends on whether the interfaces between the individual epitaxial layers are abrupt or not, and on how the dopant concentration changes at the interfaces. Hence, the measurement of dopant profiles in heterostructures by SIMS requires very good depth resolution as well as a low detection limit.

Optimization of SIMS parameters

Due to their relatively low diffusion coefficients, silicon and magnesium are used for the n- and p-doping respectively of III-V semiconductors. The instrument parameters of a Cameca IMS 3F secondary ion mass spectrometer were optimized for the analysis of GaAs-(Ga,Al)As layer structures [9-63]: because of its electrochemical properties, magnesium (Mg^+) is detected with high secondary ion yield when O_2^+ primary ions are used; the same is true for silicon

(Si^-), when Cs^+ primary ions are used (chemically-enhanced ion emission, section 2.6.5). To obtain high depth resolution, primary-ion energies should be kept as low as possible and a shallow angle of incidence should be selected (section 2.6.6). However, in a Cameca SIMS instrument it is not possible to adjust these variables independently: the angle of incidence becomes larger or smaller as ion energy is decreased, depending on the polarity of the ions (see Fig. 7-13 and Table 7-1). Experiment is therefore necessary to find the optimum settings. Fig. 9-60 shows values achieved for depth resolution: at a crater depth of between 1 and 2 μm, depth resolution is roughly the same with 3 keV O_2^+ and 14.5 keV Cs^+, i.e. approx. 10 nm. This will provide sufficient resolution for the analysis of layer structures some 100 nm in thickness.

Under the experimental conditions used for the measurements shown in Fig. 9-60, the detection limit for magnesium is $1 \cdot 10^{15}$ cm^{-3}. The detection limit of silicon is matrix-dependent: it has a value of $5 \cdot 10^{15}$ cm^{-3} in GaAs and $1 \cdot 10^{16}$ cm^{-3} in $Ga_{0.66}Al_{0.34}As$. In the latter instance, the poorer detection limit is due to the fact that $^{28}(AlH)$ and $^{29}(AlH_2)$ (with hydrogen from the residual gas) overlap the ^{28}Si and ^{29}Si mass lines. For this reason, high mass resolution ($M/\Delta M \approx 3000$) or a voltage offset (section 7.3.2) is required for the measurements. The signal-to-noise ratio is also good when the rare isotope ^{30}Si is recorded. In any event, signal yield is low and hence the poorer detection limit of silicon in (Ga,Al)As (in comparison with silicon in GaAs) must suffice.

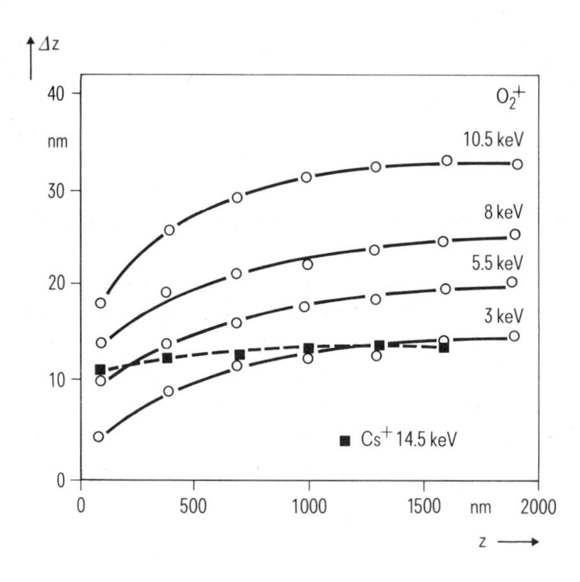

Fig. 9-60.
Depth resolution Δz as a function of sputtering depth z for O_2^+ primary ions of varying energy and 14.5 keV Cs^+ ions. Δz was measured at GaAs-$Ga_{0.66}Al_{0.34}As$ interfaces using the 16% and 84% values of the maximum aluminum signal (see Fig. 2-68). Measuring conditions: CAMECA IMS 3F; crater 300 × 300 μm^2; optical gate 60 μm diameter. [9-63]

Magnesium- and silicon-doped heterostructures

Fig. 9-61 shows the depth profiles of a double heterostructure laser grown by MOVPE measured at the experimentally optimized conditions described above. The 100 nm thick active, light-emitting zone in the middle of the structure consists of nominally undoped GaAs that is, however, slightly n-doped by residual contaminants. The adjacent (Ga,Al)As layer on the left is p-doped with magnesium, the layer on the right is n-doped with silicon. The (Ga,Al)As

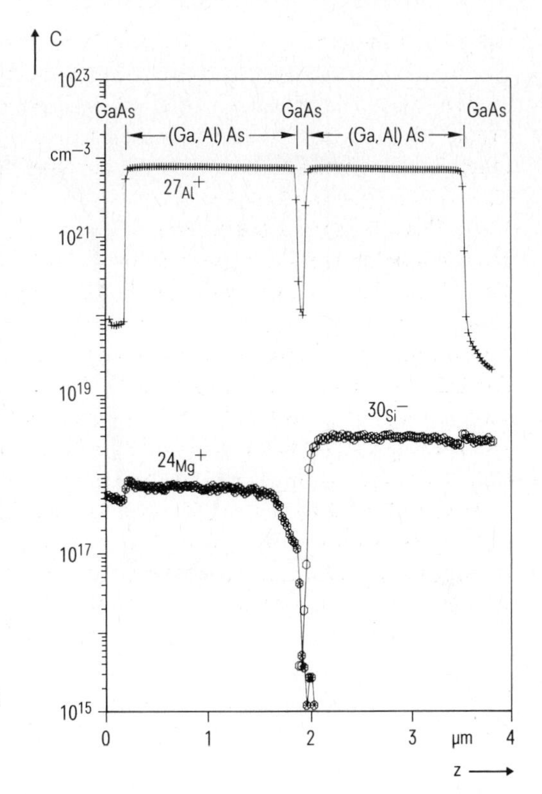

Fig. 9-61.
SIMS depth profiles of aluminum, magnesium and silicon in a GaAs-(Ga,Al)As double heterostructure laser diode grown by MOVPE [9-63]. Measuring conditions: CAMECA IMS 3F; O_2^+ primary ions, E_0 = 3 keV for magnesium and aluminum; Cs^+ primary ions, E_0 = 14.5 keV for silicon; crater 300 × 300 µm^2; optical gate 60 µm diameter. Quantification with ion-implanted reference specimens.

layers form two heterojunctions with the active GaAs layer. The upper GaAs layer and the substrate are also doped and serve as contact layers.

To permit localization of the interfaces of the active zone, the aluminum signal was recorded and is shown in the upper part of Fig. 9-61. The interface position is assumed to be where the signal is 50% of the maximum. As the figure shows, the magnesium concentration in the upper (Ga,Al)As layer reaches its specified value during the deposition process only after some delay (growth direction in all depth profiles is from right to left). In contrast, the silicon doping in the lower (Ga,Al)As layer of this specimen drops off abruptly at the interface with the active zone. However, there is a danger of silicon migrating into the active zone during deposition. Both the delayed incorporation of magnesium and the migration of silicon affect device characteristics and require investigation.

Silicon migration during vapor phase epitaxy

Fig. 9-62 shows the aluminum and silicon depth profiles of GaAs-(Ga,Al)As heterostructures on an enlarged depth scale. The specimens (Nos. 1 to 4) were deposited by MOVPE with silane as the doping gas for silicon. Although not quantified, the profiles were measured under the same conditions and are thus comparable. The signal intensities were normalized to the same level of constant concentration in the lower (Ga,Al)As layer (not contained in the part of the

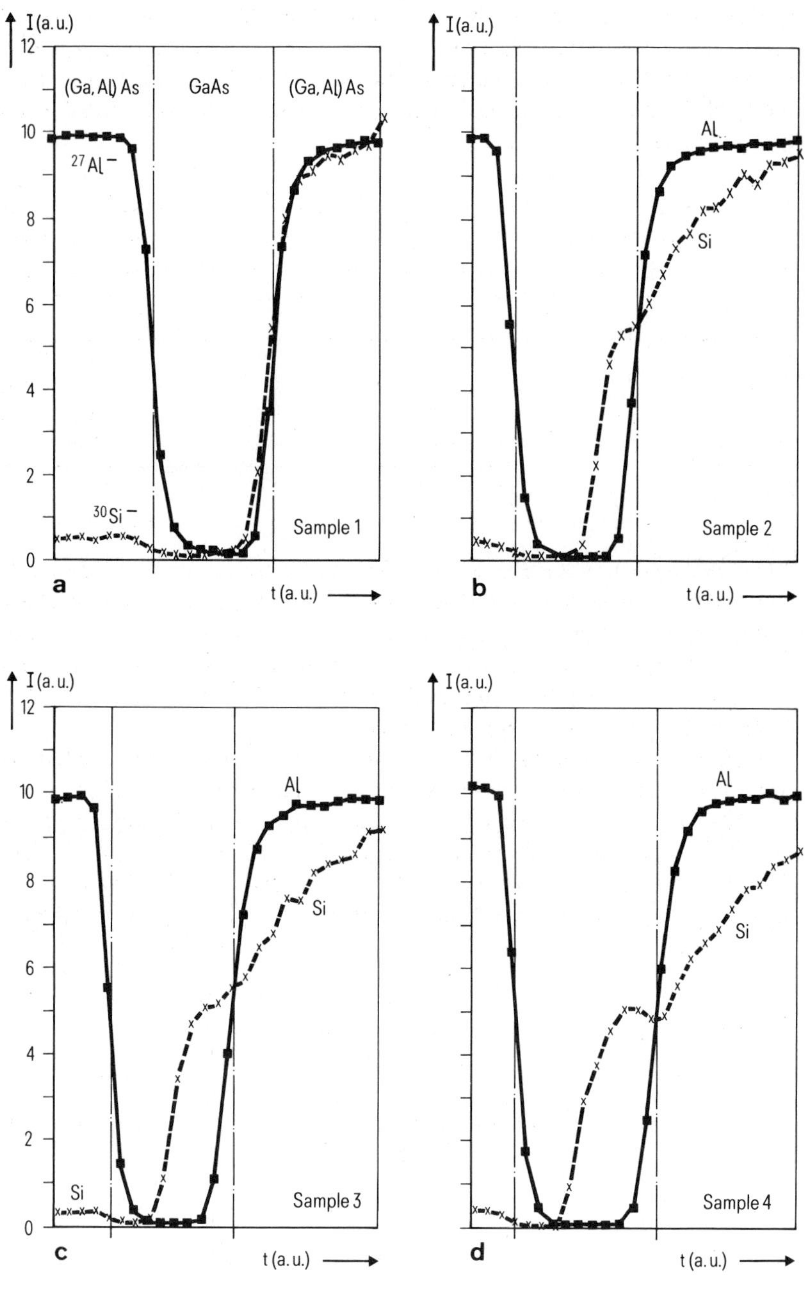

Fig. 9-62. SIMS depth profiles of silicon and aluminum in GaAs-(Ga,Al)As heterolayer structures grown by MOVPE at various conditions (Table 9-2): The normalized secondary-ion intensities I for aluminum and silicon are plotted in *linear* scale versus sputter time t (arbitrary units). Growth conditions: a) 700 °C, b) 750 °C, c) 770 °C, d) 750 °C and higher silane flow ratio than for Fig. a to c. SIMS measuring conditions as for Fig. 9-61. The silicon concentrations measured in the doped (Ga,Al)As layers are given in Table 9-2.

profiles shown in Fig. 9-62). The quantified values for silicon are summarized in Table 9-2. Specimens 1 to 3 (Fig. 9-62a to c) differ in their growth temperatures (as well as in the width of the GaAs layer). As the temperature increases, the silicon front moves into the nominally undoped GaAs layer. At the same time, the concentration of silicon in the doped (Ga,Al)As layer increases (see Table 9-2).

Table 9-2. Silicon concentrations C_{Si} in the doped (Ga,Al)As-layers of the heterostructures shown in Fig. 9-62a to d. The heterostructures were grown by MOVPE at atmospheric pressure and different temperatures T using silane as doping gas. The silicon concentrations were measured by SIMS. [9-64]

Sample No.	T	C_{Si}
1	700 °C	$3.0 \cdot 10^{18}$ cm^{-3}
2	750 °C	$5.1 \cdot 10^{18}$ cm^{-3}
3	770 °C	$5.7 \cdot 10^{18}$ cm^{-3}
4*)	750 °C	$7.4 \cdot 10^{18}$ cm^{-3}

*) Higher silane flow rate than for specimens 1 to 3

Specimen No. 4 (Fig. 9-62d) was grown at the same temperature as specimen 2 (750 °C) but with a higher silane flow ratio. The concentration of silicon is correspondingly higher than that of specimen 2 (Table 9-2). Although the growth temperature was the same, the silicon front in specimen 4 has advanced considerably further than that of specimen 2. Consequently, it may be assumed that the concentration of silicon rather than the growth temperature is the determining factor for the migration effect [9-64].

Specimens with an inverse structure (doped (Ga,Al)As layer in the middle between two undoped GaAs layers) were prepared so that the dependence of this effect on the growth direction could be studied. In accordance with specimen 1, the SIMS profile in Fig. 9-63a indicates that there is no significant migration at a silicon concentration of $< 3 \cdot 10^{18}$ cm^{-3}. At a higher concentration ($5 \cdot 10^{18}$ cm^{-3} in Fig. 9-63b), in contrast, migration can be clearly observed, although it occurs mainly in the growth direction. The process is therefore one which,

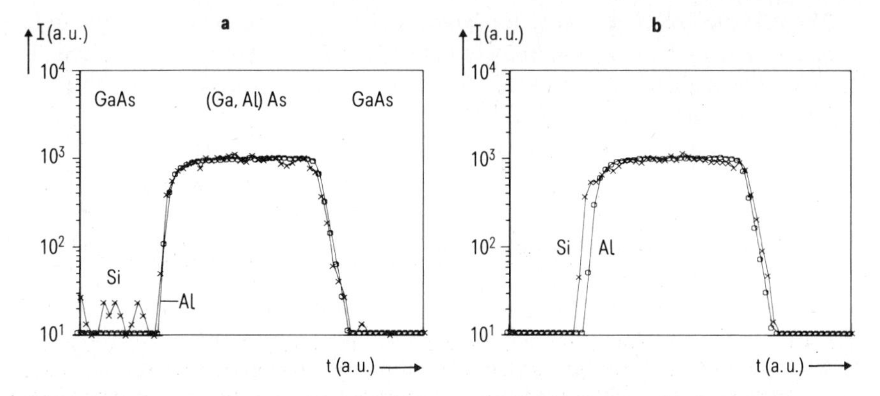

Fig. 9-63. SIMS depth profiles of silicon and aluminum in inverse layer structures (doped (Ga,Al)As layer in the middle of undoped GaAs layers) grown by MOVPE at 750 °C using a low (a) and a high (b) silane flow ratio. Plotted as in Fig. 9-62, but with *logarithmic* scale of intensity. [9-64]

rather than depending merely on diffusion, appears to be primarily the result of segregation at the growth front: when the silicon source is switched off, segregated silicon on the surface is incorporated in the crystal with some delay. This process takes place above a critical concentration of approx. $3 \cdot 10^{18}$ cm^{-3} [9-64]. It must therefore be ensured that silicon is kept to a sufficiently low concentration during the growth process or that the silicon source is switched off early to prevent silicon penetrating into the active zone.

Magnesium memory effect during vapor phase epitaxy

Fig. 9-64 shows SIMS depth profiles of the emitter-base region of a hetero-bipolar transistor. The emitter, base and collector of a transistor of this type differ not only by their doping (as is the case with homo-bipolar transistors) but also by their layer composition and thus by their band gap. This results in an extra degree of freedom that can be used advantageously in dimensioning very-high-frequency transistors [9-48].

The transistor structure shown in Fig. 9-64 has a GaAs base and a (Ga,Al)As emitter. The base is doped with magnesium. Once again, the aluminum profile is used to localize the heterojunctions. However, the ^{27}Al signal normally used would yield an excessively high intensity under the measuring conditions used here (chosen so that the comparatively low magnesium concentration could be determined). For this reason, the $^{54}(Al_2)$ ion was measured instead.

When the magnesium depth profile is evaluated, it must be borne in mind that with O_2^+ primary-ion bombardment, the signal yield in (Ga,Al)As is greater than in GaAs. This effect is due to a higher steady-state oxygen coverage resulting from the aluminum content and, in turn, to a higher ionization probability (section 2.6.5) [9-65]. When the various matrix effects are taken into account by quantifying the layers separately (with the aid of Mg-implanted GaAs and (Ga,Al)As reference specimens), the result is the profile shown in Fig. 9-64b. The data points at the interfaces were omitted, because they would give the impression of a discontinuity. This is principally due to the fact that for the purposes of quantification, the transitions from one matrix to the next are presumed to be abrupt, whereas in reality, the matrix changes gradually at the interface because of atomic mixing (compositional modifications, section 2.6.3).

The reference specimens used for quantification were implanted with the isotope ^{24}Mg, whereas the natural mixture of isotopes is always present in the MOVPE process. This is also true of other methods of deposition and must be taken into account when implanted standard specimens are used for quantification.

The emitter in Fig. 9-64b exhibits a relatively high magnesium concentration decreasing gradually in the growth direction. This finding was obtained even though the magnesium source was switched off at the juncture of the MOVPE process indicated by the arrow. To study the cause of this effect, experiments were carried out with a GaAs test specimen. As can be seen in Fig. 9-65, instead of an abrupt rise when the magnesium flow ratio is increased suddenly, a considerable delay elapses before the concentration increases (as in Fig. 9-61). On the other hand, the concentration falls only gradually when the flow of doping gas is reduced. These effects are explained by the fact that after the source has been switched on, magnesium is adsorbed onto the walls of the MOVPE system before being incorporated into the specimen. When the source is switched off, this magnesium is desorbed from the walls, producing the memory effect observed in the experiments.

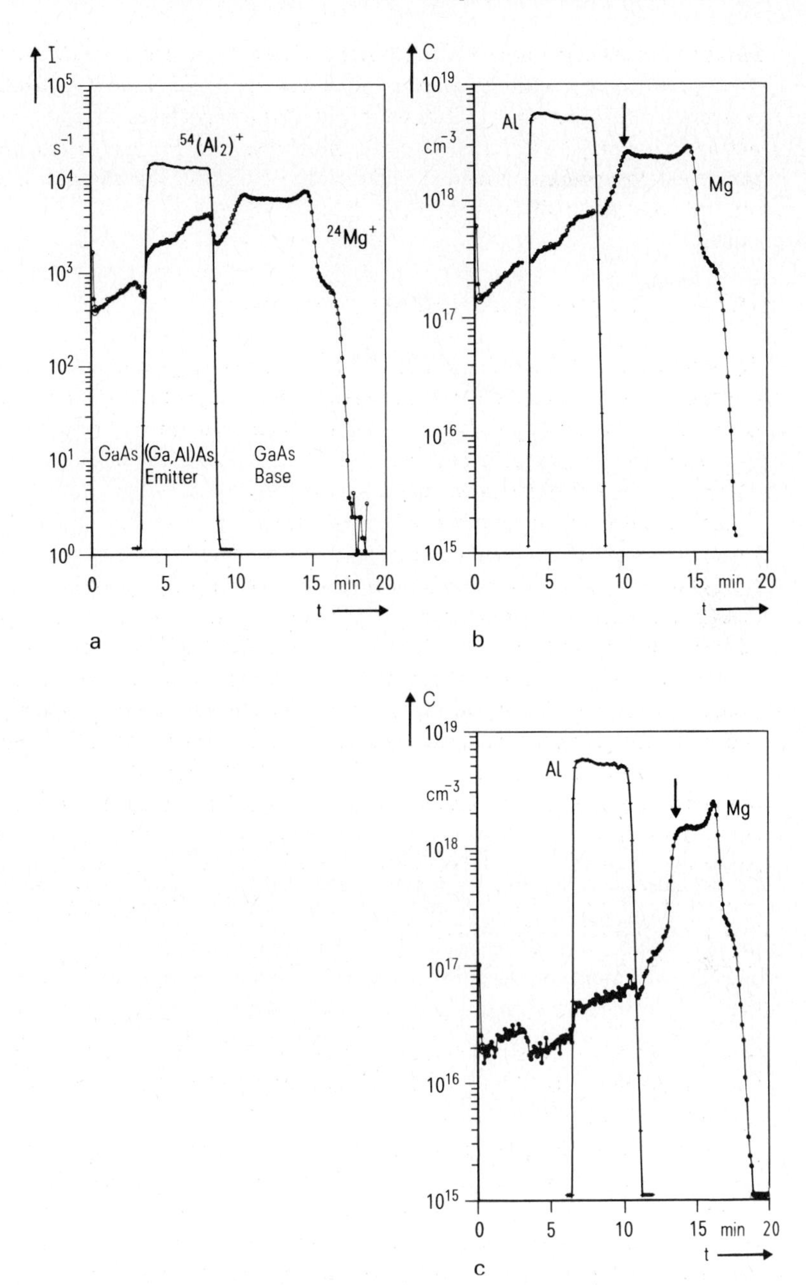

Fig. 9-64. SIMS depth profiles of magnesium and aluminum in GaAs-(Ga,Al)As hetero-bipolar transistor grown by MOVPE. Measuring conditions: CAMECA IMS 3F; O_2^+ primary ions, $E_0 = 3$ keV, $I_0 = 2 \cdot 10^{-7}$ A; crater 300 × 300 µm²; optical gate 60 µm diameter. a) Secondary ion intensity I as a function of sputtering time t; b) same measurement as (a), but showing concentration after quantification; c) specimen with 10 min growth interruption at arrow (see text). Quantification of (b) and (c) with ion-implanted reference specimens (see text).

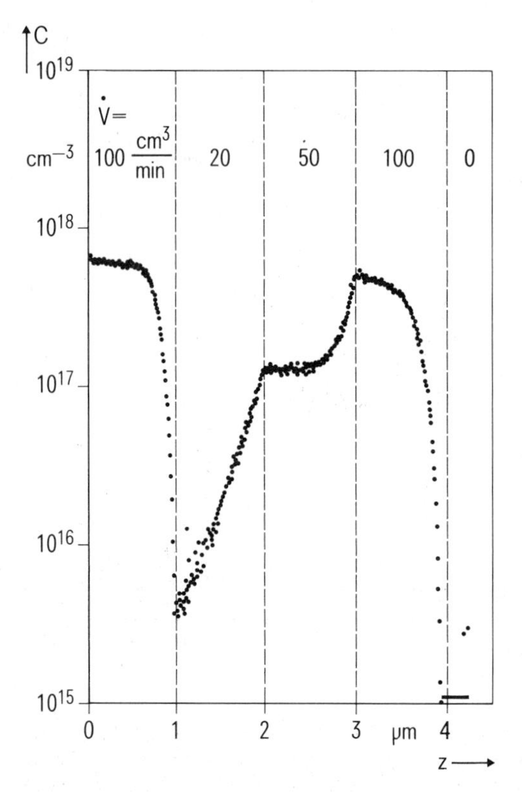

Fig. 9-65.
SIMS depth profile of magnesium in GaAs test specimen grown by MOVPE with varied doping ($\dot{V}$ = volume flow of doping gas). Same measuring conditions and quantification as for Fig. 9-64.

When the magnesium source was switched off, the growth of the GaAs layer of the transistor structure shown in Fig. 9-64c was interrupted for 10 minutes (arrow). During this time, most of the magnesium desorbed from the walls was removed from the reactor. Once removed, the magnesium is no longer available for incorporation in the layers subsequently deposited and the magnesium concentration in the emitter is lower by more than one order of magnitude than is the case in the structure shown in Fig. 9-64b. Interrupting growth in this way is of major assistance in avoiding the detrimental influence that the memory effect exerts on device properties. Using the results of these and other experiments it is now possible to produce steep magnesium doping profiles [9-66].

Silicon contamination during vapor phase epitaxy

Fig. 9-66 shows mass spectra and depth profiles of an (In,Al)As-(In,Ga)As heterostructure for field-effect transistors. The gallium signal was recorded in Fig. 9-66c to localize the heterojunction. The lower region of the (In,Ga)As layer is undoped; the upper region is doped with silicon. The (In,Al)As layer is nominally undoped. However, electrical measurements showed results indicative of n-doping in this layer. Silicon contamination is primarily suspected to be the cause of this unwanted doping.

In the mass spectrum, the lines of other ions such as ^{28}(AlH), ^{28}Co and $^{28}N_2$ overlap the ^{28}Si line. Fig. 9-66a shows a mass spectrum of an (In,Al)As reference specimen implanted

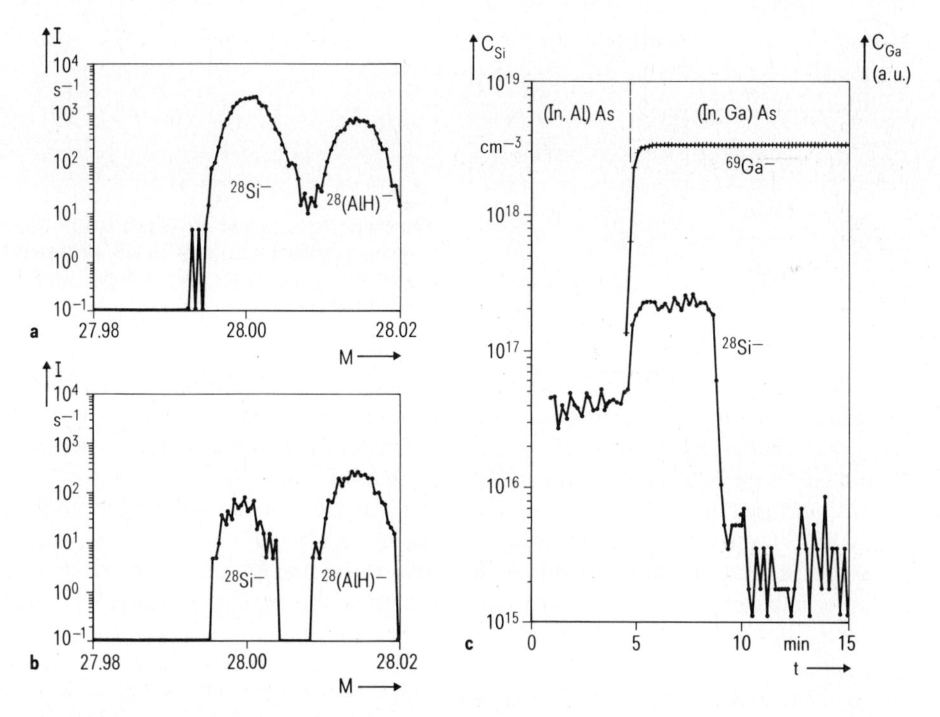

Fig. 9-66. SIMS analysis of silicon contamination in (In,Al)As-(In,Ga)As heterostructure. a) Mass spectrum of ^{28}Si-implanted reference specimen; b) mass spectrum of the (In,Al)As layer of the heterostructure showing clear separation of $^{28}Si^-$ and $^{28}(AlH)^-$ mass lines; c) depth profiles of silicon contamination and gallium in the heterostructure. Measuring conditions: CAMECA IMS 3F; Cs^+ primary ions, $E_0 = 14.5$ keV, $I_0 = 10^{-7}$ A; mass resolution $M/\Delta M = 3000$; crater 300×300 µm^2; optical gate 60 µm diameter. Quantification with ion-implanted (In,Al)As and (In,Ga)As reference specimens.

with ^{28}Si measured with a high mass resolution of $M/\Delta M = 3000$. The $^{28}Si^-$ and $^{28}(AlH)^-$ mass lines are clearly separated. The rest of the lines listed above are outside the mass range shown and are thus safely eliminated. Fig. 9-66b shows the corresponding section of the mass spectrum of the (In,Al)As layer to be analyzed. The spectrum clearly verifies that this layer is in fact contaminated with silicon. The depth profile recorded with an equally high mass resolution is shown in Fig. 9-66c: the silicon concentration drops off steeply at the heterojunction and assumes a constant value of $4 \cdot 10^{16}$ cm^{-3} in the (In,Al)As layer. This excludes a migration or memory effect such as was observed in the case of the GaAs-(Ga,Al)As heterostructures described above. Instead, the cause of the problem here was a poor quality of the aluminum gas source in the MOVPE system.

Replacement of magnesium during zinc diffusion

Bipolar npn-transistors based on (In,Ga)As-InP heterostructures can also be fabricated by MOVPE (Fig. 9-67) [9-67]. The p-doped base is contacted by converting parts of the emitter layer from n- to p-type by means of an acceptor diffusion (hatched areas in Fig. 9-67). To

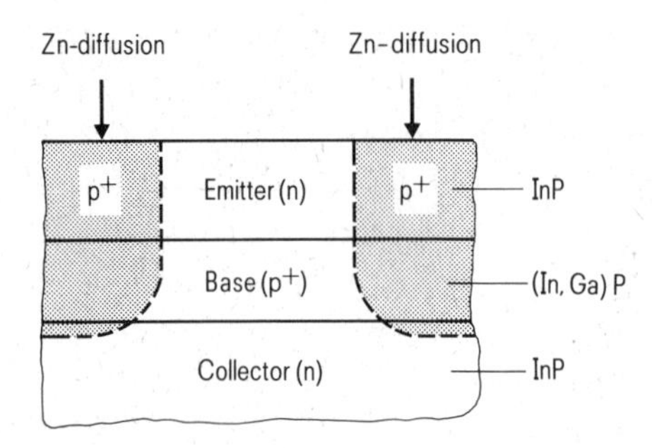

Fig. 9-67.
Schematic cross section of (In,Ga)As-InP heterojunction bipolar transistor with zinc diffusion for planar base contact; the base is doped with magnesium [9-67].

avoid out-diffusion of the initial p-dopant in the base during in-diffusion, two dopants with significantly different diffusion coefficients (D) are required. Magnesium ($D = 5 \cdot 10^{-12}$ cm^2/s) was chosen as the initial dopant during growth of the (In,Ga)As base, whereas zinc ($D = 6 \cdot 10^{-11}$ cm^2/s) is used for the in-diffusion that follows.

Zinc is diffused from spin-on Zn-doped SiO_2 films at 550 °C under a flowing hydrogen gas atmosphere. The resulting dopant distributions in full-surface-diffused specimens were measured with SIMS. Fig. 9-68 shows the depth profiles before and after diffusion. As described at the beginning of this section, oxygen should be used as primary ions to achieve a low detec-

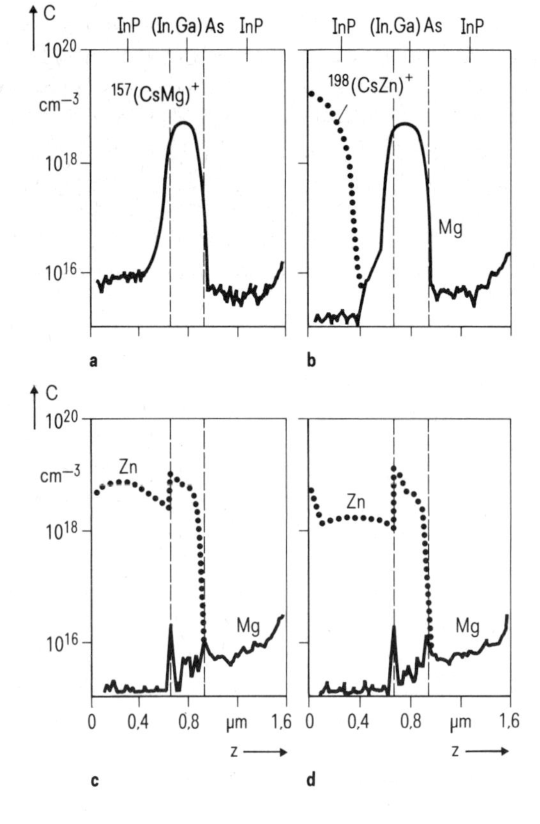

Fig. 9-68.
SIMS depth profiles of magnesium and zinc in Mg-doped (In,Ga)As-InP heterostructure before and after zinc diffusion for various times t from spin-on films at 550 °C (spin-on film removed prior to SIMS profiling) [9-67]: a) $t = 0$ (as-grown); b) $t = 10$ s; c) $t = 60$ s; d) $t = 210$ s. Measuring conditions: CAMECA IMS 3 F; Cs^+ primary ions, $E_0 = 14.5$ keV; crater 300 × 300 μm^2; optical gate 60 µm diameter. Quantification with ion-implanted reference specimen.

tion limit for magnesium. In the case of zinc, however, the detection limit is then poor because the ^{64}Zn and the $^{64}(O_4)$ ions overlap in the mass spectrum. Therefore, the profiles in Fig. 9-68 were recorded with Cs^+ primary ions in order to allow magnesium and zinc to be recorded in a single measurement. Due to the better values for secondary-ion yield, the complex ions $CsMg^+$ and $CsZn^+$ were recorded instead of the atomic ions Mg^+ and Zn^+. The interfaces of the heterolayers marked as broken lines were determined from arsenic profiles that have been left out of Fig. 9-68 for the sake of clarity.

The SIMS depth profile for the as-grown structure (after removal of the spin-on SiO_2 layer) is shown in Fig. 9-68a. Due to the memory effect of the MOVPE reactor, magnesium has penetrated slightly into the upper InP layer (emitter). Nevertheless, the dopant gradient is high on both sides.

After a diffusion time of 10 s, the Zn front reaches the middle of the upper InP layer and shows a profile typical of sole Zn diffusion into InP (Fig. 9-68b). After a diffusion time of 60 s, the Zn front has completely crossed both the emitter and the base layers (Fig. 9-68c). The most striking fact, however, is the complete removal of the initial Mg dopant from both layers. In the ternary layer in particular, the concentration of Mg is reduced by more than three orders of magnitude. Extending the diffusion time to 210 s (Fig. 9-68d) changes the shape of the Zn profile slightly but yields no further penetration of the Zn front (some out-diffusion of zinc from the emitter layer and the segregation of zinc at the emitter-base heterojunction will not be discussed here).

For comparison, diffusion experiments were carried out with a similar heterostructure but without Mg doping. The results of diffusion times of 60 s and 210 s at 550° C are displayed in Fig. 9-69 and reveal an appreciably slower Zn diffusion than is the case for the structures doped with magnesium. Even after a diffusion time of 210 s, the Zn front has not yet advanced halfway across the base layer (Fig. 9-69b), since the diffusion coefficient of zinc is smaller in (In,Ga)As than in InP. In the case of an Mg-doped layer, however, a diffusion time of 60 s is sufficient for the diffusion front to advance right across the base layer (Fig. 9-68c).

Two types of diffusion process must be involved if the Zn depth profiles are to be explained:

– a normal diffusion process operating in Fig. 9-69b and resulting in a concentration discontinuity at the InP-(In,Ga)As interface in Fig. 9-68d because of different diffusion coefficients in both materials,

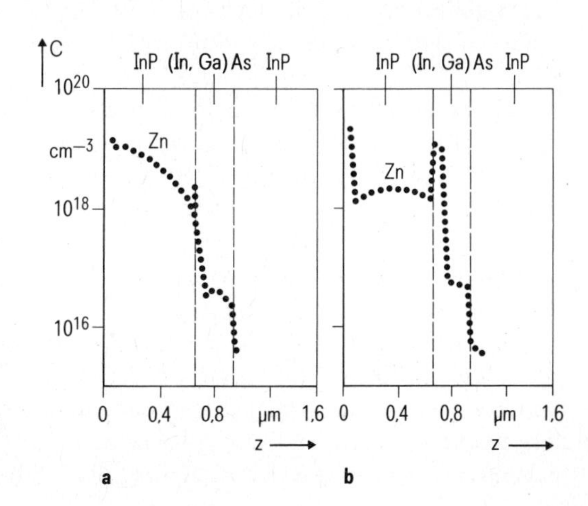

Fig. 9-69.
SIMS depth profiles of the same heterostructure as in Fig. 9-68, but without magnesium doping of the (In,Ga)As layer, after zinc diffusion for various times t from spin-on films at 550°C [9-67]: a) t = 60 s; b) t = 210 s. Same measuring conditions as for Fig. 9-68.

– a fast diffusion process coupled to the presence of magnesium, since in both Figs. 9-68c and d the Zn profile in the (In,Ga)As layer corresponds to the original Mg distribution.

The results of Figs. 9-68 and 9-69 show that in addition to the complete replacement of the initial magnesium dopant by zinc, the zinc diffusivity is distinctly enhanced in the presence of magnesium. These two observations reveal that there is a strong interaction between magnesium and zinc during diffusion.

In order to confirm this anomalous diffusion behavior and to exclude erroneous interpretations, several additional measurements were performed [9-67]. Within the detection limit of SIMS, magnesium was found to be replaced completely by zinc and enriched in the spin-on SiO_2 film. The replacement of substitutional magnesium by the diffusing zinc is assumed to be the relevant physical mechanism. This "kick-off" mechanism results in immobile substitutional zinc and mobile interstitial magnesium. Since all substitutional sites will be occupied by zinc, magnesium will diffuse rapidly in the opposite direction to zinc. Qualitatively, this is the reason why magnesium is enriched in the spin-on film and no change in the Mg distribution is observed before the zinc front has passed. Nonetheless, further experiments will be required before the diffusion behaviour of zinc described above is fully understood and this knowledge can be used for a better control to the fabrication process.

9.3.4 Ohmic contacts to gallium arsenide

High-quality, reliable contacts to n-GaAs are important for the fabrication of high-speed GaAs integrated circuits. In contrast to silicon, complex metallization systems are needed to attain ohmic contacts to GaAs. The Au-Ni-Ge system is the most widely used contact metallization for this purpose. Although very low contact resistances can be obtained, the alloying process is crucial to the contact properties. Another disadvantage of this metallization system is the deterioration found after annealing at elevated temperatures [9-68].

Another metallization system made up of a Ge-Au-Cr-Au sequence of layers that has been used successfully for some years is described below [9-69]. Contacts of this type exhibit a rather low contact resistance together with high thermal stability [9-70,71]. TEM cross sections were examined with the aid of analytical electron microscopy and compared with Auger depth profiles to clarify the complex metallurgical reactions that occur during the alloying process [9-70].

After ion implantation to form the conductive channel and etching to remove the native oxide, layers of 10 nm Ge, 140 nm Au, 40 nm Cr and 200 nm Au are vapor-deposited. Fig. 9-70 shows the Auger depth profiles of the as-deposited film structure. It is noticeable that the Cr layer contains a few percent of oxygen and that a small amount of germanium has diffused into the lower Au layer. The decrease in the slope of the profiles with increasing sputter time (i.e. with depth) is a result of a decrease in depth resolution due to the polycrystalline microstructure and some interface roughness of the various films. In the TEM cross sections of Fig. 9-71, the two Au layers are clearly separated by the Cr film. Both Au films and the Cr film exhibit a columnar microstructure with many grains extending from the bottom to the top of the films (Figs. 9-71a,b). The Cr film, however, has a much smaller grain size and contains a large number of vertically oriented pores. Instead of forming a separate layer, the germanium was incorporated in the lower Au film. This probably occurred during deposition of

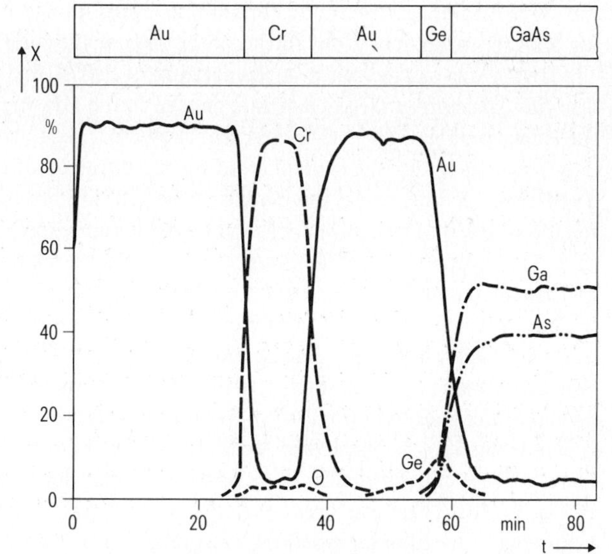

Fig. 9-70.
Auger depth profiles of as-deposited Ge-Au-Cr-Au film structure for ohmic metallization on n-GaAs. Standard sputtering conditions were used for all depth profiles: 2 keV argon ions with 55° incidence angle to the surface normal. [9-70]

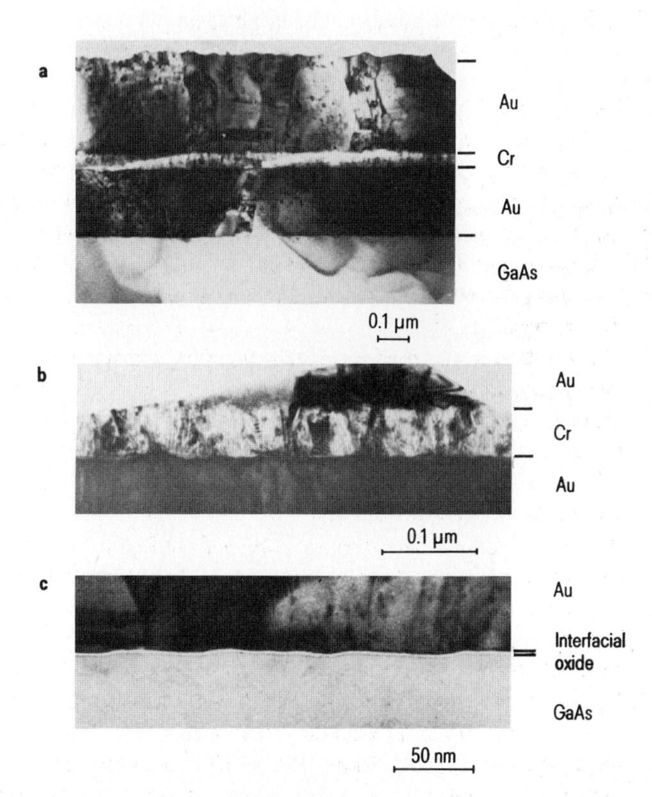

Fig. 9-71. TEM cross sections of as-deposited Ge-Au-Cr-Au film structure. a) Overall film structure, b) Cr layer, c) Residual oxide at metal-GaAs interface. [9-70]

the lower Au film. A continuous bright line representing the residual oxide formed after etching and approximately 2 nm thick can be seen at the interface to the GaAs (Fig. 9-71c).

Alloying was carried out at a temperature of 390 °C for 12 min in a nitrogen atmosphere. The film structure undergoes drastic change in the alloying process, as is evident from both the Auger profiles of Fig. 9-72 and the TEM cross sections of Fig. 9-73. Whereas the Auger depth profiles reflect the general trend for the diffusion of the various elements involved, the TEM cross sections reveal the details of the alloying reaction. Starting the description from the top of the film structure, there is a Cr oxide film approximately 5 nm thick at the top surface. The morphology of the upper Au layer did not change during alloying, but grain size has increased and the layer contains a few percent Ga in solid solution which will be discussed in more detail below. The Cr layer has decreased from 40 nm to between 35 and 25 nm in thickness and it contains a considerable amount of oxygen (Fig. 9-72). The fact that this film consists of Cr oxide or a mixture of Cr and Cr oxide rather than the pure metal is reflected in its very small grain size (Fig. 9-73b). After alloying, the lower Au film is split into two layers separated by a Cr-As phase (Fig. 9-72). In the TEM cross sections the position of the original GaAs surface can be identified by a row of bright dots representing oxide particles (arrows in Fig. 9-73b). These particles were formed when the interfacial oxide film balled up at the beginning of the alloying reaction. The grains in the vicinity of the original GaAs surface appear brighter than the Au grains that, apart from diffraction effects, are usually darker due to their high atomic number. The brighter grains correspond to the Cr-As phase indicated in the Auger profiles by the coincidence of the Cr and As peaks. Gold penetrates the GaAs and forms most of the contact area with the substrate. The depth of the alloying zone is about 120 nm.

Individual grains were subjected to X-ray microanalysis by energy-dispersive spectrometry in a STEM in order to identify the various phases in the alloying zone (Fig. 9-74). Only chromium and arsenic could be detected in grains of the Cr-As phase (Fig. 9-74a). Quantitative evaluation showed the ratio of the Cr and As mole fractions to deviate from unity by no more

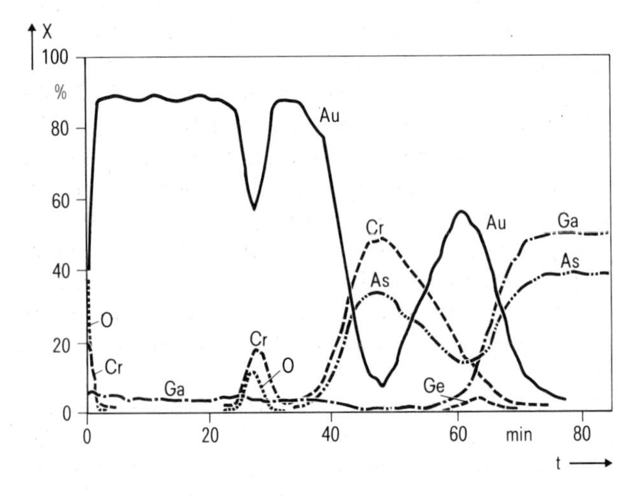

Fig. 9-72. Auger depth profiles of alloyed film structure. The coincidence of the Cr and As peaks between the two parts of the lower Au layer indicates the formation of a Cr-As phase. [9-70]

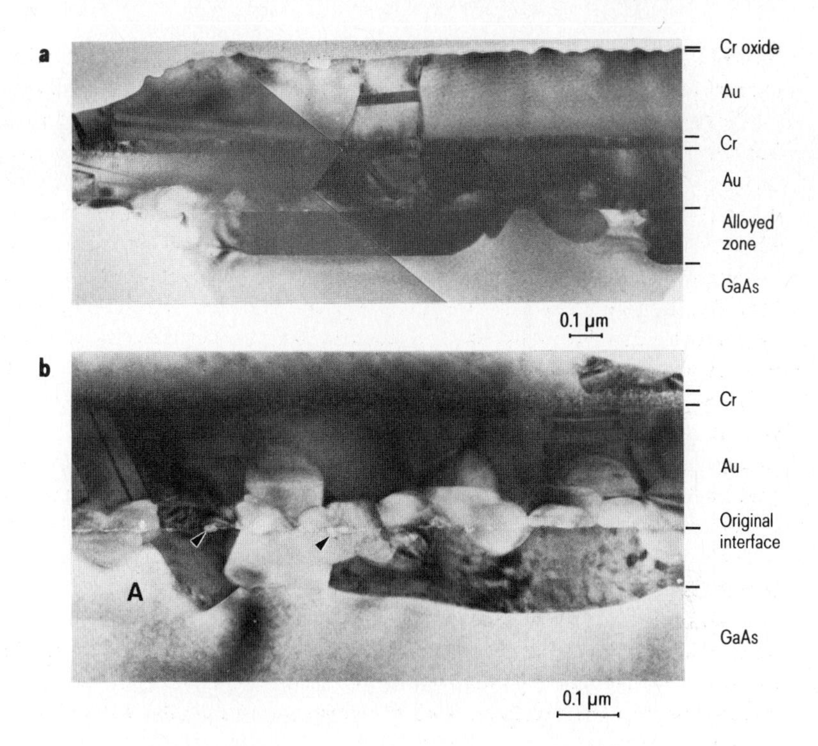

Fig. 9-73. TEM cross sections of alloyed film structure. a) Total structure, b) alloying zone. The dark grains in the alloying zone correspond to the Au(Ga) phase, the bright ones to the CrAs phase. Grain A consists of pure Ge grown epitaxially onto the GaAs substrate. In Fig. (a) two micrographs are mounted together. [9-70]

than a few percent. The conclusion that this phase is indeed CrAs was also supported by selected area diffraction. Some grains in the alloying zone that exhibited only a slight difference in contrast from the GaAs substrate were found to consist wholly of pure germanium (Fig. 9-74b). These grains precipitated epitaxially onto the GaAs substrate (at A in Fig. 9-73b) and gave rise to a Ge peak at the metallization-GaAs interface in the Auger profile (Fig. 9-72).

No significant variation was found in the composition of the Au grains in the various layers. Apart from gold, some gallium and arsenic were measured (Fig. 9-74c). To some extent at least, however, the X-ray intensities of As and Ga stem from secondary excitation (e.g. by electrons scattered in the specimen). Since the intensity ratio of Ga to As was about twice that in the GaAs substrate, we conclude that the grains of gold do in fact contain several percent Ga. Ge could not be found in these grains. The detection of Ge, however, was impeded by the fact that only the Ge Kß line could be used for analysis since the Ge Kα line overlaps the Au Lα line.

To obtain better quantitative results for the composition of the Au grains, the upper Au layer was analyzed by X-ray microanalysis using beam voltages low enough to confine the excitation volume to within this Au layer. The resulting mole fractions measured with two different beam voltages (5 and 10 kV) are listed in Table 9-3. The major part of chromium stems from the Cr oxide covering the Au layer, because the Cr fraction increased when a zone of smaller

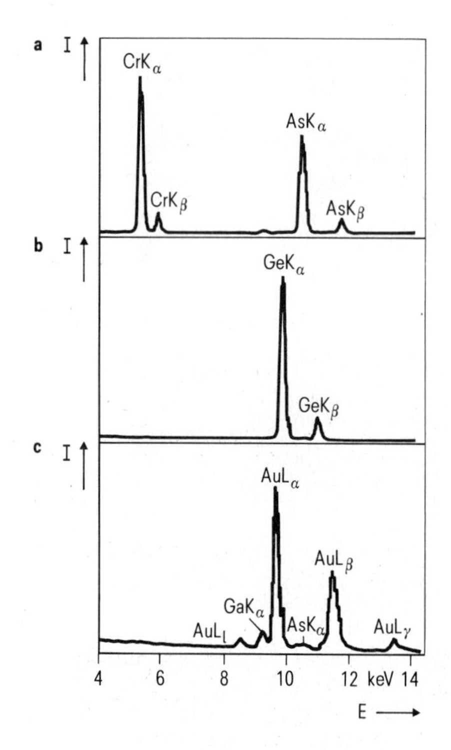

Fig. 9-74.
X-ray spectra of the various phases in the alloying zone recorded by energy-dispersive spectrometry in a STEM. a) CrAs phase, b) epitaxial Ge grain, c) Au grain from the lower Au layer. [9-70]

Table 9-3. Composition of upper gold layer determined by X-ray microanalysis

Beam voltage	Penetration depth*)	Mole fractions (%)				
		Au	Cr	Ga	As	Ge
5 kV	20 nm	87.0	6.1	6.5	–	0.5
10 kV	100 nm	90.5	2.2	6.7	–	0.7

*) The penetration depth is well below the 200 nm thickness of the top gold layer.

depth was analyzed at a lower beam voltage. No arsenic could be detected. The high Ga content is also visible in the Auger profile (Fig. 9-72) that also shows that the Ga contents of both the upper and lower Au layers are similar. Furthermore, it is worth noting that the Ge mole fraction is in the order of 0.5% – 1%, which could not be detected by X-ray analysis in the STEM or by AES.

Finally, the film structure was studied after prolonged aging for 10 h at the alloying temperature of 390 °C. Comparison of the Auger profiles (Fig. 9-75) with those of the alloyed specimen (Fig. 9-72) shows that the Cr peak of the CrAs phase has broadened as a result of aging and that consequently the splitting of the lower Au film into two parts is less pronounced. The TEM cross section of Fig. 9-76 demonstrates that the CrAs grains have increased in size and now extend throughout the lower gold film and the alloying zone. This explains the broadening of the CrAs peak. The "chromium" layer below the upper Au film decreased in thickness to about 15 nm but is still a continuous fine-grained film (Fig. 9-76). Accordingly, the

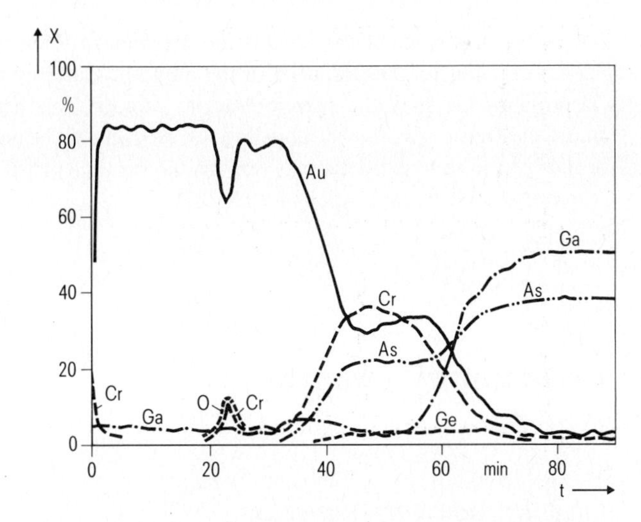

Fig. 9-75. Auger depth profiles of aged film structure. Compared to the alloyed film structure (Fig. 9-72) the peaks of the CrAs phase are broader due to an increased grain size. [9-70]

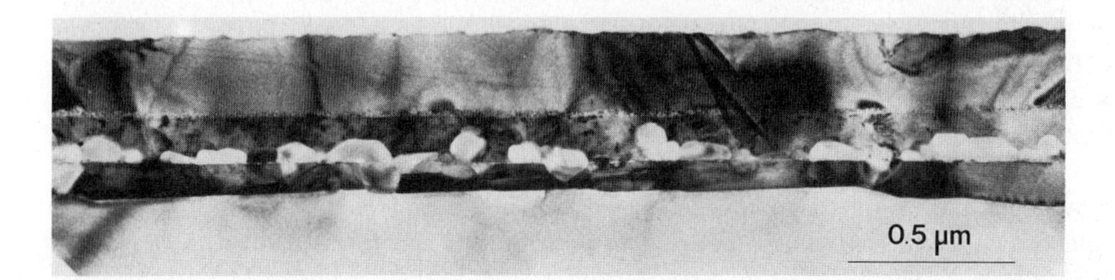

Fig. 9-76. TEM cross section of aged film structure. The metal-GaAs interface is smooth and Au(Ga) grains form the major part of the interface. [9-70,71]

Cr peak in the Auger profile remained sharp but was reduced in height (Fig. 9-75). The oxygen peak stayed about the same during aging and Cr oxide is probably all that is left of this layer. No longer concentrated at the metallization-GaAs interface, germanium has spread over the entire alloying zone and the lower Au layer. No epitaxial Ge grains could be found in the TEM cross sections. The depth of the alloying zone did not increase during aging and, as in the case of the non-aged specimen, Au grains form the major part of the contact area. Furthermore, the interface between the alloyed zone and the substrate becomes smoother during aging.

The line of oxide particles at the original metallization-GaAs interface (Fig. 9-73b) indicates that no melting occurred and alloying took place by solid-state diffusion. Gold certainly provides the major driving force for the alloying reaction, supported by chromium. Gold penetrates the metal-GaAs interface and dissolves gallium, while arsenic reacts with chromium to form the CrAs phase. No compounds of germanium with other elements were found here, in contrast to the Ni_2GeAs phase described in [9-68]. Germanium forms a separate phase but is also dissolved in the Au(Ga) grains. Precise knowledge of the alloying reaction allows conclusions to be drawn about the conduction mechanism. During the formation of the Au(Ga) phase, Ga vacancies are produced in the adjacent GaAs crystal. The in-diffusing Ge atoms will populate Ga sites and therefore act as donors. This highly doped n^+-zone allows enhanced electron transport via tunneling. Consequently, the Au(Ga) phase that forms the major part of the contact interface is responsible for the current transport, to which neither the CrAs nor the Ge phases contribute [9-70,71].

9.4 Electronic ceramics

9.4.1 Introductory remarks

In structural ceramics, the mechanical and high-temperature properties are major factors, whereas in electronic ceramics the electrical characteristics are important. Electronic ceramics include $BaTiO_3$ ceramics for capacitors and positive temperature coefficient resistors (PTCRs), ZnO varistor ceramics, piezoelectric ceramics on the basis of $Pb(Zr,Ti)O_3$ (PZT) and, most recently, high-temperature superconductors such as $YBa_2Cu_3O_{7-\delta}$. On the one hand, the electrical properties of the ceramic materials depend on their composition, including dopants which generally range from a few tenths of one percent to several percent. On the other hand, these properties are also determined by the microstructure that results from the composition, the powder morphology of the raw products and the ceramic processing, especially sintering. The mixed-oxide method is often used in manufacturing [9-72]. This technique comprises the following steps: mixing and grinding of the oxide powders of the various elements, calcination to form the compound, grinding, pressing with a binder and finally sintering.

Various methods of analysis are needed to characterize ceramic materials. The particle size of the powders and the shape of the particles can be determined in the SEM. After sintering, the first step is an accurate analysis of the chemical composition because the nominal composition can change during the sintering process if volatile elements are present (e.g. lead). Optical emission spectrometry with inductively coupled plasma (ICP-OES) can be used for this purpose, providing quantitative analysis with an error of $< 1\%$ (for the main constituents) of small specimen quantities (≤ 1 mg). The phases that form during sintering can be identified with the aid of X-ray powder diffractometry. If the grain size exceeds 1 μm, optical microscopy of polished and etched sections will suffice to display the grain structure. Besides the size and shape of the matrix grains, characterization of the second phases is of particular interest. If their portion is less than about 0.5%, second phases are not detectable by X-ray diffractometry. However, these phases are visible in the sections and can be analyzed qualitatively or quantitatively by X-ray microanalysis if the grains are larger than 1 μm (Fig. 5-16). Analytical electron microscopy in the TEM is needed if the grains are smaller than 1 μm or if an analysis of the crystal structure is also required. In addition, TEM allows lattice defects within the grains and grain boundaries to be studied in detail.

9.4.2 Second phases in yttrium-doped barium titanate ceramics

In undoped $BaTiO_3$ ceramic, individual grains in a fine-grained matrix (grain size < 1 μm) will grow anomalously large to sizes of 100 μm even below the eutectic temperature of 1320 °C in the $BaTiO_3$-TiO_2 system [9-73]. This anomalous grain growth can be suppressed by doping if a concentration of a few tenths of a percent is exceeded. Sintering will then produce a fine-grained ceramic suitable for use as a capacitor dielectric. One analytical objective was to identify the second phases produced by sintering yttrium-doped specimens. In addition, the effect of the sintering atmosphere (reducing or oxidizing) on the stoichiometry of the various phases was studied [9-74].

Fig. 9-77 shows a backscattered-electron image of $BaTiO_3$ ceramics doped with a mass fraction of approximately 1% yttrium. Besides the large, bright grains of the $BaTiO_3$ perovskite phase, the image also reveals two different, darker second phases situated at multiple grain junctions or forming inclusions in the $BaTiO_3$ matrix. These phases appear dark in the BSE image because their mean atomic number is lower than that of $BaTiO_3$ (material contrast, section 3.5.1). In Fig. 5-16b, the elemental map of yttrium is superimposed on the BSE image shown in Fig. 9-77. Comparison of the two images shows that the darker of the two second phases contains a large amount of yttrium. Since the brighter of the second phases contains no yttrium but shows up darker than $BaTiO_3$ in the BSE image, it must contain a greater fraction of the lighter element titanium than does $BaTiO_3$.

Fig. 9-77.
Microstructure of $BaTiO_3$ ceramic containing 0.1% yttrium (backscattered electron image of polished and etched section). The $BaTiO_3$ matrix appears bright, the second phases are dark. [9-74]

The grains, including those of the second phases, are large enough to permit quantitative analysis of the individual phases by X-ray microanalysis. For such measurements, it is of advantage to record the X-rays of the elements Ba, Ti, O and Y simultaneously with the four crystal spectrometers, as all the X-ray intensities will then originate from the same specimen volume of approximately 1 μm^3. Matched standards were used to minimize the error for ZAF correction (section 5.4.4): a polished $BaTiO_3$ single crystal as the standard for the elements Ba, Ti and O, and a Y_2O_3 single crystal for Y. The results of quantitative analysis of two specimens are summarized in Table 9-4: one of these specimens was sintered in a reducing atmosphere and the other in an oxidizing atmosphere. Since all the elements (including oxygen) were measured in both cases, the sum of the mass fractions can be formed. This sum allows the quality of the ZAF correction to be assessed and should deviate only slightly from 100%. The column beside the mass fractions lists their standard deviations resulting from counting statistics so that the error in analysis can be evaluated.

When the results of sintering in reducing or oxidizing atmospheres are compared (Table 9-4), it becomes apparent that both specimens contain small fractions of Y_2O_3 dissolved in the $BaTiO_3$ matrix. After sintering in a reducing atmosphere, however, this fraction is more than twice that found after sintering in an oxidizing ambient. The portion of the second phase containing yttrium was correspondingly higher after sintering in an oxidizing atmosphere. However, the sintering atmosphere has no effect on the oxygen content of the $BaTiO_3$ perovskite phase: the specimen sintered in a reducing atmosphere does not exhibit an oxygen deficiency. The yttrium-rich phase contains only a small fraction of barium and again, there is no difference in the oxygen content, regardless of the atmosphere in which the specimens were sintered (Table 9-4). Within the limits of measuring accuracy, the composition corresponds to a $Y_2Ti_2O_7$ phase. The titanium-rich phase contains no yttrium (Table 9-4). Its composition corresponds to a $Ba_6Ti_{17}O_{40-x}$ phase. In the specimen sintered in a reducing atmosphere, however, this phase exhibits a noticeable oxygen deficiency.

These measurements allowed the various second phases occurring in the barium titanate ceramics to be identified, and in addition showed that the sintering atmosphere has an effect on the dopant solubility in the $BaTiO_3$ matrix and on the oxygen content of the Ti-rich phase [9-74].

Table 9-4. Results of quantitative X-ray microanalysis of the various phases in yttrium-doped $BaTiO_3$ ceramics sintered in reducing and oxidizing atmospheres, respectively. Together with the elements, the X-ray lines used for analysis are given. The sum of the mass fractions indicates the quality of the ZAF correction. The mole fractions were normalized so that their sum is always 100%. The standard deviation of the mass fractions results from the counting statistics and is a measure for the analysis error. [9-74]

Phase	Element and X-ray line	Sintering in reducing atmosphere: Mass fraction %	Standard deviation %	Mole fraction %	Sintering in oxidizing atmosphere: Mass fraction %	Standard deviation %	Mole fraction %
$BaTiO_3$ matrix	Ti Kα	20.74	0.89	19.96	20.06	0.99	19.90
	Ba Lα	56.92	0.86	19.11	58.10	0.99	20.11
	O K	20.90	3.40	60.24	20.09	3.31	59.68
	Y Lα	1.34	2.11	0.69	0.59	5.41	0.31
	Sum	99.90		100.00	98.84		100.00
Y-rich phase	Ti Kα	24.44	0.89	19.05	24.03	0.97	18.69
	Ba Lα	0.86	4.05	0.24	0.75	7.25	0.20
	O K	26.85	3.48	62.66	27.28	3.41	63.52
	Y Lα	42.97	0.67	18.05	41.97	0.76	17.59
	Sum	95.12		100.00	94.03		100.00
Ti-rich phase	Ti Kα	35.21	0.86	29.39	33.99	0.92	25.98
	Ba Lα	35.89	0.91	10.45	37.07	1.09	9.88
	O K	24.07	3.43	60.15	28.01	3.25	64.11
	Y Lα	0.02*)	64.84	0.01	0.08*)	26.24	0.03
	Sum	95.19		100.00	99.15		100.00

*) below detection limit

9.4.3 Bismuth oxide phases in ZnO varistor ceramics

ZnO varistor devices are used to suppress transient voltages in electrical supply systems. Apart from ZnO, these varistor ceramics contain Bi_2O_3 and various other metal oxides (e.g. Sb_2O_3, Cr_2O_3, Co_3O_4, MnO_2). The mole fraction for each dopant is in the order of 1%. X-ray powder diffraction showed that besides ZnO, a $Zn_7Sb_2O_{12}$ spinel phase and a small amount of Bi_2O_3 were present in our samples. X-ray microanalysis revealed that the ZnO phase contains Co and Mn as doping elements, while the spinel phase contains the dopants Cr, Co, Mn and Ni [9-75]. Bi_2O_3 forms a liquid phase during sintering and solidifies at multiple-grain junctions during cooling.

The homojunction of the ZnO-ZnO grain boundary is responsible for the non-ohmic behavior of the varistor. Experiments with microvaristors have shown that these homojunctions are electrically bypassed by the Bi_2O_3 phases, resulting in additional leakage currents [9-76]. Four different modifications of Bi_2O_3 are known. Besides the stable room-temperature modi-

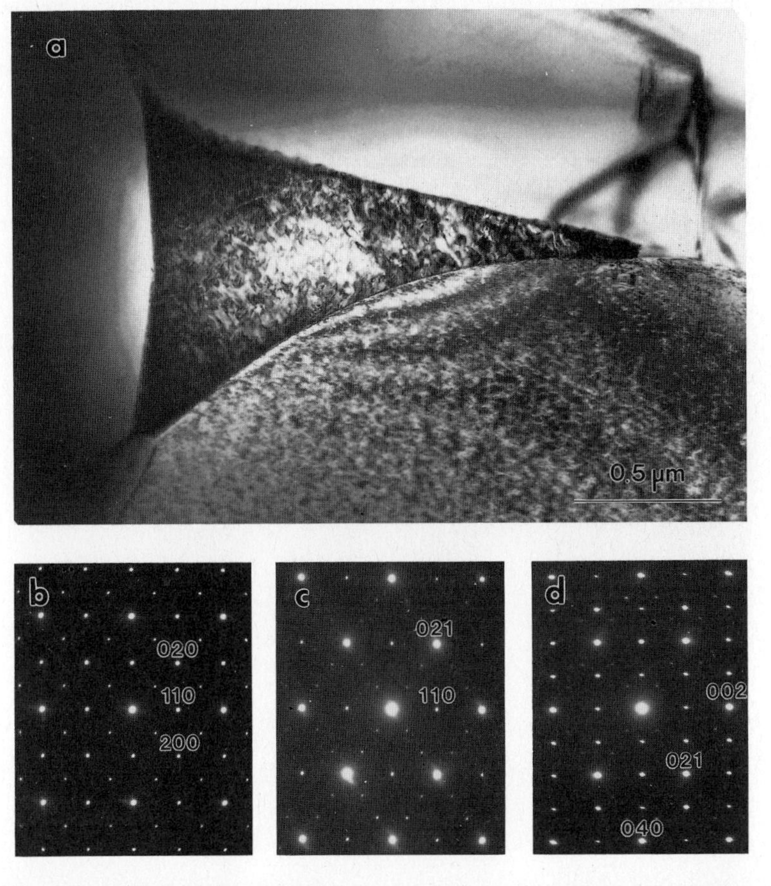

Fig. 9-78. a) Single grain of Bi_2O_3 at triple grain junction in ZnO varistor ceramic (TEM bright-field image). b), c), d) Selected area electron diffraction patterns of β-Bi_2O_3 grains in various low-index poles: b) [001], c) [$1\bar{1}2$], d) [100]. [9-77]

fication α-Bi_2O_3, various high-temperature phases (β, γ, δ) occur. However, they can be stabilized to room temperature by doping with various metal ions. Some of these phases exhibit high ionic conductivity. Since the additional leakage currents caused by these Bi_2O_3 phases are undesirable, the microstructure and crystal structure of the Bi_2O_3 grains in the multiple-grain junctions were studied in detail [9-77]. Because of their small fraction and coincidence of reflections, the various Bi_2O_3 phases cannot be clearly distinguished by X-ray powder diffraction. This investigation therefore resorted to TEM including selected area- and micro-diffraction as well as high-resolution imaging. Dimple grinding (section 4.3.2, Fig. 4-15) down to a thickness of about 10 μm prior to ion milling was an essential step in specimen preparation for the TEM. Without dimpling, in other words if ion milling is used to thin a specimen some 50 μm thick, the higher sputter rate leads to preferential removal of the small Bi_2O_3 grains in the multiple-grain junctions.

After sintering (at 1200 °C), most of the multiple-grain junctions are filled by single Bi_2O_3 grains (Fig. 9-78a). The mottled contrast is due to small strain centers produced by a high concentration of microdefects. At typically 1 μm, grain size is large enough to yield selected area diffraction (SAD) patterns. The majority of these grains were found to consist of the tetragonal β-Bi_2O_3 phase. Some SAD patterns of low-index poles are shown in Figs. 9-78b, c and d. All the strong spots in the patterns can be identified as β-Bi_2O_3 reflections. However, there are also weak reflections (e.g. {100} in Fig. 9-78b) that are forbidden in the nominal space group (No. 114) of β-Bi_2O_3. Therefore, it would appear that not all the symmetry elements of this space group are present in the structure of these grains. On the other hand, the additional weak reflections (1/3{021} in Fig. 9-78c) indicate a superstructure.

In the [100] pole, two types of domain can be distinguished in the β-Bi_2O_3 grains (Fig. 9-79). The domains A reveal a mottled contrast due to strain centers caused by microdefects, whereas the domains B exhibit a more uniform bright contrast. High-resolution lattice imaging in the [100] projection shows that the lattice fringes pass continuously through both domains without any change in spacing (Fig. 9-80a). If the same area of the crystal is imaged

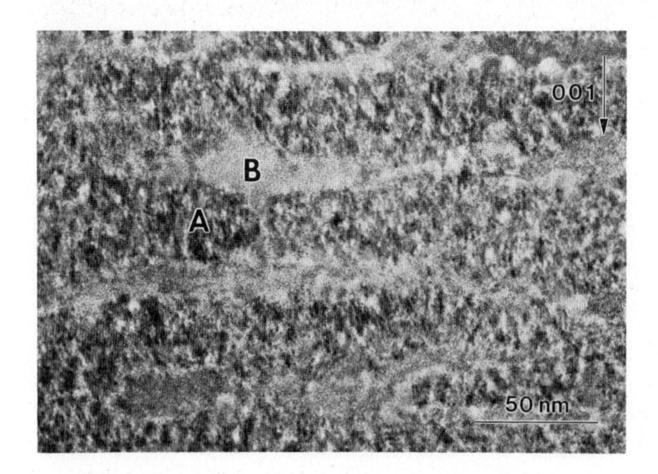

Fig. 9-79. Bright-field image of β-Bi_2O_3 grain in [100] orientation showing two types of domain. Domains A exhibit a mottled contrast due to microdefects, domains B show a more uniform contrast. [9-77]

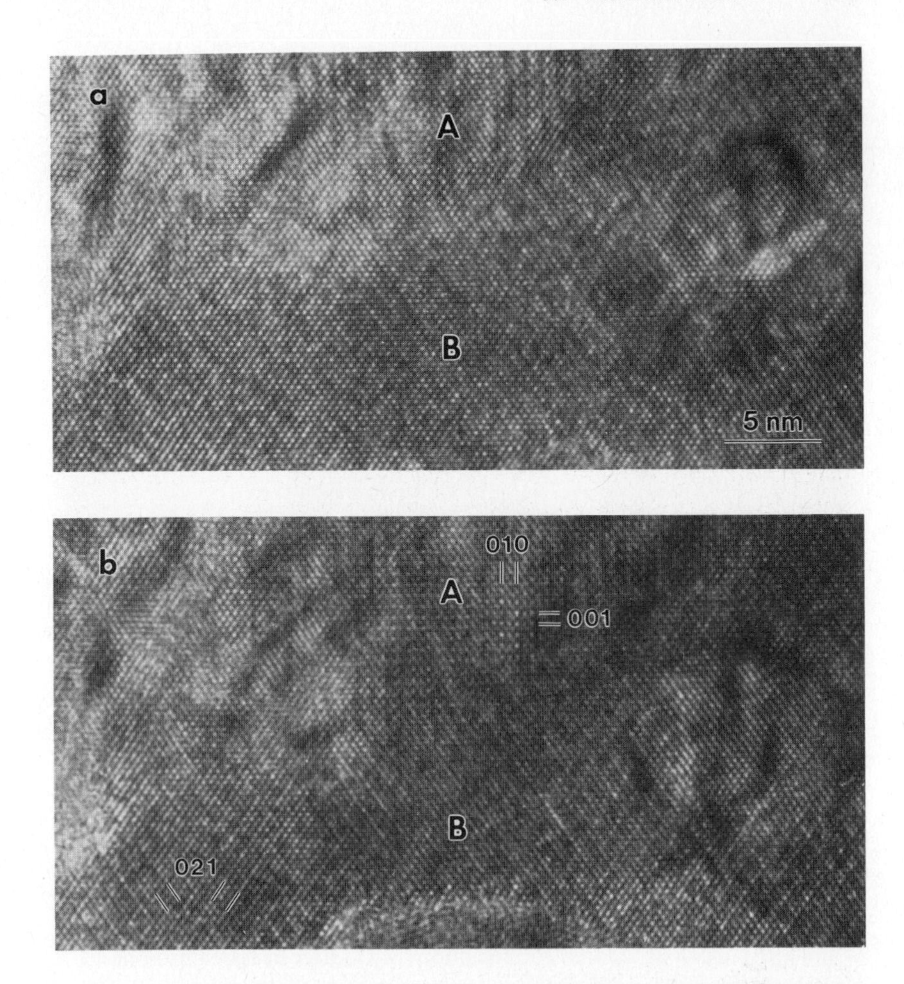

Fig. 9-80. High-resolution images of β-Bi_2O_3 grain in [100] projection recorded with different defocusing. a) The lattice fringes pass continuously through both domains A and B; b) Different ordering appears in domains A and B: parallel (010) and (001) planes in domain A, and parallel to the {021} planes in domain B. [9-77]

using another defocus setting, the dot pattern of the lattice image changes slightly in a different way for each domain (Fig. 9-80b). The pronounced bright-dot contrast shows a superlattice of the (010) and (001) planes in domain A, whereas ordering is perpendicular to the two sets of {021} planes in domain B and every third lattice fringe appears bright. The difference in ordering in domains A and B can be verified by optical diffractograms of the corresponding regions in the HREM micrographs (Fig. 9-81a,b). These optical diffractograms show that (001) and (010) spots appear in addition to the (002) and (021) spots in domain A (arrowed in Fig. 9-81a). In domain B, two superlattice spots of type 1/3{021} spots are visible between the origin and the {021} spots (Fig. 9-81b). The same result is obtained when electron microdiffraction patterns of both types of domain are recorded with a 20 nm diameter beam (Fig. 9-81c,d).

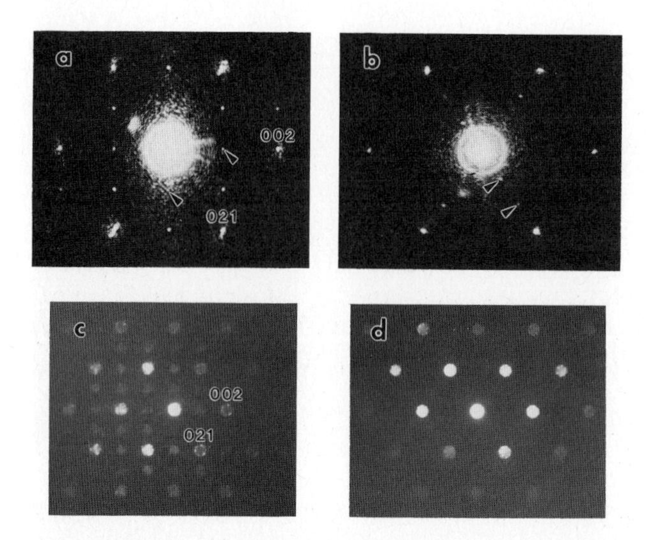

Fig. 9-81. Diffraction information of the domains in β-Bi_2O_3. a), b) Optical diffractograms of domains A (a) and B (b) by selecting specific regions of the HREM image in Fig. 9-80b; c), d) Electron microdiffraction patterns of domains A (c) and B (d) recorded with an electron beam of 20 nm diameter. The arrows indicate the superlattice spots which are only weakly visible in domains B. [9-77]

Since beam divergence is larger than in SAD patterns, the diffraction spots appear as disks (section 4.4.4). The (001) and (010) disks are then clearly visible in the A domains. In domain B, however, the 1/3{021} disks are scarcely visible after reproduction because of the low intensity of the fine electron beam and the weak intensity of the 1/3{021} reflection itself. As discussed above, both types of ordering show up clearly in the HREM image in Fig. 9-80b.

Besides β-Bi_2O_3, the as-sintered specimen was also found to contain α-Bi_2O_3 grains. Both α- and β-Bi_2O_3 phases are transformed into γ-Bi_2O_3 after heat treatment at 600 °C [9-77]. In general, the properties of the bismuth oxide phases depend on the metal dopants, sintering and cooling temperatures and the sintering ambient.

9.4.4 High-temperature superconductors

Since the first discovery of superconductivity at high temperatures ($T_c \approx 40$ K) in $(La_{1-x}Sr_x)_2CuO_{4-\delta}$ compounds by Bednorz and Müller, other compounds, especially in the Y-Ba-Cu-O and Bi-Sr-Ca-Cu-O systems have been found with transition temperatures in excess of 77 K. All these compounds are cuprates and have an oxygen-deficient perovskite-like structure. Since they are layered structures, their physical properties are highly anisotropic.

Materials development aims at achieving a high transition temperature in conjunction with a high critical current density. Thin films of $YBa_2Cu_3O_{7-\delta}$ hold promise for applications in microelectronics, whereas Bi-Sr-Ca-Cu-O compounds and related materials are interesting as bulk materials on account of the high transition temperatures that can be obtained [9-78].

$YBa_2Cu_3O_7$ thin films

In the orthorhombic crystal structure of $YBa_2Cu_3O_{7-\delta}$, the superconducting transport of current is severely anisotropic, being 10 times higher in directions parallel to the (001) plane than in the [001] direction normal to these. Hence, when $YBa_2Cu_3O_{7-\delta}$ films are produced, the aim is to grow layers with the *c*-axis normal to the film plane. Substrate temperatures in excess of 700 °C and substrate crystals with a perovskite-related structure (e.g. $SrTiO_3$) are essential for epitaxial growth. At sputtering temperatures lower than 600 °C, the films deposited on (100) $SrTiO_3$ grow amorphously; the films grown at less than 700 °C are polycrystalline. Sputtering [9-79] or laser evaporation [9-80] are suitable methods of deposition. Again, the chemical composition of the films (except oxygen) can be controlled by optical emission spectrometry (ICP-OES). Cross-sectional TEM is used for the detailed characterization of the microstructure and the interface with the substrate. Fig. 9-82 is a cross section through a $YBa_2Cu_3O_{7-\delta}$ film deposited by DC sputtering at a substrate temperature of 450 °C and annealed at 900 °C [9-81]. The film is polycrystalline and contains numerous pores measuring approximately 100 nm in diameter. Numerous dislocations (10^8 cm^{-2}) can be seen in the $SrTiO_3$ substrate. Annealing has produced a reaction zone approximately 150 nm in thickness between the substrate and the film. Fig. 9-83 shows X-ray spectra recorded in the STEM at the various specimen points. A small quantity of strontium is dissolved in the $YBa_2Cu_3O_{7-\delta}$ (Fig. 9-83a). The Ba Lα and Ti Kα lines overlap in the X-ray spectrum of the reaction zone (Fig. 9-83b). Proof that the phase in question is indeed the specified $(Bi_{1-x}Sr_x)_2TiO_4$ was furnished by electron diffraction patterns obtained by carefully tilting individual crystallites. The reaction zone contains a small fraction of copper. In contrast, it may be assumed that the small Cu peak in the $SrTiO_3$ is due to spurious radiation from the specimen holder (Fig. 9-83c). The results suggest that the reaction zone was formed by the diffusion of barium from the amorphously deposited film into the substrate and the diffusion of strontium from the substrate into the film [9-81].

A $YBa_2Cu_3O_{7-\delta}$ film deposited on (100) $SrTiO_3$ by RF magnetron sputtering at room temperature and annealed at 900 °C is shown in Fig. 9-84 [9-82]. The process has produced columnar grains with a diameter of approximately 100 nm and extending throughout the film which

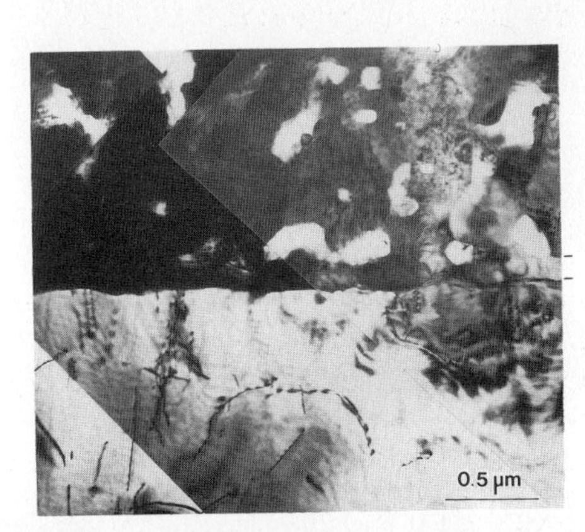

Fig. 9-82. TEM cross section of $YBa_2Cu_3O_7$ film deposited by DC sputtering onto a $SrTiO_3$ substrate. During annealing at 900 °C a reaction zone has formed at the interface. [9-81]

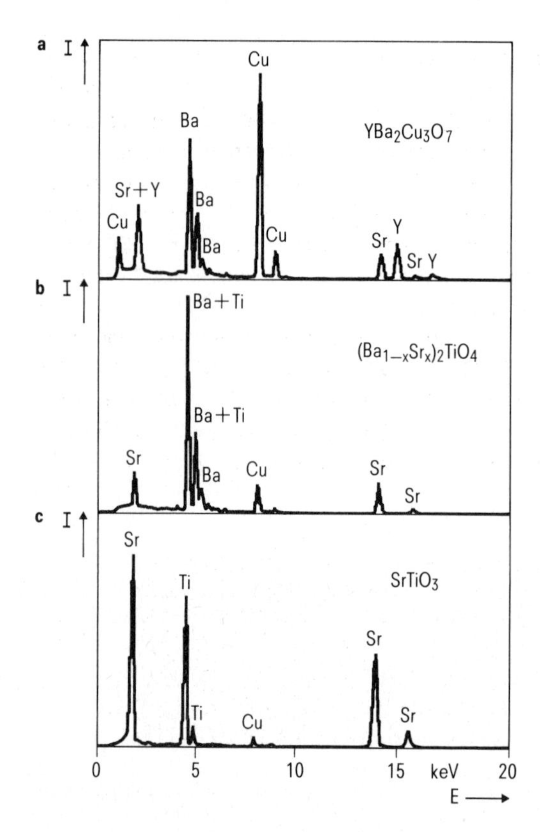

Fig. 9-83.
X-ray spectra of various regions of the TEM cross section of Fig. 9-82: a) $YBa_2Cu_3O_7$ film, b) reaction zone at the interface, c) $SrTi0_3$ substrate. The phase $(Ba_{1-x}Sr_x)_2TiO_4$ of the reaction zone was determined by electron diffraction and X-ray microanalysis.

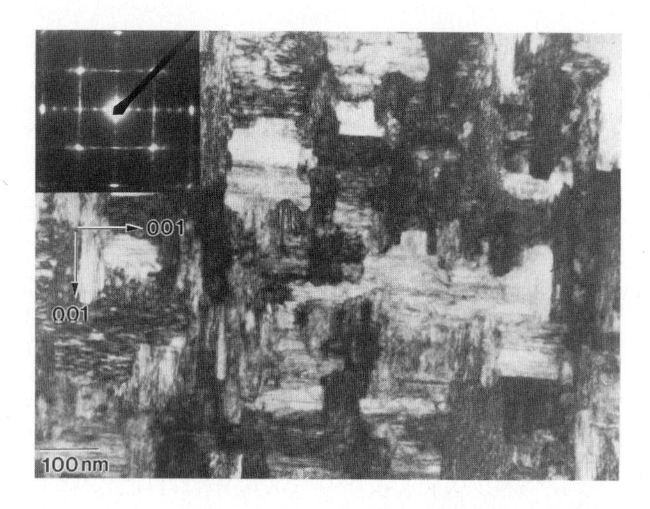

Fig. 9-84. $YBa_2Cu_3O_7$ film deposited by RF magnetron sputtering onto a $SrTiO_3$ substrate after annealing at 900 °C (TEM bright-field image of plan-view specimen). Two types of grains appear. Their *c*-axes [001] lie in the film plane and are perpendicular to each other. In the electron diffraction pattern (insert) the patterns of both types of grains are superimposed. [9-82]

is about 1 μm thick. These grains contain a high density of planar defects. Oriented crystallization has progressed through the film during annealing. However, instead of being normal to the film plane, the *c*-axes of the grains are parallel to the two ⟨100⟩ directions along the surface of the (100) oriented $SrTiO_3$ substrate. There are, therefore, two types of grain having *c*-axes perpendicular to each other. The orientation of the grains is indicated by the fringes in the grains which result from planar defects on (001) planes. The *c*-axes in both types of grain are perpendicular to these fringes. The two orientations are superimposed in the diffraction pattern (insert in Fig. 9-84). The streaks between the diffraction spots are produced by the numerous planar defects lying parallel to the electron beam.

Fig. 9-85 shows a cross section through a film grown epitaxially by laser evaporation at a substrate temperature of approximately 720 °C. The lattice fringes of the (001) planes in the $YBa_2Cu_3O_{7-\delta}$ film are clearly seen to be parallel to the interface. Very high critical currents $j_c > 10^6$ A/cm² at 77 K can be obtained with these epitaxial films in contrast to the polycrystalline film in Fig. 9-82 in which the critical current density j_c was no more than 10^4 A/cm² (cf. [9-80]).

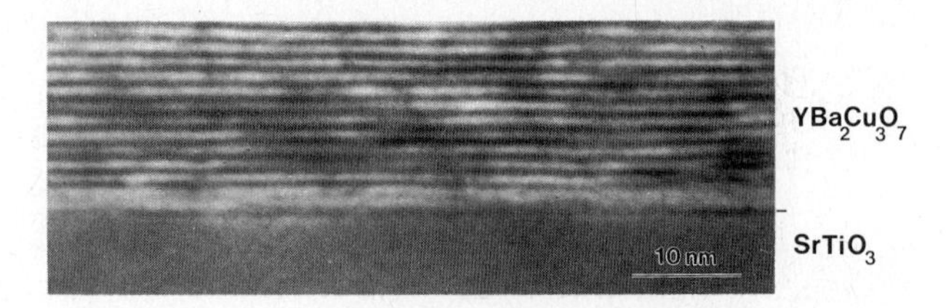

Fig. 9-85. (001) lattice fringe image of $YBa_2Cu_3O_7$ film on $SrTiO_3$ (TEM cross section). Laser evaporation at high substrate temperatures (720 °C) results in epitaxial growth with the *c*-axis normal to the film plane.

Bi-Sr-Ca-Cu-O ceramics

There are several phases with different superconducting transition temperatures in the Bi-Sr-Ca-Cu-O system. The crystal structures of the various phases are similar to each other: there is a variable number *n* of perovskite-like units inserted between the Bi-O double layers. The HREM image in Fig. 9-86 shows the phase with $n = 2$ and a composition $Bi_2Sr_2CaCu_2O_8$ in [110] projection [9-83]. Since the lattice spacings are larger than 0.18 nm – the point resolution of the microscope used – and the specimen is very thin, the image can be interpreted directly in terms of the projected crystal potential (section 4.6.1, equation (4-24): the atomic columns of the cations ($Z \geq 20$) appear as black dots. The rows of Bi atoms in the Bi-O double layers are marked by arrows. The square pattern of the perovskite units is clearly visible between these double layers. The insert shows the projection of cations in two half unit cells. In the lower half, two perovskite units are inserted between the Bi-O double layers as is characteristic for the $n = 2$ phase. In the upper half, however, three perovskite units are inserted indicating an extrinsic stacking fault. A lamella of the $n = 3$ phase is produced by this stacking fault.

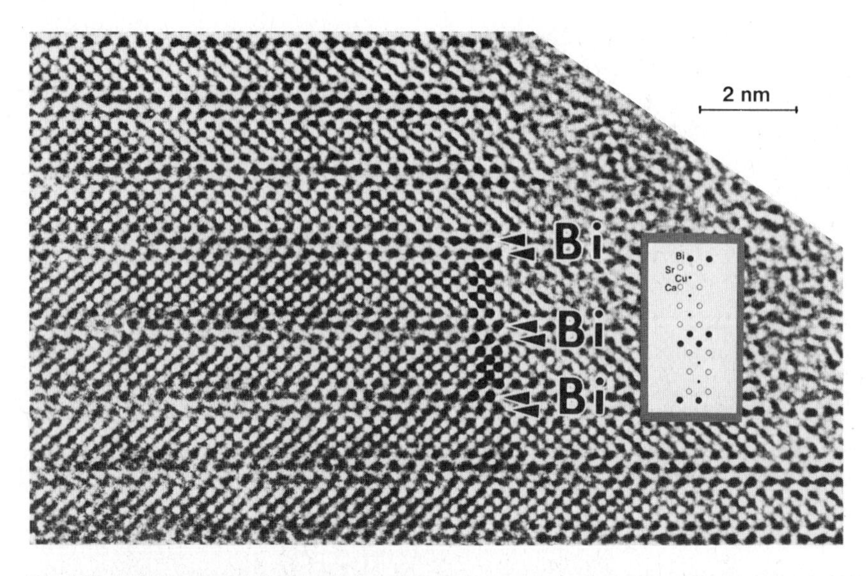

Fig. 9-86. High-resolution image of the $Bi_2Sr_2CaCu_2O_8$ phase in [110] projection. Between Bi-O double layers (arrows) perovskite units producing a square pattern are inserted. In the middle, a lamella of the $n = 3$ phase is formed (by an extrinsic stacking fault). [9-83]

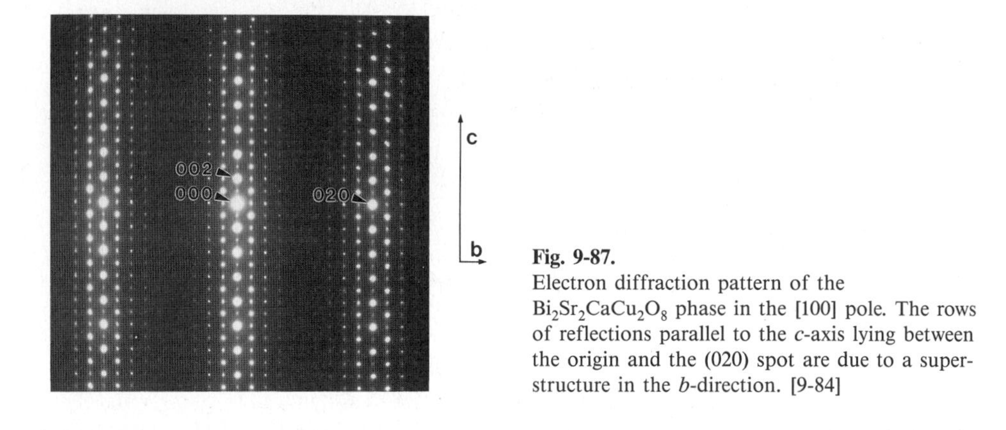

Fig. 9-87. Electron diffraction pattern of the $Bi_2Sr_2CaCu_2O_8$ phase in the [100] pole. The rows of reflections parallel to the *c*-axis lying between the origin and the (020) spot are due to a superstructure in the *b*-direction. [9-84]

Fig. 9-87 shows an electron diffraction pattern of the orthorhombic $n = 2$ phase $Bi_2Sr_2CaCu_2O_8$ in the [100] pole [9-84]. At $c = 3.05$ nm, the lattice constant is relatively large and so the reflections are closely spaced in the *c*-direction. There are several rows of reflections between the origin and the (020) reflection, caused by a superstructure in the *b*-direction. This superstructure is incommensurable, in other words its periodicity does not accord with any lattice periodicity. In the diffraction pattern, this becomes apparent in the fact that the distance of the (020) reflection from the (000) beam is not an integral multiple of the spacing of the superlattice spots. Electron diffraction patterns of various orientations yield valuable information about the symmetry elements of the structure [9-84].

The $Bi_2Sr_2CaCu_2O_8$ phase ($n = 2$) has a transition temperature of 80 K and can be produced as a primarily single-phase material by the mixed-oxide technique. The composition of the phase with $n = 3$ is $Bi_2Sr_2Ca_2Cu_3O_{10}$ and the transition temperature is 110 K. The even higher transition temperature of this compound renders it extremely interesting, but as yet it cannot be produced as a single-phase material. Fig. 9-88 shows a HREM image of a specimen containing large regions of the $n = 3$ phase [9-85]. The lattice fringes are due to the (002) planes. The perfect regions of this phase are identified by the spacing $c = 3.7$ nm of the (001) planes. However, these regions are very narrow because there is a high density of planar defects on (001) planes. These defects are lamellae of the phase with $n = 2$ having a (001) spacing of $c = 3.05$ nm. Many of these lamellae are only one lattice spacing in thickness, although some are thicker. Differences in the local chemical composition can be accommodated within regions of a few nanometers by the intergrowth of the phases with $n = 2$ and $n = 3$.

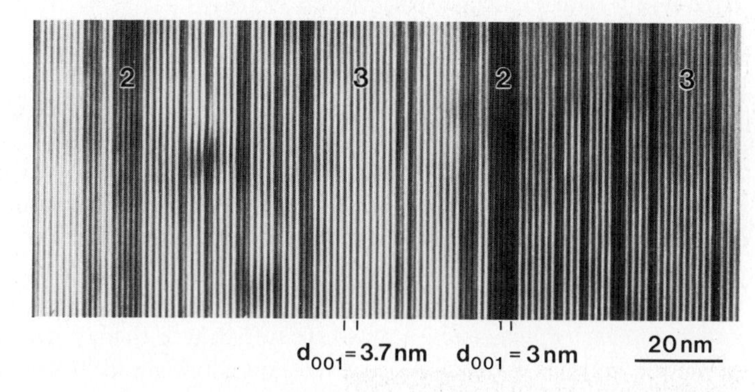

Fig. 9-88. (002) lattice fringe image of the $Bi_2Sr_2Ca_2Cu_3O_{10}$ phase ($n = 3$) with intergrown lamellae of the $n = 2$ phase. The two phases can be distinguished by their (001) spacing: $c = 3.7$ nm for the $n = 3$ phase, and $c = 3.05$ nm for the $n = 2$ phase. [9-85]

9.5 Electronic device testing

9.5.1 Electron beam test strategies

The electron beam methods described in Chapter 8 are used mainly for testing integrated circuits. The reasons are twofold:

a) Design verification: in the development phase, the electrical functions of a circuit must be checked and voltages at internal circuit nodes compared with previously simulated values.
b) Failure analysis: after stressing a device, any faults that may arise must be localized and characterized.

These purposes require different test strategies and the use of a number of methods [8-3,7]:

Design verification

In a circuit development phase, the aim is to check the circuit design and reveal design weaknesses. As a rule, only circuits already recognized as faulty in a prior test with an automatic test system are examined by electron beam testing. Since the function test takes place during the development phase, the circuits are mostly mounted in open packages. If they are already passivated, the uppermost interconnection level must be exposed, for instance by plasma etching. The device operating conditions must be selected to correspond to those used in the automatic test system and must reveal the detected fault.

The fault is conveniently localized by *logic-state mapping* or by recording a *timing diagram* (section 8.2.3) at constant frequency, starting with a low supply voltage. In most cases, the fault will occur at a specific voltage as the voltage is increased.

At the nodes in a circuit recognized as defective, *waveform measurements* (section 8.2.2.) are finally performed. The cause of the defect can then be determined by comparing the measured results with simulation calculations [9-86].

IC failure analysis

Often, a number of methods have to be used to clarify faults that have led to the failure of a device [8-7]. In this case, too, a start is made with electrical measurements with or without an automatic test system. After the housing has been opened, the circuit is examined for visible faults with an optical microscope or a scanning electron microscope.

As a rule, qualitative methods such as *stroboscopic voltage contrast imaging* (section 8.2.1) and *logic-state mapping* (section 8.2.3) are sufficient to characterize the specimen in the electron beam tester. In the case of devices of external origin for which no design data are available, a comparison with a perfect sample ("golden device") is of advantage. Finally, further analyses using scanning electron microscopy and X-ray microprobe analysis as well as special preparation techniques are needed to determine the physical causes of the failure [9-87] (section 9.6). If possible, passivation layers should not be removed prior to electron beam testing as this process can change the character of the fault. In such a case, an attempt must be made to observe the voltage contrast through the passivation layer (section 8.3.1).

Other applications

In principle, the applications for other electronic components such as microwave devices and surface acoustic wave components are the same: checking the functions during the development phase and analyzing faulty devices. The test procedure and the selected methods must be adapted to the problem in hand. It is therefore not possible to cite general guidelines such as those for testing VLSI devices.

9.5.2 Circuit verification of a 4-Mbit DRAM

Fig. 9-89 is a schematic view of a 4-Mbit dynamic random access memory (DRAM) [9-88, 89]. The 2^{22} = 4 194 304 memory cells are selected by address lines $A_0 \ldots A_{10}$ via word and bit decoders. Each addressed cell (C_{ik}) is at the intersection of the selected pair of word and bit lines (WL_i, BL_k). The signal $\overline{WE}$ (write enable) determines whether the operation in question will be a read or a write process. In a write operation, the information to be stored is present

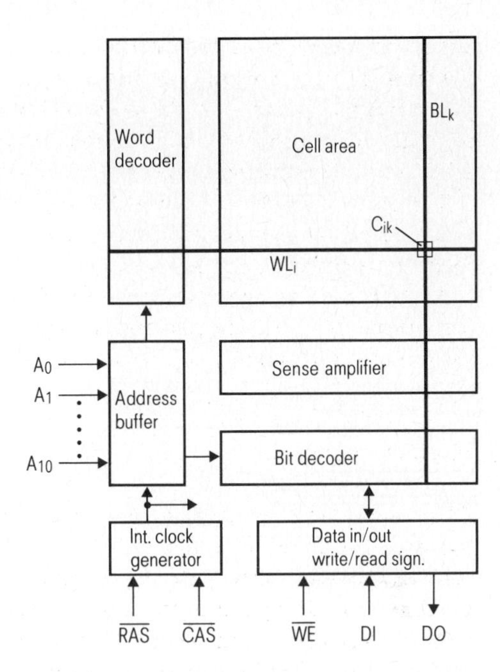

Fig. 9-89.
Block diagram of a 4-Mbit dynamic random access memory (DRAM). The cell area is surrounded by logic circuits that process the information. $A_0 \ldots A_{10}$ are the address lines, WL_i and BL_k are word and bit lines, C_{ik} is a memory cell, $\overline{WE}$ is the write/read control, $\overline{RAS}$ and $\overline{CAS}$ are external clocks, DI is the data input and DO is the data output.

at the data input (DI) and is written into the cell via the input stage and the selected bit line. In a read operation, the stored data are amplified by the read amplifier before being transferred to the data output (DO) via the output stage. The chip operates with two external clocks designated $\overline{RAS}$ and $\overline{CAS}$ (row access strobe and column access strobe).

The memory chip will not work perfectly unless certain signals of specified levels arrive at given times at given points in the circuit (timing). The signals in question are external control signals or internal signals derived from these. In the circuit design phase, computer modeling is used to simulate the switching response in normal and worst-case conditions. However, it is impossible to model all possible cases. As a result, design weaknesses requiring experimental clarification may occur.

Logic state testing

Fig. 9-90a shows a subcircuit in the address buffer of the 4-Mbit DRAM. To obtain a rapid overview of the internal switching processes (the logic), timing diagrams were recorded at five circuit nodes indicated by the signal names [9-90,91]. A distinction is made only between "low" and "high" levels. Fig. 9-90b shows the result: the threshold for evaluation was set to 50% of the specified voltage range (5 V). In other words, low is <2.5 V and high is >2.5 V. Comparison with the results of simulation soon shows whether the sequence and duration of the logical states in this subcircuit are correct (as they are in this case) or not.

Logic-state mapping and waveform measurements yield more detailed information [9-90, 91]. Fig. 9-91a shows a different part from the peripheral circuit of the chip. Preceding electrical testing with the automatic tester showed that certain memory cells in this specimen fail at specific supply voltages. For this reason, the write/read cycle of this circuit was exam-

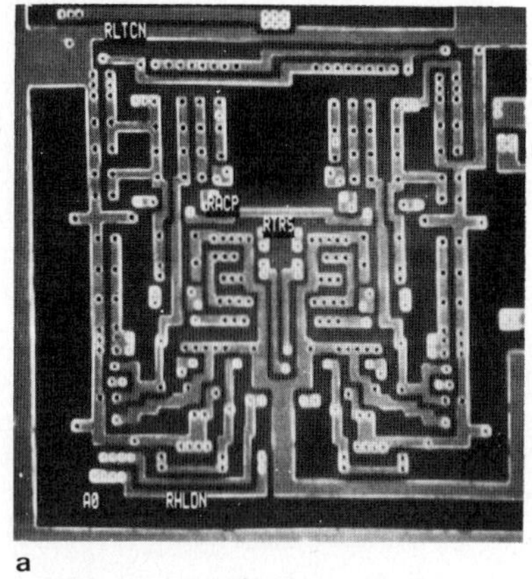

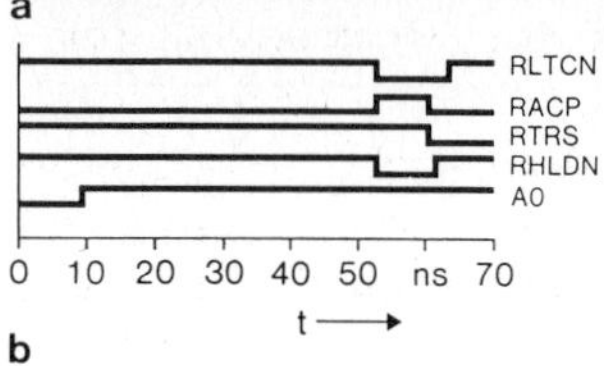

Fig. 9-90.
Subcircuit from the periphery of a 4-Mbit DRAM [9-90,91]; a) SE micrograph; b) timing diagram, measured at the nodes marked in Fig. a. Measuring conditions: $E_0 = 1$ keV; $I_0 = 2 \cdot 10^{-9}$ A; pulse frequency $f_p = 1$ MHz; pulse width $t_p = 3$ ns.

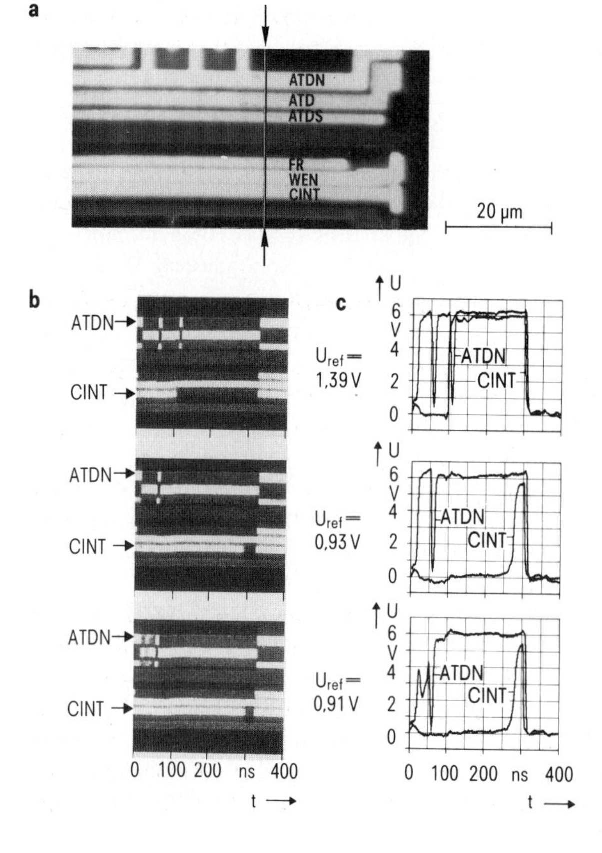

Fig. 9-91.
Checking the write/read cycle in a subcircuit of the 4-Mbit DRAM for various values of the reference voltage U_{ref} [9-90,91]. a) Stroboscopic voltage contrast micrograph; b) logic state pattern, measured along the line marked in Fig. a; c) waveform measurements of the ATDN and CINT signals. Measuring conditions: $E_0 = 1$ keV; $I_0 = 2 \cdot 10^{-9}$ A; $f_p = 1$ MHz; $t_p = 3$ ns.

ined under critical conditions with an excessive supply voltage (6 V instead of 5 V) and at various reference voltages U_{ref}. The logic-state mappings in Fig. 9-91b were recorded along the line indicated by the arrows in Fig. 9-91a (corresponding to BE in Fig. 8-8a). The line scan extends over six interconnections, the lowest of which is identified by CINT and the uppermost by ATDN. The waveforms shown in Fig. 9-91c were measured on these two interconnections. In accordance with the generation mechanism of the voltage contrast (section 8.2.1), a high potential ($U \approx 6$ V) appears dark in the logic-state pattern while a low potential ($U \approx 0$ V) appears bright.

When the chip is operating under normal conditions, U_{ref} is generated internally. Different values were externally impressed for U_{ref} in the experiment shown in Fig. 9-91. At U_{ref} = 1.39 V, the ATDN signal contains two voltage pulses from high to low (bright "spikes" in Fig. 9-91b, minima in c). The pulses switch transistors and thus produce signals in other parts of the circuit. As one result of the second pulse, the CINT signal is raised from 0 V to 6 V. This response is correct for this circuit during the recorded 400 ns of the write/read cycle.

At U_{ref} = 0.93 V, the internal switching processes are too slow. The second pulse of the ATDN signal no longer occurs and the CINT signal does not rise at the right time. The switching response becomes even worse at U_{ref} = 0.91 V: the CINT signal is only a weak delayed pulse. Moreover, the pulse on the ATDN line is no longer clear and changes over to incipient oscillations.

It follows from these results that the potential U_{ref} should not be allowed to fall below a specific lower limit. This could not be guaranteed for the device investigated here. A resimulation was then necessary to simulate this weakness. In the next step, the circuit was redesigned in order to ensure correct operation.

Measuring the read cycle

Fig. 9-92a shows the circuit diagram of a memory cell. Information is stored as a charge in the memory capacitor C_{ik}. Transfer transistor T and the bit line BL_k connect this capacitor to the sense amplifier that outputs to the bit line $\overline{BL_k}$ ("folded bit line" concept [9-88]).

Figs. 9-92b and c are schematic plots of the potential changes on the bit lines during the read cycle. Before reading, BL_k and $\overline{BL_k}$ float at the same mean voltage, in other words the potential difference at the read amplifier is zero. If the capacitor contains a "1" (Fig. 9-92b), its charge is transferred to the bit line BL_k once the read operation is initiated. A slight potential difference ΔU_1 is established between BL_k and $\overline{BL_k}$. This difference is then amplified to the specified value ΔU_2 by the sense amplifier. If the capacitor contains a "0" (Fig.9-92c), the signs of ΔU_1 and ΔU_2 are reversed.

High demands are placed on the measuring accuracy and the spatial and voltage resolutions in order to measure the read signal on the capacitively sensitive interconnections measuring only 1.1 μm in width (section 8.4.1). Fig. 9-92d shows the word-line and bit-line signals of memory cell C_{00} measured with the sampling method: before the read operation BL_0 and $\overline{BL_0}$ have the same mean potential U = 2.5 V. The read operation starts when the voltage of the selected word line WL_0 begins to rise from 0 V. At approx. 5 V, the transfer transistor switches through and the charge stored in the capacitor flows to the bit line BL_0. Within ap-

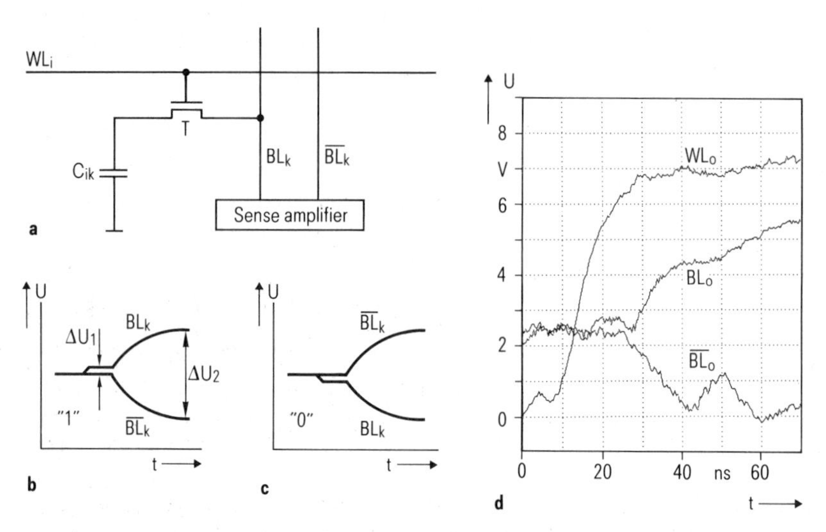

Fig. 9-92. Measuring the read cycle of a 4-Mbit DRAM [9-90,91]. a) Circuit diagram of the C_{ik} memory cell with transfer transistor T and sense amplifier; WL_i, BL_k, $\overline{BL_k}$ are word and bit lines. b) Schematic waveforms if the cell contains a "1" and c) if the cell contains a "0". d) Measured waveforms; measuring conditions: $E_0 = 1$ keV; $I_0 = 2 \cdot 10^{-9}$ A; $f_p = 1$ MHz; $t_p = 1$ ns.

prox. 10 ns, a potential difference $\Delta U_1 \approx 200$ mV is set up between BL_0 and $\overline{BL_0}$. After amplification, $\Delta U_2 = +5$ V is measured. The information stored was a "1".

The potential difference ΔU_1 set up between BL_0 and $\overline{BL_0}$ when the transistor switches through determines whether the stored information can be evaluated with certainty as a "1" or a "0". This critical value depends on the capacitances of the memory cell (approx. 40 fF) and bit line (approx. 350 fF). The measurement shows that the specified value of $\Delta U_2 = +5$ V was reached in the second phase of the read cycle and hence the ratio of capacitances suffices to reliably evaluate the information in the memory cell.

9.5.3 Examination of surface acoustic wave devices

Piezoelectric crystals are deformed by the application of an electric field. High-frequency electric fields produce mechanical oscillations of different modes. Surface acoustic waves (SAW) are of particular interest for technical applications. Fig. 9-93 shows how a wave of this nature is produced by deformation of the crystal surface. The amplitude is less than 1 nm. The velocity of propagation is in the order of 10^3 m/s. This means that the wavelength of a wave of, say, 10 MHz is only a few 100 μm. This allows to produce miniaturized high-frequency components such as frequency filters and resonators [9-92].

Mechanical surface acoustic waves are linked with electric potential waves of the same wavelength and frequency by way of the piezoelectric properties of the crystal (Fig. 9-93). The potential amplitudes vary between some mV and a few V. Hence, the SAW can be visualized by the voltage-contrast methods, using static voltage contrast mapping to view stationary waves [9-93] and the stroboscopic method to image travelling waves with frequencies of up to

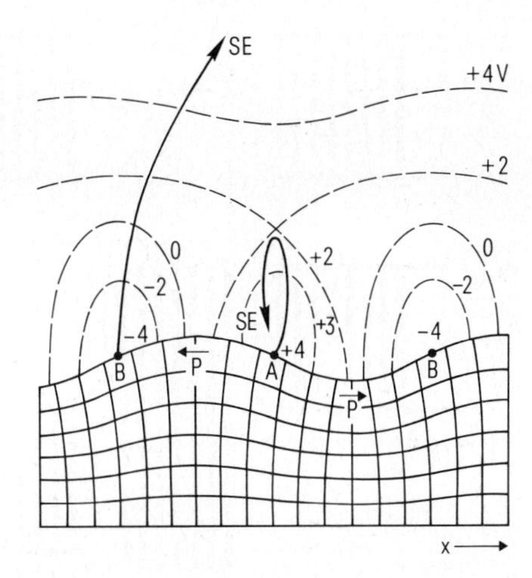

Fig. 9-93. Voltage contrast of surface acoustic waves (SAW): The deformations at the surface of a piezoelectric crystal produce an associated electrical potential distribution. Regions of positive potential (A) appear dark in a voltage contrast image: the secondary electrons (SE) are retarded and return to the surface; regions of negative potential (B) appear bright: the SEs are accelerated and reach the detector (compare Fig. 8-3). ***P*** is the vector of polarization.

500 MHz [9-94,95]. The attenuation of a SAW along the crystal surface can be measured quantitively with the aid of a special feedback loop [9-96].

Design of SAW devices

The piezoelectric crystal is the heart of a SAW device. Quartz and lithium niobate ($LiNbO_3$) are materials of technological significance here. An interdigital transducer on the surface of the crystal excites the SAW. This transducer consists of finger-like interlocking aluminum thin-film electrodes (Fig. 9-94a). An alternating voltage applied to the electrodes produces a periodic electric field between them and causes periodic deformations. These then propagate as travelling SAWs. A second interdigital transducer detects the SAWs by extracting parts of the SAW energy. Stroboscopic voltage contrast imaging was used to visualize the emission of a continuous SAW or of SAW packets emitted from a transducer (Fig. 9-95).

The arrangement shown in Fig. 9-94a also allows bulk waves to reach the detecting transducer. These waves propagate through the interior of the crystal at a higher propagation speed than the SAWs and degrade the characteristic of the device. To suppress this effect, the transducers can be arranged with a mutual shift. The SAW is then coupled to the track of the detecting transducer by a multi-strip coupler (Fig. 9-94b). This coupler consists of a series of aluminum strips insulated from each other.

The propagation of the SAW can be selectively influenced by placing other metal structures on the surface of the crystal. The properties of a device depend critically on the dimensions of these structures. Thus the number, spacing and overlap length of the transducer fingers determine the transfer function of the device, for example its characteristic as a frequency filter. But secondary effects also influence the device characteristic, for example diffraction phenom-

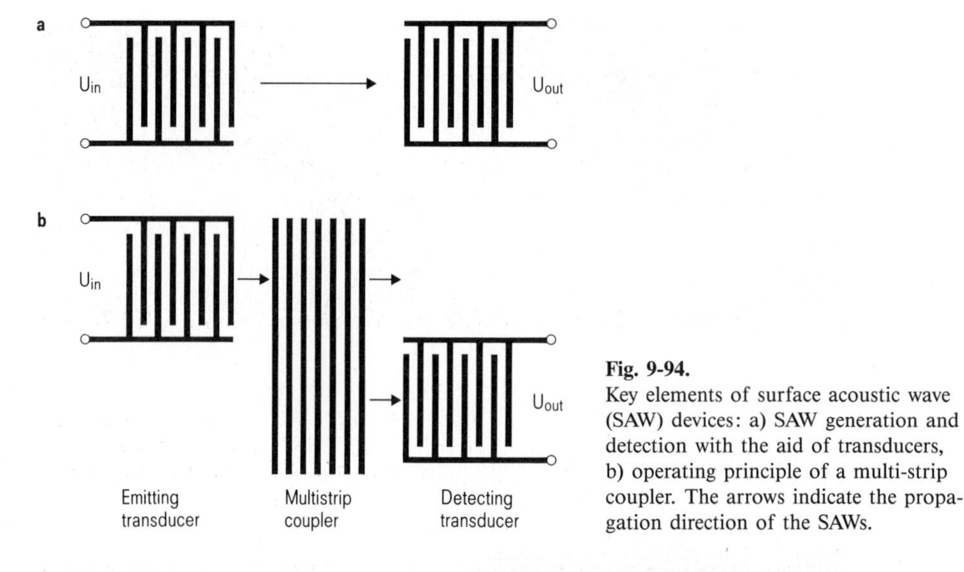

Fig. 9-94. Key elements of surface acoustic wave (SAW) devices: a) SAW generation and detection with the aid of transducers, b) operating principle of a multi-strip coupler. The arrows indicate the propagation direction of the SAWs.

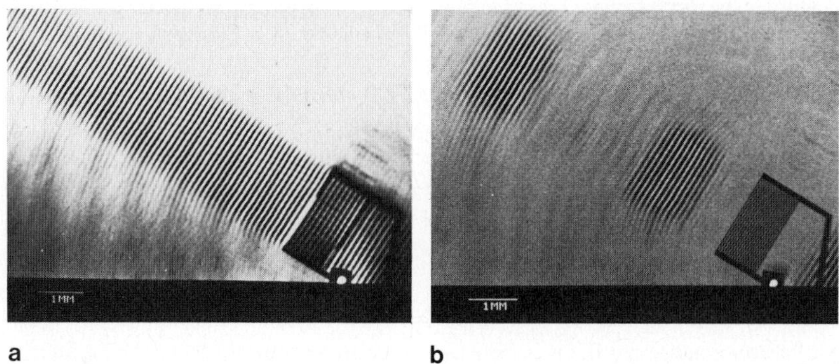

Fig. 9-95. Stroboscopic voltage contrast micrographs of travelling SAW emitted by a transducer (bottom right) on lithium niobate [9-96]. a) Continuous emission of SAWs; b) emission of SAW packets. Measuring conditions: $U_0 = 2.5$ kV; $I_0 = 10^{-7}$ A; wave frequency $f_w = 36$ MHz; pulse frequency $f_p = f_w = 36$ MHz; duty cycle $c = 10$ for (a); $c = 10^3$ for (b).

ena and reflections at the metal structures, at the edges of the crystal and at crystal defects. Effects like these are not always foreseeable. However, they can be visualized with the aid of stroboscopic voltage-contrast imaging.

Examination of bandpass filters

Fig. 9-96 shows voltage-contrast micrographs of a bandpass filter. These filters are designed to select signals of a particular frequency range. In addition, they have to provide high rejection of out-of-band signals. The structure shown in Fig. 9-96 consists of two transducers and a multi-strip coupler and is located on a 2 inch $LiNbO_3$ wafer to allow the propagation of

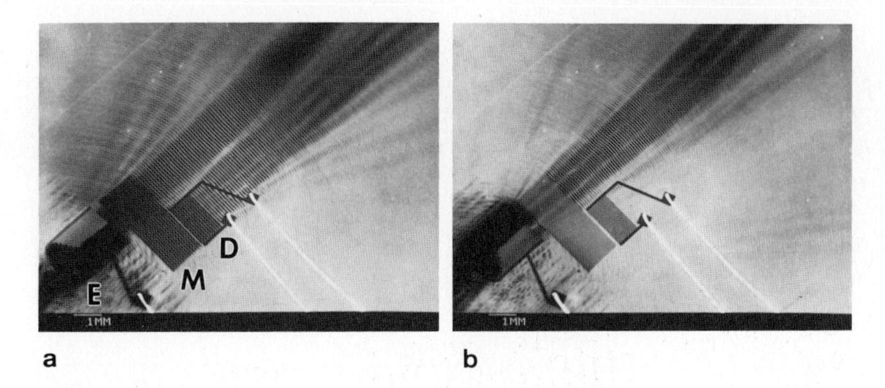

Fig. 9-96. Stroboscopic voltage contrast micrographs showing SAW propagation in a TV intermediate-frequency filter [9-96]: a) at center frequency of the filter (f_w = 36 MHz), b) out of band (f_w = 40.4 MHz). E, D are the emitting and detecting transducer and M the multistrip coupler. Measuring conditions: U_0 = 2.5 kV, $I_0 = 10^{-7}$ A; $f_p = f_w$; c = 10.

the SAW to be traced over large distances. The structure is optimized for a center frequency of 36 MHz and is used as a frequency filter in the intermediate frequency stage of television sets.

The filter in Fig. 9-96a is operated at the center frequency. In this case the multi-strip coupler (M) transfers part of the SAW energy from the emitting transducer (E) to the track of the detecting transducer (D). This can be seen in the micrograph from the fact that two beams propagate behind the multi-strip coupler and the detecting transducer. This is not so in Fig. 9-96b, where the filter was operated in the out-of-band mode (at 40.4 MHz). In this case, the detecting transducer does not receive a signal via the multi-strip coupler.

Detection of spurious waves

Secondary effects are also clearly to be seen in Figs. 9-96a and b: the exit of the emitting transducer acts like an aperture. As a result, several orders of diffraction maxima are visible. However, this scarcely affects the passband characteristics, although it does influence the filter's ability to reject out-of-band frequencies, e.g. at 40 MHz (Fig. 9-96b). At this frequency, a 180° phase shift occurs in the emitted SAW beam and consequently the multi-strip coupler does not couple energy to the detecting transducer. Nonetheless, transmission at this fequency may occur because of the diffracted waves travelling diagonally from the emitting to the detecting transducer. Voltage contrast investigations of this kind have led to improved design programs that can now compensate for the effects of diffraction and provide filters with higher rejection.

Bulk waves emitted in a direction normal to SAW propagation are another cause of spurious transmission. This can be seen at the bottom left of Figs. 9-96a and b. Due to their higher propagation speed, bulk waves have longer wavelengths than the SAWs. By reflection at the edges of the crystal, they can reach the detecting transducer and impair the filter characteristic.

The SAWs, too, are reflected at the edges of the crystal. This phenomenon can be most usefully investigated by the wave-packet mode. Figs. 9-97a and b show a wave packet before and

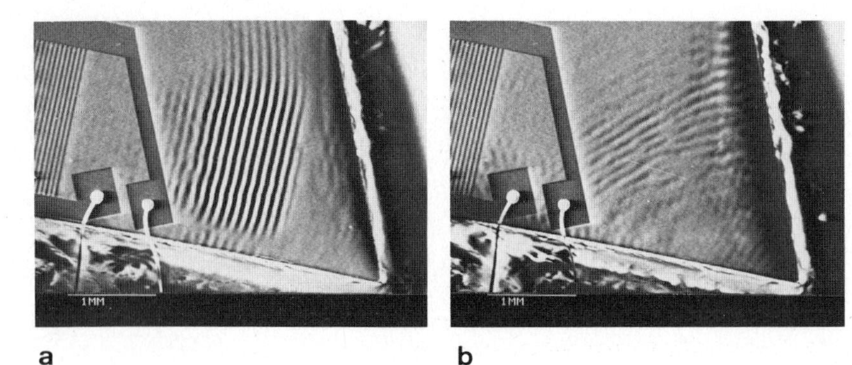

Fig. 9-97. Stroboscopic voltage contrast micrographs showing the propagation of a wave packet in a TV intermediate-frequency filter: a) before and b) after reflection at a crystal edge [9-96]. Same measuring conditions as Fig. 9-95b.

after reflection. To avoid undesirable reflections like this, an adhesive that absorbs the SAW energy is applied at critical points on the edges of the crystal.

9.6 Failure analysis

9.6.1 Problems and procedures of failure analysis

The applications dealt with so far have been concerned almost exclusively with analyses that support technology development. The objectives were to characterize features and parameters *typical* for the process in question and to uncover any typical defects.

Failure analysis attempts to localize and analyze *specific*, isolated faults occurring on a wafer or in a chip. The methodology of IC failure analysis was described above in section 9.5.1 in conjunction with electron beam testing: the initial step involves electrical characterization, to be followed by clarification of the technological cause of the fault. The complexity and fineness of the structures in modern VLSI devices makes this procedure difficult [9-97]. For example, an optical microscope does not offer a sufficiently high resolution to permit visual inspection. With its highly superior spatial resolution and depth of focus, the scanning electron microscope (SEM) has become an indispensable tool. Fitted with an X-ray spectrometer attachment, the SEM can also serve as an X-ray microprobe.

In many instances, examination in the SEM must be preceded by specimen preparation [9-98]. Preparation is needed to expose defects concealed beneath the surface of the device. Similarly, faults localized beforehand by electrical measurements (e.g. in one of the 4 million memory cells in a 4 Mbit memory chip) have to be made accessible for analysis. Polishing and etching techniques (section 3.3.4) are used providing an uncertainty for localization of less than 1 μm.

As a rule, a fault cannot be localized by electrical measurement unless it is within the cell array of a memory chip. In this case, a specific address can be given for the defective cell, which can then be sought with the aid of a precision specimen stage (computer-controlled if possible) under a high-resolution optical microscope and marked for subsequent SEM analysis. The nature of the marking must be such that the defect is not destroyed and the steps involved in preparation do not remove the marker. One way of marking a defect is to make an indent with an ultra micro-hardness tester [9-97], although laser or ion bombardment are also suitable.

A fault cannot be addressed electrically if the fault is in the periphery of a memory chip or if the device in question is not a memory chip at all. In cases like this, searching for hot spots with the aid of liquid crystal thermography [9-99] or IR emission microscopy [9-100] may be of assistance as the first step in fault localization. Alternatively, the fault can be localized by electron beam testing (section 9.5).

9.6.2 Failure analysis of dynamic memory devices

Fig. 9-98 shows secondary-electron micrographs of cross sections through the cell arrays of two DRAMs (dynamic random access memories). The memory capacitor of the 1 Mbit DRAM lies parallel to the surface (Fig. 9-98a). The 4 Mbit DRAM makes use of the third dimension and the capacitor is located at the side-walls of a trench (Fig. 9-98b, cf. Fig. 3-9). This approach allowed packing density to be increased by a factor of 2.5 for a virtually unchanged cell capacitance.

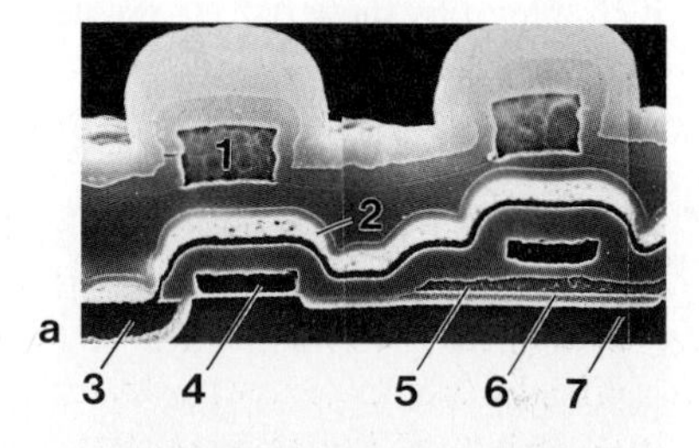

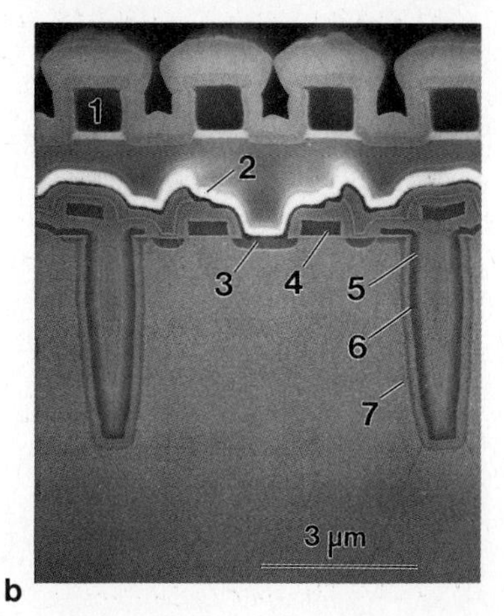

Fig. 9-98.
EM micrographs showing cross sections through memory cells of dynamic random access memories (DRAM) [9-97]. 1 wordline (aluminum), 2 bitline (polysilicon-$MoSi_2$ double layer), 3 bitline contact (P-doped substrate), 4 wordline (P-doped polysilicon), 5 second capacitor electrode (P-doped polysilicon), 6 capacitor dielectric (SiO_2), 7 first capacitor electrode (As-doped substrate). a) 1 Mbit DRAM with planar memory capacitor; b) 4 Mbit DRAM with trench capacitor. The polished cross sections were selectively etched and sputtered with 6 nm gold-palladium (as were those in the following SEM images).

Dynamic memories can be produced with higher packing densities than static memories. The drawback, however, is that the charged state in the memory capacitor can be maintained only for a limited time. The information stored must therefore be refreshed at regular intervals. The 1 Mbit DRAM has an 8 ms refresh cycle, 16 ms being the cycle time for the 4 Mbit DRAM.

Refresh failures

Memory cells that lose their information prematurely are localized electrically by the automatic test system and marked in a fail-bit map. The cell in question then has to be exposed before the technological cause for the failure can be found [9-97]. Fig. 9-99a shows part of the cell array in a 1 Mbit chip following removal of all technology layers: the critical spot is marked by a circle. Following defect etching in a specific etching solution [9-101], three etch pits have become visible (arrowed). Further etching (Fig. 9-99b) reveals that rather than isolated defects, the fault in question here is a single dislocation line. The dislocation extends beyond the memory cell found to be defective and after longer holding time would have led to the failure of all the memory cells affected. Fig. 9-100a shows the etch pit of a dislocation imaged at high magnification in the SEM. The pit lies between the memory capacitor (C) and the transfer transistor (T) (cf. Fig. 9-92a) and was the cause of failure in this cell. Dislocations were found to be most frequently the cause of refresh failures in the memory cells examined.

Dislocations were also found when refresh failures in 4 Mbit DRAMs were analyzed [9-97, 102]. Typically, these defects extend from the silicon surface to the trench wall as shown in Fig. 9-100b.

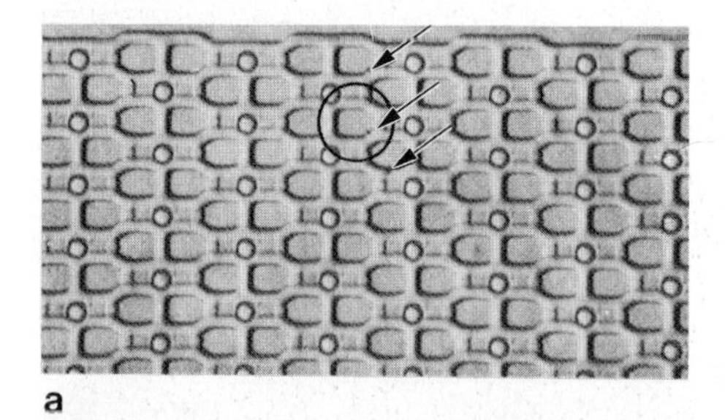
a

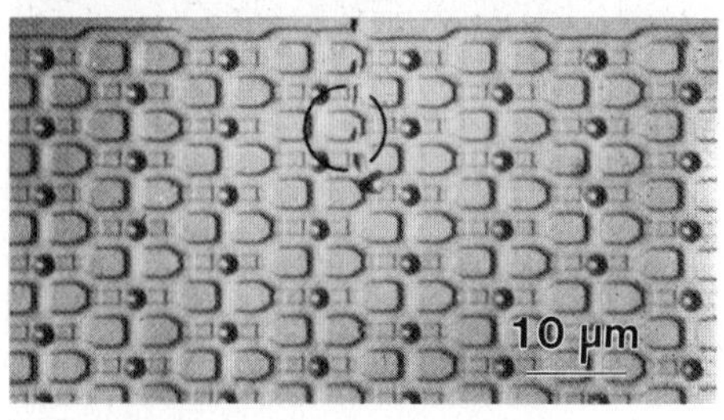

b

Fig. 9-99.
Optical micrographs illustrating refresh failures: part of the cell array of a 1 Mbit DRAM after removal of all technology layers [9-97]. The circle marks the first memory cell to fail. a) The arrows show etch pits after 0.4 µm defect etching; b) a dislocation line becomes visible at this point after 3.4 µm defect etching.

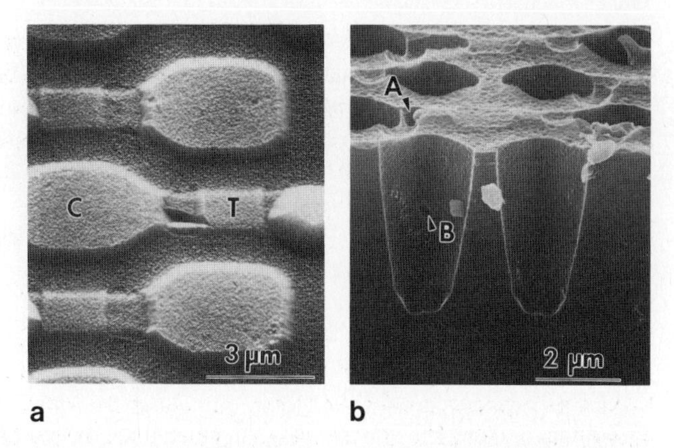

Fig. 9-100. SEM micrographs of memory cells after removal of all technology layers and defect etching: a) dislocation penetrating the surface between capacitor (C) and transfer transistor (T) of a 1 Mbit DRAM [9-97]; b) dislocation extending from the silicon surface (A) to the trench wall (B) of a 4 Mbit DRAM [9-102].

Single bit failures

Some memory cells lose their charge immediately after the write cycle. The charge state of a cell like this will always be the same, regardless of what was written into it before the electrical measurement. This effect is known as a single bit failure and is indicative of a short between the two capacitor electrodes (i.e. between the polysilicon layer (5) and the substrate (7) in Fig. 9-98a). There are two ways of verifying this assumption by analytical means [9-97]: one technique is to remove the technology layers down into the polysilicon layer (5) by face lapping. The residual polysilicon is then selectively etched to expose the oxide layer beneath it

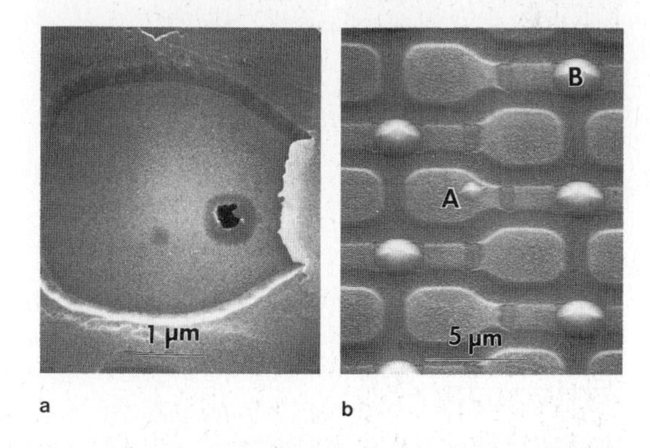

Fig. 9-101. SEM micrographs illustrating single bit failures in 1 Mbit DRAMs [9-97]. a) Memory cell after face lapping and selective etching: there is a hole in the capacitor dielectric (6 in Fig. 9-98 a). b) Part of the cell array after removal of all technology layers and defect etching: there is a hillock A due to phosphorus doping of the substrate through a hole in the oxide; B phosphorus-doped bitline contact.

(this oxide layer forms the capacitor dielectric 6). This process reveals holes in the oxide (Fig. 101a). The alternative method is to etch off all the technology layers down to the substrate by etching. Defect etching [9-101] reveals characteristic pyramid-shaped hillocks in the defective memory cells (A in Fig. 9-101b). These hillocks are shaped like the heavily phosphorus-doped bitline contacts (B). It may thus be concluded that phosphorus has diffused from the phosphorus-doped polysilicon of the upper capacitor electrode through the holes in the oxide and into the substrate. The result of these analyses yields defects in the oxide layer as the cause of the single bit failures.

Trench oxide failure

The capacitor dielectric of a trench cell is much more difficult to examine than that of a planar capacitor. The trench capacitors can be exposed by selective etching of the substrate from the back up to the oxide forming the dielectric [9-103]. Fig. 9-102a shows a trench capacitor prepared in this way. The oxide layer (6 in Fig. 9-98b) coats the polysilicon electrode (5). A defect in the capacitor oxide allows the etching solution to penetrate to the interior, where it dissolves the polysilicon capacitor electrode. This effect can be seen through the trench dielectric (Fig. 9-102b) and produces a contrast differing from that of the fault-free capacitors.

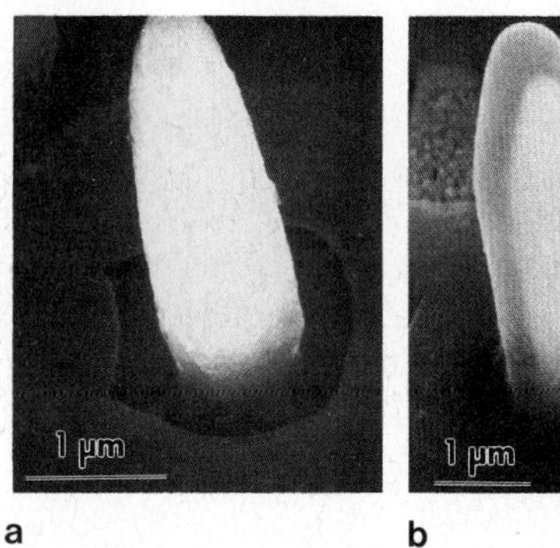

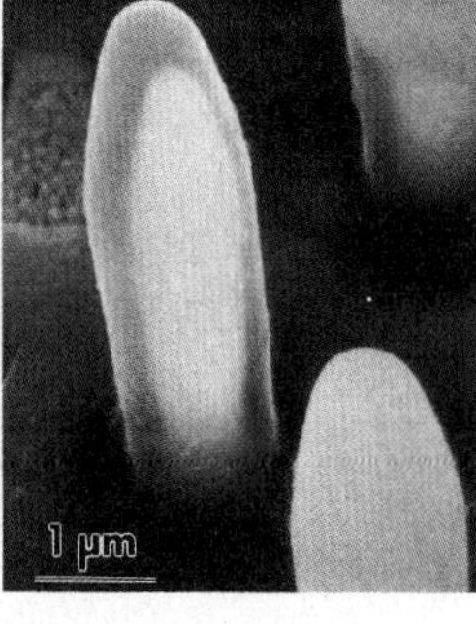

Fig. 9-102.
SEM micrographs of 4 Mbit DRAM [9-97]. The trench capacitors have been exposed down to the capacitor dielectric (6 in Fig. 9-98b) by selective etching from the back. a) Perfect cell; b) defective cell (see text).

Substrate leakage

4 Mbit DRAMs with increased substrate leakage currents were examined using the hotspot detection method. The problem was localized in the wordline decoder areas that contain a large number of contact holes from the aluminum layer to the substrate. Figs. 9-103a,b show typical SEM cross sections of wafers with normal and high leakage respectively. Whereas all the contact holes in wafers with low leakage currents looked like that shown in Fig. 9-103a, some holes in wafers with high leakage showed aluminum spiking (section 9.2.6) occurring frequently in the corner at the bottom of the contact hole (4 in Fig. 9-103b, compare Fig. 9-34). A titanium-titanium nitride double layer (2) was used as a barrier layer to prevent interdiffusion of this kind. A crack in the barrier layer was found to be the cause for spiking [9-97].

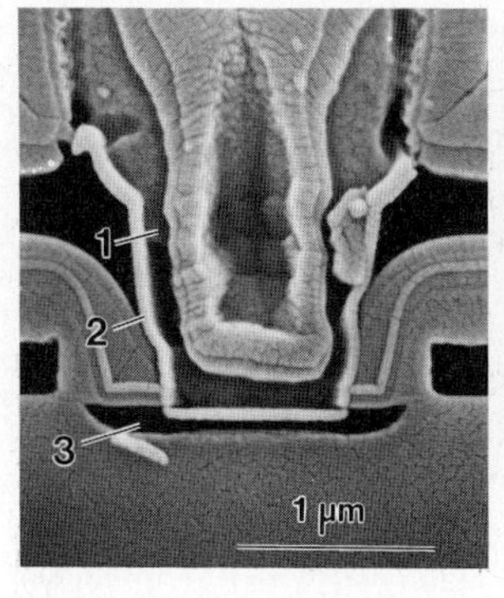

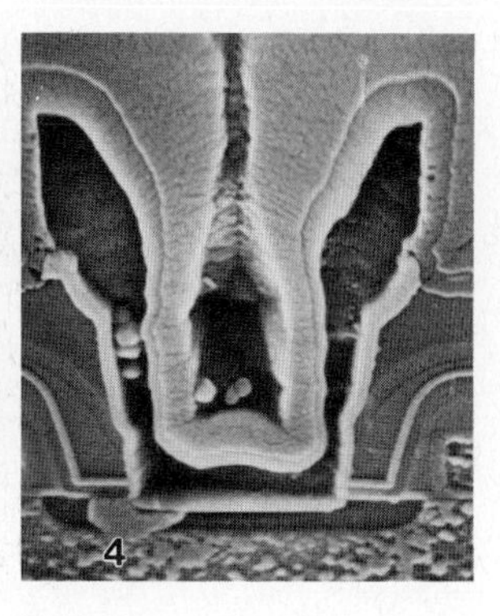

Fig. 9-103. Cross sections of contact holes in the wordline decoder area of 4 Mbit DRAM [9-97]. 1 aluminum layer, 2 Ti-TiN barrier layer, 3 As-doped region, 4 aluminum spike. a) SEM micrograph of a normal contact hole, b) of a contact hole with aluminum spiking (4).

9.7 References

[9-1] R. Singh, *J. Appl. Phys. 63,* R59 (1988)

[9-2] T. Sands, J. Washburn, R. Gronsky, W. Maszara, D.K. Sadana, G.A. Rozgonyi, *Proc. 13th. Int. Conf. on Defects in Semiconductors 9,* 531 (1985)

[9-3] R. v. Criegern, H. Zeininger, S. Röhl, in [7-6], 419

[9-4] H. Zeininger, R. v. Criegern, S. Röhl, *Surface and Interface Analysis 12,* 324 (1988)

[9-5] R. v. Criegern, I. Weitzel, H. Rehme, *Siemens Forsch.- u. Entwickl.-Ber. 14,* 208 (1985)

[9-6] V. Probst, H.J. Böhm, H. Schaber, H. Oppolzer, I. Weitzel, *J. Electrochem. Soc. 135,* 671 (1988)

[9-7] T.T. Sheng, R.B. Marcus, *J. Electrochem. Soc. 128,* 881 (1981)

[9-8] H. Rehme, H. Oppolzer, *Siemens Forsch.- u. Entwickl.-Ber. 14,* 193 (1985)

[9-9] P. Vitanov, U. Schwabe, I. Eisele, *IEEE Trans. ED31,* 96 (1984)

[9-10] U. Schwabe, E.P. Jacobs, D. Takacs, J. Winnerl, E. Lange, *Proc. IEEE Int. Electr. Device Meeting IEDM 1984,* p. 410

[9-11] H. Schink, *Thesis,* Techn. Univ. München 1984

[9-12] H. Kabza, K. Ehinger, I. Kerner, H.-W. Meul, D. Hartwig, I. Weng, M. Miura-Mattausch, T.F. Meister, P. Weger, M. Ohnemus, H. Klose, R. Köpl, H. Schaber, L. Treitinger, *IEEE-EDL 10,* 344 (1989)

[9-13] B. Benna, T.F. Meister, H. Schaber, *Solid-State Electronics 30,* 1153 (1987)

[9-14] H. Schaber, R. v. Criegern, I. Weitzel, *J. Appl. Phys. 58,* 4036 (1985)

[9-15] W.M. Werner, A. Wieder, H. Schaber, H. Kaiser, K. Wiesinger, A. Glasl, *Siemens Forsch.-u. Entwickl.-Ber. 17,* 221 (1988)

[9-16] H.J. Böhm, H. Wendt, H. Oppolzer, K. Masseli, R. Kassing, *J. Appl. Phys. 62,* 2784 (1987)

[9-17] H. Oppolzer, W. Eckers, H. Schaber, *J. de Physique 46,* Colloque C4, Suppl. No.4, C4-523 (1985)

[9-18] C.R.M. Grovenor, P.E. Batson, D.A.Smith, C. Wong, *Phil. Mag. A50,* 409 (1984)

[9-19] S.P. Murarka, *Silicides for VLSI Application.* Academic Press, New York 1983

[9-20] D. Pawlik, H. Oppolzer, T. Hillmer, *J. Vac. Sci. Technol. B5,* 492 (1984)

[9-21] R. v. Criegern, T. Hillmer, V. Huber, H. Oppolzer, I. Weitzel, *Fresenius Z. Anal. Chem. 319,* 861 (1984)

[9-22] H. Oppolzer, *J. de Physique 48,* Colloque C5, Suppl. No.11, C5-65 (1987)

[9-23] S.P. Murarka, D.B. Fraser, W.S. Lindenberg, A.K. Sinha, *J. Appl. Phys. 51,* 3241 (1980)

[9-24] C.K. Lau, Y.C. See, D.B. Scott, J.M. Bridges, S.N. Perna, R.D. Davies, *IEDM Tech. Dig. 82,* 714 (1982)
[9-25] M.E. Alperin, T.C. Holloway, R.A. Haken, C.D. Gosmeyer, R.V. Karnaugh, W.D. Paramantie, *IEEE Trans. Electron Devices ED-32,* 141 (1985)
[9-26] T. Okamoto, K. Tsukamoto, M. Shimizu, T. Matsukawa, *J. Appl. Phys. 57,* 5251 (1985)
[9-27] A.E. Morgan, E.K. Broadbent, K.N. Ritz, D.K. Sadana, B.J. Burrow, *J. Appl. Phys. 64,* 344 (1988)
[9-28] W. Pamler, K. Wangemann, W. Bensch, E. Bußmann, A. Mitwalsky, *Fresenius Z. Anal. Chem. 333,* 569 (1989)
[9-29] C. Ernsberger, J. Nickerson, A.E. Miller, J. Moulder, *J. Vac. Sci. Technol. A3,* 2415 (1985)
[9-30] J.W. Coburn, H.F. Winters, *J. Vac. Sci. Technol. 16,* 391 (1979)
[9-31] J. Segner, E.-G. Mohr, T. Hillmer, *Les vides – les couches minces, Suppl., 229,* 85 (1985)
[9-32] H. Cerva, E.-G. Mohr, H. Oppolzer, *J. Vac. Sci. Technol. B5,* 590 (1987)
[9-33] H.P. Strunk, H. Cerva, E.-G. Mohr, *J. Electrochem. Soc. 135,* 2876 (1988)
[9-34] S.W. Pang, D.D. Rathmann, D.J. Silversmith, R.W. Mountain, P.D. DeGraff, *J. Appl. Phys. 54,* 3272 (1983)
[9-35] H. Oppolzer, in P.J. Goodhew, H.G. Dickinson (Eds.): *Proc. 9th Europ. Congr. Electron Microscopy 1988.* Inst. Phys. Conf. Ser. No. 93: Vol. 2. Institute of Physics Publ., Bristol 1988, 73
[9-36] H. Oppolzer, F. Neppl, K. Hieber, V. Huber, *J. Vac. Sci. Technol. B2,* 630 (1984)
[9-37] M. Nicolet, *Thin Solid Films 52,* 415 (1978)
[9-38] M. Wittmer, *J. Vac. Sci. Technol. A3,* 1797 (1985)
[9-39] W. Pamler, *Surf. Interface Anal. 13,* 55 (1988)
[9-40] H. Oppolzer, H. Cerva, C. Fruth, V. Huber, S. Schild, in A.G. Cullis, P.D. Augustus (Eds.): *Microscopy of Semiconducting Materials 1987.* Inst. Phys. Conf. Ser. No. 87: Sect. 6. Institute of Physics Publ., Bristol 1987, 433
[9-41] B.O. Kolbesen, W. Pamler, *Fresenius Z. Anal. Chem. 333,* 561 (1989)
[9-42] H. Cerva, T. Hillmer, H. Oppolzer, R. v. Criegern, see [9-40], 445
[9-43] H. Oppolzer, in A.G. Cullis, D.B. Holt (Eds.): *Microscopy of Semiconducting Materials 1985.* Inst. Phys. Conf. Ser. No. 76: Sect. 11. Adam Hilger, Bristol 1985, 461
[9-44] T.T. Sheng, R.B. Marcus, *J. Electrochem. Soc. 125,* 432 (1978)
[9-45] L.O. Wilsen, *J. Electrochem. Soc. 129,* 831 (1982)
[9-46] R.B. Marcus, T.T. Sheng, *J. Electrochem. Soc. 129,* 1278 (1982)
[9-47] H.C. Casey, M.B. Panish, *Heterostructure Lasers,* Academic Press, New York, London 1978
[9-48] T. Ishibashi, Y. Yamauchi, *IEEE Trans. Electron. Dev. 35,* 401 (1988)
[9-49] W.T. Tsang, in R. Dingle (Ed.): *Semiconductors and Semimetals, Vol. 24,* Academic Press, San Diego, 1987, p. 397
[9-50] W.J. Bartels, *J. Vac. Sci. Technol. B1,* 338 (1983)
[9-51] P.M. Petroff, *J. Vac. Sci. Technol. 14,* 973 (1977)
[9-52) K.-H. Küsters, B.C. Cooman, J.R. Shealy, C.B. Carter, *J. Cryst. Growth 71,* 514 (1985)
[9-53] K. Alavi, P.M. Petroff, W.R. Wagner, A.Y. Cho, *J. Vac. Sci. Technol. B 1,* 146 (1983)
[9-54] W.D. Laidig et.al., *Appl. Phys. Lett. 38,* 776 (1981)
[9-55] K. Ishida, T. Ohta, S. Semura, H. Nakashima, *Jap. J. Appl. Phys. 24,* L620 (1985)
[9-56] W. Stolz, K. Fujiwara, L. Tapfer, H. Oppolzer, K. Ploog, *Inst. Phys. Conf. Ser. No. 74:* Chapt. 3. Adam Hilger, Bristol 1985, 139
[9-57] R. Gessner, M. Druminski, M.B. Schorner, *Electron. Lett. 25,* 516 (1989)
[9-58] H. Kakibayashi, F. Nagata, *Jap. J. Appl. Phys. 24,* L905 (1985), and *25,* 1644 (1986)
[9-59] A.F. de Jong, K.T.F. Janssen, *J. Mater. Res. 5,* 578 (1990)
[9-60] A.F. de Jong, H. Bender, W. Coene, *Ultramicroscopy 21,* 373 (1987)
[9-61] Y. Suzuki, H. Okamoto, *J. Appl. Phys. 58,* 3456 (1985)
[9-62] C.J.D. Hetherington, J.C. Barry, J.M. Bi, C.J. Humphreys, J. Grange, C. Wood, in *Materials Research Society Symp. Vol. 37.* Mat. Res. Soc., Pittsburgh 1985, 41
[9-63] R. Treichler, L. Korte, R. v. Criegern, in [7-6], 469
[9-64] E. Veuhoff, H. Baumeister, R. Treichler, *J. Crystal Growth 93,* 650 (1988)
[9-65] Ch. Meyer, M. Maier, D. Bimberg, *J. Appl. Phys. 54,* 2672 (1983)
[9-66] H. Tews, R. Neumann, T. Humer-Hager, R. Treichler, *J. Appl. Phys. 68,* 1318 (1990)
[9-67] F. Dildey, R. Treichler, M.-C. Amann, M. Schicr, G. Ebbinghaus, *Appl. Phys. Lett. 55,* 878 (1989)
[9-68] T.S. Kuan, P.E. Batson, T.N. Jackson, H. Rupprecht, E. Wilkie, *J. Appl. Phys. 54,* 6952 (1983)

[9-69] W. Kellner, *Siemens Forsch.- u. Entwickl.-Ber. 4,* 137 (1975)
[9-70] J. Willer, H. Oppolzer, *Thin Solid Films 147,* 117 (1987)
[9-71] J. Willer, D. Ristow, W. Kellner, H. Oppolzer, *J. Electrochem. Soc. 135,* 179 (1988)
[9-72] W.D. Kingery, H.K. Bowen, D.R. Uhlmann, *Introduction to Ceramics.* John Wiley, New York 1976
[9-73] H. Schmelz, A. Meyer, *cfi/Ber.DKG 59,* 436 (1982)
[9-74] A. Meyer, A. Papp, *Siemens Forsch.- u. Entwickl.-Ber. 14,* 306 (1985)
[9-75] M. Matsuoka, *Jap. J. Appl. Phys. 10,* 736 (1971)
[9-76] R. Einzinger, in L.M. Levinson (Ed.): *Advances in Ceramics, Vol. 1.* Amer. Ceramic Soc., Columbus 1981, 359
[9-77] H. Cerva, W. Russwurm, *J. Am. Ceram. Soc. 71,* 522 (1988)
[9-78] J. Müller, J.L. Olson (Eds.): *Conf. Proc. on High Temperature Superconductors and Materials and Mechanisms of Superconductivity. Physics C 153-155,* North-Holland, Amsterdam 1988
[9-79] H.C. Li, G. Linker, F. Ratzel, R. Smithey, J. Geerk, *Appl. Phys. Lett. 52,* 1098 (1988)
[9-80] B. Roas, L. Schultz, G. Endres, *Appl. Phys. Lett. 53,* 1557 (1988)
[9-81] O. Eibl, G. Gieres, H. Behner, in G. W. Bailey (Ed.): *Proc. 47th Annual EMSA Meeting.* San Francisco Press Inc., San Francisco 1989, 172
[9-82] O. Eibl, H. Schmid, *Inst. Phys. Conf. Ser. No. 93,* Vol. 2, Chapt. 6, IOP Publishing, Bristol 1988, 253
[9-83] O. Eibl, in G. W. Bailey (Ed.): *Proc. 47th Annual EMSA Meeting.* San Francisco Press Inc., San Francisco 1989, 159
[9-84] O. Eibl, *Solid State Communications 69,* 509 (1989)
[9-85] O. Eibl, *Solid State Communications 67,* 1049 (1988)
[9-86] H.P. Feuerbaum, K. Hernaut, *Scanning Electron Microscopy 1978/I,* SEM Inc., AMF O'Hare 1978, 795
[9-87] A. Dallmann, G. Menzel, R. Weyl, F. Fox, *Proc. 23rd Reliability Physics Conf.,* IEEE 1985, 224
[9-88] W. Pribyl, J. Harter, W. Müller, *Siemens Forsch.- u. Entwickl.-Ber. 16,* 253 (1987)
[9-89] J. Harter, W. Pribyl, M. Bähring, A. Lill, H. Mattes, W. Müller, L. Risch, D. Sommer, R. Strunz, W. Weber, K. Hofmann, *Digest of the IEEE Int. Solid State Circuits Conf.,* San Francisco 1988, p. 244
[9-90] D. Sommer, J. Kölzer, F. Bonner, F. Fox, M. Killian, W. Pribyl, *ITG-Fachbericht 102,* VDE-Verlag GmbH, Berlin 1988, 277
[9-91] F. Fox, J. Kölzer, J. Otto, E. Plies, *IBM J. Res. Develop. 34,* 215 (1990)
[9-92] D.P. Morgan, *Surface-Wave Devices for Signal Processing,* Elsevier, Amsterdam 1985
[9-93] P. Hiesinger, *Ultrasonics Symposium, Proceedings,* IEEE 1978, 611
[9-94] H.P. Feuerbaum, G. Eberharter, G. Tobolka, *Scanning Electron Microscopy 1980/I,* SEM Inc., AMF O'Hare 1980, 503
[9-95] R. Veith, G. Eberharter, H.P. Feuerbaum, U. Knauer, *1980 Ulrasonics Symposium, Proceedings,* IEEE 1980, 348
[9-96] H. P. Feuerbaum, H. P. Grassl, U. Knauer, R. Veith, *Scanning Electron Microscopy 1983/I,* SEM Inc., AMF O'Hare 1983, 55
[9-97] R. Lemme, M. Gentsch, R. Kutzner, *Proc. Intern. Symp. on Testing and Failure Analysis,* ASM International 1988, 31
[9-98] F. Beck, *Integrierte Halbleiterschaltungen. Präparationstechniken in der Fehler- und Konstruktionsanalyse.* VCH-Verlagsgesellschaft, Weinheim 1988
[9-99] R. Weyl, B. Lischke, F. Beck, R. Kappelmeyer, *NTG-Fachberichte 87,* 116 (1985)
[9-100) C.-L. Chiang, N. Khurana, *Proc. Intern. Electron. Devices Meeting IEDM 86,* IEEE, New York 1986, 672
[9-101] M. Wright-Jenkins, *J. Electrochem. Soc. 124,* 757 (1977)
[9-102] H. Wendt, S. Sauter, *J. Electrochem. Soc. 136,* 1568 (1989)
[9-103] T. Kure, T. Konoda, H. Sunami, A. Sato, Y. Kawamoto, T. Hayashida, *VLSI Symposium, Ext. Abstracts SSDM* (1987), p. 307
[9-104] H. Zeininger, R. v. Criegern, in [7-59], 419

Index